Lecture Notes in Computer Science 14228

Founding Editors

Gerhard Goos
Juris Hartmanis

Editorial Board Members

The series Lecture Notes in Computer Science (LNCS), including its subseries Lecture Notes in Artificial Intelligence (LNAI) and Lecture Notes in Bioinformatics (LNBI), has established itself as a medium for the publication of new developments in computer science and information technology research, teaching, and education.

LNCS enjoys close cooperation with the computer science R & D community, the series counts many renowned academics among its volume editors and paper authors, and collaborates with prestigious societies. Its mission is to serve this international community by providing an invaluable service, mainly focused on the publication of conference and workshop proceedings and postproceedings. LNCS commenced publication in 1973.

Hayit Greenspan · Anant Madabhushi ·
Parvin Mousavi · Septimiu Salcudean ·
James Duncan · Tanveer Syeda-Mahmood ·
Russell Taylor
Editors

Medical Image Computing and Computer Assisted Intervention – MICCAI 2023

26th International Conference
Vancouver, BC, Canada, October 8–12, 2023
Proceedings, Part IX

 Springer

Editors
Hayit Greenspan
Icahn School of Medicine, Mount Sinai,
NYC, NY, USA

Tel Aviv University
Tel Aviv, Israel

Parvin Mousavi
Queen's University
Kingston, ON, Canada

James Duncan ⓘ
Yale University
New Haven, CT, USA

Russell Taylor ⓘ
Johns Hopkins University
Baltimore, MD, USA

Anant Madabhushi ⓘ
Emory University
Atlanta, GA, USA

Septimiu Salcudean ⓘ
The University of British Columbia
Vancouver, BC, Canada

Tanveer Syeda-Mahmood ⓘ
IBM Research
San Jose, CA, USA

ISSN 0302-9743   ISSN 1611-3349 (electronic)
Lecture Notes in Computer Science
ISBN 978-3-031-43995-7   ISBN 978-3-031-43996-4 (eBook)
https://doi.org/10.1007/978-3-031-43996-4

This Springer imprint is published by the registered company Springer Nature Switzerland AG
The registered company address is: Gewerbestrasse 11, 6330 Cham, Switzerland

Paper in this product is recyclable.

Preface

We are pleased to present the proceedings for the 26th International Conference on Medical Image Computing and Computer-Assisted Intervention (MICCAI). After several difficult years of virtual conferences, this edition was held in a mainly in-person format with a hybrid component at the Vancouver Convention Centre, in Vancouver, BC, Canada October 8–12, 2023. The conference featured 33 physical workshops, 15 online workshops, 15 tutorials, and 29 challenges held on October 8 and October 12. Co-located with the conference was also the 3rd Conference on Clinical Translation on Medical Image Computing and Computer-Assisted Intervention (CLINICCAI) on October 10.

MICCAI 2023 received the largest number of submissions so far, with an approximately 30% increase compared to 2022. We received 2365 full submissions of which 2250 were subjected to full review. To keep the acceptance ratios around 32% as in previous years, there was a corresponding increase in accepted papers leading to 730 papers accepted, with 68 orals and the remaining presented in poster form. These papers comprise ten volumes of Lecture Notes in Computer Science (LNCS) proceedings as follows:

- Part I, LNCS Volume 14220: Machine Learning with Limited Supervision and Machine Learning – Transfer Learning
- Part II, LNCS Volume 14221: Machine Learning – Learning Strategies and Machine Learning – Explainability, Bias, and Uncertainty I
- Part III, LNCS Volume 14222: Machine Learning – Explainability, Bias, and Uncertainty II and Image Segmentation I
- Part IV, LNCS Volume 14223: Image Segmentation II
- Part V, LNCS Volume 14224: Computer-Aided Diagnosis I
- Part VI, LNCS Volume 14225: Computer-Aided Diagnosis II and Computational Pathology
- Part VII, LNCS Volume 14226: Clinical Applications – Abdomen, Clinical Applications – Breast, Clinical Applications – Cardiac, Clinical Applications – Dermatology, Clinical Applications – Fetal Imaging, Clinical Applications – Lung, Clinical Applications – Musculoskeletal, Clinical Applications – Oncology, Clinical Applications – Ophthalmology, and Clinical Applications – Vascular
- Part VIII, LNCS Volume 14227: Clinical Applications – Neuroimaging and Microscopy
- Part IX, LNCS Volume 14228: Image-Guided Intervention, Surgical Planning, and Data Science
- Part X, LNCS Volume 14229: Image Reconstruction and Image Registration

The papers for the proceedings were selected after a rigorous double-blind peer-review process. The MICCAI 2023 Program Committee consisted of 133 area chairs and over 1600 reviewers, with representation from several countries across all major continents. It also maintained a gender balance with 31% of scientists who self-identified as women. With an increase in the number of area chairs and reviewers, the reviewer load on the experts was reduced this year, keeping to 16–18 papers per area chair and about 4–6 papers per reviewer. Based on the double-blinded reviews, area chairs' recommendations, and program chairs' global adjustments, 308 papers (14%) were provisionally accepted, 1196 papers (53%) were provisionally rejected, and 746 papers (33%) proceeded to the rebuttal stage. As in previous years, Microsoft's Conference Management Toolkit (CMT) was used for paper management and organizing the overall review process. Similarly, the Toronto paper matching system (TPMS) was employed to ensure knowledgeable experts were assigned to review appropriate papers. Area chairs and reviewers were selected following public calls to the community, and were vetted by the program chairs.

Among the new features this year was the emphasis on clinical translation, moving Medical Image Computing (MIC) and Computer-Assisted Interventions (CAI) research from theory to practice by featuring two clinical translational sessions reflecting the real-world impact of the field in the clinical workflows and clinical evaluations. For the first time, clinicians were appointed as Clinical Chairs to select papers for the clinical translational sessions. The philosophy behind the dedicated clinical translational sessions was to maintain the high scientific and technical standard of MICCAI papers in terms of methodology development, while at the same time showcasing the strong focus on clinical applications. This was an opportunity to expose the MICCAI community to the clinical challenges and for ideation of novel solutions to address these unmet needs. Consequently, during paper submission, in addition to MIC and CAI a new category of "Clinical Applications" was introduced for authors to self-declare.

MICCAI 2023 for the first time in its history also featured dual parallel tracks that allowed the conference to keep the same proportion of oral presentations as in previous years, despite the 30% increase in submitted and accepted papers.

We also introduced two new sessions this year focusing on young and emerging scientists through their Ph.D. thesis presentations, and another with experienced researchers commenting on the state of the field through a fireside chat format.

The organization of the final program by grouping the papers into topics and sessions was aided by the latest advancements in generative AI models. Specifically, Open AI's GPT-4 large language model was used to group the papers into initial topics which were then manually curated and organized. This resulted in fresh titles for sessions that are more reflective of the technical advancements of our field.

Although not reflected in the proceedings, the conference also benefited from keynote talks from experts in their respective fields including Turing Award winner Yann LeCun and leading experts Jocelyne Troccaz and Mihaela van der Schaar.

We extend our sincere gratitude to everyone who contributed to the success of MIC-CAI 2023 and the quality of its proceedings. In particular, we would like to express our profound thanks to the MICCAI Submission System Manager Kitty Wong whose meticulous support throughout the paper submission, review, program planning, and proceeding preparation process was invaluable. We are especially appreciative of the effort and dedication of our Satellite Events Chair, Bennett Landman, who tirelessly coordinated the organization of over 90 satellite events consisting of workshops, challenges and tutorials. Our workshop chairs Hongzhi Wang, Alistair Young, tutorial chairs Islem

Rekik, Guoyan Zheng, and challenge chairs, Lena Maier-Hein, Jayashree Kalpathy-Kramer, Alexander Seitel, worked hard to assemble a strong program for the satellite events. Special mention this year also goes to our first-time Clinical Chairs, Drs. Curtis Langlotz, Charles Kahn, and Masaru Ishii who helped us select papers for the clinical sessions and organized the clinical sessions.

We acknowledge the contributions of our Keynote Chairs, William Wells and Alejandro Frangi, who secured our keynote speakers. Our publication chairs, Kevin Zhou and Ron Summers, helped in our efforts to get the MICCAI papers indexed in PubMed. It was a challenging year for fundraising for the conference due to the recovery of the economy after the COVID pandemic. Despite this situation, our industrial sponsorship chairs, Mohammad Yaqub, Le Lu and Yanwu Xu, along with Dekon's Mehmet Eldegez, worked tirelessly to secure sponsors in innovative ways, for which we are grateful.

An active body of the MICCAI Student Board led by Camila Gonzalez and our 2023 student representatives Nathaniel Braman and Vaishnavi Subramanian helped put together student-run networking and social events including a novel Ph.D. thesis 3-minute madness event to spotlight new graduates for their careers. Similarly, Women in MICCAI chairs Xiaoxiao Li and Jayanthi Sivaswamy and RISE chairs, Islem Rekik, Pingkun Yan, and Andrea Lara further strengthened the quality of our technical program through their organized events. Local arrangements logistics including the recruiting of University of British Columbia students and invitation letters to attendees, was ably looked after by our local arrangement chairs Purang Abolmaesumi and Mehdi Moradi. They also helped coordinate the visits to the local sites in Vancouver both during the selection of the site and organization of our local activities during the conference. Our Young Investigator chairs Marius Linguraru, Archana Venkataraman, Antonio Porras Perez put forward the startup village and helped secure funding from NIH for early career scientist participation in the conference. Our communications chair, Ehsan Adeli, and Diana Cunningham were active in making the conference visible on social media platforms and circulating the newsletters. Niharika D'Souza was our cross-committee liaison providing note-taking support for all our meetings. We are grateful to all these organization committee members for their active contributions that made the conference successful.

We would like to thank the MICCAI society chair, Caroline Essert, and the MICCAI board for their approvals, support and feedback, which provided clarity on various aspects of running the conference. Behind the scenes, we acknowledge the contributions of the MICCAI secretariat personnel, Janette Wallace, and Johanne Langford, who kept a close eye on logistics and budgets, and Diana Cunningham and Anna Van Vliet for including our conference announcements in a timely manner in the MICCAI society newsletters. This year, when the existing virtual platform provider indicated that they would discontinue their service, a new virtual platform provider Conference Catalysts was chosen after due diligence by John Baxter. John also handled the setup and coordination with CMT and consultation with program chairs on features, for which we are very grateful. The physical organization of the conference at the site, budget financials, fund-raising, and the smooth running of events would not have been possible without our Professional Conference Organization team from Dekon Congress & Tourism led by Mehmet Eldegez. The model of having a PCO run the conference, which we used at

MICCAI, significantly reduces the work of general chairs for which we are particularly grateful.

Finally, we are especially grateful to all members of the Program Committee for their diligent work in the reviewer assignments and final paper selection, as well as the reviewers for their support during the entire process. Lastly, and most importantly, we thank all authors, co-authors, students/postdocs, and supervisors for submitting and presenting their high-quality work, which played a pivotal role in making MICCAI 2023 a resounding success.

With a successful MICCAI 2023, we now look forward to seeing you next year in Marrakesh, Morocco when MICCAI 2024 goes to the African continent for the first time.

October 2023

Tanveer Syeda-Mahmood
James Duncan
Russ Taylor
General Chairs

Hayit Greenspan
Anant Madabhushi
Parvin Mousavi
Septimiu Salcudean
Program Chairs

Organization

General Chairs

Tanveer Syeda-Mahmood   IBM Research, USA
James Duncan   Yale University, USA
Russ Taylor   Johns Hopkins University, USA

Program Committee Chairs

Hayit Greenspan   Tel-Aviv University, Israel and Icahn School of
     Medicine at Mount Sinai, USA
Anant Madabhushi   Emory University, USA
Parvin Mousavi   Queen's University, Canada
Septimiu Salcudean   University of British Columbia, Canada

Satellite Events Chair

Bennett Landman   Vanderbilt University, USA

Workshop Chairs

Hongzhi Wang   IBM Research, USA
Alistair Young   King's College, London, UK

Challenges Chairs

Jayashree Kalpathy-Kramer   Harvard University, USA
Alexander Seitel   German Cancer Research Center, Germany
Lena Maier-Hein   German Cancer Research Center, Germany

Tutorial Chairs

Islem Rekik   Imperial College London, UK
Guoyan Zheng   Shanghai Jiao Tong University, China

Clinical Chairs

Curtis Langlotz   Stanford University, USA
Charles Kahn   University of Pennsylvania, USA
Masaru Ishii   Johns Hopkins University, USA

Local Arrangements Chairs

Purang Abolmaesumi   University of British Columbia, Canada
Mehdi Moradi   McMaster University, Canada

Keynote Chairs

William Wells   Harvard University, USA
Alejandro Frangi   University of Manchester, UK

Industrial Sponsorship Chairs

Mohammad Yaqub   MBZ University of Artificial Intelligence,
                                        Abu Dhabi
Le Lu   DAMO Academy, Alibaba Group, USA
Yanwu Xu   Baidu, China

Communication Chair

Ehsan Adeli   Stanford University, USA

Publication Chairs

Ron Summers   National Institutes of Health, USA
Kevin Zhou   University of Science and Technology of China,
                               China

Young Investigator Chairs

Marius Linguraru   Children's National Institute, USA
Archana Venkataraman   Boston University, USA
Antonio Porras   University of Colorado Anschutz Medical
                               Campus, USA

Student Activities Chairs

Nathaniel Braman   Picture Health, USA
Vaishnavi Subramanian   EPFL, France

Women in MICCAI Chairs

Jayanthi Sivaswamy   IIIT, Hyderabad, India
Xiaoxiao Li   University of British Columbia, Canada

RISE Committee Chairs

Islem Rekik   Imperial College London, UK
Pingkun Yan   Rensselaer Polytechnic Institute, USA
Andrea Lara   Universidad Galileo, Guatemala

Submission Platform Manager

Kitty Wong   The MICCAI Society, Canada

Virtual Platform Manager

John Baxter   INSERM, Université de Rennes 1, France

Cross-Committee Liaison

Niharika D'Souza   IBM Research, USA

Program Committee

Sahar Ahmad   University of North Carolina at Chapel Hill, USA
Shadi Albarqouni   University of Bonn and Helmholtz Munich,
                                           Germany
Angelica Aviles-Rivero   University of Cambridge, UK
Shekoofeh Azizi   Google, Google Brain, USA
Ulas Bagci   Northwestern University, USA
Wenjia Bai   Imperial College London, UK
Sophia Bano   University College London, UK
Kayhan Batmanghelich   University of Pittsburgh and Boston University,
                                           USA
Ismail Ben Ayed   ETS Montreal, Canada
Katharina Breininger   Friedrich-Alexander-Universität
                                           Erlangen-Nürnberg, Germany
Weidong Cai   University of Sydney, Australia
Geng Chen   Northwestern Polytechnical University, China
Hao Chen   Hong Kong University of Science and
                                           Technology, China
Jun Cheng   Institute for Infocomm Research, A*STAR,
                                           Singapore
Li Cheng   University of Alberta, Canada
Albert C. S. Chung   University of Exeter, UK
Toby Collins   Ircad, France
Adrian Dalca   Massachusetts Institute of Technology and
                                           Harvard Medical School, USA
Jose Dolz   ETS Montreal, Canada
Qi Dou   Chinese University of Hong Kong, China
Nicha Dvornek   Yale University, USA
Shireen Elhabian   University of Utah, USA
Sandy Engelhardt   Heidelberg University Hospital, Germany
Ruogu Fang   University of Florida, USA

Aasa Feragen   Technical University of Denmark, Denmark
Moti Freiman   Technion - Israel Institute of Technology, Israel
Huazhu Fu   IHPC, A*STAR, Singapore
Adrian Galdran   Universitat Pompeu Fabra, Barcelona, Spain
Zhifan Gao   Sun Yat-sen University, China
Zongyuan Ge   Monash University, Australia
Stamatia Giannarou   Imperial College London, UK
Yun Gu   Shanghai Jiao Tong University, China
Hu Han   Institute of Computing Technology, Chinese
                                Academy of Sciences, China
Daniel Hashimoto   University of Pennsylvania, USA
Mattias Heinrich   University of Lübeck, Germany
Heng Huang   University of Pittsburgh, USA
Yuankai Huo   Vanderbilt University, USA
Mobarakol Islam   University College London, UK
Jayender Jagadeesan   Harvard Medical School, USA
Won-Ki Jeong   Korea University, South Korea
Xi Jiang   University of Electronic Science and Technology
                                of China, China
Yueming Jin   National University of Singapore, Singapore
Anand Joshi   University of Southern California, USA
Shantanu Joshi   UCLA, USA
Leo Joskowicz   Hebrew University of Jerusalem, Israel
Samuel Kadoury   Polytechnique Montreal, Canada
Bernhard Kainz   Friedrich-Alexander-Universität
                                Erlangen-Nürnberg, Germany and Imperial
                                College London, UK
Davood Karimi   Harvard University, USA
Anees Kazi   Massachusetts General Hospital, USA
Marta Kersten-Oertel   Concordia University, Canada
Fahmi Khalifa   Mansoura University, Egypt
Minjeong Kim   University of North Carolina, Greensboro, USA
Seong Tae Kim   Kyung Hee University, South Korea
Pavitra Krishnaswamy   Institute for Infocomm Research, Agency for
                                Science Technology and Research (A*STAR),
                                Singapore
Jin Tae Kwak   Korea University, South Korea
Baiying Lei   Shenzhen University, China
Xiang Li   Massachusetts General Hospital, USA
Xiaoxiao Li   University of British Columbia, Canada
Yuexiang Li   Tencent Jarvis Lab, China
Chunfeng Lian   Xi'an Jiaotong University, China

Jianming Liang	Arizona State University, USA
Jianfei Liu	National Institutes of Health Clinical Center, USA
Mingxia Liu	University of North Carolina at Chapel Hill, USA
Xiaofeng Liu	Harvard Medical School and MGH, USA
Herve Lombaert	École de technologie supérieure, Canada
Ismini Lourentzou	Virginia Tech, USA
Le Lu	Damo Academy USA, Alibaba Group, USA
Dwarikanath Mahapatra	Inception Institute of Artificial Intelligence, United Arab Emirates
Saad Nadeem	Memorial Sloan Kettering Cancer Center, USA
Dong Nie	Alibaba (US), USA
Yoshito Otake	Nara Institute of Science and Technology, Japan
Sang Hyun Park	Daegu Gyeongbuk Institute of Science and Technology, South Korea
Magdalini Paschali	Stanford University, USA
Tingying Peng	Helmholtz Munich, Germany
Caroline Petitjean	LITIS Université de Rouen Normandie, France
Esther Puyol Anton	King's College London, UK
Chen Qin	Imperial College London, UK
Daniel Racoceanu	Sorbonne Université, France
Hedyeh Rafii-Tari	Auris Health, USA
Hongliang Ren	Chinese University of Hong Kong, China and National University of Singapore, Singapore
Tammy Riklin Raviv	Ben-Gurion University, Israel
Hassan Rivaz	Concordia University, Canada
Mirabela Rusu	Stanford University, USA
Thomas Schultz	University of Bonn, Germany
Feng Shi	Shanghai United Imaging Intelligence, China
Yang Song	University of New South Wales, Australia
Aristeidis Sotiras	Washington University in St. Louis, USA
Rachel Sparks	King's College London, UK
Yao Sui	Peking University, China
Kenji Suzuki	Tokyo Institute of Technology, Japan
Qian Tao	Delft University of Technology, Netherlands
Mathias Unberath	Johns Hopkins University, USA
Martin Urschler	Medical University Graz, Austria
Maria Vakalopoulou	CentraleSupelec, University Paris Saclay, France
Erdem Varol	New York University, USA
Francisco Vasconcelos	University College London, UK
Harini Veeraraghavan	Memorial Sloan Kettering Cancer Center, USA
Satish Viswanath	Case Western Reserve University, USA
Christian Wachinger	Technical University of Munich, Germany

Hua Wang   Colorado School of Mines, USA
Qian Wang   ShanghaiTech University, China
Shanshan Wang   Paul C. Lauterbur Research Center, SIAT, China
Yalin Wang   Arizona State University, USA
Bryan Williams   Lancaster University, UK
Matthias Wilms   University of Calgary, Canada
Jelmer Wolterink   University of Twente, Netherlands
Ken C. L. Wong   IBM Research Almaden, USA
Jonghye Woo   Massachusetts General Hospital and Harvard
                            Medical School, USA
Shandong Wu   University of Pittsburgh, USA
Yutong Xie   University of Adelaide, Australia
Fuyong Xing   University of Colorado, Denver, USA
Daguang Xu   NVIDIA, USA
Yan Xu   Beihang University, China
Yanwu Xu   Baidu, China
Pingkun Yan   Rensselaer Polytechnic Institute, USA
Guang Yang   Imperial College London, UK
Jianhua Yao   Tencent, China
Chuyang Ye   Beijing Institute of Technology, China
Lequan Yu   University of Hong Kong, China
Ghada Zamzmi   National Institutes of Health, USA
Liang Zhan   University of Pittsburgh, USA
Fan Zhang   Harvard Medical School, USA
Ling Zhang   Alibaba Group, China
Miaomiao Zhang   University of Virginia, USA
Shu Zhang   Northwestern Polytechnical University, China
Rongchang Zhao   Central South University, China
Yitian Zhao   Chinese Academy of Sciences, China
Tao Zhou   Nanjing University of Science and Technology,
                            USA
Yuyin Zhou   UC Santa Cruz, USA
Dajiang Zhu   University of Texas at Arlington, USA
Lei Zhu   ROAS Thrust HKUST (GZ), and ECE HKUST,
                            China
Xiahai Zhuang   Fudan University, China
Veronika Zimmer   Technical University of Munich, Germany

Reviewers

Alaa Eldin Abdelaal
John Abel
Kumar Abhishek
Shahira Abousamra
Mazdak Abulnaga
Burak Acar
Abdoljalil Addeh
Ehsan Adeli
Sukesh Adiga Vasudeva
Seyed-Ahmad Ahmadi
Euijoon Ahn
Faranak Akbarifar
Alireza Akhondi-asl
Saad Ullah Akram
Daniel Alexander
Hanan Alghamdi
Hassan Alhajj
Omar Al-Kadi
Max Allan
Andre Altmann
Pablo Alvarez
Charlems Alvarez-Jimenez
Jennifer Alvén
Lidia Al-Zogbi
Kimberly Amador
Tamaz Amiranashvili
Amine Amyar
Wangpeng An
Vincent Andrearczyk
Manon Ansart
Sameer Antani
Jacob Antunes
Michel Antunes
Guilherme Aresta
Mohammad Ali Armin
Kasra Arnavaz
Corey Arnold
Janan Arslan
Marius Arvinte
Muhammad Asad
John Ashburner
Md Ashikuzzaman
Shahab Aslani

Mehdi Astaraki
Angélica Atehortúa
Benjamin Aubert
Marc Aubreville
Paolo Avesani
Sana Ayromlou
Reza Azad
Mohammad Farid
  Azampour
Qinle Ba
Meritxell Bach Cuadra
Hyeon-Min Bae
Matheus Baffa
Cagla Bahadir
Fan Bai
Jun Bai
Long Bai
Pradeep Bajracharya
Shafa Balaram
Yaël Balbastre
Yutong Ban
Abhirup Banerjee
Soumyanil Banerjee
Sreya Banerjee
Shunxing Bao
Omri Bar
Adrian Barbu
Joao Barreto
Adrian Basarab
Berke Basaran
Michael Baumgartner
Siming Bayer
Roza Bayrak
Aicha BenTaieb
Guy Ben-Yosef
Sutanu Bera
Cosmin Bercea
Jorge Bernal
Jose Bernal
Gabriel Bernardino
Riddhish Bhalodia
Jignesh Bhatt
Indrani Bhattacharya

Binod Bhattarai
Lei Bi
Qi Bi
Cheng Bian
Gui-Bin Bian
Carlo Biffi
Alexander Bigalke
Benjamin Billot
Manuel Birlo
Ryoma Bise
Daniel Blezek
Stefano Blumberg
Sebastian Bodenstedt
Federico Bolelli
Bhushan Borotikar
Ilaria Boscolo Galazzo
Alexandre Bousse
Nicolas Boutry
Joseph Boyd
Behzad Bozorgtabar
Nadia Brancati
Clara Brémond Martin
Stéphanie Bricq
Christopher Bridge
Coleman Broaddus
Rupert Brooks
Tom Brosch
Mikael Brudfors
Ninon Burgos
Nikolay Burlutskiy
Michal Byra
Ryan Cabeen
Mariano Cabezas
Hongmin Cai
Tongan Cai
Zongyou Cai
Liane Canas
Bing Cao
Guogang Cao
Weiguo Cao
Xu Cao
Yankun Cao
Zhenjie Cao

Jaime Cardoso
M. Jorge Cardoso
Owen Carmichael
Jacob Carse
Adrià Casamitjana
Alessandro Casella
Angela Castillo
Kate Cevora
Krishna Chaitanya
Satrajit Chakrabarty
Yi Hao Chan
Shekhar Chandra
Ming-Ching Chang
Peng Chang
Qi Chang
Yuchou Chang
Hanqing Chao
Simon Chatelin
Soumick Chatterjee
Sudhanya Chatterjee
Muhammad Faizyab Ali
    Chaudhary
Antong Chen
Bingzhi Chen
Chen Chen
Cheng Chen
Chengkuan Chen
Eric Chen
Fang Chen
Haomin Chen
Jianan Chen
Jianxu Chen
Jiazhou Chen
Jie Chen
Jintai Chen
Jun Chen
Junxiang Chen
Junyu Chen
Li Chen
Liyun Chen
Nenglun Chen
Pingjun Chen
Pingyi Chen
Qi Chen
Qiang Chen

Runnan Chen
Shengcong Chen
Sihao Chen
Tingting Chen
Wenting Chen
Xi Chen
Xiang Chen
Xiaoran Chen
Xin Chen
Xiongchao Chen
Yanxi Chen
Yixiong Chen
Yixuan Chen
Yuanyuan Chen
Yuqian Chen
Zhaolin Chen
Zhen Chen
Zhenghao Chen
Zhennong Chen
Zhihao Chen
Zhineng Chen
Zhixiang Chen
Chang-Chieh Cheng
Jiale Cheng
Jianhong Cheng
Jun Cheng
Xuelian Cheng
Yupeng Cheng
Mark Chiew
Philip Chikontwe
Eleni Chiou
Jungchan Cho
Jang-Hwan Choi
Min-Kook Choi
Wookjin Choi
Jaegul Choo
Yu-Cheng Chou
Daan Christiaens
Argyrios Christodoulidis
Stergios Christodoulidis
Kai-Cheng Chuang
Hyungjin Chung
Matthew Clarkson
Michaël Clément
Dana Cobzas

Jaume Coll-Font
Olivier Colliot
Runmin Cong
Yulai Cong
Laura Connolly
William Consagra
Pierre-Henri Conze
Tim Cootes
Teresa Correia
Baris Coskunuzer
Alex Crimi
Can Cui
Hejie Cui
Hui Cui
Lei Cui
Wenhui Cui
Tolga Cukur
Tobias Czempiel
Javid Dadashkarimi
Haixing Dai
Tingting Dan
Kang Dang
Salman Ul Hassan Dar
Eleonora D'Arnese
Dhritiman Das
Neda Davoudi
Tareen Dawood
Sandro De Zanet
Farah Deeba
Charles Delahunt
Herve Delingette
Ugur Demir
Liang-Jian Deng
Ruining Deng
Wenlong Deng
Felix Denzinger
Adrien Depeursinge
Mohammad Mahdi
    Derakhshani
Hrishikesh Deshpande
Adrien Desjardins
Christian Desrosiers
Blake Dewey
Neel Dey
Rohan Dhamdhere

Maxime Di Folco
Songhui Diao
Alina Dima
Hao Ding
Li Ding
Ying Ding
Zhipeng Ding
Nicola Dinsdale
Konstantin Dmitriev
Ines Domingues
Bo Dong
Liang Dong
Nanqing Dong
Siyuan Dong
Reuben Dorent
Gianfranco Doretto
Sven Dorkenwald
Haoran Dou
Mitchell Doughty
Jason Dowling
Niharika D'Souza
Guodong Du
Jie Du
Shiyi Du
Hongyi Duanmu
Benoit Dufumier
James Duncan
Joshua Durso-Finley
Dmitry V. Dylov
Oleh Dzyubachyk
Mahdi (Elias) Ebnali
Philip Edwards
Jan Egger
Gudmundur Einarsson
Mostafa El Habib Daho
Ahmed Elazab
Idris El-Feghi
David Ellis
Mohammed Elmogy
Amr Elsawy
Okyaz Eminaga
Ertunc Erdil
Lauren Erdman
Marius Erdt
Maria Escobar

Hooman Esfandiari
Nazila Esmaeili
Ivan Ezhov
Alessio Fagioli
Deng-Ping Fan
Lei Fan
Xin Fan
Yubo Fan
Huihui Fang
Jiansheng Fang
Xi Fang
Zhenghan Fang
Mohammad Farazi
Azade Farshad
Mohsen Farzi
Hamid Fehri
Lina Felsner
Chaolu Feng
Chun-Mei Feng
Jianjiang Feng
Mengling Feng
Ruibin Feng
Zishun Feng
Alvaro Fernandez-Quilez
Ricardo Ferrari
Lucas Fidon
Lukas Fischer
Madalina Fiterau
Antonio
        Foncubierta-Rodríguez
Fahimeh Fooladgar
Germain Forestier
Nils Daniel Forkert
Jean-Rassaire Fouefack
Kevin François-Bouaou
Wolfgang Freysinger
Bianca Freytag
Guanghui Fu
Kexue Fu
Lan Fu
Yunguan Fu
Pedro Furtado
Ryo Furukawa
Jin Kyu Gahm
Mélanie Gaillochet

Francesca Galassi
Jiangzhang Gan
Yu Gan
Yulu Gan
Alireza Ganjdanesh
Chang Gao
Cong Gao
Linlin Gao
Zeyu Gao
Zhongpai Gao
Sara Garbarino
Alain Garcia
Beatriz Garcia Santa Cruz
Rongjun Ge
Shiv Gehlot
Manuela Geiss
Salah Ghamizi
Negin Ghamsarian
Ramtin Gharleghi
Ghazal Ghazaei
Florin Ghesu
Sayan Ghosal
Syed Zulqarnain Gilani
Mahdi Gilany
Yannik Glaser
Ben Glocker
Bharti Goel
Jacob Goldberger
Polina Golland
Alberto Gomez
Catalina Gomez
Estibaliz
        Gómez-de-Mariscal
Haifan Gong
Kuang Gong
Xun Gong
Ricardo Gonzales
Camila Gonzalez
German Gonzalez
Vanessa Gonzalez Duque
Sharath Gopal
Karthik Gopinath
Pietro Gori
Michael Götz
Shuiping Gou

Maged Goubran
Sobhan Goudarzi
Mark Graham
Alejandro Granados
Mara Graziani
Thomas Grenier
Radu Grosu
Michal Grzeszczyk
Feng Gu
Pengfei Gu
Qiangqiang Gu
Ran Gu
Shi Gu
Wenhao Gu
Xianfeng Gu
Yiwen Gu
Zaiwang Gu
Hao Guan
Jayavardhana Gubbi
Houssem-Eddine Gueziri
Dazhou Guo
Hengtao Guo
Jixiang Guo
Jun Guo
Pengfei Guo
Wenzhangzhi Guo
Xiaoqing Guo
Xueqi Guo
Yi Guo
Vikash Gupta
Praveen Gurunath Bharathi
Prashnna Gyawali
Sung Min Ha
Mohamad Habes
Ilker Hacihaliloglu
Stathis Hadjidemetriou
Fatemeh Haghighi
Justin Haldar
Noura Hamze
Liang Han
Luyi Han
Seungjae Han
Tianyu Han
Zhongyi Han
Jonny Hancox

Lasse Hansen
Degan Hao
Huaying Hao
Jinkui Hao
Nazim Haouchine
Michael Hardisty
Stefan Harrer
Jeffry Hartanto
Charles Hatt
Huiguang He
Kelei He
Qi He
Shenghua He
Xinwei He
Stefan Heldmann
Nicholas Heller
Edward Henderson
Alessa Hering
Monica Hernandez
Kilian Hett
Amogh Hiremath
David Ho
Malte Hoffmann
Matthew Holden
Qingqi Hong
Yoonmi Hong
Mohammad Reza
    Hosseinzadeh Taher
William Hsu
Chuanfei Hu
Dan Hu
Kai Hu
Rongyao Hu
Shishuai Hu
Xiaoling Hu
Xinrong Hu
Yan Hu
Yang Hu
Chaoqin Huang
Junzhou Huang
Ling Huang
Luojie Huang
Qinwen Huang
Sharon Xiaolei Huang
Weijian Huang

Xiaoyang Huang
Yi-Jie Huang
Yongsong Huang
Yongxiang Huang
Yuhao Huang
Zhe Huang
Zhi-An Huang
Ziyi Huang
Arnaud Huaulmé
Henkjan Huisman
Alex Hung
Jiayu Huo
Andreas Husch
Mohammad Arafat
    Hussain
Sarfaraz Hussein
Jana Hutter
Khoi Huynh
Ilknur Icke
Kay Igwe
Abdullah Al Zubaer Imran
Muhammad Imran
Samra Irshad
Nahid Ul Islam
Koichi Ito
Hayato Itoh
Yuji Iwahori
Krithika Iyer
Mohammad Jafari
Srikrishna Jaganathan
Hassan Jahanandish
Andras Jakab
Amir Jamaludin
Amoon Jamzad
Ananya Jana
Se-In Jang
Pierre Jannin
Vincent Jaouen
Uditha Jarayathne
Ronnachai Jaroensri
Guillaume Jaume
Syed Ashar Javed
Rachid Jennane
Debesh Jha
Ge-Peng Ji

Luping Ji
Zexuan Ji
Zhanghexuan Ji
Haozhe Jia
Hongchao Jiang
Jue Jiang
Meirui Jiang
Tingting Jiang
Xiajun Jiang
Zekun Jiang
Zhifan Jiang
Ziyu Jiang
Jianbo Jiao
Zhicheng Jiao
Chen Jin
Dakai Jin
Qiangguo Jin
Qiuye Jin
Weina Jin
Baoyu Jing
Bin Jing
Yaqub Jonmohamadi
Lie Ju
Yohan Jun
Dinkar Juyal
Manjunath K N
Ali Kafaei Zad Tehrani
John Kalafut
Niveditha Kalavakonda
Megha Kalia
Anil Kamat
Qingbo Kang
Po-Yu Kao
Anuradha Kar
Neerav Karani
Turkay Kart
Satyananda Kashyap
Alexander Katzmann
Lisa Kausch
Maxime Kayser
Salome Kazeminia
Wenchi Ke
Youngwook Kee
Matthias Keicher
Erwan Kerrien

Afifa Khaled
Nadieh Khalili
Farzad Khalvati
Bidur Khanal
Bishesh Khanal
Pulkit Khandelwal
Maksim Kholiavchenko
Ron Kikinis
Benjamin Killeen
Daeseung Kim
Heejong Kim
Jaeil Kim
Jinhee Kim
Jinman Kim
Junsik Kim
Minkyung Kim
Namkug Kim
Sangwook Kim
Tae Soo Kim
Younghoon Kim
Young-Min Kim
Andrew King
Miranda Kirby
Gabriel Kiss
Andreas Kist
Yoshiro Kitamura
Stefan Klein
Tobias Klinder
Kazuma Kobayashi
Lisa Koch
Satoshi Kondo
Fanwei Kong
Tomasz Konopczynski
Ender Konukoglu
Aishik Konwer
Thijs Kooi
Ivica Kopriva
Avinash Kori
Kivanc Kose
Suraj Kothawade
Anna Kreshuk
AnithaPriya Krishnan
Florian Kromp
Frithjof Kruggel
Thomas Kuestner

Levin Kuhlmann
Abhay Kumar
Kuldeep Kumar
Sayantan Kumar
Manuela Kunz
Holger Kunze
Tahsin Kurc
Anvar Kurmukov
Yoshihiro Kuroda
Yusuke Kurose
Hyuksool Kwon
Aymen Laadhari
Jorma Laaksonen
Dmitrii Lachinov
Alain Lalande
Rodney LaLonde
Bennett Landman
Daniel Lang
Carole Lartizien
Shlomi Laufer
Max-Heinrich Laves
William Le
Loic Le Folgoc
Christian Ledig
Eung-Joo Lee
Ho Hin Lee
Hyekyoung Lee
John Lee
Kisuk Lee
Kyungsu Lee
Soochahn Lee
Woonghee Lee
Étienne Léger
Wen Hui Lei
Yiming Lei
George Leifman
Rogers Jeffrey Leo John
Juan Leon
Bo Li
Caizi Li
Chao Li
Chen Li
Cheng Li
Chenxin Li
Chnegyin Li

Dawei Li
Fuhai Li
Gang Li
Guang Li
Hao Li
Haofeng Li
Haojia Li
Heng Li
Hongming Li
Hongwei Li
Huiqi Li
Jian Li
Jieyu Li
Kang Li
Lin Li
Mengzhang Li
Ming Li
Qing Li
Quanzheng Li
Shaohua Li
Shulong Li
Tengfei Li
Weijian Li
Wen Li
Xiaomeng Li
Xingyu Li
Xinhui Li
Xuelu Li
Xueshen Li
Yamin Li
Yang Li
Yi Li
Yuemeng Li
Yunxiang Li
Zeju Li
Zhaoshuo Li
Zhe Li
Zhen Li
Zhenqiang Li
Zhiyuan Li
Zhjin Li
Zi Li
Hao Liang
Libin Liang
Peixian Liang

Yuan Liang
Yudong Liang
Haofu Liao
Hongen Liao
Wei Liao
Zehui Liao
Gilbert Lim
Hongxiang Lin
Li Lin
Manxi Lin
Mingquan Lin
Tiancheng Lin
Yi Lin
Zudi Lin
Claudia Lindner
Simone Lionetti
Chi Liu
Chuanbin Liu
Daochang Liu
Dongnan Liu
Feihong Liu
Fenglin Liu
Han Liu
Huiye Liu
Jiang Liu
Jie Liu
Jinduo Liu
Jing Liu
Jingya Liu
Jundong Liu
Lihao Liu
Mengting Liu
Mingyuan Liu
Peirong Liu
Peng Liu
Qin Liu
Quan Liu
Rui Liu
Shengfeng Liu
Shuangjun Liu
Sidong Liu
Siyuan Liu
Weide Liu
Xiao Liu
Xiaoyu Liu

Xingtong Liu
Xinwen Liu
Xinyang Liu
Xinyu Liu
Yan Liu
Yi Liu
Yihao Liu
Yikang Liu
Yilin Liu
Yilong Liu
Yiqiao Liu
Yong Liu
Yuhang Liu
Zelong Liu
Zhe Liu
Zhiyuan Liu
Zuozhu Liu
Lisette Lockhart
Andrea Loddo
Nicolas Loménie
Yonghao Long
Daniel Lopes
Ange Lou
Brian Lovell
Nicolas Loy Rodas
Charles Lu
Chun-Shien Lu
Donghuan Lu
Guangming Lu
Huanxiang Lu
Jingpei Lu
Yao Lu
Oeslle Lucena
Jie Luo
Luyang Luo
Ma Luo
Mingyuan Luo
Wenhan Luo
Xiangde Luo
Xinzhe Luo
Jinxin Lv
Tianxu Lv
Fei Lyu
Ilwoo Lyu
Mengye Lyu

Qing Lyu
Yanjun Lyu
Yuanyuan Lyu
Benteng Ma
Chunwei Ma
Hehuan Ma
Jun Ma
Junbo Ma
Wenao Ma
Yuhui Ma
Pedro Macias Gordaliza
Anant Madabhushi
Derek Magee
S. Sara Mahdavi
Andreas Maier
Klaus H. Maier-Hein
Sokratis Makrogiannis
Danial Maleki
Michail Mamalakis
Zhehua Mao
Jan Margeta
Brett Marinelli
Zdravko Marinov
Viktoria Markova
Carsten Marr
Yassine Marrakchi
Anne Martel
Martin Maška
Tejas Sudharshan Mathai
Petr Matula
Dimitrios Mavroeidis
Evangelos Mazomenos
Amarachi Mbakwe
Adam McCarthy
Stephen McKenna
Raghav Mehta
Xueyan Mei
Felix Meissen
Felix Meister
Afaque Memon
Mingyuan Meng
Qingjie Meng
Xiangzhu Meng
Yanda Meng
Zhu Meng

Martin Menten
Odyssée Merveille
Mikhail Milchenko
Leo Milecki
Fausto Milletari
Hyun-Seok Min
Zhe Min
Song Ming
Duy Minh Ho Nguyen
Deepak Mishra
Suraj Mishra
Virendra Mishra
Tadashi Miyamoto
Sara Moccia
Marc Modat
Omid Mohareri
Tony C. W. Mok
Javier Montoya
Rodrigo Moreno
Stefano Moriconi
Lia Morra
Ana Mota
Lei Mou
Dana Moukheiber
Lama Moukheiber
Daniel Moyer
Pritam Mukherjee
Anirban Mukhopadhyay
Henning Müller
Ana Murillo
Gowtham Krishnan
    Murugesan
Ahmed Naglah
Karthik Nandakumar
Venkatesh
    Narasimhamurthy
Raja Narayan
Dominik Narnhofer
Vishwesh Nath
Rodrigo Nava
Abdullah Nazib
Ahmed Nebli
Peter Neher
Amin Nejatbakhsh
Trong-Thuan Nguyen

Truong Nguyen
Dong Ni
Haomiao Ni
Xiuyan Ni
Hannes Nickisch
Weizhi Nie
Aditya Nigam
Lipeng Ning
Xia Ning
Kazuya Nishimura
Chuang Niu
Sijie Niu
Vincent Noblet
Narges Norouzi
Alexey Novikov
Jorge Novo
Gilberto Ochoa-Ruiz
Masahiro Oda
Benjamin Odry
Hugo Oliveira
Sara Oliveira
Arnau Oliver
Jimena Olveres
John Onofrey
Marcos Ortega
Mauricio Alberto
    Ortega-Ruíz
Yusuf Osmanlioglu
Chubin Ou
Cheng Ouyang
Jiahong Ouyang
Xi Ouyang
Cristina Oyarzun Laura
Utku Ozbulak
Ece Ozkan
Ege Özsoy
Batu Ozturkler
Harshith Padigela
Johannes Paetzold
José Blas Pagador
    Carrasco
Daniel Pak
Sourabh Palande
Chengwei Pan
Jiazhen Pan

Jin Pan
Yongsheng Pan
Egor Panfilov
Jiaxuan Pang
Joao Papa
Constantin Pape
Bartlomiej Papiez
Nripesh Parajuli
Hyunjin Park
Akash Parvatikar
Tiziano Passerini
Diego Patiño Cortés
Mayank Patwari
Angshuman Paul
Rasmus Paulsen
Yuchen Pei
Yuru Pei
Tao Peng
Wei Peng
Yige Peng
Yunsong Peng
Matteo Pennisi
Antonio Pepe
Oscar Perdomo
Sérgio Pereira
Jose-Antonio
    Pérez-Carrasco
Mehran Pesteie
Terry Peters
Eike Petersen
Jens Petersen
Micha Pfeiffer
Dzung Pham
Hieu Pham
Ashish Phophalia
Tomasz Pieciak
Antonio Pinheiro
Pramod Pisharady
Theodoros Pissas
Szymon Płotka
Kilian Pohl
Sebastian Pölsterl
Alison Pouch
Tim Prangemeier
Prateek Prasanna

Raphael Prevost
Juan Prieto
Federica Proietto Salanitri
Sergi Pujades
Elodie Puybareau
Talha Qaiser
Buyue Qian
Mengyun Qiao
Yuchuan Qiao
Zhi Qiao
Chenchen Qin
Fangbo Qin
Wenjian Qin
Yulei Qin
Jie Qiu
Jielin Qiu
Peijie Qiu
Shi Qiu
Wu Qiu
Liangqiong Qu
Linhao Qu
Quan Quan
Tran Minh Quan
Sandro Queirós
Prashanth R
Febrian Rachmadi
Daniel Racoceanu
Mehdi Rahim
Jagath Rajapakse
Kashif Rajpoot
Keerthi Ram
Dhanesh Ramachandram
João Ramalhinho
Xuming Ran
Aneesh Rangnekar
Hatem Rashwan
Keerthi Sravan Ravi
Daniele Ravì
Sadhana Ravikumar
Harish Raviprakash
Surreerat Reaungamornrat
Samuel Remedios
Mengwei Ren
Sucheng Ren
Elton Rexhepaj

Mauricio Reyes
Constantino
    Reyes-Aldasoro
Abel Reyes-Angulo
Hadrien Reynaud
Razieh Rezaei
Anne-Marie Rickmann
Laurent Risser
Dominik Rivoir
Emma Robinson
Robert Robinson
Jessica Rodgers
Ranga Rodrigo
Rafael Rodrigues
Robert Rohling
Margherita Rosnati
Łukasz Roszkowiak
Holger Roth
José Rouco
Dan Ruan
Jiacheng Ruan
Daniel Rueckert
Danny Ruijters
Kanghyun Ryu
Ario Sadafi
Numan Saeed
Monjoy Saha
Pramit Saha
Farhang Sahba
Pranjal Sahu
Simone Saitta
Md Sirajus Salekin
Abbas Samani
Pedro Sanchez
Luis Sanchez Giraldo
Yudi Sang
Gerard Sanroma-Guell
Rodrigo Santa Cruz
Alice Santilli
Rachana Sathish
Olivier Saut
Mattia Savardi
Nico Scherf
Alexander Schlaefer
Jerome Schmid

Adam Schmidt
Julia Schnabel
Lawrence Schobs
Julian Schön
Peter Schueffler
Andreas Schuh
Christina
    Schwarz-Gsaxner
Michaël Sdika
Suman Sedai
Lalithkumar Seenivasan
Matthias Seibold
Sourya Sengupta
Lama Seoud
Ana Sequeira
Sharmishtaa Seshamani
Ahmed Shaffie
Jay Shah
Keyur Shah
Ahmed Shahin
Mohammad Abuzar
    Shaikh
S. Shailja
Hongming Shan
Wei Shao
Mostafa Sharifzadeh
Anuja Sharma
Gregory Sharp
Hailan Shen
Li Shen
Linlin Shen
Mali Shen
Mingren Shen
Yiqing Shen
Zhengyang Shen
Jun Shi
Xiaoshuang Shi
Yiyu Shi
Yonggang Shi
Hoo-Chang Shin
Jitae Shin
Keewon Shin
Boris Shirokikh
Suzanne Shontz
Yucheng Shu

Hanna Siebert
Alberto Signoroni
Wilson Silva
Julio Silva-Rodríguez
Margarida Silveira
Walter Simson
Praveer Singh
Vivek Singh
Nitin Singhal
Elena Sizikova
Gregory Slabaugh
Dane Smith
Kevin Smith
Tiffany So
Rajath Soans
Roger Soberanis-Mukul
Hessam Sokooti
Jingwei Song
Weinan Song
Xinhang Song
Xinrui Song
Mazen Soufi
Georgia Sovatzidi
Bella Specktor Fadida
William Speier
Ziga Spiclin
Dominik Spinczyk
Jon Sporring
Pradeeba Sridar
Chetan L. Srinidhi
Abhishek Srivastava
Lawrence Staib
Marc Stamminger
Justin Strait
Hai Su
Ruisheng Su
Zhe Su
Vaishnavi Subramanian
Gérard Subsol
Carole Sudre
Dong Sui
Heung-Il Suk
Shipra Suman
He Sun
Hongfu Sun

Jian Sun
Li Sun
Liyan Sun
Shanlin Sun
Kyung Sung
Yannick Suter
Swapna T. R.
Amir Tahmasebi
Pablo Tahoces
Sirine Taleb
Bingyao Tan
Chaowei Tan
Wenjun Tan
Hao Tang
Siyi Tang
Xiaoying Tang
Yucheng Tang
Zihao Tang
Michael Tanzer
Austin Tapp
Elias Tappeiner
Mickael Tardy
Giacomo Tarroni
Athena Taymourtash
Kaveri Thakoor
Elina Thibeau-Sutre
Paul Thienphrapa
Sarina Thomas
Stephen Thompson
Karl Thurnhofer-Hemsi
Cristiana Tiago
Lin Tian
Lixia Tian
Yapeng Tian
Yu Tian
Yun Tian
Aleksei Tiulpin
Hamid Tizhoosh
Minh Nguyen Nhat To
Matthew Toews
Maryam Toloubidokhti
Minh Tran
Quoc-Huy Trinh
Jocelyne Troccaz
Roger Trullo

Chialing Tsai
Apostolia Tsirikoglou
Puxun Tu
Samyakh Tukra
Sudhakar Tummala
Georgios Tziritas
Vladimír Ulman
Tamas Ungi
Régis Vaillant
Jeya Maria Jose Valanarasu
Vanya Valindria
Juan Miguel Valverde
Fons van der Sommen
Maureen van Eijnatten
Tom van Sonsbeek
Gijs van Tulder
Yogatheesan Varatharajah
Madhurima Vardhan
Thomas Varsavsky
Hooman Vaseli
Serge Vasylechko
S. Swaroop Vedula
Sanketh Vedula
Gonzalo Vegas
    Sanchez-Ferrero
Matthew Velazquez
Archana Venkataraman
Sulaiman Vesal
Mitko Veta
Barbara Villarini
Athanasios Vlontzos
Wolf-Dieter Vogl
Ingmar Voigt
Sandrine Voros
Vibashan VS
Trinh Thi Le Vuong
An Wang
Bo Wang
Ce Wang
Changmiao Wang
Ching-Wei Wang
Dadong Wang
Dong Wang
Fakai Wang
Guotai Wang

Haifeng Wang
Haoran Wang
Hong Wang
Hongxiao Wang
Hongyu Wang
Jiacheng Wang
Jing Wang
Jue Wang
Kang Wang
Ke Wang
Lei Wang
Li Wang
Liansheng Wang
Lin Wang
Ling Wang
Linwei Wang
Manning Wang
Mingliang Wang
Puyang Wang
Qiuli Wang
Renzhen Wang
Ruixuan Wang
Shaoyu Wang
Sheng Wang
Shujun Wang
Shuo Wang
Shuqiang Wang
Tao Wang
Tianchen Wang
Tianyu Wang
Wenzhe Wang
Xi Wang
Xiangdong Wang
Xiaoqing Wang
Xiaosong Wang
Yan Wang
Yangang Wang
Yaping Wang
Yi Wang
Yirui Wang
Yixin Wang
Zeyi Wang
Zhao Wang
Zichen Wang
Ziqin Wang

Ziyi Wang
Zuhui Wang
Dong Wei
Donglai Wei
Hao Wei
Jia Wei
Leihao Wei
Ruofeng Wei
Shuwen Wei
Martin Weigert
Wolfgang Wein
Michael Wels
Cédric Wemmert
Thomas Wendler
Markus Wenzel
Rhydian Windsor
Adam Wittek
Marek Wodzinski
Ivo Wolf
Julia Wolleb
Ka-Chun Wong
Jonghye Woo
Chongruo Wu
Chunpeng Wu
Fuping Wu
Huaqian Wu
Ji Wu
Jiangjie Wu
Jiong Wu
Junde Wu
Linshan Wu
Qing Wu
Weiwen Wu
Wenjun Wu
Xiyin Wu
Yawen Wu
Ye Wu
Yicheng Wu
Yongfei Wu
Zhengwang Wu
Pengcheng Xi
Chao Xia
Siyu Xia
Wenjun Xia
Lei Xiang

Tiange Xiang
Deqiang Xiao
Li Xiao
Xiaojiao Xiao
Yiming Xiao
Zeyu Xiao
Hongtao Xie
Huidong Xie
Jianyang Xie
Long Xie
Weidi Xie
Fangxu Xing
Shuwei Xing
Xiaodan Xing
Xiaohan Xing
Haoyi Xiong
Yujian Xiong
Di Xu
Feng Xu
Haozheng Xu
Hongming Xu
Jiangchang Xu
Jiaqi Xu
Junshen Xu
Kele Xu
Lijian Xu
Min Xu
Moucheng Xu
Rui Xu
Xiaowei Xu
Xuanang Xu
Yanwu Xu
Yanyu Xu
Yongchao Xu
Yunqiu Xu
Zhe Xu
Zhoubing Xu
Ziyue Xu
Kai Xuan
Cheng Xue
Jie Xue
Tengfei Xue
Wufeng Xue
Yuan Xue
Zhong Xue

Ts Faridah Yahya
Chaochao Yan
Jiangpeng Yan
Ming Yan
Qingsen Yan
Xiangyi Yan
Yuguang Yan
Zengqiang Yan
Baoyao Yang
Carl Yang
Changchun Yang
Chen Yang
Feng Yang
Fengting Yang
Ge Yang
Guanyu Yang
Heran Yang
Huijuan Yang
Jiancheng Yang
Jiewen Yang
Peng Yang
Qi Yang
Qiushi Yang
Wei Yang
Xin Yang
Xuan Yang
Yan Yang
Yanwu Yang
Yifan Yang
Yingyu Yang
Zhicheng Yang
Zhijian Yang
Jiangchao Yao
Jiawen Yao
Lanhong Yao
Linlin Yao
Qingsong Yao
Tianyuan Yao
Xiaohui Yao
Zhao Yao
Dong Hye Ye
Menglong Ye
Yousef Yeganeh
Jirong Yi
Xin Yi

Chong Yin
Pengshuai Yin
Yi Yin
Zhaozheng Yin
Chunwei Ying
Youngjin Yoo
Jihun Yoon
Chenyu You
Hanchao Yu
Heng Yu
Jinhua Yu
Jinze Yu
Ke Yu
Qi Yu
Qian Yu
Thomas Yu
Weimin Yu
Yang Yu
Chenxi Yuan
Kun Yuan
Wu Yuan
Yixuan Yuan
Paul Yushkevich
Fatemeh Zabihollahy
Samira Zare
Ramy Zeineldin
Dong Zeng
Qi Zeng
Tianyi Zeng
Wei Zeng
Kilian Zepf
Kun Zhan
Bokai Zhang
Daoqiang Zhang
Dong Zhang
Fa Zhang
Hang Zhang
Hanxiao Zhang
Hao Zhang
Haopeng Zhang
Haoyue Zhang
Hongrun Zhang
Jiadong Zhang
Jiajin Zhang
Jianpeng Zhang

Jiawei Zhang
Jingqing Zhang
Jingyang Zhang
Jinwei Zhang
Jiong Zhang
Jiping Zhang
Ke Zhang
Lefei Zhang
Lei Zhang
Li Zhang
Lichi Zhang
Lu Zhang
Minghui Zhang
Molin Zhang
Ning Zhang
Rongzhao Zhang
Ruipeng Zhang
Ruisi Zhang
Shichuan Zhang
Shihao Zhang
Shuai Zhang
Tuo Zhang
Wei Zhang
Weihang Zhang
Wen Zhang
Wenhua Zhang
Wenqiang Zhang
Xiaodan Zhang
Xiaoran Zhang
Xin Zhang
Xukun Zhang
Xuzhe Zhang
Ya Zhang
Yanbo Zhang
Yanfu Zhang
Yao Zhang
Yi Zhang
Yifan Zhang
Yixiao Zhang
Yongqin Zhang
You Zhang
Youshan Zhang

Yu Zhang
Yubo Zhang
Yue Zhang
Yuhan Zhang
Yulun Zhang
Yundong Zhang
Yunlong Zhang
Yuyao Zhang
Zheng Zhang
Zhenxi Zhang
Ziqi Zhang
Can Zhao
Chongyue Zhao
Fenqiang Zhao
Gangming Zhao
He Zhao
Jianfeng Zhao
Jun Zhao
Li Zhao
Liang Zhao
Lin Zhao
Mengliu Zhao
Mingbo Zhao
Qingyu Zhao
Shang Zhao
Shijie Zhao
Tengda Zhao
Tianyi Zhao
Wei Zhao
Yidong Zhao
Yiyuan Zhao
Yu Zhao
Zhihe Zhao
Ziyuan Zhao
Haiyong Zheng
Hao Zheng
Jiannan Zheng
Kang Zheng
Meng Zheng
Sisi Zheng
Tianshu Zheng
Yalin Zheng

Yefeng Zheng
Yinqiang Zheng
Yushan Zheng
Aoxiao Zhong
Jia-Xing Zhong
Tao Zhong
Zichun Zhong
Hong-Yu Zhou
Houliang Zhou
Huiyu Zhou
Kang Zhou
Qin Zhou
Ran Zhou
S. Kevin Zhou
Tianfei Zhou
Wei Zhou
Xiao-Hu Zhou
Xiao-Yun Zhou
Yi Zhou
Youjia Zhou
Yukun Zhou
Zongwei Zhou
Chenglu Zhu
Dongxiao Zhu
Heqin Zhu
Jiayi Zhu
Meilu Zhu
Wei Zhu
Wenhui Zhu
Xiaofeng Zhu
Xin Zhu
Yonghua Zhu
Yongpei Zhu
Yuemin Zhu
Yan Zhuang
David Zimmerer
Yongshuo Zong
Ke Zou
Yukai Zou
Lianrui Zuo
Gerald Zwettler

Outstanding Area Chairs

Mingxia Liu	University of North Carolina at Chapel Hill, USA
Matthias Wilms	University of Calgary, Canada
Veronika Zimmer	Technical University Munich, Germany

Outstanding Reviewers

Kimberly Amador	University of Calgary, Canada
Angela Castillo	Universidad de los Andes, Colombia
Chen Chen	Imperial College London, UK
Laura Connolly	Queen's University, Canada
Pierre-Henri Conze	IMT Atlantique, France
Niharika D'Souza	IBM Research, USA
Michael Götz	University Hospital Ulm, Germany
Meirui Jiang	Chinese University of Hong Kong, China
Manuela Kunz	National Research Council Canada, Canada
Zdravko Marinov	Karlsruhe Institute of Technology, Germany
Sérgio Pereira	Lunit, South Korea
Lalithkumar Seenivasan	National University of Singapore, Singapore

Honorable Mentions (Reviewers)

Kumar Abhishek	Simon Fraser University, Canada
Guilherme Aresta	Medical University of Vienna, Austria
Shahab Aslani	University College London, UK
Marc Aubreville	Technische Hochschule Ingolstadt, Germany
Yaël Balbastre	Massachusetts General Hospital, USA
Omri Bar	Theator, Israel
Aicha Ben Taieb	Simon Fraser University, Canada
Cosmin Bercea	Technical University Munich and Helmholtz AI and Helmholtz Center Munich, Germany
Benjamin Billot	Massachusetts Institute of Technology, USA
Michal Byra	RIKEN Center for Brain Science, Japan
Mariano Cabezas	University of Sydney, Australia
Alessandro Casella	Italian Institute of Technology and Politecnico di Milano, Italy
Junyu Chen	Johns Hopkins University, USA
Argyrios Christodoulidis	Pfizer, Greece
Olivier Colliot	CNRS, France

Lei Cui   Northwest University, China
Neel Dey   Massachusetts Institute of Technology, USA
Alessio Fagioli   Sapienza University, Italy
Yannik Glaser   University of Hawaii at Manoa, USA
Haifan Gong   Chinese University of Hong Kong, Shenzhen,
                                   China
Ricardo Gonzales   University of Oxford, UK
Sobhan Goudarzi   Sunnybrook Research Institute, Canada
Michal Grzeszczyk   Sano Centre for Computational Medicine, Poland
Fatemeh Haghighi   Arizona State University, USA
Edward Henderson   University of Manchester, UK
Qingqi Hong   Xiamen University, China
Mohammad R. H. Taher   Arizona State University, USA
Henkjan Huisman   Radboud University Medical Center,
                                   the Netherlands
Ronnachai Jaroensri   Google, USA
Qiangguo Jin   Northwestern Polytechnical University, China
Neerav Karani   Massachusetts Institute of Technology, USA
Benjamin Killeen   Johns Hopkins University, USA
Daniel Lang   Helmholtz Center Munich, Germany
Max-Heinrich Laves   Philips Research and ImFusion GmbH, Germany
Gilbert Lim   SingHealth, Singapore
Mingquan Lin   Weill Cornell Medicine, USA
Charles Lu   Massachusetts Institute of Technology, USA
Yuhui Ma   Chinese Academy of Sciences, China
Tejas Sudharshan Mathai   National Institutes of Health, USA
Felix Meissen   Technische Universität München, Germany
Mingyuan Meng   University of Sydney, Australia
Leo Milecki   CentraleSupelec, France
Marc Modat   King's College London, UK
Tiziano Passerini   Siemens Healthineers, USA
Tomasz Pieciak   Universidad de Valladolid, Spain
Daniel Rueckert   Imperial College London, UK
Julio Silva-Rodríguez   ETS Montreal, Canada
Bingyao Tan   Nanyang Technological University, Singapore
Elias Tappeiner   UMIT - Private University for Health Sciences,
                                   Medical Informatics and Technology, Austria
Jocelyne Troccaz   TIMC Lab, Grenoble Alpes University-CNRS,
                                   France
Chialing Tsai   Queens College, City University New York, USA
Juan Miguel Valverde   University of Eastern Finland, Finland
Sulaiman Vesal   Stanford University, USA

Wolf-Dieter Vogl	RetInSight GmbH, Austria
Vibashan VS	Johns Hopkins University, USA
Lin Wang	Harbin Engineering University, China
Yan Wang	Sichuan University, China
Rhydian Windsor	University of Oxford, UK
Ivo Wolf	University of Applied Sciences Mannheim, Germany
Linshan Wu	Hunan University, China
Xin Yang	Chinese University of Hong Kong, China

Contents – Part IX

Image-Guided Intervention, Surgical Planning, and Data Science

UXDiff: Synthesis of X-Ray Image from Ultrasound Coronal Image of Spine with Diffusion Probabilistic Network

Yihao Zhou[1], Chonglin Wu[1], Xinyi Wang[1], and Yongping Zheng[1,2(✉)]

[1] Department of Biomedical Engineering, The Hong Kong Polytechnic University,
Hong Kong SAR, China
`yongping.zheng@polyu.edu.hk`
[2] Research Institute for Smart Ageing, The Hong Kong Polytechnic University,
Hong Kong SAR, China

Abstract. X-ray radiography with measurement of the Cobb angle is the gold standard for scoliosis diagnosis. However, cumulative exposure to ionizing radiation risks the health of patients. As a radiation-free alternative, imaging of scoliosis using 3D ultrasound scanning has recently been developed for the assessment of spinal deformity. Although these coronal ultrasound images of the spine can provide angle measurement comparable to X-rays, not all spinal bone features are visible. Diffusion probabilistic models (DPMs) have recently emerged as high-fidelity image generation models in medical imaging. To enhance the visualization of bony structures in coronal ultrasound images, we proposed UX-Diffusion, the first diffusion-based model for translating ultrasound coronal images to X-ray-like images of the human spine. To mitigate the underestimation in angle measurement, we first explored using ultrasound curve angle (UCA) to approximate the distribution of X-ray under Cobb angle condition in the reverse process. We then presented an angle embedding transformer module, establishing the angular variability conditions in the sampling stage. The quantitative results on the ultrasound and X-ray pair dataset achieved the state-of-the-art performance of high-quality X-ray generation and showed superior results in comparison with other reported methods. This study demonstrated that the proposed UX-diffusion method has the potential to convert coronal ultrasound image of spine into X-ray image for better visualization.

Keywords: Diffusion model · Image-to-image translation · Image synthesis · Ultrasound imaging · Cobb angle · Scoliosis

1 Introduction

Adolescent idiopathic scoliosis (AIS), the most prevalent form of spinal deformity among children, and some patients tend to worsen over time, ultimately leading to surgical treatment if not being treated timely [1]. Determining the

H. Greenspan et al. (Eds.): MICCAI 2023, LNCS 14228, pp. 3–12, 2023.
https://doi.org/10.1007/978-3-031-43996-4_1

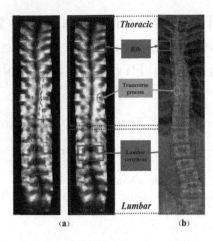

Fig. 1. Paired coronal ultrasound image with ultrasound curve angle (UCA) and X-ray image with Cobb angle. (a) Ultrasound images; (b) The corresponding X-ray image

observation interval is crucial for monitoring the likelihood of curve progression. Some scoliosis patients may progress within a short period. The gold standard for clinical diagnosis of scoliosis is the Cobb angle measured with X-ray imaging. However, the follow-up observation using X-ray requires an interval of 3–12 months because of the potential oncogenic effect of radiation exposure [2,3]. To reduce X-ray exposure and assess scoliosis frequently, coronal ultrasound imaging of spine formed by 3D ultrasound scanning has recently been developed and commercialized for scoliosis assessment [4]. This technique used a volume projection imaging (VPI) method to form the coronal ultrasound image, which contains the information of lateral curvature of spine (Fig. 1(a)) [5]. The ultrasound curve angle (UCA) can be obtained using the coronal ultrasound image of spine and has been demonstrated to be comparable with the radiographic Cobb angle [6]. However, clinicians hesitate to adopt this image modality since spinal images formed by VPI method are new to users, and the bone features look different from those in X-ray images (Fig. 1). If these bony features can be presented similarly to X-ray images, this radiation-free spine image will be more accepted. Moreover, such a conversion from ultrasound coronal images to X-ray-like images can not only help clinicians understand bone structure without any barrier, but also indirectly minimizes the patient's exposure to cumulative radiation.

In previous studies, generative adversarial networks (GANs) have been widely used in medical image translation applications. Long et al. proposed an enhanced Cycle-GAN for integrated translation in ultrasound, which introduced a perceptual constraint to increasing the quality of synthetic ultrasound texture [7,8]. For scoliosis assessment, UXGAN has been added with an attention mechanism into the Cycle-GAN model to focus on the spine feature during modal transformation for translating coronal ultrasound image to X-ray image [9]. However, it worked well on mapping local texture but was not so good for the translation with more extensive geometric changes. Besides, it was only tested on patients with less than 20° of scoliosis, thus limiting its application for scoliosis assessment.

Diffusion and score-matching models, which have emerged as high-fidelity image generation models, have achieved impressive performance in the medical field [10,11]. Pinaya et al. adapted denoising diffusion probabilistic models(DDPM) for high-resolution 3D brain image generation [12]. Qing et al. investigated the performance of the DPM-based model with different sampling strategies for the conversion between CT and MRI [13]. Despite the achievements of all these previous works, there is no research on the diffusion model for converting coronal ultrasound images to X-ray-like images. So far, it is still a challenging topic because the difference in texture and shape between the two modalities are substantial.

In this study, based on the conditional diffusion model, we design an ultrasound-to-X-ray synthesis network, which incorporates an angle-embedding transformer module into the noise prediction model. We have found that the guidance on angle information can rectify for offsets in generated images with different amounts of the inclination of the vertebrae. To learn the posterior probability of X-ray in actual Cobb angle conditions, we present a conditional consistency function to utilize UCA as the prior knowledge to approximate the objective distribution. Our contributions are summarized as follows:

- We propose a novel method for the Us-to-X-ray translation using probabilistic denoising diffusion probabilistic model and a new angle embedding attention module, which takes UCA and Cobb angle as the prerequisite for image generation.
- Our attention module facilitates the model to close the objective X-ray distribution by incorporating the angle and source domain information. The conditional consistency function ensures that the angle guidance flexibly controls the curvature of the spine during the reverse process.
- correlation with authentic Cobb angle indicates its high potential in scoliosis evaluation.

2 Method

2.1 Conditional Diffusion Models for U2X Translation

Diffusion models are the latent variable model that attempts to learn the data distribution, followed by a Markovian process. The objective is to optimize the usual variational bound on negative log likelihood denoted as:

$$\mathbb{E}_{q(x_0)}[\log p_\theta(x_0)] \leq \mathbb{E}_{q(x_{0:T})}[\log p_\theta(x_{0:T}) - \log q(x_{1:T}|x_0)] \tag{1}$$

$p_\theta(x_{0:T})$ is the joint distribution called the reverse process. $q(x_{1:T}|x_0)$ is the diffusion process, progressively adding Gaussian noise to the previous state of the system. Given $x \sim p(x) \in \mathbb{R}^{W \times H}$ be an X-ray image, conditional diffusion models tend to learn the data distribution in condition y. In this work, we partition y

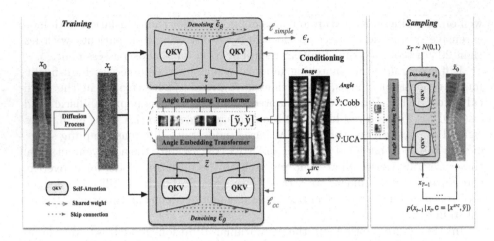

Fig. 2. *Training* The prediction models $\breve{\epsilon}_\theta$ and $\tilde{\epsilon}_\theta$ receive the corrupted image at the moment t of the diffusion process and output the noise at moment $t-1$ with an attention module for the introduction of the Cobb angle and UCA conditions, respectively; *Sampling* The embedded ultrasound images and its UCA are fed to the trained prediction model $\tilde{\epsilon}_\theta$ to generate X-ray-like images iteratively.

into the set of Cobb angle $\breve{y}$ and paired ultrasound image $x^{src} \sim p'(x) \in \mathbb{R}^{W \times H}$. Then the training objective can be reparameterized as the prediction of mean of noising data distribution with y at all timestep t:

$$\mu(x_t, t, y) = \frac{1}{\sqrt{a_t}}(x_t - \frac{\beta_t}{\sqrt{1-\bar{\alpha}_t}}\epsilon(x_t, t, y)) \tag{2}$$

$$\ell_{simple}(\breve{\epsilon}_\theta) = \sum_{t=1}^{T} \mathbb{E}_{x_0 \sim q(x_0), \epsilon_t \sim \mathcal{N}(0,I)}[\left\| \breve{\epsilon}_\theta^{(t)}(\sqrt{a_t}x_0 + \sqrt{1-\alpha_t}\epsilon_t, y) - \epsilon_t \right\|_2^2] \tag{3}$$

2.2   Attention Module with Angle Embedding

Theoretically, if the denoising $\breve{\epsilon}_\theta$ is correct, then as $T \to \infty$, we can obtain X-ray images that the sample paths are distributed as $p_\theta(x_{t-1}|x_t, c = [\breve{y}, x^{src}])$. However, acquiring Cobb angles from real X-ray images is not feasible during the sampling stage. Accordingly, we propose an auxiliary model $\tilde{\epsilon}_\theta$, taking estimated UCA as the prior knowledge, to approximate the objective distribution. Specifically, let $\tilde{y}$ be the UCA of ultrasound images (Fig. 2). This transformer-based module establishes the relationship between $\tilde{y}$ and $\breve{y}$. We run a linear projection of dimension $\mathbb{R}^{dim}$ over the two scalars of angle, and a learnable 1D position embedding is employed to them, representing their location information. Then we reshape them to the same dimension as image tokens. A shared weight attention module compute the self-attention after taking all image tokens, and the projection of $\tilde{y}$ or $\breve{y}$ as input.

Algorithm 1. UXDiffusion Training

Require: c: Condition information
Optimization parameters: $\breve{\theta}$, $\tilde{\theta}$, Angle-Embedding module $\theta_{\mathcal{A}}$
1: **repeat**
2:    $x_0 \sim q(x_0)$
3:    $x^{src}, \breve{y}, \tilde{y} \Longleftarrow$ Paired source image, Cobb angle, UCA retrieval
4:    $t \sim Uniform(\{1, ..., T\})$
5:    $x_t \Longleftarrow$ Corrupt target image x_0
6:    $\epsilon \sim \mathcal{N}(0, I)$
7:    Taking gradient descent step on
$$\nabla_{\breve{\theta}, \theta_{\mathcal{A}}} \left\| \epsilon - \breve{\epsilon}_\theta^{(t)}(x_t, \mathbf{c} = [x^{src}, \breve{y}]) \right\|^2 \;\; // \text{ Eq. 3}$$
8:    $\breve{\theta}$ Freeze
9:    Taking gradient descent step on
$$\nabla_{\tilde{\theta}, \theta_{\mathcal{A}}} \left\| \breve{\epsilon}_\theta^{(t)}(x_t, \mathbf{c} = [x^{src}, \breve{y}]) - \tilde{\epsilon}_\theta^{(t)}(x_t, \mathbf{c} = [x^{src}, \tilde{y}]) \right\|^2 \;\; // \text{ Eq. 5}$$
10: **until** converged

$$z = \begin{cases} \mathcal{A}_\theta([E(\tilde{y}); E(-1); x_{src}^1 E; \cdots x_{src}^N E;] + E_{pos}) & \text{if angle is UCA} \\ \mathcal{A}_\theta([E(-1); E(\breve{y}); x_{src}^1 E; \cdots x_{src}^N E;] + E_{pos}) & \text{else} \end{cases} \tag{4}$$

where $\mathcal{A}$ is the operation of a standard transformer encoder [14]. N is the patch number of the source image. Since the transformer can record the indexes of each token, we set the value of the patch of $\breve{y}$ to -1 when predicting the $\tilde{\epsilon}(c = [x^{src}, \tilde{y}, \breve{y} = \phi])$, and vice versa. The dimension of output z matches the noise prediction model's input for the self-attention mechanism. Thus, the sequence of the patch of the angle and image can be entered point-wise into the denoising model.

2.3 Consistent Conditional Loss

Training UXDiffusion. As described in Algorithm 1, the input to the denoising model $\tilde{\epsilon}_\theta$ and $\breve{\epsilon}_\theta$ are the corrupt image, source image and the list of angle. Rather than learning in the direction of the gradient of the Cobb angle, the auxiliary denoising model instead predicts the score estimates of the posterior probability in UCA conditions for approximating the X-ray distribution under the actual Cobb angle condition. The bound can be denoted as $KL(p_\theta(x_{t-1}|x_t, \tilde{y}, x^{src})||p_\theta(x_{t-1}|x_t, \breve{y}, x^{src}))$, where KL denotes Kullback-Leibler divergence. The reverse process mean comes from an estimate $x_\theta(\tilde{y}) \approx x_\theta(\breve{y})$ plugged into $q(x_{t-1}|x_t, y)$. Note that the parameterization of the mean of distribution can be simplified to the noise prediction [15]. The posterior probability of X-ray in the condition of Cobb angle can be calculated by minimizing the conditional consistency loss defined as:

$$\ell_{cc} = \sum_{t=1}^{T} \mathbb{E}[\left\| \epsilon_\theta^{(t)}(x_t, x^{src}, \breve{y}) - \epsilon_\theta^{(t)}(x_t, x^{src}, \tilde{y}) \right\|_2^2] \tag{5}$$

Sampling with UCA. We follow the sampling scheme in [16] to speed up the sampling process. The embedded ultrasound image and corresponding UCA are fed into the trained noise estimation network $\tilde{\epsilon}$ to sample the X-ray-like image iteratively.

3   Experiments

3.1   Dataset

The dataset consists of 150 paired coronal ultrasound and X-ray images. Each patient took X-ray radiography and ultrasound scanning at the same day. Patients with BMI indices greater than $25.0\,\mathrm{kg/m^2}$ and patients with scoliosis angles exceeding $60°$ were excluded from this study, as the 7.5 MHz ultrasound transducer could not penetrate well for those fatty body and the spine would deform and rotate too much with large Cobb angle, thus affecting ultrasound image quality. The Cobb angle was acquired from an expert with 15 years of experience on scoliosis radiographs, while the UCA was acquired by two raters with at least 5 years of evaluating scoliosis using ultrasound. We manually aligned and cropped the paired images to the same reference space. The criterion for registration was to align the spatial positions of the transverse and spinous processes and the ribs. We resized them into 256×512, and grouped the data by 90, 30 and 30 for training, validation and test, respectively.

3.2   Implementation Details

We followed the same architecture of DDPM, using a U-Net network with attention blocks to predict ϵ. For the angle embedding attention module, we transformed the token in ViT [14] for classification into the tokens for the angle list. The transformer encoder blocks were used to compute the self-attention on the embedding angle and image tokens. We set $T = 2000$ and forward variances from 10^{-4} to 0.02 linearly. We apply the exponential moving average strategy to update the model weights, the decay rate is 0.999. All the experiments were conducted on a 48GB NVIDIA RTX A6000 GPU.

3.3   Synthesis Performance

In this section, three different types of generated models were chosen for comparison: 1) GAN-based model for paired image-to-image translation [17], because of the paired data we use; 2) GAN-based model for unpaired image-to-image translation [9], the first model to synthesize X-ray image from coronal ultrasound image, is based on CycleGAN; 3) Classifier-free conditional diffusion-based model [18], which proposed a conditional diffusion model toward medical image segmentation (Table 1). We could transform the model into our U2X task by replacing the segmentation image with the ultrasound image. The visualization comparison is presented in Fig. 4, and the quantitative comparison is demonstrated. Our proposed model has a higher quality in the generation of vertebral

Table 1. Quantitative comparison. The average SSIM and PSNR scores of X-ray-like images using different methods.

Method	SSIM	PSNR (dB)
Pix2pix [17]	0.673 ± 0.05	18.44 ± 3.27
UXGAN [9]	0.715 ± 0.04	20.86 ± 2.85
MedSegDiff [18]	0.729 ± 0.04	21.02 ± 1.77
UXDiff (Ours)	0.740 ± 0.04	21.55 ± 1.69

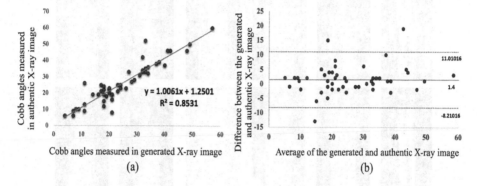

Fig. 3. (a) Correlation between Cobb angles measured with the synthesized image and ground-truth X-ray images. The coefficient of determination $R^2 = 0.8531$; (b) Bland-Altman plot of Cobb angles for the synthesized image and ground-truth X-ray image.

contour edges compared to the baselines. Also, the synthesis images have the same spine curvature orientation as ground-true images, with high fidelity of structural information. Then, we measured the structural similarity (SSIM) and peak signal-to-noise ratio (PSNR) to evaluate the performance of the synthesized images. The results show that our model outperforms paired and unpaired GAN-based methods by 2.5–6.7 points on SSIM and is also better than the reference diffusion-based method. Our model achieves the highest value of PSNR along with the lowest standard deviation, showing the stability of the model. We believe that the diffusion-based baseline only considers the source images and disregards the scoliosis offset of the generated X-ray images, which can be addressed using angle-based guidance. Experimental results demonstrate that the performance of predicting the conditional distribution with angle guidance is superior to simply using the source image.

Ultrasound	Ground-Truth	UXDiff(Ours)	MedSegDiff	UXGAN	Pix2pix

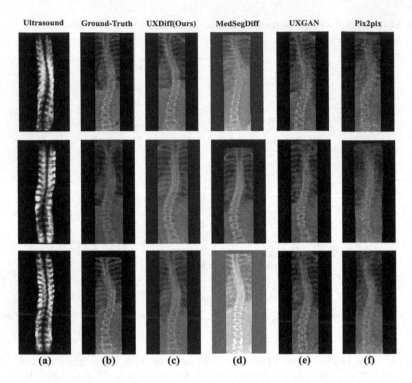

Fig. 4. Qualitative comparison of Us-to-X-ray image translation. The first two columns show the ultrasound (**a**) and ground-truth X-ray images (**b**), respectively. Our model generate a more realistic X-ray-like image that reveals the anatomical structure of spine curvature, with higher visualization quality of vertebrae than baselines (**c–f**).

3.4 Comparison with Cobb Angle

Since the Cobb angle is widely used as the gold standard for scoliosis assessment, our objective is to synthesize an X-ray-like image that can be applied to the measurement of the Cobb angle. As depicted in Fig. 3, we measure the Cobb angle difference between the synthesized image and the original X-ray image. Then we use linear regression to study the correlation between the Cobb angles measured using the generated and original images and the Bland-Altman plot to analyze the agreement between the two Cobb angles. The result demonstrates that our model can generate images maintaining a high consistency in restoring the overall curvature of the bones. The coefficient of determination is $R^2 = 0.8531(p < 0.001)$. The slope of the regression line is $45.17°$ for all parameters, which is close to the ideal value of $45°$. Bland-Altman plots demonstrate the measured angle value difference between the generated and GT X-rays. The mean difference is $1.4°$ for all parameters indicating that it is comparable with the ground-truth Cobb angle. The experiments demonstrate that the proposed model for calculating the Cobb angle is practical and can be used to evaluate scoliosis.

4   Conclusion

This paper developed a synthesized model for translating coronal ultrasound images to X-ray-like images using a probabilistic diffusion network. Our purpose is to use a single network to parameterize the X-ray distribution, and the generated images can be applied to the Cobb angle measurement. We achieved this by introducing the angular information corresponding to ultrasound and X-ray image to the model for noise prediction. An attention module was proposed to guide the model for generating high-quality images based on embedded image and angle information. Furthermore, to overcome the unavailability of the Cobb angle in the sampling process, we presented a conditional consistency function to train the model to learn the gradient according to the UCA for approximating the X-ray distribution in the condition of Cobb angle. Experiments on paired ultrasound and X-ray coronal images demonstrated that our diffusion-based method advanced the state-of-the-art significantly. In summary, this new model has great potential to facilitate 3D ultrasound imaging to be used for scoliosis assessment with accurate Cobb angle measurement and X-ray-like images obtained without any radiation.

Acknowledgement. This study was support by The Research Grant Council of Hong Kong (R5017-18).

References

1. Reamy, B.V., Slakey, J.B.: Adolescent idiopathic scoliosis: review and current concepts. Am. Family Phys. **64**(1), 111–117 (2001)
2. Yamamoto, Y., et al.: How do we follow-up patients with adolescent idiopathic scoliosis? Recommendations based on a multicenter study on the distal radius and ulna classification. Eur. Spine J. **29**, 2064–2074 (2020)
3. Knott, P., et al.: SOSORT 2012 consensus paper: reducing X-ray exposure in pediatric patients with scoliosis. Scoliosis **9**(1), 4 (2014)
4. Zheng, Y.-P., et al.: A reliability and validity study for Scolioscan: a radiation-free scoliosis assessment system using 3D ultrasound imaging. Scoliosis Spinal Disord. **11**, 1–15 (2016)
5. Cheung, C.-W.J., Zhou, G.-Q., Law, S.-Y., Mak, T.-M., Lai, K.-L., Zheng, Y.-P.: Ultrasound volume projection imaging for assessment of scoliosis. IEEE Trans. Med. Imaging **34**(8), 1760–1768 (2015)
6. Lee, T.T.-Y., Lai, K.K.-L., Cheng, J.C.-Y., Castelein, R.M., Lam, T.-P., Zheng, Y.-P.: 3D ultrasound imaging provides reliable angle measurement with validity comparable to x-ray in patients with adolescent idiopathic scoliosis. J. Orthop. Transl. **29**, 51–59 (2021)
7. Teng, L., Fu, Z., Yao, Y.: Interactive translation in echocardiography training system with enhanced cycle-GAN. IEEE Access **8**, 106147–106156 (2020)
8. Zhu, J.-Y., Park, T., Isola, P., Efros, A.A.: Unpaired image-to-image translation using cycle-consistent adversarial networks. In: Proceedings of the IEEE International Conference on Computer Vision, pp. 2223–2232 (2017)

9. Jiang, W., Yu, C., Chen, X., Zheng, Y., Bai, C.: Ultrasound to X-ray synthesis generative attentional network (UXGAN) for adolescent idiopathic scoliosis. Ultrasonics **126**, 106819 (2022)

10. Ho, J., Jain, A., Abbeel, P.: Denoising diffusion probabilistic models. Adv. Neural. Inf. Process. Syst. **33**, 6840–6851 (2020)

11. Song, Y., Sohl-Dickstein, J., Kingma, D.P., Kumar, A., Ermon, S., Poole, B.: Score-based generative modeling through stochastic differential equations. arXiv preprint arXiv:2011.13456 (2020)

12. Pinaya, W.H.L., et al.: Brain imaging generation with latent diffusion models. In: Mukhopadhyay, A., Oksuz, I., Engelhardt, S., Zhu, D., Yuan, Y. (eds.) DGM4MICCAI 2022. LNCS, vol. 13609, pp. 117–126. Springer, Cham (2022). https://doi.org/10.1007/978-3-031-18576-2_12

13. Lyu, Q., Wang, G.: Conversion between CT and MRI images using diffusion and score-matching models. arXiv preprint arXiv:2209.12104 (2022)

14. Dosovitskiy, A., et al.: An image is worth 16×16 words: transformers for image recognition at scale. arXiv preprint arXiv:2010.11929 (2020)

15. Nichol, A.Q., Dhariwal, P.: Improved denoising diffusion probabilistic models. In: International Conference on Machine Learning, pp. 8162–8171. PMLR (2021)

16. Song, J., Meng, C., Ermon, S.: Denoising diffusion implicit models. arXiv preprint arXiv:2010.02502 (2020)

17. Isola, P., Zhu, J.-Y., Zhou, T., Efros, A.A.: Image-to-image translation with conditional adversarial networks. In: Proceedings of the IEEE Conference on Computer Vision and Pattern Recognition, pp. 1125–1134 (2017)

18. Wu, J., Fang, H., Zhang, Y., Yang, Y., Xu, Y.: MedSegDiff: medical image segmentation with diffusion probabilistic model. arXiv preprint arXiv:2211.00611 (2022)

EndoSurf: Neural Surface Reconstruction of Deformable Tissues with Stereo Endoscope Videos

Ruyi Zha[1](✉), Xuelian Cheng[2,4,5], Hongdong Li[1], Mehrtash Harandi[2], and Zongyuan Ge[2,3,4,5,6]

[1] Australian National University, Canberra, Australia
ruyi.zha@anu.edu.au
[2] Faculty of Engineering, Monash University, Melbourne, Australia
[3] Faculty of IT, Monash University, Melbourne, Australia
[4] AIM for Health Lab, Monash University, Melbourne, Australia
[5] Monash Medical AI, Monash University, Melbourne, Australia
[6] Airdoc-Monash Research Lab, Monash University, Melbourne, Australia

Abstract. Reconstructing soft tissues from stereo endoscope videos is an essential prerequisite for many medical applications. Previous methods struggle to produce high-quality geometry and appearance due to their inadequate representations of 3D scenes. To address this issue, we propose a novel neural-field-based method, called *EndoSurf*, which effectively learns to represent a deforming surface from an RGBD sequence. In EndoSurf, we model surface dynamics, shape, and texture with three neural fields. First, 3D points are transformed from the observed space to the canonical space using the deformation field. The signed distance function (SDF) field and radiance field then predict their SDFs and colors, respectively, with which RGBD images can be synthesized via differentiable volume rendering. We constrain the learned shape by tailoring multiple regularization strategies and disentangling geometry and appearance. Experiments on public endoscope datasets demonstrate that EndoSurf significantly outperforms existing solutions, particularly in reconstructing high-fidelity shapes. Code is available at https://github.com/Ruyi-Zha/endosurf.git.

Keywords: 3D Reconstructon · Neural Fields · Robotic Surgery

1 Introduction

Surgical scene reconstruction using stereo endoscopes is crucial to Robotic-Assisted Minimally Invasive Surgery (RAMIS). It aims to recover a 3D model of

R. Zha and X. Cheng—Equal contribution.

Supplementary Information The online version contains supplementary material available at https://doi.org/10.1007/978-3-031-43996-4_2.

H. Greenspan et al. (Eds.): MICCAI 2023, LNCS 14228, pp. 13–23, 2023.
https://doi.org/10.1007/978-3-031-43996-4_2

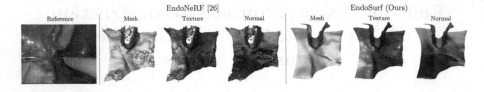

Fig. 1. 3D meshes extracted from EndoNeRF [26] and our method. EndoNeRF cannot recover a smooth and accurate surface even with post-processing filters.

the observed tissues from a stereo endoscope video. Compared with traditional 2D monitoring, 3D reconstruction offers notable advantages because it allows users to observe the surgical site from any viewpoint. Therefore, it dramatically benefits downstream medical applications such as surgical navigation [21], surgeon-centered augmented reality [18], and virtual reality [7]. General reconstruction pipelines first estimate depth maps with stereo-matching [5,6,13] and then fuse RGBD images into a 3D model [12,14,17,23,28]. Our work focuses on the latter, *i.e.*, how to accurately reconstruct the shape and appearance of deforming surfaces from RGBD sequences.

Existing approaches represent a 3D scene in two ways: discretely or continuously. Discrete representations include point clouds [12,14,23,28] and mesh grids [17]. Additional warp fields [9] are usually utilized to compensate for tissue deformation. Discrete representation methods produce surfaces efficiently due to their sparsity property. However, this property also limits their ability to handle complex high-dimensional changes, *e.g.*, non-topology deformation and color alteration resulting from cutting or pulling tissues.

Recently, continuous representations have become popular with the blossoming of neural fields, *i.e.*, neural networks that take space-time inputs and return the required quantities. Neural-field-based methods [19,20,22,25–27] exploit deep neural networks to implicitly model complex geometry and appearance, outperforming discrete-representation-based methods. A good representative is EndoNeRF [26]. It trains two neural fields: one for tissue deformation and the other for canonical density and color. EndoNeRF can synthesize reasonable RGBD images with post-processing filters. However, the ill-constrained properties of the density field deter the network from learning a solid surface shape. Figure 1 shows that EndoNeRF can not accurately recover the surface even with filters. While there have been attempts to parameterize other geometry fields, *e.g.*, occupancy fields [19,20] and signed distance function (SDF) fields [25,27], they hypothesize static scenes and diverse viewpoints. Adapting them to surgical scenarios where surfaces undergo deformation and camera movement is confined is non-trivial.

We propose EndoSurf: neural implicit fields for **Endo**scope-based **Surf**ace reconstruction, a novel neural-field-based method that effectively learns to represent dynamic scenes. Specifically, we model deformation, geometry, and appearance with three separate multi-layer perceptrons (MLP). The deformation network transforms points from the observation space to the canonical space. The

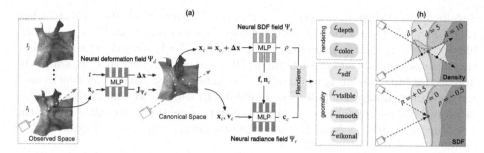

Fig. 2. (a) The overall pipeline of EndoSurf. (b) Density field *v.s.* SDF field. Red lines represent surfaces. The density field is depth ambiguous, while the SDF field clearly defines the surface as the zero-level set. (Color figure online)

geometry network represents the canonical scene as an SDF field. Compared with density, SDF is more self-contained as it explicitly defines the surface as the zero-level set. We enforce the geometry network to learn a solid surface by designing various regularization strategies. Regarding the appearance network, we involve positions and normals as extra clues to disentangle the appearance from the geometry. Following [25], we adopt unbiased volume rendering to synthesize color images and depth maps. The network is optimized with gradient descent by minimizing the error between the real and rendered results. We evaluate EndoSurf quantitatively and qualitatively on public endoscope datasets. Our work demonstrates superior performance over existing solutions, especially in reconstructing smooth and accurate shapes.

2 Method

2.1 Overview

Problem Setting. Given a stereo video of deforming tissues, we aim to reconstruct the surface shape S and texture C. Similar to EndoNeRF [26], we take as inputs a sequence of frame data $\{(\mathbf{I}_i, \mathbf{D}_i, \mathbf{M}_i, \mathbf{P}_i, t_i)\}_{i=1}^{T}$. Here T stands for the total number of frames. $\mathbf{I}_i \in \mathbb{R}^{H \times W \times 3}$ and $\mathbf{D}_i \in \mathbb{R}^{H \times W}$ refer to the i-th left RGB image and depth map with height H and width W. Foreground mask $\mathbf{M}_i \in \mathbb{R}^{H \times W}$ is utilized to exclude unwanted pixels, such as surgical tools, blood, and smoke. Projection matrix $\mathbf{P}_i \in \mathbb{R}^{4 \times 4}$ maps 3D coordinates to 2D pixels. $t_i = i/T$ is each frame's timestamp normalized to $[0, 1]$. While stereo matching, surgical tool tracking, and pose estimation are also practical clinical concerns, in this work we prioritize 3D reconstruction and thus take depth maps, foreground masks, and projection matrices as provided by software or hardware solutions.

Pipeline. Figure 2(a) illustrates the overall pipeline of our approach. Similar to [25,26], we incorporate our EndoSurf network into a volume rendering scheme.

Specifically, we begin by adopting a mask-guided sampling strategy [26] to select valuable pixels from a video frame. We then cast 3D rays from these pixels and hierarchically sample points along the rays [25]. The EndoSurf network utilizes these sampled points and predicts their SDFs and colors. After that, we adopt the unbiased volume rendering method [25] to synthesize pixel colors and depths used for network training. We tailor loss functions to enhance the network's learning of geometry and appearance. In the following subsections, we will describe the EndoSurf network (cf. Sect. 2.2) and the optimization process (cf. Sect. 2.3) in detail.

2.2 EndoSurf: Representing Scenes as Deformable Neural Fields

We represent a dynamic scene as canonical neural fields warped to an observed pose. Separating the learning of deformation and canonical shapes has been proven more effective than directly modeling dynamic shapes [22]. Particularly, we propose a neural deformation field $\mathbf{\Psi}_d$ to transform 3D points from the observed space to the canonical space. The geometry and appearance of the canonical scene are described by a neural SDF field $\mathbf{\Psi}_s$ and a neural radiance field $\mathbf{\Psi}_r$, respectively. All neural fields are modeled with MLPs with position encoding [16, 24].

Neural Deformation Field. Provided a 3D point $\mathbf{x}_o \in \mathbb{R}^3$ in the observed space at time $t \in [0, 1]$, the neural deformation field $\mathbf{\Psi}_d(\mathbf{x}_o, t) \mapsto \mathbf{\Delta x}$ returns the displacement $\mathbf{\Delta x} \in \mathbb{R}^3$ that transforms $\mathbf{x}_o$ to its canonical position $\mathbf{x}_c = \mathbf{x}_o + \mathbf{\Delta x}$. The canonical view direction $\mathbf{v}_c \in \mathbb{R}^3$ of point $\mathbf{x}_c$ can be obtained by transforming the raw view direction $\mathbf{v}_o$ with the Jacobian of the deformation field $\mathbf{J}_{\mathbf{\Psi}_d}(\mathbf{x}_o) = \partial \mathbf{\Psi}_d / \partial \mathbf{x}_o$, i.e., $\mathbf{v}_c = (\mathbf{I} + \mathbf{J}_{\mathbf{\Psi}_d}(\mathbf{x}_o))\mathbf{v}_o$.

Neural SDF Field. The shape of the canonical scene is represented by a neural field $\mathbf{\Psi}_s(\mathbf{x}_c) \mapsto (\rho, \mathbf{f})$ that maps a spatial position $\mathbf{x}_c \in \mathbb{R}^3$ to its signed distance function $\rho \in \mathbb{R}$ and a geometry feature vector $\mathbf{f} \in \mathbb{R}^F$ with feature size F.

In 3D vision, SDF is the orthogonal distance of a point $\mathbf{x}$ to a watertight object's surface, with the sign determined by whether or not $\mathbf{x}$ is outside the object. In our case, we slightly abuse the term SDF since we are interested in a segment of an object rather than the whole thing. We extend the definition of SDF by imagining that the surface of interest divides the surrounding space into two distinct regions, as shown in Fig. 2 (b). SDF is positive if $\mathbf{x}$ falls into the region which includes the camera and negative if it is in the other. As $\mathbf{x}$ approaches the surface, the SDF value gets smaller until it reaches zero at the surface. Therefore, the surface of interest $\mathcal{S}$ is the zero-level set of SDF, i.e., $\mathcal{S} = \{\mathbf{p} \in \mathbb{R}^3 | \mathbf{\Psi}_s(\mathbf{p}) = 0\}$.

Compared with the density field used in [26], the SDF field provides a more precise representation of surface geometry because it explicitly defines the surface as the zero-level set. The density field, however, encodes the probability of an object occupying a position, making it unclear which iso-surface defines the object's boundary. As a result, density-field-based methods [16, 26] can not directly identify a depth via ray marching but rather render it by integrating the depths of sampled points with density-related weights. Such a rendering method can lead to potential depth ambiguity, *i.e.*, camera rays pointing to the same surface produce different surface positions (Fig. 2(b)).

Given a surface point $\mathbf{p}_c \in \mathbb{R}^3$ in the canonical space, the surface normal $\mathbf{n}_c \in \mathbb{R}^3$ is the gradient of the neural SDF field $\boldsymbol{\Psi}_s$: $\mathbf{n}_c = \nabla_{\boldsymbol{\Psi}_s}(\mathbf{p}_c)$. Normal $\mathbf{n}_o$ of the deformed surface point $\mathbf{p}_o$ can also be obtained with the chain rule.

Neural Radiance Field. We model the appearance of the canonical scene as a neural radiance field $\boldsymbol{\Psi}_r(\mathbf{x}_c, \mathbf{v}_c, \mathbf{n}_c, \mathbf{f}) \mapsto \mathbf{c}_c$ that returns the color $\mathbf{c}_c \in \mathbb{R}^3$ of a viewpoint $(\mathbf{x}_c, \mathbf{v}_c)$. Unlike [16, 26], which only take the view direction $\mathbf{v}_c$ and feature vector $\mathbf{f}$ as inputs, we also feed the normal $\mathbf{n}_c$ and position $\mathbf{x}_c$ to the radiance field as extra geometric clues. Although the feature vector implies the normal and position information, it is validated that directly incorporating them benefits the disentanglement of geometry, *i.e.*, allowing the network to learn appearance independently from the geometry [25, 27].

2.3 Optimization

Unbiased Volume Rendering. Given a camera ray $\mathbf{r}(h) = \mathbf{o}_o + h\mathbf{v}_o$ at time t in the observed space, we sample N points $\mathbf{x}_i$ in a hierarchical manner along this ray [25] and predict their SDFs ρ_i and colors $\mathbf{c}_i$ via EndoSurf. The color $\hat{\mathbf{C}}$ and depth $\hat{\mathbf{D}}$ of the ray can be approximated by unbiased volume rendering [25]:

$$\hat{\mathbf{C}}(\mathbf{r}(h)) = \sum_{i=1}^{N} T_i \alpha_i \mathbf{c}_i, \quad \hat{\mathbf{D}}(\mathbf{r}(h)) = \sum_{i=1}^{N} T_i \alpha_i h_i, \tag{1}$$

where $T_i = \prod_{j=1}^{i-1}(1 - \alpha_j)$, $\alpha_i = \max((\phi(\rho_i) - \phi(\rho_{i+1}))/\phi(\rho_i), 0)$ and $\phi(\rho) = (1 + e^{-\rho/s})^{-1}$. Note that s is a trainable standard deviation, which approaches zero as the network training converges.

Loss. We train the network with two objectives: 1) to minimize the difference between the actual and rendered results and 2) to impose constraints on the neural SDF field such that it aligns with its definition. Accordingly, we design two categories of losses: rendering constraints and geometry constraints:

$$\mathcal{L} = \underbrace{\left(\lambda_1 \mathcal{L}_{\text{color}} + \lambda_2 \mathcal{L}_{\text{depth}}\right)}_{\text{rendering}} + \underbrace{\left(\lambda_3 \mathcal{L}_{\text{eikonal}} + \lambda_4 \mathcal{L}_{\text{sdf}} + \lambda_5 \mathcal{L}_{\text{visible}} + \lambda_6 \mathcal{L}_{\text{smooth}}\right)}_{\text{geometry}},$$

$$\tag{2}$$

where $\lambda_{i=1,\cdots,6}$ are balancing weights. The rendering constraints include the color reconstruction loss $\mathcal{L}_{\text{color}}$ and depth reconstruction loss $\mathcal{L}_{\text{depth}}$:

$$\mathcal{L}_{\text{color}} = \sum_{\mathbf{r} \in \mathcal{R}} \|M(\mathbf{r})(\hat{\mathbf{C}}(\mathbf{r}) - \mathbf{C}(\mathbf{r}))\|_1, \mathcal{L}_{\text{depth}} = \sum_{\mathbf{r} \in \mathcal{R}} \|M(\mathbf{r})(\hat{\mathbf{D}}(\mathbf{r}) - \mathbf{D}(\mathbf{r}))\|_1, \tag{3}$$

where $M(\mathbf{r})$, $\{\hat{\mathbf{C}}, \hat{\mathbf{D}}\}$, $\{\mathbf{C}, \mathbf{D}\}$ and $\mathcal{R}$ are ray masks, rendered colors and depths, real colors and depths, and ray batch, respectively.

We regularize the neural SDF field $\boldsymbol{\Psi}_s$ with four losses: Eikonal loss $\mathcal{L}_{\text{eikonal}}$, SDF loss $\mathcal{L}_{\text{sdf}}$, visibility loss $\mathcal{L}_{\text{visible}}$, and smoothness loss $\mathcal{L}_{\text{smooth}}$.

$$\mathcal{L}_{\text{eikonal}} = \sum_{\mathbf{x} \in \mathcal{X}} (\|\nabla_{\boldsymbol{\Psi}_s}(\mathbf{x})\|_2 - 1)^2, \mathcal{L}_{\text{sdf}} = \sum_{\mathbf{p} \in \mathcal{D}} \|\boldsymbol{\Psi}_s(\mathbf{p})\|_1,$$

$$\mathcal{L}_{\text{visible}} = \sum_{\mathbf{p} \in \mathcal{D}} \max(\langle \nabla_{\boldsymbol{\Psi}_s}(\mathbf{p}), \mathbf{v}_c \rangle, 0), \mathcal{L}_{\text{smooth}} = \sum_{\mathbf{p} \in \mathcal{D}} \|\nabla_{\boldsymbol{\Psi}_s}(\mathbf{p}) - \nabla_{\boldsymbol{\Psi}_s}(\mathbf{p} + \epsilon)\|_1.$$

$$(4)$$

Here the Eikonal loss $\mathcal{L}_{\text{eikonal}}$ [10] encourages $\boldsymbol{\Psi}_s$ to satisfy the Eikonal equation [8]. Points $\mathbf{x}$ are sampled from the canonical space $\mathcal{X}$. The SDF loss $\mathcal{L}_{\text{sdf}}$ restricts the SDF value of points lying on the ground truth depths $\mathcal{D}$ to zero. The visibility loss $\mathcal{L}_{\text{visible}}$ limits the angle between the canonical surface normal and the viewing direction $\mathbf{v}_c$ to be greater than $90°$. The smoothness loss $\mathcal{L}_{\text{smooth}}$ encourages a surface point and its neighbor to be similar, where ϵ is a random uniform perturbation.

3   Experiments

3.1   Experiment Settings

Datasets and Evaluation. We conduct experiments on two public endoscope datasets, namely ENDONERF [26] and SCARED [1] (See statistical details in the supplementary material). ENDONERF provides two cases of in-vivo prostatectomy data with estimated depth maps [13] and manually labeled tool masks. SCARED [1] collects the ground truth RGBD images of five porcine cadaver abdominal anatomies. We pre-process the datasets by normalizing the scene into a unit sphere and splitting the frame data into 7:1 training and test sets.

Our approach is compared with EndoNeRF [26], the state-of-the-art neural-field-based method. There are three outputs for test frames: RGB images, depth maps, and 3D meshes. The first two outputs are rendered the same way as the training process. We use marching cubes [15] to extract 3D meshes from the density and SDF fields. The threshold is set to 5 for the density field and 0 for the SDF field. See the supplementary material for the validation of threshold selection. Five evaluation metrics are used: PSNR, SSIM, LPIPS, RMSE, and point cloud distance (PCD). The first three metrics assess the similarity between the actual and rendered RGB images [26], while RMSE and PCD measure depth map [5,6,14] and 3D mesh [3,4] reconstruction quality, respectively.

Implementation Details. We train neural networks per scene, *i.e.*, one model for each case. All neural fields consist of 8-layer 256-channel MLPs with a skip connection at the 4th layer. Position encoding frequencies in all fields are 6,

Table 1. Quantitative metrics of appearance (PSNR/SSIM/LPIPS) and geometry (RMSE/PCD) on two datasets. The unit for RMSE/PCD is millimeter.

Methods	EndoNeRF [26]					EndoSurf (Ours)				
Metrics	PSNR↑	SSIM↑	LPIPS↓	RMSE↓	PCD↓	PSNR↑	SSIM↑	LPIPS↓	RMSE↓	PCD↓
ENDONERF-cutting	34.186	0.932	0.151	0.930	1.030	**34.981**	**0.953**	**0.106**	**0.835**	**0.559**
ENDONERF-pulling	34.212	0.938	0.161	1.485	2.260	**35.004**	**0.956**	**0.120**	**1.165**	**0.841**
SCARED-d1k1	24.365	0.763	0.326	0.697	2.982	**24.395**	**0.769**	**0.319**	**0.522**	**0.741**
SCARED-d2k1	25.733	0.828	**0.240**	0.583	1.788	**26.237**	**0.829**	0.254	**0.352**	**0.515**
SCARED-d3k1	19.004	0.599	0.467	1.809	3.244	**20.041**	**0.649**	**0.441**	**1.576**	**1.091**
SCARED-d6k1	24.041	0.833	0.464	1.194	3.268	**24.094**	**0.866**	**0.461**	**1.065**	**1.331**
SCARED-d7k1	22.637	0.813	0.312	2.272	3.465	**23.421**	**0.861**	**0.282**	**2.123**	**1.580**
Average	26.311	0.815	0.303	1.281	2.577	**26.882**	**0.840**	**0.283**	**1.091**	**0.952**

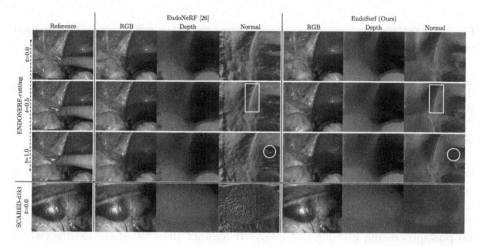

Fig. 3. 2D rendering results on the dynamic case "ENDONERF-cutting" and static case "SCARED-d1k1". Our method yields high-quality depth and normal maps, whereas those of EndoNeRF exhibit jagged noise, over-smoothed edges (white boxes), and noticeable artifacts (white rings).

except those in the radiance field are 10 and 4 for location and direction, respectively. The SDF network is initialized [2] for better training convergence. We use Adam optimizer [11] with a learning rate of 0.0005, which warms up for 5k iterations and then decays with a rate of 0.05. We sample 1024 rays per batch and 64 points per ray. The initial standard deviation s is 0.3. The weights in Eq. 2 are $\lambda_1 = 1.0$, $\lambda_2 = 1.0$, $\lambda_3 = 0.1$, $\lambda_4 = 1.0$, $\lambda_5 = 0.1$ and $\lambda_6 = 0.1$. We train our model with $100K$ iterations for 9 h on an NVIDIA RTX 3090 GPU.

3.2 Qualitative and Quantitative Results

As listed in Table 1, EndoSurf yields superior results against EndoNeRF. On the one hand, EndoSurf produces better appearance quality than EndoNeRF by ↑ 0.571 PSNR, ↑ 0.025 SSIM, and ↑ 0.020 LPIPS. On the other hand, EndoSurf

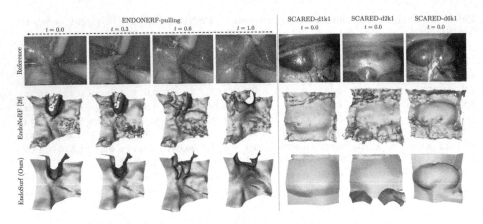

Fig. 4. Extracted meshes on one dynamic scene and three static scenes. Our method produces accurate and smooth surfaces.

dramatically outperforms EndoNeRF in terms of geometry recovery by ↓ 0.190 RMSE and ↓ 1.625 PCD. Note that both methods perform better on ENDON-ERF than on SCARED. This is because ENDONERF fixes the camera pose, leading to easier network fitting.

Figure 3 shows the 2D rendering results. While both methods synthesize high-fidelity RGB images, only EndoSurf succeeds in recovering depth maps with a smoother shape, more details, and fewer artifacts. First, the geometry constraints in EndoSurf prevent the network from overfitting depth supervision, suppressing rough surfaces as observed in the EndoNeRF's normal maps. Second, the brutal post-processing filtering in EndoNeRF cannot preserve sharp details (white boxes in Fig. 3). Moreover, the texture and shape of EndoNeRF are not disentangled, causing depth artifacts in some color change areas (white rings in Fig. 3).

Figure 4 depicts the shapes of extracted 3D meshes. Surfaces reconstructed by EndoSurf are accurate and smooth, while those from EndoNeRF are quite noisy. There are two reasons for the poor quality of EndoNeRF's meshes. First, the density field without regularization tends to describe the scene as a volumetric fog rather than a solid surface. Second, the traditional volume rendering causes discernible depth bias [25]. In contrast, we force the neural SDF field to conform to its definition via multiple geometry constraints. Furthermore, we use unbiased volume rendering to prevent depth ambiguity [25].

We present a qualitative ablation study on how geometry constraints can influence the reconstruction quality in Fig. 5. The Eikonal loss $\mathcal{L}_{\mathrm{eikonal}}$ and SDF loss $\mathcal{L}_{\mathrm{sdf}}$ play important roles in improving geometry recovery, while the visibility loss $\mathcal{L}_{\mathrm{visible}}$ and smoothness loss $\mathcal{L}_{\mathrm{smooth}}$ help refine the surface.

| Reference
PSNR/PCD | Without $\mathcal{L}_{\text{visible}}$
34.630/0.874 | Without $\mathcal{L}_{\text{eikonal}}$
34.573/1.434 | Without $\mathcal{L}_{\text{sdf}}$
34.368/3.711 | Without $\mathcal{L}_{\text{smooth}}$
34.767/0.853 | Complete model
35.004/0.841 |

Fig. 5. Ablation study on four geometry constraints, *i.e.*, visibility loss $\mathcal{L}_{\text{visible}}$, Eikonal loss $\mathcal{L}_{\text{eikonal}}$, SDF loss $\mathcal{L}_{\text{sdf}}$, and smoothness loss $\mathcal{L}_{\text{smooth}}$.

4   Conclusion

This paper presents a novel neural-field-based approach, called EndoSurf, to reconstruct the deforming surgical sites from stereo endoscope videos. Our approach overcomes the geometry limitations of prior work by utilizing a neural SDF field to represent the shape, which is constrained by customized regularization techniques. In addition, we employ neural deformation and radiance fields to model surface dynamics and appearance. To disentangle the appearance learning from geometry, we incorporate normals and locations as extra clues for the radiance field. Experiments on public datasets demonstrate that our method achieves state-of-the-art results compared with existing solutions, particularly in retrieving high-fidelity shapes.

Acknowledgments. This research is funded in part via an ARC Discovery project research grant (DP220100800).

References

1. Allan, M., et al.: Stereo correspondence and reconstruction of endoscopic data challenge. arXiv preprint arXiv:2101.01133 (2021)
2. Atzmon, M., Lipman, Y.: SAL: sign agnostic learning of shapes from raw data. In: Proceedings of the IEEE/CVF Conference on Computer Vision and Pattern Recognition, pp. 2565–2574 (2020)
3. Bozic, A., Zollhofer, M., Theobalt, C., Nießner, M.: DeepDeform: learning non-rigid rgb-d reconstruction with semi-supervised data. In: Proceedings of the IEEE/CVF Conference on Computer Vision and Pattern Recognition, pp. 7002–7012 (2020)
4. Cai, H., Feng, W., Feng, X., Wang, Y., Zhang, J.: Neural surface reconstruction of dynamic scenes with monocular RGB-D camera. arXiv preprint arXiv:2206.15258 (2022)
5. Cheng, X., et al.: Hierarchical neural architecture search for deep stereo matching. Adv. Neural. Inf. Process. Syst. **33**, 22158–22169 (2020)
6. Cheng, X., Zhong, Y., Harandi, M., Drummond, T., Wang, Z., Ge, Z.: Deep laparoscopic stereo matching with transformers. In: Wang, L., Dou, Q., Fletcher, P.T., Speidel, S., Li, S. (eds.) MICCAI 2022, Part VII. LNCS, vol. 13437, pp. 464–474. Springer, Cham (2022). https://doi.org/10.1007/978-3-031-16449-1_44

7. Chong, N., Si, Y., Zhao, W., Zhang, Q., Yin, B., Zhao, Y.: Virtual reality application for laparoscope in clinical surgery based on Siamese network and census transformation. In: Su, R., Zhang, Y.-D., Liu, H. (eds.) MICAD 2021. LNEE, vol. 784, pp. 59–70. Springer, Singapore (2022). https://doi.org/10.1007/978-981-16-3880-0_7

8. Crandall, M.G., Lions, P.L.: Viscosity solutions of Hamilton-Jacobi equations. Trans. Am. Math. Soc. **277**(1), 1–42 (1983)

9. Gao, W., Tedrake, R.: SurfelWarp: efficient non-volumetric single view dynamic reconstruction. arXiv preprint arXiv:1904.13073 (2019)

10. Gropp, A., Yariv, L., Haim, N., Atzmon, M., Lipman, Y.: Implicit geometric regularization for learning shapes. arXiv preprint arXiv:2002.10099 (2020)

11. Kingma, D.P., Ba, J.: Adam: a method for stochastic optimization. arXiv preprint arXiv:1412.6980 (2014)

12. Li, Y., et al.: SuPer: a surgical perception framework for endoscopic tissue manipulation with surgical robotics. IEEE Robot. Autom. Lett. **5**(2), 2294–2301 (2020)

13. Li, Z., et al.: Revisiting stereo depth estimation from a sequence-to-sequence perspective with transformers. In: Proceedings of the IEEE/CVF International Conference on Computer Vision, pp. 6197–6206 (2021)

14. Long, Y., et al.: E-DSSR: efficient dynamic surgical scene reconstruction with transformer-based stereoscopic depth perception. In: de Bruijne, M., et al. (eds.) MICCAI 2021, Part IV. LNCS, vol. 12904, pp. 415–425. Springer, Cham (2021). https://doi.org/10.1007/978-3-030-87202-1_40

15. Lorensen, W.E., Cline, H.E.: Marching cubes: a high resolution 3D surface construction algorithm. ACM SIGGRAPH Comput. Graph. **21**(4), 163–169 (1987)

16. Mildenhall, B., Srinivasan, P.P., Tancik, M., Barron, J.T., Ramamoorthi, R., Ng, R.: NeRF: representing scenes as neural radiance fields for view synthesis. In: Vedaldi, A., Bischof, H., Brox, T., Frahm, J.-M. (eds.) ECCV 2020, Part I. LNCS, vol. 12346, pp. 405–421. Springer, Cham (2020). https://doi.org/10.1007/978-3-030-58452-8_24

17. Newcombe, R.A., Fox, D., Seitz, S.M.: DynamicFusion: reconstruction and tracking of non-rigid scenes in real-time. In: Proceedings of the IEEE Conference on Computer Vision and Pattern Recognition, pp. 343–352 (2015)

18. Nicolau, S., Soler, L., Mutter, D., Marescaux, J.: Augmented reality in laparoscopic surgical oncology. Surg. Oncol. **20**(3), 189–201 (2011)

19. Niemeyer, M., Mescheder, L., Oechsle, M., Geiger, A.: Differentiable volumetric rendering: learning implicit 3D representations without 3D supervision. In: Proceedings of the IEEE/CVF Conference on Computer Vision and Pattern Recognition, pp. 3504–3515 (2020)

20. Oechsle, M., Peng, S., Geiger, A.: UNISURF: unifying neural implicit surfaces and radiance fields for multi-view reconstruction. In: Proceedings of the IEEE/CVF International Conference on Computer Vision, pp. 5589–5599 (2021)

21. Overley, S.C., Cho, S.K., Mehta, A.I., Arnold, P.M.: Navigation and robotics in spinal surgery: where are we now? Neurosurgery **80**(3S), S86–S99 (2017)

22. Pumarola, A., Corona, E., Pons-Moll, G., Moreno-Noguer, F.: D-NeRF: neural radiance fields for dynamic scenes. In: Proceedings of the IEEE/CVF Conference on Computer Vision and Pattern Recognition, pp. 10318–10327 (2021)

23. Song, J., Wang, J., Zhao, L., Huang, S., Dissanayake, G.: Dynamic reconstruction of deformable soft-tissue with stereo scope in minimal invasive surgery. IEEE Robot. Autom. Lett. **3**(1), 155–162 (2017)

24. Tancik, M., et al.: Fourier features let networks learn high frequency functions in low dimensional domains. Adv. Neural. Inf. Process. Syst. **33**, 7537–7547 (2020)

25. Wang, P., Liu, L., Liu, Y., Theobalt, C., Komura, T., Wang, W.: NeuS: learning neural implicit surfaces by volume rendering for multi-view reconstruction. arXiv preprint arXiv:2106.10689 (2021)
26. Wang, Y., Long, Y., Fan, S.H., Dou, Q.: Neural rendering for stereo 3D reconstruction of deformable tissues in robotic surgery. In: Wang, L., Dou, Q., Fletcher, P.T., Speidel, S., Li, S. (eds.) MICCAI 2022, Part VII. LNCS, vol. 13437, pp. 431–441. Springer, Cham (2022). https://doi.org/10.1007/978-3-031-16449-1_41
27. Yariv, L., et al.: Multiview neural surface reconstruction by disentangling geometry and appearance. Adv. Neural. Inf. Process. Syst. **33**, 2492–2502 (2020)
28. Zhou, H., Jayender, J.: EMDQ-SLAM: real-time high-resolution reconstruction of soft tissue surface from stereo laparoscopy videos. In: de Bruijne, M., Cattin, P.C., Cotin, S., Padoy, N., Speidel, S., Zheng, Y., Essert, C. (eds.) MICCAI 2021, Part IV. LNCS, vol. 12904, pp. 331–340. Springer, Cham (2021). https://doi.org/10.1007/978-3-030-87202-1_32

Surgical Video Captioning
with Mutual-Modal Concept Alignment

Zhen Chen[1], Qingyu Guo[1], Leo K. T. Yeung[2], Danny T. M. Chan[2],
Zhen Lei[1,3], Hongbin Liu[1,3], and Jinqiao Wang[1,3,4,5](✉)

[1] Centre for Artificial Intelligence and Robotics (CAIR), Hong Kong Institute of
Science and Innovation, Chinese Academy of Sciences, Beijing, China
zhen.chen@cair-cas.org.hk, jqwang@nlpr.ia.ac.cn
[2] Department of Surgery, The Chinese University of Hong Kong, Shatin, Hong Kong
[3] Institute of Automation, Chinese Academy of Sciences, Beijing, China
[4] Wuhan AI Research, Wuhan, China
[5] ObjectEye Inc., Beijing, China

Abstract. Automatic surgical video captioning is critical to understanding surgical procedures, and can provide the intra-operative guidance and the post-operative report generation. As the overlap of surgical workflow and vision-language learning, this cross-modal task expects
precise text descriptions of complex surgical videos. However, current
captioning algorithms neither fully leverage the inherent patterns of
surgery, nor coordinate the knowledge of visual and text modalities
well. To address these problems, we introduce the surgical concepts
into captioning, and propose the Surgical Concept Alignment Network
(SCA-Net) to bridge the visual and text modalities via surgical concepts. Specifically, to enable the captioning network to accurately perceive surgical concepts, we first devise the Surgical Concept Learning
(SCL) to predict the presence of surgical concepts with the representations of visual and text modalities, respectively. Moreover, to mitigate the semantic gap between visual and text modalities of captioning, we propose the Mutual-Modality Concept Alignment (MC-Align)
to mutually coordinate the encoded features with surgical concept representations of the other modality. In this way, the proposed SCA-Net
achieves the surgical concept alignment between visual and text modalities, thereby producing more accurate captions with aligned multi-modal
knowledge. Extensive experiments on neurosurgery videos and nephrectomy images confirm the effectiveness of our SCA-Net, which outperforms the state-of-the-arts by a large margin. The source code is available
at https://github.com/franciszchen/SCA-Net.

Keywords: Neurosurgery · Video caption · Surgical concept

1 Introduction

Automatic surgical video captioning is critical to understanding the surgery
with complicated operations, and can produce the natural language description

H. Greenspan et al. (Eds.): MICCAI 2023, LNCS 14228, pp. 24–34, 2023.
https://doi.org/10.1007/978-3-031-43996-4_3

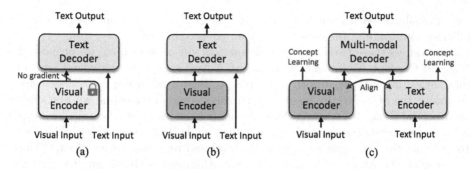

Fig. 1. Differences between existing captioning works (a) and (b), and our method (c). Different from directly mapping from visual input to text output in (a) and (b), we perform the surgical concept learning of two modality-specific encoders and mutually align two modalities for better multi-modal decoding.

with given surgical videos [24, 26]. In this way, these techniques can reduce the workload of surgeons with multiple applications, such as providing the intra-operative surgical guidance [17], generating the post-operative surgical report [4], and even training junior surgeons [8].

To generate text descriptions from input videos, existing captioning works [5, 9, 21, 24, 26] mostly consist of a visual encoder for visual representations and a text decoder for text generation. Some early works [9, 11, 21, 23] adopted a fixed object detector as the visual encoder to capture object representations for text decoding. This paradigm in Fig. 1(a) requires auxiliary annotations (*e.g.*, bounding box) to pre-train the visual encoder, and cannot adequately train the entire network for captioning. To improve performance with high efficiency in practice, recent works [13, 24, 25] followed the detector-free strategy, and opened up the joint optimization of visual encoder and text decoder towards captioning, as shown in Fig. 1(b). Despite great progress in this field, these works can be further improved with two limitations of surgical video captioning.

First, existing surgical captioning works [23, 24, 26] did not fully consider the inherent patterns of surgery to facilitate captioning. Due to the variability of lesions and surgical operations, surgical videos contain complex visual contents, and thus it is difficult to directly learn the mapping from the visual input to the text output. In fact, the same type of surgery has relatively fixed semantic patterns, such as using specific surgical instruments for a certain surgical action. Therefore, we introduce the **surgical concepts** (*e.g.*, surgical instruments, operated targets and surgical actions) from a semantic perspective, and guide the surgical captioning network to perceive these surgical concepts in the input video to generate more accurate surgical descriptions. Second, existing studies [9, 24, 26] simply processed visual and text modalities in sequential, while ignoring the semantic gap between these two modalities. This restricts the integration of visual and text modality knowledge, thereby damaging the captioning performance. Considering that both visual and text modalities revolve around the same set of surgical concepts, we aim to align the features in the visual and text modalities with each other through surgical concepts, and achieve more efficient multi-modal fusion for accurate text predictions.

To address these two limitations in surgical video captioning, we propose the Surgical Concept Alignment Network (SCA-Net) to bridge the visual and text modalities through the surgical concepts, as illustrated in Fig. 1(c). Specifically, to enable the SCA-Net to accurately perceive surgical concepts, we first devise the Surgical Concept Learning (SCL) to predict the presence of surgical concepts with the representations of visual and text modalities, respectively. Moreover, to mitigate the semantic gap between visual and text modalities of captioning, we propose the Mutual-Modality Concept Alignment (MC-Align) to mutually coordinate the encoded features with surgical concept representations of the other modality. In this way, the proposed SCA-Net achieves the surgical concept alignment between visual and text modalities, thereby producing more accurate captions with aligned multi-modal knowledge. To the best of our knowledge, this work represents the first effort to introduce the surgical concepts for the surgical video captioning. Extensive experiments are performed on neurosurgery video and nephrectomy image datasets, and demonstrate the effectiveness of our SCA-Net by remarkably outperforming the state-of-the-art captioning works.

2   Surgical Concept Alignment Network

2.1   Overview of SCA-Net

As illustrated in Fig. 2, the Surgical Concept Alignment Network (SCA-Net) follows the advanced captioning architecture [25], and consists of visual and text encoders, and a multi-modal decoder. We implement the visual encoder with VideoSwin [15] to capture the discriminative spatial and temporal representations from input videos, and utilize the Vision Transformer (ViT) [7] with causal mask [6] as the text encoder to exploit text semantics with merely previous text tokens. The multi-modal decoder with ViT structure takes both visual and text tokens as input, and finally generates the caption of the input video. Moreover, to accurately perceive surgical concepts in SCL (Sect. 2.2), the SCA-Net learns from surgical concept labels using separate projection heads after the visual and text encoders. In the MC-Align (Sect. 2.3), the visual and text tokens from two encoders are mutually aligned with the concept representations of the other modality for better multi-modal decoding.

2.2   Surgical Concept Learning

Previous surgical captioning works [23,24,26] generated surgical descriptions directly from input surgical videos. Considering the variability of lesions and surgical operations, these methods may struggle to understand complex visual contents and generate erroneous surgical descriptions, thereby hindering performance to meet clinical requirements. In fact, both the surgical video and surgical caption represent the same surgical semantics in different modalities. Therefore, we decompose surgical operations into surgical concepts, and guide these two modalities to accurately perceive the presence of surgical concepts, so as to better complete this cross-modal task.

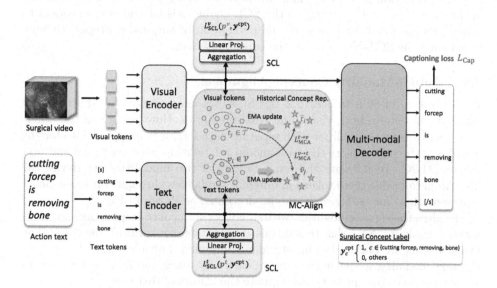

Fig. 2. Surgical Concept Alignment Network (SCA-Net) includes visual and text encoders, and a multi-modal decoder. The SCL supervises two encoders with projection heads by surgical concept labels, and the MC-Align mutually coordinates two modalities with concept representations for better multi-modal decoding.

Given a type of surgery, we regard the surgical instruments, surgical actions and the operated targets used in surgical videos as surgical concepts. Considering that both the visual input and the shifted text input contain the same set of surgical concepts, we find out which surgical concepts appear in the input surgical video by parsing the caption label. In this way, the presence of surgical concepts can be represented in a multi-hot surgical concept label $\boldsymbol{y}^{\mathrm{cpt}} \in \{0, 1\}^C$, where the surgical concepts that appear in the video are marked as 1 and the rest are marked as 0, and C is the number of possible surgical concepts. For example, the surgical video in Fig. 2 contains the instrument *cutting forcep*, the action *removing* and the target *bone*, and thus the surgical concept label $\boldsymbol{y}^{\mathrm{cpt}}$ represents these surgical concepts in corresponding dimensions.

To guide the visual modality to perceive surgical concepts, we aggregate visual tokens generated by the visual encoder in average, and add a linear layer to predict the surgical concepts of input videos, where the normalized output $\boldsymbol{p}^v \in [0, 1]^C$ estimates the probability of each surgical concept. We perform the multi-label classification using binary sigmoid cross-entropy loss, as follows:

$$L_{\mathrm{SCL}}^v = -\sum_{c=1}^{C} \boldsymbol{y}_c^{\mathrm{cpt}} \log\left(\boldsymbol{p}_c^v\right) + \left(1 - \boldsymbol{y}_c^{\mathrm{cpt}}\right) \log\left(1 - \boldsymbol{p}_c^v\right). \tag{1}$$

In this way, the visual tokens are supervised to contain discriminative semantics related to valid surgical concepts, which can reduce prediction errors in surgical descriptions. For the text modality, we also perform SCL for surgical concept prediction $\boldsymbol{p}_c^t \in [0,1]^C$ and calculate the loss L_{SCL}^t in the same way. By optimizing $L_{\text{SCL}} = L_{\text{SCL}}^v + L_{\text{SCL}}^t$, the SCL enables visual and text encoders to exploit multi-modal features with the perception of surgical concepts, thereby facilitating the SCA-Net towards the captioning task.

2.3 Mutual-Modality Concept Alignment

With the help of SCL in Sect. 2.2, our SCA-Net can perceive the shared set of surgical concepts in both visual and text modalities. However, given the differences between two modalities with separate encoders, it is inappropriate for the decoder to directly explore the cross-modal relationship between visual and text tokens [24,26]. To mitigate the semantic gap of two modalities, we devise the MC-Align to bridge these tokens in different modalities through surgical concept representations for better multi-modal decoding, as shown in Fig. 2.

To align these two modalities, we first collect surgical concept representations for each modality. Note that text tokens are separable for surgical concepts, while visual tokens are part of the input video containing multiple surgical concepts. For text modality, we parse the label of each text token and average text tokens of each surgical concept as $\boldsymbol{t}_c$, and update the historical text concept representations $\{\bar{\boldsymbol{t}}_c\}_{c=1}^C$ using Exponential Moving Average (EMA), as $\bar{\boldsymbol{t}}_c \leftarrow \gamma\bar{\boldsymbol{t}}_c + (1-\gamma)\boldsymbol{t}_c$, where the coefficient γ controls the updating for stable training and is empirically set as 0.9. For visual modality, we average visual tokens as the representation of each surgical concept present in the input video (i.e., $\boldsymbol{v}_c$ if surgical concept label $\boldsymbol{y}_c^{\text{cpt}} = 1$), and update the historical visual concept representations $\{\bar{\boldsymbol{v}}_c\}_{c=1}^C$ with EMA, as $\bar{\boldsymbol{v}}_c \leftarrow \gamma\bar{\boldsymbol{v}}_c + (1-\gamma)\boldsymbol{v}_c$. In this way, we obtain the text and visual concept representations with tailored strategies for the alignment.

Then, we mutually align visual and text concept representations with corresponding historical ones in another modality. For visual-to-text alignment, visual concept representations are expected to be similar to corresponding text concept representations, while differing from other text concept representations as possible. Thus, we calculate the alignment objective $L_{\text{MCA}}^{v \rightarrow t}$ with regard to surgical concepts [10], and the visual encoder can be optimized with the gradients of visual concept representations in backward, thereby gradually aligning visual modality to text modality. Similarly, text concept representations are also aligned to the historical visual ones, as text-to-visual alignment $L_{\text{MCA}}^{t \rightarrow v}$. The MC-Align is summarized as follows:

$$L_{\text{MCA}} = -\underbrace{\sum_{v_i \in \mathcal{V}} \log \frac{\exp\left(\boldsymbol{v}_i \cdot \bar{\boldsymbol{t}}_i\right)}{\sum_{c=1}^C \exp\left(\boldsymbol{v}_i \cdot \bar{\boldsymbol{t}}_c\right)}}_{L_{\text{MCA}}^{v \rightarrow t}} - \underbrace{\sum_{t_j \in \mathcal{T}} \log \frac{\exp\left(\boldsymbol{t}_j \cdot \bar{\boldsymbol{v}}_j\right)}{\sum_{c=1}^C \exp\left(\boldsymbol{t}_j \cdot \bar{\boldsymbol{v}}_c\right)}}_{L_{\text{MCA}}^{t \rightarrow v}}, \quad (2)$$

where $\mathcal{V}$ and $\mathcal{T}$ denote all visual and text representations respectively, and $\cdot$ is the inner product of vectors. In this way, the MC-Align aligns visual and text representations with each other modality according to the surgical concept, thus benefiting multi-modal decoding for captioning.

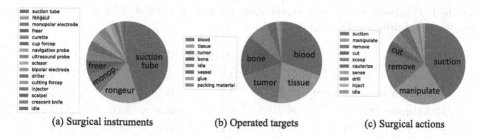

(a) Surgical instruments   (b) Operated targets   (c) Surgical actions

Fig. 3. Surgical concepts and proportions in neurosurgery video captioning dataset.

2.4 Optimization

For the surgical captioning task, we adopt standard captioning loss L_{Cap} to optimize the cross-entropy of each predicted word based on previous words $y_{<t}$ and input video x, as follows:

$$L_{\text{Cap}} = -\sum_{t=1}^{T} \log p\left(y_t \mid y_{<t}, x\right), \tag{3}$$

where T is the length of caption prediction. Overall, the final objective of SCA-Net is summarized as $L = L_{\text{Cap}} + \lambda_1 L_{\text{SCL}} + \lambda_2 L_{\text{MCA}}$, where loss coefficients λ_1 and λ_2 control the trade-off of SCL and MC-Align. By optimizing this final objective L, the proposed SCA-Net can achieve multi-modal concept alignment, and generate superior descriptions for the surgical video captioning.

3 Experiment

3.1 Dataset and Implementation Details

Neurosurgery Video Captioning Dataset. To evaluate the effectiveness of surgical video captioning, we collect a large-scale dataset with 41 surgical videos of endonasal skull base neurosurgery. These surgical videos are recorded at the Prince of Wales Hospital, Chinese University of Hong Kong, where surgeons remove pituitary tumors through the endonasal corridor to the skull base. After necessary data cleaning, we divide these surgical videos with resolution of $1,920 \times 1,080$ into $11,004$ thirty-second video clips with clear surgical purposes. These video clips are annotated under Tool-Tissue Interaction (TTI) principle [18], and include a total of 16 instruments, 8 targets, and 10 surgical actions. The annotation preprocessing follows [26] using NLTK [16] toolkit. The proportion of surgical concepts is illustrated in Fig. 3. We split these video clips at patient-level, where the video clips of 31 patients are used for training and the rest of 10 patients are utilized for test.

EndoVis Image Captioning Dataset. We further compare our method with state-of-the-arts on the public EndoVis-2018 Image Captioning Dataset [1,23].

Table 1. Comparison on neurosurgery video captioning dataset. Best and second best results are **highlighted** and <u>underlined</u>.

Method	BLEU@4	METEOR	SPICE	ROUGE	CIDEr
VideoSwin + Self-Seq [21]	34.4	24.8	39.7	53.5	183.8
VideoSwin + AOANet [9]	41.5	29.7	46.5	58.0	288.1
SIG-Former [26]	36.2	30.1	35.8	52.0	181.7
SwinMLP-TranCAP [24]	39.8	28.7	39.2	51.9	195.9
M^2Transformer [5]	43.2	30.9	46.6	57.8	317.8
Ours *w/o* SCL, MC-Align	40.3	29.6	46.4	55.7	279.8
Ours *w/o* MC-Align	44.3	32.4	52.6	61.8	298.9
Ours *w/o* SCL	<u>45.8</u>	<u>32.9</u>	<u>53.2</u>	<u>62.7</u>	<u>325.1</u>
Ours	**48.1**	**35.1**	**56.1**	**64.9**	**368.4**

This dataset reveals robotic nephrectomy procedures acquired by the da Vinci X or Xi system, and is annotated with surgical actions between 9 possible tools and surgical targets [23]. We follow the official split in [24] with 11 sequences for training and 3 sequences for test. In this way, these two datasets can comprehensively evaluate the captioning tasks under both surgical videos and images.

Implementation Details. We implement our SCA-Net and state-of-the-art captioning methods [5,9,21,24,26] in PyTorch [20]. We optimize the SCA-Net and compared captioning methods using Adam with the batch size of 12 for both captioning datasets. All models are trained for 20 and 50 epochs in neurosurgery and EndoVis datasets, respectively. We adopt the step-wise learning rate decay strategy to facilitate training convergence, where the learning rate is initialized as 1×10^{-2} and halved after every 5 epochs. The loss coefficients λ_1 of L_{SCL} and λ_2 of of L_{MCA} are empirically set to 0.1 and 0.01, respectively. All experiments are performed on a single NVIDIA A100 GPU.

Evaluation Metrics. To evaluate the captioning performance, we adopt standard metrics, including BLEU@4 [19], METEOR [3], SPICE [2], ROUGE [12] and CIDEr [22]. Specifically, BLEU@4 [19] evaluates the 4-gram precision of the predicted caption, and CIDEr [22] is based on the n-gram similarity with TF-IDF weights. METEOR [3] considers both precision and recall. ROUGE [12] and SPICE [2] measure the matching between predictions and ground truth. The higher scores of these metrics indicate better performance in surgical captioning.

3.2   Comparison on Neurosurgery Video Captioning

To evaluate the performance of our SCA-Net, we perform a comprehensive comparison with the state-of-the-art captioning methods, including Self-Seq [21], AOANet [9], SIG-Former [26], M^2Transformer [5], and SwinMLP-TranCAP [24]. As illustrated in Table 1, our SCA-Net achieves the best performance, with the overwhelming BLEU@4 of 48.1%, METEOR of 35.1% and CIDEr of 368.4%.

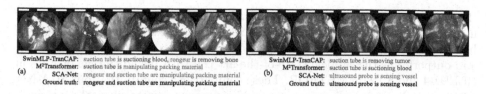

(a) SwinMLP-TranCAP: suction tube is suctioning blood, rongeur is removing bone
M²Transformer: suction tube is manipulating packing material
SCA-Net: rongeur and suction tube are manipulating packing material
Ground truth: rongeur and suction tube are manipulating packing material

(b) SwinMLP-TranCAP: suction tube is removing tumor
M²Transformer: suction tube is suctioning blood
SCA-Net: ultrasound probe is sensing vessel
Ground truth: ultrasound probe is sensing vessel

Fig. 4. The qualitative comparison between our SCA-Net and state-of-the-arts. With surgical concept alignment, our SCA-Net generates more accurate surgical descriptions.

Noticeably, our SCA-Net outperforms the surgical captioning work, SwinMLP-TranCAP [24], by a large margin, e.g., 16.9% in SPICE and 13.0% in ROUGE. This advantage confirms that the proposed surgical concept alignment can alleviate the modalities gap in surgical captioning. Moreover, compared with the second-best M²Transformer [5] with meshed attention between the visual encoder and the text decoder, our SCA-Net obtains superior performance with a remarkable increase of 9.5% in SPICE and 7.1% in ROUGE. These experimental results demonstrate the performance advantage of our SCA-Net over state-of-the-arts in the neurosurgery video captioning.

Ablation Study. To further validate the effectiveness of SCL and MC-Align, we perform the detailed ablation study in Table 1. Specifically, we implement three ablative baselines of the proposed SCA-Net, by removing the MC-Align (denoted as w/o MC-Align) and the SCL (denoted as w/o SCL) individually, as well as removing both (denoted as w/o SCL, MC-Align). As illustrated in Table 1, the proposed SCL and MC-Align can bring an individual improvement of 4.0% and 5.5% in BLEU@4, respectively, to the baseline of 40.3%. Furthermore, the SCL and MC-Align can work together to facilitate the captioning, with a BLEU@4 gain of 7.8%. These ablation experiments confirm that the proposed SCL and MC-Align play an important role in solving the modality gap in surgical video captioning, resulting in the performance advantage of our SCA-Net.

Qualitative Analysis. We present qualitative results of our SCA-Net and state-of-the-arts [5, 24] on neurosurgery video captioning. In Fig. 4(a), SwinMLP-TranCAP [24] and M²Transformer [5] incorrectly predict the operated targets and ignore important surgical instruments, respectively, and both methods [5, 24] cannot recognize the rare instrument *ultrasound probe* as well as the corresponding surgical action in Fig. 4(b). With the help of surgical concept alignment, our SCA-Net can perceive the surgical concepts present in the surgical videos and thus generate correct descriptions in these two complex videos.

3.3 Comparison on EndoVis Image Captioning

To further confirm the effectiveness of surgical captioning, we perform the comparison on the public EndoVis image captioning dataset. As shown in Table 2, the end-to-end captioning methods [24, 26] outperform the detector-based works using instrument bounding box as auxiliary annotations [9, 21], by optimizing the visual encoder to meet the requirement of the captioning task. In particular, our SCA-Net with Swin Transformer [14] as visual encoder achieves the best performance of four metrics (*e.g.*, 47.6% in BLEU@4 and 58.4% in SPICE), and outperforms the surgical state-of-the-art [24] with the advantage of 7.3% in BLEU@4 and 5.1% in METEOR. These comparisons confirm that our SCA-Net with surgical concept alignment can produce more accurate surgical captions.

Table 2. Comparison on EndoVis-2018 image captioning dataset. Best and second best results are **highlighted** and underlined.

Method	Aux. Anno	BLEU@4	METEOR	SPICE	CIDEr
FasterRCNN + Self-seq [21]	✓	29.5	28.3	49.6	180.1
FasterRCNN + AOANet [9]	✓	37.7	32.4	58.0	181.1
SIG-Former [26]	✗	42.6	33.5	52.4	282.6
SwinMLP-TranCAP [24]	✗	40.3	31.3	54.7	250.4
M^{2}Transformer [5]	✗	43.0	32.5	55.3	245.2
Ours	✗	**47.6**	**36.4**	**58.4**	**300.8**

4   Conclusion

To achieve accurate surgical video captioning, we propose the SCA-Net to mitigate the semantic gap of visual and text modalities with surgical concepts. Specifically, we devise the SCL to enable the SCA-Net with the perception of surgical concepts in visual and text modalities, respectively. Moreover, we propose the MC-Align to mutually coordinate visual and text representations with surgical concept representations of the other modality for multi-modal decoding, thereby generating more accurate captions with aligned multi-modal knowledge. Extensive experiments on neurosurgery and nephrectomy datasets confirm the advantage of our SCA-Net over state-of-the-arts on the surgical captioning.

Acknowledgments. This work is supported by National Key R&D Program of China under Grant No. 2021YFE0205700, National Natural Science Foundation of China (No. 62276260, 62076235, 62176254, 61976210, 62002356, 62006230), sponsored by Zhejiang Lab (No. 2021KH0AB07) and the InnoHK program.

References

1. Allan, M., et al.: 2018 robotic scene segmentation challenge. arXiv preprint arXiv:2001.11190 (2020)
2. Anderson, P., Fernando, B., Johnson, M., Gould, S.: SPICE: semantic propositional image caption evaluation. In: Leibe, B., Matas, J., Sebe, N., Welling, M. (eds.) ECCV 2016. LNCS, vol. 9909, pp. 382–398. Springer, Cham (2016). https://doi.org/10.1007/978-3-319-46454-1_24

3. Banerjee, S., Lavie, A.: METEOR: an automatic metric for MT evaluation with improved correlation with human judgments. In: ACL Workshop, pp. 65–72 (2005)
4. Bieck, R., et al.: Generation of surgical reports using keyword-augmented next sequence prediction. Curr. Direct. Biomed. Eng. **7**(2), 387–390 (2021)
5. Cornia, M., Stefanini, M., Baraldi, L., Cucchiara, R.: Meshed-memory transformer for image captioning. In: CVPR, pp. 10578–10587 (2020)
6. Czempiel, T., Paschali, M., Ostler, D., Kim, S.T., Busam, B., Navab, N.: OperA: attention-regularized transformers for surgical phase recognition. In: de Bruijne, M., et al. (eds.) MICCAI 2021. LNCS, vol. 12904, pp. 604–614. Springer, Cham (2021). https://doi.org/10.1007/978-3-030-87202-1_58
7. Dosovitskiy, A., et al.: An image is worth 16×16 words: transformers for image recognition at scale. In: ICLR (2021)
8. Elnikety, S., Badr, E., Abdelaal, A.: Surgical training fit for the future: the need for a change. Postgrad. Med. J. **98**(1165), 820–823 (2022)
9. Huang, L., Wang, W., Chen, J., Wei, X.Y.: Attention on attention for image captioning. In: ICCV, pp. 4634–4643 (2019)
10. Khosla, P., et al.: Supervised contrastive learning. In: NeurIPS, vol. 33, pp. 18661–18673 (2020)
11. Lin, C., Zheng, S., Liu, Z., Li, Y., Zhu, Z., Zhao, Y.: SGT: scene graph-guided transformer for surgical report generation. In: Wang, L., Dou, Q., Fletcher, P.T., Speidel, S., Li, S. (eds.) MICCAI 2022. LNCS, vol. 13437, pp. 507–518. Springer, Cham (2022). https://doi.org/10.1007/978-3-031-16449-1_48
12. Lin, C.Y.: ROUGE: a package for automatic evaluation of summaries. In: Text Summarization Branches Out, pp. 74–81 (2004)
13. Lin, K., et al.: SwinBERT: end-to-end transformers with sparse attention for video captioning. In: CVPR, pp. 17949–17958 (2022)
14. Liu, Z., et al.: Swin transformer: hierarchical vision transformer using shifted windows. In: ICCV, pp. 10012–10022 (2021)
15. Liu, Z., et al.: Video swin transformer. In: CVPR, pp. 3202–3211 (2022)
16. Loper, E., Bird, S.: NLTK: the natural language toolkit. arXiv preprint cs/0205028 (2002)
17. Madani, A., et al.: Artificial intelligence for intraoperative guidance: using semantic segmentation to identify surgical anatomy during laparoscopic cholecystectomy. Ann. Surg. (2020)
18. Nwoye, C.I., et al.: CholecTriplet 2021: a benchmark challenge for surgical action triplet recognition. Med. Image Anal. **86**, 102803 (2023)
19. Papineni, K., Roukos, S., Ward, T., Zhu, W.J.: BLEU: a method for automatic evaluation of machine translation. In: ACL, pp. 311–318 (2002)
20. Paszke, A., et al.: PyTorch: an imperative style, high-performance deep learning library. arXiv preprint arXiv:1912.01703 (2019)
21. Rennie, S.J., Marcheret, E., Mroueh, Y., Ross, J., Goel, V.: Self-critical sequence training for image captioning. In: CVPR, pp. 7008–7024 (2017)
22. Vedantam, R., Lawrence Zitnick, C., Parikh, D.: CIDEr: consensus-based image description evaluation. In: CVPR, pp. 4566–4575 (2015)
23. Xu, M., Islam, M., Lim, C.M., Ren, H.: Class-incremental domain adaptation with smoothing and calibration for surgical report generation. In: de Bruijne, M., et al. (eds.) MICCAI 2021. LNCS, vol. 12904, pp. 269–278. Springer, Cham (2021). https://doi.org/10.1007/978-3-030-87202-1_26

24. Xu, M., Islam, M., Ren, H.: Rethinking surgical captioning: end-to-end window-based MLP transformer using patches. In: Wang, L., Dou, Q., Fletcher, P.T., Speidel, S., Li, S. (eds.) MICCAI 2022. LNCS, vol. 13437, pp. 376–386. Springer, Cham (2022). https://doi.org/10.1007/978-3-031-16449-1_36
25. Yu, J., Wang, Z., Vasudevan, V., Yeung, L., Seyedhosseini, M., Wu, Y.: CoCa: contrastive captioners are image-text foundation models. Trans. Mach. Learn. Res. (2022)
26. Zhang, J., Nie, Y., Chang, J., Zhang, J.J.: Surgical instruction generation with transformers. In: de Bruijne, M., et al. (eds.) MICCAI 2021. LNCS, vol. 12904, pp. 290–299. Springer, Cham (2021). https://doi.org/10.1007/978-3-030-87202-1_28

SEDSkill: Surgical Events Driven Method for Skill Assessment from Thoracoscopic Surgical Videos

Xinpeng Ding[1], Xiaowei Xu[2(✉)], and Xiaomeng Li[1(✉)]

[1] The Hong Kong University of Science and Technology, Hong Kong SAR, China
eexmli@ust.hk
[2] Guangdong Cardiovascular Institute, Guangdong Provincial People's Hospital
(Guangdong Academy of Medical Sciences), Southern Medical University,
Guangzhou, China
xiao.wei.xu@foxmail.com

Abstract. Thoracoscopy-assisted mitral valve replacement (MVR) is a crucial treatment for patients with mitral regurgitation and demands exceptional surgical skills to prevent complications and enhance patient outcomes. Consequently, surgical skill assessment (SKA) for MVR is essential for certifying novice surgeons and training purposes. However, current automatic SKA approaches have inherent limitations that include the absence of public thoracoscopy-assisted surgery datasets, exclusion of inter-video relationships, and limited to SKA of a single short surgical action. This paper introduces a novel clinical dataset for MVR, which is the first thoracoscopy-assisted long-form surgery dataset to the best of our knowledge. Our dataset, unlike existing short video clips that contain single surgical action, includes videos of the whole MVR procedure that capture multiple complex skill-related surgical events. To tackle the challenges posed by MVR, we propose a novel method called **S**urgical **E**vents **D**riven **S**kill assessment (SEDSkill). Our key idea is to develop a long-form surgical events-driven method for skill assessment, which is based on the insight that the skill level of a surgeon is closely tied to the occurrence of inappropriate operations such as excessively long suture repairing times. SEDSkill incorporates an event-aware module that automatically localizes skill-related events, thus extracting local semantics from long-form videos. Additionally, we introduce a difference regression block to learn imperceptible discrepancies, which enables precise and accurate surgical skills assessment. Extensive experiments demonstrate that our proposed method outperforms state-of-the-art approaches. Our code is available at https://github.com/xmed-lab/SEDSkill.

Keywords: Surgical skill assessment · Long-form video · Thoracoscopy-assisted surgery

H. Greenspan et al. (Eds.): MICCAI 2023, LNCS 14228, pp. 35–45, 2023.
https://doi.org/10.1007/978-3-031-43996-4_4

1   Introduction

Thoracoscopy-assisted mitral valve replacement (MVR) has become routine for the treatment of mitral valve regurgitation [4]. Compared to other surgeries such as laparoscopic operations, thoracoscopy-assisted MVR requires higher surgical skills due to the intricate structure of the heart, the mitral valve's proximity to other vital cardiac structures, and the geometric limitations of the surgical field [11]. Improving surgical skills can prevent avoidable complications [9], leading to better patient outcomes, such as improved long-term survival and reduced postoperative complications [2,3]. Therefore, surgical skill assessment (SKA), *i.e.*, evaluating the skill level of surgeons, is essential in the training and certification of novice surgeons [19,20,26].

Traditionally, SKA has been reliant on manual observation by experienced surgeons either in the operating room or via recorded videos, as described by Reznick in his work on teaching [19]. However, this method is subjective, time-consuming, and not very efficient for use in surgical education. To address these limitations, researchers have increasingly focused on developing automatic SKA tools. While current automatic SKA approaches [10,12,16,17,22] have demonstrated success on simulated and laparoscopic datasets, their application to thoracoscopy-assisted MVR poses several challenges. First, to the best of our knowledge, there are no publicly available clinical datasets for thoracoscopy-assisted surgery. Second, most existing methods [5–7,10,12,16,17] focus solely on the global information within a single video to perform SKA, such as regressing a singular skill score from the video. However, these methods disregard the inter-video information, such as subtle differences between various videos, that could be critical in predicting surgical skill scores [13]. For instance, differences in haemorrhage loss, suture repairing times, and thread twining times among videos can have a significant impact on the final scores. Generally, more thread winding, haemorrhage, and suture repairing can indicate a lower skill level; see Fig. 1(a).

To address the above challenges, we collect a new dataset for SKA, which is the first-ever long-form thoracoscopy-assisted MVR video dataset. Our dataset offers longer video duration and more surgical events with corresponding labels in comparison to the currently available public datasets such as JIGSAWS [8] or HeiChole [24]; see Fig. 1(b). Then, we present a novel **S**urgical **E**vents **D**riven **Skill** assessment (**SEDSkill**) method to address the limitations of current automatic methods for MVR assessment. Unlike prior work [10,12,17], our key idea is to develop a long-form surgical events-driven method for skill assessment, which is based on the crucial insight that the skill level of a surgeon is closely tied to the occurrence of inappropriate operations such as excessively long suture repairing times. To achieve it, we propose a **novel local-global difference method** that can *learn inter-video relations between both the global long-form and local surgical events correlated semantics*. The method includes an event-aware module and a difference regression module. The event-aware module can automatically localize skill-related surgical events and extract their corresponding features to represent the local event semantics. As surgical skill is highly correlated with

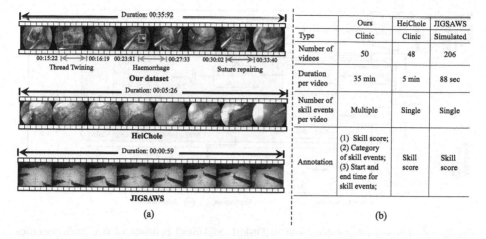

Fig. 1. (a) Visualization of different surgical skill datasets: Our collected dataset, HeiChole [24] and JIGSAWS [8]. (b) Comparison of different datasets. Unlike other datasets where each video typically contains only one surgical action (e.g., dissection or suturing), our MVR dataset provides long-form videos with multiple skill-related events and their corresponding labels.

the occurrence of inappropriate events, this module is crucial for precise SKA. To enable the accurate detection of slight differences between videos, our difference regression module captures the relationships among videos and enhances the model's ability to detect subtle variations. By incorporating video-wise and event-wise difference learning, our framework can capture both local and global inter-video relations, thereby enabling precise SKA.

In summary, our contributions are three-fold: (1) We introduce a novel SED-Skill method that aims to design a long-form, surgical events-driven approach for SKA, and it is the first method designed specifically for SKA in thoracoscopic surgical videos. (2) We propose a local-global difference framework that can learn inter-video relations between both the global long-form and local surgical events correlated semantics, thereby enabling enhanced SKA performance. (3) Experimental results demonstrate that our method outperforms existing SKA methods, as well as methods designed for video quality assessment in computer vision. This indicates the great potential of our method for use in clinical practice. Our code will be publicly released upon paper acceptance.

2   Method

Figure 2 illustrates our SEDSkill framework for SKA, which takes a surgical video as input and regresses a surgical skill score. Our proposed framework consists of two main modules: (a) a basic regression module to output the skill score for each input, (b) a local-global difference module to learn both video-level and event-level inter-video differences for precise assessment.

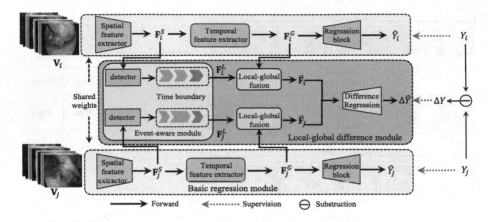

Fig. 2. Illustration of our proposed SEDSkill. SEDSkill consists of two main components: a basic regression module and a local-global difference module.

2.1 Basic Regression Module

The basic regression module aims to regress a surgical skill score, *i.e.*, Y_i, for a raw surgical video input $\mathbf{V}_i \in \mathbb{R}^{T_i \times H \times W}$, where T_i is the duration, H and W are the height and width of each frame. As shown in Fig. 2, the basic regression module consists of three components: a spatial feature extractor, a temporal feature extractor, and a regression block. Specifically, the video $\mathbf{V}_i$ is first fed into the spatial feature extractor to obtain the spatial feature, denoted as $\mathbf{F}_i^S \in \mathbb{R}^{T_i \times D_s}$, where D_s is the dimension of features, followed by a temporal feature extractor to model the intra-video relations to generate the global video feature $\mathbf{F}_i^G \in \mathbf{R}^{T \times D_t}$. Finally, a regression block consisting of several convolutional, max-pooling layers and a fully connected layer is applied to map $\mathbf{F}_i^G$ to the skill score $\hat{Y}_i$. Then, the loss function is to minimize the differences between predicted $\hat{Y}_i$ and the ground-truth Y_i as follows:

$$\mathcal{L}_{reg} = 1/N \sum_{i=1}^{N} (\hat{Y}_i - Y_i)^2, \tag{1}$$

where N is the number of videos.

2.2 Local-Global Difference Module

Surgical Event-Aware Module for Local Information. Unlike prior datasets used for skill assessment, our MVR video dataset is much longer, ranging from 30 min to 1 h, and consists of multiple skill-related events such as thread twining, haemorrhage, and suture repairing; see Fig. 1. As surgical skill is highly correlated to the qualities of surgical events, directly using the basic regression module to predict a score from the long video would consider too many irrelevant parts, thus degrading the regression performance. To encourage the model to focus on the skill-related parts and remove the less informative ones, we devise an event-aware module to localize the skill-related events, i.e., haemorrhage, surgical thread twining and surgical suture repair, from the long videos and extract the local event-level features, as shown in Fig. 2(b).

Specifically, given the spatial feature, e.g., $\mathbf{F}_i^S$, we introduce an event detector, which is a transformer-like network that maps the video features to the classified logits and regressed start/end time. The detailed architecture can refer to [28]. Formally, the prediction of the detector is a set for each time t which can be formulated as:

$$\mathcal{G}_t = \{\mathbf{p}_t, d_t^s, d_t^e\}, \tag{2}$$

where $\mathbf{p}_t \in \mathbb{R}^4$ consists of 4 values (including the background), which indicates the probability of event category, $d_t^s > 0$ and $d_t^e > 0$ denote the distance between time t to start and end time of events. Note that if p_t equals to zero, $d_t^s > 0$ and $d_t^e > 0$ are not defined. Then, following [28] the loss function for the detector is defined as:

$$\mathcal{L}_{det} = \sum_t \left(\mathcal{L}_{cls} + \lambda_{loc} \mathbf{1}_{c_t} \mathcal{L}_{loc}\right) / N_+, \tag{3}$$

where N_+ is the number of positive frames, $\mathcal{L}_{cls}$ is a focal loss [15] and $\mathcal{L}_{loc}$ is a DIoU loss [29]. $\mathbf{1}_{c_t}$ is the indicator function to identify where the time t is within a event. λ_{loc} is set to 1 following [28]. Note that the detector is pre-trained and fixed during the training of the basic regression module and the local-global difference module. After obtaining the pre-trained detector, we generate the event confidence map for each video denoted by $\mathbf{A}_i = [a_t]_{t=1}^{T_i}$, where $\mathbf{A}_i \in \mathbb{R}^{T_i \times 1}$, $a_t = \max \mathbf{p}_t$ is the confidence for each time t and T_i is the duration for the $\mathbf{V}_i$. Then, the local event-level feature is obtained by the multiplication of the confidence values and the global video feature, i.e., $\mathbf{F}_i^L = \mathbf{A}_i \circ \mathbf{F}_i^S$, where $\mathbf{F}_i^L \in \mathbb{R}^{T_i \times D_s}$ and $\circ$ is the element-wise multiplication.

Local-Global Fusion. We introduce the local-global fusion module to aggregate the local (i.e.event-level) and global (long-form video) semantics. Formally, we can define the local-global fusion as $\overline{\mathbf{F}}_i = \text{Fusion}(\mathbf{F}_i^G, \mathbf{F}_i^L)$, where $\overline{\mathbf{F}}_i \in \mathbb{R}^{T_i \times (D_t + D_s)}$. This module can be implemented by different types and we will conduct an ablation study to analyze the effect of this module in Table 3.

Difference Regression Block. Most surgeries of the same type are performed in similar scenes, leading to subtle differences among surgical videos. For example, in MVR, the surgeon first stitches two lines on one side using a needle, and then passes one of the lines through to the other side, connecting it to the extracorporeal circulation tube. Although these procedures are performed in a similar way, the imperceptible discrepancies are very important for accurately assessing surgical skills. Hence, we first leverage the relation block to capture the inter-video semantics. We use the features of the pairwise videos, i.e., $\overline{\mathbf{F}}_i$ and $\overline{\mathbf{F}}_j$, for clarity. Since attention [5,23] is widely used for capturing relations, we formulate the detailed relation block in the attention manner as follows:

$$\overline{\mathbf{F}}_{j \to i} = \text{Attention}\left(\mathbf{Q}_j; \mathbf{K}_i; \mathbf{V}_i\right) = \text{softmax}\left(\frac{\mathbf{Q}_j \mathbf{K}_i^\top}{\sqrt{D}}\right) \mathbf{V}_i, \tag{4}$$

where $\mathbf{Q}_j = \overline{\mathbf{F}}_j \mathbf{W}^q$, $\mathbf{K}_i = \overline{\mathbf{F}}_i \mathbf{W}^k$ and $\mathbf{V}_i = \overline{\mathbf{F}}_i \mathbf{W}^v$ are linear layers, $\sqrt{D}$ controls the effect of growing magnitude of dot-product with larger D [23]. Since $\overline{\mathbf{F}}_{j \to i}$ only learn the attentive relation from $\mathbf{F}_j$ to $\mathbf{F}_i$. We then learn the bi-direction attentive relation by $\overline{\mathbf{F}}_{i-j} = \text{Relation}(\overline{\mathbf{F}}_i, \overline{\mathbf{F}}_j) = \overline{\mathbf{F}}_{j \to i} + \overline{\mathbf{F}}_{i \to j}$, where $\overline{\mathbf{F}}_{i \to j} = \text{Attention}(\mathbf{Q}_i; \mathbf{K}_j; \mathbf{V}_j)$.

After that, we use the difference regression block to map $\overline{\mathbf{F}}_{i-j}$ to the difference scores $\Delta\hat{Y}$. Then, we minimize the error as follows:

$$\mathcal{L}_{diff} = (\Delta\hat{Y} - \Delta Y)^2, \tag{5}$$

where ΔY is the ground-truth of the difference scores between the pair videos, which can be computed by $|Y_i - Y_j|$. By optimizing $\mathcal{L}_{diff}$, the model would be able to distinguish differences between videos for precise SKA.

Finally, the overall loss function of our proposed method is as follows:

$$\mathcal{L} = \mathcal{L}_{reg} + \lambda_{diff}\mathcal{L}_{diff}, \tag{6}$$

where λ_{diff} is the hyper-parameter to control the weight between two loss functions (set to 1 empirically).

3   Experiments

Datasets. We collect the data from our collaborating hospitals. The data collection process follows the same protocol in a well-established study [2]. The whole procedure of the surgery is recorded by a surgeon's view camera. Each surgeon will submit videotapes when performing thoracoscopy-assisted MVR in the operating rooms. We have collected 50 high-resolution videos of thoracoscopy-assisted MVR from surgeons and patients, with a resolution of 1920×1080 and 25 frames per second. Each collected video lasts 30 min - 1 h. 50 videos are randomly divided into training and testing subsets containing 38, and 12 videos, respectively. To evaluate skill level, each video will be rated along various dimensions of technical skill on a scale of 1 to 9 (with higher scores indicating more advanced skill) by at least ten authoritative surgeons who are unaware of the identity of the operating surgeon. Furthermore, we also provide the annotations (including the category and corresponding start and end time) for three skill-related events, *i.e.*, haemorrhage, surgical thread twining and surgical suture repair times. The detailed annotation examples are illustrated in Fig. 1(a).

Implementation Details. Our model is implemented on an NVIDIA GeForce RTX 3090 GPU. We use a pre-trained inception-v3 [21] and MS-TCN [5] as the spatial and temporal feature extractors, respectively. For each video, we sample one frame per second. As the durations of different videos vary, we resample all videos to 1000 frames. We trained our model using an Adam optimizer with learning rates initialized at $1e - 3$. The total number of epochs is 200, and the batch size is 4.

Table 1. Results on our MVR dataset. †: we implement existing action quality assessment methods on our dataset. ⋆: we run existing surgical skill assessment methods on our dataset.

Method	MAE	Corr
CoRe† [27]	2.89	0.40
TPT† [1]	2.68	0.42
C3D-LSTM⋆ [18]	3.01	0.20
C3D-SVR⋆ [18]	2.92	0.14
MTL-VF [25]	2.35	0.31
ViSA⋆ [14]	2.31	0.33
Ours	**1.83**	**0.54**

Table 2. Ablation study of the local-global difference module. "Base" refers to using the basic regression module containing only global information. "EAM" refers to the event-aware module capturing local information. "DRB" indicates the difference regression block extracting inter-video information.

Method	Base	EAM	DRB	MAE	Corr
Global	✓			2.82	0.25
Local		✓		2.49	0.33
Local-global	✓	✓		2.45	0.37
Global-difference	✓		✓	2.15	0.39
Local-difference		✓	✓	2.11	0.45
Full (ours)	✓	✓	✓	**1.83**	**0.54**

Evaluation Metrics. Following previous works [12,14], we measured the performance of our model using Spearman's Rank Correlation (Corr) and Mean Absolute Error (MAE). Lower values of Corr and MAE indicate better results.

3.1 Comparison with the State-of-the-Art Methods

We compare our method with existing state-of-the-art methods in action quality assessment (AQA) [1,27] and surgical skill assessment (SKA) [14,18]. Note that the spatial and temporal feature extractors for ViSA, CoRe and TPT as the same as our method. As shown in Table 1, our method achieved the best performance with an MAE score of 1.83 and a Corr score of 0.54. The comparison demonstrates that our method not only outperformed existing SKA methods but also outshined existing AQA methods by a clear margin in surgical skill assessment.

3.2 Ablation Study

Effectiveness of Proposed Modules. Table 2 shows the effectiveness of our proposed local-global difference module. We can see that using the local features from the event-aware module (EAM) can outperform the global ones, which indicates the importance of skill-related events. Furthermore, incorporating the difference regression module (DRB) can benefit both local and global features, *e.g.*, improving the MAE of Local from 2.49 to 2.11. Finally, the combination of all proposed modules can achieve the best performance, *i.e.*, the MAE of 1.83.

Analysis of Local-Global Fusion. We explore the effect of different types of local-global fusion in Table 3. "Concatenation" indicates concatenating the two features in the feature dimension. "Multiplication" indicates the element-wise multiplication of the two features. The results show that the different fusion

Table 3. Analysis of local-global fusion.

	MAE	Corr
Concatenation	**1.83**	**0.54**
Multiplication	1.98	0.45
Attention	1.85	0.51

Table 4. Anlysis of the attention in the difference block. "Unidirectional" and "Bidirectional" indicate the unidirectional and bidirectional attentive relations (See Eq. 4).

Attention	MAE	Corr
Unidirectional ($\overline{\mathbf{F}}_{i\rightarrow j}$)	1.91	0.47
Bidirectional ($\overline{\mathbf{F}}_{i-j}$)	**1.83**	**0.54**

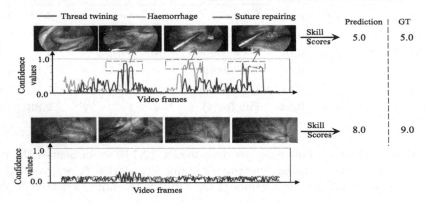

Fig. 3. Qualitative results of sampled pairwise videos. For each video, we visualize its confidence values along video frames for skill-related events. Specifically, the green, orange, and blue lines indicate the confidences scores of thread twining ($\mathbf{A}_0$), haemorrhage ($\mathbf{A}_1$) and suture repairing ($\mathbf{A}_2$) along video frames. It is worth noting that the occurrence of inappropriate surgical events such as haemorrhage and more thread twining times is highly correlated with the surgical skill level. Therefore, a lower confidence value indicates a lower probability of an event occurring, leading to a higher skill score.

methods can achieve comparable performance, indicating that different fusion methods can effectively aggregate local and global information. In this paper, we select concatenation as our default fusion method.

Effect of the Attention in the Difference Block. In Sect. 2.2, we implement the difference block by the attention, shown in Eq. 4. Here, we conduct the ablation study on the effect of different attentions in Table 4. The results indicate that using bidirectional attention, *i.e.*, $\overline{\mathbf{F}}_{i-j}$, can achieve better performance, compared with the uidirectional one.

Qualitative Results. Figure 3 shows the qualitative results of sampled videos to analyze the effectiveness of our method. The upper video presents a low surgical skill score, *i.e.*, 5.0, while the score for the lower video is higher, *i.e.*, 9.0. By comparing the two videos, the confidence lines generated by our model can find several factors that lower the skill score for the upper video, such as haemorrhage, multiple rewinds, and needle threading. Hence the upper video only obtains a skill score of 5.0, while the lower one achieves the better score, *i.e.*, 8.0.

4   Conclusion

This paper introduces a new surgical video dataset for evaluating thoracoscopy-assisted surgical skills. This dataset constitutes the first-ever collection of long surgical videos used for skill assessment from real operating rooms. To address the challenges posed by long-range videos and multiple complex surgical actions in videos, we propose a novel SEDSkill method that incorporates a local-global difference framework. In contrast to current methods that solely rely on intra-video information, our proposed framework leverages local and global difference learning to enhance the model's ability to use inter-video relations for accurate SKA in the MVR scenario.

Acknowledgement. This work was supported in part by a research grant from HKUST-BICI Exploratory Fund (HCIC-004) and in part by a grant from the Research Grants Council of the Hong Kong Special Administrative Region, China (Project Reference Number: T45-401/22-N).

References

1. Bai, Y., Zhou, D., Zhang, S., Wang, J., Ding, E., Guan, Y., Long, Y., Wang, J.: Action quality assessment with temporal parsing transformer. In: Avidan, S., Brostow, G., Cissé, M., Farinella, G.M., Hassner, T. (eds.) ECCV 2022, Part IV. LNCS, vol. 13664, pp. 422–438. Springer, Cham (2022). https://doi.org/10.1007/978-3-031-19772-7_25
2. Birkmeyer, J.D., et al.: Surgical skill and complication rates after bariatric surgery. N. Engl. J. Med. **369**(15), 1434–1442 (2013)
3. Brajcich, B.C., et al.: Association between surgical technical skill and long-term survival for colon cancer. JAMA Oncol. **7**(1), 127–129 (2021)
4. Carbello, B.: Mitral valve disease. Curr. Probl. Cardiol. **18**(7), 425–478 (1993)
5. Ding, X., Li, X.: Exploiting segment-level semantics for online phase recognition from surgical videos. arXiv preprint arXiv:2111.11044 (2021)
6. Ding, X., Wang, N., Gao, X., Li, J., Wang, X., Liu, T.: KFC: an efficient framework for semi-supervised temporal action localization. IEEE Trans. Image Process. **30**, 6869–6878 (2021)
7. Ding, X., et al.: Support-set based cross-supervision for video grounding. In: Proceedings of the IEEE/CVF International Conference on Computer Vision, pp. 11573–11582 (2021)
8. Gao, Y., et al.: JHU-ISI gesture and skill assessment working set (JIGSAWS): a surgical activity dataset for human motion modeling. In: MICCAI Workshop: M2cai, vol. 3 (2014)
9. Healey, M.A., Shackford, S.R., Osler, T.M., Rogers, F.B., Burns, E.: Complications in surgical patients. Arch. Surg. **137**(5), 611–618 (2002)
10. Jin, A., et al.: Tool detection and operative skill assessment in surgical videos using region-based convolutional neural networks. In: 2018 IEEE Winter Conference on Applications of Computer Vision (WACV), pp. 691–699. IEEE (2018)

11. Kunisaki, C., et al.: Significance of thoracoscopy-assisted surgery with a minithoracotomy and hand-assisted laparoscopic surgery for esophageal cancer: the experience of a single surgeon. J. Gastrointest. Surg. **15**, 1939–1951 (2011)
12. Lavanchy, J., et al.: Automation of surgical skill assessment using a three-stage machine learning algorithm. Sci. Rep. **11**(1), 5197 (2021)
13. Li, M., Zhang, H.B., Lei, Q., Fan, Z., Liu, J., Du, J.X.: Pairwise contrastive learning network for action quality assessment. In: Avidan, S., Brostow, G., Cissé, M., Farinella, G.M., Hassner, T. (eds.) ECCV 2022. LNCS, vol. 13664, pp. 457–473. Springer, Cham (2022). https://doi.org/10.1007/978-3-031-19772-7_27
14. Li, Z., Gu, L., Wang, W., Nakamura, R., Sato, Y.: Surgical skill assessment via video semantic aggregation. In: Wang, L., Dou, Q., Fletcher, P.T., Speidel, S., Li, S. (eds.) MICCAI 2022, Part VII. LNCS, vol. 13437, pp. 410–420. Springer, Cham (2022). https://doi.org/10.1007/978-3-031-16449-1_39
15. Lin, T.Y., Goyal, P., Girshick, R., He, K., Dollár, P.: Focal loss for dense object detection. In: Proceedings of the IEEE International Conference on Computer Vision, pp. 2980–2988 (2017)
16. Liu, D., Jiang, T., Wang, Y., Miao, R., Shan, F., Li, Z.: Surgical skill assessment on in-vivo clinical data via the clearness of operating field. In: Shen, D., et al. (eds.) MICCAI 2019, Part V. LNCS, vol. 11768, pp. 476–484. Springer, Cham (2019). https://doi.org/10.1007/978-3-030-32254-0_53
17. Mason, J.D., Ansell, J., Warren, N., Torkington, J.: Is motion analysis a valid tool for assessing laparoscopic skill? Surg. Endosc. **27**, 1468–1477 (2013)
18. Parmar, P., Tran Morris, B.: Learning to score olympic events. In: Proceedings of the IEEE Conference on Computer Vision and Pattern Recognition Workshops, pp. 20–28 (2017)
19. Reznick, R.K.: Teaching and testing technical skills. Am. J. Surg. **165**(3), 358–361 (1993)
20. Strasberg, S.M., Linehan, D.C., Hawkins, W.G.: The accordion severity grading system of surgical complications. Ann. Surg. **250**(2), 177–186 (2009)
21. Szegedy, C., Vanhoucke, V., Ioffe, S., Shlens, J., Wojna, Z.: Rethinking the inception architecture for computer vision. In: Proceedings of the IEEE Conference on Computer Vision and Pattern Recognition, pp. 2818–2826 (2016)
22. Uemura, M., et al.: Procedural surgical skill assessment in laparoscopic training environments. Int. J. Comput. Assist. Radiol. Surg. **11**, 543–552 (2016)
23. Vaswani, A., et al.: Attention is all you need. In: Advances in Neural Information Processing Systems, vol. 30 (2017)
24. Wagner, M., et al.: Comparative validation of machine learning algorithms for surgical workflow and skill analysis with the Heichole benchmark. arXiv preprint arXiv:2109.14956 (2021)
25. Wang, Tianyu, Wang, Yijie, Li, Mian: Towards accurate and interpretable surgical skill assessment: a video-based method incorporating recognized surgical gestures and skill levels. In: Martel, A.L., et al. (eds.) MICCAI 2020, Part III. LNCS, vol. 12263, pp. 668–678. Springer, Cham (2020). https://doi.org/10.1007/978-3-030-59716-0_64
26. Wanzel, K.R., Ward, M., Reznick, R.K.: Teaching the surgical craft: from selection to certification. Curr. Probl. Surg. **39**(6), 583–659 (2002)
27. Yu, X., Rao, Y., Zhao, W., Lu, J., Zhou, J.: Group-aware contrastive regression for action quality assessment. In: Proceedings of the IEEE/CVF International Conference on Computer Vision, pp. 7919–7928 (2021)

28. Zhang, C.L., Wu, J., Li, Y.: ActionFormer: localizing moments of actions with transformers. In: Avidan, S., Brostow, G., Cissé, M., Farinella, G.M., Hassner, T. (eds.) ECCV 2022, Part IV. LNCS, vol. 13664, pp. 492–510. Springer, Cham (2022). https://doi.org/10.1007/978-3-031-19772-7_29
29. Zheng, Z., Wang, P., Liu, W., Li, J., Ye, R., Ren, D.: Distance-IoU loss: faster and better learning for bounding box regression. In: Proceedings of the AAAI Conference on Artificial Intelligence, vol. 34, pp. 12993–13000 (2020)

Neural LerPlane Representations for Fast 4D Reconstruction of Deformable Tissues

Chen Yang[1], Kailing Wang[1], Yuehao Wang[2], Xiaokang Yang[1], and Wei Shen[1(✉)]

[1] MoE Key Lab of Artificial Intelligence, AI Institute,
Shanghai Jiao Tong University, Shanghai, China
`wei.shen@sjtu.edu.cn`
[2] Department of Computer Science and Engineering,
The Chinese University of Hong Kong, Sha Tin, Hong Kong

Abstract. Reconstructing deformable tissues from endoscopic stereo videos in robotic surgery is crucial for various clinical applications. However, existing methods relying only on implicit representations are computationally expensive and require dozens of hours, which limits further practical applications. To address this challenge, we introduce LerPlane, a novel method for fast and accurate reconstruction of surgical scenes under a single-viewpoint setting. LerPlane treats surgical procedures as 4D volumes and factorizes them into explicit 2D planes of static and dynamic fields, leading to a compact memory footprint and significantly accelerated optimization. The efficient factorization is accomplished by fusing features obtained through linear interpolation of each plane and enables using lightweight neural networks to model surgical scenes. Besides, LerPlane shares static fields, significantly reducing the workload of dynamic tissue modeling. We also propose a novel sample scheme to boost optimization and improve performance in regions with tool occlusion and large motions. Experiments on DaVinci robotic surgery videos demonstrate that LerPlane accelerates optimization by over 100× while maintaining high quality across various non-rigid deformations, showing significant promise for future intraoperative surgery applications.

Keywords: Fast 3D Reconstruction · Neural Rendering · Robotic Surgery

1 Introduction

Reconstructing deformable tissues in surgical scenes accurately and efficiently from endoscope stereo videos is a challenging and active research topic. Such techniques can facilitate constructing virtual surgery environments for surgery

Supplementary Information The online version contains supplementary material available at https://doi.org/10.1007/978-3-031-43996-4_5.

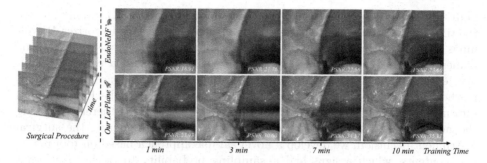

Fig. 1. Performance along with training time. We show the results of EndoNeRF (top) and LerPlane (bottom) with the same training time. LerPlane exhibits remarkable restoration of surgical scenes with just a few minutes of optimization.

robot learning and AR/VR surgery training and provide vivid and specific training for medics on human tissues. Moreover, real-time reconstruction further expands its applications to intraoperative use, allowing surgeons to navigate and precisely control surgical instruments while having a complete view of the surgical scene. This capability could reduce the need for invasive follow-up procedures and address the challenge of operating within a confined field of view.

Neural Radiance Fields (NeRFs) [16], a promising approach for 3D reconstruction, have demonstrated strong potential in accurately reconstructing deformable tissues in dynamic surgical scenes from endoscope stereo videos. EndoNeRF [26], a recent representative approach, represents deformable surgical scenes using a canonical neural radiance field and a time-dependent neural displacement field, achieving impressive reconstruction of deformable tissues. However, the optimization for dynamic NeRFs is computationally intensive, often taking dozens of hours, as each generated pixel requires hundreds of neural network calls. This computational bottleneck significantly constrains the widespread application of these methods in surgical procedures.

Recently, explicit and hybrid methods have been developed for modeling static scenes, achieving significant speedups over NeRF by employing explicit spatial data structures [8,9,27] or features decoded by small MLPs [5,17,25]. Nevertheless, these methods have only been applied to static scenes thus far. Adopting these methods to surgical scenes presents significant challenges for two primary reasons. Firstly, encoding temporal information is essential for modeling surgical scenes while naively adding a temporal dimension to the explicit data structure can significantly increase memory and computational requirements. Secondly, dynamic surgical scene reconstruction suffers from limited viewpoints, often providing only one view per timestep, as opposed to static scenes, which can fully use multi-view consistency for further regularization. This condition requires sharing information across disjoint timesteps for better reconstruction.

To address the aforementioned challenges, we propose a novel method for fast and accurate reconstruction of deformable tissues in surgical procedures, Neural LerPlane (Linear Interpolation Plane), by leveraging explicitly represented multi-plane fields. Specifically, we treat surgical procedures as 4D volumes, where the time axis is orthogonal to 3D spatial coordinates. LerPlane factorizes 4D volumes into 2D planes and uses space planes to form static fields and space-time planes to form dynamic fields. This factorization results in a compact memory footprint and significantly accelerates optimization compared to previous methods [6,26], which rely on pure MLPs, as shown in Fig. 1. LerPlane enables information sharing across timesteps within the static field, thereby reducing the negative impact of limited viewpoints. Moreover, considering the surgical instrument occlusion, we develop a novel sample approach based on tool masks and contents, which assigns higher sampling probability to tissue pixels that have been occluded by tools or have a more extensive motion range. By targeting these regions, our approach allows for more efficient sampling during the training, leading to higher-quality results and faster optimization.

We summarize our contributions:

1. A fast deformable tissue reconstruction method, with rendering quality comparable to or better than the previous method in just 3 min, which is over 100× faster.
2. An efficient representation of surgical scenes, which includes static and dynamic fields, enabling fast optimization and high reconstruction quality.
3. A novel sampling method that boosts optimization and improves the rendering quality. Compared to previous methods, our LerPlane, achieves much faster optimization with superior quantitative and qualitative performance on 3D reconstruction and deformation tracking of surgical scenes, providing significant promise for further applications.

2   Method

2.1   Overview

LerPlane represents surgical procedures using static and dynamic fields, each of which is made up of three orthogonal planes (Sect. 2.3). It starts by using spatiotemporal importance sampling to identify high-priority tissue pixels and build corresponding rays (Sect. 2.4). Then we sample points along each ray and query features using linear interpolation to construct fused features. The fused features and encoded coordinate-time information are input to a lightweight MLP, which predicts color and density for each point (Sect. 2.5). To better optimize LerPlane, we introduce some training schemes, including sample-net, various regularizers, and a warm-up training strategy (Sect. 2.6). Finally, we apply volume rendering to produce predicted color and depth values for each chosen ray. The overall framework is illustrated in Fig. 2.

2.2   Preliminaries

Neural Radiance Field (NeRF) [16] is a coordinate-based neural scene representation optimized through a differentiable rendering loss. NeRF maps the 3D coordinate and view direction of each point in the space into its color values c and volume density σ via neural networks Φ_r.

$$c, \sigma = \Phi_r(x, y, z, \theta, \phi). \tag{1}$$

It calculates the expected color $\hat{C}(\mathbf{r})$ and the expected depth $\hat{D}(\mathbf{r})$ of a pixel in an image captured by a camera by tracing a ray $\mathbf{r}(t) = \mathbf{o} + t\mathbf{d}$ from the camera center to the pixel. Here, $\mathbf{o}$ is the ray origin, $\mathbf{d}$ is the ray direction, and t is the distance from a point to the $\mathbf{o}$, ranging from a pre-defined near bound t_n to a far bound t_f. $w(t)$ represents a weight function that accounts for absorption and scattering during the propagation of light rays. The pixel color is obtained by classical volume rendering techniques [10], which involve sampling a series of points along the ray.

$$\hat{C}(\mathbf{r}) = \int_{t_n}^{t_f} w(t)\mathbf{c}(t)dt, \hat{D}(\mathbf{r}) = \int_{t_n}^{t_f} w(t)tdt, w(t) = \exp\left(-\int_{t_n}^{t} \sigma(s)ds\right)\sigma(t). \tag{2}$$

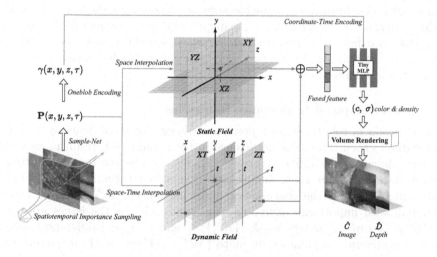

Fig. 2. Illustration of our fast 4D reconstruction method, LerPlane.

2.3 Neural LerPlane Representations for Deformable Tissues

A surgical procedure can be represented as a 4D volume, and we factorize the volume into 2D planes. Specifically, We represent a surgical scene using six orthogonal feature planes, consisting of three space planes (*i.e.*, *XY*, *YZ*, and *XZ*) for the static field and three space-time planes (*i.e.*, *XT*, *YT*, and *ZT*) for the dynamic field. Each space plane has a shape of $N \times N \times D$, and each space-time plane owns a shape of $N \times M \times D$, where N and M represent spatial and temporal resolution, respectively, and D is the size of the feature.

To extract features from an image pixel p_{ij} with color C at a specific timestep τ, we first cast a ray $\mathbf{r}(t)$ from o to the pixel. We then sample spatial-temporal points along the ray, obtaining their 4D coordinates. We acquire a feature vector for a point $\mathbf{P}(x, y, z, \tau)$ by projecting it onto each plane and using bilinear interpolation $\mathcal{B}$ to query features from the six feature planes.

$$\mathbf{v}(x, y, z, \tau) = \mathcal{B}(F_{XY}, x, y) \odot \mathcal{B}(F_{YZ}, y, z) \dots \mathcal{B}(F_{YT}, y, \tau) \odot \mathcal{B}(F_{ZT}, z, \tau), \quad (3)$$

where the $\odot$ represents element-wise multiplication, inspired by [4,22]. The fused feature vector $\mathbf{v}$ is then passed to a tiny MLP Θ, which predicts the color c and density σ of the point. Finally, we leverage the Eq. 2 to get the predicted color $\hat{C}(\mathbf{r})$. Inspired by [14,17], we build the feature planes with multi-resolution planes, e.g. F_{XY} is represented by planes with $N = 128$ and 256.

Existing methods [6,26] for reconstructing surgical procedures using pure implicit representations, requires traversing all possible positions in space-time, which is highly computationally and time-intensive. In contrast, LerPlane decomposes the surgical scene into six explicitly posed planes, resulting in a significant reduction in complexity and a much more efficient representation. This reduces the computational cost from $O(N^4)$ to $O(N^2)$ and enables the use of smaller neural networks, leading to a considerable acceleration in the training period. Besides, methods [7,12,19,20,23,26] using a single displacement field to supplement the static field struggle with handling variations in scene topology, such as non-linear deformations. In contrast, LerPlane can naturally model these situations using a dynamic field.

2.4 Spatiotemporal Importance Sampling

Tool occlusion in robotic surgery poses a challenge for reconstructing occluded tissues due to their infrequent occurrence in the training set, resulting in varied learning difficulties for different pixels. Besides, we observe that many tissues remain stationary over time, and therefore repeated training on these pixels contributes minor to the convergence, reducing efficiency. We design a novel spatiotemporal importance sampling strategy to address the issues above. In particular, we utilize binary masks $\{M_i\}_{i=1}^T$ and temporal differences among frames to generate sampling weight maps $\{\mathbf{W}\}_{i=1}^T$. These weight maps represent the sampling probabilities for each pixel/ray, drawing inspiration from [12,26]. One sampling weight map $\mathbf{W}_i$ can be determined by:

$$\mathbf{W}_i = \min(\max_{\substack{i-n<j \\ <i+n}} (\| \mathbf{I}_i M_i - \mathbf{I}_j M_j \|_1)/3, \alpha) \cdot \Omega_i, \quad \Omega_i = \beta(M_i T / \sum_{i=1}^T M_i), \quad (4)$$

where α is a lower-bound to avoid zero weight among unchanged pixels, Ω_i specifies higher importance scaling for those tissue areas with higher occlusion frequencies, and β is a hyper-parameter for balancing augmentation among frequently occluded areas and time-variant areas. By unitizing spatiotemporal importance sampling, LerPlane concentrates on tissue areas and speeds up training, improving the rendering quality of occluded areas and prioritizing tissue areas with higher occlusion frequencies and temporal variability.

2.5 Coordinate-Time Encoding

Previous methods [6,26] apply positional encoded view direction $\gamma(\mathbf{d})$ to model view-dependent appearance. However, during endoscopic operations, camera movements are restricted. The view direction changes are typically minimal. Instead of view encoding, we propose using $\gamma(x, y, z, \tau)$ to enhance the spatiotemporal information. Specifically, the encoding along with the fused features $\mathbf{v}$ from feature planes is input to the MLP Θ, which predicts σ and c of each point. Then we utilize Eq. 2 to render the expected color $\hat{C}$ and depth $\hat{D}$ of one specific ray.

2.6 Optimization

We adopt a joint supervision approach to optimize the tiny MLP Θ and feature planes using rendered color and depth. To further improve the optimization process, we propose several optimization schemes, including a sample-net for better-sampled points, a warm-up strategy to address outliers, and several regularizers.

Sample-Net. The sampling of spatiotemporal points is crucial for volume rendering, with a particular focus on sampling around tissue regions for optimal performance. We replaced the conventional two-stage time-consuming sampling strategy with a single sample-net and train it using histogram loss [3]. The sample-net is a lightweight single-resolution LerPlane model that provides more accurate sampling points for the full model.

Regularizers. We apply some regularization to address the limited information available in surgical scene reconstruction. We adopt 2D total variation (TV) loss for space planes in [5,8,24] and 1D TV loss on the space axis for space-time planes and a similar smooth loss on the time axis. Additionally, we introduce a minor time-invariant loss to separate the static and dynamic fields as much as possible, encouraging the features in space-time planes to remain unchanged.

Warm-Up Training Strategy. Since single-view captures cannot provide valid scale information, we leverage pseudo ground truth depth maps $D(\mathbf{r})$ generated by STTR-light [13] from stereo images to guide the optimization. Specifically, we apply a Huber loss for depth regularization:

$$\mathcal{L}_D = \begin{cases} 0.5 \Delta D(\mathbf{r})^2, & \text{if } |\Delta D(\mathbf{r})| < \delta \\ \delta \cdot (\Delta D(\mathbf{r}) - 0.5 \cdot \delta), & \text{otherwise} \end{cases} \tag{5}$$

where $\Delta D(\mathbf{r}) = |\hat{D}(\mathbf{r}) - D(\mathbf{r})|$ represents the absolute depth difference among valid depth values, δ is a threshold at which to change loss type. Considering that the predicted depth maps encounter a lot of unreliable depth values and missing areas [26], we design a simple by effective warm-up training strategy. Specifically, we apply the $\mathcal{L}_D$ to depths from both the sample-net and the full model during the first half of the training. In the remaining iterations, we disable the $\mathcal{L}_D$ and use other regularization to refine unreliable depths.

3 Experiments

3.1 Dataset and Evaluation Metrics

We evaluate our proposed method on the EndoNeRF dataset [26], a collection of typical robotic surgery stereo videos captured from stereo cameras at a single viewpoint during in-house DaVinci robotic prostatectomy procedures, which is designed to capture challenging surgical scenes with non-rigid deformation and tool occlusion. We evaluate our proposed method by comparing it to existing methods [15,26] using standard image quality metrics following [26], including PSNR, SSIM, and LPIPS. Additionally, to measure the consistency of the underlying 3D scene, we supplement these metrics using the FLIP metric [1,2]. For qualitative evaluation, we follow the exhibition method from [26].

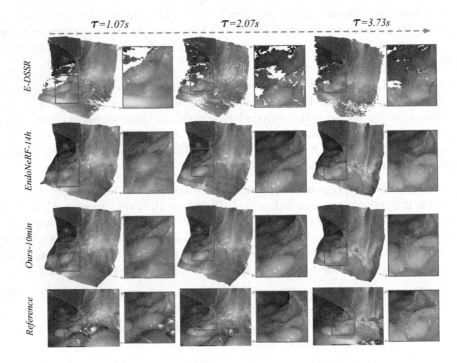

Fig. 3. Qualitative results on scene "traction" from different timesteps τ.

3.2 Implementation Details

We normalize the scene into device coordinates (NDC) to handle single-view endoscopy videos and then project rays within the NDC space. The video duration is normalized to $[-1, 1]$. We use a two-stage sampling network with 128 and 256 dimensional plane features for the sample-net. Oneblob encoding [18] is applied to encode the spatiotemporal information. The full model consists of four resolutions, 64, 128, 256, and 512 dimensions among space. Hyperparameters include $D = 32$ for feature planes, $j = 25$ for spatiotemporal importance sampling, $\xi = 16$ for Oneblob encoding dimensionality, and $\delta = 0.2$ for depth loss across all experiments. An Adam [11] optimizer is used with default values for optimization. In each iteration, 2048 rays are randomly sampled from the whole dataset to form a batch. The initial learning rate is set to 0.01. We apply a cosine schedule with a 512 iterations warming-up stage. We train all scenes with $9k$ and $32k$ iterations, which take around 3 and 10 minutes, respectively, on a single RTX 3090 GPU running the Ubuntu 20.04. Our LerPlane is implemented with pure Pytorch [21]. The code is available at https://github.com/Loping151/LerPlane.

3.3   Evaluation

We compare our proposed method, LerPlane, against two existing SOTA methods: the surfel warping-based method, E-DSSR [15] and NeRF-based method EndoNeRF [26]. We find that E-DSSR struggles to completely reconstruct surgical scenes, resulting in many holes and noisy points (see Fig. 3), which leads to poor numerical performance. In contrast, EndoNeRF achieves high-fidelity reconstruction of deformable tissues but requires around 14 h of optimization, which is computationally expensive and constrains intraoperative use. LerPlane, on the other hand, achieves comparable results to EndoNeRF with only 3 min of optimization, providing nearly 280-fold acceleration. Moreover, with a longer optimization time of 10 min, LerPlane outperforms both E-DSSR and EndoNeRF in terms of all metrics, as shown in Table 1. Our novel importance sampling and encoding strategies further enhance the ability of LerPlane to preserve details and produce accurate visualizations of deformable tissues, as demonstrated in Fig. 3. Our results demonstrate that LerPlane achieves significantly faster optimization without compromising reconstruction quality, showing great potential for future clinical applications in robotic surgery.

Table 1. Quantitative results on the EndoNeRF dataset. Please refer to Sect. 3.4 for explanations of the acronyms used.

Methods	PSNR↑	SSIM↑	LPIPS↓	FLIP↓	Time↓
E-DSSR [15]	13.398 ± 1.387	0.630 ± 0.057	0.423 ± 0.047	\	\
EndoNeRF [26]	29.272 ± 2.836	0.921 ± 0.022	0.088 ± 0.020	0.085 ± 0.018	14 h
Ours-NS	31.532 ± 1.665	0.886 ± 0.021	0.142 ± 0.020	0.112 ± 0.016	3 min
Ours-TS	31.544 ± 1.669	0.886 ± 0.021	0.142 ± 0.020	0.111 ± 0.015	3 min
Ours-VE	32.353 ± 1.742	0.897 ± 0.022	0.131 ± 0.024	0.103 ± 0.012	3 min
Ours-NE	32.230 ± 1.655	0.895 ± 0.023	0.131 ± 0.020	0.102 ± 0.012	3 min
Ours-9k	32.589 ± 1.451	0.901 ± 0.021	0.126 ± 0.028	0.103 ± 0.014	3 min
Ours-32k	35.504 ± 3.076	0.935 ± 0.026	0.083 ± 0.022	0.075 ± 0.031	10 min

3.4 Ablation Study

We conduct ablation studies on the EndoNeRF dataset to understand the key components and demonstrate their effectiveness. Table 1 shows the performance of all experiments.

1. *Sampling Strategy.* We compare with two different methods: naively avoiding tool masks, assigning equal weights to other pixels (Ours-NS), and assigning higher probabilities to highly occluded areas (Ours-TS), as in [26]. Our method effectively prioritizes time-variant and highly occluded areas, significantly improving convergence speed.
2. *Encoding Strategy.* Experiments showed that coordinate-time encoding achieves better performance in all metrics compared to no encoding (Ours-NE) or direction encoding (Ours-VE), showing the effectiveness of the proposed encoding.

Further analysis of the optimization schemes is available in the Supplementary Materials.

4   Conclusion and Future Work

In this paper, we introduced LerPlane, a fast and accurate method for reconstructing deformable tissues from endoscopic videos. By utilizing multi-plane fields and spatiotemporal importance sampling, we can handle tool occlusion and large motion while significantly accelerating optimization. Our experiments show that LerPlane achieves rendering quality comparable to or better than EndoNeRF in just three minutes, which is over 100× faster. We believe that LerPlane could improve robotic surgery scene understanding, benefiting various clinical-oriented tasks and intraoperative surgery applications.

Currently, the inference speed of our Lerplane is slow, In our future research endeavors, our main emphasis will revolve around improving the inference time of our approach, with the primary goal of efficiently supporting intraoperative operations. Moreover, we will dedicate our efforts to reducing the input data requirements, thereby aiming to broaden the applicability of LerPlane to a wider range of surgical scenarios.

Data Use Declaration and Acknowledgment. This work was supported in part by the National Key R&D Program of China 2022YFF1202600, in part by the National Natural Science Foundation of China under Grant 62176159, in part by the Natural Science Foundation of Shanghai 21ZR1432200, and in part by the Shanghai Municipal Science and Technology Major Project 2021SHZDZX0102. This paper uses the EndoNeRF dataset, which is supported by Multi-Scale Medical Robotics Centre InnoHK, CUHK Shun Hing Institute of Advanced Engineering, and Shenzhen-HK Collaborative Development Zone.

References

1. Andersson, P., Nilsson, J., Akenine-Möller, T., Oskarsson, M., Åström, K., Fairchild, M.D.: FLIP: a difference evaluator for alternating images. Proc. ACM Comput. Graph. Interact. Tech. **3**(2), 15-1 (2020)
2. Andersson, P., Nilsson, J., Shirley, P., Akenine-Möller, T.: Visualizing errors in rendered high dynamic range images. Eurographics (2021)
3. Barron, J.T., Mildenhall, B., Verbin, D., Srinivasan, P.P., Hedman, P.: Mip-NeRF 360: unbounded anti-aliased neural radiance fields. In: Proceedings of the IEEE/CVF Conference on Computer Vision and Pattern Recognition, pp. 5470–5479 (2022)
4. Chan, E.R., et al.: Efficient geometry-aware 3D generative adversarial networks. In: Proceedings of the IEEE/CVF Conference on Computer Vision and Pattern Recognition, pp. 16123–16133 (2022)
5. Chen, A., Xu, Z., Geiger, A., Yu, J., Su, H.: Tensorf: tensorial radiance fields. In: Avidan, S., Brostow, G., Cissé, M., Farinella, G.M., Hassner, T. (eds.) ECCV 2022, Part XXXII. LNCS, vol. 13692, pp. 333–350. Springer, Cham (2022). https://doi.org/10.1007/978-3-031-19824-3_20
6. Corona-Figueroa, A., Frawley, J., Bond-Taylor, S., Bethapudi, S., Shum, H.P., Willcocks, C.G.: MedNeRF: medical neural radiance fields for reconstructing 3D-aware CT-projections from a single x-ray. In: 2022 44th Annual International Conference of the IEEE Engineering in Medicine & Biology Society (EMBC), pp. 3843–3848. IEEE (2022)
7. Fang, J., et al.: Fast dynamic radiance fields with time-aware neural voxels. In: SIGGRAPH Asia 2022 Conference Papers, pp. 1–9 (2022)
8. Fridovich-Keil, S., Yu, A., Tancik, M., Chen, Q., Recht, B., Kanazawa, A.: Plenoxels: radiance fields without neural networks. In: Proceedings of the IEEE/CVF Conference on Computer Vision and Pattern Recognition, pp. 5501–5510 (2022)
9. Hedman, P., Srinivasan, P.P., Mildenhall, B., Barron, J.T., Debevec, P.: Baking neural radiance fields for real-time view synthesis. In: Proceedings of the IEEE/CVF International Conference on Computer Vision, pp. 5875–5884 (2021)
10. Kajiya, J.T., Von Herzen, B.P.: Ray tracing volume densities. ACM SIGGRAPH Comput. Graph. **18**, 165–174 (1984)
11. Kingma, D.P., Ba, J.: Adam: a method for stochastic optimization. arXiv preprint arXiv:1412.6980 (2014)
12. Li, T., et al.: Neural 3d video synthesis from multi-view video. In: Proceedings of the IEEE/CVF Conference on Computer Vision and Pattern Recognition, pp. 5521–5531 (2022)
13. Li, Z., et al.: Revisiting stereo depth estimation from a sequence-to-sequence perspective with transformers. In: Proceedings of the IEEE/CVF International Conference on Computer Vision, pp. 6197–6206 (2021)
14. Liu, L., Gu, J., Zaw Lin, K., Chua, T.S., Theobalt, C.: Neural sparse voxel fields. Adv. Neural. Inf. Process. Syst. **33**, 15651–15663 (2020)
15. Long, Y., et al.: E-DSSR: efficient dynamic surgical scene reconstruction with transformer-based stereoscopic depth perception. In: de Bruijne, M., et al. (eds.) MICCAI 2021, Part IV. LNCS, vol. 12904, pp. 415–425. Springer, Cham (2021). https://doi.org/10.1007/978-3-030-87202-1_40
16. Mildenhall, B., Srinivasan, P.P., Tancik, M., Barron, J.T., Ramamoorthi, R., Ng, R.: NeRF: representing scenes as neural radiance fields for view synthesis. Commun. ACM **65**(1), 99–106 (2021)

17. Müller, T., Evans, A., Schied, C., Keller, A.: Instant neural graphics primitives with a multiresolution hash encoding. ACM Trans. Graph. (ToG) **41**(4), 1–15 (2022)
18. Müller, T., McWilliams, B., Rousselle, F., Gross, M., Novák, J.: Neural importance sampling. ACM Trans. Graph. (ToG) **38**(5), 1–19 (2019)
19. Park, K., et al.: NeRFies: deformable neural radiance fields. In: Proceedings of the IEEE/CVF International Conference on Computer Vision, pp. 5865–5874 (2021)
20. Park, K., et al.: HyperNeRF: a higher-dimensional representation for topologically varying neural radiance fields. arXiv preprint arXiv:2106.13228 (2021)
21. Paszke, A., et al.: PyTorch: an imperative style, high-performance deep learning library. In: Wallach, H., Larochelle, H., Beygelzimer, A., d'Alché Buc, F., Fox, E., Garnett, R. (eds.) Advances in Neural Information Processing Systems, vol. 32, pp. 8024–8035. Curran Associates, Inc. (2019)
22. Peng, S., Niemeyer, M., Mescheder, L., Pollefeys, M., Geiger, A.: Convolutional occupancy networks. In: Vedaldi, A., Bischof, H., Brox, T., Frahm, J.-M. (eds.) ECCV 2020, Part III. LNCS, vol. 12348, pp. 523–540. Springer, Cham (2020). https://doi.org/10.1007/978-3-030-58580-8_31
23. Pumarola, A., Corona, E., Pons-Moll, G., Moreno-Noguer, F.: D-NeRF: neural radiance fields for dynamic scenes. In: Proceedings of the IEEE/CVF Conference on Computer Vision and Pattern Recognition, pp. 10318–10327 (2021)
24. Schwarz, K., Sauer, A., Niemeyer, M., Liao, Y., Geiger, A.: VoxGRAF: fast 3D-aware image synthesis with sparse voxel grids. arXiv preprint arXiv:2206.07695 (2022)
25. Sun, C., Sun, M., Chen, H.T.: Direct voxel grid optimization: super-fast convergence for radiance fields reconstruction. In: Proceedings of the IEEE/CVF Conference on Computer Vision and Pattern Recognition, pp. 5459–5469 (2022)
26. Wang, Y., Long, Y., Fan, S.H., Dou, Q.: Neural rendering for stereo 3D reconstruction of deformable tissues in robotic surgery. In: Wang, L., Dou, Q., Fletcher, P.T., Speidel, S., Li, S. (eds.) MICCAI 2022, Part VII. LNCS, vol. 13437, pp. 431–441. Springer, Cham (2022). https://doi.org/10.1007/978-3-031-16449-1_41
27. Yu, A., Li, R., Tancik, M., Li, H., Ng, R., Kanazawa, A.: PlenOctrees for real-time rendering of neural radiance fields. In: Proceedings of the IEEE/CVF International Conference on Computer Vision, pp. 5752–5761 (2021)

SegmentOR: Obtaining Efficient Operating Room Semantics Through Temporal Propagation

Lennart Bastian[✉], Daniel Derkacz-Bogner, Tony D. Wang, Benjamin Busam,
and Nassir Navab

Computer Aided Medical Procedures,
Technical University Munich, Munich, Germany
`lennart.bastian@tum.de`

Abstract. The digitization of surgical operating rooms (OR) has gained
significant traction in the scientific and medical communities. However,
existing deep-learning methods for operating room recognition tasks still
require substantial quantities of annotated data. In this paper, we introduce a method for weakly-supervised semantic segmentation for surgical operating rooms. Our method operates directly on 4D point cloud
sequences from multiple ceiling-mounted RGB-D sensors and requires
less than 0.01% of annotated data. This is achieved by incorporating
a self-supervised temporal prior, enforcing semantic consistency in 4D
point cloud video recordings. We show how refining these priors with
learned semantic features can increase segmentation mIoU to 10% above
existing works, achieving higher segmentation scores than baselines that
use four times the number of labels. Furthermore, the 3D semantic predictions from our method can be projected back into 2D images; we
establish that these 2D predictions can be used to improve the performance of existing surgical phase recognition methods. Our method
shows promise in automating 3D OR segmentation with a 20 times lower
annotation cost than existing methods, demonstrating the potential to
improve surgical scene understanding systems.

Keywords: Surgical Scene Understanding · Context-aware Systems ·
Surgical Phase Recognition · Surgical Data Science

1 Introduction

Automating systems to interpret complex behaviors in surgical operating rooms
(OR) has seen a surge of interest in recent years [13, 20]. Robot-assisted surgeries

---

Lennart Bastian and Daniel Derkacz-Bogner contributed equally to this work.

Supplementary Information The online version contains supplementary material
available at https://doi.org/10.1007/978-3-031-43996-4_6.

have improved patient outcomes by reducing blood loss, recovery periods, and hospitalization times [26, 27]. For robotic systems to autonomously interact with hospital staff, surgical tools, or patients, they must attain a sophisticated and detailed understanding of a highly complex environment. To achieve this, robotic detection systems must comprehend high-level surgical phases and granular 3D object semantics and interactions [16, 25].

Recent works have established the necessity of combining data from multiple cameras to obtain better coverage of surgical procedures [25, 26], as frequent occlusions and obstructions due to personnel and medical equipment obscure important events for individual cameras. Furthermore, a metric 3D semantic understanding is crucial for robotic systems operating and interacting with objects in an environment [31]. In surgical operating rooms, 3D segmentation has previously been approached by fusing semantic RGB predictions from multiple views in 3D under the full supervision of dense 2D labels [16].

However, to adequately represent the distribution of possible events in the surgical domain, large volumes of data must be acquired [27]. Training deep-learning models for automated recognition tasks thus induces an enormous annotation burden, particularly as the privacy-sensitive nature of such materials can prevent annotation outsourcing. Therefore, the surgical data science community actively seeks methods to alleviate this burden, particularly through means of domain-adaptation [24], as well as unsupervised and self-supervised learning [12]. While progress has been made in surgical workflow recognition, methods for 3D surgical scene understanding still require fine-grained labels [25], particularly for semantic segmentation [16].

To this end, we propose SegmentOR, a weakly-supervised indoor semantic segmentation method for 4D multi-view OR datasets. By leveraging the innate temporal consistency of 4D point cloud sequences, we reduce the annotation burden to only a single click per class (about 0.005% of points), decreasing average annotation time per surgical phase from 3 h to 9.6 min while achieving a higher segmentation mIoU than existing methods that use four times the amount of labels. Furthermore, we establish the soundness of semantic predictions from our model for surgical scene understanding by showing that surgical phase recognition performance can be improved using our segmentation predictions as input. Our main contributions can thus be summarized as follows:

- We propose the first 3D weakly-supervised semantic segmentation method for operating room environments and validate it on a manually annotated dataset of 3D point clouds from real surgical acquisitions.
- We demonstrate that various temporal priors can be used to exploit consistency in weakly-supervised semantic segmentation, improving performance to 10% mIoU above baseline methods.
- We show that the semantic outputs from our model can improve the performance of downstream surgical phase recognition methods, formally establishing the link between these two previously disjoint tasks.

– Finally, we release all code and tools, as well as 2577 anonymized and annotated point clouds from the dataset, to advance progress in surgical scene
understanding. https://bastianlb.github.io/segmentOR/

2   Related Work

Surgical Scene Understanding. Workflow recognition is pivotal for contextual awareness in operating room (OR) intelligent systems. Activity recognition has been achieved for single-frame [27,29] and multi-view acquisitions [26],
including laparoscopic views and ceiling-mounted cameras [6,7].

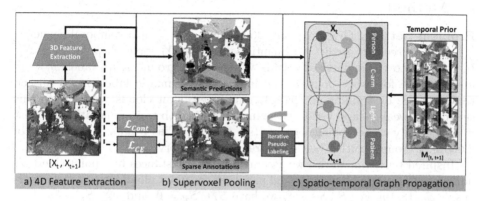

Fig. 1. Proposed architecture. SegmentOR extracts semantic and class-specific features from a sequence of point clouds (a). Sparse labels (in red) are expanded to their
nearest supervoxel cluster (b) and used as supervision for learning the segmentation
task. In contrast to previous works, we propose to incorporate a prior to establish a
temporal consistency between the pooled semantic features in a point cloud sequence
(c), enabling spatiotemporal pseudo-label propagation across timesteps. (Color figure
online)

Modeling OR activities also requires semantic understanding [16]. Semantic scene graphs offer a detailed approach to surgical procedure modeling [25].
While future intelligent OR systems will employ semantics, manual labeling is
time-consuming. However, recent progress in weakly-supervised semantic segmentation, such as one-thing-one-click (OTOC) [19], can decrease this burden,
requiring only a single annotation per semantic class [18,19,30].

The effectiveness of such weakly-supervised segmentation methods in
dynamic OR environments remains unclear. Unlike static indoor datasets like
ScanNet [8], dynamic ORs blur the geometric class separation due to human-
object interaction. Moreover, 3D surgical acquisitions, unlike static indoor reconstructions, are fragmented due to static sensor positions and severe occlusions.
Imprecise 3D registration or temporal synchronization creates artifacts that further complicate 3D modeling, worsened by dynamic non-rigid movements (see
suppl. for examples).

Temporal Modeling. Optical flow effectively extracts movement from image sequences [1,9]. Self-supervised methods have recently offered scene flow extraction from LiDAR point clouds [15,23], but few ground-truth flow annotated datasets exist, leading to potential generalization issues [21,22].

Few have combined temporal consistency with 3D semantic segmentation for dynamic point cloud sequences outside autonomous driving settings [4,10,14,28]. These methods typically rely on dense ground truth labels or require multi-stage label propagation methods and pre-training. Our approach guides temporal label propagation with unsupervised priors, resolving this cold-start problem cost-effectively.

3 Method

Problem Setting. Given a point cloud $X_t \in \mathbb{R}^{N \times 3}$ in a temporal sequence $t \in [T]$, we seek to predict a semantic label $\hat{y}_i \in Y_t$ for each point $p_i \in X_t$. In contrast to the supervised setting where dense ground truth labels $y_i \in Y_t$ are available for every point p_i, we infer dense semantic labels in an unseen test sequence by training on sparsely annotated point clouds. We refer to this setting as "weakly-supervised", meaning ground truth train annotations consist of a randomly chosen point per class (see Fig. 1b). This results in an average of 0.005% of ground truth labels compared to full supervision.

Inspired by recent works, we assume semantic instances in a point cloud X_t adhere to geometric boundaries and partition the point cloud into a set of super-voxels $\mathcal{S}_t$ [18,19]. For $S_j, S_{j'} \in \mathcal{S}_t$ we have $S_j \cap S_{j'} = \emptyset$, and $\cup_{j=1}^N S_j = X_t$. This allows us to represent a group of points with a single feature, increasing the level of supervision obtained from a single "click", and enabling efficient label propagation. Due to the sparse nature of the click annotations, most supervoxels in each point cloud are unlabeled (see Fig. 1c). By propagating learned information into unlabeled supervoxels, the level of supervision can be drastically increased through pseudo-labeling. For clarity, we use indices i for points and j for supervoxels [19]. Indices t are used to indicate timesteps.

Proposed Method. Following the approach of OTOC [19], we train a sparse 3D-UNet [5] F_Θ to model the semantic class of each point p_i, $Y_t = F_\Theta(X_t)$ with a cross-entropy loss $\mathcal{L}_{CE}$, and an identically structured relation network R_Ψ to predict a category specific embedding $R_t = R_\Psi(X_t)$ with a contrastive loss $\mathcal{L}_{Cont}$. The point-wise embeddings obtained from both networks are accumulated using mean-pooling over each supervoxel S_j. The pooled feature embeddings f_j, r_j can then be used to construct a fully-connected graph G_i to propagate labels to unlabeled supervoxels. This is achieved by maximizing the expectation E of unlabeled supervoxels given the complete supervoxel set $\mathcal{S}$:

$$E(Y|\mathcal{S}) = \sum_{j'} \psi_u(y_{j'}|\mathcal{S}, F_\Theta) + \sum_{j'} \psi_p(y_j, y_{j'}|\mathcal{S}, R_\Psi, F_\Theta) \tag{1}$$

where ψ_u represents the pooled class predictions, and ψ_p is a pairwise similarity between supervoxels $S_j, S_{j'}$ [19]. By measuring the similarity between supervoxels in this manner, semantic information in the two networks F_Θ and R_Ψ can be propagated to unlabeled supervoxels iteratively through pseudo-labeling, using a likelihood threshold of, e.g., $E(Y|S_j) \geq 0.90$. Setting a high confidence threshold reduces incorrect pseudo-labels, which would negatively impact subsequent training iterations.

OTOC [19] considers all supervoxel pairs in a single static acquisition as candidates during this expansion. An intuitive way to extend the graph propagation in the temporal dimension would be to pool all supervoxels over a pair of frames X_t and X_{t+1}, creating a fully connected graph over both supervoxel sets. We refer to this method as $OTOC+T$, as the original method uses only spatial context. Aside from being computationally expensive, this naive approach does not consider that the nearest supervoxel in an adjacent timestep is highly likely to describe a similar region in the point cloud.

To further improve upon this idea, we propose to enforce temporal consistency through the use of a supervoxel matching matrix $M_{[t,t+1]} \in \mathbf{R}^{m \times n}$ where $|S_t| = m$ and $|S_{t+1}| = n$ are the dimensions of the respective supervoxel sets, reducing computational complexity to an additional comparison per supervoxel instead of n as with the $OTOC+T$. An entry $m_{j,j'} \in M_{[t,t+1]}$ indicates the probability that supervoxels S_j and $S_{j'}$ describe a similar region across time steps. Intuitively, this can establish consistency between the pair of point clouds by considering matched supervoxels from a different timestep $X_{t'}$ as pseudo-label candidates. To initialize the matching matrix, we explore how nearest neighbor, unsupervised optical [9] and scene flow [23] priors can improve temporal pseudo-label propagation (see suppl. sec. 1 for mathematical details).

After initialization through any of these priors, we propose to update the matching iteratively during training, establishing a link between temporal consistency and semantic understanding. To strengthen this dynamic and account for potentially incorrect matches, we additionally incorporate relation net R_Ψ features to refine the supervoxel matching matrix M. Formally, we can define the matching update for a single entry $m_{j,j'} \in M_{[t,t+1]}$ as follows:

$$p(S_{j'}|S_j) = \lambda\, p(S_{j'}|S_j, M_{[t,t+1]}) + (1 - \lambda)\, p(S_{j'}|S_j, R_\Psi) \tag{2}$$

The updated matching is the supervoxel with the highest matching probability, i.e., $\hat{m}_{j,j'} = \arg\max_{S_j \in Y_{t+1}} p(S_{j'}|S_j, M_{[t,t+1]})$. We can then additionally regularize the graph propagation (Eq. 1) using the updated matching matrix for the merged supervoxel sets $\hat{S} = S_t \cup S_{t+1}$:

$$E(Y|\hat{S}) = \sum_{j'} \psi_u(y_{j'}|\hat{S}, F_\Theta) + \sum_{j'} \psi_p(y_j, y_{j'}|\hat{S}, M_{[t,t+1]}) \tag{3}$$

where ψ_u describes the probability of S_j being assigned label $y_{j'}$ based on the prediction $f_j = F_\Theta(S_j)$. ψ_p describes the pairwise similarity between two supervoxels additionally based on $r_j = R_\Psi(S_j)$, mean supervoxel color c_j, and mean coordinate p_j.

$$\psi_u(y_{j'}|S_j, F_\Theta) = -\log P(y_{j'}|S_j, F_\Theta) \tag{4}$$

$$\psi_p(y_j, y_{j'} | S_j, M_{[t,t+1]}) = m_{j,j'} \cdot e^{\{-\|c_j - c_{j'}\|_2^2 - \|p_j - p_{j'}\|_2^2 - \|f_j - f_{j'}\|_2^2 - \|r_j - r_{j'}\|_2^2\}}$$

(5)

4   Experiments

Dataset Description. All experiments are carried out on an existing dataset of 18 laparoscopic surgeries [2,3]. The full dataset consists of RGB-D video acquisitions from four co-registered ceiling-mounted Azure Kinect cameras, containing 582,000 images per camera, with video annotations for 8 surgical workflow phases. We uniformly split a subset of the fused point cloud data into training and validation sets, ensuring similar distributions of non-overlapping surgeries, camera calibrations, and class distributions. We use 1500 point clouds for training and 1077 for validation. We additionally manually annotate these point clouds with segmentation labels covering 12 medically relevant classes.

Each training split contains around 500 sparsely annotated point clouds, with 5436 click annotations on average per split. The densely labeled validation annotations comprise approximately 93% of the on average one million points per point cloud. To reduce the bias from a subjective annotation, we employed four different data annotators for a total annotation time of ~115 h. To measure the robustness of our method, we split the train and validation sets into 3 non-overlapping splits, referring to this as "cross-validation" despite the train and validation splits having different types of labels (sparse click and dense, respectively). For more details on data annotation and splits, as well as qualitative examples, please refer to the supplementary materials.

Experimental Setup. In contrast to OTOC [19], which relies on additional ground truth instance segmentation as input [8], we use a heuristic over-segmentation approach to generate supervoxels [17]. All experiments use the same hyperparameters unless otherwise noted. The relation and feature networks were pre-trained on ScanNet [8]. We then fine-tune on our dataset for four iterations, with pseudo-label propagation occurring after each iteration. We report standardized segmentation metrics. The average training time on an NVIDIA A40 GPU was 7.45 h, using PyTorch 1.13.1 with CUDA 11.7.

5   Results and Discussion

Experiment 1: Baseline Comparisons and Color Ablation. To quantify the overall impact of temporal guidance in the weakly-supervised setting, we compare SegmentOR with the baseline OTOC [19] over three-fold cross-validation. As color information may be unavailable in OR datasets due to privacy concerns, we perform additional experiments without RGB features to contextualize the performance in such a setting. Furthermore, we explore how different unsupervised priors can improve semantic predictions, namely (i) nearest neighbor matching, (ii) optical flow matching, and (iii) scene flow matching. For each prior, the matching matrix $M_{[t,t+1]}$ is then initialized based on obtained

Table 1. Main Results. Comparison of our best proposed model, using a nearest neighbor temporal prior with a learned update, with OTOC [19] in a three-fold cross-validation.

Method	Color	mIoU (%)	F1 (%)	Recall (%)	Precision (%)	Accuracy (%)
OTOC	✗	63.29 ± 3.49	75.51 ± 3.76	75.88 ± 4.07	76.80 ± 3.88	75.88 ± 4.07
	✓	62.40 ± 0.01	74.59 ± 0.01	75.02 ± 0.54	76.19 ± 0.20	75.02 ± 0.54
SegmentOR	✗	69.32 ± 2.05	70.57 ± 1.06	79.88 ± 2.06	79.96 ± 2.49	79.88 ± 2.05
	✓	**72.99 ± 0.19**	**82.77 ± 0.13**	**83.25 ± 0.71**	**83.31 ± 1.08**	**83.25 ± 0.71**

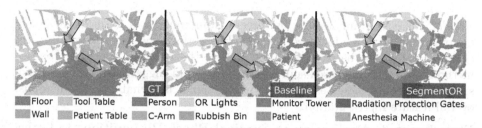

■ Floor	■ Tool Table	■ Person	■ OR Lights	■ Monitor Tower	■ Radiation Protection Gates
■ Wall	■ Patient Table	■ C-Arm	■ Rubbish Bin	■ Patient	■ Anesthesia Machine

Fig. 2. Qualitative Results. Comparison between the ground truth annotation (GT), the OTOC baseline, and SegmentOR. SegmentOR demonstrates improved segmentation mIOU for moving classes such as Person (red arrow) but also for static classes such as Patient Table (orange arrow). Best viewed digitally.

flow. Interestingly, initializing the matching matrix with the nearest neighbor prior outperformed optical and scene flow initialization. Using any single prior resulted in improvements over $OTOC+T$. We refer to the supplementary materials for a detailed description and ablation of the flow priors.

Results: Table 1 shows that SegmentOR consistently outperforms the baseline both with and without RGB features when initialized with a nearest neighbor temporal prior and a learned update according to Eq. 2. The addition of RGB features marginally impacts the baseline, with a difference of only 0.9% mIoU. Without color, SegmentOR outperforms OTOC by ∼6% mIoU, making it more suitable for privacy-constrained setups. This improves to ∼10% for colored point clouds, indicating that the proposed learned temporal matching can leverage color similarities across time steps more effectively. The most significantly moving entities in ORs are surgical staff, who tend to wear gowns with specific and consistent colors. The presence of color features could influence the ability to distinguish humans from other objects more consistently (Fig. 2). This is supported by a segmentation mIoU increase of over 15% concerning the human class for SegmentOR (see supplementary for all class distributions).

Experiment 2: Number of Clicks. A model's performance should theoretically increase with the level of supervision. We thus quantify the impact of adding up to three additional clicks on OTOC's performance. This experiment is performed on the first training and validation split, using a varying number of click annotations per class.

Table 2. To assess the trade-off between annotation % and performance, we train the baseline OTOC method with up to four times the amount of "click" annotations.

# clicks	annotated (%)	mIoU (%)	F1 (%)	Recall (%)	Precision (%)	Accuracy (%)
1	0.0054	62.60	74.91	74.45	77.86	74.40
2	0.0108	69.01	79.92	79.86	81.13	79.82
3	0.0163	**70.83**	**81.45**	**81.57**	**82.12**	**81.56**
4	0.0217	70.56	80.01	79.13	81.75	79.17

Results: Increasing annotations by three or four times lead to an improvement of approximately 8% mIoU (see Table 2). The performance saturates with three clicks. Notably, the baseline does not achieve the 73% mIoU of the proposed method, even with increased supervision. This could suggest that temporal consistency not only increases the supervision signal but enables a more robust overall feature representation.

Experiment 3: Application to Surgical Phase Recognition. To further assess the quality of our semantic predictions, we evaluate their impact on surgical workflow analysis (see Table 3). We use a ResNet50 [11] backbone and perform four-fold cross-validation using random splits over different surgeries of the complete, larger RGB-D video dataset. We use our best-performing segmentation model to infer the semantic predictions for each of the 582k fused point clouds (inference @12.5 fps), projecting them back into the two cameras. We then compare the performance of raw RGB, depth, and semantic labels inputs against fusing the latter two via late fusion [29].

Table 3. Downstream Surgical Phase Recognition. We evaluate the capability of a ResNet50 [11] as a surgical phase recognition backbone based on 3 modalities. Accuracy and mAP are reported for two differently placed surgical workflow cameras

Input	Camera 01		Camera 02	
	Accuracy	mAP	Accuracy	mAP
Semantic	61.98 ± 7.89	70.19 ± 8.44	52.48 ± 5.19	58.34 ± 6.74
Depth	69.19 ± 1.46	77.95 ± 2.82	63.62 ± 6.45	67.56 ± 5.98
RGB	$\mathbf{76.04 \pm 4.27}$	$\mathbf{88.71 \pm 3.61}$	$\mathbf{73.74 \pm 1.85}$	$\mathbf{85.44 \pm 3.84}$
Depth + Semantic	74.98 ± 5.62	83.34 ± 4.08	64.29 ± 5.35	71.76 ± 4.76

Results: Consistent with previous works [2,29], RGB achieves the best performance across both cameras. Semantic predictions alone yield a performance well below RGB or depth. However, the fact that noisy semantic predictions from our network (achieving 73% mIoU) can be used for this challenging task demonstrates the benefits of our segmentation outputs for surgical scene understanding. Furthermore, when combined with depth through late fusion, results are improved by nearly 6% and 0.6% accuracy over depth for the surgical and workflow cameras, respectively. This suggests that segmentation maps could substitute RGB features when unavailable due to privacy reasons [2,27].

Conclusion. This work presents a novel semantic segmentation method for surgical scene understanding that significantly reduces the annotation burden by leveraging the temporal consistency of point cloud sequences. We demonstrate the effectiveness of our approach on point clouds from a surgical phase recognition dataset, which we enrich with manual 3D annotations. By incorporating self-supervised temporal priors, our method achieves a high segmentation mIoU of 73.10% using only 0.005% of annotated points. Furthermore, we establish a formal link between semantic segmentation and workflow analysis by demonstrating that our semantic predictions benefit downstream surgical phase recognition methods. Finally, we release all anonymized point clouds, annotations, and code used to ease the deployment of context-aware systems in surgical environments.

Acknowledgements. This work was funded by the German Federal Ministry of Education and Research (BMBF), No.: 16SV8088 and 13GW0236B. We additionally thank the J&J Robotics & Digital Solutions team for their support. Furthermore, we thank Ruiyang Li for supporting the point cloud annotation. Code and data can be found at: https://bastianlb.github.io/segmentOR/.

References

1. Baker, S., Matthews, I.: Lucas-Kanade 20 years on: a unifying framework. Int. J. Comput. Vision **56**, 221–255 (2004)
2. Bastian, L., et al.: Know your sensors-a modality study for surgical action classification. Comput. Methods Biomech. Biomed. Eng.: Imaging Vis. **11**, 1–9 (2022)
3. Bastian, L., Wang, T.D., Czempiel, T., Busam, B., Navab, N.: DisguisOR: holistic face anonymization for the operating room. Int. J. Comput. Assist. Radiol. Surg. 1–7 (2023)
4. Choy, C., Gwak, J., Savarese, S.: 4D spatio-temporal convnets: Minkowski convolutional neural networks. In: Proceedings of the IEEE/CVF Conference on Computer Vision and Pattern Recognition, pp. 3075–3084 (2019)
5. Spconv Contributors: Spconv: spatially sparse convolution library (2022). https://github.com/traveller59/spconv
6. Czempiel, T., et al.: TeCNO: surgical phase recognition with multi-stage temporal convolutional networks. In: Martel, A.L., et al. (eds.) MICCAI 2020. LNCS, vol. 12263, pp. 343–352. Springer, Cham (2020). https://doi.org/10.1007/978-3-030-59716-0_33
7. Czempiel, T., Sharghi, A., Paschali, M., Navab, N., Mohareri, O.: Surgical workflow recognition: from analysis of challenges to architectural study. In: Karlinsky, L., Michaeli, T., Nishino, K. (eds.) ECCV 2022, Part III. LNCS, vol. 13803, pp. 556–568. Springer, Cham (2023). https://doi.org/10.1007/978-3-031-25066-8_32
8. Dai, A., Chang, A.X., Savva, M., Halber, M., Funkhouser, T., Nießner, M.: ScanNet: richly-annotated 3D reconstructions of indoor scenes. In: CVPR, pp. 5828–5839 (2017)

9. Farnebäck, G.: Two-frame motion estimation based on polynomial expansion. In: Bigun, J., Gustavsson, T. (eds.) SCIA 2003. LNCS, vol. 2749, pp. 363–370. Springer, Heidelberg (2003). https://doi.org/10.1007/3-540-45103-x_50

10. Hanyu, S., Jiacheng, W., Hao, W., Fayao, L., Guosheng, L.: Learning spatial and temporal variations for 4D point cloud segmentation. arXiv preprint arXiv:2207.04673 (2022)

11. He, K., Zhang, X., Ren, S., Sun, J.: Deep residual learning for image recognition. In: CVPR, pp. 770–778 (2016)

12. Liu, M., Zhou, Y., Qi, C.R., Gong, B., Su, H., Anguelov, D.: Less: Label-efficient semantic segmentation for lidar point clouds. In: Avidan, S., Brostow, G., Cissé, M., Farinella, G.M., Hassner, T. (eds.) ECCV 2022, Part VII. LNCS, vol. 13699, pp. 70–89. Springer, Cham (2022). https://doi.org/10.1007/978-3-031-19842-7_5

13. Kennedy-Metz, L.R., et al.: Computer vision in the operating room: opportunities and caveats. IEEE Trans. Med. Robot. Bionics 3(1), 2–10 (2020)

14. Kochanov, D., Ošep, A., Stückler, J., Leibe, B.: Scene flow propagation for semantic mapping and object discovery in dynamic street scenes. In: IROS, pp. 1785–1792. IEEE (2016)

15. Li, R., Zhang, C., Lin, G., Wang, Z., Shen, C.: RigidFlow: self-supervised scene flow learning on point clouds by local rigidity prior. In: Proceedings of the IEEE/CVF Conference on Computer Vision and Pattern Recognition, pp. 16959–16968 (2022)

16. Li, Z., Shaban, A., Simard, J.G., Rabindran, D., DiMaio, S., Mohareri, O.: A robotic 3D perception system for operating room environment awareness. arXiv:2003.09487 [cs] (2020)

17. Lin, Y., Wang, C., Zhai, D., Li, W., Li, J.: Toward better boundary preserved supervoxel segmentation for 3D point clouds. ISPRS J. Photogram. Remote Sens. 143, 39–47 (2018). https://www.sciencedirect.com/science/article/pii/S0924271618301370. iSPRS Journal of Photogrammetry and Remote Sensing Theme Issue "Point Cloud Processing"

18. Liu, M., Zhou, Y., Qi, C.R., Gong, B., Su, H., Anguelov, D.: LESS: label-efficient semantic segmentation for lidar point clouds. In: Avidan, S., Brostow, G., Cissé, M., Farinella, G.M., Hassner, T. (eds.) ECCV 2022. LNCS, vol. 13699, pp. 70–89. Springer, Cham (2022). https://doi.org/10.1007/978-3-031-19842-7_5

19. Liu, Z., Qi, X., Fu, C.W.: One thing one click: a self-training approach for weakly supervised 3d semantic segmentation. In: CVPR, pp. 1726–1736 (2021)

20. Maier-Hein, L., et al.: Surgical data science-from concepts toward clinical translation. Med. Image Anal. 76, 102306 (2022)

21. Mayer, N., et al.: A large dataset to train convolutional networks for disparity, optical flow, and scene flow estimation. In: CVPR, pp. 4040–4048 (2016)

22. Menze, M., Geiger, A.: Object scene flow for autonomous vehicles. In: CVPR (2015)

23. Mittal, H., Okorn, B., Held, D.: Just go with the flow: self-supervised scene flow estimation. In: CVPR, pp. 11177–11185 (2020)

24. Mottaghi, A., Sharghi, A., Yeung, S., Mohareri, O.: Adaptation of surgical activity recognition models across operating rooms. In: Wang, L., Dou, Q., Fletcher, P.T., Speidel, S., Li, S. (eds.) MICCAI 2022, Part VII. LNCS, vol. 13437, pp. 530–540. Springer, Cham (2022). https://doi.org/10.1007/978-3-031-16449-1_51

25. Özsoy, E., Örnek, E.P., Eck, U., Czempiel, T., Tombari, F., Navab, N.: 4D-OR: semantic scene graphs for or domain modeling. In: Wang, L., Dou, Q., Fletcher, P.T., Speidel, S., Li, S. (eds.) MICCAI 2022. LNCS, vol. 13437, pp. 475–485. Springer, Cham (2022). https://doi.org/10.1007/978-3-031-16449-1_45

26. Schmidt, A., Sharghi, A., Haugerud, H., Oh, D., Mohareri, O.: Multi-view surgical video action detection via mixed global view attention. In: de Bruijne, M., et al. (eds.) MICCAI 2021. LNCS, vol. 12904, pp. 626–635. Springer, Cham (2021). https://doi.org/10.1007/978-3-030-87202-1_60
27. Sharghi, A., Haugerud, H., Oh, D., Mohareri, O.: Automatic operating room surgical activity recognition for robot-assisted surgery. In: Martel, A.L., et al. (eds.) MICCAI 2020. LNCS, vol. 12263, pp. 385–395. Springer, Cham (2020). https://doi.org/10.1007/978-3-030-59716-0_37
28. Shi, H., Wei, J., Li, R., Liu, F., Lin, G.: Weakly supervised segmentation on outdoor 4D point clouds with temporal matching and spatial graph propagation. In: CVPR, pp. 11840–11849 (2022)
29. Twinanda, A.P., Winata, P., Gangi, A., Mathelin, M., Padoy, N.: Multi-stream deep architecture for surgical phase recognition on multi-view RGBD videos. In: Proceedings of the M2CAI Workshop MICCAI, pp. 1–8 (2016)
30. Yang, C.K., Wu, J.J., Chen, K.S., Chuang, Y.Y., Lin, Y.Y.: An mil-derived transformer for weakly supervised point cloud segmentation. In: CVPR, pp. 11830–11839 (2022)
31. Yousif, K., Bab-Hadiashar, A., Hoseinnezhad, R.: An overview to visual odometry and visual SLAM: applications to mobile robotics. Intell. Industr. Syst. 1(4), 289–311 (2015)

Revisiting Distillation for Continual Learning on Visual Question Localized-Answering in Robotic Surgery

Long Bai[1], Mobarakol Islam[2], and Hongliang Ren[1,3(✉)]

[1] Department of Electronic Engineering, The Chinese University of Hong Kong (CUHK), Hong Kong SAR, China
b.long@link.cuhk.edu.hk
[2] Wellcome/EPSRC Centre for Interventional and Surgical Sciences (WEISS), University College London, London, UK
mobarakol.islam@ucl.ac.uk
[3] Shun Hing Institute of Advanced Engineering, CUHK, Hong Kong SAR, China
hlren@ee.cuhk.edu.hk

Abstract. The visual-question localized-answering (VQLA) system can serve as a knowledgeable assistant in surgical education. Except for providing text-based answers, the VQLA system can highlight the interested region for better surgical scene understanding. However, deep neural networks (DNNs) suffer from catastrophic forgetting when learning new knowledge. Specifically, when DNNs learn on incremental classes or tasks, their performance on old tasks drops dramatically. Furthermore, due to medical data privacy and licensing issues, it is often difficult to access old data when updating continual learning (CL) models. Therefore, we develop a non-exemplar continual surgical VQLA framework, to explore and balance the rigidity-plasticity trade-off of DNNs in a sequential learning paradigm. We revisit the distillation loss in CL tasks, and propose rigidity-plasticity-aware distillation (RP-Dist) and self-calibrated heterogeneous distillation (SH-Dist) to preserve the old knowledge. The weight aligning (WA) technique is also integrated to adjust the weight bias between old and new tasks. We further establish a CL framework on three public surgical datasets in the context of surgical settings that consist of overlapping classes between old and new surgical VQLA tasks. With extensive experiments, we demonstrate that our proposed method excellently reconciles learning and forgetting on the continual surgical VQLA over conventional CL methods. Our code is publicly accessible at github.com/longbai1006/CS-VQLA.

L. Bai and M. Islam—Co-first authors.

Supplementary Information The online version contains supplementary material available at https://doi.org/10.1007/978-3-031-43996-4_7.

1 Introduction

Trustworthy and reliable visual question-answering (VQA) models have proved their potential in the medical domain [17,22]. A deep learning (DL)-based surgical VQA system [22] has been developed as a surgical training and popularization tool for junior surgeons, medical students, and patients. However, one pivotal problem with surgical VQA is the lack of localized answers. VQA can provide the answer to the question, but cannot relate the answers to its localization at an instance level. Surgical scenarios with various similar instruments and actions may further confuse the learners. Answers with localization can further assist learners in dealing with confusion. In this case, a surgical visual-question localized-answering (VQLA) system can thereby be established for effective surgical training and scene understanding [3].

Meanwhile, catastrophic forgetting has become a largely discussed topic in deep neural networks. Deep neural networks (DNNs) shall abruptly and drastically forget old knowledge when learning new [16]. Various continual learning (CL) methods have been proposed to mitigate catastrophic forgetting and study the balance of rigidity and plasticity in deep models [16,20]. Rigidity refers to the ability of the model not to diverge and remember old knowledge, while plasticity represents the acquisition of new knowledge by DNNs [4]. Some pioneering works have attempted to tackle the CL problem in the medical domain [6]. Catastrophic forgetting may occur in various real-world medical scenarios, e.g., data collected (i) over time, and (ii) across devices/institutions. More seriously, due to issues of data privacy, storage, and licensing, old data may not be accessible anymore [13]. Therefore, it is necessary to develop a non-exemplar CL method for surgical VQLA tasks to resist catastrophic forgetting in clinical applications.

Furthermore, most medical decision-making tasks shall include classes overlapping with the old tasks and newly appeared classes, as shown in Fig. 1. We should not distillate the entire previous model when we deal with CL with overlapping classes. Firstly, the model will not emphasize new classes and have a high bias toward overlapping classes rather than new classes. Overlapping classes will dominate the model prediction if we naively follow the distillation from existing CL models. Secondly, catastrophic forgetting will be severe in old non-overlapping classes and the overlapping classes will dominate in the model prediction, and forget the old classes. For this purpose, we revisit distillation methods in CL and design a Continual Surgical VQLA (CS-VQLA) framework for learning incremental classes by balancing the performance of the old overlapping and non-overlapping classes. CS-VQLA has the following attributes: (i) it is a multi-task model including answering and localization, (ii) domain shift and class increment problems both exist, (iii) there may be overlapping classes between old and new tasks. These points shall further complicate the CL tasks.

In this work, (**1**) We establish a non-exemplar CS-VQLA framework. While being applied to surgical education and scene understanding, the framework can learn data in a streaming manner and effectively resist catastrophic forgetting. (**2**) We revisit the distillation method for CL, and propose rigidity-plasticity-aware distillation (RP-Dist) and self-calibrated heterogeneous distilla-

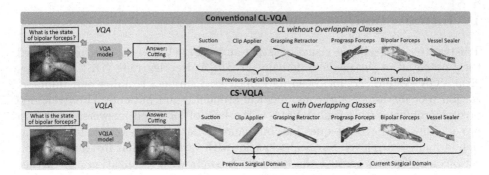

Fig. 1. Comparison between conventional CL-VQA and our CS-VQLA. Besides providing localized answers, our CS-VQLA framework also pays attention to the issue of overlapping and non-overlapping classes in sequential surgical domains.

tion (SH-Dist) for the output logits and intermediate feature maps, respectively. The weight aligning (WA) technique is further integrated to adjust model bias between old and new data. **(3)** Extensive comparison and ablation studies prove the outstanding performance of our method in mitigating catastrophic forgetting, demonstrating its potential in real-world applications.

2 Methodology

2.1 Preliminaries

Problem Definition. We define the continual learning sequence with $\mathcal{TP}$ time periods, and $t \in \{1, ..., \mathcal{TP}\}$ means the current time period. $\mathcal{D}_t$ denotes the training dataset at time period t, with x representing a sample of the input question and image pair in $\mathcal{D}_t$. $\mathcal{C}_{old}$ denotes the classes appearing in previous time period $\{1, ..., t-1\}$, and $\mathcal{C}_{new}$ represents the classes appearing in current time period t. Furthermore, we define the classes existing in both $\mathcal{C}_{old}$ and $\mathcal{C}_{new}$ as *overlapping classes* $\mathcal{C}_{op}$, and define unique classes in $\mathcal{C}_{old}$ as *old non-overlapping classes* $\mathcal{C}_{no}$. F stands for the output feature map from the network backbone.

Knowledge Distillation (KD) [9,26] on output logits [16] or intermediate feature map [19] is a widely used approach to retain knowledge on old tasks. With z^o and z^{cl} denote the output logits from the old and CL model, respectively, we can formulate the logits distillation loss [16] as:

$$\mathcal{L}_{LKD} = \sum_{c=0}^{\mathcal{C}_{old}} -p_T^o(x) \log \left(p_T^{cl}(x) \right) \tag{1}$$

in which $p_T^o(x) = SM(z^o/T)$ and $p_T^{cl}(x) = SM(z^{cl}/T)$ represent the probabilities. SM means Softmax. T is temperature normalization for all old classes.

Weight Aligning (WA) [27] is a simple technique to align the weight bias in the classifier layer. We use $\mathbf{W}_{new}$ to represent the weights for newly appeared classes in the classifier, and $\mathbf{W}_{old}$ to denote those of old classes, then we have:

$$\hat{\mathbf{W}}_{new} = \frac{Mean[Norm(\mathbf{W}_{old})]}{Mean[Norm(\mathbf{W}_{new})]} \cdot \mathbf{W}_{new} \tag{2}$$

where *norm* means normalizing all the elements in the vector. In class-incremental learning, WA can effectively avoid the model bias towards new classes.

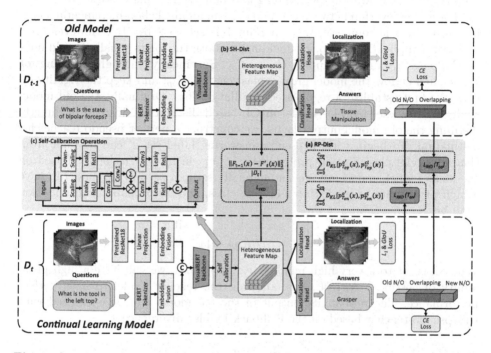

Fig. 2. Overview of our CS-VQLA network. The VQLA model is used to process bimodal input (image and text) and provide predictions for two tasks (answering and localization). The proposed RP-Dist and SH-Dist are designed to help the CL model retain old knowledge from the old model and trade-off model rigidity-plasticity. 'N/O' means non-overlapping classes.

2.2 Continual Surgical VQLA (CS-VQLA)

Visual-Question Localized Answering. We define our VQLA framework following [3], by building a parallel detector on top of the VQA-based classification model. Therefore, the VLQA model includes the following components: a ResNet18 [8] pre-trained on ImageNet [5] as a prior image feature extractor, a BERT tokenizer [7], the VisualBERT [15] as the backbone (it can also be called as the deep feature extractor), a fully-connected layer as the classifier, and a 3-layer MLP as the detector. The classification task is optimized via the cross-entropy loss $\mathcal{L}_{CE}$, and the bounding box regression is optimized by the sum of $\mathcal{L}_1$ and $GIoU$ loss [21]. Thus, the VQLA loss can be formulated as: $\mathcal{L}_{VQLA} = \mu \cdot \mathcal{L}_{CE} + (\mathcal{L}_1 + \mathcal{L}_{GIoU})$, where μ is set as 100 to balance the optimization progress of the two tasks.

Rigidity-Plasticity-Aware Distillation (RP-Dist). The current rigidity-plasticity trade-off is towards the entire model. However, we shall make the rigidity-plasticity aware in overlapping and non-overlapping classes.

There is no overlap between $\mathcal{C}_{old}$ and $\mathcal{C}_{new}$ in an ideal class-incremental learning setup, so the temperature T in Equ. 1 is set to 2 by [16]. However, in a real-world application setup, T should not smooth the logits equally for old non-overlapping $\mathcal{C}_{on}$ and overlapping classes $\mathcal{C}_{op}$. Specifically, through adjusting for T, we shall endow the model greater plasticity on $\mathcal{C}_{op}$, and keep the rigidity on $\mathcal{C}_{on}$. We first establish a regularized distillation loss. Originally, the old model shall serve as the 'teacher' model in CL-based distillation. Instead of directly distilling the old model output logits, we construct a perfect pseudo teacher for distillation. To begin with, a pseudo answering label set a' can be built from the old model classification probabilities $p^o(x)$ via $a' = Max[p(x)]$. Based on the idea of label smoothing, we can manually setup a pseudo old model to have a high probability of predicting a correct class, and its probability distribution shall be:

$$p^{o'}(x) = \begin{cases} \lambda & \text{if } x = a' \\ (1 - \lambda)/(\mathcal{C}_{old} - 1) & \text{if } x \neq a' \end{cases} \tag{3}$$

When λ is set to a very high number (e.g., $\lambda \geq 0.9$), we will have a high probability of getting a correct class, allowing the teacher model to have perfect performance. The probability output of the CL model can be optimized with this pseudo-teacher based on the Kullback-Leibler divergence D_{KL}:

$$\mathcal{L}_{RKD} = \sum_{c=0}^{\mathcal{C}_{old}} D_{KL} \left[p_T^{o'}(x), p_T^{cl}(x) \right] \tag{4}$$

T is the KD temperature used to generate soft probabilities for the pseudo old model. As discussed above, this naive setting of T is not suitable for general CL scenarios. Therefore, we treat T_{op} and T_{on} differently to strengthen the plasticity on $\mathcal{C}_{op}$ and the rigidity on $\mathcal{C}_{on}$ respectively. $\mathcal{L}_{RKD}$ can thereby be rewritten as:

$$\mathcal{L}_{RKD} = \sum_{c=0}^{\mathcal{C}_{op}} D_{KL} \left[p_{T_{op}}^{o'}(x), p_{T_{op}}^{cl}(x) \right] + \sum_{c=0}^{\mathcal{C}_{on}} D_{KL} \left[p_{T_{on}}^{o'}(x), p_{T_{on}}^{cl}(x) \right] \tag{5}$$

We keep $T_{op} > T_{on}$ to balance the rigidity and plasticity trade-off in the CL model, and set $T_{op} = 25$, $T_{on} = 20$ empirically in our implementation.

Self-Calibrated Heterogeneous Distillation (SH-Dist). Works have discussed the use of self-calibration to improve model performance [18,32]. However, assuming we obtain an old model and we would like to conduct CL training on it, we can hardly modify the old model itself directly. Therefore, we perform a self-calibration operation on the heterogeneous output features F_t from the VisualBERT backbone to get self-calibrated feature F_t'. The details can be referred to at the bottom of Fig. 2. Without engaging more learnable parameters, we endow the heterogeneous features with adaptively modeled long-range context information. Therefore, we can construct our feature distillation using the self-calibrated feature map F_t' and the old model feature map F_{t-1}. $\mathcal{L}_2$ loss is used to minimize the distance between F_t' and F_{t-1} empirically by following [19]:

$$\mathcal{L}_{FKD} = \frac{\|F_{t-1}(x) - F_t'(x)\|_2^2}{|\mathcal{D}_t|} \tag{6}$$

Subsequently, the self-calibrated feature map F_t' shall be propagated through the parallel classifier and detector for the multi-task prediction.

Overall Framework. Figure 2 shows the overview of our CS-VQLA framework. The given image and question input are respectively processed as feature embedding by pre-trained ResNet18 and BERT tokenier, and fed to the VisualBERT backbone after embedding fusion. Then the output heterogeneous feature map is used to train the parallel predictors. The loss functions establish the essential components of our CS-VQLA framework. In the initial time period $t = 0$, the model is only trained on the VQLA loss. When $t > 0$, we combine the VQLA loss for training on the current dataset D_t, with the RP-Dist & SH-Dist loss to retain the old knowledge. We can summarize our final loss function as follows:

$$\mathcal{L} = \begin{cases} \mathcal{L}_{VQLA} & t = 0 \\ \alpha \cdot \mathcal{L}_{VQLA} + \beta \cdot \mathcal{L}_{RKD} + \gamma \cdot \mathcal{L}_{FKD} & t > 0 \end{cases} \tag{7}$$

We set $\alpha = \beta = 1$ and $\gamma = 5$ in our implementation. Furthermore, WA is deployed after training on each time period, to balance the weight bias of new classes on the classification layer. Through the combination of multiple distillation paradigms and model weight adjustment, we successfully realize the general continual learning framework in the VQLA scenario.

3 Experiments

3.1 Dataset and Setup

Dataset. We construct our continual procedure as follows: when $t = 0$, we train on EndoVis18 Dataset, $t = 1$ on EndoVis17 Dataset, and $t = 2$ on M2CAI Dataset. Therefore, we can establish our CS-VQLA framework with a large initial step, and several smaller sequential steps. When splitting the dataset, we isolate the training and test sets in different sequences to avoid information leakage.

EndoVis18 Dataset is a public dataset with 14 videos on robotic surgery [1]. The question-answer (QA) pairs are accessible in [22], and the bounding box annotations are from [10]. The answers are in single-word form with three categories (organ, interaction, and locations). We further extend the QA pairs and include cases when the answer is a surgical tool. Besides, if the answer is regarding the organ-tool interaction, the bounding box shall contain both the organ and the tool. Statistically, the training set contains 1560 frames with 12741 QA pairs, and the test set contains 447 frames with 3930 QA pairs.

EndoVis17 Dataset is a public dataset with 10 videos on robotic surgery [2]. We randomly select frames and manually annotate the QA pairs and bounding boxes. The training set contains 167 frames with 1034 QA pairs, and the test set contains 40 frames with 201 QA pairs.

M2CAI Dataset is also a public robotic surgery dataset [24,25], and the location bounding box is publicly accessible in [11]. Similarly, we randomly select 167

Table 1. Comparison experiments from the time period $t = 0$ to $t = 1$. **Bold** and underlined represent best and second best, respectively. 'W/O' denotes 'without', and 'W/I' denotes 'within'. 'N/O' means non-overlapping. 'Old N/O' represents the classes that exist in $t = 0$ but do not exist in $t = 1$, and 'New N/O' represents the opposite. 'Overlapping' denotes the classes that exist in both $t = 0, 1$.

$t = 1$	Methods	Old N/O		Overlapping		New N/O		EndoVis18		EndoVis17		Average	
		Acc	mIoU	Acc	mIoU	Acc	mIoU	Acc	mIoU	Acc	mIoU	Acc	mIoU
W/O CL	Base ($t = 0$)	6.11	62.12	64.60	75.68	✗	✗	62.65	75.23	✗	✗	✗	✗
	FT	0.00	62.55	38.07	71.88	86.96	80.41	34.86	71.29	81.59	78.30	58.23	74.79
W/I CL	LwF [16]	0.00	**63.40**	54.36	69.94	<u>73.91</u>	77.47	53.20	69.53	43.79	74.64	48.50	72.08
	WA [27]	<u>0.76</u>	60.70	55.11	71.61	52.17	78.10	52.85	71.02	63.68	76.71	58.26	73.86
	iCaRL [20]	<u>0.76</u>	<u>62.23</u>	<u>55.87</u>	72.36	43.48	**79.51**	<u>53.85</u>	71.76	58.21	**78.41**	56.03	<u>75.08</u>
	IL2A [29]	0.00	57.74	53.00	69.88	56.52	78.20	51.48	69.23	48.76	75.75	50.12	72.49
	PASS [30]	0.00	56.49	54.01	70.08	69.57	77.60	51.70	69.46	65.67	74.56	58.69	72.01
	SSRE [31]	0.00	60.29	54.04	70.34	65.22	76.07	51.76	69.78	64.68	75.44	58.22	72.61
	CLVQA [14]	0.00	59.87	51.83	72.98	65.22	78.36	49.14	72.40	72.14	76.40	60.64	74.40
	CLiMB [23]	0.00	60.16	52.88	<u>72.99</u>	69.57	77.37	50.13	<u>72.44</u>	<u>74.13</u>	75.87	<u>62.13</u>	74.16
	Ours	**1.53**	61.08	**56.98**	**74.57**	**78.26**	<u>78.59</u>	**54.33**	74.02	**75.12**	<u>77.02</u>	**64.73**	**75.52**

Table 2. Comparison experiments from the time period $t = 0, 1$ to $t = 2$. 'Old N/O' represents the classes that exist in $t = 0, 1$ but do not exist in $t = 2$, and 'New N/O' represents the opposite. 'Overlapping' denotes the classes that exist in $t = 0, 1, 2$.

$t = 2$	Methods	Old N/O		Overlapping		New N/O		EndoVis18		EndoVis17		M2CAI16		Average	
		Acc	mIoU	Acc	mIoU	Acc	mIoU	Acc	mIoU	Acc	mIoU	Acc	mIoU	Acc	mIoU
W/O CL	FT	4.00	60.90	19.08	58.69	55.56	70.83	15.57	58.75	41.79	60.18	51.06	69.48	36.14	62.80
W/I CL	LwF [16]	4.20	<u>62.91</u>	**42.75**	63.34	27.78	72.68	<u>38.04</u>	63.25	41.29	**62.71**	31.91	69.65	37.08	<u>65.20</u>
	WA [27]	6.80	59.32	40.62	61.55	**55.56**	73.06	36.67	61.20	36.32	60.86	40.43	69.80	37.81	63.96
	iCaRL [20]	2.00	58.55	38.06	58.59	41.67	73.42	33.46	58.30	38.81	61.12	38.30	**71.01**	36.85	63.48
	IL2A [29]	11.00	57.25	33.80	58.50	27.78	72.71	30.61	58.28	37.31	58.29	36.17	67.10	34.70	61.22
	PASS [30]	22.60	56.57	24.80	58.39	30.56	72.07	23.52	58.07	37.81	58.91	**41.49**	66.68	34.27	61.22
	SSRE [31]	13.80	58.12	19.49	57.02	<u>47.22</u>	73.31	18.12	57.00	27.36	58.09	40.43	67.47	28.64	60.85
	CLVQA [14]	21.80	58.01	36.54	62.92	25.00	<u>74.09</u>	34.40	62.35	39.80	61.05	36.17	68.89	36.79	64.10
	CLiMB [23]	<u>23.00</u>	57.03	38.90	<u>64.30</u>	33.33	73.45	36.62	<u>63.50</u>	<u>42.29</u>	61.35	40.43	69.20	<u>39.78</u>	64.68
	Ours	**28.20**	**68.14**	<u>41.04</u>	**65.74**	44.44	**74.41**	**39.13**	**66.21**	**46.77**	<u>62.17</u>	41.49	<u>70.04</u>	**42.46**	**66.14**

frames and annotate 449 QA pairs for the training set, and 40 frames with 94 QA pairs in different videos for the test set.

Implementation Details. We compare our solution against the fine-tuning (FT) baseline and state-of-the-art (SOTA) methods, including LwF [16], WA [27], iCaRL [20], IL2A [29], PASS [30], SSRE [31], CLVQA [14], and CLiMB [23]. All the methods are implemented using [28][1], with PyTorch and on NVIDIA RTX 3090 GPU. We removed the exemplars in all methods for a non-exemplar comparison. All methods are firstly trained on EndoVis18 ($t = 0$) for 60 epochs with a learning rate of 1×10^{-5}, and then trained on EndoVis17 ($t = 2$) and M2CAI ($t = 2$) for 30 epochs with a learning rate of 5×10^{-5}. We use Adam optimizer [12] and a batch size of 64. Answering and localization performance are evaluated by Accuracy (Acc) and mean intersection over union (mIoU), respectively.

Table 3. Ablation experiments from the time period $t = 0$ to $t = 1$, and from $t = 1$ to $t = 2$. To observe the contribution of each component, we degenerate the proposed RP-Dist and SH-Dist to the normal distillation paradigm, and remove the WA module.

Methods			$t = 0$ to $t = 1$								$t = 1$ to $t = 2$							
			Old N/O		Overlapping		New N/O		Average		Old N/O		Overlapping		New N/O		Average	
RP	SH	WA	Acc	mIoU	Acc	mIoU	Acc	mIoU	Acc	mIoU	Acc	mIoU	Acc	mIoU	Acc	mIoU	Acc	mIoU
✓	✗	✗	0.00	60.58	55.30	72.94	73.91	77.62	62.66	74.37	11.40	58.29	40.96	64.72	30.56	73.31	40.80	64.85
✗	✓	✗	0.00	59.94	56.23	73.66	**82.61**	78.04	64.33	74.90	9.80	57.17	38.14	65.30	38.89	74.15	40.26	64.35
✗	✗	✓	0.00	60.33	53.08	72.45	52.17	76.18	61.94	74.30	12.20	59.49	39.75	64.81	41.67	72.12	39.07	64.81
✗	✓	✓	0.00	60.97	54.59	74.12	73.91	78.52	61.83	74.89	11.00	65.16	40.69	64.45	41.67	74.18	39.63	65.45
✓	✗	✓	0.00	61.04	56.08	74.11	60.87	78.91	61.60	74.88	11.00	59.10	39.25	62.60	22.22	71.66	39.16	64.55
✓	✓	✗	0.00	60.07	54.56	74.15	73.91	**79.79**	61.81	75.00	10.40	63.32	39.14	64.40	27.78	73.89	40.21	65.65
✓	✓	✓	**1.53**	**61.08**	**56.98**	**74.57**	78.26	78.59	**64.73**	**75.52**	**28.20**	**68.14**	41.04	**65.74**	**44.44**	**74.41**	**42.46**	**66.14**

3.2 Results

Except for testing on three datasets separately, we set three specific categories in our continual learning setup: *old non-overlapping* (old N/O) classes, *overlapping* classes, and *new non-overlapping* (new N/O) classes. By measuring the performance in these three categories, we can easily observe the catastrophic forgetting phenomenon and the performance of mitigating catastrophic forgetting.

As shown in Table 1 & 2, firstly, catastrophic forgetting can be apparently observed in the performance of FT. Then, among all baselines, iCaRL achieves the best performance when the model learns from $t = 0$ to $t = 1$, and gets to forget when there are more time periods. On the contrary, LwF exhibits a strong retention of old knowledge, but a lack of ability to learn new. Our proposed methods demonstrate superior performance in almost all metrics and classes. In classification tasks, the overall average of our methods outperforms the second best with 2.60% accuracy improvement at $t = 1$ and 2.68% at $t = 2$. In localization tasks, our method is 0.44 mIoU higher than the second best at $t = 1$

---

[1] github.com/G-U-N/PyCIL.

and 0.94 mIoU higher at $t = 2$. The results prove the remarkable ability of our method to balance the rigidity-plasticity trade-off. Furthermore, an ablation study is conducted to demonstrate the effectiveness of each component in our proposed method. We (i) degenerate the RP-Dist to original logits distillation [16], (ii) degenerate the SH-Dist to normal feature distillation [19], and (iii) remove the WA module, as shown in Table 3. Experimental results show that each component we propose or integrate plays an essential role in the final rigidity-plasticity trade-off. Therefore, we demonstrate that each of our components is indispensable. More evaluation and ablation studies can be found in the supplementary materials.

4   Conclusion

This paper introduces CS-VQLA, a general continual learning framework on surgical VQLA tasks. This is a significant attempt to continue learning under complicated clinical tasks. Specifically, we propose the RP-Dist on output logits, and the SH-Dist on the intermediate feature space, respectively. The WA technique is further integrated for model weight bias adjustment. Superior performance on VQLA tasks demonstrates that our method has an excellent ability to deal with CL-based surgical scenarios. Except for giving localized answers for better surgical scene understanding, our solution can conduct continual learning in any questions in surgical applications to solve the problem of class increment, domain shift, and overlapping/non-overlapping classes. Our framework can also be applied when adapting a vision-language foundation model in the surgical domain. Therefore, our solution holds promise for deploying auxiliary surgical education tools across time/institutions. Potential future works also include combining various surgical training systems (e.g., mixed reality-based training, surgical skill assessment) to develop an effective and comprehensive virtual teaching system.

Acknowledgements. This work was funded by Hong Kong RGC CRF C4063-18G, CRF C4026-21GF, RIF R4020-22, GRF 14203323, GRF 14216022, GRF 14211420, NSFC/RGC JRS N_CUHK420/22; Shenzhen-Hong Kong-Macau Technology Research Programme (Type C 202108233000303); Guangdong GBABF #2021B1515120035. M. Islam was funded by EPSRC grant [EP/W00805X/1].

References

1. Allan, M., et al.: 2018 robotic scene segmentation challenge. arXiv preprint arXiv:2001.11190 (2020)
2. Allan, M., et al.: 2017 robotic instrument segmentation challenge. arXiv preprint arXiv:1902.06426 (2019)
3. Bai, L., Islam, M., Seenivasan, L., Ren, H.: Surgical-VQLA: transformer with gated vision-language embedding for visual question localized-answering in robotic surgery. arXiv preprint arXiv:2305.11692 (2023)

4. De Lange, M., et al.: A continual learning survey. defying forgetting in classification tasks. IEEE Trans. Pattern Anal. Mach. Intell. **44**(7), 3366–3385 (2021)

5. Deng, J., Dong, W., Socher, R., Li, L.J., Li, K., Fei-Fei, L.: Imagenet: a large-scale hierarchical image database. In: 2009 IEEE Conference on Computer Vision and Pattern Recognition, pp. 248–255. IEEE (2009)

6. Derakhshani, M.M., et al.: Lifelonger: a benchmark for continual disease classification. In: Wang, L., Dou, Q., Fletcher, P.T., Speidel, S., Li, S. (eds.) Medical Image Computing and Computer Assisted Intervention – MICCAI 2022. MICCAI 2022. LNCS, vol. 13432, pp. 314–324. Springer, Cham (2022). https://doi.org/10.1007/978-3-031-16434-7_31

7. Devlin, J., Chang, M.W., Lee, K., Toutanova, K.: Bert: Pre-training of deep bidirectional transformers for language understanding. arXiv preprint arXiv:1810.04805 (2018)

8. He, K., Zhang, X., Ren, S., Sun, J.: Deep residual learning for image recognition. In: Proceedings of the IEEE Conference on Computer Vision and Pattern Recognition, pp. 770–778 (2016)

9. Hinton, G., Vinyals, O., Dean, J.: Distilling the knowledge in a neural network. arXiv preprint arXiv:1503.02531 (2015)

10. Islam, M., Seenivasan, L., Ming, L.C., Ren, H.: Learning and reasoning with the graph structure representation in robotic surgery. In: Martel, A.L., et al. (eds.) MICCAI 2020. LNCS, vol. 12263, pp. 627–636. Springer, Cham (2020). https://doi.org/10.1007/978-3-030-59716-0_60

11. Jin, A., et al.: Tool detection and operative skill assessment in surgical videos using region-based convolutional neural networks. In: IEEE Winter Conference on Applications of Computer Vision (2018)

12. Kingma, D.P., Ba, J.: Adam: a method for stochastic optimization. arXiv preprint arXiv:1412.6980 (2014)

13. Lee, C.S., Lee, A.Y.: Clinical applications of continual learning machine learning. Lancet Digit. Health **2**(6), e279–e281 (2020)

14. Lei, S.W., et al.: Symbolic replay: scene graph as prompt for continual learning on VQA task. In: Proceedings of the AAAI Conference on Artificial Intelligence, vol. 37, pp. 1250–1259 (2023)

15. Li, L.H., Yatskar, M., Yin, D., Hsieh, C.J., Chang, K.W.: Visualbert: a simple and performant baseline for vision and language. arXiv preprint arXiv:1908.03557 (2019)

16. Li, Z., Hoiem, D.: Learning without forgetting. IEEE Trans. Pattern Anal. Mach. Intell. **40**(12), 2935–2947 (2017)

17. Lin, Z., et al.: Medical visual question answering: a survey. arXiv preprint arXiv:2111.10056 (2021)

18. Liu, J.J., Hou, Q., Cheng, M.M., Wang, C., Feng, J.: Improving convolutional networks with self-calibrated convolutions. In: Proceedings of the IEEE/CVF Conference on Computer Vision and Pattern Recognition, pp. 10096–10105 (2020)

19. Michieli, U., Zanuttigh, P.: Incremental learning techniques for semantic segmentation. In: Proceedings of the IEEE/CVF International Conference on Computer Vision Workshops (2019)

20. Rebuffi, S.A., Kolesnikov, A., Sperl, G., Lampert, C.H.: iCaRL: incremental classifier and representation learning. In: Proceedings of the IEEE Conference on Computer Vision and Pattern Recognition, pp. 2001–2010 (2017)

21. Rezatofighi, H., Tsoi, N., Gwak, J., Sadeghian, A., Reid, I., Savarese, S.: Generalized intersection over union: a metric and a loss for bounding box regression.

In: Proceedings of the IEEE/CVF Conference on Computer Vision and Pattern Recognition, pp. 658–666 (2019)

22. Seenivasan, L., Islam, M., Krishna, A., Ren, H.: Surgical-VQA: visual question answering in surgical scenes using transformer. In: Wang, L., Dou, Q., Fletcher, P.T., Speidel, S., Li, S. (eds.) Medical Image Computing and Computer Assisted Intervention – MICCAI 2022. MICCAI 2022. LNCS, vol. 13437, pp. 33–43. Springer, Cham (2022). https://doi.org/10.1007/978-3-031-16449-1_4

23. Srinivasan, T., Chang, T.Y., Pinto Alva, L., Chochlakis, G., Rostami, M., Thomason, J.: Climb: a continual learning benchmark for vision-and-language tasks. Adv. Neural Inf. Process. Syst. **35**, 29440–29453 (2022)

24. Stauder, R., Ostler, D., Kranzfelder, M., Koller, S., Feußner, H., Navab, N.: The tum lapchole dataset for the m2cai 2016 workflow challenge. arXiv preprint arXiv:1610.09278 (2016)

25. Twinanda, A.P., Shehata, S., Mutter, D., Marescaux, J., De Mathelin, M., Padoy, N.: EndoNet: a deep architecture for recognition tasks on laparoscopic videos. IEEE Trans. Med. Imaging **36**(1), 86–97 (2016)

26. Yuan, L., Tay, F.E., Li, G., Wang, T., Feng, J.: Revisiting knowledge distillation via label smoothing regularization. In: Proceedings of the IEEE/CVF Conference on Computer Vision and Pattern Recognition, pp. 3903–3911 (2020)

27. Zhao, B., Xiao, X., Gan, G., Zhang, B., Xia, S.T.: Maintaining discrimination and fairness in class incremental learning. In: Proceedings of the IEEE/CVF Conference on Computer Vision and Pattern Recognition, pp. 13208–13217 (2020)

28. Zhou, D.W., Wang, F.Y., Ye, H.J., Zhan, D.C.: PyCIL: a python toolbox for class-incremental learning (2021)

29. Zhu, F., Cheng, Z., Zhang, X.Y., Liu, C.L.: Class-incremental learning via dual augmentation. Adv. Neural Inf. Process. Syst. **34**, 14306–14318 (2021)

30. Zhu, F., Zhang, X.Y., Wang, C., Yin, F., Liu, C.L.: Prototype augmentation and self-supervision for incremental learning. In: Proceedings of the IEEE/CVF Conference on Computer Vision and Pattern Recognition, pp. 5871–5880 (2021)

31. Zhu, K., Zhai, W., Cao, Y., Luo, J., Zha, Z.J.: Self-sustaining representation expansion for non-exemplar class-incremental learning. In: Proceedings of the IEEE/CVF Conference on Computer Vision and Pattern Recognition, pp. 9296–9305 (2022)

32. Zou, W., Ye, T., Zheng, W., Zhang, Y., Chen, L., Wu, Y.: Self-calibrated efficient transformer for lightweight super-resolution. In: Proceedings of the IEEE/CVF Conference on Computer Vision and Pattern Recognition, pp. 930–939 (2022)

Intra-operative Forecasting of Standing Spine Shape with Articulated Neural Kernel Fields

Sylvain Thibeault[1], Stefan Parent[2], and Samuel Kadoury[1,2](✉)

[1] MedICAL, Polytechnique Montreal, Montreal, QC, Canada
samuel.kadoury@polymtl.ca
[2] Sainte-Justine Hospital Research Center, Montreal, QC, Canada

Abstract. Minimally invasive spine surgery such as anterior vertebral tethering (AVT), enables the treatment of spinal deformities while seeking to preserve lower back mobility. However the intra-operative positioning and posture of the spine affects surgical outcomes. Forecasting the standing shape from adolescent patients with growing spines remains challenging with many factors influencing corrective spine surgery, but can allow spine surgeons to better prepare and position the spine prior to surgery. We propose a novel intra-operative framework anticipating the standing posture of the spine immediately after surgery from patients with idiopathic scoliosis. The method is based on implicit neural representations, which uses a backbone network to train kernels based on neural splines and estimate network parameters from intra-pose data, by regressing the standing shape on-the-fly using a simple positive definite linear system. To accommodate with the variance in spine appearance, we use a Signed Distance Function for articulated structures (A-SDF) to capture the articulation vectors in a disentangled latent space, using distinct encoding vectors to represent both shape and articulation parameters. The network's loss function incorporates a term regularizing outputs from a pre-trained population growth trajectory to ensure transformations are smooth with respect to the variations seen on first-erect exams. The model was trained on 735 3D spine models and tested on a separate set of 81 patients using pre- and intra-operative models used as inputs. The neural kernel field framework forecasted standing shape outcomes with a mean average error of 1.6 ± 0.6 mm in vertebral points, and generated shapes with IoU scores of 94.0 compared to follow-up models.

Keywords: Spine surgery · Prediction model · Neural kernel fields · Upright posture · Disentangled latent representation

1 Introduction

Surgical treatment remains to this day the main approach to correct severe spinal deformities, which consists of rectifying the alignment of the spine over several

Supported by the Canada Research Chairs and NSERC Discovery Grant RGPIN-2020-06558.

vertebral bodies with intrinsic forces. However major limiting factors remain the loss in spinal mobility, as well as the increased risk of pain and osteoarthritis in years after surgery [1]. Furthermore, prone positioning during spinal surgery combined with hip extension can greatly affect post-operative outcomes of sagittal alignment and balance, particularly in degenerative surgical cases [2]. Recent literature has demonstrated the importance to preserve the lumbar positioning, as well as the normal sagittal alignment once surgery has been completed [5,7]. Therefore, by optimizing the intra-op posture, surgeons can help to reduce long-term complications such as lower back pain and improve spinal mobility. A predictive model anticipating the upright 3D sagittal alignment of patients during spine surgery [15], such as anterior vertebral tethering (AVT), may not only have an impact on patients to preserve their post-operative balance, but also provide information about patients at risk of degenerative problems in the lumbar region of the spine following surgery [24].

To interpret the positioning effects during surgery, identifying features leading to reliable predictions of shape representations in postural changes is a deciding factor as shown in [14], which estimated correlations with geometrical parameters. However, with the substantial changes during surgery such as the prone positioning of the spine and global orientation of the pelvic area, this becomes very challenging. Previous classification methods were proposed in the literature to estimate 3D articulated shape based on variational auto-encoders [19] or with spatio-temporal neural networks [10], combining soft tissue information of the pre-operative models. The objective here is to address these important obstacles by presenting a forecasting framework used in the surgical workflow, describing the spine shape changes within an implicit neural representation, with a clinical target of estimating Cobb angles within 5° [14].

Recently, Neural Radiance Fields (NeRF) [12] have been exploited for synthetic view generation [23]. The implicit representation provided by the parameters of the neural network captures properties of the objects such as radiance field and density, but can be extended to 3D in order to generate volumetric views and shapes from images [4], meshes [16] or clouds of points [9]. As opposed to traditional non-rigid and articulated registration methods based on deep learning for aligning pre-op to intra-op 2D X-rays [3,26], kernel methods allow to map input data into a different space, where subsequently simpler models can be trained on the new feature space, instead of the original space. They can also integrate prior knowledge [22], but have been mostly focused on partial shapes [21]. However, due to the difficulty to apply neural fields in organ rendering applications in medical imaging, and the important variations in the environment and spine posture/shape which can impact the accuracy of the synthesis process, their adoption has been limited.

This paper presents an intra-operative predictive framework for spine surgery, allowing to forecast on-the-fly the standing geometry of the spine following corrective AVT procedures. The proposed forecasting framework uses as input the intra-operative 3D spine model generated from a multi-view Transformer network which integrates a pre-op model (Sect. 2.1). The predictive model is based

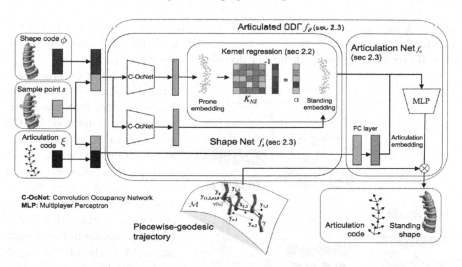

Fig. 1. The proposed forecasting architecture of the standing spine posture following surgery. During training, randomly selected spine shapes in the prone and first-erect standing posture are used to regress a neural spline kernel (K_{NS}). The generated shape in latent space (f_s) is then used as input to an articulated signed distance function (A-SDF) capturing the inter-vertebral shape constellation (f_a), transforming the signed distance values and infer the new instances from the trained kernel. Identical shape instances can yield multiple articulated instances, regardless of the shape code. At inference time, the shape and articulation code are inferred jointly by f_θ from backpropagation based on the input prone shape. A regularization term is added based on a pre-trained trajectory model, capturing the natural changes of spinal curvature.

on neural field representations, capturing the articulation variations in a disentangled latent space intra-operative positioning in the prone position. The method regresses a kernel function (Sect. 2.2) based on neural splines to estimate an implicit function warping the spine geometry between various poses. The inference process is regularized with a piecewise geodesic term, capturing the natural evolution of spine morphology (Sect. 2.3). As an output, the framework produces the geometry of the spine at the first-erect examination in a standing posture following surgery, as illustrated in Fig. 1.

2   Methods

2.1   Intra-operative 3D Model from Multi-view X-Rays

The first step of the pipeline consists of inferring a 3D model of the spine in the OR prior to instrumentation, using as input pair of orthogonal C-arm acquisitions $\mathcal{I} = \{\mathbf{I}^1, \mathbf{I}^2\}$ and previously generated pre-op 3D model [6], capturing the spine geometry in a lying prone position on the OR table. We use a multi-view 3D reconstruction approach based on Transformers [20]. The Transformer-based framework integrates both a bi-planar Transformer-based encoder, which combines information and relationships between the multiple calibrated 2D views, and a 3D reconstruction Transformer decoder, as shown in Fig. 2.

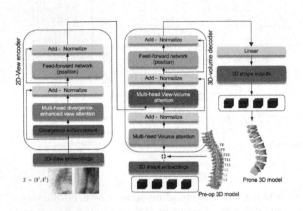

The 3D-reconstruction Transformer decodes and combines features from the biplanar views generated by the encoder, in order to produce a 3D articulated mesh model with probabilistic outputs for every spatial query token. The decoder's attention layers captures the 2D-3D relationship between the resulting grid node of the mesh and input 2D X-rays. Attention layers in the 3D network on the other hand analyzes correlations between 3D mesh landmarks to learn a 3D representation. A conditional prior using the pre-op model is integrated at the input of

Fig. 2. Schematic illustration of the 3D shape inference of the spine from bi-planar X-rays in the prone position using multi-view Transformers. The model incorporates a view-divergence mechanisms to enhance features from the 2D view embeddings.

the 3D module to inject knowledge about the general geometry. By combining these modules (2D-2D, 2D-3D, 3D-3D) into a unified framework, feature correlations can be processed simultaneously using the attention layers in the encoder/decoder networks, generating as an output shape model $S = \{\mathbf{s}_1, \ldots, \mathbf{s}_m\}$ with $\mathbf{s}_m$ as a vertebral mesh with a specific topology for the vertebral level m with an associated articulation vector:

$$\mathbf{y}_i = [T_1, ; T_1 \circ T_2, , \ldots, \\ T_1 \circ T_2 \circ \ldots \circ T_m] \tag{1}$$

representing the m inter-vertebral transformations T_i between consecutive vertebrae i and $i+1$, each consisting of 6 DoF with translation and rotation, expressed with recursive compositions. To accommodate with the Transformer's inability to explore and synthesize multi-view associations at deeper levels, the divergence decay is slowed down within the self-attention layers by augmenting the discrepancy in multi-view embeddings.

2.2   Learnable Shape Deformation Kernel

We use a training set $\mathcal{S}$ of N spine models from a surgical population, where each spine model is articulated into pre-operative, prone and first-erect poses ($K = 3$), leading to a training set of $L = N \times K$ spine models from a population of scoliotic patients. We denote $S_{n,k}$ as the spine model n given a particular shape articulation k, where $n \in \{1, \ldots, N\}$, and $k \in \{1, \ldots, K\}$. We also define $s_i \in \mathbb{R}^3$ as a sample 3D point from a spine shape in $\mathcal{S}$.

We first train a kernel based on neural splines using the embeddings generated with a neural network [21], in order to map spine input points from the prone to the first-erect position. Input points $s_i \in \mathcal{S}$, associated to a shape code $\phi \in \mathbb{R}^D$ in latent space, with D indicating the dimensionality, are associated with a feature vector $\rho(s_i|\mathcal{S}, \Omega)$, where Ω conditions the parameters of the neural network ρ using the dataset $\mathcal{S}$. Based on the learned feature points of prone and upright spine models, the data-related kernel is given by:

$$K_{\mathcal{S},\Omega}(s_i, s_j) = K_{NS}([s_i : \rho(s_i|\mathcal{S}, \Omega)], [s_j : \rho(s_j|\mathcal{S}, \Omega)]) \tag{2}$$

with $[c : d]$ concatenating feature vectors c and d, representing the features of sample points i and j taken from a mesh model, respectively, and K_{NS} representing Neural Spline kernel function. A Convolutional Occupancy Network [17] is used to generate the features from sampled points in both spaces. Once the space around each input point is discretized into a volumetric grid space, PointNet [18] is used in every grid cell which contains points, to extract non-zero features from each 3D cell. These are then fed to a 3D FCN, yielding output features of same size.

The neural network's implicit function is determined by finding coefficients α_j associated for each point s_j, such that:

$$\alpha = [\alpha_j]_{j=1}^{2L} = (\mathcal{G}(\mathcal{S}, \Omega) + \lambda \mathbf{I})^{-1}\mathbf{y}_j \tag{3}$$

which solves a $2L \times 2L$ linear system to obtain the standing articulations ($\mathbf{y}_j$), with $\mathcal{G}(\mathcal{S}, \Omega)$ as the gram matrix, such that $\mathcal{G}(\mathcal{S}, \Omega)_{ij} = K_{\mathcal{S},\Omega}(s_i, s_j)$, with s_i and s_j are the input sample points of the spine models, and regularized by the λ parameter using the identity matrix $\mathbf{I}$. For new feature points s, the regressed function is evaluated using α in the following term:

$$f_s(s, \phi) = \sum_{s_j \in S, \phi} \alpha_j K_{\mathcal{S},\Omega}(s, s_j) \tag{4}$$

which represents the ShapeNet function and maps feature points s from the original prone embedding onto the standing embedding.

2.3 Articulation SDF

Each training shape $S_{n,k}$ is also associated with an articulation vector field, $\xi \in \mathbb{R}^d$. The shape embedding ϕ_n is common across all instances n with various articulation vectors. We then train ArticulationNet, where for each instance, the shape code is maintained and each individual code for shape is updated. The articulation vector $\mathbf{y}_i$ defined in Eq.(1) is used to define ξ_i.

The shape function f_s trained from kernel regression is integrated into an articulation network [13] using the signed distance function (A-SDF) which is determined by using $s \in \mathbb{R}^3$ as a shape's sample point. We recall that ϕ and ξ are codes representing the shape and inter-vertebral articulation, respectively.

The A-SDF for the spine shape provided by f_θ is represented with an auto-decoder model, which includes the trained shape kernel function f_s, while f_a describes ArticulationNet:

$$f_\theta(s, \phi, \xi) = f_a[f_s(s, \phi), s, \xi] = v \tag{5}$$

with $v \in \mathbb{R}$ representing a value of the SDF on the output mesh, where the sign specifies if the sampled spine point falls within or outside the generated 3D surface model, described implicitly with the zero level-set $f_\theta(.) = 0$.

At training, the ground-truth articulation vectors ξ are used to train the parameters of the model and the shape code. Sample points s are combined with the shape code ϕ to create a feature vector which is used as input to the shape embedding network. A similar process is used for the articulation code ξ, which are also concatenated with sample points s. This vector is used as input to the articulation network in order to predict for each 3D point s the value of the SDF. Hence, the network uses a fully connected layer at the beginning to map the vector in the latent space, while a classification module is added in the final hidden layer to output the series of vertebrae. We define the loss function using the L1 distance term, regressing the SDF values for the M number of 3D points describing each spine shape using the function f_θ:

$$\mathcal{L}^s = \frac{1}{M} \sum_{m=1}^{M} \| f_\theta(s_m, \phi, \xi) - v_m \|_1 \tag{6}$$

where $s_m \in \mathcal{S}$ is a sample point from the shape space, and v_m is the SDF value used as ground-truth for $m \in \{1, \ldots, M\}$. The second loss term evaluates the vertebral constellation and alignment with the inter-vertebral transformations:

$$\mathcal{L}^p = \frac{1}{M} \sum_{m=1}^{M} [CE(f_\theta(s_m, \phi, \xi), p_m)] \tag{7}$$

with p_m identifying the vertebral level for sample s_m and CE represents the cross-entropy loss. This classification loss enables to separate the constellation of vertebral shapes $\mathbf{s}_m$.

To ensure the predicted shape is anatomically consistent, a regularization term $\mathcal{L}^r$ is used to ensure generated models fall near a continuous piecewise-geodesic trajectory defined in a spatio-temporal domain. The regularization term integrates pre-generated embeddings capturing the spatiotemporal changes of the spine geometry following surgery [11]. The piecewise manifold is obtained from nearest points with analogue features in the training data of surgical spine patients in the low-dimensional embedding, producing local regions $\mathcal{N}(s_i)$ with samples lying within the shape change spectrum. A Riemannian domain with samples $\mathbf{y}_i$ is generated, with i a patient model acquired at regular time intervals. Assuming the latent space describes the overall variations in a surgical population and the geodesic trajectory covers the temporal aspects, new labeled points can be regressed within the domain $\mathbb{R}^D$, producing continuous curves in the shape/time domain. The overall loss function is estimated as:

$$\mathcal{L}(\mathcal{S}, \phi, \xi) = \mathcal{L}^s(\mathcal{S}, \phi, \xi) + \beta_p \mathcal{L}^p(\mathcal{S}, \phi, \xi) + \beta_\theta \mathcal{L}^r \tag{8}$$

with β_p and β_θ weighting the parts and regularization terms, respectively. During training, a random initialization of the shape codes based on Gaussian distributions is used. An optimization procedure is then applied during training for each shape code, which are used for all instances of articulations. Hence the global objective seeks to minimize for all $N \times K$ shapes, the following energy:

$$\arg\min_{\phi, \xi} \sum_{n=1}^{N} \sum_{k=1}^{K} \mathcal{L}(S_{n,k}, \phi_n, \xi_k). \tag{9}$$

At test time, the intra-operative spine model (obtained in 2.1) is given as input, generating the articulated standing pose, with shape and articulation vectors, using back-propagation. The shape and articulation codes ϕ and ξ are randomly initialized while the network parameters stay fixed. To overcome divergence problems and avoid a local minima for Eq. (9), a sequential process is applied where first articulated codes are estimated while shape outputs are ignored, allowing to capture the global appearance. Then, using the fixed articulated representation, the shape code is re-initialized and optimized only for ϕ.

3 Experiments

A mono-centric dataset of 735 spine models was used to train the articulated neural kernel field forecasting model. The dataset included pre-, intra- and post-operative thoraco/lumbar 3D models. Pre- and post-op models were generated from bi-planar X-rays using a stereo-reconstruction method (EOS system, Paris, France), integrating a semi-supervised approach generating vertebral landmarks [6]. Each patient in the cohort underwent corrective spine surgery, with a main angulation range of [35°–65°]. Vertebral landmarks were validated by a surgeon.

During spine surgery, a pair of C-arm images at frontal and 90° angulations were acquired after prone positioning of the patient on the OR table and before starting the surgical instrumentation. The C-arm image pair was used to generate the input 3D model of the instrumented spine segment. For the Transformer model, batch size was set at 64 and image sizes for training were set at 225×225. For the neural field model, batch size was set at 32, with a dropout rate of 0.4, a learning rate of 0.003, $\beta_p = 0.4$, $\beta_\theta = 0.5$, $\lambda = 0.25$. The AdamW optimizer was used for both models [8]. Inference time was of 1.2 s on a NVIDIA A100 GPU. We assessed the intra-operative modeling of the spine in a prone position using the Transformer based framework on a separate set of 20 operative patients. Ground-truth models with manually annotated landmarks on pairs of C-arm images were used as basis of comparison, yielding a 3D RMS error of 0.9 ± 0.4 mm, a Dice score (based on overlap of GT and generated vertebral meshes) of 0.94 ± 0.3 and a difference of $0.9° \pm 0.3$ in the main spine angulation. These are in the clinical acceptable ranges to work in the field.

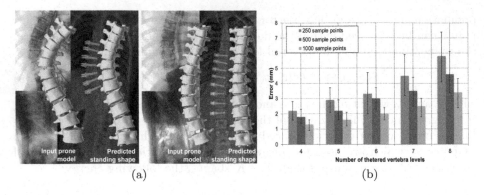

(a)   (b)

Fig. 3. (a) Sample predicted 3D geometric models of the thoraco/lumbar spine in standing posture following corrective procedures. Both examples show the input intra-operative model on the left, and the predicted geometric shape on the right, alongside the first-erect X-ray images at follow-up. (b) Effect of increasing number of tethered segments and number sampled points per model on the overall accuracy.

Finally, the predicted standing spine shape accuracy was evaluated on a hold-out set of 81 surgical patients who underwent minimally invasive corrective spine surgery. The cohort had an initial mean angulation of 48°, and immediate follow-up exam at 2 weeks. Results are presented in Table 1. For each predicted standing spine model, errors in 3D vertebral landmarks, IoU measures and Chamfer distance, as well as differences in Cobb and lordosis angles were computed for the inferred shapes, and were compared to state-of-the art spatio-temporal and neural fields models. Ablation experiments were also conducted, demonstrating

Table 1. 3D RMS errors (mm), IoU (%) Chamfer distance, Cobb angle difference (°) and lordosis angle difference (°) for the proposed articulated neural kernel field (A-NKF) method, compared to ST-ResNet [25], Convolutional Occupancy Network [17], DenseSDF [16], NeuralPull [9] and ST-Manifold [10]. The ground-truth standing 3D spine models obtained at follow-up from bi-planar X-rays were used as basis for comparison. Bottom rows shows the ablation experiments.

	3D RMS ↓	IoU ↑	Chamfer ↓	Cobb ↓	Lordosis ↓
ST-ResNet [25]	6.3 ± 3.8	78.5 ± 3.6	8.3 ± 4.2	5.6 ± 3.9	5.9 ± 4.2
C-OccNet [17]	4.6 ± 2.8	81.2 ± 3.4	7.0 ± 3.6	4.3 ± 2.9	4.8 ± 4.4
DeepSDF [16]	3.8 ± 2.1	84.5 ± 2.8	6.1 ± 2.2	3.9 ± 2.4	4.2 ± 3.9
NeuralPull [9]	3.1 ± 1.4	86.9 ± 1.9	5.3 ± 1.8	3.7 ± 2.0	3.9 ± 3.6
ST-Manifold [10]	2.8 ± 1.2	88.2 ± 1.5	4.9 ± 1.6	3.1 ± 1.8	3.5 ± 3.0
A-NKF (no kernel + no regular.)	2.7 ± 1.0	87.4 ± 1.3	5.0 ± 1.8	3.3 ± 2.0	3.5 ± 2.9
A-NKF (no kernel)	2.1 ± 0.7	91.7 ± 2.5	4.3 ± 0.6	2.6 ± 1.8	2.9 ± 2.4
A-NKF (no regular.)	1.9 ± 0.7	92.2 ± 2.3	4.1 ± 0.7	2.5 ± 1.7	2.7 ± 2.0
Proposed A-NKF	1.6 ± 0.6	94.0 ± 2.0	3.7 ± 0.5	2.1 ± 1.2	2.3 ± 1.5

statistically significant improvements ($p < 0.05$) with the kernel and regularization modules to the overall accuracy. We finally evaluated the performance of the model by measuring the 3D RMS errors versus the input size of points sampled from each shape and tethered vertebral levels. We can observe limitations when using a sparser set of samples points at 250 points with a mean error of 2.8 mm, but a significant increase in performance of the kernel-based neural fields using denser inputs for each model (n=1000).

4 Conclusion

In this paper, we proposed an online forecasting model predicting the first-erect spine shape based on intra-operative positioning in the OR, capturing the articulated shape constellation changes between the prone and the standing posture with neural kernel fields and an articulation network. Geometric consistency is integrated with the network's training with a pre-trained spine correction geodesic trajectory model used to regularize outputs of ArticulationNet. The model yielded results comparable to ground-truth first-erect 3D geometries in upright positions, based on statistical tests. The neural field network implicitly captures the physiological changes in pose, which can be helpful for planning the optimal posture during spine surgery. Future work will entail evaluating the model in a multi-center study to evaluate the predictive robustness and integrate the tool for real-time applications.

References

1. Cheng, J.C., et al.: Adolescent idiopathic scoliosis. Nat. Rev. Dis. Primers **1**(1), 1–21 (2015)
2. Elysee, J.C., et al.: Supine imaging is a superior predictor of long-term alignment following adult spinal deformity surgery. Glob. Spine J. **12**(4), 631–637 (2022)
3. Esfandiari, H., Anglin, C., Guy, P., Street, J., Weidert, S., Hodgson, A.J.: A comparative analysis of intensity-based 2D–3D registration for intraoperative use in pedicle screw insertion surgeries. Int. J. Comput. Assist. Radiol. Surg. **14**, 1725–1739 (2019)
4. Ge, L., et al.: 3D hand shape and pose estimation from a single RGB image. In: Proceedings of the IEEE/CVF Conference on Computer Vision and Pattern Recognition, pp. 10833–10842 (2019)
5. Harimaya, K., Lenke, L.G., Mishiro, T., Bridwell, K.H., Koester, L.A., Sides, B.A.: Increasing lumbar lordosis of adult spinal deformity patients via intraoperative prone positioning. Spine **34**(22), 2406–2412 (2009)
6. Humbert, L., de Guise, J., Aubert, B., Godbout, B., Skalli, W.: 3D reconstruction of the spine from biplanar X-rays using parametric models based on transversal and longitudinal inferences. Med. Eng. Phys. **31**(6), 681–87 (2009)
7. Karikari, I.O., et al.: Key role of preoperative recumbent films in the treatment of severe sagittal malalignment. Spine Deform. **6**, 568–575 (2018)
8. Loshchilov, I., Hutter, F.: Decoupled weight decay regularization. ICLR, 2019 (2017)

9. Ma, B., Han, Z., Liu, Y.S., Zwicker, M.: Neural-pull: learning signed distance functions from point clouds by learning to pull space onto surfaces. arXiv preprint arXiv:2011.13495 (2020)
10. Mandel, W., Oulbacha, R., Roy-Beaudry, M., Parent, S., Kadoury, S.: Image-guided tethering spine surgery with outcome prediction using Spatio-temporal dynamic networks. IEEE Trans. Med. Imaging 40(2), 491–502 (2020)
11. Mandel, W., Turcot, O., Knez, D., Parent, S., Kadoury, S.: Prediction outcomes for anterior vertebral body growth modulation surgery from discriminant spatiotemporal manifolds. IJCARS 14(9), 1565–1575 (2019)
12. Mildenhall, B., Srinivasan, P.P., Tancik, M., Barron, J.T., Ramamoorthi, R., Ng, R.: Nerf: representing scenes as neural radiance fields for view synthesis. Commun. ACM 65(1), 99–106 (2021)
13. Mu, J., Qiu, W., Kortylewski, A., Yuille, A., Vasconcelos, N., Wang, X.: A-SDF: learning disentangled signed distance functions for articulated shape representation. In: Proceedings of the IEEE/CVF International Conference on Computer Vision, pp. 13001–13011 (2021)
14. Nault, M.L., Mac-Thiong, J.M., Roy-Beaudry, M., Labelle, H., Parent, S., et al.: Three-dimensional spine parameters can differentiate between progressive and non-progressive patients with AIS at the initial visit: a retrospective analysis. J. Pediatr. Orthop. 33(6), 618–623 (2013)
15. Oren, J.H., et al.: Measurement of spinopelvic angles on prone intraoperative long-cassette lateral radiographs predicts postoperative standing global alignment in adult spinal deformity surgery. Spine Deform. 7(2), 325–330 (2019)
16. Park, J.J., Florence, P., Straub, J., Newcombe, R., Lovegrove, S.: DeepSDF: learning continuous signed distance functions for shape representation. In: Proceedings of the IEEE/CVF Conference on Computer Vision and Pattern Recognition, pp. 165–174 (2019)
17. Peng, S., Niemeyer, M., Mescheder, L., Pollefeys, M., Geiger, A.: Convolutional occupancy networks. In: Vedaldi, A., Bischof, H., Brox, T., Frahm, J.-M. (eds.) ECCV 2020. LNCS, vol. 12348, pp. 523–540. Springer, Cham (2020). https://doi.org/10.1007/978-3-030-58580-8_31
18. Qi, C.R., Su, H., Mo, K., Guibas, L.J.: PointNet: deep learning on point sets for 3D classification and segmentation. In: Proceedings of the IEEE Conference on Computer Vision and Pattern Recognition, pp. 652–660 (2017)
19. Thong, W., Parent, S., Wu, J., Aubin, C.E., Labelle, H., Kadoury, S.: Three-dimensional morphology study of surgical adolescent idiopathic scoliosis patient from encoded geometric models. Eur. Spine J. 25, 3104–3113 (2016)
20. Wang, D., et al.: Multi-view 3D reconstruction with transformers. In: Proceedings of the IEEE/CVF International Conference on Computer Vision, pp. 5722–5731 (2021)
21. Williams, F., et al.: Neural fields as learnable kernels for 3D reconstruction. In: Proceedings of the IEEE/CVF Conference on Computer Vision and Pattern Recognition, pp. 18500–18510 (2022)
22. Williams, F., Trager, M., Bruna, J., Zorin, D.: Neural splines: fitting 3D surfaces with infinitely-wide neural networks. In: Proceedings of the IEEE/CVF Conference on Computer Vision and Pattern Recognition, pp. 9949–9958 (2021)
23. Xie, Y., et al.: Neural fields in visual computing and beyond. In: Computer Graphics Forum, vol. 41, pp. 641–676. Wiley Online Library (2022)
24. Yuan, L., Zeng, Y., Chen, Z., Li, W., Zhang, X., Ni, J.: Risk factors associated with failure to reach minimal clinically important difference after correction surgery in patients with degenerative lumbar scoliosis. Spine 45(24), E1669–E1676 (2020)

25. Zhang, J., Zheng, Y., Qi, D.: Deep spatio-temporal residual networks for citywide crowd flows prediction. In: Proceedings of the Thirty-First AAAI Conference on Artificial Intelligence, pp. 1655–1661 (2017)
26. Zhao, L., et al.: Spineregnet: spine registration network for volumetric MR and CT image by the joint estimation of an affine-elastic deformation field. Med. Image Anal. **86**, 102786 (2023)

Rectifying Noisy Labels with Sequential Prior: Multi-scale Temporal Feature Affinity Learning for Robust Video Segmentation

Beilei Cui[1][iD], Minqing Zhang[2][iD], Mengya Xu[3][iD], An Wang[1][iD], Wu Yuan[2(✉)][iD], and Hongliang Ren[1,3(✉)][iD]

[1] Department of Electronic Engineering, The Chinese University of Hong Kong, Hong Kong SAR, China
{beileicui,wa09,lars.zhang}@link.cuhk.edu.hk
[2] Department of Biomedical Engineering, The Chinese University of Hong Kong, Hong Kong SAR, China
wyuan@cuhk.edu.hk
[3] Department of Biomedical Engineering, National University of Singapore, Singapore, Singapore
mengya@u.nus.edu, ren@nus.edu.sg

Abstract. Noisy label problems are inevitably in existence within medical image segmentation causing severe performance degradation. Previous segmentation methods for noisy label problems only utilize a single image while the potential of leveraging the correlation between images has been overlooked. Especially for video segmentation, adjacent frames contain rich contextual information beneficial in cognizing noisy labels. Based on two insights, we propose a Multi-Scale Temporal Feature Affinity Learning (MS-TFAL) framework to resolve noisy-labeled medical video segmentation issues. First, we argue the sequential prior of videos is an effective reference, i.e., pixel-level features from adjacent frames are close in distance for the same class and far in distance otherwise. Therefore, Temporal Feature Affinity Learning (TFAL) is devised to indicate possible noisy labels by evaluating the affinity between pixels in two adjacent frames. We also notice that the noise distribution exhibits considerable variations across video, image, and pixel levels. In this way, we introduce Multi-Scale Supervision (MSS) to supervise the network from three different perspectives by re-weighting and refining the samples. This design enables the network to concentrate on clean samples in a coarse-to-fine manner. Experiments with both synthetic and real-world label noise demonstrate that our method outperforms recent state-of-the-art robust segmentation approaches. Code is available at https://github.com/BeileiCui/MS-TFAL.

B. Cui, M. Xu and W. Yuan—Authors contributed equally to this work.

Supplementary Information The online version contains supplementary material available at https://doi.org/10.1007/978-3-031-43996-4_9.

Keywords: Noisy label learning · Feature affinity · Semantic segmentation

1   Introduction

Video segmentation, which refers to assigning pixel-wise annotation to each frame in a video, is one of the most vital tasks in medical image analysis. Thanks to the advance in deep learning algorithms based on Convolutional Neural Networks, medical video segmentation has achieved great progress over recent years [9]. But a major problem of deep learning methods is that they are largely dependent on both the quantity and quality of training data [13]. Datasets annotated by non-expert humans or automated systems with little supervision typically suffer from very high label noise and are extremely time-consuming. Even expert annotators could generate different labels based on their cognitive bias [6]. Based on the above, noisy labels are inevitably in existence within medical video datasets causing misguidance to the network and resulting in severe performance degradation. Hence, it is of great importance to design medical video segmentation methods that are robust to noisy labels within training data [4,18].

Most of the previous noisy label methods mainly focus on classification tasks. Only in recent years, the problem of noise labels in segmentation tasks has been more explored, but still less involved in medical image analysis. Previous techniques for solving noisy label problems in medical segmentation tasks can be categorized in three directions. The first type of method aims at deriving and modeling the general distribution of noisy labels in the form of Noise Transition Matrix (NTM) [3,8]. Secondly, some researchers develop special training strategies to re-weight or re-sample the data such that the model could focus on more dependable samples. Zhang et al. [19] concurrently train three networks and each network is trained with pixels filtered by the other two networks. Shi et al. [14] use stable characteristics of clean labels to estimate samples' uncertainty map which is used to further guide the network. Thirdly, label refinement is implemented to renovate noisy labels. Li et al. [7] represent the image with superpixels to exploit more advanced information in an image and refine the labels accordingly. Liu et al. [10] use two different networks to jointly determine the error sample, and use each other to refine the labels to prevent error accumulation. Xu et al. [15] utilize the mean-teacher model and Confident learning to refine the low-quality annotated samples.

Despite the amazing performance in tackling noisy label issues for medical image segmentation, almost all existing techniques only make use of the information within a single image. *To this end, we make the effort in exploring the feature affinity relation between pixels from consecutive frames.* The motivation is that the embedding features of pixels from adjacent frames should be close if they belong to the same class, and should be far if they belong to different classes. Hence, if a pixel's feature is far from the pixels of the same class in the adjacent frame and close to the ones of different classes, its label is more likely to be incorrect. Meanwhile, the distribution of noisy labels may vary among different videos and frames, which also motivates us to supervise the network from multiple perspectives.

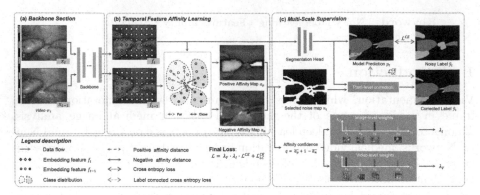

Fig. 1. Illustration of proposed Multi-Scale Temporal Feature Affinity Learning framework. We acquire the embedding feature maps of adjacent frames in the Backbone Section. Then, the temporal affinity is calculated for each pixel in current frame to obtain the positive and negative affinity map indicating possible noisy labels. The affinity maps are then utilized to supervise the network in a multi-scale manner.

Inspired by the motivation above and to better resolve noisy label problems with temporal consistency, we propose Multi-Scale Temporal Feature Affinity Learning (MS-TFAL) framework. Our contributions can be summarized as the following points:

1. In this work, we first propose a novel Temporal Feature Affinity Learning (TFAL) method to evaluate the temporal feature affinity map of an image by calculating the similarity between the same and different classes' features of adjacent frames, therefore indicating possible noisy labels.
2. We further develop a Multi-Scale Supervision (MSS) framework based on TFAL by supervising the network through video, image, and pixel levels. Such a coarse-to-fine learning process enables the network to focus more on correct samples at each stage and rectify the noisy labels, thus improving the generalization ability.
3. Our method is validated on a publicly available dataset with synthetic noisy labels and a real-world label noise dataset and obtained superior performance over other state-of-the-art noisy label techniques.
4. To the best of our knowledge, we are the first to tackle noisy label problems using inter-frame information and discover the superior ability of sequential prior information to resolve noisy label issues.

2   Method

The proposed Multi-Scale Temporal Feature Affinity Learning Framework is illustrated in Fig. 1. We aim to exploit the information from adjacent frames to identify the possible noisy labels, thereby learning a segmentation network robust to label noises by re-weighting and refining the samples. Formally, given an input training image $x_t \subset \mathbb{R}^{H \times W \times 3}$, and its adjacent frame x_{t-1}, two feature maps $f_t, f_{t-1} \in \mathbb{R}^{h \times w \times C_f}$ are first generated by a CNN backbone, where h, w and C_f represent the height, width and channel number. Intuitively, for each pair of features from f_t and f_{t-1}, their distance should be close if they belong to the same class and far otherwise. Therefore for each pixel in f_t, we calculate two affinity relations with f_{t-1}. The first one is called positive affinity, computed by averaging the cosine similarity between one pixel $f_t(i)$ in the current frame and all the same class' pixels as $f_t(i)$ in previous frame. The second one is called negative affinity, computed by averaging the cosine similarity between one pixel $f_t(i)$ in current frame and all the different class' pixels as $f_t(i)$ in previous frame. Then through up-sampling, the Positive Affinity Map a_p and Negative Affinity Map a_n can be obtained, where $a_p, a_n \in \mathbb{R}^{H \times W}$, denote the affinity relation between x_t and x_{t-1}. The positive affinity of clean labels should be high while the negative affinity of clean labels should be low. Therefore, the black areas in a_p and the white areas in a_n are more likely to be noisy labels.

Then we use two affinity maps a_p, a_n to conduct Multi-Scale Supervision training. Multi-scale refers to video, image, and pixel levels. Specifically, for pixel-level supervision, we first obtain thresholds t_p and t_n by calculating the average positive and negative affinity over the entire dataset. The thresholds are used to determine the possible noisy label sets based on positive and negative affinity separately. The intersection of two sets is selected as the final noisy set and relabeled with the model prediction p_t. The affinity maps are also used to estimate the image-level weights λ_I and video-level weights λ_V. The weights enable the network to concentrate on videos and images with higher affinity confidence. Our method is a plug-in module that is not dependent on backbone type and can be applied to both image-based backbones and video-based backbones by modifying the shape of inputs and feature maps.

2.1   Temporal Feature Affinity Learning

The purpose of this section is to estimate the affinity between pixels in the current frame and previous frame, thus indicating possible noisy labels. Specifically, in addition to the aforementioned feature map $f_t, f_{t-1} \in \mathbb{R}^{h \times w \times C_f}$, we obtain the down-sampled labels with the same size of feature map $\tilde{y}'_t, \tilde{y}'_{t-1} \in \mathbb{R}^{h \times w \times \mathcal{C}}$, where $\mathcal{C}$ means the total class number. We derive the positive and negative label maps with binary variables: $M_p, M_n \subseteq \{0, 1\}^{hw \times hw}$. The value corresponds to pixel (i, j) is determined by the label as:

$$M_p(i, j) = \mathbb{1}\left[\tilde{y}'_t(i) = \tilde{y}'_{t-1}(j)\right], \qquad M_n(i, j) = \mathbb{1}\left[\tilde{y}'_t(i) \neq \tilde{y}'_{t-1}(j)\right] \quad (1)$$

where $\mathbb{1}(\cdot)$ is the indicator function. $M_p(i, j) = 1$ when ith label in $\tilde{y}'_t$ and jth label in $\tilde{y}'_{t-1}$ are the same class, while $M_p(i, j) = 0$ otherwise; and M_n vise versa. The value of cosine similarity map $S \in \mathbb{R}^{hw \times hw}$ corresponds to pixel (i, j) is determined by: $S(i, j) = \frac{f_t(i) \cdot f_{t-1}(j)}{\|f_t(i)\| \times \|f_{t-1}(j)\|}$. We then use the average cosine similarity of a pixel with all pixels in the previous frame belonging to the same or different class to represent its positive or negative affinity:

$$a_{p,f}(i) = \frac{\sum_{j=1}^{hw} S(i,j) M_p(i,j)}{\sum_{j=1}^{hw} M_p(i,j)}, \quad a_{n,f}(i) = \frac{\sum_{j=1}^{hw} S(i,j) M_n(i,j)}{\sum_{j=1}^{hw} M_n(i,j)} \quad (2)$$

where $a_{p,f}, a_{n,f} \in \mathbb{R}^{h \times w}$ means the positive and negative map with the same size as the feature map. With simple up-sampling, we could obtain the final affinity maps $a_p, a_n \in \mathbb{R}^{H \times W}$, indicating the positive and negative affinity of pixels in the current frame.

2.2 Multi-scale Supervision

The feature map is first connected with a segmentation head generating the prediction p. Besides the standard cross entropy loss $\mathcal{L}^{CE}(p, \tilde{y}) = -\sum_i^{HW} \tilde{y}(i) log p(i)$, we applied a label corrected cross entropy loss $\mathcal{L}_{LC}^{CE}(p, \hat{y}) = -\sum_i^{HW} \hat{y}(i) log p(i)$ to train the network with pixel-level corrected labels. We further use two weight factors λ_I and λ_V to supervise the network in image and video levels. The specific descriptions are explained in the following sections.

Pixel-Level Supervision. Inspired by the principle in Confident Learning [12], we use affinity maps to denote the confidence of labels. if a pixel $x(i)$ in an image has both small enough positive affinity $a_p(i) \leqslant t_p$ and large enough negative affinity $a_n(i) \geqslant t_n$, then its label $\tilde{y}(i)$ can be suspected as noisy. The threshold t_p, t_n are obtained empirically by calculating the average positive and negative affinity, formulated as $t_p = \frac{1}{|A_p|} \sum_{a_p \in A_p} \overline{a_p}$, $t_n = \frac{1}{|A_n|} \sum_{a_n \in A_n} \overline{a_n}$, where $\overline{a_p}, \overline{a_n}$ means the average value of positive and negative affinity over an image. The noisy pixels set can therefore be defined by:

$$\tilde{x} := \{x(i) \in x : a_p(i) \leqslant t_p\} \bigcap \{x(i) \in x : a_n(i) \geqslant t_n\}. \quad (3)$$

Then we update the pixel-level label map $\hat{y}$ as:

$$\hat{y}(i) = \mathbb{1}(x(i) \in \tilde{x}) p(i) + \mathbb{1}(x(i) \notin \tilde{x}) \tilde{y}(i), \quad (4)$$

where $p(i)$ is the prediction of network. Through this process, we only replace those pixels with both low positive affinity and large negative affinity.

Image-Level Supervision. Even in the same video, different frames may contain different amounts of noisy labels. Hence, we first define the affinity confidence value as: $q = \overline{a_p} + 1 - \overline{a_n}$. The average affinity confidence value is therefore denoted as: $\bar{q} = t_p + 1 - t_n$. Finally, we define the image-level weight as:

$$\lambda_I = e^{2(q - \bar{q})}. \quad (5)$$

$\lambda_I > 1$ if the sample has large affinity confidence and $\lambda_I < 1$ otherwise, therefore enabling the network to concentrate more on the clean samples.

Video Level Supervision. We assign different weights to different videos such that the network can learn from more correct videos in the early stage. We first define the video affinity confidence as the average affinity confidence of all the frames: $q_v = \frac{1}{|V|}\sum_{x \in V} q_x$. Supposing there are N videos in total, we use $k \in \{1, 2, \cdots, N\}$ to represent the ranking of video affinity confidence from small to large, which means $k = 1$ and $k = N$ denote the video with lowest and highest affinity confidence separately. Video-level weight is thus formulated as:

$$\lambda_V = \begin{cases} \theta_l, & \text{if } k < \frac{N}{3} \\ \theta_l + \frac{3k-N}{N}(\theta_u - \theta_l), & \text{if } \frac{N}{3} \leqslant k \leqslant \frac{2N}{3} \\ \theta_u, & \text{if } k > \frac{2N}{3} \end{cases} \tag{6}$$

where θ_l and θ_u are the preseted lower-bound and upper-bound of weight.

Combining the above-defined losses and weights, we obtain the final loss as: $\mathcal{L} = \lambda_V \lambda_I \mathcal{L}^{CE} + \mathcal{L}_{LC}^{CE}$, which supervise the network in a multi-scale manner. These losses and weights are enrolled in training after initialization in an order of video, image, and pixel enabling the network to enhance the robustness and generalization ability by concentrating on clean samples from rough to subtle.

3 Experiments

3.1 Dataset Description and Experiment Settings

EndoVis 2018 Dataset and Noise Patterns. EndoVis 2018 Dataset is from the MICCAI robotic instrument segmentation dataset[1] of endoscopic vision challenge 2018 [1]. It is officially divided into 15 videos with 2235 frames for training and 4 videos with 997 frames for testing separately. The dataset contains 12 classes including different anatomy and robotic instruments. Each image is resized into 256×320 in pre-processing. To better simulate manual noisy annotations within a video, we first randomly select a ratio of α of videos and in each selected video, we divide all frames into several groups in a group of 3–6 consecutive frames. Then for each group of frames, we randomly apply dilation, erosion, affine transformation, or polygon noise to each class [7,16,18,19]. We investigated our algorithms in several noisy settings with α being $\{0.3, 0.5, 0.8\}$. Some examples of data and noisy labels are shown in supplementary.

Rat Colon Dataset. For real-world noisy dataset, we have collected rat colon OCT images using 800nm ultra-high resolution endoscopic spectral domain OCT. We refer readers to [17] for more details. Each centimeter of rat colon imaged corresponds to 500 images with 6 class layers of interest. We select 8 sections with 2525 images for training and 3 sections with 1352 images for testing. The labels of test set were annotated by professional endoscopists as ground truth while the training set was annotated by non-experts. Each image is resized into 256×256 in pre-processing. Some dataset examples are shown in supplementary.

---

[1] https://endovissub2018-roboticscenesegmentation.grand-challenge.org/.

Implementation Details. We adopt Deeplabv3+ [2] as our backbone network for fair comparison. The framework is implemented with PyTorch on two Nvidia 3090 GPUs. We adopt the Adam optimizer with an initial learning rate of $1e-4$ and weight decay of $1e-4$. Batch size is set to 4 with a maximum of 100 epochs for both Datasets. θ_l and θ_u are set to 0.4 and 1 separately. The video, image, and pixel level supervision are involved from the 16th, 24th, and 40th epoch respectively. The segmentation performance is assessed by $mIOU$ and $Dice$ scores.

Data	Method	$mIOU$ (%)	Sequence $mIOU$ (%)				$Dice$ (%)
			Seq 1	Seq 2	Seq 3	Seq 4	
Clean	Deeplabv3+ [2]	53.98	54.07	51.46	72.35	38.02	64.30
Noisy, $\alpha = 0.3$	Deeplabv3+ [2]	50.42	49.90	48.50	67.18	36.10	60.60
	STswin (22′) [5]	50.29	**49.96**	48.67	66.52	35.99	60.62
	RAUNet (19′) [11]	50.36	44.97	48.06	68.90	**39.53**	60.61
	JCAS (22′) [4]	48.65	48.77	46.60	64.83	34.39	58.97
	VolMin (21′) [8]	47.64	45.60	45.31	64.01	35.63	57.42
	MS-TFAL (Ours)	**52.91**	49.48	**51.60**	**71.08**	39.52	**62.91**
Noisy, $\alpha = 0.5$	Deeplabv3+ [2]	42.87	41.72	42.96	59.54	27.27	53.02
	STswin (22′) [5]	44.48	40.78	45.22	60.50	31.45	54.99
	RAUNet (19′) [11]	46.74	46.16	43.08	63.00	**34.73**	57.44
	JCAS (22′) [4]	45.24	41.90	44.06	61.13	33.90	55.22
	VolMin (21′) [8]	44.02	42.68	46.26	59.67	27.47	53.59
	MS-TFAL (Ours)	**50.34**	**49.15**	**50.17**	**67.37**	34.67	**60.50**
Noisy, $\alpha = 0.8$	Deeplabv3+ [2]	33.35	27.57	35.69	45.30	24.86	42.22
	STswin (22′) [5]	32.27	28.92	34.48	42.97	22.72	42.61
	RAUNet (19′) [11]	33.25	30.23	34.95	44.99	22.88	43.67
	JCAS (22′) [4]	35.99	28.29	38.06	51.00	26.66	44.75
	VolMin (21′) [8]	33.85	28.40	39.38	43.76	23.90	42.63
	MS-TFAL (Ours)	**41.36**	**36.33**	**41.65**	**59.57**	**27.88**	**51.01**
Noisy, $\alpha = 0.5$	w/V	47.80	44.73	48.71	66.87	30.91	57.45
	w/V & I	48.72	43.20	48.44	66.34	**36.93**	58.54
	Same frame	48.99	46.77	49.58	64.70	34.94	59.31
	Any frame	48.69	46.25	48.92	65.56	34.08	58.89
	MS-TFAL (Ours)	**50.34**	**49.15**	**50.17**	**67.37**	34.67	**60.50**

Table 1. Comparison of other methods and our models on EndoVis 2018 Dataset under different ratios of noise. The best results are **highlighted**.

3.2 Experiment Results on EndoVis 2018 Dataset

Table 1 presents the comparison results under different ratios of label noises. We evaluate the performance of backbone trained with clean labels, two state-of-the-art instrument segmentation network [5,11], two noisy label learning techniques [4,8], backbone [2] and the proposed MS-TFAL. We re-implement [4,8] with the same backbone [2] for a fair comparison. Compared with all other methods, MS-TFAL shows the minimum performance gap with the upper bound (Clean) for both *mIOU* and *Dice* scores under all ratios of noises demonstrating the robustness of our method. As noise increases, the performance of all baselines decreases significantly indicating the huge negative effect of noisy labels. It is noteworthy that when the noise ratio rises from 0.3 to 0.5 and from 0.5 to 0.8, our method only drops 2.57% *mIOU* with 2.41% *Dice* and 8.98% *mIOU* with 9.49% *Dice*, both are the minimal performance degradation, which further demonstrates the robustness of our method against label noise. In the extreme noise setting ($\alpha = 0.8$), our method achieves 41.36% *mIOU* and 51.01% *Dice* and outperforms second best method 5.37% *mIOU* and 6.26% *Dice*. As shown in Fig. 2, we provide partial qualitative results indicating the superiority of MS-TFAL over other methods in the qualitative aspect. More qualitative results are shown in supplementary.

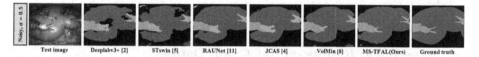

Fig. 2. Comparison of qualitative segmentation results on EndoVis18 Dataset.

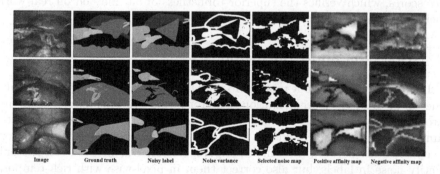

Fig. 3. Illustration of Noise variance and feature affinity. Selected noisy label (Fifth column) means the noise map selected with Eq. (3).

Ablation Studies. We further conduct two ablation studies on our multi-scale components and choice of frame for feature affinity under noisy dataset with $\alpha = 0.5$. With only video-level supervision (w/V), *mIOU* and *Dice* are increased by 4.93% and 4.43% compared with backbone only. Then we apply both video

and image level supervision (w/V & I) and gain an increase of 0.92% *mIOU* and
1.09% *Dice*. Pixel-level supervision is added at last forming the complete Multi-
Scale Supervision results in another improvement of 1.62% *mIOU* and 1.96%
Dice verifying the effectiveness in attenuating noisy label issues of individual
components. For the ablation study of the choice of frame, we compared two
different attempts with ours: conduct TFAL with the same frame and any frame
in the dataset (Ours is adjacent frame). Results show that using adjacent frame
has the best performance compared to the other two choices.

Visualization of Temporal Affinity. To prove the effectiveness of using affinity relation we defined to represent the confidence of label, we display comparisons between noise variance and selected noise map in Fig. 3. Noise variance
(Fourth column) represents the incorrect label map and the Selected noise map
(Fifth column) denotes the noise map we select with Eq. (3). We can observe
that the noisy labels we affirm have a high overlap degree with the true noise
labels, which demonstrates the validity of our TFAL module.

Method	Deeplabv3+ [2]	STswin [5]	RAUNet [11]	JCAS [4]	VolMin [8]	MS-TFAL (Ours)
mIOU (%)	68.46	68.21	68.24	68.15	68.81	**71.05**
Dice (%)	75.25	77.70	77.39	77.50	77.89	**80.17**

Table 2. Comparison of other methods and our models on Rat Colon Dataset.

3.3 Experiment Results on Rat Colon Dataset

The comparison results on real-world noisy Rat Colon Dataset are presented in
Table 2. Our method outperforms other methods consistently on both *mIOU* and
Dice scores, which verifies the superior robustness of our method on real-world
label noise issues. Qualitative results are shown in supplementary.

4   Discussion and Conclusion

In this paper, we propose a robust MS-TFAL framework to resolve noisy label
issues in medical video segmentation. Different from previous methods, we first
introduce the novel TFAL module to use affinity between pixels from adjacent
frames to represent the confidence of label. We further design MSS framework
to supervise the network from multiple perspectives. Our method can not only
identify noise in labels, but also correct them in pixel-wise with rich temporal
consistency. Extensive experiments under both synthetic and real-world label
noise data demonstrate the excellent noise resilience of MS-TFAL.

Acknowledgements. This work was supported by Hong Kong Research Grants
Council (RGC) Collaborative Research Fund (C4026-21G), General Research Fund
(GRF 14211420 & 14203323), Shenzhen-Hong Kong-Macau Technology Research Pro-
gramme (Type C) STIC Grant SGDX20210823103535014 (202108233000303).

References

1. Allan, M., et al.: 2018 robotic scene segmentation challenge. arXiv preprint arXiv:2001.11190 (2020)
2. Chen, L.C., Zhu, Y., Papandreou, G., Schroff, F., Adam, H.: Encoder-decoder with atrous separable convolution for semantic image segmentation. In: Proceedings of the European conference on computer vision (ECCV), pp. 801–818 (2018)
3. Guo, X., Yang, C., Li, B., Yuan, Y.: MetaCorrection: domain-aware meta loss correction for unsupervised domain adaptation in semantic segmentation. In: Proceedings of the IEEE/CVF Conference on Computer Vision and Pattern Recognition, pp. 3927–3936 (2021)
4. Guo, X., Yuan, Y.: Joint class-affinity loss correction for robust medical image segmentation with noisy labels. In: Wang, L., Dou, Q., Fletcher, P.T., Speidel, S., Li, S. (eds.) MICCAI 2022, Part IV. LNCS, vol. 13434, pp. 588–598. Springer, Cham (2022). https://doi.org/10.1007/978-3-031-16440-8_56
5. Jin, Y., Yu, Y., Chen, C., Zhao, Z., Heng, P.A., Stoyanov, D.: Exploring intra-and inter-video relation for surgical semantic scene segmentation. IEEE Trans. Med. Imaging **41**(11), 2991–3002 (2022)
6. Karimi, D., Dou, H., Warfield, S.K., Gholipour, A.: Deep learning with noisy labels: exploring techniques and remedies in medical image analysis. Med. Image Anal. **65**, 101759 (2020)
7. Li, S., Gao, Z., He, X.: Superpixel-guided iterative learning from noisy labels for medical image segmentation. In: de Bruijne, M., et al. (eds.) MICCAI 2021, Part I. LNCS, vol. 12901, pp. 525–535. Springer, Cham (2021). https://doi.org/10.1007/978-3-030-87193-2_50
8. Li, X., Liu, T., Han, B., Niu, G., Sugiyama, M.: Provably end-to-end label-noise learning without anchor points. In: Meila, M., Zhang, T. (eds.) Proceedings of the 38th International Conference on Machine Learning. Proceedings of Machine Learning Research, vol. 139, pp. 6403–6413. PMLR (2021). https://proceedings.mlr.press/v139/li21l.html
9. Litjens, G., et al.: A survey on deep learning in medical image analysis. Med. Image Anal. **42**, 60–88 (2017). https://doi.org/10.1016/j.media.2017.07.005, https://www.sciencedirect.com/science/article/pii/S1361841517301135
10. Liu, L., Zhang, Z., Li, S., Ma, K., Zheng, Y.: S-CUDA: self-cleansing unsupervised domain adaptation for medical image segmentation. Med. Image Anal. **74**, 102214 (2021)
11. Ni, Z.-L., et al.: RAUNet: residual attention U-net for semantic segmentation of cataract surgical instruments. In: Gedeon, Tom, Wong, Kok Wai, Lee, Minho (eds.) ICONIP 2019. LNCS, vol. 11954, pp. 139–149. Springer, Cham (2019). https://doi.org/10.1007/978-3-030-36711-4_13
12. Northcutt, C., Jiang, L., Chuang, I.: Confident learning: estimating uncertainty in dataset labels. J. Artif. Intell. Res. **70**, 1373–1411 (2021)
13. Shen, D., Wu, G., Suk, H.I.: Deep learning in medical image analysis. Annu. Rev. Biomed. Eng. **19**(1), 221–248 (2017). https://doi.org/10.1146/annurev-bioeng-071516-044442. pMID: 28301734
14. Shi, J., Wu, J.: Distilling effective supervision for robust medical image segmentation with noisy labels. In: de Bruijne, M., et al. (eds.) MICCAI 2021, Part I. LNCS, vol. 12901, pp. 668–677. Springer, Cham (2021). https://doi.org/10.1007/978-3-030-87193-2_63

15. Xu, Z., et al.: Anti-interference from noisy labels: mean-teacher-assisted confident learning for medical image segmentation. IEEE Trans. Med. Imaging **41**(11), 3062–3073 (2022)

16. Xue, C., Deng, Q., Li, X., Dou, Q., Heng, P.-A.: Cascaded robust learning at imperfect labels for chest X-ray segmentation. In: Martel, A.L., Abolmaesumi, P., Stoyanov, D., Mateus, D., Zuluaga, M.A., Zhou, S.K., Racoceanu, D., Joskowicz, L. (eds.) MICCAI 2020, Part VI. LNCS, vol. 12266, pp. 579–588. Springer, Cham (2020). https://doi.org/10.1007/978-3-030-59725-2_56

17. Yuan, W., et al.: In vivo assessment of inflammatory bowel disease in rats with ultrahigh-resolution colonoscopic oct. Biomed. Opt. Express **13**(4), 2091–2102 (2022). https://doi.org/10.1364/BOE.453396, https://opg.optica.org/boe/abstract.cfm?URI=boe-13-4-2091

18. Zhang, M., et al.: Characterizing label errors: confident learning for noisy-labeled image segmentation. In: Martel, A.L., et al. (eds.) MICCAI 2020, Part I. LNCS, vol. 12261, pp. 721–730. Springer, Cham (2020). https://doi.org/10.1007/978-3-030-59710-8_70

19. Zhang, T., Yu, L., Hu, N., Lv, S., Gu, S.: Robust medical image segmentation from non-expert annotations with tri-network. In: Martel, A.L., et al. (eds.) MICCAI 2020, Part IV. LNCS, vol. 12264, pp. 249–258. Springer, Cham (2020). https://doi.org/10.1007/978-3-030-59719-1_25

Foundation Model for Endoscopy Video Analysis via Large-Scale Self-supervised Pre-train

Zhao Wang[1], Chang Liu[2], Shaoting Zhang[3(✉)], and Qi Dou[1(✉)]

[1] The Chinese University of Hong Kong, Hong Kong, China
qidou@cuhk.edu.hk
[2] SenseTime Research, Shanghai, China
[3] Shanghai Artificial Intelligence Laboratory, Shanghai, China

Abstract. Foundation models have exhibited remarkable success in various applications, such as disease diagnosis and text report generation. To date, a foundation model for endoscopic video analysis is still lacking. In this paper, we propose Endo-FM, a foundation model specifically developed using massive endoscopic video data. First, we build a video transformer, which captures both local and global long-range dependencies across spatial and temporal dimensions. Second, we pre-train our transformer model using global and local views via a self-supervised manner, aiming to make it robust to spatial-temporal variations and discriminative across different scenes. To develop the foundation model, we construct a large-scale endoscopy video dataset by combining 9 publicly available datasets and a privately collected dataset from Baoshan Branch of Renji Hospital in Shanghai, China. Our dataset overall consists of over 33K video clips with up to 5 million frames, encompassing various protocols, target organs, and disease types. Our pre-trained Endo-FM can be easily adopted for a given downstream task via fine-tuning by serving as the backbone. With experiments on 3 different types of downstream tasks, including classification, segmentation, and detection, our Endo-FM surpasses the current state-of-the-art (SOTA) self-supervised pre-training and adapter-based transfer learning methods by a significant margin, such as VCL (3.1% F1, 4.8% Dice, and 5.5% F1 for classification, segmentation, and detection) and ST-Adapter (5.9% F1, 9.6% Dice, and 9.9% F1 for classification, segmentation, and detection). Code, datasets, and models are released at https://github.com/med-air/Endo-FM.

Keywords: Foundation model · Endoscopy video · Pre-train

1 Introduction

Foundation models pre-trained on large-scale data have recently showed success in various downstream tasks on medical images including classification [9],

---

Z. Wang and C. Liu—Equal contributions.

H. Greenspan et al. (Eds.): MICCAI 2023, LNCS 14228, pp. 101–111, 2023.
https://doi.org/10.1007/978-3-031-43996-4_10

detection [33], and segmentation [31]. However, medical data have various imaging modalities, and clinical data collection is expensive. It is arguable that a specific foundation model trained on some certain type of data is useful at the moment. In this paper, we focus on endoscopic video, which is a routine imaging modality and increasingly studied in gastrointestinal disease diagnosis, minimally invasive procedure and robotic surgery. Having an effective foundation model is promising to facilitate downstream tasks that necessitate endoscopic video analysis.

Existing work on foundation models for medical tasks, such as X-ray diagnosis [4] and radiology report generation [20, 21], involves pre-training on large-scale image-text pairs and relies on large language models to learn cross-modality features. However, since clinical routines for endoscopy videos typically do not involve text data, a pure image-based foundation model is currently more feasible. To this end, we develop a video transformer, based on ViT B/16 [8], containing 121M parameters, which serves as the foundation model backbone for our video data. We note that a similarly scaled foundation model in recent work [33] based on Swin UNETR [11] with 62M parameters has been successfully employed for CT scans. This would indicate that our video transformer could have sufficient capacity to model the rich spatial-temporal information of endoscopy videos.

To learn rich spatial-temporal information from endoscopy video data [12], our Endo-FM is pre-trained via a self-supervised manner by narrowing the gap between feature representations from different spatial-temporal views of the same video. These views are generated to address the variety of context information and motions of endoscopy videos. Drawing inspiration from self-supervised vision transformers [6, 29], we propose to pre-train the model via a teacher-student scheme. Under this scheme, the student is trained to predict (match) the teacher's output in the latent feature space. In other words, given two spatial-temporal aware views from the same video, one view processed by the teacher is predicted by another one processed by the student to learn the spatial-temporal information. Therefore, designing effective and suitable matching strategies for different spatial-temporal views from the same endoscopy video is important.

In this paper, we propose Endo-FM, a novel foundation model designed for endoscopic video analysis. First, we build a video transformer based on ViT [8] to capture long-range spatial and temporal dependencies, together with dynamic spatial-temporal positional encoding designed for tackling input data with diverse spatial sizes and temporal frame rates. Second, Endo-FM is pre-trained under a teacher-student scheme via spatial-temporal matching on diverse video views. Specifically, we create various spatial-temporal aware views differing in spatial sizes and frame rates for an input video clip. Both teacher and student models process these views of a video and predict one view from another in the latent feature space. This enables Endo-FM to learn *spatial-temporal invariant* (to view, scale, and motion) features that are transferable across different endoscopic domains and disease types while retaining discriminative features that are specific to each context. We construct a large-scale endoscopic video dataset by

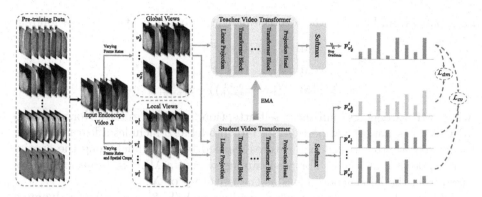

Fig. 1. Illustration of our proposed Endo-FM. We build a video transformer model and design a self-supervised pre-training approach.

combining 9 public and a new private collected dataset from Baoshan Branch of Renji Hospital in Shanghai, China, with over 33K video clips with up to 5 million frames. Our pre-trained Endo-FM can be easily applied to various downstream tasks by serving as the backbone. Experimental results on 3 different types of downstream tasks demonstrate the effectiveness of Endo-FM, surpassing the current state-of-the-art self-supervised pre-training and adapter-based transfer learning methods by a significant margin, such as VCL (3.1% F1, 4.8% Dice, and 5.5% F1 for classification, segmentation, and detection) and ST-Adapter (5.9% F1, 9.6% Dice, and 9.9% F1 for classification, segmentation, and detection).

2 Method

To begin with, we build a video transformer as the architecture of our Endo-FM (Sect. 2.1). Then, we propose a novel self-supervised spatial-temporal matching scheme (Sect. 2.2). Finally, we describe the overall training objective and specifics in Sect. 2.3. An overview of our method is shown in Fig. 1.

2.1 Video Transformer for Spatial-Temporal Encoding

We build a video transformer to encode input endoscopic video. The spatial and temporal attention mechanisms in our model capture long-range dependencies across both spatial and temporal dimensions, with a larger receptive field than conventional convolutional kernels [22]. Our model is built using 12 encoder blocks, equipped with space-time attention [3]. Specifically, given an endoscopic video clip $X \in \mathbb{R}^{T \times 3 \times H \times W}$ as input, consisting of T frames with size $H \times W$, each frame in X is divided into $N = HW/P^2$ patches of size $P \times P$, and these patches are then mapped into N patch tokens. Thus, each encoder block processes N patch (spatial) and T temporal tokens. Given the intermediate token $z^m \in \mathbb{R}^D$

for a patch from block m, the token computation in the next block is as follows:

$$z_{\text{time}}^{m+1} = \text{MHSA}_{\text{time}}\left(\text{LN}\left(z^m\right)\right) + z^m,$$
$$z_{\text{space}}^{m+1} = \text{MHSA}_{\text{space}}\left(\text{LN}\left(z_{\text{time}}^{m+1}\right)\right) + z_{\text{time}}^{m+1}, \tag{1}$$
$$z^{m+1} = \text{MLP}\left(\text{LN}\left(z_{\text{space}}^{m+1}\right)\right) + z_{\text{space}}^{m+1},$$

where MHSA denotes multi-head self-attention, LN denotes LayerNorm [1], and MLP denotes multi-layer perceptron. Our model also includes a learnable class token, representing the global features learned by the model along the spatial and temporal dimensions. For pre-training, we use a MLP to project the class token from the last encoder block as the feature f of X.

Different from static positional encoding in ViT [8], we design a dynamic spatial-temporal encoding strategy to help our model tackle various spatial-temporal views with different spatial sizes and frame rates (Sect. 2.2). Specifically, We fix the spatial and temporal positional encoding vectors to the highest resolution of the input view for each dimension, making it easy to interpolate for views with smaller spatial size or lower temporal frame rate. These spatial and temporal positional encoding vectors are added to the corresponding spatial and temporal tokens. Such dynamic strategy ensures that the learned positional encoding is suitable for downstream tasks with diverse input sizes.

2.2　Self-supervised Pre-train via Spatial-Temporal Matching

Considering the difficulties of tackling the context information related with lesions, tissues, and dynamic scenes in endoscopic data, we pre-train Endo-FM to be robust to such spatial-temporal characteristics. Inspired by self-supervised vision transformers [6], the pre-training is designed in a teacher-student scheme, where the student is trained to match the teacher's output. To achieve this, given an input video X, we create two types of spatial-temporal views serving as the model inputs: global and local views, as shown in Fig. 1. The global views $\{v_g^i \in \mathbb{R}^{T_g^i \times 3 \times H_g \times W_g}\}_{i=1}^G$ are generated by uniformly sampling X with different frame rates, and the local ones $\{v_l^j \in \mathbb{R}^{T_l^j \times 3 \times H_l \times W_l}\}_{j=1}^L$ are generated by uniformly sampling video frames with different frame rates from a randomly cropped region of X ($T_l \leq T_g$). During pre-training, the global views are fed into both teacher and student, and the local ones are only fed into the student. The model output f is then normalized by a softmax function with a temperature τ to obtain the probability distribution $p = \text{softmax}(f/\tau)$. In the following, we design two matching schemes with respect to the difficulties of tackling endoscopy videos.

Cross-View Matching. Different from image-based pre-training [33], our video-oriented pre-training is designed to capture the relationships between different spatial-temporal variations. Specifically, the context information presented in different frames of the same endoscope video can vary under two key factors: 1) the proportion of tissue and lesions within the frame, and 2) the presence or absence of lesion areas. To address these, we employ a *cross-view matching* approach where the target global views processed by the teacher ($\{\boldsymbol{p}^t_{v^i_g}\}^C_{i=1}$) are predicted from the online local views processed by the student ($\{\boldsymbol{p}^s_{v^j_l}\}^L_{j=1}$). By adopting this strategy, our model learns high-level context information from two perspectives: 1) spatial context in terms of the possible neighboring tissue and lesions within a local spatial crop, and 2) temporal context in terms of the possible presence of lesions in the previous or future frames of a local temporal crop. Thus, our method effectively addresses the proportion and existence issues that may be encountered. We minimize the following loss for cross-view matching:

$$\mathcal{L}_{cv} = \sum_{i=1}^{G}\sum_{j=1}^{L} -\boldsymbol{p}^t_{v^i_g} \cdot \log \boldsymbol{p}^s_{v^j_l}. \qquad (2)$$

Dynamic Motion Matching. In addition to the proportion and existence issues of lesions, a further challenge arises from the inherent dynamic nature of the scenes captured in endoscopy videos. The speeds and ranges of motion can vary greatly across different videos, making it difficult to train a model that is effective across a wide range of dynamic scenarios. The previous model [27] learned from clips with fixed frame rate can not tackle this issue, as clips sampled with various frame rates contain different motion context information (e.g., fast v.s. slow scene changing) and differ in nuanced tissue and lesions. To address this challenge, our approach involves motion modeling during pre-training under dynamic endoscope scenes by predicting a target global view ($\boldsymbol{p}^t_{v^i_g}$) processed by the teacher from another online global view ($\boldsymbol{p}^s_{v^k_g}$) processed by the student. Moreover, by predicting the nuanced differences of tissue and lesions in a view with a high frame rate from another with a low frame rate, the model is encouraged to learn more comprehensive motion-related contextual information. The dynamic motion difference among global view pairs is minimized by

$$\mathcal{L}_{dm} = \sum_{i=1}^{G}\sum_{k=1}^{G} -\mathbb{1}_{[i\neq k]}\boldsymbol{p}^t_{v^i_g} \cdot \log \boldsymbol{p}^s_{v^k_g}, \qquad (3)$$

where $\mathbb{1}[\cdot]$ is an indicator function.

2.3   Overall Optimization Objective and Pre-training Specifics

The overall training objective for Endo-FM is $\mathcal{L}_{pre\text{-}train} = \mathcal{L}_{cv} + \mathcal{L}_{dm}$. Centering and sharpening schemes [6] are incorporated to the teacher outputs. To prevent the problem of the teacher and student models constantly outputting the same value during pre-training, we update the student model θ through backpropagation, while the teacher model ϕ is updated through exponential moving average (EMA) using the student's weights. This is achieved by updating the teacher's weights as $\phi_t \leftarrow \alpha\phi_{t-1} + (1 - \alpha)\theta_t$ at each training iteration t. Here, α is a momentum hyper-parameter that determines the updating rate.

Except for the challenges posed by the issues of size proportion, existence, and dynamic scenes in Sect. 2.2, we have also observed that the appearance of

Table 1. Details of all pre-train and downstream datasets used in this work.

Phase	Dataset	Provider	Videos	Frames	Protocol	Disease
Pre-train	Colonoscopic [19]	CNRS	210	36534	colonoscope	adenoma, hyperplasia
	SUN-SEG [13]	ANU	1018	159400	colonoscope	SSL, adenoma, hyperplasia, T1b
	LDPolypVideo [18]	USTC	237	40186	colonoscope	polyp
	Hyper-Kvasir [5]	Simula	5704	875940	gastroscope	barrett's oesophagus, polyp, cancer
	Kvasir-Capsule [30]	Simula	1000	158892	gastroscope	erosion, erythema, etc
	CholecTriplet [23]	BIDMC	580	90444	laparoscope	cholecystectomy
	Ours	Baoshan Branch	16494	2491952	colonoscope	polyp, erosion, etc.
		of Renji Hospital	7653	1170753	gastroscope	
	Summary	6 providers	32896	5024101	3 protocols	10+ diseases
Downstream	PolypDiag [32]	Adelaide	253	485561	gastroscope	polyp, cancer
	CVC-12k [2]	UAB	29	612	colonoscope	polyp
	KUMC [15]	Kansas	53	19832	colonoscope	adenoma, hyperplasia
	Summary	3 providers	335	506005	2 protocols	4 diseases

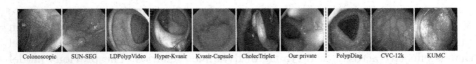

Colonoscopic   SUN-SEG   LDPolypVideo   Hyper-Kvasir   Kvasir-Capsule   CholecTriplet   Our private   PolypDiag   CVC-12k   KUMC

Fig. 2. Example frames of the 10 pre-train and downstream datasets used in this work.

endoscope videos is highly diverse. These videos are captured using different surgical systems and in a wide range of environmental conditions [10]. To address this variability, we apply temporally consistent spatial augmentations [27] to all frames within a single view. Our augmentation approach includes random horizontal flips, color jitter, Gaussian blur, solarization, and so on, which enhances the robustness and generalizability of Endo-FM.

For Endo-FM, we set the patch size P as 16 and embedding dimension D as 768. We create $G=2$ global views and $L=8$ local views for every input endoscopy video, where $T_g \in [8, 16]$ and $T_l \in [2, 4, 8, 16]$. The spatial sizes of global and local views are 224×224 and 96×96, respectively. The MLP head projects the dimension of class token to 65536. The temperature hyper-parameters are set as $\tau_t = 0.04$ and $\tau_s = 0.07$. The EMA update momentum α is 0.996. The training batch size is 12 with AdamW [17] optimizer (learning rate $2e{-}5$, weight decay $4e{-}2$). The pre-training is finished with 30 epochs with a cosine schedule [16].

3   Experiment

3.1   Datasets and Downstream Setup

We collect all possible public endoscope video datasets and a new one from Baoshan Branch of Renji Hospital for pre-training. As shown in Table 1, these public datasets are provided by world-wide research groups [5, 13, 18, 19, 30] and previous EndoVis challenge [23], covering 3 endoscopy protocols and 10+ types of diseases. We process the original videos into 30fps short clips with a duration

Table 2. Comparison with other latest SOTA methods on 3 downstream tasks. We report F1 score (%) for PolypDiag, Dice (%) for CVC-12k, and F1 score (%) for KUMC.

Method	Venue	Pre-training Time (h)	PolypDiag (Classification)	CVC-12k (Segmentation)	KUMC (Detection)
Scratch (Rand. init.)		N/A	83.5±1.3	53.2±3.2	73.5±4.3
TimeSformer [3]	ICML'21	104.0	84.2±0.8	56.3±1.5	75.8±2.1
CORP [12]	ICCV'21	65.4	87.1±0.6	68.4±1.1	78.2±1.4
FAME [7]	CVPR'22	48.9	85.4±0.8	67.2±1.3	76.9±1.2
ProViCo [25]	CVPR'22	71.2	86.9±0.5	69.0±1.5	78.6±1.7
VCL [26]	ECCV'22	74.9	87.6±0.6	69.1±1.2	78.1±1.9
ST-Adapter [24]	NeurIPS'22	8.1	84.8±0.7	64.3±1.9	74.9±2.9
Endo-FM (Ours)		20.4	**90.7±0.4**	**73.9±1.2**	**84.1±1.3**

of 5 s on average. We evaluate our pre-trained Endo-FM on three downstream tasks: disease diagnosis (PolypDiag [32]), polyp segmentation (CVC-12k [2]), and detection (KUMC [15]). The detailed information of three downstream datasets is shown in Table 1. The example frames of the 10 datasets are shown in Fig. 2.

For downstream fine-tuning, we utilize the following setup: 1) PolypDiag: A randomly initialized linear layer is appended to our pre-trained Endo-FM. We sample 8 frames with spatial size 224×224 for every video as the input and train for 20 epochs. 2) CVC-12k: A TransUNet equipped with Endo-FM as the backbone is implemented. We resize the spatial size as 224×224 and train for 150 epochs. 3) KUMC: we implement a STFT [34] with our pre-trained model as backbone for generating feature pyramid. We resize the spatial size as 640×640 and train for 24k iterations. We report F1 score for PolypDiag, Dice for CVC-12k, and F1 score for KUMC.

3.2 Comparison with State-of-the-Art Methods

We compare our method with recent SOTA video-based pre-training methods, including the **TimeSformer** [3] introduces spatial-temporal attention for video processing, the **CORP** [12] presents a self-supervised contrast-and-order representation framework, the **FAME** [7] proposes a foreground-background merging scheme, the **ProViCo** [25] applies a self-supervised probabilistic video contrastive learning strategy, the **VCL** [26] learns the static and dynamic visual concepts, and the **ST-Adapter** [24] adapts the CLIP [28] by a depth-wise convolution. We also train our model from scratch to serve as a baseline. The same experimental setup is applied to all the experiments for fair comparisons.

Quantitative comparison results are shown in Table 2. We can observe that the scratch model shows low performance on all 3 downstream tasks, especially for segmentation. Compared with training from scratch, our Endo-FM achieves +7.2% F1, +20.7% Dice, and +10.6% F1 improvements for classification, segmentation, and detection tasks, respectively, indicating the high effectiveness of our proposed pre-training approach. Moreover, our Endo-FM outperforms all SOTA methods, with +3.1% F1, +4.8% Dice, and +5.5% F1 boosts for the 3 downstream tasks over the second-best. Such significant improvements are benefited from our specific spatial-temporal pre-training designed for endoscopy videos to tackle the complex context information and dynamic scenes. Meanwhile, Endo-FM requires less pre-training time than SOTA pre-training methods, except the lighter but much worse ST-Adapter [24].

3.3   Analytical Studies

Without loss of generality, we conduct ablation studies on polyp diagnosis task from 3 aspects: 1) components analysis of our pre-training method; 2) varying combinations of global and local views in spatial-temporal matching; 3) varying the construction of global and local views.

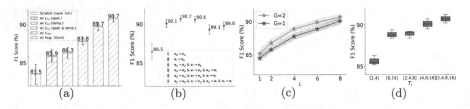

Fig. 3. Ablations on PolypDiag: (a) components analysis; (b) different combinations of views; (c) number of global and local views; (d) length of local views.

Components Analysis. We first study each component in our approach, as shown in Fig. 3(a). Here, "w/ $\mathcal{L}_{cv}$ (spat.)" and "w/ $\mathcal{L}_{cv}$ (temp.)" indicate that only spatial and temporal sampling are used for generating the local views. We can learn that both spatial and temporal sampling for local views can help improve the performance and their combination produces a plus, yielding +4.3% F1 improvement. Furthermore, our proposed dynamic matching scheme boosts the performance to 89.7%, demonstrating the importance of capturing the motion related context information from dynamic scenes. Additionally, the performance is further improved with video augmentations from 89.7% to 90.7%.

Spatial-Temporal Matching Combinations. We further investigate the effects of combinations of global and local views in spatial-temporal matching, as depicted in Fig. 3(b). Here, the notation $v_l \rightarrow v_g$ represents the prediction of v_g from v_l, and vice versa. It indicates that joint prediction scenarios, where we predict v_g from both v_l (cross-view matching) and v_g (dynamic motion matching), result in optimal performance. This trend can be attributed to the fact that joint prediction scenarios allow for a more comprehensive understanding of the context in complex endoscopy videos, which is lacking in individual cases.

Construction of Global and Local Views. We conduct a further analysis of the strategies for constructing global ($G \in [1, 2]$) and local views ($L \in [1, 2, 4, 6, 8]$). We vary the number of global and local views, and the length of local views (T_l), as depicted in Fig. 3(c), and Fig. 3(d). We find that incorporating more views and increasing the length variations of local views yields better performance. For "$G = 1$", we still create 2 global views for $\mathcal{L}_{dm}$ but only consider the longer one for $\mathcal{L}_{cv}$. These improvements stem from the spatial-temporal change invariant and cross-video discriminative features learned from the diverse endoscopy videos.

4   Conclusion and Discussion

To the best of our knowledge, we develop the first foundation model, Endo-FM, Which is specifically designed for analyzing endoscopy videos. Endo-FM is built upon a video transformer to capture rich spatial-temporal information and pre-trained to be robust to diverse spatial-temporal variations. A large-scale endoscope video dataset with over 33K video clips is constructed. Extensive experimental results on 3 downstream tasks demonstrate the effectiveness of Endo-FM, significantly outperforming other state-of-the-art video-based pre-training methods, and showcasing its potential for clinical application.

Regarding the recent SAM [14] model, which is developed for segmentation task, we try to apply SAM for our downstream task CVC-12k with the same fine-tuning scheme as Endo-FM. The experimental results show that SAM can achieve comparable performance with our Endo-FM for the downstream segmentation task. Considering that SAM is trained with 10x samples, our domain Endo-FM is considered to be powerful for endoscopy scenarios. Moreover, besides segmentation, Endo-FM can also be easily applied to other types of tasks including classification and detection. Therefore, we envision that, despite existence of general-purpose foundation models, Endo-FM or similar domain-specific foundation models will be helpful for medical applications.

Acknowledgements. This work was supported in part by Shenzhen Portion of Shenzhen-Hong Kong Science and Technology Innovation Cooperation Zone under HZQB-KCZYB-20200089, in part by Science, Technology and Innovation Commission of Shenzhen Municipality Project No. SGDX20220530111201008, in part by Hong Kong Innovation and Technology Commission Project No. ITS/237/21FP, in part by Hong Kong Research Grants Council Project No. T45-401/22-N, and in part by the Action Plan of Shanghai Science and Technology Commission [21SQBS02300].

References

1. Ba, J.L., Kiros, J.R., Hinton, G.E.: Layer normalization. Arxiv (2016)
2. Bernal, J., Sánchez, F.J., Fernández-Esparrach, G., Gil, D., Rodríguez, C., Vilariño, F.: WM-DOVA maps for accurate polyp highlighting in colonoscopy: validation vs. saliency maps from physicians. Comput. Med. Imaging Graph. **43**, 99–111 (2015)

3. Bertasius, G., Wang, H., Torresani, L.: Is space-time attention all you need for video understanding? In: ICML, vol. 2, p. 4 (2021)
4. Boecking, B., et al.: Making the most of text semantics to improve biomedical vision-language processing. In: Avidan, S., Brostow, G., Cisse, M., Farinella, G.M., Hassner, T. (eds.) Computer Vision – ECCV 2022. ECCV 2022. LNCS, vol. 13696, pp. 1–21. Springer, Cham (2022). https://doi.org/10.1007/978-3-031-20059-5_1
5. Borgli, H., et al.: Hyperkvasir, a comprehensive multi-class image and video dataset for gastrointestinal endoscopy. Sci. Data **7**(1), 1–14 (2020)
6. Caron, M., et al.: Emerging properties in self-supervised vision transformers. In: ICCV (2021)
7. Ding, S., et al.: Motion-aware contrastive video representation learning via foreground-background merging. In: CVPR, pp. 9716 9726 (2022)
8. Dosovitskiy, A., et al.: An image is worth 16×16 words: transformers for image recognition at scale. In: ICLR (2021)
9. Fu, Z., Jiao, J., Yasrab, R., Drukker, L., Papageorghiou, A.T., Noble, J.A.: Anatomy-aware contrastive representation learning for fetal ultrasound. In: Karlinsky, L., Michaeli, T., Nishino, K. (eds.) Computer Vision – ECCV 2022 Workshops. ECCV 2022. LNCS, vol. 13803, pp. 422–436. Springer, Cham (2023). https://doi.org/10.1007/978-3-031-25066-8_23
10. Goodman, E.D., et al.: A real-time spatiotemporal AI model analyzes skill in open surgical videos. arXiv:2112.07219 (2021)
11. Hatamizadeh, A., Nath, V., Tang, Y., Yang, D., Roth, H.R., Xu, D.: Swin UNETR: swin transformers for semantic segmentation of brain tumors in MRI images. In: Crimi, A., Bakas, S. (eds.) Brainlesion: Glioma, Multiple Sclerosis, Stroke and Traumatic Brain Injuries. BrainLes 2021. LNCS, vol. 12962, pp. 272–284. Springer, Cham (2022). https://doi.org/10.1007/978-3-031-08999-2_22
12. Hu, K., Shao, J., Liu, Y., Raj, B., Savvides, M., Shen, Z.: Contrast and order representations for video self-supervised learning. In: Proceedings of the IEEE/CVF International Conference on Computer Vision, pp. 7939–7949 (2021)
13. Ji, G.P., et al.: Video polyp segmentation: a deep learning perspective. Mach. Intell. Res. **19**(6), 531–549 (2022)
14. Kirillov, A., et al.: Segment anything. arXiv:2304.02643 (2023)
15. Li, K., et al.: Colonoscopy polyp detection and classification: dataset creation and comparative evaluations. PLoS ONE **16**(8), e0255809 (2021)
16. Loshchilov, I., Hutter, F.: SGDR: stochastic gradient descent with warm restarts. In: ICLR (2017)
17. Loshchilov, I., Hutter, F.: Decoupled weight decay regularization. In: ICLR (2019)
18. Ma, Yiting, Chen, Xuejin, Cheng, Kai, Li, Yang, Sun, Bin: LDPolypVideo benchmark: a large-scale colonoscopy video dataset of diverse polyps. In: de Bruijne, M., et al. (eds.) MICCAI 2021. LNCS, vol. 12905, pp. 387–396. Springer, Cham (2021). https://doi.org/10.1007/978-3-030-87240-3_37
19. Mesejo, P., et al.: Computer-aided classification of gastrointestinal lesions in regular colonoscopy. IEEE TMI **35**(9), 2051–2063 (2016)
20. Moon, J.H., Lee, H., Shin, W., Kim, Y.H., Choi, E.: Multi-modal understanding and generation for medical images and text via vision-language pre-training. IEEE JBHI **26**(12), 6070–6080 (2022)
21. Moor, M., et al.: Foundation models for generalist medical artificial intelligence. Nature **616**(7956), 259–265 (2023)
22. Naseer, M.M., Ranasinghe, K., Khan, S.H., Hayat, M., Shahbaz Khan, F., Yang, M.H.: Intriguing properties of vision transformers. NeurIPS (2021)

23. Nwoye, C.I., et al.. Rendezvous: attention mechanisms for the recognition of surgical action triplets in endoscopic videos. Media **78**, 102433 (2022)
24. Pan, J., Lin, Z., Zhu, X., Shao, J., Li, H.: St-adapter: parameter-efficient image-to-video transfer learning for action recognition. NeurIPS (2022)
25. Park, J., Lee, J., Kim, I.J., Sohn, K.: Probabilistic representations for video contrastive learning. In: CVPR, pp. 14711–14721 (2022)
26. Qian, R., Ding, S., Liu, X., Lin, D.: Static and dynamic concepts for self-supervised video representation learning. In: Avidan, S., Brostow, G., Cisse, M., Farinella, G.M., Hassner, T. (eds.) Computer Vision – ECCV 2022. ECCV 2022. LNCS, vol. 13686, pp. 145–164. Springer, Cham (2022). https://doi.org/10.1007/978-3-031-19809-0_9
27. Qian, R., et al.: Spatiotemporal contrastive video representation learning. In: CVPR (2021)
28. Radford, A., et al.: Learning transferable visual models from natural language supervision. In: ICML, pp. 8748–8763. PMLR (2021)
29. Ranasinghe, K., Naseer, M., Khan, S., Khan, F.S., Ryoo, M.: Self-supervised video transformer. In: CVPR, June 2022
30. Smedsrud, P.H., et al.: Kvasir-capsule, a video capsule endoscopy dataset. Sci. Data **8**(1), 1–10 (2021)
31. Tang, Y., et al.: Self-supervised pre-training of Swin transformers for 3D medical image analysis. In: CVPR, pp. 20730–20740 (2022)
32. Tian, Y., et al.: Contrastive transformer-based multiple instance learning for weakly supervised polyp frame detection. In: Wang, L., Dou, Q., Fletcher, P.T., Speidel, S., Li, S. (eds.) Medical Image Computing and Computer Assisted Intervention – MICCAI 2022. MICCAI 2022. LNCS, vol. 13433, pp. 88–98. Springer, Cham (2022). https://doi.org/10.1007/978-3-031-16437-8_9
33. Willemink, M.J., Roth, H.R., Sandfort, V.: Toward foundational deep learning models for medical imaging in the new era of transformer networks. Radiol. Artif. Intell. **4**(6), e210284 (2022)
34. Wu, L., Hu, Z., Ji, Y., Luo, P., Zhang, S.: Multi-frame collaboration for effective endoscopic video polyp detection via spatial-temporal feature transformation. In: de Bruijne, M., et al. (eds.) MICCAI 2021. LNCS, vol. 12905, pp. 302–312. Springer, Cham (2021). https://doi.org/10.1007/978-3-030-87240-3_29

Point Cloud Diffusion Models for Automatic Implant Generation

Paul Friedrich[1]([✉]), Julia Wolleb[1], Florentin Bieder[1], Florian M. Thieringer[1,2], and Philippe C. Cattin[1]

[1] Department of Biomedical Engineering, University of Basel, Allschwil, Switzerland
paul.friedrich@unibas.ch
[2] Department of Oral and Cranio-Maxillofacial Surgery,
University Hospital Basel, Basel, Switzerland

Abstract. Advances in 3D printing of biocompatible materials make patient-specific implants increasingly popular. The design of these implants is, however, still a tedious and largely manual process. Existing approaches to automate implant generation are mainly based on 3D U-Net architectures on downsampled or patch-wise data, which can result in a loss of detail or contextual information. Following the recent success of Diffusion Probabilistic Models, we propose a novel approach for implant generation based on a combination of 3D point cloud diffusion models and voxelization networks. Due to the stochastic sampling process in our diffusion model, we can propose an ensemble of different implants per defect, from which the physicians can choose the most suitable one. We evaluate our method on the SkullBreak and Skull-Fix datasets, generating high-quality implants and achieving competitive evaluation scores. The project page can be found at https://pfriedri. github.io/pcdiff-implant-io.

Keywords: Automatic Implant Generation · Diffusion Models · Point Clouds · Voxelization

1 Introduction

The design of 3D-printed patient-specific implants, commonly used in cranioplasty and maxillofacial surgery, is a challenging and time-consuming task that is usually performed manually. To speed up the design process and enable point-of-care implant generation, approaches for automatically deriving suitable implant designs from medical images are needed. This paper presents a novel approach based on a Denoising Diffusion Probabilistic Model for 3D point clouds that reconstructs complete anatomical structures S_c from segmented CT images of subjects showing bone defects S_d. An overview of the proposed method is shown in Fig. 1.

Supplementary Information The online version contains supplementary material available at https://doi.org/10.1007/978-3-031-43996-4_11.

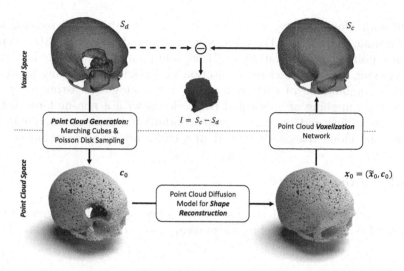

Fig. 1. Proposed implant generation method with shape reconstruction in point cloud space. The implant geometry I is defined as the Boolean subtraction between completed output S_c and defective input S_d in voxel space.

Since performing the anatomy reconstruction task in a high resolution voxel space would be memory inefficient and computationally expensive, we propose a method that builds upon a sparse surface point cloud representation of the input anatomy. This point cloud c_0, which can be obtained from the defective segmentation mask S_d, serves as input to a Denoising Diffusion Probabilistic Model [37] that, conditioned on this input c_0, reconstructs the complete anatomy x_0 by generating missing points $\tilde{x}_0$. The second network transforms the point cloud x_0 back into voxel space using a Differentiable Poisson Solver [23]. The final implant I is generated by the Boolean subtraction of the completed and defective anatomical structure. We thereby ensure a good fit at the junction between implant and skull. Our main contributions are:

- We employ 3D point cloud diffusion models for an automatic patient-specific implant generation task. The stochastic sampling process of diffusion models allows for the generation of multiple anatomically reasonable implant designs per subject, from which physicians can choose the most suitable one.
- We evaluate our method on the SkullBreak and SkullFix datasets, generating high-quality implants and achieving competitive evaluation scores.

Related Work. Previous work on automatic implant generation methods mainly derived from the AutoImplant challenges [11–13] at the 2020/21 MICCAI conferences. Most of the proposed methods were based on 2D slice-wise U-Nets [27] and 3D U-Nets on downsampled [8,20,31] or patch-wise data [4,14,22]. Other approaches were based on Statistical Shape Models [34], Generative Adversarial Networks [24] or Variational Autoencoders [30]. This work is based on Diffusion Models [3, 28], which achieved good results in 2D reconstruction tasks like image inpainting [17, 26] and were also already applied to 3D generative tasks like point cloud generation [18, 21, 36] and point cloud completion [19, 37]. For retrieving a dense voxel representation of a point cloud, many approaches rely on a combination of surface meshing [1, 2, 7] and ray casting algorithms. Since surface meshing of unoriented point clouds with a non-uniform distribution is challenging and ray casting implies additional computational effort, we look at point cloud voxelization based on a Differentiable Poisson Solver [23].

2   Methods

As presented in Fig. 1, we propose a multi-step approach for generating an implant I from a binary voxel representation S_d of a defective anatomical structure.

Point Cloud Generation. Since the proposed method for shape reconstruction works in the point cloud space, we first need to derive a point cloud $c_0 \in \mathbb{R}^{N \times 3}$ from S_d. We therefore create a surface mesh of S_d using Marching Cubes [16]. Then we sample N points from this surface mesh using Poisson Disk Sampling [35]. During training, we generate the ground truth point cloud $\tilde{x}_0 \in \mathbb{R}^{M \times 3}$ by sampling M points from the ground truth implant using the same approach.

Diffusion Model for Shape Reconstruction. Reconstructing the shape of an anatomical structure can be seen as a conditional generation process. We train a diffusion model ϵ_θ to reconstruct the point cloud $x_0 = (\tilde{x}_0, c_0)$ that describes the complete anatomical structure S_c. The generation process is conditioned on the points c_0 belonging to the known defective anatomical structure S_d. An overview is given in Fig. 2. For describing the diffusion model, we follow the formulations in [37]. Starting from x_0, we first define the forward diffusion process, that gradually adds small amounts of noise to $\tilde{x}_0$, while keeping c_0 unchanged and thus produces a series of point clouds $\{x_0 = (\tilde{x}_0, c_0), x_1 = (\tilde{x}_1, c_0), ..., x_T = (\tilde{x}_T, c_0)\}$.

This *conditional forward diffusion process* can be modeled as a Markov chain with a defined number of timesteps T and transfers $\tilde{x}_0$ into a noise distribution:

$$q(\tilde{x}_{0:T}) = q(\tilde{x}_0) \prod_{t=1}^{T} q(\tilde{x}_t | \tilde{x}_{t-1}). \tag{1}$$

Each transition is modeled as a parameterized Gaussian and follows a predefined variance schedule $\beta_1, ..., \beta_T$ that controls the diffusion rate of the process:

$$q(\tilde{x}_t | \tilde{x}_{t-1}) := \mathcal{N}(\sqrt{1 - \beta_t} \tilde{x}_{t-1}, \beta_t I). \tag{2}$$

The goal of the diffusion model is to learn the *reverse diffusion process* that is able to gradually remove noise from $\tilde{x}_T \sim \mathcal{N}(0, I)$. This reverse process is also

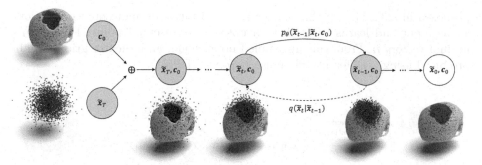

Fig. 2. Conditional diffusion model for anatomy reconstruction in point cloud space. Points belonging to the defective anatomical structure c_0 are shown in color and remain unchanged throughout the whole process. The forward and reverse diffusion processes therefore only affect the gray points $\tilde{x}_{0:T}$ belonging to the implant.

modeled as a Markov chain

$$p_\theta(\tilde{\boldsymbol{x}}_{0:T}) = p(\tilde{\boldsymbol{x}}_T) \prod_{t=1}^{T} p_\theta(\tilde{\boldsymbol{x}}_{t-1}|\tilde{\boldsymbol{x}}_t, \boldsymbol{c}_0), \qquad (3)$$

with each transition being defined as a Gaussian, with the estimated mean μ_θ:

$$p_\theta(\tilde{\boldsymbol{x}}_{t-1}|\tilde{\boldsymbol{x}}_t, \boldsymbol{c}_0) := \mathcal{N}(\mu_\theta(\tilde{\boldsymbol{x}}_t, \boldsymbol{c}_0, t), \beta_t \boldsymbol{I}). \qquad (4)$$

As derived in [37], the network ϵ_θ can be adapted to predict the noise $\epsilon_\theta(\tilde{\boldsymbol{x}}_t, \boldsymbol{c}_0, t)$ to be removed from a noisy point cloud $\tilde{\boldsymbol{x}}_t$. During training, we compute $\tilde{\boldsymbol{x}}_t$ at a random timestep $t \in \{1, ..., T\}$ and optimize a Mean Squared Error (MSE) loss

$$\mathcal{L}_t = \|\epsilon - \epsilon_\theta(\tilde{\boldsymbol{x}}_t, \boldsymbol{c}_0, t)\|^2, \quad \text{where} \quad \tilde{\boldsymbol{x}}_t = \sqrt{\tilde{\alpha}_t}\tilde{\boldsymbol{x}}_0 + \sqrt{1 - \tilde{\alpha}_t}\epsilon, \qquad (5)$$

with $\epsilon \sim \mathcal{N}(0, \boldsymbol{I})$, $\alpha_t = 1 - \beta_t$, and $\tilde{\alpha}_t = \prod_{s=1}^{t} \alpha_s$. To perform shape reconstruction with the trained network, we start with a point cloud $\boldsymbol{x}_T = (\tilde{\boldsymbol{x}}_T, \boldsymbol{c}_0)$ with $\tilde{\boldsymbol{x}}_T \sim \mathcal{N}(0, \boldsymbol{I})$. This point cloud is then passed through the reverse diffusion process

$$\tilde{\boldsymbol{x}}_{t-1} = \frac{1}{\sqrt{\alpha_t}} \left(\tilde{\boldsymbol{x}}_t - \frac{1 - \alpha_t}{\sqrt{1 - \tilde{\alpha}_t}} \epsilon_\theta(\tilde{\boldsymbol{x}}_t, \boldsymbol{c}_0, t) \right) + \sqrt{\beta_t}\boldsymbol{z}, \qquad (6)$$

with $\boldsymbol{z} \sim \mathcal{N}(0, \boldsymbol{I})$, for $t = T, ..., 1$. While this reverse diffusion process gradually removes noise from $\tilde{\boldsymbol{x}}_T$, the points belonging to the defective anatomical structure $\boldsymbol{c}_0$ remain unchanged. As proposed in [37], our network ϵ_θ is based on a PointNet++ [25] architecture with Point-Voxel Convolutions [15]. Details on the network architecture can be found in the supplementary material.

Voxelization. To create an implant, the point cloud of the restored anatomy must be converted back to voxel space. We follow a learning-based pipeline proposed in [23]. The pipeline shown in Fig. 3 takes an unoriented point cloud x_0 as input and learns to predict an upsampled, oriented point cloud $\hat{x}$ with normal vectors $\hat{n}$. From this upsampled point cloud, an indicator grid $\hat{\chi}$ can be produced using a Differentiable Poisson Solver (DPSR).

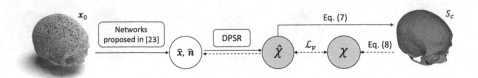

Fig. 3. Voxelization pipeline based on a trainable network and a Differentiable Poisson Solver (DPSR). The trainable part learns to produce an upsampled point cloud $\hat{x}$ with estimated normal vectors $\hat{n}$. The DPSR transforms $(\hat{x}, \hat{n})$ into an indicator function $\hat{\chi}$ from which a voxel representation can be derived.

The complete voxel representation S_c can then be obtained by evaluating the following equation for every voxel position i:

$$S_c(i) = \begin{cases} 1, & \text{if } \hat{\chi}(i) \leq 0 \\ 0, & \text{if } \hat{\chi}(i) > 0 \end{cases}. \tag{7}$$

During training, the ground truth indicator grid χ can be obtained directly from the ground truth voxel representation S_c and, as described in [6], is defined as:

$$\chi(i) = \begin{cases} 0.5, & \text{if } S_c(i) = 0 \\ -0.5, & \text{if } S_c(i) = 1 \end{cases}. \tag{8}$$

Due to the differentiability of the used Poisson solver, the networks can be trained with an MSE loss between the estimated and ground truth indicator grid:

$$\mathcal{L}_v = \|\hat{\chi} - \chi\|^2. \tag{9}$$

For further information on the used network architectures, we refer to [23].

Implant Generation. With the predicted complete anatomical structure S_c and the defective input structure S_d, an implant geometry I can be derived by the Boolean subtraction between S_c and S_d:

$$I = S_c - S_d. \tag{10}$$

To further improve the implant quality and remove noise, we apply a median filter as well as binary opening to the generated implant.

Ensembling. As the proposed point cloud diffusion model features a stochastic generation process, we can sample multiple anatomically reasonable implant designs for each defect. This offers physicians the opportunity of selecting from various possible implants and allows the determination of a mean implant from a previously generated ensemble of n different implants. As presented in [32], the ensembling strategy can also be used to create voxel-wise variance over an ensemble. These variance maps highlight areas with high differences between multiple anatomically reasonable implants.

3   Experiments

We evaluated our method on the publicly available parts of the SkullBreak and SkullFix datasets. For the point cloud diffusion model we chose a total number of 30 720 points ($N = 27\,648, M = 3072$), set $T = 1000$, followed a linear variance schedule between $\beta_0 = 10^{-4}$ and $\beta_T = 0.02$, used the Adam optimizer with a learning rate of 2×10^{-4}, a batch size of 8, and trained the network for 15 000 epochs. This took about 20 d/4 d for the SkullBreak/SkullFix dataset. For training the voxelization network, we used the Adam optimizer with a learning rate of 5×10^{-4}, a batch size of 2 and trained the networks for 1300/500 epochs on the SkullBreak/SkullFix dataset. This took about 72 h/5 h. All experiments were performed on an NVIDIA A100 GPU using PyTorch as the framework.

SkullBreak/SkullFix. Both datasets, SkullBreak and SkullFix [9], contain binary segmentation masks of head CT images with artificially created skull defects. While the SkullFix dataset mainly features rectangular defect patterns with additional craniotomy drill holes, SkullBreak offers more diverse defect patterns. The SkullFix dataset was resampled to an isotropic voxel size of 0.45 mm, zero padded to a volume size of $512 \times 512 \times 512$, and split into a training set with 75 and a test set with 25 volumes. The SkullBreak dataset already has an isotropic voxel size of 0.4 mm and a volume size of $512 \times 512 \times 512$. We split the SkullBreak dataset into a training set with 430 and a test set with 140 volumes. All point clouds sampled from these datasets were normalized to a range between $[-3, 3]$ in all spatial dimensions. The SkullBreak and SkullFix datasets were both adapted from the publicly available head CT dataset CQ500, which is licensed under a CC-BY-NC-SA 4.0 and End User License Agreement (EULA). The SkullBreak and SkullFix datasets were adapted and published under the same licenses.

4   Results and Discussion

For evaluating our approach, we compared it to three methods from AutoImplant 2021 challenge: the winning 3D U-Net based approach [31], a 3D U-Net based approach with sparse convolutions [10] and a slice-wise 2D U-Net approach [33]. We also evaluated the mean implant produced by the proposed ensembling

Table 1. Mean evaluation scores on SkullBreak and SkullFix test sets.

Model	SkullBreak			SkullFix		
	DSC ↑	bDSC ↑	HD95 ↓	DSC ↑	bDSC ↑	HD95 ↓
3D U-Net [31]	0.87	0.91	2.32	0.91	0.95	1.79
3D U-Net (sparse) [10]	0.71	0.80	4.60	0.81	0.87	3.04
2D U-Net [33]	0.87	0.89	2.13	0.89	0.92	1.98
Ours	0.86	0.88	2.51	0.90	0.92	1.73
Ours (n=5)	0.87	0.89	2.45	0.90	0.93	1.69

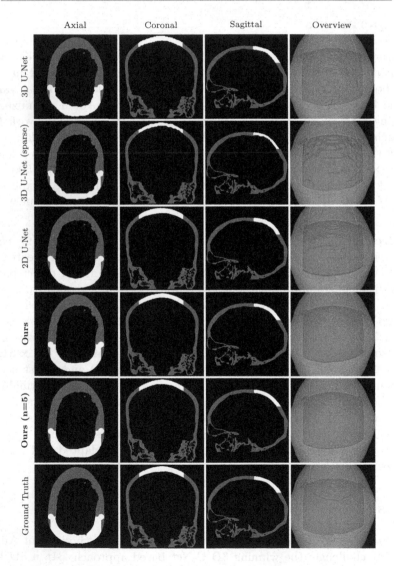

Fig. 4. Predicted implants for a skull defect from the SkullFix dataset.

strategy ($n = 5$). The implant generation time ranges from ~1000 s for Skull-Break ($n = 1$) to ~1200 s for SkullFix ($n = 5$), with the diffusion model requiring most of this time. In Table 1, the Dice score (DSC), the 10 mm boundary DSC (bDSC), as well as the 95 percentile Hausdorff Distance (HD95) are presented as mean values over the respective test sets.

Qualitative results of the different implant generation methods are shown in Fig. 4 and Fig. 5, as well as in the supplementary material. Implementation detail for the comparing methods, more detailed runtime information, as well as the used code can be found at https://github.com/pfriedri/pcdiff-implant.

We outperformed the sparse 3D U-Net, while achieving comparable results to the challenge winning 3D U-Net and the 2D U-Net based approach. By visually comparing the results in Fig. 4, our method produces significantly smoother surfaces that are more similar to the ground truth implant. In Fig. 5, we show that our method reliably reconstructs defects of various sizes, as well as complicated geometric structures. For large defects, however, we achieve lower evaluation scores, which can be explained with multiple anatomically reasonable implant solutions. This was also reported in [31] and [34].

| Fronto-Orbital | Bilateral | Parieto-Temporal | Random 1 | Random 2 |

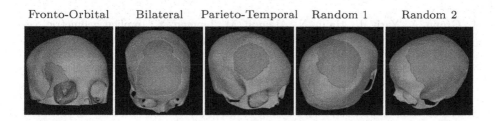

Fig. 5. Results of our method for the five defect classes of the SkullBreak dataset.

In Fig. 6, we present the proposed ensembling strategy for an exemplary defect. Apart from generating multiple implants, we can compute the mean implant and the variance map. To the best of our knowledge, we are the first to combine such an approach with an automatic implant generation method. Not only do we offer a choice of implants, but we can also provide physicians with information about implant areas with more anatomical variation.

| Implant 1 | Implant 2 | Implant 5 | Mean | Variance Map |

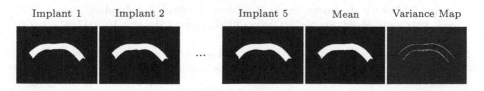

Fig. 6. Different implants, mean implant and variance map for a single skull defect.

5   Conclusion

We present a novel approach for automatic implant generation based on a combination of a point cloud diffusion model and a voxelization network. Due to the sparse point cloud representation of the anatomical structure, the proposed approach can directly handle high resolution input images without losing context information. We achieve competitive evaluation scores, while producing smoother, more realistic surfaces. Furthermore, our method is capable of producing different implants per defect, accounting for the anatomical variation seen in the training population. Thereby, we can propose several solutions to the physicians, from which they can choose the most suitable one. For future work, we plan to speed up the sampling process by using different sampling schemes, as proposed in [5,29].

Acknowledgments. This work was financially supported by the Werner Siemens Foundation through the MIRACLE II project.

References

1. Gropp, A., Yariv, L., Haim, N., Atzmon, M., Lipman, Y.: Implicit geometric regularization for learning shapes. In: Proceedings of Machine Learning and Systems, pp. 3569–3579 (2020)
2. Hanocka, R., Metzer, G., Giryes, R., Cohen-Or, D.: Point2mesh: a self-prior for deformable meshes. ACM Trans. Graph. **39**(4), 126:1–126:12 (2020). Article-Nr. 126
3. Ho, J., Jain, A., Abbeel, P.: Denoising diffusion probabilistic models. In: Advances in Neural Information Processing Systems (2020)
4. Jin, Y., Li, J., Egger, J.: High-resolution cranial implant prediction via patchwise training. In: Li, J., Egger, J. (eds.) AutoImplant 2020. LNCS, vol. 12439, pp. 94–103. Springer, Cham (2020). https://doi.org/10.1007/978-3-030-64327-0_11
5. Karras, T., Aittala, M., Aila, T., Laine, S.: Elucidating the design space of diffusion-based generative models. In: Advances in Neural Information Processing Systems (2022)
6. Kazhdan, M., Bolitho, M., Hoppe, H.: Poisson surface reconstruction. In: Eurographics Symposium on Geometry Processing (2006)
7. Kazhdan, M., Hoppe, H.: Screened poisson surface reconstruction. ACM Trans. Graph. **32**, 1–13 (2013)
8. Kodym, O., Španěl, M., Herout, A.: Cranial defect reconstruction using cascaded CNN with alignment. In: Li, J., Egger, J. (eds.) AutoImplant 2020. LNCS, vol. 12439, pp. 56–64. Springer, Cham (2020). https://doi.org/10.1007/978-3-030-64327-0_7
9. Kodym, O., et al.: Skullbreak/skullfix - dataset for automatic cranial implant design and a benchmark for volumetric shape learning tasks. Data Brief **35**, 106902 (2021)
10. Kroviakov, A., Li, J., Egger, J.: Sparse convolutional neural network for skull reconstruction. In: Li, J., Egger, J. (eds.) AutoImplant 2021. LNCS, vol. 13123, pp. 80–94. Springer, Cham (2021). https://doi.org/10.1007/978-3-030-92652-6_7

11. Li, J., Egger, J. (eds.): Towards the Automatization of Cranial Implant Design in Cranioplasty, vol. 12439. Springer International Publishing, Cham (2020). https://doi.org/10.1007/978-3-030-64327-0

12. Li, J., Egger, J. (eds.): Towards the Automatization of Cranial Implant Design in Cranioplasty II, vol. 13123. Springer International Publishing, Cham (2021). https://doi.org/10.1007/978-3-030-92652-6

13. Li, J., et al.: Autoimplant 2020-first MICCAI challenge on automatic cranial implant design. IEEE Trans. Med. Imaging 40(9), 2329–2342 (2021)

14. Li, J., et al.: Automatic skull defect restoration and cranial implant generation for cranioplasty. Med. Image Anal. 73, 102171 (2021)

15. Liu, Z., Tang, H., Lin, Y., Han, S.: Point-voxel CNN for efficient 3D deep learning. In: Advances in Neural Information Processing Systems (2019)

16. Lorensen, W.E., Cline, H.E.: Marching cubes: a high resolution 3D surface construction algorithm. ACM SIGGRAPH Comput. Graph. 21, 163–169 (1987)

17. Lugmayr, A., Danelljan, M., Romero, A., Yu, F., Timofte, R., Gool, L.V.: Repaint: inpainting using denoising diffusion probabilistic models. In: Proceedings of the IEEE/CVF Conference on Computer Vision and Pattern Recognition (2022)

18. Luo, S., Hu, W.: Diffusion probabilistic models for 3D point cloud generation. In: Proceedings of the IEEE/CVF Conference on Computer Vision and Pattern Recognition (2021)

19. Lyu, Z., Kong, Z., Xu, X., Pan, L., Lin, D.: A conditional point diffusion-refinement paradigm for 3D point cloud completion. In: International Conference on Learning Representations (2022)

20. Matzkin, F., Newcombe, V., Glocker, B., Ferrante, E.: Cranial implant design via virtual craniectomy with shape priors. In: Li, J., Egger, J. (eds.) AutoImplant 2020. LNCS, vol. 12439, pp. 37–46. Springer, Cham (2020). https://doi.org/10.1007/978-3-030-64327-0_5

21. Nichol, A., Jun, H., Dhariwal, P., Mishkin, P., Chen, M.: Point-e: a system for generating 3D point clouds from complex prompts. arXiv preprint arXiv:2212.08751 (2022)

22. Pathak, S., Sindhura, C., Gorthi, R.K.S.S., Kiran, D.V., Gorthi, S.: Cranial implant design using V-Net based region of interest reconstruction. In: Li, J., Egger, J. (eds.) AutoImplant 2021. LNCS, vol. 13123, pp. 116–128. Springer, Cham (2021). https://doi.org/10.1007/978-3-030-92652-6_10

23. Peng, S., Jiang, C.M., Liao, Y., Niemeyer, M., Pollefeys, M., Geiger, A.: Shape as points: a differentiable Poisson solver. In: Advances in Neural Information Processing Systems (2021)

24. Pimentel, P., et al.: Automated virtual reconstruction of large skull defects using statistical shape models and generative adversarial networks. In: Li, J., Egger, J. (eds.) AutoImplant 2020. LNCS, vol. 12439, pp. 16–27. Springer, Cham (2020). https://doi.org/10.1007/978-3-030-64327-0_3

25. Qi, C.R., Yi, L., Su, H., Guibas, L.J.: Pointnet++: deep hierarchical feature learning on point sets in a metric space. In: Advances in Neural Information Processing Systems (2017)

26. Saharia, C., et al.: Palette: image-to-image diffusion models. In: ACM SIGGRAPH 2022 Conference Proceedings (2021)

27. Shi, H., Chen, X.: Cranial implant design through multiaxial slice inpainting using deep learning. In: Li, J., Egger, J. (eds.) AutoImplant 2020. LNCS, vol. 12439, pp. 28–36. Springer, Cham (2020). https://doi.org/10.1007/978-3-030-64327-0_4

28. Sohl-Dickstein, J., Weiss, E.A., Maheswaranathan, N., Ganguli, S.: Deep unsupervised learning using nonequilibrium thermodynamics. In: Proceedings of the 32nd International Conference on Machine Learning (2015)

29. Song, J., Meng, C., Ermon, S.: Denoising diffusion implicit models. In: International Conference on Learning Representations (2020)

30. Wang, B., et al.: Cranial implant design using a deep learning method with anatomical regularization. In: Li, J., Egger, J. (eds.) AutoImplant 2020. LNCS, vol. 12439, pp. 85–93. Springer, Cham (2020). https://doi.org/10.1007/978-3-030-64327-0_10

31. Wodzinski, M., Daniol, M., Hemmerling, D.: Improving the automatic cranial implant design in cranioplasty by linking different datasets. In: Li, J., Egger, J. (eds.) AutoImplant 2021. LNCS, vol. 13123, pp. 29–44. Springer, Cham (2021). https://doi.org/10.1007/978-3-030-92652-6_4

32. Wolleb, J., Sandkühler, R., Bieder, F., Valmaggia, P., Cattin, P.C.: Diffusion models for implicit image segmentation ensembles. In: Proceedings of Machine Learning Research, pp. 1–13 (2022)

33. Yang, B., Fang, K., Li, X.: Cranial implant prediction by learning an ensemble of slice-based skull completion networks. In: Li, J., Egger, J. (eds.) AutoImplant 2021. LNCS, vol. 13123, pp. 95–104. Springer, Cham (2021). https://doi.org/10.1007/978-3-030-92652-6_8

34. Yu, L., Li, J., Egger, J.: PCA-Skull: 3D skull shape modelling using principal component analysis. In: Li, J., Egger, J. (eds.) AutoImplant 2021. LNCS, vol. 13123, pp. 105–115. Springer, Cham (2021). https://doi.org/10.1007/978-3-030-92652-6_9

35. Yuksel, C.: Sample elimination for generating poisson disk sample sets. Comput. Graph. Forum **34**, 25–32 (2015)

36. Zeng, X., et al.: LION: latent point diffusion models for 3D shape generation. In: Advances in Neural Information Processing Systems (2022)

37. Zhou, L., Du, Y., Wu, J.: 3D shape generation and completion through point-voxel diffusion. In: Proceedings of the IEEE/CVF International Conference on Computer Vision, pp. 5826–5835 (2021)

A Novel Video-CTU Registration Method with Structural Point Similarity for FURS Navigation

Mingxian Yang[1]([✉]), Yinran Chen[1]([✉]), Bei Li[1], Zhiyuan Liu[1], Song Zheng[3]([✉]), Jianhui Chen[3], and Xiongbiao Luo[1,2]([✉])

[1] Department of Computer Science and Technology,
Xiamen University, Xiamen, China
`yangmingxian@stu.xmu.edu.cn`, `yinran_chen@xmu.edu.cn`
[2] National Institute for Data Science in Health and Medicine,
Xiamen University, Xiamen, China
`xiongbiao.luo@gmail.com`
[3] Fujian Medical University Union Hospital, Fuzhou, China
`zhengwu_99@aliyun.com`

Abstract. Flexible ureteroscopy (FURS) navigation remains challenging since ureteroscopic images are poor quality with artifacts such as water and floating matters, leading to a difficulty in directly registering these images to preoperative images. This paper presents a novel 2D-3D registration method with structure point similarity for robust vision-based flexible ureteroscopic navigation without using any external positional sensors. Specifically, this new method first uses vision transformers to extract structural regions of the internal surface of the kidneys in real FURS video images and then generates virtual depth maps by the ray-casting algorithm from preoperative computed tomography urogram (CTU) images. After that, a novel similarity function without using pixel intensity is defined as an intersection of point sets from the extracted structural regions and virtual depth maps for the video-CTU registration optimization. We evaluate our video-CTU registration method on in-house ureteroscopic data acquired from the operating room, with the experimental results showing that our method attains higher accuracy than current methods. Particularly, it can reduce the position and orientation errors from (11.28 mm, 10.8°) to (5.39 mm, 8.13°).

Keywords: Surgical navigation · Flexible Ureteroscopy · Vision transformer · Computed tomography urogram · Ureteroscopic lithotripsy

1 Introduction

Flexible ureteroscopy (FURS) is a routinely performed surgical procedure for renal lithotripsy. This procedure inserts a flexible ureteroscope through the blad-

---

M .Yang and Y. Chen—Shows the equally contributed authors.
X. Luo and S. Zheng is the corresponding author.

© The Author(s), under exclusive license to Springer Nature Switzerland AG 2023
H. Greenspan et al. (Eds.): MICCAI 2023, LNCS 14228, pp. 123–132, 2023.
https://doi.org/10.1007/978-3-031-43996-4_12

der and ureters to get inside the kidneys for diagnosis and treatment of stones and tumors. Unfortunately, such an examination and treatment depends on skills and experiences of surgeons. On the other hand, surgeons may miss stones and tumors and unsuccessfully orientate the ureteroscope inside the kidneys due to limited field of views, just 2D images without depth information, and the complex anatomical structure of the kidneys. To this end, ureteroscope tracking and navigation is increasingly developed as a promising tool to solve these issues.

Many researchers have developed various methods to boost endoscopic navigation. These methods generally consist of vision- and sensor-based tracking. Han et al. [3] utilized the porous structures in renal video images to develop a vision-based navigation method for ureteroscopic holmium laser lithotripsy. Zhao et al. [15] designed a master-slave robotic system to navigate the flexible ureteroscope. Luo et al. [7] reported a discriminative structural similarity measure driven 2D-3D registration for vision-based bronchoscope tracking. More recently, Huang et al. [4] developed an image-matching navigation system using shape context for robotic ureteroscopy. Additionally, sensor-based methods are widely sued in surgical navigation [1,6]. Zhang et al. [14] employed electromagnetic sensors to estimate the ureteroscope shape for navigation.

Although these methods mentioned above work well, ureteroscopic navigation is still a challenging problem. Compared to other endoscopes such as colonoscope and bronchoscope, the diameter of the ureteroscope is smaller, resulting in more limited lighting source and field of view. Particularly, ureteroscopy involves much solids (impurities) and fluids (liquids), making ureteroscopic video images low-quality, as well as these solids and fluids inside the kidneys cannot be regularly observed in computed tomography (CT) images. On the other hand, the complex internal structures such as calyx, papilla, and pyramids of the kidneys are difficult to be observed in CT images. These issues introduce a difficulty in directly aligning ureteroscopic video sequences to CT images, leading to a challenge of image-based continuous ureteroscopic navigation.

This work aims to explore an accurate and robust vision-based navigation method for FURS procedures without using any external positional sensors. Based on ureteroscopic video images and preoperative computed tomography urogram (CTU) images, we propose a novel video-CTU registration method to precisely locate the flexible ureteroscope in the CTU space. Several highlights of this work are clarified as follows. To the best of our knowledge, this work shows the first study to continuously track the flexible ureteroscope in preoperative data using a vision-based method. Technically, we propose a novel 2D-3D (video-CTU) registration method that introduces a structural point similarity measure without using image pixel intensity information to characterize the difference between the structural regions in real video images and CTU-driven virtual image depth maps. Additionally, our proposed method can successfully deal with solid and fluid ureteroscopic video images and attains higher navigation accuracy than intensity-based 2D-3D registration methods.

2   Video-CTU Registration

Our proposed video-CTU registration method consists of several steps: (1) ureteroscopic structure extraction, (2) virtual depth map generation, and (3) structural point similarity and optimization. Figure 1 illustrates the flowchart of our method.

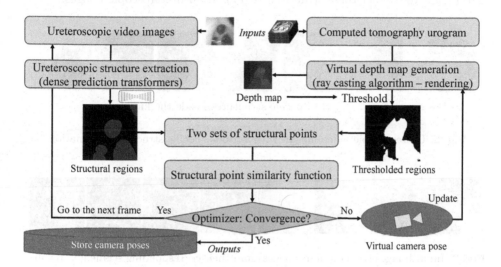

Fig. 1. Processing flowchart of our ureteroscopic navigation method.

2.1   Ureteroscopic Image Characteristics

The internal kidneys consist of complex anatomical structures such as pelvis and calyx and also contain solid particles (e.g., stones and impurities) floating in fluids (e.g., water, urine and small blood), resulting in poor image quality during ureteroscopy. Therefore, it is a challenging task to extract meaningful features from these low-quality images for achieving accurate 2D-3D registration.

Our idea is to introduce specific structures inside the kidneys to boost the video-CTU registration since these structural regions are meaningful features that can facilitate the similarity computation. During ureteroscopy, various anatomical structures observed in ureteroscopic video images indicate different poses of the ureteroscope inside the kidneys. While some structural features such as capillary texture and striations at the tip of the renal pyramids are observed ureteroscopic images, they are not discernible in CT or other preoperative data. Typical structural or texture regions (Columns 1~3 in Fig. 2 (b)) observed both in ureteroscopic video and CTU images are the renal papilla when the uretero-scope gets into the kidneys through the ureter and renal pelvis to reach the major and minor calyxes. Additionally, we also find that the renal pelvis (dark or ultra-low light) regions (Columns 4~6 in Fig. 2 (b)) are also useful to enhance the registration. Hence, this work employs these interior renal structural char-acteristics to calculate the similarity between ureteroscopic images and CTU.

2.2   Ureteroscopic Structure Extraction

Deep learning is widely used for medical image segmentation. Lazo et al. [5] used spatial-temporal ensembles to segment lumen structures in ureteroscopic images. More recently, vision transformers show the potential to precisely segment various medical images [8,12]. This work employs the dense prediction transformer (DPT) [11] to extract these structural regions from ureteroscopic images.

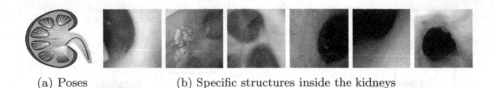

(a) Poses   (b) Specific structures inside the kidneys

Fig. 2. Ureteroscopic video images with various specific structure information

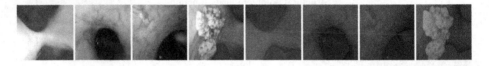

Fig. 3. Input images (the first four images) and their corresponding segmented results: Red indicates structural feature regions and green denotes stones (Color figure online)

DPT is a general deep learning framework for dense prediction tasks such as semantic segmentation and has three versions of DPT-Base, DPT-Large, and DPT-Hybrid. This work use DPT-Base since it only requires a small number of parameters but provides a high inference speed. DPT-Base consists of a transformer encoder and a convolutional decoder. Its backbone is vision transformers [2], where input images are transformed into tokens by non-overlapping patches extraction, followed by a linear projection of their flattened representation. The conventional decoder employs a reassemble operation [11] to assemble a set of tokens into image-like feature representations at various resolutions:

$$\text{Reassemble}_s^{\hat{D}}(t) = (\text{Resample}_s \circ \text{Concatenate} \circ \text{Read})(t) \tag{1}$$

where s is the output size ratio of the feature representation and $\hat{D}$ is the output feature dimension. Tokens from layers $l = \{6, 12, 18, 24\}$ are reassembled in DPT-Base. These feature representations are subsequently fused into the final dense prediction. In the structure extraction, we define three classes: Non-structural regions (background), structural regions, and stones. We manually select and annotate ureteroscopic video images for training and testing. Vision transformers require large datasets for training, so we initialize the encoder with weights pretrained on ImageNet and further train it on our in-house database. Figure 3 displays some segmentation results of ureteroscopic video images.

2.3   Virtual CTU Depth Map Generation and Thresholding

This step is to compute depth maps of 2D virtual images generated from CTU images by volume rendering [13]. Depth maps can represent structural information of virtual images. Note that this work uses CTU to create virtual images since non-contrast CT images cannot capture certain internal structures of the kidneys, although these structures can be observed in ureteroscopic video images.

We introduce a ray casting algorithm in volume rendering to generate depth maps of virtual rending images [13]. The ray casting algorithm is to trace a ray that starts from the viewpoint and passes through a pixel on the screen. When the tracing ray intersects with a voxel, the properties of that voxel will affect the value of corresponding pixel in the final image. For a 3D point (x_0, y_0, z_0) and a normalized direction vector (r_x, r_y, r_z) of its casting ray, a corresponding point (x, y, z) at any distance d on the tracing ray is:

$$(x, y, z) = (x_0 + dr_x, \ y_0 + dr_y, \ z_0 + dr_z). \tag{2}$$

The tracing ray $\mathbf{R}(x, y, z, r_x, r_y, r_z)$ will stop when it encounters opaque voxels. For 3D point (x,y,z), its depth value V can be calculated by projecting the ray $\mathbf{R}(x, y, z, r_x, r_y, r_z)$ onto the normal vector $\mathbf{N}(x, y, z)$ of the image plane:

$$V(x, y, z) = \mathbf{R}(x, y, z, r_x, r_y, r_z) \bullet \mathbf{N}(x, y, z), \tag{3}$$

where symbol $\bullet$ denotes the dot product.

To obtain the depth map with structural regions, we define two thresholds t_u and t_v. Only CT intensity values within $[t_u, t_v]$ are opaque voxels that the casting rays cannot pass through in the ray-casting algorithm. According to CTU characteristics [10], this work uses our excretory-phase data to generate virtual images and set $[t_u, t_v]$ to $[-1000, 120]$, where -1000 represents air and 120 was determined by the physician's experience and characteristics of contrast agents.

Unfortunately, the accuracy of thresholded structural regions suffers from inaccurate depth maps caused by renal stones and contrast agents. Stones and agents are usually high intensity in CTU images, which result in incorrect depth information of structural regions (e.g., renal papilla). To deal with these issues, we use the segmented stones as a mask to remove these regions with wrong depth. On the other hand, the structural regions usually have larger depth values than the agent-contrasted regions. Therefore, we sort the depth values outside the mask and only use the thresholded structural regions with the largest depth values for the structural point similarity computation.

2.4   Structural Point Similarity and Optimization

We define a point similarity measure between DPT-base segmented structural regions in ureteroscopic images and thresholded structural regions in virtual depth maps generated by the volume rendering ray casting algorithm.

The structural point similarity function (cost function) is defined as an intersection of point sets from the extracted real and virtual structural regions:

$$\mathcal{F}(\mathbf{I}_i, \mathbf{D}_v) = \frac{2|\mathcal{P}_i \bigcap \mathcal{P}_v|}{|\mathcal{P}_i| + |\mathcal{P}_v|}, \mathcal{P}_i \in E_i(x, y), \mathcal{P}_v \in E_v(x, y), \tag{4}$$

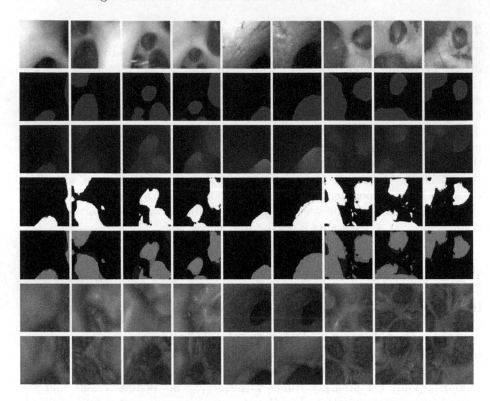

Fig. 4. Visual navigation results: Rows 1∼5 show the real video images, segmented structural regions, depth maps, thresholded structural regions, and overlapped regions for similarity computation, while Rows 6∼7 illustrate the tracking results (2D virtual rendering images) of using Luo et al. [7] and our method.

where $\mathbf{I}_i$ is the ureteroscopic video image at frame i, point sets $\mathcal{P}_i$ and $\mathcal{P}_v$ are from the ureteroscopic image extracted structural region $E_i(a, b)$ and the thresholded structural region $E_v(a, b)$ ((a, b) denotes a point)from the depth map $\mathbf{D}_v$ of the 2D virtual rendering image $\mathbf{I}_v(\mathbf{p}_i, \mathbf{q}_i)$, respectively:

$$\mathbf{D}_v \propto \mathbf{I}_v(\mathbf{p}_i, \mathbf{q}_i), \tag{5}$$

where $(\mathbf{p}_i, \mathbf{q}_i)$ is the endoscope position and orientation in the CTU space.

Eventually, the optimal pose $(\tilde{\mathbf{p}}_i, \tilde{\mathbf{q}}_i)$ of the ureteroscope in the CTU space can be estimated by maximizing the structural point similarity:

$$(\tilde{\mathbf{p}}_i, \tilde{\mathbf{q}}_i) = \arg \max_{\mathbf{p}_i, \mathbf{q}_i} \mathcal{F}(\mathbf{I}_i, \mathbf{D}_v), \qquad \mathbf{D}_v \in \mathbf{I}_v(\mathbf{p}_i, \mathbf{q}_i), \tag{6}$$

where Powell method [9] is used as an optimizer to run this procedure.

Table 1. DPT-base segmented results of intersection over union (IoU), accuracy (Acc), and dice similarity coefficient (DSC), and estimated ureteroscope (position, orientation) errors of using the two vision-based navigation methods

Classes	IoU%	Acc%	DSC%	Methods	Luo et al. [7]	Our method
Background	92.14	98.20	95.91	Case A	(18.13 mm, 11.78°)	(5.66 mm, 7.05°)
Structures	80.09	83.17	88.95	Case B	(11.04 mm, 4.87°)	(2.59 mm, 3.66°)
Renal stones	92.80	95.40	96.27	Case C	(8.60 mm, 12.37°)	(6.20 mm, 10.04°)
Average	88.34	92.26	93.71	Average	(11.28 mm, 10.82°)	(5.39 mm, 8.13°)

Table 2. Sensitivity analysis results of threshold in virtual CTU depth map generation. In this work, the threshold range was set to [−1000, 120].

Threshold	[−1000, 70]	[−1000, 95]	[−1000, 120]	[−1000, 145]	[−1000, 170]
Position Errors	11.08 mm	8.48 mm	5.39 mm	6.70 mm	5.59 mm
Orientation Errors	15.55°	10.52°	8.13°	8.92°	11.10°

3 Results and Discussion

We validate our method on clinical ureteroscopic lithotripsy data with video sequences and CTU volumes. Ureteroscopic video images were a size of 400 × 400 pixels, while the space parameters of CTU volumes were 512 × 512 pixels, 361∼665 slices, 0.625∼1.25 mm slice thickness. Three ureteroscopic videos more than 30000 frames were acquired from three ureteroscopic procedures for experiments. While we manually annotated ureteroscopic video images for DPT-base segmentation, three experts also manually generated ureteroscope pose ground-truth data by our developed software, which can manually adjust position and direction parameters of the virtual camera to visually align endoscopic real images to virtual images, evaluating the navigation accuracy of the different methods.

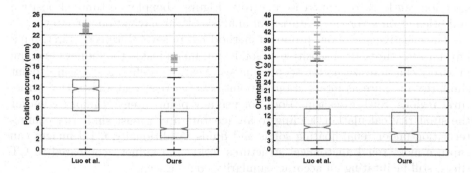

Fig. 5. Boxplotted position and orientation errors of using the two methods

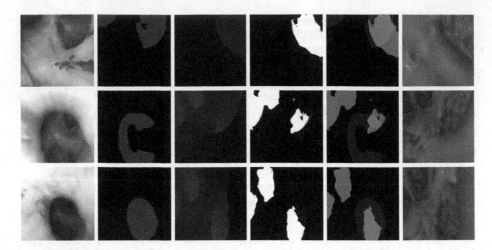

Fig. 6. Our method fails to track the ureteroscope in CTU: Columns 1∼6 correspond to input images, segmented structural regions, depth maps, thresholded structural regions, overlapped regions, and generated virtual images.

Figure 4 illustrates the navigation results of segmentation, depth maps, extracted structural regions for similarity calculation, and generated 2D virtual images corresponding to estimated ureteroscope poses. Structural regions can be extracted from ureteroscopic images and virtual depth maps. Particularly, we can see that our method generated virtual images (Row 7 in Fig. 4) resemble real video images (Row 1 in Fig. 4) much better than Luo et al. [7] generated ones (Row 6 in Fig. 4). This implies that our method can estimate the ureteroscope pose much more accurate than Luo et al. [7]. Table 1 summarizes quantitative segmentation results and position and orientation errors. DPT-Base can achieve average segmentation IoU 88.34%, accuracy 92.26%, and DSC 93.71%. The average position and orientation errors of our method were 5.39 mm and 8.14°, which much outperform the compared method. Table 2 shows the results of sensitivity analysis for the threshold values. It can be seen that inappropriate threshold selection can lead to an increase in errors. Figure 5 boxplots estimated position and orientation errors for a statistical analysis of our navigation accuracy.

The effectiveness of our proposed method lies in several aspects. First, renal interior structures are insensitive to solids and fluids inside the kidneys and can precisely characterize ureteroscopic images. Next, we define a structural point similarity measure as intersection of point sets between real and virtual structural regions. Such a measure does not use any point intensity information for the similarity calculation, leading to an accurate and robust similarity characterization under renal floating solids and fluids. Additionally, CTU images can capture more renal anatomical structures inside the kidneys compared to CT slices, still facilitating an accurate similarity computation.

Our method still suffers from certain limitations. Figure 6 displays some ureteroscopic video images our method fails to track. This is because that the segmentation method cannot successfully extract structural regions, while the ray casting algorithm cannot correctly generate virtual depth maps with structural regions. Both unsuccessfully extracted real and virtual structural regions collapse the similarity characterization. We will improve the segmentation of ureteroscopic video images, while generating more ground-truth data for training and testing.

4 Conclusion

This paper proposes a new 2D-3D registration approach for vision-based FURS navigation. Specifically, such an approach can align 2D ureteroscopic video sequences to 3D CTU volumes and successfully locate an ureteroscope into CTU space. Different from intensity-based cost function, a novel structural point similarity measure is proposed to effectively and robustly characterize ureteroscopic video images. The experimental results demonstrate that our proposed method can reduce the navigation errors from $(11.28\,\text{mm}, 10.8°)$ to $(5.39\,\text{mm}, 8.13°)$.

Acknowledgements. This work was supported in part by the National Natural Science Foundation of China under Grants 61971367, 82272133, and 62001403, in part by the Natural Science Foundation of Fujian Province of China under Grants 2020J01004 and 2020J05003, and in part by the Fujian Provincial Technology Innovation Joint Funds under Grant 2019Y9091.

References

1. Attivissimo, F., Lanzolla, A.M.L., Carlone, S., Larizza, P., Brunetti, G.: A novel electromagnetic tracking system for surgery navigation. Comput. Assist. Surg. **23**(1), 42–52 (2018)
2. Dosovitskiy, A., et al.: An image is worth 16×16 words: transformers for image recognition at scale. arXiv preprint arXiv:2010.11929 (2020)
3. Han, M., Dai, Y., Zhang, J.: Endoscopic navigation based on three-dimensional structure registration. In: 2020 IEEE/RSJ International Conference on Intelligent Robots and Systems (IROS), pp. 2900–2905. IEEE (2020)
4. Huang, Z.: Image-matching based navigation system for robotic ureteroscopy in kidney exploration. Master's thesis, Delft University of Technology, Netherlands (2022)
5. Lazo, J.F., et al.: Using spatial-temporal ensembles of convolutional neural networks for lumen segmentation in ureteroscopy. Int. J. Comput. Assist. Radiol. Surg. **16**(6), 915–922 (2021). https://doi.org/10.1007/s11548-021-02376-3
6. Luo, X.: A new electromagnetic-video endoscope tracking method via anatomical constraints and historically observed differential evolution. In: Martel, A.L., et al. (eds.) MICCAI 2020. LNCS, vol. 12263, pp. 96–104. Springer, Cham (2020). https://doi.org/10.1007/978-3-030-59716-0_10
7. Luo, X.: Accurate multiscale selective fusion of CT and video images for real-time endoscopic camera 3D tracking in robotic surgery. In: IEEE International Conference on Acoustics, Speech and Signal Processing (ICASSP), vol. 33, pp. 1386–1390. IEEE (2022)

8. Matsoukas, C., Haslum, J.F., Söderberg, M., Smith, K.: Is it time to replace CNNs with transformers for medical images? arXiv preprint arXiv:2108.09038 (2021)
9. Nocedal, J., Wright, S.J.: Numerical Optimization, 2nd ed. Springer, New York (2006). https://doi.org/10.1007/978-0-387-40065-5
10. Noorbakhsh, A., Aganovic, L., Vahdat, N., Fazeli, S., Chung, R., Cassidy, F.: What a difference a delay makes! CT urogram: a pictorial essay. Abdom. Radiol. **44**(12), 3919–3934 (2019)
11. Ranftl, R., Bochkovskiy, A., Koltun, V.: Vision transformers for dense prediction. In: Proceedings of the IEEE/CVF International Conference on Computer Vision, pp. 12179–12188 (2021)
12. Shamshad, F., et al.: Transformers in medical imaging: a survey. arXiv preprint arXiv:2201.09873 (2022)
13. Wrenninge, M.: Production Volume Rendering: Design and Implementation, vol. 5031 (2020)
14. Zhang, C., et al.: Shape estimation of the anterior part of a flexible ureteroscope for intraoperative navigation. Int. J. Comput. Assist. Radiol. Surg. **17**(10), 1787–1799 (2022)
15. Zhao, J., Li, J., Cui, L., Shi, C., Wei, G., et al.: Design and performance investigation of a robot-assisted flexible ureteroscopy system. Appl. Bionics Biomech. **2021** (2021)

Pelphix: Surgical Phase Recognition from X-Ray Images in Percutaneous Pelvic Fixation

Benjamin D. Killeen(✉)🆔, Han Zhang, Jan Mangulabnan🆔,
Mehran Armand🆔, Russell H. Taylor🆔, Greg Osgood🆔,
and Mathias Unberath🆔

Johns Hopkins University, Baltimore, MD, USA
{killeen,hzhan206,jmangul1,rht,unberath}@jhu.edu,
{marmand2,gosgood2}@jhmi.edu

Abstract. Surgical phase recognition (SPR) is a crucial element in the digital transformation of the modern operating theater. While SPR based on video sources is well-established, incorporation of interventional X-ray sequences has not yet been explored. This paper presents Pelphix, a first approach to SPR for X-ray-guided percutaneous pelvic fracture fixation, which models the procedure at four levels of granularity – corridor, activity, view, and frame value – simulating the pelvic fracture fixation workflow as a Markov process to provide fully annotated training data. Using added supervision from detection of bony corridors, tools, and anatomy, we learn image representations that are fed into a transformer model to regress surgical phases at the four granularity levels. Our approach demonstrates the feasibility of X-ray-based SPR, achieving an average accuracy of 99.2% on simulated sequences and 71.7% in cadaver across all granularity levels, with up to 84% accuracy for the target corridor in real data. This work constitutes the first step toward SPR for the X-ray domain, establishing an approach to categorizing phases in X-ray-guided surgery, simulating realistic image sequences to enable machine learning model development, and demonstrating that this approach is feasible for the analysis of real procedures. As X-ray-based SPR continues to mature, it will benefit procedures in orthopedic surgery, angiography, and interventional radiology by equipping intelligent surgical systems with situational awareness in the operating room. Code and data available at https://github.com/benjamindkilleen/pelphix.

Keywords: Activity recognition · fluoroscopy · orthopedic surgery · surgical data science

---

Supplementary Information The online version contains supplementary material available at https://doi.org/10.1007/978-3-031-43996-4_13.

H. Greenspan et al. (Eds.): MICCAI 2023, LNCS 14228, pp. 133–143, 2023.
https://doi.org/10.1007/978-3-031-43996-4_13

1 Introduction

In some ways, surgical data is like the expanding universe: 95% of it is dark and unobservable [2]. The vast majority of intra-operative X-ray images, for example, are "dark", in that they are not further analyzed to gain quantitative insights into routine practice, simply because the human-hours required would drastically outweigh the benefits. As a consequence, much of this data not only goes un-analyzed but is discarded directly from the imaging modality after inspection. Fortunately, machine learning algorithms for automated intra-operative image analysis are emerging as an opportunity to leverage these data streams. A popular application is surgical phase recognition (SPR), a way to obtain quantitative analysis of surgical workflows and equip automated systems with situational awareness in the operating room (OR). SPR can inform estimates of surgery duration to maximize OR throughput [7] and augment intelligent surgical systems, *e.g.* for suturing [20] or image acquisition [4,10,11], enabling smooth transitions from one specialized subsystem to the next. Finally, SPR provides the backbone for automated skill analysis to produce immediate, granular feedback based on a specific surgeon's performance [5,21].

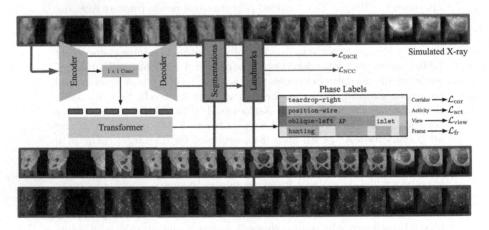

Fig. 1. Our model architecture incorporates frame-level spatial annotations using a U-Net encoder-decoder variant. Anatomical landmarks and segmentation maps provide added supervision to the image encoder for a transformer, which predicts the surgical phase. The images shown here are the result of Markov-based simulation of percutaneous fixation, used for training.

The possibilities described above have motivated the development of algorithms for surgical phase recognition based on the various video sources in the OR [15,19,22,23]. However, surgical phase recognition based on interventional X-ray sequences remains largely unexplored. Although X-ray guidance informs more than 17 million procedures across the United States (as of 2006) [13], the unique challenges of processing X-ray sequences compared to visible or structured light imaging have so far hindered research in this area. Video cameras collect many images per second from relatively stationary viewpoints. By contrast, C-arm X-ray imaging often features consecutive images from vastly different viewpoints, resulting in highly varied object appearance due to the transmissive nature of X-rays. X-ray images are also acquired irregularly, usually amounting to several hundred frames in a procedure of several hours, limiting the availability of training data for machine learning algorithms.

Following recent work that enables sim-to-real transfer in the X-ray domain [6], we now have the capability to train generalizable deep neural networks (DNNs) using simulated images, where rich annotations are freely available. *This paper represents the first step in breaking open SPR for the X-ray domain, establishing an approach to categorizing phases, simulating realistic image sequences, and analyzing real procedures.* We focus our efforts on percutaneous pelvic fracture fixation, which involves the acquisition of standard views and the alignment of Kirschner wires (K-wires) and orthopedic screws with bony corridors [17]. We model the procedure at four levels, the current target corridor, activity (`position-wire`, `insert-wire`, and `insert-screw`), C-arm view (`AP`, `lateral`, etc.), and frame-level clinical value. Because of radiation exposure for both patients and clinicians, it is relevant to determine which X-ray images are acquired in the process of "fluoro-hunting" (`hunting`) versus those used for clinical `assessment`. Each of these levels is modeled as a Markov process in a stochastic simulation, which provides fully annotated training data for a transformer architecture.

2   Related Work

SPR from video sources is a popular topic, and has benefited from the advent of transformer architectures for analyzing image sequences. The use of convolutional layers as an image encoder has proven effective for recognizing surgical phases in endoscopic video [22] and laparoscopic video [3,19]. These works especially demonstrate the effectiveness of transformers for dealing with long image sequences [3], while added spatial annotations improve both the precision and information provided by phase recognition [19]. Although some work explores activity recognition in orthopedic procedures [8,9] they rely on head-mounted cameras with no way to assess tool-to-tissue relationships in percutaneous procedures. The inclusion of X-ray image data in this space recenters phase recognition on patient-centric data and makes possible the recognition of surgical phases which are otherwise invisible.

3   Method

The Pelphix pipeline consists of stochastic simulation of X-ray image sequences, based on a large database of annotated CT images, and a transformer architecture for phase recognition with additional task-aware supervision. A statistical shape model is used to propagate landmark and corridor annotations over 337

CTs, as shown in Fig. 2a. The simulation proceeds by randomly aligning virtual K-wires and screws with the annotated corridors (Sect. 3.1). In Sect. 3.2, we describe a transformer architecture with a U-Net style encoder-decoder structure enables sim-to-real transfer for SPR in X-ray.

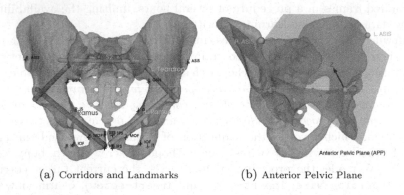

(a) Corridors and Landmarks   (b) Anterior Pelvic Plane

Fig. 2. (a) The ramus, teardrop and S2 bony corridors, as well as 16 anatomical landmarks with added supervision for phase recognition. (b) The anterior pelvic plane (APP) coordinate system is used to define principle ray directions for standard views of the pelvis, enabling realistic simulation of image sequences for percutaneous fixation.

3.1   Image Sequence Simulation for Percutaneous Fixation

Unlike sequences collected from real surgery [15] or human-driven simulation [14], our workflow simulator must capture the procedural workflow while also maintaining enough variation to allow algorithms to generalize. We accomplish this by modeling the procedural state as a Markov process, in which the transitions depend on evaluations of the projected state, as well as an adjustment factor $\lambda_{adj} \in [0, 1]$ that affects the number of images required for a given task. A low adjustment factor decreases the probability of excess acquisitions for the simulated procedure. In our experiments, we sample $\lambda_{adj} \in \mathcal{U}(0.6, 0.8)$ at the beginning of each sequence.

Figure 3 provides an overview of this process. Given a CT image with annotated corridors, we first sample a target corridor with start and endpoints $\mathbf{a}, \mathbf{b} \in \mathbb{R}^3$. For the ramus corridors, we randomly swap the start and endpoints to simulate the retrograde and antegrade approaches. We then uniformly sample the initial wire tip position within 5 mm of $\mathbf{a}$ and the direction within 15° of $\mathbf{b} - \mathbf{a}$.

Sample Desired View. The desired view is sampled from views appropriate for the current target corridor. For example, appropriate views for evaluating wire placement in the superior ramus corridor are typically the inlet and obturator oblique views, and other views are sampled with a smaller probability. We refer to the "oblique left" and "oblique right" view independent of the

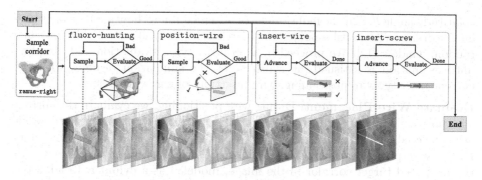

Fig. 3. The image sequence simulation pipeline for Pelphix. We model the procedure as a Markov random process, where transition probabilities depend on realistic evaluation of the current frame.

affected patient side, so that for the right pubic ramus, the obturator oblique is the "oblique left" view, and the iliac oblique is "oblique right." We define the "ideal" principle ray direction $\hat{\mathbf{r}}^*$ for each standard view in the anterior pelvic plane (APP) coordinate system (see supplement) and the ideal viewing point $\mathbf{p}^*$ as the midpoint of the target corridor. At the beginning of each sequence, we sample the intrinsic camera matrix of the virtual C-arm with sensor width $w_{\mathrm{s}} \sim \mathcal{U}(300, 400)$ mm, $d_{\mathrm{sd}} \sim \mathcal{U}(900, 1200)$, and an image size of 384×384. Given a viewing point and direction $(\mathbf{p}, \hat{\mathbf{r}})$, the camera projection matrix $\mathbf{P}$ is computed with the X-ray source (or camera center) at $\mathbf{p} - d_{\mathrm{sp}}\hat{\mathbf{r}}$ and principle ray $\hat{\mathbf{r}}$, where d_{sp} $\mathcal{U}(0.65\,d_{\mathrm{sd}}, 0.75\,d_{\mathrm{sd}})$ is the source-to-viewpoint distance, and d_{sd} is the source-to-detector distance (or focal length) of the virtual C-arm.

Evaluate View. Given a current view $(\mathbf{p}, \hat{\mathbf{r}})$ and desired view $(\mathbf{p}^*, \hat{\mathbf{r}}^*)$, we first evaluate whether the current view is acceptable and, if it is not, make a random adjustment. View evaluation considers the principle ray alignment and whether the viewing point is reasonably centered in the image, computing,

$$\hat{\mathbf{r}} \cdot \hat{\mathbf{r}}^* < \cos(\theta_t) \text{ AND } \left\| \mathbf{P}\mathbf{p}^* - \left[\tfrac{H}{2} \ \tfrac{W}{2} \ 1 \right]^T \right\| < \frac{2}{5} \min(H, W) \qquad (1)$$

where the angular tolerance $\theta_t \in [3°, 10°]$ depends on the desired view, ranging from teardrop views (low) to lateral (high tolerance).

Sample View. If Eq. 1 is not satisfied, then we sample a new view $(\mathbf{p}, \hat{\mathbf{r}})$ uniformly within a uniform window that shrinks every iteration by the adjustment factor, according to

$$\mathbf{p} \sim \mathcal{U}_{\circ} \left(\mathbf{p}^*, \mathrm{clip}(\lambda_{\mathrm{adj}} \|\mathbf{p}^* - \mathbf{p}\|, 5\,\mathrm{mm}, 100\,\mathrm{mm})\right) \qquad (2)$$

$$\hat{\mathbf{r}} \sim \mathcal{U}_{\angle} \left(\hat{\mathbf{r}}^*, \mathrm{clip}(\lambda_{\mathrm{adj}} \arccos(\hat{\mathbf{r}}^* \cdot \hat{\mathbf{r}}), 1°, 45°)\right), \qquad (3)$$

where $\mathcal{U}_{\circ}(\mathbf{c}, r)$ is the uniform distribution in the sphere with center $\mathbf{c}$ and radius r, and $\mathcal{U}_{\angle}(\hat{\mathbf{r}}, \theta)$ is the uniform distribution on the solid angle centered on $\hat{\mathbf{r}}$ with colatitude angle θ. This formula emulates observed fluoro-hunting by converging on the desired view until a point, when further adjustments are within the same random window [12]. We proceed by alternating view evaluation and sampling until evaluation is satisfied, at which point the simulation resumes with the current activity: wire positioning, wire insertion, or screw insertion.

Evaluate Wire Placement. During wire positioning, we evaluate the current wire position and make adjustments from the current view, iterating until evaluation succeeds. Given the current wire tip $\mathbf{x}$, direction $\hat{\mathbf{v}}$, and projection matrix $\mathbf{P}$, the wire placement is considered "aligned" if it *appears* to be aligned with the projected target corridor in the image, modeled as a cylinder. In addition, we include a small likelihood of a false positive evaluation, which diminishes as the wire is inserted.

Sample Wire Placement. If the wire evaluation determines the current placement is unsuitable, then a new wire placement is sampled. For the down-the-barrel views, this is done similarly to Eq. 2, by bringing the wire closer to the corridor in 3D. For orthogonal views, repositioning consists of a small random adjustment to $\mathbf{x}$, a rotation about the principle ray (the in-plane component), and a minor perturbation orthogonal to the ray (out-of-plane). This strategy emulates real repositioning by only adjusting the degree of freedom visible in the image, i.e. the projection onto the image plane:

$$\mathbf{x} \sim \mathcal{U}_\circ(\mathbf{x}, \text{clip}(\lambda_{\text{adj}}||\mathbf{x} - \mathbf{a}||, 5\,\text{mm}, 10\,\text{mm})) \tag{4}$$

$$\hat{\mathbf{v}} \leftarrow \text{Rot}\left(\hat{\mathbf{v}} \times \hat{\mathbf{r}}, \theta_\perp\right)\text{Rot}\left(\hat{\mathbf{r}}, \theta^* + \theta_\parallel\right), \text{ where } \theta_\perp \sim \mathcal{U}(-0.1\,\theta^*, 0.1\,\theta^*), \tag{5}$$

$$\theta_\parallel \sim \mathcal{U}(-\text{clip}(\lambda_{\text{adj}}\theta^*, 3°, 10°), \text{ clip}(\lambda_{\text{adj}}\theta^*, 3°, 10°)), \tag{6}$$

and θ^* is the angle between the wire and the target corridor in the image plane. If the algorithm returns "Good," the sequence either selects a new view to acquire (and stays in the `position-wire` activity) or proceeds to `insert-wire` or `insert-screw`, according to random transitions.

In our experiments, we used 337 CT images: 10 for validation, and 327 for generating the training set. (Training images were collected continuously during development, after setting aside a validation set.) A DRR was acquired at every decision point in the simulation, with a maximum of 1000 images per sequence, and stored along with segmentations and anatomical landmarks. We modeled a K-wire with 2 mm diameter and orthopedic screws with lengths from 30 to 130 mm and a 16 mm thread, with up to eight instances of each in a given sequence. Using a customized version of DeepDRR [18], we parallelized image generation across 4 RTX 3090 GPUs with an observed GPU memory footprint of ~ 13 GB per worker, including segmentation projections. Over approximately five days, this resulted in a training set of 677 sequences totaling 279,709 images and 22 validation sequences with 8,515 images.

3.2   Transformer Architecture for X-ray-based SPR

Figure 1 shows the transformer architecture used to predict surgical phases based on embedding tokens for each frame. To encourage local temporal features in each

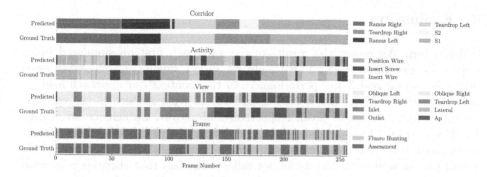

Fig. 4. Results of surgical phase recognition for a cadaveric procedure. We observe varying performance based on the target corridor, either because of the associated views or due to the accumulated orthopedic hardware.

embedding token, we cross-pollinate adjacent frames in the channel dimension, so that each $(3, H, W)$ encoder input contains the previous, current, and next frame. The image encoder is a U-Net [16] encoder-decoder variant with 5 Down and Up blocks and 33 spatial output channels, consisting of (a) 7 segmentation masks of the left hip, right hip, left femur, right femur, sacrum, L5 vertebra, and pelvis; (b) 8 segmentation masks of bony corridors, including the ramus (2), teardrop (2) and sacrum corridors (4), as in Fig. 2a; (c) 2 segmentation masks for wires and screws; and (d) 16 heatmaps corresponding to the anatomical landmarks in Fig. 2a. These spatial annotations provide additional supervision, trained with DICE loss $\mathcal{L}_{\mathrm{DICE}}$ for segmentation channels and normalized cross correlation $\mathcal{L}_{\mathrm{NCC}}$ for heatmap channels as in [1,6]. To compute tokens for input to the transformer, we apply a 1×1 Conv + BatchNorm + ReLU block with kernel size 512 to the encoder output, followed by global average pooling. The transformer has 6 layers with 8 attention heads and a feedforward dimension of 2048. During training and inference, we apply forward masking so that only previous frames are considered. The output of the transformer are vectors in $\mathbb{R}^{21}$ with phase predictions for each frame, corresponding to (a) the 8 target corridors; (b) 3 activities (`position-wire`, `insert-wire`, or `insert-screw`); (c) 8 standard views (see Sect. 3.1); and (d) 2 frame values (`hunting` or `assessment`). We compute the cross entropy loss separately for the corridor $\mathcal{L}_{\mathrm{cor}}$, activity $\mathcal{L}_{\mathrm{act}}$, view $\mathcal{L}_{\mathrm{view}}$, and frame $\mathcal{L}_{\mathrm{fr}}$ phases, and take the mean.

Training Details. Following [6,11], we use a pipeline of heavy domain randomization techniques to enable sim-to-real transfer. In our experiments, we trained the transformer for 200 epochs on 2 RTX 3090 GPUs with 24 GB of memory each, with a sequence length of 48 and a batch size of 4. The initial learning rate was 0.0001, reduced by a factor of 10 at epoch 150 and 180.

4   Evaluation

Simulation. We report the results of our approach first on simulated image sequences, generated from the withheld set of CT images, which serves as an upper bound on real X-ray performance. In this context our approach achieves an accuracy of 99.7%, 98.2%, 99.8%, and 99.0% with respect to the corridor, activity, view, and frame level, respectively. Moreover, we achieve an average DICE score of 0.72 and landmark detection error of 1.01 ± 0.153 pixels in simulation, indicating that these features provide a meaningful signal. That the model generalizes so well to the validation set reflects the fact that these sequences are sampled using the same Markov-based simulation as the training data.

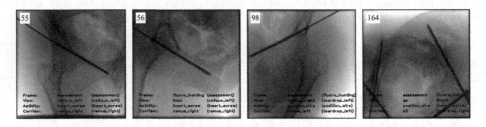

Fig. 5. Exemplary cases from the cadaver study. Image 55 is an ideal prediction, whereas image 56 merely confirms the final placement. Image 98 illustrates an ambiguity that can arise between the teardrop and oblique views during fluoro-hunting, while image 164 is interpreted as belonging to an S2 insertion.

Cadaver Study. We evaluate our approach on cadaveric image sequences with five screw insertions. An attending orthopedic surgeon performed percutaneous fixation on a lower torso specimen, taking the antegrade approach for the left and right pubic ramus corridors, followed by the left and right teardrop and S1 screws. An investigator acted as the radiological technician, positioning a mobile C-arm according to the surgeon's direction. A total of 257 images were acquired during these fixations, with phase labels based on the surgeon's narration.

Our results, shown in Fig. 4 demonstrate the potential for Pelphix as a viable approach to SPR in X-ray. We achieve an overall accuracy of 84%, 60%, 65%, and 77% with respect to the corridor, activity, view, and frame levels, respectively. Figure 5 shows exemplary success and failure modes for our approach, which struggles with ambiguities that may arise due to the similarity of certain views. For instance, image 98 was acquired during fluoro-hunting for the teardrop left view, but our approach associates this sequence with verification of the left ramus screw, which was just finished. Similarly, after the right teardrop wire was inserted, our approach anticipated the insertion of an S2 wire. This was a valid transition in our simulation, so surgeon preferences may be needed to resolve the ambiguity. At the same time, we observe significantly higher accuracy for the pubic ramus corridors (97.7, 76.9, 98.3, and 84.4% respectively) than the teardrop (60.2, 56.9, 64.2, 73.7%) and S1 corridors (100%, 40.6%, 23%, 71%), which may reflect the challenges of interpreting associated images or simply the accumulation of orthopedic hardware.

5   Discussion and Conclusion

As our results show, Pelphix is a potentially viable approach to robust SPR based on X-ray images. We showed that stochastic simulation of percutaneous fracture fixation, despite having no access to real image sequences, is a sufficiently realistic data source to enable sim-to-real transfer. While we expect adjustments to the simulation approach will close the gap even further, truly performative SPR algorithms for X-ray may rely on Pelphix-style simulation for pretraining, before fine-tuning on real image sequences to account for human-like behavior. Extending this approach to other procedures in orthopedic surgery, angiography, and interventional radiology will require task-specific simulation capable of modeling possibly more complex tool-tissue interactions and human-in-the-loop workflows. Nevertheless, Pelphix provides a viable first route toward X-ray-based surgical phase recognition, which we hope will motivate routine collection and interpretation of these data, in order to enable advances in surgical data science that ultimately improve the standard of care for patients.

Acknowledgements. This work was supported by NIH Grant No. R21EB028505 and Johns Hopkins University internal funds. Thank you to Demetries Boston, Henry Phalen, and Justin Ma for assistance with cadaver experiments.

References

1. Bier, B., et al.: X-ray-transform invariant anatomical landmark detection for pelvic trauma surgery. In: Frangi, A.F., Schnabel, J.A., Davatzikos, C., Alberola-López, C., Fichtinger, G. (eds.) MICCAI 2018. LNCS, vol. 11073, pp. 55–63. Springer, Cham (2018). https://doi.org/10.1007/978-3-030-00937-3_7
2. Caldwell, R., Kamionkowski, M.: Dark matter and dark energy. Nature **458**(7238), 587–589 (2009). https://doi.org/10.1038/458587a
3. Czempiel, T., Paschali, M., Ostler, D., Kim, S.T., Busam, B., Navab, N.: OperA: attention-regularized transformers for surgical phase recognition. In: de Bruijne, M., et al. (eds.) MICCAI 2021, Part IV. LNCS, vol. 12904, pp. 604–614. Springer, Cham (2021). https://doi.org/10.1007/978-3-030-87202-1_58
4. Da Col, T., Mariani, A., Deguet, A., Menciassi, A., Kazanzides, P., De Momi, E.: SCAN: system for camera autonomous navigation in robotic-assisted surgery. In: 2020 IEEE/RSJ International Conference on Intelligent Robots and Systems (IROS), pp. 2996–3002. IEEE (2021). https://doi.org/10.1109/IROS45743.2020. 9341548
5. DiPietro, R., et al.: Segmenting and classifying activities in robot-assisted surgery with recurrent neural networks. Int. J. Comput. Assist. Radiol. Surg. **14**(11), 2005–2020 (2019). https://doi.org/10.1007/s11548-019-01953-x
6. Gao, C., et al.: SyntheX: scaling up learning-based X-ray image analysis through in silico experiments. arXiv (2022). https://doi.org/10.48550/arXiv.2206.06127
7. Guédon, A.C.P.: Deep learning for surgical phase recognition using endoscopic videos. Surg. Endosc. **35**(11), 6150–6157 (2020). https://doi.org/10.1007/s00464-020-08110-5

8. Hossain, M., Nishio, S., Hiranaka, T., Kobashi, S.: Real-time surgical tools recognition in total knee arthroplasty using deep neural networks. In: 2018 Joint 7th International Conference on Informatics, Electronics & Vision (ICIEV) and 2018 2nd International Conference on Imaging, Vision & Pattern Recognition (icIVPR), pp. 470–474. IEEE (2018). https://doi.org/10.1109/ICIEV.2018.8641074

9. Kadkhodamohammadi, A., et al.: Towards video-based surgical workflow understanding in open orthopaedic surgery. Comput. Meth. Biomech. Biomed. Eng. Imaging Vis. 9(3), 286–293 (2021). https://doi.org/10.1080/21681163.2020.1835552

10. Kausch, L., et al.: C-Arm positioning for spinal standard projections in different intra-operative settings. In: de Bruijne, M., et al. (eds.) MICCAI 2021. LNCS, vol. 12904, pp. 352–362. Springer, Cham (2021). https://doi.org/10.1007/978-3-030-87202-1_34

11. Killeen, B.D., et al.: An autonomous X-ray image acquisition and interpretation system for assisting percutaneous pelvic fracture fixation. Int. J. CARS 18, 1–8 (2023). https://doi.org/10.1007/s11548-023-02941-y

12. Killeen, B.D., et al.: Mixed reality interfaces for achieving desired views with robotic X-ray systems. Comput. Meth. Biomech. Biomed. Eng. Imaging Vis. 11, 1–6 (2022). https://doi.org/10.1080/21681163.2022.2154272

13. Kim, K.P., et al.: Occupational radiation doses to operators performing fluoroscopically-guided procedures. Health Phys. 103(1), 80 (2012). https://doi.org/10.1097/HP.0b013e31824dae76

14. Munawar, A., et al.: Virtual reality for synergistic surgical training and data generation. Comput. Meth. Biomech. Biomed. Eng. Imaging Vis. 10, 1–9 (2021)

15. Padoy, N.: Machine and deep learning for workflow recognition during surgery. Minim. Invasive Therapy Allied Technol. 28(2), 82–90 (2019). https://doi.org/10.1080/13645706.2019.1584116

16. Ronneberger, O., Fischer, P., Brox, T.: U-Net: convolutional networks for biomedical image segmentation. In: Navab, N., Hornegger, J., Wells, W.M., Frangi, A.F. (eds.) MICCAI 2015. LNCS, vol. 9351, pp. 234–241. Springer, Cham (2015). https://doi.org/10.1007/978-3-319-24574-4_28

17. Simonian, P.T., Routt Jr, M.L.C., Harrington, R.M., Tencer, A.F.: Internal fixation of the unstable anterior pelvic ring a biomechanical comparison of standard plating techniques and the retrograde medullary superior pubic ramus screw. J. Orthop. Trauma 8(6), 476 (1994)

18. Unberath, M., et al.: DeepDRR – a catalyst for machine learning in fluoroscopy-guided procedures. In: Frangi, A.F., Schnabel, J.A., Davatzikos, C., Alberola-López, C., Fichtinger, G. (eds.) MICCAI 2018. LNCS, vol. 11073, pp. 98–106. Springer, Cham (2018). https://doi.org/10.1007/978-3-030-00937-3_12

19. Valderrama, N., et al.: Towards holistic surgical scene understanding. In: Wang, L., Dou, Q., Fletcher, P.T., Speidel, S., Li, S. (eds.) MICCAI 2022. Lecture Notes in Computer Science, vol. 13437, pp. 442–452. Springer, Cham (2022). https://doi.org/10.1007/978-3-031-16449-1_42

20. Varier, V.M., Rajamani, D.K., Goldfarb, N., Tavakkolmoghaddam, F., Munawar, A., Fischer, G.S.: Collaborative suturing: a reinforcement learning approach to automate hand-off task in suturing for surgical robots. In: 2020 29th IEEE International Conference on Robot and Human Interactive Communication (RO-MAN), pp. 1380–1386. IEEE (2020). https://doi.org/10.1109/RO-MAN47096.2020.9223543

21. Wu, J.Y., Tamhane, A., Kazanzides, P., Unberath, M.: Cross-modal self-supervised representation learning for gesture and skill recognition in robotic surgery. Int. J. Comput. Assist. Radiol. Surg. **16**(5), 779–787 (2021). https://doi.org/10.1007/s11548-021-02343-y
22. Zhang, B., et al.: Towards accurate surgical workflow recognition with convolutional networks and transformers. Comput. Meth. Biomech. Biomed. Eng. Imaging Vis. **10**(4), 349–356 (2022). https://doi.org/10.1080/21681163.2021.2002191
23. Zisimopoulos, O., et al.: DeepPhase: surgical phase recognition in CATARACTS videos. In: Frangi, A.F., Schnabel, J.A., Davatzikos, C., Alberola-López, C., Fichtinger, G. (eds.) MICCAI 2018. LNCS, vol. 11073, pp. 265–272. Springer, Cham (2018). https://doi.org/10.1007/978-3-030-00937-3_31

TCL: Triplet Consistent Learning for Odometry Estimation of Monocular Endoscope

Hao Yue[1,2,3] and Yun Gu[1,2,3(✉)]

[1] Institute of Image Processing and Pattern Recognition,
Shanghai Jiao Tong University, Shanghai, China
{yuehao6,geron762}@sjtu.edu.cn
[2] Department of Automation, Shanghai Jiao Tong University, Shanghai, China
[3] Institute of Medical Robotics, Shanghai Jiao Tong University, Shanghai, China

Abstract. The depth and pose estimations from monocular images are essential for computer-aided navigation. Since the ground truth of depth and pose are difficult to obtain, the unsupervised training method has a broad prospect in endoscopic scenes. However, endoscopic datasets lack sufficient diversity of visual variations, and appearance inconsistency is also frequently observed in *image triplets*. In this paper, we propose a triplet-consistency-learning framework (TCL) consisting of two modules: Geometric Consistency module(GC) and Appearance Inconsistency module(AiC). To enrich the diversity of endoscopic datasets, the GC module generates synthesis triplets and enforces geometric consistency via specific losses. To reduce the appearance inconsistency in the *image triplets*, the AiC module introduces a triplet-masking strategy to act on photometric loss. TCL can be easily embedded into various unsupervised methods without adding extra model parameters. Experiments on public datasets demonstrate that TCL effectively improves the accuracy of unsupervised methods even with limited number of training samples. Code is available at https://github.com/EndoluminalSurgicalVision-IMR/TCL.

Keywords: Self-supervised monocular pose estimation · Endoscopic images · Data augmentation · Appearance inconsistency

1 Introduction

The technical advances in endoscopes have extended the diagnostic and therapeutic value of endoluminal interventions in a wide range of clinical applications. Due to the restricted field of view, it is challenging to control the flexible endoscopes inside the lumen. Therefore, the development of navigation systems, which

---

Supplementary Information The online version contains supplementary material available at https://doi.org/10.1007/978-3-031-43996-4_14.

locates the position of the end tip of endoscopes and enables the depth-wise visualization, is essential to assisting the endoluminal interventions. A typical task is to visualize the depth-wise information and estimate the six-degree-of-freedom (6DoF) pose of endoscopic camera based on monocular imaging.

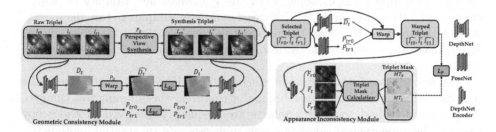

Fig. 1. Overview of the proposed triplet-consistency-learning framework (TCL) with two modules: Geometric Consistency module(GC) and Appearance Inconsistency module(AiC). The GC module utilizes the Perspective View Synthesis technique to produce Synthesis Triplets, while enforcing geometric consistency through the Depth Consistent Loss $\mathcal{L}_{dc}$ and the Pose Consistent Loss $\mathcal{L}_{pc}$. The AiC module generates Triplet Masks based on the Warped Triplet to apply to the photometric loss $\mathcal{L}_p$. TCL can be easily embedded into unsupervised SfM methods without adding extra model parameters.

Due to the clinical limitations, the ground truth of depth and pose trajectory of endoscope imaging is difficult to acquire. Previous works jointly estimated the depth and the pose via the unsupervised frameworks [2,8,10,12,17]. The main idea is modeling the differences of video frames with the Structure-from-Motion (SfM) mechanisms. In this framework [17], the *image triplet*, including a specific frame (i.e. *target frame*) and its temporal neighborhood (i.e. *reference frames*), is fed into the model to estimate the pose and depth. The model is optimized by minimizing the warping loss between the target frame and reference frames. Following the basic SfM method [17], scale-consistency [2], auto-masking [5], cost-volume [14] and optical flows [15] are also introduced to further improve the performance. These methods have also been applied to endoscopic scenarios with specific designs of attention modules [10] and priors from sparse depth [8]. Although the performance of SfM methods is promising, the intrinsic challenges of endoscopic data still require further consideration.

The first issue is the insufficient visual diversity of endoscopic datasets. Compared with the large-scaled KITTI dataset [4] for general vision tasks, the collection of endoscopic datasets is challenged by the limited freedom of instruments and the scenarios. In this case, the visual variations of lumen structures and texture appearances cannot be fully explored. While road datasets mostly exhibit 3DOF motion(2DOF translation and 1DOF rotation in the road plane), endoscopy involves 6DOF motion within 3D anatomical structures. Therefore,

SfM algorithms for endoscopic imaging are designed to estimate complicated trajectories with limited data diversity. General data augmentation methods, including random flipping and cropping, cannot generate synthesis with sufficient variance of camera views, and the realistic camera motion cannot be fully guaranteed. Recent works have tried to mimic the camera motion to improve the diversity of samples. For example, PDA [16] generated new samples for supervised depth estimation, and 3DCC [6] used synthetic samples to test the robustness of the models. However, the perspective view synthesis for unsupervised SfM framework, especially the transformation on *image triplets* is still underexplored.

The second issue is the appearance inconsistency in *image triplets*. Due to the complicated environment of endoluminal structures, the illumination changes, motion blurness and specular artefacts are frequently observed in endoscopy images. The appearance-inconsistent area may generate substantial photometric losses even in the well-aligned adjacent frames. These photometric losses caused by the inconsistent appearance impede the training process and remain unable to optimize. To handle this problem, AF-SfMLearner [12] adopted the flow network to predict appearance flow to correct inconsistency between consecutive frames. RNNSLAM [9] used an encoder network to predict masks with supervised HSV signals from the original images. Consequently, these methods adopted auxiliary modules to handle the visual inconsistency, involving more parameters to learn.

In this paper, we propose a triplet-consistency-learning framework (TCL) for unsupervised depth and pose estimation of monocular endoscopes. To improve the visual diversity of *image triplets*, the perspective view synthesis is introduced, considering the geometric consistency of camera motion. Specifically, the depth-consistency and pose-consistency are preserved via specific losses. To reduce the appearance inconsistency in the *image triplets*, a triplet-masking strategy is proposed by measuring the differences between the triplet-level and the frame-level representations. The proposed framework does not involve additional model parameters, which can be easily embedded into previous SfM methods. Experiments on public datasets demonstrate that TCL can effectively improve the accuracy of depth and pose estimation even with small amounts of training samples.

2   Methodology

2.1   Unsupervised SfM with Triplet Consistency Learning

Unsupervised SfM Method. The unsupervised SfM methods adopt Depth-Net and PoseNet to predict the depth and the pose, respectively. With depth and pose prediction, the reference frames $I_{ri}(i = 0, 1)$ are warped to the warped frames $I_{ti}(i = 0, 1)$. The photometric loss, denoted by $\mathcal{L}_P$, is introduced to measure the differences between $I_{ti}(i = 0, 1)$ and the target frame I_t. The loss can be implemented by L_1 norm [17] and further improved with SSIM metrics [13].

In addition to $\mathcal{L}_P$, auxiliary regularization loss functions, such as depth map smoothing loss [5] and depth scale consistency loss [2], are also introduced to improve the performance. In this work, these regularization terms are denoted by $\mathcal{L}_{Reg}$. Therefore, the final loss functions $\mathcal{L}$ of unsupervised methods can be summarized as follows:

$$\mathcal{L} = \mathcal{L}_P + \mathcal{L}_{Reg} \tag{1}$$

Previous unsupervised methods [2,5,17] have achieved excellent results on realistic road datasets such as KITTI dataset [4]. However, endoscopic datasets lack sufficient diversity of visual variations, and appearance inconsistency is also frequently observed in *image triplets*. Therefore, the unsupervised SfM methods based on $\mathcal{L}_p$ require further considerations to address the issues above.

Framework Architecture. As shown in Fig. 1, we propose a triplet-consistency-learning framework (TCL) based on unsupervised SfM with *image triplets*. TCL can be easily embedded into SfM variants without adding model parameters. To enrich the diversity of endoscopic samples, the Geometric Consistency module (GC) performs the perspective view synthesis method to generate synthesis triplets. Additionally, we introduce the Depth Consistent Loss $\mathcal{L}_{dc}$ and the Pose Consistent Loss $\mathcal{L}_{pc}$ to preserve the depth-consistency and the pose-consistency between raw and synthesis triplets. To reduce the affect of appearance inconsistency in the triplet, we propose Appearance Inconsistency module (AiC) where the Triplet Masks of reference frames are generated to reduce the inconsistent warping in the photometric loss.

2.2   Learning with Geometric Consistency

Since the general data augmentation methods cannot generate synthesis with sufficient variance of camera views, 3DCC [6] and PDA [16] mimic the camera motion to generate the samples by applying perspective view synthesis. However, raw and novel samples are used separately in the previous works. To enrich the diversity of endoscopic datasets, we perform the synthesis on triplets. Furthermore, the loss functions are introduced to preserve the depth-consistency and the pose-consistency.

Synthesis Triplet Generation. The perspective view synthesis method aims to warp the original image I to generate a new image I'. The warping process is based on the camera intrinsic matrix K, the depth map D of the original image and perturbation pose P_0. For any point q in I, its depth value is denoted as z in D. The corresponding point q' on the new image I' is calculated by Eq.(2):

$$q' \sim K P_0 z K^{-1} q \tag{2}$$

Given the depth maps generated from a pre-trained model, we perform perspective view synthesis with the same pose transformation P_0 on the three frames of raw triplet respectively. Then we obtain the synthesis triplet $[I'_{r0}, I'_t, I'_{r1}]$. Selected triplet $[\widehat{I_{r0}}, \widehat{I_t}, \widehat{I_{r1}}]$ is randomly selected from raw and synthesis triplets as the unsupervised training triplet to calculate $\mathcal{L}$ in Eq.(1).

Depth Consistent Loss. Figure 1 illustrates that the depth prediction of raw target frame I_t is D_t, and the depth prediction of synthesis target frame I'_t is D'_t. The relative pose between the raw and synthesis triplets is randomly generated as P_0. With P_0, we can warp the depth prediction D_t to $\widehat{D'_t}$. To preserve depth-consistency between target frames of raw and synthesis triplet, we propose the Depth Consistent Loss function $\mathcal{L}_{dc}$, defined as the L1 loss function of D'_t and $\widehat{D'_t}$, written as

$$\mathcal{L}_{dc} = |D'_t - \widehat{D'_t}| \tag{3}$$

Pose Consistent Loss. As shown in Fig. 1, the inner pose predictions of the adjacent frames of the raw and synthesis triplet are $P_{tri}(i = 0, 1)$ and $P'_{tri}(i = 0, 1)$. Since the three frames of the synthesis triplet are warped from the same pose perturbation P_0, the inner poses of the synthesis triplet remain the same as the raw triplet. To preserve inner pose-consistency between raw and synthesis triplets, we propose Pose Consistent Loss $\mathcal{L}_{pc}$, defined as the weighted L1 loss function of the inner pose prediction of the raw and synthesis triplets. Specifically, we use the translational $t_{tri}, t'_{tri} \in \mathcal{R}^{3 \times 1}(i = 0, 1)$ and rotational $R_{tri}, R'_{tri} \in \mathcal{R}^{3 \times 3}(i = 0, 1)$ components of the pose transformation matrix to calculate $\mathcal{L}_{pc}$. We weight the translation component by λ_{pt}, Pose Consistent Loss is written as Eq.(4).

$$\mathcal{L}_{pc} = \sum_{i=0,1} (|R_{tri} - R'_{tri}| + \lambda_{pt}|t_{tri} - t'_{tri}|) \tag{4}$$

2.3   Learning with Appearance Inconsistency

The complicated environment of endoluminal structures may cause appearance inconsistency in *image triplets*, leading to the misalignment of reference and target frames. Previous works [9,12] proposed auxiliary modules to handle the appearance inconsistency, involving more parameters of models.

Since the movement of camera is normally slow, the same appearance inconsistency is unlikely to exist multiple times within an endoscopic *image triplet*. Therefore, we can measure the differences between the triplet-level and the frame-level representations to eliminate the appearance inconsistency.

Specifically, for the selected triplet $[\widehat{I_{r0}}, \widehat{I_t}, \widehat{I_{r1}}]$, two warped frames $[\widehat{I_{t0}}, \widehat{I_{t1}}]$ are generated from two reference frames. To obtain the frame-level representations, we used the encoder of DepthNet to extract the feature maps of the three frames $[\widehat{I_{t0}}, \widehat{I_t}, \widehat{I_{t1}}]$ respectively. The feature maps are upsampled to the size of the original image, denoted by $[F_{r0}, F_t, F_{r1}]$. As in Eq.(5), the triplet-level representations F_R are generated by the weighted aggregation of the feature maps, which is dominated by the feature of the target frame with weight λ_t. To measure the differences between the triplet-level and the frame-level representations, we calculate feature difference maps $Df_i(i = 0, 1)$ by weighting direct subtraction

Table 1. Quantitative results on the SCARED and SERV-CT Dataset. The best results among all methods are in **bold**. The best results among each series are underlined.

Series	Methods	SCARED Dataset									
		Pose Metrics			Depth Metrics						
		ATE(mm)↓	t_{RPE}(mm)↓	r_{RPE}(deg)↓	Abs Rel↓	Sq Rel↓	RMSE↓	RMSE log↓	$\delta<1.25$↑	$\delta<1.25^2$↑	$\delta<1.25^3$↑
	SfMLearner	6.308	$0.356_{\pm0.18}$	$0.217_{\pm0.12}$	$0.472_{\pm0.07}$	$7.870_{\pm5.52}$	$14.024_{\pm5.55}$	$0.499_{\pm0.05}$	$0.365_{\pm0.10}$	$0.636_{\pm0.10}$	$0.810_{\pm0.06}$
MonoDepth2-Based	MonoDepth2	4.848	$0.479_{\pm0.20}$	$0.459_{\pm0.22}$	$0.454_{\pm0.06}$	$7.311_{\pm2.87}$	$13.583_{\pm5.02}$	$0.487_{\pm0.05}$	$0.368_{\pm0.10}$	$0.650_{\pm0.11}$	$0.818_{\pm0.06}$
	AF-SfMLearner	3.506	$\underline{0.161}_{\pm0.10}$	$0.265_{\pm0.14}$	$\underline{0.446}_{\pm0.06}$	$\underline{7.153}_{\pm3.02}$	$13.517_{\pm5.16}$	$0.481_{\pm0.05}$	$0.371_{\pm0.10}$	$0.651_{\pm0.11}$	$0.825_{\pm0.06}$
	MonoDepth2+Ours	$\underline{3.110}$	$\underline{0.161}_{\pm0.11}$	$\underline{0.250}_{\pm0.15}$	$\underline{0.446}_{\pm0.06}$	$7.185_{\pm3.30}$	$\underline{13.405}_{\pm5.20}$	$\underline{0.480}_{\pm0.05}$	$\underline{0.373}_{\pm0.11}$	$\underline{0.655}_{\pm0.12}$	$\underline{0.828}_{\pm0.06}$
SC-SfMLearner-Based	SC-SfMLearner	4.743	$\underline{0.478}_{\pm0.21}$	$0.466_{\pm0.24}$	$0.442_{\pm0.06}$	$7.044_{\pm2.96}$	$13.580_{\pm5.42}$	$0.479_{\pm0.04}$	$\underline{0.368}_{\pm0.11}$	$0.650_{\pm0.12}$	$0.826_{\pm0.06}$
	Endo-SfMLearner	5.013	$0.494_{\pm0.22}$	$\underline{0.461}_{\pm0.24}$	$0.438_{\pm0.06}$	$6.969_{\pm3.34}$	$13.592_{\pm5.46}$	$0.478_{\pm0.05}$	$0.365_{\pm0.10}$	$0.650_{\pm0.11}$	$0.826_{\pm0.06}$
	SC-SfMLearner+Ours	$\underline{4.601}$	$0.490_{\pm0.22}$	$0.464_{\pm0.24}$	$\mathbf{\underline{0.437}}_{\pm0.05}$	$\mathbf{\underline{6.865}}_{\pm2.93}$	$\underline{13.471}_{\pm5.32}$	$\mathbf{\underline{0.475}}_{\pm0.04}$	$\underline{0.368}_{\pm0.11}$	$\underline{0.653}_{\pm0.12}$	$\mathbf{\underline{0.831}}_{\pm0.05}$

Series	Methods	SERV-CT Dataset									
		Pose Metrics			Depth Metrics						
		ATE(mm)↓	t_{RPE}(mm)↓	r_{RPE}(deg)↓	Abs Rel↓	Sq Rel↓	RMSE↓	RMSE log↓	$\delta<1.25$↑	$\delta<1.25^2$↑	$\delta<1.25^3$↑
	SfMLearner	-	-	-	$0.114_{\pm0.04}$	$2.005_{\pm1.42}$	$12.632_{\pm6.05}$	$0.149_{\pm0.05}$	$0.864_{\pm0.12}$	$0.985_{\pm0.02}$	$\mathbf{1.000}_{\pm0.00}$
MonoDepth2-Based	MonoDepth2	-	-	-	$0.124_{\pm0.04}$	$2.298_{\pm1.38}$	$13.639_{\pm8.78}$	$0.165_{\pm0.04}$	$0.843_{\pm0.10}$	$0.976_{\pm0.04}$	$0.998_{\pm0.01}$
	AF-SfMLearner	-	-	-	$0.115_{\pm0.04}$	$2.101_{\pm1.56}$	$13.136_{\pm6.74}$	$0.156_{\pm0.05}$	$0.860_{\pm0.10}$	$0.979_{\pm0.03}$	$0.998_{\pm0.01}$
	MonoDepth2+Ours	-	-	-	$\underline{0.103}_{\pm0.03}$	$\mathbf{\underline{1.694}}_{\pm1.20}$	$\mathbf{\underline{11.711}}_{\pm5.89}$	$\mathbf{\underline{0.139}}_{\pm0.04}$	$\underline{0.886}_{\pm0.09}$	$\mathbf{\underline{0.986}}_{\pm0.02}$	$\mathbf{1.000}_{\pm0.00}$
SC-SfMLearner-Based	SC-SfMLearner	-	-	-	$0.113_{\pm0.04}$	$2.417_{\pm2.21}$	$13.719_{\pm6.34}$	$0.177_{\pm0.08}$	$0.872_{\pm0.09}$	$0.959_{\pm0.05}$	$0.980_{\pm0.03}$
	Endo-SfMLearner	-	-	-	$0.133_{\pm0.05}$	$3.295_{\pm2.88}$	$15.974_{\pm9.23}$	$0.224_{\pm0.10}$	$0.829_{\pm0.15}$	$0.928_{\pm0.07}$	$0.961_{\pm0.05}$
	SC-SfMLearner+Ours	-	-	-	$\underline{0.103}_{\pm0.04}$	$\underline{1.868}_{\pm1.50}$	$\underline{12.199}_{\pm6.71}$	$\underline{0.150}_{\pm0.06}$	$\mathbf{\underline{0.888}}_{\pm0.09}$	$\underline{0.970}_{\pm0.04}$	$\underline{0.996}_{\pm0.01}$

and SSIM similarity [13] with weight λ_{sub}. The Triplet Mask of each reference frame is generated by reverse normalizing the difference map to $[\beta,1]$.

$$F_R = \lambda_t F_t + \frac{1}{2}(1 - \lambda_t)(F_{r0} + F_{r1})$$

$$Df_i = \lambda_{sub} N_{(0,1)}(|F_R - F_{ri}|)$$
$$+ (1 - \lambda_{sub}) N_{(0,1)}(|1 - \text{SSIM}(F_R, F_{ri})|), (i = 0, 1) \tag{5}$$

$$MT_i = N_{(\beta,1)}(1 - Df_i), (i = 0, 1)$$

where $N_{(a,b)}(\cdot)$ normalizes the input to the range $[a,b]$.

2.4 Overall Loss

The final loss of TCL $\mathcal{L}_t$ is formulated as follows:

$$\mathcal{L}_t = MT \odot \mathcal{L}_P + \mathcal{L}_{Reg} + \lambda_d \mathcal{L}_{dc} + \lambda_p \mathcal{L}_{pc} \tag{6}$$

where $\odot$ denotes that the Triplet Mask $MT_i(i = 0, 1)$ is applied to the photometric loss calculation of the two reference frames respectively. The final photometric loss is obtained by averaging the photometric losses of the two reference frames after applying $MT_i(i = 0, 1)$. λ_d, λ_p are weights of $\mathcal{L}_{dc}$ and $\mathcal{L}_{pc}$. Since the early adoption of $\mathcal{L}_{dc}$ and $\mathcal{L}_{pc}$ may lead to overfitting, the DepthNet and PoseNet are warmed up with N_w epochs before adding the two loss functions. The synthesis method may inherently generate invalid (black) areas in the augmented samples. This arises from the single-image-based augmentation process, which lacks the additional information to fill the new areas generated from the viewpoint transformation. The invalid regions should be masked in the related loss functions.

3 Experiments

Dataset and Implementation Details. The public datasets, including SCARED [1] with ground truth of both depth and pose and SERV-CT [3] with only depth ground truth, were used to evaluate the proposed method. Following the settings in [12], we trained on SCARED and tested on SCARED and SERV-CT. The depth metrics (*Abs Rel, Sq Rel, RMSE, RMSE log and* δ), and pose metrics (*ATE,* t_{RPE}, r_{RPE}) were used to measure the difference of predictions and the ground truth[1]. ATE is noted as the weighted average of the RMSE of sub-trajectories, and the rest metrics are noted as Mean $\pm$ Standard Deviation of error.

We implemented networks using PyTorch [11] and trained the networks on 1 NVIDIA RTX 3090 GPU with Adam [7] for 100 epochs with a learning rate of $1e^{-4}$, dropped by a scale factor of 10 after 10 epochs. Given the SCARED dataset, we divided 5/2/2 subsets for training/validation/testing. We finally obtained $\sim$ 8k frames for training. The batch size was 12 and all images were downsampled to 256×320. $\lambda_{pt}, \lambda_t, \lambda_{sub}, \beta, \lambda_d, \lambda_p, N_w$ in the loss function were empirically set to $1, 0.5, 0.2, 0.5, 0.001, 0.5, 5$ which were tuned on validation set. For more details of the experiments, please refer to the supplementary material.

Results. To evaluate the effectiveness of the proposed method, TCL was applied to MonoDepth2 [5] and SC-SfMLearner [2] by exactly using the same architectures of DepthNet and PoseNet. For comparisons, SfMLearner [17], MonoDepth2 [5], AF-SfMLearner [12], SC-SfMLearner [2] and Endo-SfMLearner [10] were adopted as baseline methods. Specifically, AF-SfMLearner improved the MonoDepth2 by predicting the appearance flow, while Endo-SfMLearner is the alternative of SC-SfMLearner with better model architectures.

Table 1 presents the quantitative results on SCARED and SERV-CT. After TCL is applied to the series baselines(MonoDepth2 and SC-SfMLearner), most depth and pose metrics are significantly improved, and most metrics achieve the best performance in their series. Our method not only outperforms the series baseline on SERV-CT, but also achieves all the best values of the MonoDepth2-Based and SC-SfMLearner-Based series. For visualization and further ablations, we present the results of TCL applied to MonoDepth2, which is denoted as Ours below. Figure 2(a) presents the comparisons of the depth prediction. In the first image, our method predicts the depth most accurately. In the second image, the ground truth reveals that the dark area in the upper right corner is closer, and only our method accurately predicts this area, resulting in a detailed and accurate full-image depth map. Despite the improvement, our prediction results still remain inaccurate compared to the ground truths. Figure 2(b) visualizes the trajectory and the projection on three planes. Our predicted trajectory is the closest to the ground truth compared with baselines. The SfMlearner predicts almost all trajectories as straight lines, a phenomenon also observed in [10], in which case a low r_{RPE} metric is ineffective.

---

[1] The detailed implementations of the metrics can be found in [2,10].

Ablations on Proposed Modules. We introduce two proposed modules to the baseline (MonoDepth2) separately. We additionally propose a simple version of AiC that computes Triplet Masks directly using two reference frames without warping, denoted as AiC*. From Table 2, GC and AiC can significantly improve the effect of the baseline when introduced separately. AiC outperforms AiC* as the warping process can provide greater benefits for pixel-level alignment. Figure 3(a) intuitively demonstrates the effect of the proposed Triplet Mask(AiC), which effectively covers pixels with apparent appearance inconsistency between reference and target frames.

Ablations on Different Dataset Amounts. To verify the effect of our proposed method on different amounts of training and validation sets, we utilize the main depth metric RMSE and pose metric ATE for comparison. In Fig. 3(b), our method achieves significant improvements over MonoDepth2 with different dataset amounts. The performance of our approach is almost optimal for depth and pose at 11k and 8k training samples, respectively. Therefore, our proposed framework has the potential to effectively enhance the performance of various unsupervised SfM methods, even with limited training data.

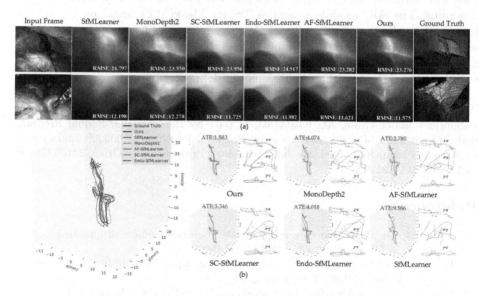

Fig. 2. Qualitative depth and pose results on the SCARED dataset.

Table 2. Ablation results of proposed modules on the SCARED dataset.

Settings	Module Ablations									
	Pose Metrics			Depth Metrics						
	ATE(mm)↓	t_{RPE}(mm)↓	r_{RPE}(deg)↓	Abs Rel↓	Sq Rel↓	RMSE↓	RMSE log↓	$\delta < 1.25$ ↑	$\delta < 1.25^2$ ↑	$\delta < 1.25^3$ ↑
MonoDepth2	4.848	0.479±0.30	0.459±0.22	0.454±0.06	7.311±2.87	13.583±5.02	0.487±0.05	0.368±0.10	0.650±0.11	0.818±0.06
AF-SfMLearner	3.506	0.161±0.10	0.265±0.14	**0.446**±0.06	**7.153**±3.02	13.517±5.16	0.481±0.05	0.371±0.10	0.651±0.11	0.825±0.06
Ours w/GC Only	3.334	**0.160**±0.11	0.266±0.15	0.449±0.06	7.314±3.25	13.519±5.84	0.485±0.05	0.370±0.10	0.651±0.12	0.824±0.06
Ours w/AiC Only	3.121	0.161±0.10	**0.244**±0.14	0.447±0.06	7.179±2.92	13.551±4.91	0.482±0.05	0.368±0.11	0.653±0.11	0.824±0.06
Ours w/AiC* Only	3.518	0.162±0.11	0.269±0.14	0.451±0.06	7.445±2.93	13.623±4.96	0.488±0.05	0.368±0.11	0.649±0.11	0.822±0.06
Ours	**3.110**	0.161±0.11	0.250±0.26	**0.446**±0.06	7.185±3.30	**13.405**±5.20	**0.480**±0.05	**0.373**±0.11	**0.655**±0.12	**0.828**±0.06

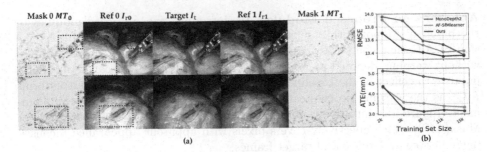

Mask 0 MT_0 Ref 0 I_{r0} Target I_t Ref 1 I_{r1} Mask 1 MT_1

(a)   (b)

Fig. 3. Qualitative results of Triplet Mask and ablation results on different dataset amounts.

4   Conclusion

We present a triplet-consistency-learning framework (TCL) to improve the effect of monocular endoscopy unsupervised depth and pose estimation. The GC module generates synthesis triplets to increase the diversity of the endoscopic samples. Furthermore, we constrain the depth and pose consistency using two loss functions. The AiC module generates Triplet Mask(MT) based on the triplet information. MT can effectively mask the appearance inconsistency in the triplet, which leads to more efficient training of the photometric loss. Extensive experiments demonstrate the effectiveness of TCL, which can be easily embedded into various SfM methods without additional model parameters.

Acknowledgement. This work is supported in part by the Open Funding of Zhejiang Laboratory under Grant 2021KH0AB03, in part by the Shanghai Sailing Program under Grant 20YF1420800, and inpart by NSFC under Grant 62003208, and in part by Shanghai Municipal of Science and Technology Project, under Grant 20JC1419500 and Grant 20DZ2220400.

References

1. Allan, M., et al.: Stereo correspondence and reconstruction of endoscopic data challenge. arXiv preprint arXiv:2101.01133 (2021)
2. Bian, J.W., et al.: Unsupervised scale-consistent depth learning from video. Int. J. Comput. Vision **129**(9), 2548–2564 (2021)
3. Edwards, P.E., Psychogyios, D., Speidel, S., Maier-Hein, L., Stoyanov, D.: SERV-CT: a disparity dataset from cone-beam CT for validation of endoscopic 3D reconstruction. Med. Image Anal. **76**, 102302 (2022)
4. Geiger, A., Lenz, P., Stiller, C., Urtasun, R.: Vision meets robotics: the kitti dataset. Int. J. Robot. Res. (IJRR) **32**, 1231–1237 (2013)
5. Godard, C., Mac Aodha, O., Firman, M., Brostow, G.J.: Digging into self-supervised monocular depth estimation. In: Proceedings of the IEEE/CVF International Conference on Computer Vision, pp. 3828–3838 (2019)
6. Kar, O.F., Yeo, T., Atanov, A., Zamir, A.: 3D common corruptions and data augmentation. In: Proceedings of the IEEE/CVF Conference on Computer Vision and Pattern Recognition, pp. 18963–18974 (2022)

7. Kingma, D.P., Ba, J.: Adam: a method for stochastic optimization. arXiv preprint arXiv:1412.6980 (2014)
8. Liu, X., et al.: Dense depth estimation in monocular endoscopy with self-supervised learning methods. IEEE Trans. Med. Imaging **39**(5), 1438–1447 (2019)
9. Ma, R., et al.: RNNSLAM: reconstructing the 3D colon to visualize missing regions during a colonoscopy. Med. Image Anal. **72**, 102100 (2021)
10. Ozyoruk, K.B., et al.: EndoSLAM dataset and an unsupervised monocular visual odometry and depth estimation approach for endoscopic videos. Med. Image Anal. **71**, 102058 (2021)
11. Paszke, A., et al.: Automatic differentiation in pytorch (2017)
12. Shao, S., et al.: Self-supervised monocular depth and ego-motion estimation in endoscopy: appearance flow to the rescue. Med. Image Anal. **77**, 102338 (2022)
13. Wang, Z., Bovik, A.C., Sheikh, H.R., Simoncelli, E.P.: Image quality assessment: from error visibility to structural similarity. IEEE Trans. Image Process. **13**(4), 600–612 (2004)
14. Watson, J., Aodha, O.M., Prisacariu, V., Brostow, G., Firman, M.: The temporal opportunist: self-supervised multi-frame monocular depth. In: Computer Vision and Pattern Recognition (CVPR) (2021)
15. Zhao, W., Liu, S., Shu, Y., Liu, Y.J.: Towards better generalization: joint depth-pose learning without posenet. In: Proceedings of the IEEE/CVF Conference on Computer Vision and Pattern Recognition, pp. 9151–9161 (2020)
16. Zhao, Y., Kong, S., Fowlkes, C.: Camera pose matters: improving depth prediction by mitigating pose distribution bias. In: Proceedings of the IEEE/CVF Conference on Computer Vision and Pattern Recognition, pp. 15759–15768 (2021)
17. Zhou, T., Brown, M., Snavely, N., Lowe, D.G.: Unsupervised learning of depth and ego-motion from video. In: Proceedings of the IEEE Conference on Computer Vision and Pattern Recognition, pp. 1851–1858 (2017)

Regressing Simulation to Real: Unsupervised Domain Adaptation for Automated Quality Assessment in Transoesophageal Echocardiography

Jialang Xu[1]([📧])(iD), Yueming Jin[4], Bruce Martin[2], Andrew Smith[2], Susan Wright[3], Danail Stoyanov[1](iD), and Evangelos B. Mazomenos[1]([📧])(iD)

[1] UCL Wellcome/EPSRC Centre for Interventional and Surgical Sciences, Department of Medical Physics and Biomedical Engineering, University College London, London, UK
{jialang.xu.22,e.mazomenos}@ucl.ac.uk
[2] St Bartholomew's Hospital, NHS Foundation Trust, London, UK
[3] St George's University Hospitals, NHS Foundation Trust, London, UK
[4] Department of Biomedical Engineering and Department of Electrical and Computer Engineering, National University of Singapore, Singapore, Singapore

Abstract. Automated quality assessment (AQA) in transoesophageal echocardiography (TEE) contributes to accurate diagnosis and echocardiographers' training, providing direct feedback for the development of dexterous skills. However, prior works only perform AQA on simulated TEE data due to the scarcity of real data, which lacks applicability in the real world. Considering the cost and limitations of collecting TEE data from real cases, exploiting the readily available simulated data for AQA in real-world TEE is desired. In this paper, we construct the first simulation-to-real TEE dataset, and propose a novel Simulation-to-Real network (SR-AQA) with unsupervised domain adaptation for this problem. It is based on uncertainty-aware feature stylization (UFS), incorporating style consistency learning (SCL) and task-specific learning (TL), to achieve high generalizability. Concretely, UFS estimates the uncertainty of feature statistics in the real domain and diversifies simulated images with style variants extracted from the real images, alleviating the domain gap. We enforce SCL and TL across different real-stylized variants to learn domain-invariant and task-specific representations. Experimental results demonstrate that our SR-AQA outperforms state-of-the-art methods with 3.02% and 4.37% performance gain in two AQA regression tasks, by using only 10% unlabelled real data. Our code and dataset are available at https://doi.org/10.5522/04/23699736.

Supplementary Information The online version contains supplementary material available at https://doi.org/10.1007/978-3-031-43996-4_15.

Keywords: Automated quality assessment · Unsupervised domain adaptation regression · Transoesophageal echocardiography · Uncertainty · Consistency learning · Style transfer

1 Introduction

Transoesophageal echocardiography (TEE) is a valuable diagnostic and monitoring imaging modality with widespread use in cardiovascular surgery for anaesthesia management and outcome assessment, as well as in emergency and intensive care medicine. The quality of TEE views is important for diagnosis and professional organisations publish guidelines for performing TEE exams [5,22]. These guidelines standardise TEE view acquisition and set benchmarks for the education of new echocardiographers. Computational methods for automated quality assessment (AQA) will have great impact, guaranteeing quality of examinations and facilitating training of new TEE operators.

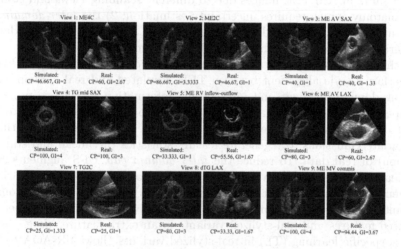

Fig. 1. Overview of our proposed dataset. The 9 TEE views in our dataset: 1: Mid-Esophageal 4-Chamber, 2: Mid-Esophageal 2-Chamber, 3: Mid-Esophageal Aortic Valve Short-Axis, 4: Transgastric Mid-Short-Axis, 5: Mid-Esophageal Right Ventricle inflow-outflow, 6: Mid-Esophageal Aortic Valve Long-Axis, 7: Transgastric 2-Chamber, 8: Deep Transgastric Long-Axis, 9: Mid-Esophageal Mitral Commissural

Deep models for AQA have shown promise in transthoracic echocardiography (TTE) and other ultrasound (US) applications [1,9,13,14]. Investigation of such methods in the real TEE domain remains underexplored and has been restricted to simulated datasets from Virtual Reality (VR) systems [15]. Although VR technology is useful for developing and retaining US scanning skills [7,10,16, 17,20], AQA methods developed on simulation settings cannot meet real-world usability without addressing the significant domain gap. As shown in Fig. 1, there are significant content differences between simulated and real TOE images. Simulated images are free of speckle noise and contain only the heart muscle, ignoring tissue in the periphery of the heart. In this work, we take the first step in exploring AQA in the real TEE domain. We propose to leverage readily accessible simulated data, and transfer knowledge learned in the simulated domain, to boost performance in real TEE space.

To alleviate the domain mismatch, a feasible solution is unsupervised domain adaptation (UDA). UDA aims to increase the performance on the target domain by using labelled source data with unlabelled target data to reduce the domain shift. For example, Mixstyle [24], performs style regularization by mixing instance-level feature statistics of training samples. The most relevant UDA work for our AQA regression task is representation subspace distance (RSD) [4], which aligns features from simulated and real domains via representation sub-spaces. Despite its effectiveness in several tasks, performing UDA on the simulation-to-real AQA task of TEE has two key challenges that need to be addressed. From Fig. 1, it is evident that: 1) there are many unknown *intra-domain* shifts in real TEE images due to different scanning views and complex heart anatomy, which requires uncertainty estimation; 2) the *inter-domain* gap (heart appearance, style, and resolution) between simulated and real data is considerable, necessitating robust, domain-invariant features.

In this paper, we propose a novel UDA regression network named SR-AQA that performs style alignment between TEE simulated and real domains while retaining domain-invariant and task-specific information to achieve promising AQA performance. To estimate the uncertainty of *intra-domain* style offsets in real data, we employ uncertainty-aware feature stylization (UFS) utilizing multivariate Gaussians to regenerate feature statistics (i.e. mean and standard deviation) of real data. To reduce the *inter-domain gap*, UFS augments simulated features to resemble diverse real styles and obtain real-stylized variants. We then design a style consistency learning (SCL) strategy to learn domain-invariant representations by minimizing the negative cosine similarity between simulated features and real-stylized variants in an extra feature space. Enforcing task-specific learning (TL) in real-stylized variants allows SR-AQA to keep task-specific information useful for AQA. Our work represents the original effort to address the TEE domain shift in AQA tasks. For method evaluation, we present the *first* simulation-to-real TEE dataset with two AQA tasks (see Fig. 1), and benchmark the proposed SR-AQA model against four state-of-the-art UDA methods. Our proposed SR-AQA outperforms other UDA methods, achieving 2.13%–5.08% and 4.37%–16.28% mean squared error (MSE) reduction for two AQA tasks, respectively.

2 Methodology

2.1 Dataset Collection

We collected a dataset of 16,192 simulated and 4,427 real TEE images from 9 standard views. From Fig. 1, it is clear that significant style differences (e.g.

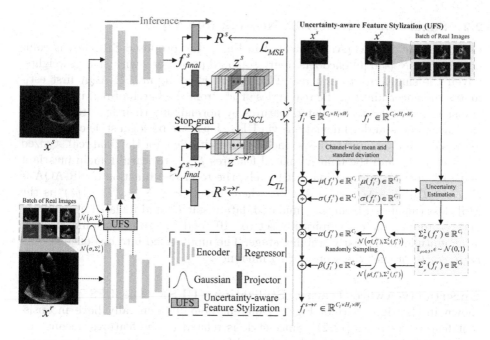

Fig. 2. Overall architecture of our proposed SR-AQA. The right part shows uncertainty-aware feature stylization (UFS) for one simulated image x^s with real images at l^{th} layer. No gradient calculation is performed on the dotted line. The red line indicates the path of model inference.

brightness, contrast, acoustic shadowing, and refraction artifact) exist between simulated and real data, posing a considerable challenge to UDA. Simulated images were collected with the HeartWorks TEE simulation platform from 38 participants of varied experience asked to image the 9 views. Fully anonymized real TEE data were collected from 10 cardiovascular procedures in 2 hospitals, with ethics for research use and collection approved by the respective Research Ethics Committees. Each image is annotated by 3 expert anaesthetists with two independent scores w.r.t. two AQA tasks for TEE. The criteria percentage (CP) score ranging from "0–100", measuring the number of essential criteria, from the checklists (provided in supplementary material) of the ASE/SCA/BSE imaging guidelines [5], met during image acquisition and a general impression (GI) score ranging from "0–4", representing overall US image quality.

As the number of criteria thus the maximum score varies for different views, we normalise CP as a percentage to provide a consistent measure across all views. Scores from the 3 raters were averaged to obtain the final score for each view. The Pearson product-moment correlation coefficients between CP and GI are 0.81 for simulated data and 0.70 for real data, indicating that these two metrics are correlated but focus on different clinical quality aspects. Inter-rater variability is assessed using the two-way mixed-effects interclass correlation coefficient (ICC) with the definition of absolute agreement. Both CP and GI, show very good agreement between the 3 annotators with ICCs of 0.959, 0.939 and 0.813, 0.758 for simulated and real data respectively.

2.2 Simulation-to-Real AQA Network (SR-AQA)

Overview of SR-AQA. Illustrated in Fig. 2, the proposed SR-AQA is composed of ResNet-50 [6] encoders, regressors and projectors, with shared weights. Given the simulated x^s and real TEE image x^r as input, SR-AQA first estimates the uncertainty in the real styles of x^s, from the batch of real images and transfers real styles to simulated features by normalizing their feature statistics (i.e. mean and standard deviation) via UFS. Then, we perform style consistency learning with $\mathcal{L}_{SCL}$ and task-specific learning with $\mathcal{L}_{TL}$ for the final real-stylized features $f^{s \to r}_{final}$ and the final simulated features f^s_{final} to learn domain-invariant and task-specific information. Ultimately, the total loss function of SR-AQA is $\mathcal{L}_{total} = \mathcal{L}_{MSE} + \lambda_1 \mathcal{L}_{SCL} + \lambda_2 \mathcal{L}_{TL}$, where $\mathcal{L}_{MSE} = \frac{1}{N} \sum_{i=1}^{N} (R^s_i - y^s_i)^2$ is the MSE loss calculated from the simulated data result R^s and its label y^s, while λ_1 and λ_2 are parameters empirically set to "10" and "1" to get a uniform order of magnitude at the early training stage. The input is fed into one encoder and regressor to predict the score during inference.

Uncertainty-Aware Feature Stylization (UFS). The UFS pipeline is shown in the right part of Fig. 2. Different domains generally have inconsistent feature statistics [12,21]. Since style is related to the features' means and standard deviations [2,8,11,24], simulated features can be augmented to resemble real styles by adjusting their mean and standard deviation with the help of unlabelled real data. Let f^s_l and f^r_l be the simulated features and real features extracted from the lth layer of the encoder, respectively. We thus can transfer the style of f^r_l to f^s_l to obtain real-stylized features $f^{s \to r}_l$ as:

$$f^{s \to r}_l = \sigma(f^r_l) \frac{f^s_l - \mu(f^s_l)}{\sigma(f^s_l)} + \mu(f^r_l), \tag{1}$$

where: $\mu(f)$ and $\sigma(f)$ denote channel-wise mean and standard deviation of feature f, respectively.

However, due to the complexity of real-world TEE, there are significant intra-domain differences, leading to uncertainties in the feature statistics of real data. To explore the potential space of unknown intra-domain shifts, instead of using fixed feature statistics, we generate multivariate Gaussian distributions to represent the uncertainty of the mean and standard deviation in the real data. Considering this, the new feature statistics of real features f^r_l, i.e. mean $\beta(f^r_l)$ and standard deviation $\alpha(f^r_l)$, are sampled from $\mathcal{N}\left(\mu(f^r_l), \Sigma^2_\mu(f^r_l)\right)$ and $\mathcal{N}\left(\sigma(f^r_l), \Sigma^2_\sigma(f^r_l)\right)$, respectively and computed as:

$$\beta(f^r_l) = \mu(f^r_l) + (\epsilon \Sigma_\mu(f^r_l)) \cdot \mathbb{I}_{\rho > 0.5}, \alpha(f^r_l) = \sigma(f^r_l) + (\epsilon \Sigma_\sigma(f^r_l)) \cdot \mathbb{I}_{\rho > 0.5}, \tag{2}$$

where: $\epsilon \sim \mathcal{N}(\mathbf{0}, \mathbf{1})^1$, variances $\Sigma^2_\mu(f^r_l) = \frac{1}{B} \sum_{b=1}^{B} (\mu(f^r_l) - \mathbb{E}_b[\mu(f^r_l)])^2$ and $\Sigma^2_\sigma(f^r_l) = \frac{1}{B} \sum_{b=1}^{B} (\sigma(f^r_l) - \mathbb{E}_b[\sigma(f^r_l)])^2$ are estimated from the mean and standard deviation of the batch B of real images, $\mathbb{I}_{\rho > 0.5}$ is an indicator function and

---

[1] Re-parameterization trick is applied here to make the sampling operation differentiable.

$\rho \sim \mathcal{U}(\mathbf{0}, \mathbf{1})$. Finally, our UFS for l^{th} layer is defined as:

$$f_l^{s \to r} = \alpha(f_l^r) \frac{f_l^s - \mu(f_l^s)}{\sigma(f_l^s)} + \beta(f_l^r). \tag{3}$$

To this end, the proposed UFS approach can close the reality gap by generating real-stylized features with sufficient variations, so that the network interprets real data as just another variation.

Style Consistency Learning (SCL). Through the proposed UFS, we obtain the final real-stylized features $f_{final}^{s \to r}$ that contain a diverse range of real styles. The $f_{final}^{s \to r}$ can be seen as style perturbations of the final simulated features f_{final}^s. We thus incorporate a SCL step, that maximizes the similarity between f_{final}^s and $f_{final}^{s \to r}$ to enforce their consistency in the feature level, allowing the encoder to learn robust representations. Specifically, the SCL adds a projector independently of the regressor to transform the f_{final}^s ($f_{final}^{s \to r}$) in an extra feature embedding, and then matches it to the other one. To prevent the Siamese encoder and Siamese projector (i.e. the top two encoders and projectors in Fig. 2) from collapsing to a constant solution, similar to [3], we adopt the stop-gradient (stopgrad) operation for the projected features z^s and $z^{s \to r}$. The SCL process is summarized as:

$$\mathcal{L}_{SCL} = \frac{1}{N} \sum_{i=1}^{N} \left(\frac{1}{2} \mathcal{D} \left(f_{final}^s, \text{stopgrad}\,(z^{s \to r}) \right) + \frac{1}{2} \mathcal{D} \left(f_{final}^{s \to r}, \text{stopgrad}\,(z^s) \right) \right), \tag{4}$$

where: $\mathcal{D}\left(f^1, f^2\right) = -\frac{f^1}{\|f^1\|_2} \cdot \frac{f^2}{\|f^2\|_2}$ is the negative cosine similarity between the input features f^1 and f^2, and $\|\cdot\|_2$ is $L2$-normalization.

The SCL guides the network to learn domain-invariant features, via various style perturbations, so that it can generalize well to the different visual appearances of the real domain.

Task-Specific Learning (TL). While alleviating the style differences between the simulated and real domain, UFS filters out some task-specific information (e.g. semantic content) encoded in the simulated features, as content and style are not orthogonal [11,19], resulting in performance deterioration. Therefore, we embed TL to retain useful representations for AQA. Specifically, $f_{final}^{s \to r}$ should retain task-specific information to allow the regressor to predict results $R^{s \to r}$ that correspond to the quality scores (CP, GI) in the simulated data. In TL, simulated labels y^s are used as the supervising signal:

$$\mathcal{L}_{TL} = \frac{1}{N} \sum_{i=1}^{N} (R_i^{s \to r} - y_i^s)^2. \tag{5}$$

The TL performs AQA tasks for style variants to complement the loss of information due to feature stylization.

3   Experiments

3.1   Experimental Settings

Experiments are implemented in PyTorch on a Tesla V100 GPU. The maximum training iteration is 40,000 with a batch size of 32. We adopted the SGD optimizer with a weight decay of 5e−4, a momentum of 0.9 and a learning rate of 1e-4. Input images are resized to 224×224, the CP and GI scores are normalized to $[0, 1]$. The MSE is adopted as the evaluation metric for both the CP and GI regression tasks. Following the standard approach for UDA [4,18,23], we use all 16,192 labelled simulated data and all 4,427 unlabelled real data for domain adaptation, and test on all 4,427 real data. To further explore the data efficiency of UDA methods on our simulation-to-real AQA tasks, we also conduct experiments with fractions (10%, 30%, and 50%) of unlabeled real data for domain adaptation, randomly selected from the 10 TEE real cases.

3.2   Comparison with State-of-the-Arts

We compare the proposed SR-AQA with MixStyle [24], MDD [23], RSD [4], and SDAT [18]. All methods are implemented based on their released code and original literature, and fine-tuned to fit our tasks to provide a basis for a fair comparison.

As shown in Table 1, all UDA methods show better AQA overall performance (lower MSE) compared to the model trained only with simulated data ("Simulated Only"), demonstrating the effectiveness of UDA methods in TEE simulation-to-real AQA. The proposed SR-AQA achieves superior performance with the lowest MSE in all CP experiments and all but one GI experiment (50%), in which is very close (0.7648 to 0.76) to the best-performing SDAT. We calculate the MSE reduction percentage between our proposed SR-AQA and the second-best method, to obtain the degree of performance improvement. Specifically, on the CP task among the five real data ratio settings, the MSE of our method dropped by 2.13%–5.08% against the suboptimal method SDAT. It is evident that even with a small amount (10%) of unlabelled real data used for UDA, our SR-AQA still achieves a significant MSE reduction, of at least 3.02% and 4.37% compared to other UDA methods, on the CP and GI tasks respectively, showcasing high data efficiency. We also conduct paired t-tests on MSE results, from multiple runs with different random seeds. The obtained p-values ($p < 0.05$ in all but one case, see supplementary material Table S3), validate that the improvements yielded by the proposed SR-AQA are statistically significant. In Table 2, we report the performance over different (low, medium, and high) score ranges with SR-AQA, obtaining promising results among all ranges[2].

3.3   Ablation Study

We first explore the impact of the amount of UFS on generalization performance. As shown in the left part of Table 3, the performance continues to improve as UFS

---

[2] Example GI and CP results are provided in the supplementary material

Table 1. MSE results for CP and GI scores, in different unlabelled real data ratios. Lower MSE means better model performance. The 'Reduction' row lists the MSE reduction percentage of our proposed SR-AQA compared to the second-best method. Top two results are in **bold** and <u>underlined</u>.

Methods	CP Task (MSE)				GI Task (MSE)			
	Unlabelled Real Data Ratio				Unlabelled Real Data Ratio			
	100%	50%	30%	10%	100%	50%	30%	10%
SR-AQA (Ours)	**411**	**411**	**413**	**417**	**0.6992**	<u>0.7648</u>	**0.7440**	**0.7696**
Reduction	−5.08%	−2.38%	−2.13%	−3.02%	−16.28%	+0.63%	−7.55%	−4.37%
SDAT [18]	<u>433</u>	<u>421</u>	<u>422</u>	<u>430</u>	0.9792	**0.7600**	<u>0.8048</u>	1.1024
RSD [4]	540	507	501	513	0.8768	0.8768	0.9392	<u>0.8048</u>
Mixstyle [24]	466	474	465	496	<u>0.8352</u>	1.0048	0.8736	1.0400
MDD [23]	766	787	755	742	1.1696	1.1360	1.1328	1.1936
Simulated Only	913				1.2848			

Table 2. MSE results on AQA tasks for different score ranges. The full real dataset (without labels) is used for unsupervised domain adaptation.

Methods	CP Task (MSE)			GI Task (MSE)		
	Low	Medium	High	Low	Medium	High
	(0,30]	(30, 60]	(60, 100]	(0,1.2]	(1.2, 2.4]	(2.4, 4]
SR-AQA (Ours)	**1668**	**403**	**286**	**2.3200**	**0.4080**	**0.4832**
Reduction	−2.68%	−2.66%	−1.38%	−13.38%	−19.56%	−1.30%
SDAT [18]	1886	441	<u>290</u>	3.2512	0.6400	<u>0.4896</u>
RSD [4]	1881	489	427	<u>2.6784</u>	0.5168	0.6912
Mixstyle [24]	1752	<u>414</u>	359	2.7152	<u>0.5072</u>	0.5632
MDD [23]	<u>1714</u>	550	771	3.1376	0.8208	0.8480
Simulated Only	2147	850	831	3.4592	0.8896	1.0336

Table 3. Ablation results on two AQA tasks via different UFS settings (left part), and for the proposed uncertainty-aware, $\mathcal{L}_{SCL}$, and $\mathcal{L}_{TL}$ (right part). Headings 1, 1-2, 1-3, 1-4, 1-5 refer to replacing the 1st, $1\text{-}2^{nd}$, $1\text{-}3^{rd}$, $1\text{-}4^{th}$, $1\text{-}5^{th}$ batch normalization layer(s) of ResNet-50 with the UFS. The full real dataset (without labels) is used for unsupervised domain adaptation.

UFS	1	1-2	1-3	1-4	1-5
CP Task (MSE)	421	<u>412</u>	**411**	422	430
GI Task (MSE)	0.7808	0.7488	**0.6992**	<u>0.7312</u>	0.7840

Feature Stylization	$\mathcal{L}_{SCL}$	$\mathcal{L}_{TL}$	CP Task (MSE)	GI Task (MSE)
w/o Uncertainty	✓		426	0.8080
	✓	✓	<u>419</u>	<u>0.7408</u>
w/ Uncertainty	✓		422	0.7840
	✓	✓	**411**	**0.6992**

is applied to more shallow layers, but decreases when UFS is added to deeper layers. This is because semantic information is more important than style in the deeper layers. Using a moderate number of UFS to enrich simulated features with real-world styles, without corrupting semantic information, improves model generalization. Secondly, we study the effect of uncertainty-aware, SCL, and TL, as shown in the right part of Table 3, removing each component leads to performance degradation.

4   Conclusion

This paper presents the first annotated TEE dataset for simulation-to-real AQA with 16,192 simulated images and 4,427 real images. Based on this, we propose a novel UDA network named SR-AQA for boosting the generalization of AQA performance. The network transfers diverse real styles to the simulated domain based on uncertainty-award feature stylization. Style consistency learning enables the encoder to learn style-independent representations while task-specific learning allows our model to naturally adapt to real styles by preserving task-specific information. Experimental results on two AQA tasks for CP and GI scores show that the proposed method outperforms state-of-the-art methods with at least 5.08% and 16.28% MSE reduction, respectively, resulting in superior TEE AQA performance. We believe that our work provides an opportunity to leverage large amounts of simulated data to improve the generalisation performance of AQA for real TEE. Future work will focus on reducing negative transfer to extend UDA methods towards simulated-to-real TEE quality assessment.

Acknowledgements. This work is supported by the Wellcome/EPSRC Centre for Interventional and Surgical Sciences (WEISS) [203145Z/16/Z and NS/A000050/1]; EPSRC-funded UCL Centre for Doctoral Training in Intelligent, Integrated Imaging in Healthcare (i4health) [EP/S021930/1]; Horizon 2020 FET [863146]; a UCL Graduate Research Scholarship; and Singapore MoE Tier 1 Start up grant (WBS: A-8001267-00-00). Danail Stoyanov is supported by a RAE Chair in Emerging Technologies [CiET1819/2/36] and an EPSRC Early Career Research Fellowship [EP/P012841/1].

References

1. Abdi, A.H., et al.: Automatic quality assessment of echocardiograms using convolutional neural networks: feasibility on the apical four-chamber view. IEEE Trans. Med. Imaging **36**(6), 1221–1230 (2017)
2. Chen, C., Li, Z., Ouyang, C., Sinclair, M., Bai, W., Rueckert, D.: MaxStyle: adversarial style composition for robust medical image segmentation. In: Wang, L., Dou, Q., Fletcher, P.T., Speidel, S., Li, S. (eds.) MICCAI 2022. LNCS, vol. 13435, pp. 151–161. Springer, Cham (2022). https://doi.org/10.1007/978-3-031-16443-9_15
3. Chen, X., He, K.: Exploring simple Siamese representation learning. In: CVPR 2021, pp. 15745–15753 (2021)
4. Chen, X., Wang, S., Wang, J., Long, M.: Representation subspace distance for domain adaptation regression. In: ICML 2021, pp. 1749–1759 (2021)

5. Hahn, R.T., et al.: Guidelines for performing a comprehensive transesophageal echocardiographic examination: recommendations from the American society of echocardiography and the society of cardiovascular anesthesiologists. J. Am. Soc. Echocardiogr. **26**(9), 921–964 (2013)

6. He, K., Zhang, X., Ren, S., Sun, J.: Deep residual learning for image recognition. In: CVPR 2016, pp. 770–778 (2016)

7. Hempel, C., et al.: Impact of simulator-based training on acquisition of transthoracic echocardiography skills in medical students. Ann. Card. Anaesth. **23**(3), 293 (2020)

8. Huang, X., Belongie, S.: Arbitrary style transfer in real-time with adaptive instance normalization. In: ICCV 2017, pp. 1501–1510 (2017)

9. Labs, R.B., Vrettos, A., Loo, J., Zolgharni, M.: Automated assessment of transthoracic echocardiogram image quality using deep neural networks. Intell. Med. (2022)

10. Le, C.K., Lewis, J., Steinmetz, P., Dyachenko, A., Oleskevich, S.: The use of ultrasound simulators to strengthen scanning skills in medical students: a randomized controlled trial. J. Ultrasound Med. **38**(5), 1249–1257 (2019)

11. Lee, S., Seong, H., Lee, S., Kim, E.: WildNet: learning domain generalized semantic segmentation from the wild. In: CVPR 2022, pp. 9926–9936 (2022)

12. Li, X., Dai, Y., Ge, Y., Liu, J., Shan, Y., Duan, L.: Uncertainty modeling for out-of-distribution generalization. In: ICLR 2022 (2022)

13. Liao, Z., et al.: On modelling label uncertainty in deep neural networks: automatic estimation of intra-observer variability in 2D echocardiography quality assessment. IEEE Trans. Med. Imaging **39**(6), 1868–1883 (2019)

14. Lin, Z., et al.: Multi-task learning for quality assessment of fetal head ultrasound images. Med. Image Anal. **58**, 101548 (2019)

15. Mazomenos, E.B., Bansal, K., Martin, B., Smith, A., Wright, S., Stoyanov, D.: Automated performance assessment in transoesophageal echocardiography with convolutional neural networks. In: Frangi, A.F., Schnabel, J.A., Davatzikos, C., Alberola-López, C., Fichtinger, G. (eds.) MICCAI 2018. LNCS, vol. 11073, pp. 256–264. Springer, Cham (2018). https://doi.org/10.1007/978-3-030-00937-3_30

16. Mazomenos, E.B., et al.: Motion-based technical skills assessment in transoesophageal echocardiography. In: Zheng, G., Liao, H., Jannin, P., Cattin, P., Lee, S.-L. (eds.) MIAR 2016. LNCS, vol. 9805, pp. 96–103. Springer, Cham (2016). https://doi.org/10.1007/978-3-319-43775-0_9

17. Montealegre-Gallegos, M., et al.: Imaging skills for transthoracic echocardiography in cardiology fellows: the value of motion metrics. Ann. Card. Anaesth. **19**(2), 245 (2016)

18. Rangwani, H., Aithal, S.K., Mishra, M., Jain, A., Radhakrishnan, V.B.: A closer look at smoothness in domain adversarial training. In: ICML 2022, pp. 18378–18399 (2022)

19. Shen, Y., Lu, Y., Jia, X., Bai, F., Meng, M.Q.H.: Task-relevant feature replenishment for cross-centre polyp segmentation. In: Wang, L., Dou, Q., Fletcher, P.T., Speidel, S., Li, S. (eds.) MICCAI 2022. LNCS, vol. 13434, pp. 599–608. Springer, Cham (2022). https://doi.org/10.1007/978-3-031-16440-8_57

20. Song, H., Peng, Y.G., Liu, J.: Innovative transesophageal echocardiography training and competency assessment for Chinese anesthesiologists: role of transesophageal echocardiography simulation training. Curr. Opin. Anaesthesiol. **25**(6), 686–691 (2012)

21. Wang, X., Long, M., Wang, J., Jordan, M.: Transferable calibration with lower bias and variance in domain adaptation. In: NeurIPS 2020, pp. 19212–19223 (2020)

22. Wheeler, R., et al.: A minimum dataset for a standard transoesphageal echocardiogram: a guideline protocol from the British society of echocardiography. Echo Res. Pract. **2**(4), G29 (2015)
23. Zhang, Y., Liu, T., Long, M., Jordan, M.: Bridging theory and algorithm for domain adaptation. In: ICML 2019, pp. 7404–7413 (2019)
24. Zhou, K., Yang, Y., Qiao, Y., Xiang, T.: Domain generalization with mixstyle. In: ICLR 2021 (2021)

Estimated Time to Surgical Procedure Completion: An Exploration of Video Analysis Methods

Barak Ariel, Yariv Colbeci, Judith Rapoport Ferman, Dotan Asselmann, and Omri Bar[✉]

Theator Inc., Palo Alto, CA, USA
{barak,yariv,judith,dotan,omri}@theator.io

Abstract. An accurate estimation of a surgical procedure's time to completion (ETC) is a valuable capability that has significant impact on operating room efficiency, and yet remains challenging to predict due to significant variability in procedure duration. This paper studies the ETC task in depth; rather than focusing on introducing a novel method or a new application, it provides a methodical exploration of key aspects relevant to training machine learning models to automatically and accurately predict ETC. We study four major elements related to training an ETC model: evaluation metrics, data, model architectures, and loss functions. The analysis was performed on a large-scale dataset of approximately 4,000 surgical videos including three surgical procedures: Cholecystectomy, Appendectomy, and Robotic-Assisted Radical Prostatectomy (RARP). This is the first demonstration of ETC performance using video datasets for Appendectomy and RARP. Even though AI-based applications are ubiquitous in many domains of our lives, some industries are still lagging behind. Specifically, today, ETC is still done by a mere average of a surgeon's past timing data without considering the visual data captured in the surgical video in real time. We hope this work will help bridge the technological gap and provide important information and experience to promote future research in this space. The source code for models and loss functions is available at: https://github.com/theator/etc.

Keywords: Surgical Intelligence · Operating Room Efficiency · ETC · RSD · Cholecystectomy · Appendectomy · Radical Prostatectomy

1 Introduction

One of the significant logistical challenges facing hospital administrations today is operating room (OR) efficiency. This parameter is determined by many fac-

---

B. Ariel and Y. Colbeci—Equal contribution.

Supplementary Information The online version contains supplementary material available at https://doi.org/10.1007/978-3-031-43996-4_16.

H. Greenspan et al. (Eds.): MICCAI 2023, LNCS 14228, pp. 165–175, 2023.
https://doi.org/10.1007/978-3-031-43996-4_16

tors, one of which is surgical procedure duration that reflects intracorporeal time, which in itself poses a challenge as, even across the same procedure type, duration can vary greatly. This variability is influenced by numerous elements, including the surgeon's experience, the patient's comorbidities, unexpected events occurring during the procedure, the procedure's complexity and more. Accurate real-time estimation of procedure duration improves scheduling efficiency because it allows administrators to dynamically reschedule before a procedure has run overtime. Another important aspect is the ability to increase patient safety and decrease complications by dosing and timing anesthetics more accurately.

Currently, methods aiming for OR workflow optimization through ETC are lacking. One study showed that surgeons underestimated surgery durations by an average of 31 min, while anesthesiologists underestimated the durations by 35 min [21]. These underestimations drive inefficiencies, causing procedures to be delayed or postponed, forcing longer waiting times for patients. For example, a large variation of waiting time (47 ± 17 min) was observed in a study assessing 157 Cholecystectomy patients [15].

As AI capabilities have evolved greatly in recent years, the field of minimally invasive procedures, which is inherently video-based, has emerged as a potent platform for the harnessing of these capabilities to improve both patient care and workflow efficiency. Consequently, ETC has become a technologically achievable and clinically beneficial task.

2   Related Work

Initially, ETC studies performed preoperative estimates based on surgeon data, patient data, or a combination of these [2,10]. Later on, intraoperative estimates were performed, with some studies requiring manual annotations or the addition of external information [7,11,12,16]. Recently, a study by Twinanda et al. [23] achieved robust ETC results, even without incorporating external information, showing that video-based ETC is better than statistical analysis of past surgeons' data. However, all these studies have evaluated ETC using limited size datasets with inherent biases, as they are usually curated from a small number of surgeons and medical centers or exclude complex cases with significant unexpected events.

In this work, we study the key elements important to the development of ETC models and perform an in-depth methodical analysis of this task. First, we suggest an adequate metric for evaluation - SMAPE, and introduce two new architectures, one based on LSTM networks and one on the transformer architecture. Then, we examine how different ETC methods perform when trained with various loss functions and show that their errors are not necessarily correlated. We then test the hypothesis that an ensemble composed of several ETC model variations can significantly improve estimation compared to any single model.

3   Methods

3.1   Evaluation Metrics

Mean Absolute Error (MAE). The evaluation metric used in prior work was MAE.

$$MAE(y, \hat{y}) = \frac{1}{T} \cdot \sum_{t=0}^{T-1} |y_t - \hat{y}_t| \tag{1}$$

where T is a video duration, y is the actual time left until completion, and $\hat{y}$ is the ETC predictions. A disadvantage of MAE is its reliance on the magnitude of values, consequently, short videos are likely to have small errors while long videos are likely to have large errors. In addition, MAE does not consider the actual video duration or the temporal location for which the predictions are made.

Symmetric Mean Absolute Percentage Error (SMAPE). SMAPE is invariant to the magnitude and keeps an equivalent scale for videos of different duration, thus better represents ETC performance [3, 4, 20].

$$SMAPE(y, \hat{y}) = \frac{1}{T} \cdot \sum_{t=0}^{T-1} \left(\frac{|y_t - \hat{y}_t|}{|y_t| + |\hat{y}_t|} \cdot 100 \right) \tag{2}$$

3.2   Datasets

We focus on three different surgical video datasets (a total of 3,993 videos) that were curated from several medical centers (MC) and include procedures performed by more than 100 surgeons. The first dataset is Laparoscopic Cholecystectomy that contains 2,400 videos (14 MC and 118 surgeons). This dataset was utilized for the development and ablation study. Additionally, we explore two other datasets: Laparoscopic Appendectomy which contains 1,364 videos (5 MC and 61 surgeons), and Robot-Assisted Radical Prostatectomy (RARP) which contains 229 videos (2 MC and 14 surgeons). The first two datasets are similar, both are relatively linear and straightforward procedures, have similar duration distribution, and are abdominal procedures with similarities in anatomical views. However, RARP is almost four times longer on average. Therefore, it is interesting to explore how methods developed on a relatively short and linear procedure will perform on a much longer procedure type such as RARP. Table 3 in the appendix provides a video duration analysis for all datasets. The duration is defined as the difference between surgery start and end times, which is the time interval between scope-in and scope-out. All datasets were randomly divided into training, validation, and test sets with a ratio of 60/15/25%.

3.3   Loss Functions

Loss values are calculated by comparing ETC predictions $(\hat{y}_t)$ for each timestamp to the actual time left until the procedure is complete (y_t). The final loss for each video is the result of averaging these values across all timestamps.

MAE Loss. The MAE loss is defined by:

$$L_{MAE}(y, \hat{y}) = \frac{1}{T} \cdot \sum_{t=0}^{T-1} |y_t - \hat{y}_t| \tag{3}$$

Smooth L1 Loss. The smooth L1 loss is less sensitive to outliers [9].

$$L(y, \hat{y}) = \frac{1}{T} \cdot \sum_{t=0}^{T-1} SmoothL1(y_t - \hat{y}_t) \tag{4}$$

in which

$$SmoothL1(x) = \begin{cases} 0.5x^2 & |x| < 1 \\ |x| - 0.5 & otherwise \end{cases} \tag{5}$$

SMAPE Loss. Based on the understanding that SMAPE (Sect. 3.1) is a good representation of the ETC problem, we also formulated it as a loss function:

$$L_{SMAPE}(y, \hat{y}) = \frac{1}{T} \cdot \sum_{t=0}^{T-1} (\frac{|y_t - \hat{y}_t|}{|y_t| + |\hat{y}_t|} \cdot 100) \tag{6}$$

Importantly, SMAPE produces higher loss values for the same absolute error as the procedure progresses, when the denominator is getting smaller. This property is valuable as the models should be more accurate as the surgery nears its end.

Corridor Loss. A key assumption overlooked in developing ETC methods is that significant and time-impacting events might occur during a procedure. For example, a prolonged procedure due to significant bleeding occurring after 30 min of surgery is information that is absent from the model when providing predictions at the 10 min timestamp. To tackle this problem, we apply the corridor loss [17] that considers both the actual progress of a procedure and the average duration in the dataset (see Fig. 2 in the appendix for a visual example). The corridor loss acts as a wrapper (π) for other loss functions:

$$Corridor(Loss(y, \hat{y})) = \frac{1}{T} \cdot \sum_{t=0}^{T-1} \pi(y, t) \cdot Loss(y_t, \hat{y}_t) \tag{7}$$

Interval L1 Loss. The losses described above focus on the error between predictions and labels for each timestamp independently. Influenced by the total variation loss, we suggest considering the video's sequential properties. The interval L1 loss focuses on jittering in predictions between timestamps in a pre-defined interval, aiming to force them to act more continuously. $\hat{y}_t$ are the predictions per timestamp, and S is an interval time span (jump) between two timestamps.

$$L_{IntervalL1}^{S}(\hat{y}_t) = \sum_{t=0}^{T-S-1} |\hat{y}_{t+S} - \hat{y}_t| \tag{8}$$

Total Variation Denoising Loss. This loss is inspired by a 1D total variation denoising loss and was modified to fit as part of the ETCouple model.

$$L_{squared\_error}(y, \hat{y}) = \frac{1}{2T} \cdot \sum_{t=1}^{T-1} ((y_t - \hat{y}_t)^2 + (y_{t-S} - \hat{y}_{t-S})^2) \qquad (9)$$

$$L_{total\_variation\_denoising}(y, \hat{y}) = L_{squared\_error}(y, \hat{y}) + \lambda \cdot L_{IntervalL1}^{S=120}(\hat{y}) \quad (10)$$

3.4   ETC Models

Feature Representation. All models and experiments described in this work are based on fixed visual features that were extracted from the surgical videos using a pre-trained model. This approach allows for shorter training cycles, less computing requirements, and benefits from a model that was trained on a different task [6,14]. Previous works showed that pre-training could be done with either progress labels or surgical steps labels and that similar performances are achieved, with a slight improvement when using the steps label pre-training [1,23]. In this work, we use a pre-trained Video Transformer Network (VTN) [13] model with a Vision Transformer (ViT) [8] backbone as a feature extraction module. It was originally trained using the same training set (Sect. 3.2) for the step recognition task with a similar protocol to the one described by [5].

Inferring ETC. Our ETC architectures end with a single shared fully connected (FC) layer and a Sigmoid that outputs two values: ETC and *progress*. ETC is inferred by averaging the predicted ETC value and the one calculated from the *progress*.

$$ETC = T - t_{el} = \frac{t_{el}}{progress} - t_{el} \qquad (11)$$

where T is the video duration and t_{el} marks the elapsed time. Inspired by [12], we also incorporate t_{max} which is defined as the expected maximum video length. t_{max} is applied to scale the elapsed time and ensures values in a range [0, 1].

ETC-LSTM. A simple architecture that consists of an LSTM layer with a hidden size of 128. Following hyperparameters tuning on the validation set, the ETC-LSTM was trained using an SGD optimizer with a constant learning rate of 0.1 and a batch size of 32 videos.

ETCouple. ETCouple is a different approach to applying LSTM networks. In contrast to ETC-LSTM and similar methods which predict ETC for a single timestamp, here we randomly select one timestamp from the input video and set it as an anchor sample. The anchor is then paired with a past timestamp using a fixed interval of S = 120 s. The model is given two inputs, the features from the beginning of the procedure up to the anchor and the features up to the pair location. Instead of processing the entire video in an end-to-end manner, we only process past information and are thus able to use a bi-directional LSTM (hidden dimension is 128). The rest of the architecture contains a dropout layer ($P = 0.5$), the shared FC layer, and a Sigmoid function. We explored various hyperparameters and the final model was trained with a batch size of 16, an AdamW optimizer with a learning rate of $5 \cdot 10^{-4}$, and a weight decay of $5 \cdot 10^{-3}$.

ETCformer. LSTM networks have been shown to struggle with capturing long-term dependencies [22]. Intuitively, ETC requires attending to events that occur in different temporal locations throughout the procedure. Thus, we propose a transformer-based network that uses attention modules [24]. The transformer encoder architecture has four self-attention heads with a hidden dimension of size 512. To allow this model to train in a framework where all the video's features are the input to the model but still maintain a causal system, we used a forward direction self-attention [18,19]. This is done by masking out future samples for each timestamp, thus relying only on past information. Best results on the validation set were achieved when training with a batch size of two videos, an AdamW optimizer with a learning rate of 10^{-4}, and a weight decay of 0.1.

Table 1. Comparing validation set results of the mean SMAPE and MAE performance when training ETC models with one or a combination of a few loss functions.

	MAE loss	SMAPE loss	Interval L1 loss (S = 1)	Corridor loss	Mean SMAPE	MAE
ETC-LSTM	✓				**20.68**	**7.67**
	✓	✓			21	7.9
	✓	✓	✓		20.81	7.96
	✓	✓		✓	20.85	7.79
	✓	✓	✓	✓	20.92	7.97
	MAE loss	SMAPE loss	Total variation denoising loss		Mean SMAPE	MAE
ETCouple	✓				21.84	**7.81**
	✓	✓			21.7	8.82
	✓	✓	✓		**21.6**	8.1
	MAE loss	SMAPE loss	Interval L1 loss (S = 1)	Corridor loss	Mean SMAPE	MAE
ETCformer	✓				21.13	7.79
	✓	✓			21.06	7.65
	✓	✓	✓		20.78	**7.38**
	✓	✓		✓	20.68	7.72
	✓	✓	✓	✓	**20.54**	7.67

4   Results

4.1   Ablation Experiments

This section provides ablation studies on the Cholecystectomy dataset.

Loss Functions. Table 1 provides a comparison on the same validation set when using one or the sum of a few loss functions. The classic approach of using LSTM produces the best results when using only MAE loss. However, ETCouple and ETCformer benefit from the combination of several loss functions.

Error Analysis. To test whether the errors of the various models are correlated, we compared the predictions made by the different models on a per-video basis. We use SMAPE and analyze the discrepancy by comparing the difference of every two model variations independently. Then, we divided the videos into a *similar* and a *dissimilar* group, by using a fixed threshold, i.e., if the SMAPE difference is smaller than the threshold the models are considered as providing *similar* results. The threshold was empirically set to 2, deduced from the ETC curves, which are almost identical when the SMAPE difference is smaller than 2 (Fig. 1(a) and appendix Fig. 4. We demonstrate these results visually in Fig. 1(b). Interestingly, there are significant differences in SMAPE between different models (disagreement in more than 50%). ETC-LSTM and ETCouple show the highest disagreement.

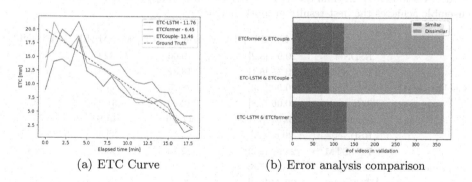

(a) ETC Curve   (b) Error analysis comparison

Fig. 1. (a) ETC per minute on a Cholecystectomy video. Each color represents a different ETC model, the numbers are the SMAPE scores. The dashed lines represents the ground truth progress. (b) An error analysis comparison was performed by measuring the difference in SMAPE per video between two independent models (each row is a pair of two models). The blue color represents the number of *similar* videos, and the orange color the number of *dissimilar* videos in the validation set (total of 366). (Color figure online)

Baseline Comparison. We reproduce RSDNet [23] on top of our extracted features and use it as a baseline for comparison. We followed all methodical details described in the original paper, only changing the learning rate reduction policy to match the same epoch proportion in our dataset. Table 2 shows that ETC-LSTM and RSDNet have similar results, ETC-LSTM achieves better SMAPE scores while RSDNet is more accurate in MAE. These differences can be the product of scaling the elapsed time using t_{max} vs. s_{norm} and shared vs. independent FC layer. The ETCformer model reaches the best SMAPE results but is still short on MAE.

Ensemble Analysis. There are many tasks in machine learning in which data can be divided into easy or hard samples. We argue that the ETC task is different in these regards. Based on the error analysis, we explored how an ensemble of models performs and if it produces better results (Table 2). In contrast to a classic use case of models ensemble, in which the same model is trained with bootstrapping on different folds of data, here we suggest an ensemble that uses different models, which essentially learn to perform differently on the same input video. Figure 3 in the appendix illustrates the MAE error graph for the ensemble, presenting the mean and SD of the MAE. All model variations' performance is also provided in the appendix in Table 4. When using more than one model, the ETC predictions for each model are averaged into a single end result.

Table 2. ETC models comparison on the test set, using mean SMAPE and standard deviation (SD), median SMAPE, 90th percentile (90p) SMAPE, and MAE. The ensemble achieves the best results in most metrics.

		Mean SMAPE [SD]	Median SMAPE	90p SMAPE	MAE (min)
Cholecystectomy	RSDNet	20.97 [8.2]	19.06	32.6	7.48
	ETC-LSTM	20.75 [8.1]	18.82	31.33	7.88
	ETCouple	20.99 [8.3]	18.75	33.18	8.15
	ETCformer	20.06 [7.8]	18.22	31.34	7.56
	Ensemble	**19.57** [7.9]	**17.56**	**30.87**	**7.33**
Appendectomy	RSDNet	20.6 [8.5]	19.18	32.55	10.98
	ETC-LSTM	21.92 [9.2]	19.73	33.98	11.88
	ETCouple	21.49 [7.6]	20.08	31.76	8.49
	ETCformer	22.74 [8.3]	20.58	33.88	12.24
	Ensemble	**19.87** [7.8]	**17.9**	**30.75**	**7.47**
RARP	RSDNet	17.03 [5.9]	15.81	25.07	**21.01**
	ETC-LSTM	16.85 [4.7]	16.07	24.26	23.8
	ETCouple	19.05 [9.4]	17.58	30.57	30.88
	ETCformer	17.55 [6.8]	15.94	29.28	26.18
	Ensemble	**15** [5.9]	**13.94**	**23.43**	21.35

4.2   Appendectomy and RARP Results

We examine the results on two additional datasets to showcase the key elements explored in this work (Table 2). In Appendectomy, the ensemble also achieves the best results on the test set with a significant drop in SMAPE and MAE scores. The ETCformer performs the worst compared to other model variations, this might be because transformers require more data for training, therefore additional data could show its potential as seen in Cholecystectomy. The RARP dataset contains fewer videos, but they are of longer duration. Here too, the ensemble achieves better SMAPE scores.

5   Discussion

In this work, we examine different architectures trained with several loss functions and show how SMAPE can be utilized as a better metric to compare ETC models. In the error analysis, we conclude that each model learns to operate differently on the same videos. This led us to explore an ensemble of models which eventually achieves the best results. Yet, this conclusion can facilitate future work, focusing on understanding the differences and commonalities of the models' predictions and developing a unified or enhanced model, potentially reducing the complexity of training several ETC models and achieving better generalizability. Future work should also incorporate information regarding the surgeon's experience, which may improve the model's performance.

This work has several limitations. First, the proposed models need to be evaluated across other procedures and specialties for their potential to be validated further. Second, the ensemble's main disadvantage is its requirement for more computing resources. In addition, there may be data biases due to variability in the time it takes surgeons to perform certain actions at different stages of their training. Finally, although our model relies on video footage only, and no annotations are required for ETC predictions, manual annotations of surgical steps are still needed for pre-training of the feature extraction model.

Real-time ETC holds great potential for surgical management. First, in optimizing OR scheduling, and second as a proxy to complications that cause unusual deviations in anticipated surgery duration. However, optimizing OR efficiency with accurate procedural ETC, based on surgical videos, has yet to be realized. We hope the information in this study will assist researchers in developing new methods and achieve robust performance across multiple surgical specialties, ultimately leading to better OR management and improved patient care.

Acknowledgements. We thank Ross Girshick for providing valuable feedback on this manuscript and for helpful suggestions on several experiments.

References

1. Aksamentov, I., Twinanda, A.P., Mutter, D., Marescaux, J., Padoy, N.: Deep neural networks predict remaining surgery duration from cholecystectomy videos. In: Descoteaux, M., Maier-Hein, L., Franz, A., Jannin, P., Collins, D.L., Duchesne, S. (eds.) MICCAI 2017. LNCS, vol. 10434, pp. 586–593. Springer, Cham (2017). https://doi.org/10.1007/978-3-319-66185-8_66
2. Ammori, B., Larvin, M., McMahon, M.: Elective laparoscopic cholecystectomy. Surg. Endosc. **15**(3), 297–300 (2001)
3. Armstrong, J.S., Collopy, F.: Error measures for generalizing about forecasting methods: empirical comparisons. Int. J. Forecast. **8**(1), 69–80 (1992)
4. Armstrong, J.S., Forecasting, L.R.: From Crystal Ball to Computer, p. 348. New York (1985)
5. Bar, O., et al.: Impact of data on generalization of AI for surgical intelligence applications. Sci. Rep. **10**(1), 1–12 (2020)
6. Colbeci, Y., Zohar, M., Neimark, D., Asselmann, D., Bar, O.: A multi instance learning approach for critical view of safety detection in laparoscopic cholecystectomy. In: Proceedings of Machine Learning Research, vol. 182, pp. 1–14 (2022)
7. Dexter, F., Epstein, R.H., Lee, J.D., Ledolter, J.: Automatic updating of times remaining in surgical cases using Bayesian analysis of historical case duration data and "instant messaging" updates from anesthesia providers. Anesth. Analg. **108**(3), 929–940 (2009)
8. Dosovitskiy, A., et al.: An image is worth 16×16 words: transformers for image recognition at scale. arXiv preprint arXiv:2010.11929 (2020)
9. Girshick, R.: Fast R-CNN. In: Proceedings of the IEEE International Conference on Computer Vision, pp. 1440–1448 (2015)
10. Macario, A., Dexter, F.: Estimating the duration of a case when the surgeon has not recently scheduled the procedure at the surgical suite. Anesth. Analg. **89**(5), 1241–1245 (1999)
11. Maktabi, M., Neumuth, T.: Online time and resource management based on surgical workflow time series analysis. Int. J. Comput. Assist. Radiol. Surg. **12**(2), 325–338 (2017)
12. Marafioti, A., et al.: CataNet: predicting remaining cataract surgery duration. In: de Bruijne, M., et al. (eds.) MICCAI 2021. LNCS, vol. 12904, pp. 426–435. Springer, Cham (2021). https://doi.org/10.1007/978-3-030-87202-1_41
13. Neimark, D., Bar, O., Zohar, M., Asselmann, D.: Video transformer network. In: Proceedings of the IEEE/CVF International Conference on Computer Vision, pp. 3163–3172 (2021)
14. Neimark, D., Bar, O., Zohar, M., Hager, G.D., Asselmann, D.: "Train one, classify one, teach one"-cross-surgery transfer learning for surgical step recognition. In: Medical Imaging with Deep Learning, pp. 532–544. PMLR (2021)
15. Paalvast, M., et al.: Real-time estimation of surgical procedure duration. In: 2015 17th International Conference on E-Health Networking, Application & Services (HealthCom), pp. 6–10. IEEE (2015)
16. Padoy, N., Blum, T., Feussner, H., Berger, M.O., Navab, N.: On-line recognition of surgical activity for monitoring in the operating room. In: AAAI, pp. 1718–1724 (2008)
17. Rivoir, D., Bodenstedt, S., von Bechtolsheim, F., Distler, M., Weitz, J., Speidel, S.: Unsupervised temporal video segmentation as an auxiliary task for predicting the remaining surgery duration. In: Zhou, L., et al. (eds.) OR 2.0/MLCN -2019. LNCS,

vol. 11796, pp. 20 37. Springer, Cham (2019). https://doi.org/10.1007/978-3-030-32695-1_4

18. Shen, T., Zhou, T., Long, G., Jiang, J., Pan, S., Zhang, C.: DiSAN: directional self-attention network for RNN/CNN-free language understanding. In: Proceedings of the AAAI Conference on Artificial Intelligence, vol. 32 (2018)

19. Shen, T., Zhou, T., Long, G., Jiang, J., Zhang, C.: Fast directional self-attention mechanism. arXiv preprint arXiv:1805.00912 (2018)

20. Tofallis, C.: A better measure of relative prediction accuracy for model selection and model estimation. J. Oper. Res. Soc. **66**(8), 1352–1362 (2015)

21. Travis, E., Woodhouse, S., Tan, R., Patel, S., Donovan, J., Brogan, K.: Operating theatre time, where does it all go? A prospective observational study. BMJ **349** (2014). https://doi.org/10.1136/bmj.g7182, https://www.bmj.com/content/349/bmj.g7182

22. Trinh, T., Dai, A., Luong, T., Le, Q.: Learning longer-term dependencies in RNNs with auxiliary losses. In: International Conference on Machine Learning, pp. 4965–4974. PMLR (2018)

23. Twinanda, A.P., Yengera, G., Mutter, D., Marescaux, J., Padoy, N.: RSDNet: learning to predict remaining surgery duration from laparoscopic videos without manual annotations. IEEE Trans. Med. Imaging **38**(4), 1069–1078 (2018)

24. Vaswani, A., et al.: Attention is all you need. In: Advances in Neural Information Processing Systems, vol. 30 (2017)

A Transfer Learning Approach to Localising a Deep Brain Stimulation Target

Ying-Qiu Zheng[1]([✉]) [iD], Harith Akram[2] [iD], Stephen Smith[1] [iD],
and Saad Jbabdi[1] [iD]

[1] Wellcome Centre for Integrative Neuroimaging, University of Oxford, Oxford, UK
ying-qiu.zheng@ndcn.ox.ac.uk
[2] UCL Institute of Neurology, Queen Square, London, UK

Abstract. The ventral intermediate nucleus of thalamus (Vim) is a well-established surgical target in magnetic resonance-guided (MR-guided) surgery for the treatment of tremor. As the structure is not identifiable from conventional MR sequences, targeting the Vim has predominantly relied on standardised Vim atlases and thus fails to model individual anatomical variability. To overcome this limitation, recent studies define the Vim using its white matter connectivity with both primary motor cortex and dentate nucleus, estimated via tractography. Although successful in accounting for individual variability, these connectivity-based methods are sensitive to variations in image acquisition and processing, and require high-quality diffusion imaging techniques which are often not available in clinical contexts. Here we propose a novel transfer learning approach to accurately target the Vim particularly on clinical-quality data. The approach transfers anatomical information from publicly available high-quality datasets to a wide range of white matter connectivity features in low-quality data, to augment inference on the Vim. We demonstrate that the approach can robustly and reliably identify the Vim despite compromised data quality, and is generalisable to different datasets, outperforming previous surgical targeting methods.

Keywords: Surgical targeting · Deep brain stimulation · MR-guided surgery · Ventral intermediate nucleus of thalamus · Transfer learning

1 Introduction

Functional neurosurgical techniques, such as deep brain stimulation (DBS), and MR-guided stereotactic ablation, have been used as effective treatments of neurological and psychiatric disorders for decades. By intervening on a target brain structure, a neurosurgical treatment typically modulates brain activity in disease-related circuits and can often mitigate the disease symptoms and restore

---

Supplementary Information The online version contains supplementary material available at https://doi.org/10.1007/978-3-031-43996-4_17.

brain function when drug treatments are ineffective. An example is the ventral intermediate nucleus of thalamus (Vim), a well-established surgical target in DBS and stereotactic ablation for the treatment of tremor in Parkinson's Disease, essential tremor, and multiple sclerosis [1]. The Vim plays a central role in tremor circuitry [2–9]. It receives efferent fibers from the dentate nucleus of the contralateral cerebellum and projects primarily to the primary motor cortex (M1) [5], as part of the dentato-thalamo-cortical pathway (DTCp).

As the structure is not readily visible on conventional MR images, targeting the Vim has relied primarily on standardised coordinates provided by stereotactic atlases, instead of directly targeting the structure based on subject-specific image-derived features. Such standardised stereotactic coordinates/atlases provide a reproducible way to identify the nucleus. However, the atlas-based Vim targeting falls short of accounting for the inter-individual and often inter-hemispheric anatomical variability, which are often substantial for thalamic nuclei [14], whilst efficacy of stereotactic operations heavily depends on accurate identification of the target nucleus.

To overcome this limitation, several recent studies proposed to localise the Vim using its anatomical connectivity features (e.g., with M1 and the dentate nucleus) *in vivo* on an individual basis [11,15,16], aided by cutting-edge diffusion-weighted imaging (DWI) and tractography techniques. DWI allows estimation of local fiber orientations, based upon which tractography generates streamline samples representative of the underlying white matter pathway, starting from a given seed. Typically, these studies identify Vim by finding the region of maximum connectivity with both M1 and contralateral dentate, and thus better capture individual variations of the nucleus, leading to improved clinical outcome [11,12].

However, the connectivity-driven Vim requires high angular resolution diffusion imaging (HARDI) techniques to allow reconstruction of Vim's connectivity features, which are often not readily available in advanced-care clinical contexts. Furthermore, even with cutting-edge HARDI and higher order diffusion modelling techniques, the connectivity-derived Vim has exhibited considerable variations across different acquisition protocol and processing pipelines, suggesting that they have to be used with caution to ensure that they adhere to the true underlying anatomical variability instead of simply reflecting the methodological confounds erroneously interpreted as variability.

Given the limitations of the standardised approaches and connectivity-based methods, we propose a novel approach, HQ-augmentation, to reliably target the Vim, particularly for clinical (or generally lower-quality) data. We utilised the publicly-available high-quality (HQ) Human Connectome Project (HCP) dataset [18,19] to augment surgical targeting of the Vim on low-quality (LQ) data. More specifically, the approach transfers the anatomical information derived from high-quality data, i.e., the approximate position of the Vim, to a wide range of low-quality white matter connectivity features in order to infer the likelihood that a given voxel is classified as part of Vim on the corresponding low-quality data. We demonstrate that the proposed approach not only yields

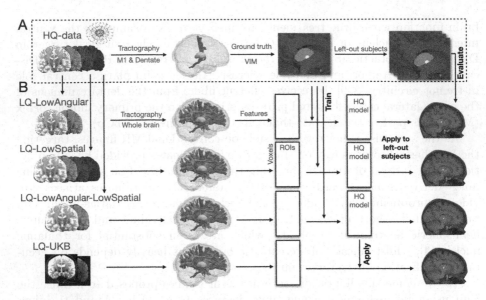

Fig. 1. Schematic illustration of the HQ-augmentation model. The process begins with the use of HQ diffusion data to establish a "ground truth" Vim within the thalamic masks, utilising white matter connectivity with the M1 and contralateral cerebellum. Following this, surrogate low-quality diffusion datasets are generated through intentional degradation of the HQ datasets. The HQ-augmentation model is then trained using the "ground truth" Vim derived from the HCP HQ data as target labels, and a broad set of HCP's low-quality connectivity profiles (in the form of voxel-by-ROI matrices) as input features. After training, the model is applied to unseen low-quality datasets, which include surrogate low-quality HCP and UK Biobank diffusion data, to predict Vim location. The performance of the model is subsequently evaluated against the corresponding HQ "ground truth" data.

consistent Vim targets despite compromised data quality, but also preserves inter-individual anatomical variability of the structure. Furthermore, the approach generalises to unseen datasets with different acquisition protocols, showing potential of translating into a reliable clinical routine for MR-guided surgical targeting.

2   Materials and Methods

HQ Dataset I: Human Connectome Project (HCP) 3T MRI Datasets. We leveraged the 3T diffusion MRI data from HCP [18] as the HQ dataset. The minimally pre-processed T1-, T2-, and diffusion-weighted MRI scans from a total of 1,062 healthy young adults were included, among which 43 subjects were scanned twice. Diffusion-weighted imaging was acquired at isotropic spatial resolution 1.25 mm, with three shells (b-values = 1000, 2000 and 3000 s/mm^2) and approximately 90 unique diffusion directions per shell, acquired twice (total scan time 60 min per subject) [17].

HQ Dataset II: UK Biobank (UKB) 3T MRI Datasets. The UKB 3T MRI datasets [24] were also used as HQ data. The T1, T2 and diffusion-weighted scans from 2,560 subjects with retest (second scan) sessions were included. The diffusion MRI was carried out at isotropic spatial resolution 2 mm with two shells (b-values = 1000 and 2000 s/mm^2), 50 diffusion directions per shell (total scan time 6min per subject).

Surrogate Low-Quality (LQ) Datasets. We considered a range of low-quality datasets representing the typical data quality in clinical contexts. This included 1) HCP surrogate low angular resolution diffusion dataset, obtained by extracting the b = 0 s/mm^2 and 32 b = 1000 s/mm^2 volumes from HQ HCP 3T diffusion MRI ("**LQ-LowAngular**"); 2) HCP surrogate low spatial resolution dataset, obtained by downsampling the HCP 3T diffusion MRI to isotropic 2 mm sptial resolution ("**LQ-LowSpatial**"); 3) HCP surrogate low angular and spatial resolution dataset, created by downsampling the surrogate low-angular-resolution dataset to isotropic 2 mm spatial resolution ("**LQ-LowAngular-LowSpatial**"); 4) UKB surrogate low angular resolution dataset, created by extracting the b = 0 s/mm^2 and 32 b = 1000 s/mm^2 volumes from the original UKB diffusion dataset ("**LQ-UKB**").

The Connectivity-Driven Approach. We followed the practice in [11] to generate connectivity-driven Vim. This approach identifies the Vim by finding the maximum probability of connection to M1 and contralateral dentate nucleus within the thalamic mask, generated via probablistic tractography. When this is applied to HQ original data, it generates the ground truth segmentations. When it is applied to LQ surrogate data, it provides results against which other methods (atlas-based and HQ-augmentation) can be compared.

The Atlas-Defined Approach. We used a previously published and validated Vim atlas [11] to find atlas-defined Vim on individual subjects. The atlas was registered into the individual T1 space (also referred to native space) via the warp fields between the corresponding individual T1 scans and the MNI152 standard brain. The warped group-average Vim atlas was thresholded at 50% percentile in the native space as the atlas-defined Vim.

The HQ-Augmentation Approach. The goal of this approach is to leverage anatomical information in HQ data to infer the likelihood of a voxel belonging to the Vim, given a wide range of tract-density features (multiple distinct tract bundles) derived from low-quality data. The HQ-augmentation model was trained on the HCP dataset for each type of low-quality dataset. Using the HCP HQ data, we first generated the connectivity-driven Vim (referred to as HQ-Vim) as the "ground truth" location of the nucleus, serving as training labels in the model. Next, for each low-quality counterpart, we generated an extended set of tract-density features, targeting a wide range of region-of-interests (ROIs), as the input features of the model. The richer set of connectivity features serves to compensate for the primary tract-density features (with M1 and dentate), when those are compromised by less sufficient spatial or angular resolution in low-quality diffusion MRI, thus making Vim identification less reliant on the primary tract-density features used in the connectivity-driven approach and more robust to variations in data quality. During training, the model learns to use the extended set of low-quality connectivity features to identify the Vim that is closest to the one that can be otherwise obtained from its HQ counterpart.

Specifically, assume $\mathbf{X} = [\mathbf{x}_1, \mathbf{x}_2, ...\mathbf{x}_V]^T$ is a $V \times d$ connectivity feature matrix for a given subject, where $\mathbf{x}_i$ is a $d \times 1$ vector representing the connectivity features in voxel i, V is the total number of voxels of the thalamus (per hemisphere) for this subject; $\mathbf{y} = [y_1, y_2, ...y_V]^T$ is a $V \times 1$ vector containing the HQ-Vim labels for the V voxels, in which y_i is the label of voxel i. Given the low-quality features $\mathbf{X}$, we seek to maximise the probability of reproducing the exact same HQ-Vim label assignment $\mathbf{y}$ on its low-quality counterparts, across the training subjects

$$\log P(\mathbf{y}|\mathbf{X}) = \log[\frac{1}{Z(\mathbf{X})}\exp(-E(\mathbf{y}|\mathbf{X}))] \tag{1}$$

Here $E(\mathbf{y}|\mathbf{X})$ is the cost of the label assignment $\mathbf{y}$ given the features $\mathbf{X}$, whilst $Z(\mathbf{X})$ serves as an image-dependent normalising term. Maximising the posterior $P(\mathbf{y}|\mathbf{X})$ across subjects is equivalent to minimising the cost of the label assignment $\mathbf{y}$ given the features $\mathbf{X}$. Suppose $\mathcal{N}_i$ is the set of voxels neighbouring voxel i, the cost $E(\mathbf{y}|\mathbf{X})$ is modelled as

$$E(\mathbf{y}|\mathbf{X}) = \sum_i \psi_u(y_i|\mathbf{x}_i) + \sum_i \sum_{j \in \mathcal{N}_i} \psi_p(y_i, y_j|\mathbf{x}_i, \mathbf{x}_j) + \lambda_1||\mathbf{W}||_1 + \lambda_2||\mathbf{W}||_2^2 \tag{2}$$

The first component $\psi_u(y_i)$ measures the cost (or inverse likelihood) of voxel i taking label y_i. Here $\psi_u(y_i)$ takes the form $\psi_u(y_i) = \mathbf{w}_{y_i}^T \phi(\mathbf{x}_i)$, where $\phi(\cdot)$ maps a feature vector $\mathbf{x}_i = [x_1, x_2, ...x_d]$ to a further expanded feature space in order to provide more flexibility for the parameterisation. $\mathbf{W} = [\mathbf{w}_1, \mathbf{w}_2]$ is the coefficient matrix, each column containing the coefficients for the given class (i.e., belonging to the HQ-Vim or not). Here we chose a series of polynomials along with the group-average Vim probability (registered into native space) to expand the feature space, i.e.,

$$\phi(\mathbf{x}_i) = [x_1, x_2, ...x_d, x_1^{p_1}, x_2^{p_1}, ...x_d^{p_1}, x_1^{p_2}, x_2^{p_2}, ...x_d^{p_2}, x_1^{p_3}, x_2^{p_3}, ...x_d^{p_3}, g_i]$$

where $p_1 = 2, p_2 = 0.5, p_3 = 0.2$ are the power of the polynomials, and g_i is the group-average probability of voxel i classified as Vim. The second pairwise cost encourages assigning similar labels to neighbouring voxels, particularly for those sharing similar connectivity features. We modelled this component as $\psi_p(\mathbf{x}_i, \mathbf{x}_j) = \mu(y_i, y_j)\rho_m k_m(\phi(\mathbf{x}_i), \phi(\mathbf{x}_j))$. Here $k_m(\phi(\mathbf{x}_i), \phi(\mathbf{x}_j)) = \exp(-\gamma_m||\phi(\mathbf{x}_i) - \phi(\mathbf{x}_j)||^2)$ is a kernel function modelling the similarity between voxel i and j in the extended feature space, with length scale γ_m, chosen via cross-validation. $\mu(\cdot)$ is a label compatibility function where $\mu(y_i, y_j) = 0$ if $y_i = y_j$ or $\mu(y_i, y_j) = 1$ if $y_i \neq y_j$.

Therefore, in a local neighbourhood, the kernel function penalises inconsistent label assignment of voxels that have similar features, thus allowing modelling local smoothness. ρ_m controls the relative strength of this pairwise cost weighted by $k_m(\cdot)$. Lastly, the $L1$ and $L2$ penalty terms serve to prevent overfitting of the model. We used a mean-field algorithm to iteratively approximate the maximum posterior $P(\mathbf{y}|\mathbf{X})$ [29] summed across the subjects. The approximated posterior is maximised via gradient descent in a mini-batch form, where the connectivity feature matrix of each subject serves as a mini-batch, demeaned and normalised, and sequentially fed into the optimisation problem.

3   Experiments and Discussions

Accuracy of the HQ-Augmentation Model on the HCP Surrogate Low-Quality Datasets. The HQ-augmentation model was trained using HQ-Vim as labels and the LQ connectivity profiles as features, and then tested on left-out subjects from the same LQ dataset. Its accuracy was evaluated against the HQ-Vim counterpart obtained from the HQ HCP data of these left-out subjects, which served as ground truth. The accuracies of the HQ-augmentation Vim were also compared with those of the atlas-defined Vim and the connectivity-driven Vim, the latter obtained using the low-quality M1 and dentate tract-density features. We considered two accuracy metrics: Dice coefficient, which measures the overlap with the ground truth, and centroid displacement, which calculates the Euclidean distance between the predicted and ground truth centers-of-mass. As the connectivity-driven approach may even fail on HQ data, resulting in unreliable HQ-Vim, the training and evaluation of HQ-augmentation model were conducted only on a subset of "good subjects", whose HQ-Vim's center-of-mass was within 4 mm from the atlas's center-of-mass in native space. For the subjects which HQ-Vim regarded as unreliable (i.e., deviating too much from the atlas), the accuracy of HQ-augmentation model was evaluated against the atlas-defined Vim. The HQ-augmentation model produced Vim predictions that are closest to the HQ-Vim (i.e., higher Dice coefficient and smaller centroid displacement) than the LQ connectivity-driven approach and the atlas-defined Vim, evaluated on the reliable subset (see Fig. 2A for the results on LQ-LowSpatial-LowAngular data; see supplementary materials for results on the other low-quality datasets). When the HQ-vim failed to serve as approximate ground truth, the HQ-augmentation model on low-quality data yielded Vim predictions that were closer to the atlas-defined Vim than the HQ-Vim derived from HQ data (Fig. 2B).

Generalisability of the HQ-Augmentation Model to UK Biobank. We also tested whether the HQ-augmentation models trained on HCP were generalisable to other datasets collected under different protocols. It is crucial for a model to be generalisable to unseen protocols, as collecting large datasets for training purposes in clinical contexts can often be impractical. We therefore applied the HQ-augmentation models trained on different HCP LQ datasets to UK Biobank surrogate low-quality data (LQ-UKB) and averaged the outputs to give a single

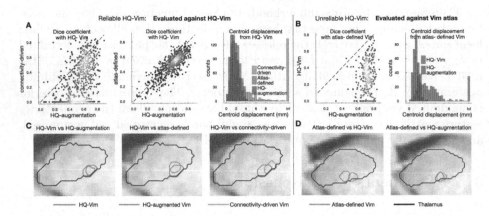

Fig. 2. Accuracy of HQ-augmentation on LQ-HCP-LowSpatial-LowAngular data. (**A**) When using the HQ-Vim as ground truth, the HQ-augmentation approach using low-quality features (here LQ-LowAngular-LowSpatial features) gives higher Dice coefficient and smaller centroid displacement with the HQ-Vim, than the atlas-defined Vim (green) and the low-quality connectivity-driven Vim (orange). (**B**) When using the atlas-defined Vim as ground truth (because, in these "unreliable" subjects, the HQ-connectivity method was considered to have failed), the HQ-augmentation model using low-quality features even gives more reliable results than the HQ-Vim (red). (**C**) and (**D**). Example contours of Vim identified by each approach. (Color figure online)

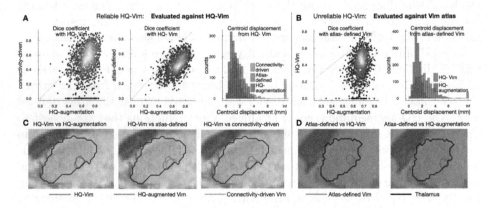

Fig. 3. Generalisability of HQ-Augmentation to UKB. The HQ-augmentation model trained on HCP was applied to the UKB low-quality features (blue) as HQ-augmented Vim. (**A**). When using the UKB HQ-Vim as ground truth, the HQ-augmentated Vim has higher Dice coefficient and smaller centroid displacement with the UKB HQ-Vim, than the atlas-defined Vim (green) and the connectivity-driven Vim using low-quality features (orange). (**B**) When using the atlas-defined Vim as ground truth, the HQ-augmentation model using low-quality features even gave more reliable Vim than the UKB HQ-Vim (red), which used HQ features to target Vim. (**C**) and (**D**). Contours of Vim, identified by each approach. (Color figure online)

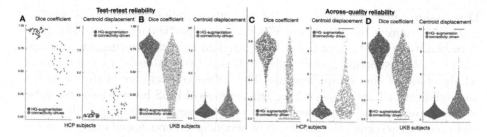

Fig. 4. The HQ-augmentation model (blue) gives more reproducible Vim identification than the connectivity-driven approach (orange), giving higher Dice coefficient and smaller centroid displacement across scanning sessions (i.e. test-retest reliability, **A** and **B**) and across datasets of different quality (across-quality reliability, **C** and **D**). (Color figure online)

ensembled Vim prediction per UKB subject in the LQ-UKB dataset. (Note that this step did not involve retraining or fine-tuning for LQ-UKB.) To evaluate its performance, similarly, we derived the HQ-Vim from the original (HQ) UK Biobank data via the connectivity-driven approach, and split it into a reliable subset, in which the UKB HQ-Vim served as the ground truth, and an unreliable subset, in which the atlas-defined Vim (warped into the UKB individual native space) served as the ground truth. On the reliable subset, the evaluations were conducted against the UKB HQ-Vim (Fig. 3A), whilst on the unreliable subset, the evaluations were conducted against the atlas-defined Vim (Fig. 3B). Despite being trained on HCP, the HQ-augmentation model produced Vim predictions that were closer to the corresponding HQ-Vim on the UKB subjects, outperforming the connectivity-driven approach on the UKB surrogate low-quality data and the atlas-defined approach. Even when using the atlas-defined Vim as the ground truth, the HQ-augmentation Vim on the UKB low-quality data showed higher accuracy than the HQ-Vim derived from the HQ data.

Reliability Analysis of the HQ-Augmentation Model. We also conducted a reliability analysis for the HQ-augmentation model and the connectivity-driven approach to assess their consistency in providing results despite variations in data quality (across-quality reliability) and across scanning sessions (test-retest reliability). To evaluate the HQ-augmentation model's across-quality reliability, we trained it using HQ tract-density maps to produce an "HQ version" of HQ-augmented Vim. The similarity between HQ-augmentation outputs using high- or low-quality features was assessed using the Dice coefficient and centroid displacement measures. Test-retest reliability was determined by comparing the outputs of the HQ-augmentation model on first-visit and repeat scanning sessions. Similarly, we assessed the across-quality reliability and test-retest reliability for the connectivity-driven approach, applied to high- and low-quality data accordingly. The HQ-augmentation model consistently provided more reliable results than the connectivity-driven approach, not only across datasets of different quality but also across different scanning sessions (Fig. 4).

Discussion. Our study presents the HQ-augmentation technique as a robust method to improve the accuracy of Vim targeting, particularly in scenarios where data quality is less than optimal. Compared to existing alternatives, our approach exhibits superior performance, indicating its potential to evolve into a reliable tool for clinical applications. The enhanced accuracy of this technique has significant clinical implications. During typical DBS procedures, one or two electrodes are strategically placed near predetermined targets, with multiple contact points to maximise the likelihood of beneficial outcomes and minimise severe side effects. Greater accuracy translates into improved overlap with the target area, which consequently increases the chances of successful surgical outcomes. Importantly, the utility of our method extends beyond Vim targeting. It can be adapted to target any area in the brain that might benefit from DBS, thus expanding its clinical relevance. As part of our ongoing efforts, we are developing a preoperative tool based on the HQ-augmentation technique. This tool aims to optimise DBS targeting tailored to individual patients' conditions, thereby enhancing therapeutic outcomes and reducing patient discomfort associated with suboptimal electrode placement.

References

1. Cury, R.G., et al.: Thalamic deep brain stimulation for tremor in Parkinson disease, essential tremor, and dystonia. Neurology **89**(13), 1416–1423 (2017)
2. Muthuraman, M., et al.: Oscillating central motor networks in pathological tremors and voluntary movements. What makes the difference? Neuroimage **60**(2), 1331–1339 (2012)
3. Baker, K.B., et al.: Deep brain stimulation of the lateral cerebellar nucleus produces frequency-specific alterations in motor evoked potentials in the rat in vivo. Exp. Neurol. **226**(2), 259–264 (2010)
4. Dum, R.P., Strick, P.L.: An unfolded map of the cerebellar dentate nucleus and its projections to the cerebral cortex. J. Neurophysiol. **89**(1), 634–639 (2003)
5. Gallay, M.N., et al.: Human pallidothalamic and cerebellothalamic tracts: anatomical basis for functional stereotactic neurosurgery. Brain Struct. Funct. **212**, 443–463 (2008)
6. Darian-Smith, C., Darian-Smith, I., Cheema, S.S.: Thalamic projections to sensorimotor cortex in the macaque monkey: use of multiple retrograde fluorescent tracers. J. Comparat. Neurol. **299**(1), 17–46 (1990)
7. Calzavara, R., et al.: Neurochemical characterization of the cerebellar-recipient motor thalamic territory in the macaque monkey. Eur. J. Neurosci. **21**(7), 1869–1894 (2005)
8. McIntyre, C.C., Hahn, P.J.: Network perspectives on the mechanisms of deep brain stimulation. Neurobiol. Dis. **38**(3), 329–337 (2010)
9. Helmich, R.C., et al.: Cerebral causes and consequences of parkinsonian resting tremor: a tale of two circuits? Brain **135**(11), 3206–3226 (2012)
10. Hirai, T., Jones, E.G.: A new parcellation of the human thalamus on the basis of histochemical staining. Brain Res. Rev. **14**(1), 1–34 (1989)
11. Akram, H., et al.: Connectivity derived thalamic segmentation in deep brain stimulation for tremor. NeuroImage: Clin. **18**, 130–142 (2018)

12. Su, J.H., et al.: Improved Vim targeting for focused ultrasound ablation treatment of essential tremor: a probabilistic and patient-specific approach. Hum. Brain Mapp. **41**(17), 4769–4788 (2020)
13. Elias, G.J.B., et al.: Probabilistic mapping of deep brain stimulation: insights from 15 years of therapy. Ann. Neurol. **89**(3), 426–443 (2021)
14. Morel, A., Magnin, M., Jeanmonod, D.: Multiarchitectonic and stereotactic atlas of the human thalamus. J. Comparat. Neurol. **387**(4), 588–630 (1997)
15. Ferreira, F., et al.: Ventralis intermedius nucleus anatomical variability assessment by MRI structural connectivity. NeuroImage **238**, 118231 (2021)
16. Bertino, S., et al.: Ventral intermediate nucleus structural connectivity-derived segmentation: anatomical reliability and variability. Neuroimage **243**, 118519 (2021)
17. Sotiropoulos, S.N., et al.: Advances in diffusion MRI acquisition and processing in the Human Connectome Project. Neuroimage **80**, 125–143 (2013)
18. Van Essen, D.C., et al.: The WU-Minn human connectome project: an overview. Neuroimage **80**, 62–79 (2013)
19. Glasser, M.F., et al.: The minimal preprocessing pipelines for the human connectome project. Neuroimage **80**, 105–124 (2013)
20. Andersson, J.L.R., Skare, S., Ashburner, J.: How to correct susceptibility distortions in spin-echo echo-planar images: application to diffusion tensor imaging. Neuroimage **20**(2), 870–888 (2003)
21. Andersson, J.L.R., Sotiropoulos, S.N.: An integrated approach to correction for off-resonance effects and subject movement in diffusion MR imaging. Neuroimage **125**, 1063–1078 (2016)
22. Jenkinson, M., et al.: Improved optimization for the robust and accurate linear registration and motion correction of brain images. Neuroimage **17**(2), 825–841 (2002)
23. Jenkinson, M., et al.: FSL. Neuroimage **62**(2), 782–790 (2012)
24. Miller, K.L., et al.: Multimodal population brain imaging in the UK Biobank prospective epidemiological study. Nat. Neurosci. **19**(11), 1523–1536 (2016)
25. Alfaro-Almagro, F., et al.: Image processing and quality control for the first 10,000 brain imaging datasets from UK Biobank. Neuroimage **166**, 400–424 (2018)
26. Destrieux, C., et al.: Automatic parcellation of human cortical gyri and sulci using standard anatomical nomenclature. Neuroimage **53**(1), 1–15 (2010)
27. Warrington, S., et al.: XTRACT-standardised protocols for automated tractography in the human and macaque brain. Neuroimage **217**, 116923 (2020)
28. Tang, Y., et al.: A probabilistic atlas of human brainstem pathways based on connectome imaging data. Neuroimage **169**, 227–239 (2018)
29. Zheng, S., et al.: Conditional random fields as recurrent neural networks. In: Proceedings of the IEEE International Conference on Computer Vision (2015)
30. LNCS Homepage. http://www.springer.com/lncs. Accessed 4 Oct 2017

Soft-Tissue Driven Craniomaxillofacial Surgical Planning

Xi Fang[1], Daeseung Kim[2(✉)], Xuanang Xu[1], Tianshu Kuang[2],
Nathan Lampen[1], Jungwook Lee[1], Hannah H. Deng[2], Jaime Gateno[2],
Michael A. K. Liebschner[3], James J. Xia[2], and Pingkun Yan[1(✉)]

[1] Department of Biomedical Engineering and Center for Biotechnology and
Interdisciplinary Studies, Rensselaer Polytechnic Institute, Troy, NY 12180, USA
yanp2@rpi.edu
[2] Department of Oral and Maxillofacial Surgery,
Houston Methodist Research Institute, Houston, TX 77030, USA
DKim@houstonmethodist.org
[3] Department of Neurosurgery, Baylor College of Medicine,
Houston, TX 77030, USA

Abstract. In CMF surgery, the planning of bony movement to achieve
a desired facial outcome is a challenging task. Current bone driven
approaches focus on normalizing the bone with the expectation that
the facial appearance will be corrected accordingly. However, due to the
complex non-linear relationship between bony structure and facial soft-
tissue, such bone-driven methods are insufficient to correct facial deformities. Despite efforts to simulate facial changes resulting from bony
movement, surgical planning still relies on iterative revisions and educated guesses. To address these issues, we propose a soft-tissue driven
framework that can automatically create and verify surgical plans. Our
framework consists of a bony planner network that estimates the bony
movements required to achieve the desired facial outcome and a facial
simulator network that can simulate the possible facial changes resulting from the estimated bony movement plans. By combining these two
models, we can verify and determine the final bony movement required
for planning. The proposed framework was evaluated using a clinical
dataset, and our experimental results demonstrate that the soft-tissue
driven approach greatly improves the accuracy and efficacy of surgical
planning when compared to the conventional bone-driven approach.

Keywords: Deep Learning · Surgical Planning · Bony Movement ·
Bony Planner · Facial Simulator

Supplementary Information The online version contains supplementary material
available at https://doi.org/10.1007/978-3-031-43996-4_18.

1   Introduction

Craniomaxillofacial (CMF) deformities can affect the skull, jaws, and midface. When the primary cause of disfigurement lies in the skeleton, surgeons cut the bones into pieces and reposition them to restore normal alignment [12]. In this context, the focus is on correcting bone deformities, as it is anticipated that the restoration of normal facial appearance will follow automatically. Consequently, it is customary to initiate the surgical planning process by estimating the positions of normal bones. The latter has given rise to bone-driven approaches [1,9,14]. For example, methods based on sparse representation [13] and deep learning [15] have been proposed to estimate the bony shape that may lead to an acceptable facial appearance.

However, the current bone-driven methods have a major limitation, subjecting to the complex and nonlinear relationship between the bones and the draping soft-tissues. Surgeons estimate the required bony movement through trial and error, while computer-aided surgical simulation (CASS) software [14] simulates the effect on facial tissues resulting from the proposed movements. Correcting the bone deformity may not completely address the soft-tissue disfigurement. The problem can be mitigated by iteratively revising the bony movement plan and simulating the corresponding soft-tissue changes. However, this iterative planning revision is inherently time-consuming, especially when the facial change simulation is performed using computationally expensive techniques such as the finite-element method (FEM) [4]. Efforts have been made to accelerate facial change prediction using deep learning algorithms [2,5,8], which, however, do not change the iterative nature of the bone-driven approaches.

To address the above challenge, this paper proposes a novel soft-tissue driven surgical planning framework. Instead of simulating facial tissue under the guessed movement, our approach directly aims at a desired facial appearance and then determines the optimal bony movements required to achieve such an appearance without the need for iterative revisions. Unlike the bone-driven methods, this soft-tissue driven framework eliminates the need for surgeons to make educated guesses about bony movement, significantly improving the efficiency of the surgical planning process. Our proposed framework consists of two main components, the Bony Planner (BP) and the Facial Simulator (FS). The BP estimates the possible bony movement plans (bony plans) required to achieve the desired facial appearance change, while the FS verifies the effectiveness of the estimated plans by simulating the facial appearance based on the bony plans. Without the intervention of clinicians, the BP automatically creates the most clinically feasible surgical plan that achieves the desired facial appearance.

The main contributions of our work are as follows. 1) This is the first soft-tissue driven approach for CMF surgical planning, which can substantially reduce the planning time by removing the need for repetitive guessing bony movement. 2) We develop a deep learning model as the bony planner, which can estimate the underlying bony movement needed for changing a facial appearance into a targeted one. 3) The developed FS module can qualitatively assess the effect of surgical plans on facial appearance, for virtual validation.

2   Method

Figure 1 shows an overview of the proposed framework, which consists of two primary modules. The first module is a BP network that plans bony movement based on the desired facial outcome and the given preoperative facial and bony surfaces. The second module is the FS network that simulates corresponding postoperative facial appearances by applying the estimated bony plans. Instead of providing one single bony plan, the BP will estimate multiple plans because not all the generated plans may result in the desired clinical effect. The FS then simulates facial appearances by using those plans and chooses the plan leading to the facial appearance closest to the desired target. The two models are deployed together to determine and confirm the final bony plan. Below we first present the details of BP and FS modules, then we introduce how they are deployed together for inference.

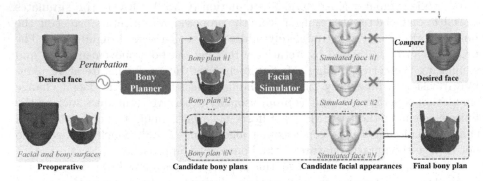

Fig. 1. Overview of the proposed soft-tissue driven framework. The framework is composed of two main components: the creation of candidate bony plans using BP, and the simulation of facial outcomes following the plans using FS. Finally, the facial outcomes are compared with the desired face to select the final bony plan.

2.1   Bony Planner (BP)

Data Preprocessing: The bony plan is created by the BP network using preoperative facial F_{pre}, bony surface B_{pre}, and desired facial surface F_{des}. The goal of the BP network is to convert the desired facial change from F_{pre} to F_{des} into rigid bony movements, denoted as T_S for each bony segment S, as required for surgical planning. However, it is very challenging to directly estimate the rigid bone transformation from the facial difference. Therefore, we first estimate the non-rigid bony movement vector field and then convert that into the rigid transformations for each bone segment. Figure 2 illustrates the BP module and the details are provided as follows. For computational efficiency, pre-facial point set $P_{F_{pre}}$, pre-bony point set $P_{B_{pre}}$, desired point set $P_{F_{des}}$ are subsampled from the pre-facial/bony and desired facial surfaces.

Non-rigid Bony Movement Vector Estimation: We adopt the Attentive Correspondence assisted Movement Transformation network (ACMT-Net) [2], which was originally proposed for facial tissue simulation, to estimate the point-wise bony displacements to acquire F_{des}. This method is capable of effectively computing the movement of individual bony points by capturing the relationship between facial points through learned affinity.

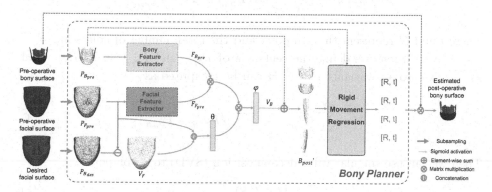

Fig. 2. Scheme of the proposed bony planner (BP). BP first estimates the non-rigidly deformed bony points based on the desired face, then regresses the rigid movements by fitting the non-rigid prediction.

In the network, point-wise facial features ($F_{F_{pre}}$) and bony features ($F_{B_{pre}}$) are extracted from $P_{F_{pre}}$ and $P_{B_{pre}}$. We used a pair of modified PointNet++ modules where the classification layers were removed and a 1D convolution layer was added at the end to project $F_{F_{pre}}$ and $F_{B_{pre}}$ into the same lower dimensions [10]. A correlation matrix R is established by computing the normalized dot product between $F_{F_{pre}}$ and $F_{B_{pre}}$ to evaluate the relationship between the facial and bony surfaces:

$$R = \frac{F_{B_{pre}}{}^{T} F_{F_{pre}}}{N_{F_{pre}}}, \qquad (1)$$

where $N_{F_{pre}}$ denotes the number of facial points $P_{F_{pre}}$. On the other hand, desired facial movement V_F is computed by subtracting $P_{F_{des}}$ and $P_{F_{pre}}$. V_F is then concatenated with $P_{F_{pre}}$ and fed into a 1D convolution layer to encode the movement information. Then the movement feature of each bony point is estimtated by the normalized summary of facial features using R. Finally, the transformed bony movement features are decoded into movement vectors after being fed into one 1D convolution layer and normalization:

$$V_B = \varphi(\theta([P_{F_{pre}}, V_F]R). \qquad (2)$$

Rigid Bony Movement Regression: Upon obtaining the estimated point-wise bony movements, they are added to the corresponding bony points, resulting in a non-rigidly transformed bony point set denoted as $P_{B_{pdt}}$. The resulting points are grouped based on their respective bony segments, with point sets $P_{S_{pre}}$ and $P_{S_{pdt}}$ representing each bony segment S before and after movement, respectively. To fit the movement between $P_{S_{pre}}$ and $P_{S_{pdt}}$, we estimate the rigid transformations $[R_S, T_S]$ by minimizing the mean square error as follows:

$$E(R_S, T_S) = \left\| \frac{1}{N} \sum_i^n (R_S P_{S_{pre}}^i + T_S - P_{S_{pdt}}^i) \right\|^2, \tag{3}$$

where i and N represent the i-th point and the total number of points in $P_{S_{pre}}$, respectively. First, we define the centroids of $P_{S_{pre}}$ and $P_{S_{pdt}}$ to be $\overline{P_{S_{pre}}}$ and $\overline{P_{S_{pdt}}}$. The cross-covariance matrix $\mathbf{H}$ can be computed as

$$\mathbf{H} = \sum_{i=1}^N (P_{S_{pre}} - \overline{P_{S_{pre}}})(P_{S_{pdt}} - \overline{P_{S_{pdt}}}) \tag{4}$$

Then we can use singular value decomposition (SVD) to decompose

$$\mathbf{H} = USV^T, \tag{5}$$

then the alignment minimizing $E(R_S, T_S)$ can be solved by

$$R_S = VU^T, T_S = -R_S \overline{P_{S_{pre}}} + \overline{P_{S_{pdt}}}. \tag{6}$$

Finally, the rigid transformation matrices are applied to their corresponding bone segments for virtual planning.

2.2 Facial Simulator (FS)

While the bony planner can estimate bony plans and we can compare them with ground truth. However, the complex relationship between bone and face makes it unknown whether adopting the bony plan will result in the desired facial outcome or not. To evaluate the effectiveness of the BP network in simulating facial soft tissue, an FS is developed to simulate the facial outcome following the estimated bony plans. For facial simulation, we employ the ACMT-Net, which takes $P_{F_{pre}}$, $P_{B_{pre}}$, and $P_{B_{pdt}}$ as input and predicts the point-wise facial movement vector V_F'. The movement vector of all vertices in the facial surface is estimated by interpolating V_F'. The simulated facial surface F_{pdt}' is then reconstructed by adding the predicted movement vectors to the vertices of F_{pre}.

2.3 Self-verified Virtual Planning

To generate a set of potential bony plans, we randomly perturbed the surfaces by flipping and translating them up to 10mm in three directions during inference. We repeated this process 10 times in our work. After estimation, the bony surfaces were re-localized to their original position prior to perturbation. Sequentially, the FS module generated a simulated facial appearance for each bony plan estimated from the BP module, serving two purposes. Firstly, it verified the feasibility of the bony plan through facial appearance. Secondly, it allowed us to evaluate the facial outcomes of different bony plans. The simulated facial surfaces were compared with the desired facial surface, and the final plan was selected based on the similarity of the resulting facial outcome. This process verified the efficacy of the selected plan for achieving the desired facial outcome.

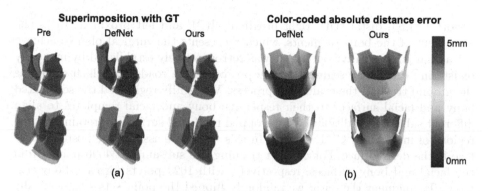

Fig. 3. Examples of the results of bony plans. (a) Visual comparison of the estimated bony surface(red) with ground truth (blue), where deeper shades of blue and red indicate shading effects. (b) Color-coded error map to display the difference between the estimated bony surface and ground truth. (Color figure online)

3   Experiments and Results

3.1   Dataset

We employed a five-fold cross-validation technique to evaluate the performance of the proposed network using 34 sets of patient CT data. We partitioned the data into five groups with {7, 7, 7, 7, 6} sets of data, respectively. During each round of validation, four of these groups (folds) were used for training and the remaining group was used for testing. The CT scans were randomly selected from our digital archive of patients who had undergone double-jaw orthognathic surgery. To obtain the necessary data, we employed a semi-automatic method to segment the facial and bony surface from the CT images [6]. Then we transformed the segmentation into surface meshes using the Marching Cube approach [7]. To retrospectively recreate the surgical plan that could "achieve" the actual postoperative outcomes, we registered the postoperative facial and bony surfaces to their respective preoperative surfaces based on surgically unaltered bony volumes, i.e., cranium, and utilized them as a roadmap [8]. To establish the surgical plan to achieve actual postoperative outcomes, virtual osteotomies were first reperformed on the preoperative bones to create bony segments, including

Table 1. Quantitative evaluation results. Prediction accuracy comparison with the state-of-the-art bone-driven methods.

Method	Mean absolute distance (mean ± std) in millimeter				
	LF	DI	RP	LP	Entire Bone
DefNet [15]	2.69 ± 1.18	6.45 ± 1.79	3.08 ± 1.15	3.23 ± 1.29	3.86 ± 0.91
BP	2.42 ± 1.13	3.16 ± 1.38	2.58 ± 1.29	2.42 ± 1.16	2.64 ± 0.97
BP + FS	2.46 ± 1.13	2.89 ± 1.64	2.56 ± 1.30	2.41 ± 1.13	2.58 ± 0.93

LeFort 1 (LF), distal (DI), right proximal (RP), and left proximal (LP). The movement of the bony segments, which represents the surgical plan to achieve the actual postoperative outcomes, was retrospectively established by manually registering each bony segment to the postoperative roadmap, which served as the ground truth in the evaluation process. We rigidly registered the segmented bony and facial surfaces to their respective bony and facial templates to align different subjects. Additionally, we cropped the facial surfaces to retain only the regions of interest for CMF surgery. In this study, we assume the postoperative face is the desired face. For efficient training, we subsampled 4096 points from the facial and bony surfaces, respectively, with 1024 points for each bony segment. To augment the data, we randomly flipped the point sets symmetrically and translated them along three directions within a range of 10 mm.

3.2   Implementation and Evaluation Methods

To compare our approach, we implemented the state-of-the-art bone-driven approach, i.e., deformation network (DefNet) [15], which takes the point coordinates and their normal vectors as input and generates the displacement vectors to deform the preoperative bones. Our method has two variations: BP and BP+FS. BP estimates only one plan, while BP+FS selects the final plan based on the simulated facial outcomes of our facial simulator. The PointNet++ networks utilized in both BP and FS are comprised of four feature-encoding blocks and four feature-decoding blocks. The output dimensions for each block are 128, 256, 512, 1024, 512, 256, 128, and 128, respectively, and the output point numbers of the modules are 1024, 512, 256, 64, 256, 512, 1024, and 4096, sequentially. We used the Adam optimizer with a learning rate of 0.001, Beta1 of 0.9, and Beta2 of 0.999 to train both the BP and FS networks. For data augmentation, both facial and bone data are subjected to the same random flipping and translation, ensuring that the relative position and scale of facial changes and bony movements remain unchanged. The models were trained for 500 epochs with a batch size of 4 and MSE loss was used, after which the models were used for evaluation. The models were trained on an NVIDIA DGX-1 deep learning server with eight V100 GPUs.

The prediction accuracy of the soft-tissue driven approach was evaluated quantitatively and qualitatively by comparing it with DefNet. Then quantitative evaluation was to assess the accuracy using the mean absolute error (MAE) between the predicted bony surfaces and the ground truth. For detailed evaluation, MAE was also separately calculated for each bony segment, including LF, DI, RP, and LP. Statistical significance was determined using the Wilcoxon signed-rank test to compare the results obtained by different methods [11]. Qualitative evaluation was carried out by directly comparing the bony surfaces and simulated facial outcomes generated by our approach and DefNet with the ground truth postoperative bony and facial surfaces.

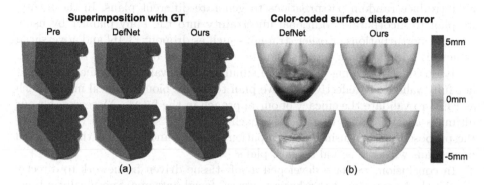

Fig. 4. Examples of the results of simulated facial outcomes. (a) Comparison of the simulated facial outcome(red) with ground truth (blue). (b) Color-coded error map of the simulated facial outcome compared with ground truth. (Color figure online)

3.3   Results

The results of the quantitative evaluation are shown in Table 1. The results of the Wilcoxon signed-rank test showed that BP outperforms DefNet on LF ($p < 0.05$), DI ($p < 0.001$), LP segments ($p < 0.01$), and the entire bone ($p < 0.001$) by statistically significant margins. In addition, BP+FS significantly outperforms BP on DI segment ($p < 0.05$). Two randomly selected patients are shown in Fig. 3. To make a clear visual comparison, we superimposed the estimated bony surfaces (Blue) with their corresponding ground truth (Red) and set the surfaces to be transparent as shown in Fig. 3(a). Figure 3(b) displays the surface distance error between the estimated bony surfaces and ground truth.

The evaluation of the bony surface showed that the proposed method successfully predicted bony surfaces that were similar to the real postoperative bones. To further assess the performance of the method, a facial simulator was used to qualitatively verify the simulated faces from the surgical plans. Figure 4 shows the comparison of the simulated facial outcomes of different methods using FS network and the desired facial appearance. The facial outcomes derived from bony plans of DefNet and our method are visualized, and the preoperative face is also superimposed with GT for reference. The results of the facial appearance simulation further validate the feasibility of the proposed learning-based framework for bony movement estimation and indicate that our method can achieve comparable simulation accuracy with the real bony plan.

4   Discussions and Conclusions

As a result of our approach that considers both the bony and soft-tissue components of the deformity, the accuracy of the estimated bony plan, especially on the DI segment, has significantly improved. Nonetheless, the non-linear relationship between the facial and bony surfaces cannot be adequately learned using only facial and bony surface data, and additional information such as tissue properties can also affect the facial outcome. To account for uncertainty, we introduce random perturbations to generate different plans. In the future, we plan to incorporate additional uncertainty into the bony planner by using stronger perturbations or other strategies such as dropout [3,16] and adversarial attacks [17,18], which could help create more diverse bony plans. Also, relying solely on a deep learning-based facial simulation to evaluate our method might not fully validate its effectiveness. We plan to utilize biomechanical models such as FEM to validate the efficacy of our approach in the future. Moreover, for our ultimate goal of translating the approach to clinical settings, we will validate the proposed method using a larger patient dataset and compare the predicted bony plans with the actual surgical plans.

In conclusion, we have developed a soft-tissue driven framework to directly predict the bony plans that achieve a desired facial outcome. Specifically, a bony planner and a facial simulator have been proposed for generating bony plans and verifying their effects on facial appearance. Evaluation results on a clinical dataset have shown that our method significantly outperforms the traditional bone-driven approach. By adopting this approach, we can create a virtual surgical plan that can be assessed and adjusted before the actual surgery, reducing the likelihood of complications and enhancing surgical outcomes. The proposed soft-tissue driven framework can potentially improve the accuracy and efficiency of CMF surgery planning by automating the process and incorporating a facial simulator to account for the complex non-linear relationship between bony structure and facial soft-tissue.

Acknowledgements. This work was partially supported by NIH under awards R01 DE022676, R01 DE027251, and R01 DE021863.

References

1. Bobek, S., Farrell, B., Choi, C., Farrell, B., Weimer, K., Tucker, M.: Virtual surgical planning for orthognathic surgery using digital data transfer and an intraoral fiducial marker: the charlotte method. J. Oral Maxillofac. Surg. **73**(6), 1143–1158 (2015)
2. Fang, X., et al.: Deep learning-based facial appearance simulation driven by surgically planned craniomaxillofacial bony movement. In: Wang, L., Dou, Q., Fletcher, P.T., Speidel, S., Li, S. (eds.) MICCAI 2022. LNCS, vol. 13437, pp. 565–574. Springer, Cham (2022). https://doi.org/10.1007/978-3-031-16449-1_54
3. Gal, Y., Ghahramani, Z.: Dropout as a Bayesian approximation: representing model uncertainty in deep learning. In: International Conference on Machine Learning, pp. 1050–1059. PMLR (2016)

4. Kim, D., et al.: A novel incremental simulation of facial changes following orthognathic surgery using fem with realistic lip sliding effect. Med. Image Anal. **72**, 102095 (2021)
5. Lampen, N., et al.: Deep learning for biomechanical modeling of facial tissue deformation in orthognathic surgical planning. Int. J. Comput. Assist. Radiol. Surg. **17**(5), 945–952 (2022)
6. Liu, Q., et al.: SkullEngine: a multi-stage CNN framework for collaborative CBCT image segmentation and landmark detection. In: Lian, C., Cao, X., Rekik, I., Xu, X., Yan, P. (eds.) MLMI 2021. LNCS, vol. 12966, pp. 606–614. Springer, Cham (2021). https://doi.org/10.1007/978-3-030-87589-3_62
7. Lorensen, W.E., Cline, H.E.: Marching cubes: a high resolution 3D surface construction algorithm. ACM Siggraph comput. graph. **21**(4), 163–169 (1987)
8. Ma, L., et al.: Deep simulation of facial appearance changes following craniomaxillofacial bony movements in orthognathic surgical planning. In: de Bruijne, M., et al. (eds.) MICCAI 2021. LNCS, vol. 12904, pp. 459–468. Springer, Cham (2021). https://doi.org/10.1007/978-3-030-87202-1_44
9. McCormick, S.U., Drew, S.J.: Virtual model surgery for efficient planning and surgical performance. J. Oral Maxillofac. Surg. **69**(3), 638–644 (2011)
10. Qi, C.R., Yi, L., Su, H., Guibas, L.J.: PointNet++: deep hierarchical feature learning on point sets in a metric space. In: Advances in Neural Information Processing Systems, vol. 30 (2017)
11. Rey, D., Neuhäuser, M.: Wilcoxon-signed-rank test. In: Lovric, M. (ed.) International Encyclopedia of Statistical Science, pp. 1658–1659. Springer, Heidelberg (2011). https://doi.org/10.1007/978-3-642-04898-2_616
12. Shafi, M., Ayoub, A., Ju, X., Khambay, B.: The accuracy of three-dimensional prediction planning for the surgical correction of facial deformities using Maxilim. Int. J. Oral Maxillofac. Surg. **42**(7), 801–806 (2013)
13. Wang, L., et al.: Estimating patient-specific and anatomically correct reference model for craniomaxillofacial deformity via sparse representation. Med. Phys. **42**(10), 5809–5816 (2015)
14. Xia, J., et al.: Algorithm for planning a double-jaw orthognathic surgery using a computer-aided surgical simulation (CASS) protocol. Part 1: planning sequence. Int. J. Oral Maxillofac. Surg. **44**(12), 1431–1440 (2015)
15. Xiao, D., et al.: Estimating reference bony shape models for orthognathic surgical planning using 3D point-cloud deep learning. IEEE J. Biomed. Health Inform. **25**(8), 2958–2966 (2021)
16. Xu, X., Sanford, T., Turkbey, B., Xu, S., Wood, B.J., Yan, P.: Shadow-consistent semi-supervised learning for prostate ultrasound segmentation. IEEE Trans. Med. Imaging **41**(6), 1331–1345 (2022)
17. Zhang, J., Chao, H., Dasegowda, G., Wang, G., Kalra, M.K., Yan, P.: Overlooked trustworthiness of saliency maps. In: Wang, L., Dou, Q., Fletcher, P.T., Speidel, S., Li, S. (eds.) MICCAI 2022. LNCS, vol. 13433, pp. 451–461. Springer, Cham (2022). https://doi.org/10.1007/978-3-031-16437-8_43
18. Zhang, J., et al.: When neural networks fail to generalize? A model sensitivity perspective. In: Proceedings of the AAAI Conference on Artificial Intelligence, vol. 37, no. 9, pp. 11219–11227 (2023)

ACT-Net: Anchor-Context Action Detection in Surgery Videos

Luoying Hao[1,2], Yan Hu[2(✉)], Wenjun Lin[2,3], Qun Wang[4], Heng Li[2], Huazhu Fu[5], Jinming Duan[1(✉)], and Jiang Liu[2(✉)]

[1] School of Computer Science, University of Birmingham, Birmingham, UK
j.duan@bham.ac.uk
[2] Research Institute of Trustworthy Autonomous Systems and Department of Computer Science and Engineering, Southern University of Science and Technology, Shenzhen, China
{huy3,liuj}@sustech.edu.cn
[3] Department of Mechanical Engineering, National University of Singapore, Singapore, Singapore
[4] Third Medical Center of Chinese PLAGH, Beijing, China
[5] Institute of High Performance Computing (IHPC), Agency for Science, Technology and Research (A*STAR), Singapore, Singapore

Abstract. Recognition and localization of surgical detailed actions is an essential component of developing a context-aware decision support system. However, most existing detection algorithms fail to provide high-accuracy action classes even having their locations, as they do not consider the surgery procedure's regularity in the whole video. This limitation hinders their application. Moreover, implementing the predictions in clinical applications seriously needs to convey model confidence to earn entrustment, which is unexplored in surgical action prediction. In this paper, to accurately detect fine-grained actions that happen at every moment, we propose an anchor-context action detection network (ACT-Net), including an anchor-context detection (ACD) module and a class conditional diffusion (CCD) module, to answer the following questions: 1) where the actions happen; 2) what actions are; 3) how confidence predictions are. Specifically, the proposed ACD module spatially and temporally highlights the regions interacting with the extracted anchor in surgery video, which outputs action location and its class distribution based on anchor-context interactions. Considering the full distribution of action classes in videos, the CCD module adopts a denoising diffusion-based generative model conditioned on our ACD estimator to further reconstruct accurately the action predictions. Moreover, we utilize the stochastic nature of the diffusion model outputs to access model confidence for each prediction. Our method reports the state-of-the-art performance, with improvements of 4.0% mAP against baseline on the surgical video dataset.

L. Hao and Y. Hu—Co-first authors.

Supplementary Information The online version contains supplementary material available at https://doi.org/10.1007/978-3-031-43996-4_19.

Keywords: Action detection · Anchor-context · Conditional diffusion · Surgical video

1   Introduction

Surgery is often an effective therapy that can alleviate disabilities and reduce the risk of death from common conditions [17]. While surgical procedures are intended to save lives, errors within the surgery may bring great risks to the patient and even cause sequelae [18], which emphasizes the development of a computer-assisted system. A context-aware assistant system for surgery can not only decrease intraoperative adverse events, and enhance the quality of interventional healthcare [28], but also contribute to surgeon training, and assist procedure planning and retrospective analysis [9].

Designing intelligent assistance systems for operating rooms requires an understanding of surgical scenes and procedures [20]. Most current works pay attention to phase and step recognition [3,27], which is to get the major types of events that occurred during the surgery. They merely provided very coarse descriptions of scenes. As the granularity of action increases, the clinical utility becomes more valuable in providing an accurate depiction of detailed motion [13,19]. Recent studies focus on fine-grained action recognition by modelling action as a group of the instrument, its role, and its target anatomy and capturing their associations [7,26]. Recognizing targets in different methods is dependent on different surgical scenarios and it also significantly increases the complexity and time consumption for anatomy annotation [30]. In addition, although most existing methods can provide accurate action positions, the predicted action class is often inaccurate. Moreover, they do not provide any information about the reliability of their output, which is a key requirement for integrating into assistance systems of surgery [11]. Thus, we propose a reliable surgical action detection method in this paper, with high-accuracy action predictions and their confidence.

Mistrust is a major barrier to deep-learning-based predictions applied to clinical implementation [14]. Existing works measuring the model uncertainty [1,8] often need several-time re-evaluations, and store multiple sets of weights. It is hard for them to apply to surgery assistance applications to get confidence for each prediction directly [10], and they are limited to improving prediction performance. Conditional diffusion-based generative models have received significant attention due to their ability to accurately recover the full distribution of data guided by conditions from the perspective of diffusion probabilistic models [24]. However, they focus on generating high-resolution photo-realistic images. Instead, after observing our surgical video dataset, our conditional diffusion model aims to reconstruct accurately class distribution. We also access the estimation of confidence with the stochastic nature of the diffusion model.

Here, to predict accurately micro-action (fine-grained action) categories happening every moment, we achieve it with two modules. Specifically, a novel anchor-context module for action detection is proposed to highlight the spatio-temporal regions that are interacted with the anchors (we extract instrument

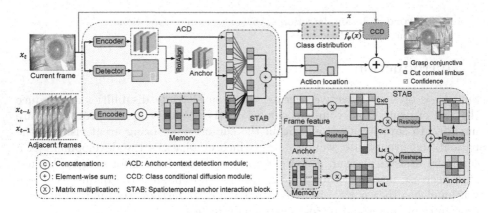

Fig. 1. The pipeline of our ACTNet includes ACD and CCD modules.

features as anchors), which includes surrounding tissues and movement information. Then, with the constraints of class distributions and the surgical videos, we propose a conditional diffusion model to cover the whole distribution of our data and to accurately reconstruct new predictions based on full learning. Furthermore, our class conditional diffusion model also accesses uncertainty for each prediction, through the stochasticity of outputs.

We summarize our main contributions as follows: 1) We develop an anchor-context action detection network (ACTNet), including an anchor-context detection (ACD) module and a class conditional diffusion (CCD) module, which combines three tasks: i) where actions locate; ii) what actions are; iii) how confident our model is about predictions. 2) For ACD module, we develop a spatio-temporal anchor interaction block (STAB) to spatially and temporally highlight the context related to the extracted anchor, which provides micro-action location and initial class. 3) By conditioning on the full distribution of action classes in the surgical videos, our proposed class conditional diffusion (CCD) model reconstructs better class prototypes in a stochastic fashion, to provide a more accurate estimations and push the assessment of the model confidence in its predictions. 4) We carry out comparison and ablation study experiments to demonstrate the effectiveness of our proposed algorithm based on cataract surgery.

2   Methodology

The overall framework of our proposed ACTNet for reliable action detection is illustrated in Fig. 1. Based on a video frame sequence, the ACD module extracts anchor features and aggregates the spatio-temporal interactions with anchor features by proposed STAB, which generates action locations and initial action class distributions. Then considering the full distribution of action classes in surgical videos, we use the CCD module to refine the action class predictions and access confidence estimations.

2.1 Our ACD Module

Anchor Extraction: Assuming a video X with T frames, denoted as $X = \{x_t\}_{t=1}^T$, where x_t is the t-th frame of the video. The task of this work is to estimate all potential locations and classes $P = \{box_n, c_n\}_{n=1}^N$ for action instances contained in video X, where box_n is the position of the n-th action happened, c_n is the action class of n-th action, and N is the number of action instances. For video representation, this work tries to encode the original videos into features based on the backbone ResNet50 [5] network to get each frame's feature $F = \{f_t\}_{t=1}^T$.

In surgical videos, the instruments, as action subjects, are significant to recognize the action. For instrument detection, it is very important but not very complicated. Existing excellent object detection method like Faster R-CNN [22] is enough to obtain results with high accuracy. After getting the detected instrument anchors, RoIAlign is applied to extract the instrument features from frame features. The instrument features are denoted as $I = \{i_t\}_{t=1}^T$. Since multiple instruments exist in surgeries, our action detection needs to solve the problem that related or disparate concurrent actions often lead to wrong predictions. Thus, in this paper, we propose to provide action location and class considering the spatio-temporal anchor interactions in the surgical videos, based on STAB.

Spatio-Temporal Action interaction Block (STAB): For several actions like pushing, pulling, and cutting, there is no difference just inferred from the local region in one frame. Thus we propose STAB to utilize spatial and temporal interactions with an anchor to improve the prediction accuracy of the action class, which finds actions with strong logical links to provide an accurate class. The structure of STAB is shown in Fig. 1. We introduce spatial and temporal interactions respectively in the following.

For spatial interaction: The instrument feature i_t acts as the anchor. In order to improve the anchor features, the module has the ability to select value features that are highly active with the anchor features and merge them. The formulation is defined as: $a_t = \frac{1}{C(f_t)} \sum_{j \in S_j} h(f_{tj}, i_t) g(f_{tj})$, where j is the index that enumerates all possible positions of f_t. A pairwise function $h(\cdot)$ computes the relationship such as affinity between i_t and all f_{tj}. In this work, dot-product is employed to compute the similarity. The unary function $g(f_{tj})$ computes a representation of the input signal at the position j. The response is normalized by a factor $C(f_t) = \sum_{j \in S_j} h(f_{tj}, i_t)$. S_j represents the set of all positions j. Through the formulation, the output a_t obtains more information from the positions related to the instrument and catches interactions in space for the actions.

For temporal interaction: We build memory features consisting of features in consecutive frames: $m_t = [f_{t-L}, ..., f_{t-1}]$. To effectively model temporal interactions of the anchor, the network offers a powerful tool for capturing the complex and dynamic dependencies that exist between elements in sequential data and anchors. Same with the spatial interaction, we take i_t as an anchor and calculate the interactions between the memory features and the anchor. The formulation

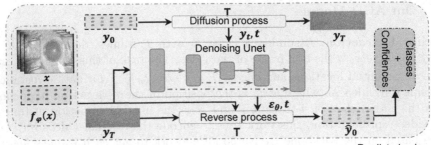

· ⇢ : Skip connection; $(f_\varphi(x), x)$: Conditional variable; T: Diffusion steps; ε_θ: Predicted noise

Fig. 2. Overview of the class conditional diffusion (CCD) model.

is defined as: $b_t = \frac{1}{C(m_t)} \sum_{j \in T_j} h(m_{tj}, i_t) g(m_{tj})$, where T_j refers to the set of all possible positions along the time series in the range of L. In this way, temporal interactions with anchors are obtained. Then a global average pooling is carried out on the spatial and temporal outputs. Action localizations and initial action class distributions are produced based on the fully-connected classifier layer.

2.2 CCD Module for Reliable Action Detection

Since the surgical procedures follow regularity, we propose a CCD module to reconstruct the action class predictions considering the full distribution of action classes in videos. The diffusion conditioned on the action classes and surgical videos is adopted in our paper. Let $y_0 \in \mathbb{R}^n$ be a sample from our data distribution. As shown in Fig. 2, a diffusion model specified in continuous time is a generative model with latent y_t, obeying a forward process $q_t(y_t|y_{t-1})$ starting at data y_0 [6]. y_0 indicates a one-hot encoded label vector. We treat each one-hot label as a class prototype, i.e., we assume a continuous data and state space, which enables us to keep the Gaussian diffusion model framework [2,6]. The forward process and reverse process of unconditional diffusion are provided in the supplementary material.

Here, for the diffusion model optimization can be better guided by meaningful information, we integrate the ACD and our surgical video data as priors or constraints in the diffusion training process. We design a conditional diffusion model $\hat{p}_\theta(y_{t-1}|y_t, x)$ that is conditioned on an additional latent variable x. Specifically, the model $\hat{p}_\theta(y_{t-1}|y_t, x)$ is built to approximate the corresponding tractable ground-truth denoising transition step $\hat{p}_t(y_{t-1}|y_t, y_0, x)$. We specify the reverse process with conditional distributions as [21]:

$$\hat{p}_t(y_{t-1}|y_t, y_0, x) = \hat{p}_t(y_{t-1}|y_t, y_0, f_\varphi(x)) = N\left(y_{t-1}; \hat{\mu}(y_t, y_0, f_\varphi(x)), \hat{\beta}_t I\right)$$

where $\hat{\mu}(y_t, y_0, f_\varphi(x))$ and $\hat{\beta}_t$ are described in supplementary material. $f_\varphi(x)$ is the prior knowledge of the relation between x and y_0, i.e., the ACD module pre-trained with our surgical video dataset. The x indicates the input surgical video frames. Since ground-truth step $\hat{p}_t(y_{t-1}|y_t, y_0, x)$ cannot get directly, the model $\hat{p}_\theta(y_{t-1}|y_t, x)$ are trained by following loss function for estimating ϵ_θ to approximate the ground truth:

$$\hat{L}(\theta) = E_{t, y_0, \epsilon, x}\left[\|\epsilon - \epsilon_\theta(x, \sqrt{\bar{\alpha}_t}y_0 + \sqrt{1 - \bar{\alpha}_t}\epsilon + (1 - \sqrt{\bar{\alpha}_t})f_\varphi(x), f_\varphi(x), t)\|^2\right]$$

where $\alpha_t := 1 - \beta_t$, $\bar{\alpha}_t := \prod_{s=1}^{t}(1 - \beta_s)$, $\epsilon \sim N(0, 1)$ and $\epsilon_\theta(\cdot)$ estimates ϵ using a time-conditional network parameterized by θ. β_t is a constant hyperparameter.

To produce model confidence for each action instance, we mainly calculate the prediction interval width (IW). Specifically, we first sample N class prototype reconstruction with the trained diffusion model. Then calculate the IW between the 2.5^{th} and 97.5^{th} percentiles of the N reconstructed values for all test classes. Compared with traditional classifiers to get deterministic outputs, the denoising diffusion model is a preferable modelling choice due to its ability to produce stochastic outputs, which enables confidence generation.

3   Experimental Results

Cataract Surgical Video Dataset: To perform reliable action detection, we build a cataract surgical video dataset. Cataract surgery is a procedure to remove the lens of the eyes and, in most cases, replace it with an artificial lens. The dataset consists of 20 videos with a frame rate of 1 fps (a total of 17511 frames and 28426 action instances). Under the direction of ophthalmologists, each video is labelled frame by frame with the categories and locations of the actions. 49 types of action bounding boxes as well as class labels are included in our dataset. The surgical video dataset is randomly split into a training set with 15 videos (13583 frames) and a testing set with 5 videos (3928 frames).

Implementation Details: The proposed architecture is implemented using the publicly available Pytorch Library. A model with ResNet50 backbone from Faster R-CNN-benchmark [23] is adopted for our instrument anchor detection. In STAB, we use ten adjacent frames. During inference, detected anchor boxes with a confidence score larger than 0.8 are used. More implementation details are listed in the supplementary material. The performances are evaluated with official metric frame level mean average precision (mAP) at IoU = 0.1, 0.3, and 0.5, respectively, obtaining figures in the following named mAP_{10}, mAP_{30} and mAP_{50} with their mean mAP_{mean}.

Method Comparison: In order to demonstrate the superiority of the proposed method for surgical action detection, we carry out a comprehensive comparison between the proposed method and the following state-of-the-art methods: 1) single-stage algorithms, including the Single Shot Detector (SSD) [16], SSDLite [25] and RetinaNet [12]. 2) two-stage algorithms, including Faster R-CNN [23], Mask R-CNN [4], Dynamic R-CNN [29] and OA-MIL [15]. The data presented in Table 1 clearly demonstrate that our method outperforms other approaches,

Table 1. Methods comparison and ablation study on cataract video dataset.

Methods	mAP_{10}	mAP_{30}	mAP_{50}	mAP_{mean}
Faster R-CNN [23]	0.388	0.384	0.371	0.381
SSD [16]	0.360	0.358	0.350	0.356
RetinaNet [12]	0.358	0.356	0.347	0.354
Mask R-CNN [4]	0.375	0.373	0.363	0.370
SSDlite [25]	0.305	0.304	0.298	0.302
Dynamic R-CNN [29]	0.315	0.310	0.296	0.307
OA-MIL [15]	0.395	0.394	0.378	0.389
backbone	0.373	0.365	0.360	0.366
+temporal	0.385	0.378	0.372	0.378
+spatial	0.394	0.385	0.377	0.385
+STAB	0.400	0.393	0.385	0.393
+CCD (ACTNet)	**0.415**	**0.406**	**0.397**	**0.406**

irrespective of the IoU threshold being set to 0.1, 0.3, 0.5, or the average values. Notably, the results obtained after incorporating diffusion even surpass Faster R-CNN by 2.5% and baseline by 4.0% in terms of average mAP. This finding provides compelling evidence for the efficacy of our method in integrating spatio-temporal interactive information under the guidance of anchors and leveraging diffusion to optimize the category distribution. The quantitative results further corroborate the effectiveness of our approach in Fig. 3, which shows that our model does not only improve the performance of the baseline models but also localizes accurately the regions of interest of the actions. More results are listed in the material.

Ablation Study: To validate the effectiveness of our ACTNet, we have done some ablation studies. We train and test the model with spatial interaction, temporal interaction, spatio-temporal interaction (STAB), and finally together with our CCD model. The testing results are shown in Fig. 3 and Table 1. For our backbone, it is achieved by concatenating the anchor features through RoIAlign and the corresponding frame features to get the detected action classes.

The results reveal that the spatial and temporal interactions for instruments can provide useful information to detect the actions. What's more, spatial interaction has slightly better performance than temporal interaction. It may be led by the number of spatially related action categories being slightly more than that of temporally related action categories. It is worth noting that spatial interaction and temporal interaction can be enhanced by each other and achieve optimal performance. After being enhanced by the diffusion model conditioned on our obtained class distributions and video frames, we get optimal performance.

Confidence Analysis: To analyze the model confidence, we take the best prediction for each instance to calculate the instance accuracy. We can observe

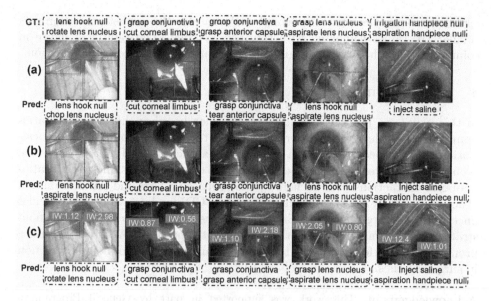

Fig. 3. Visualization on cataract dataset. We choose different actions to show the results of (a) Faster R-CNN, (b) OA_MIL, and (c) our ACTNet. For each example, we show the ground-truth (Blue), right predictions (Green) and wrong predictions (Red). The actions are labelled from left to right. IW values (multiplied by 100) mean prediction interval width to show the level of confidence. (Color figure online)

from Table 2 across the test set, the mean_IW of the class label among correctly classified instances by ACTNet is significantly narrower compared to that of incorrectly classified instances. This observation indicates that the model is more confident in its accurate predictions and is more likely to make errors when its predictions are vague. Furthermore, upon comparing the mean_IW at the true class level, we find that a more precise class tends to exhibit a larger disparity between the correct and incorrect predictions. Figure 3 also shows the

Table 2. The mean_IW (multiplied by 100) results from our CCD module.

Class	Instance	Accuracy	Mean_IW (Correct)	Mean_IW (Incorrect)
grasp conjunctiva	487	0.702	0.91 (342)	9.78 (145)
aspirate lens cortex	168	0.613	1.37 (103)	15.82 (65)
chop lens nucleus	652	0.607	0.54 (396)	9.73 (256)
polish intraocular lens	222	0.572	0.90 (127)	8.48 (95)
aspirate lens nucleus	621	0.554	0.76 (344)	10.30 (277)
inject viscoelastic	112	0.536	2.17 (60)	9.14 (52)
Remove lens cortex	174	0.471	0.42 (82)	5.84 (92)
forceps null	280	0.464	2.67 (130)	8.38 (150)

confidence estimations for some samples. We can see the correct prediction gets smaller IW values compared with the incorrect one (The rightmost figure in column (c)), which means it has more uncertainty for the incorrect prediction.

4 Conclusions

In this paper, we propose a conditional diffusion-based anchor-context spatio-temporal action detection network (ACTNet) to achieve recognition and localization of every occurring action in the surgical scenes. ACTNet improves the accuracy of the predicted action class from two considerations, including spatio-temporal interactions with anchors by the proposed STAB and full distribution of action classes by class conditional diffusion (CCD) module, which also provides uncertainty in surgical scenes. Experiments based on cataract surgery demonstrate the effectiveness of our method. Overall, the proposed ACTNet presents a promising avenue for improving the accuracy and reliability of action detection in surgical scenes.

Acknowledgement. This work was supported in part by General Program of National Natural Science Foundation of China (82272086 and 82102189), Guangdong Basic and Applied Basic Research Foundation (2021A1515012195), Shenzhen Stable Support Plan Program (20220815111736001 and 20200925174052004), and Agency for Science, Technology and Research (A*STAR) Advanced Manufacturing and Engineering (AME) Programmatic Fund (A20H4b0141) and Central Research Fund (CRF).

References

1. Gal, Y., Ghahramani, Z.: Dropout as a Bayesian approximation: representing model uncertainty in deep learning. In: International Conference on Machine Learning, pp. 1050–1059. PMLR (2016)
2. Han, X., Zheng, H., Zhou, M.: CARD: classification and regression diffusion models. arXiv preprint arXiv:2206.07275 (2022)
3. Hashimoto, D.A., et al.: Computer vision analysis of intraoperative video: automated recognition of operative steps in laparoscopic sleeve gastrectomy. Ann. Surg. **270**(3), 414 (2019)
4. He, K., Gkioxari, G., Dollár, P., Girshick, R.: Mask R-CNN. In: Proceedings of the IEEE International Conference on Computer Vision, pp. 2961–2969 (2017)
5. He, K., Zhang, X., Ren, S., Sun, J.: Deep residual learning for image recognition. In: Proceedings of the IEEE Conference on Computer Vision and Pattern Recognition, pp. 770–778 (2016)
6. Ho, J., Jain, A., Abbeel, P.: Denoising diffusion probabilistic models. In: Advances in Neural Information Processing Systems, vol. 33, pp. 6840–6851 (2020)
7. Islam, M., Seenivasan, L., Ming, L.C., Ren, H.: Learning and reasoning with the graph structure representation in robotic surgery. In: Martel, A.L., et al. (eds.) MICCAI 2020, Part III. LNCS, vol. 12263, pp. 627–636. Springer, Cham (2020). https://doi.org/10.1007/978-3-030-59716-0_60
8. Lakshminarayanan, B., Pritzel, A., Blundell, C.: Simple and scalable predictive uncertainty estimation using deep ensembles. In: Advances in Neural Information Processing Systems, vol. 30 (2017)

9. Lalys, F., Jannin, P.: Surgical process modelling: a review. Int. J. Comput. Assist. Radiol. Surg. **9**, 495–511 (2014). https://doi.org/10.1007/s11548-013-0940-5

10. Lee, Y., et al.: Localization uncertainty estimation for anchor-free object detection. In: Karlinsky, L., Michaeli, T., Nishino, K. (eds.) ECCV 2022, Part VIII. LNCS, vol. 13808, pp. 27–42. Springer, Cham (2023). https://doi.org/10.1007/978-3-031-25085-9_2

11. Leibig, C., Allken, V., Ayhan, M.S., Berens, P., Wahl, S.: Leveraging uncertainty information from deep neural networks for disease detection. Sci. Rep. **7**(1), 1–14 (2017)

12. Lin, T.Y., Goyal, P., Girshick, R., He, K., Dollár, P.: Focal loss for dense object detection. In: Proceedings of the IEEE International Conference on Computer Vision, pp. 2980–2988 (2017)

13. Lin, W., et al.: Instrument-tissue interaction quintuple detection in surgery videos. In: Wang, L., Dou, Q., Fletcher, P.T., Speidel, S., Li, S. (eds.) MICCAI 2022, Part VII. LNCS, vol. 13437, pp. 399–409. Springer, Cham (2022). https://doi.org/10.1007/978-3-031-16449-1_38

14. Linegang, M.P., et al.: Human-automation collaboration in dynamic mission planning: a challenge requiring an ecological approach. In: Proceedings of the Human Factors and Ergonomics Society Annual Meeting, vol. 50, pp. 2482–2486. SAGE Publications Sage, Los Angeles (2006)

15. Liu, C., Wang, K., Lu, H., Cao, Z., Zhang, Z.: Robust object detection with inaccurate bounding boxes. In: Avidan, S., Brostow, G., Cissé, M., Farinella, G.M., Hassner, T. (eds.) ECCV 2022, Part X. LNCS, vol. 13670, pp. 53–69. Springer, Cham (2022). https://doi.org/10.1007/978-3-031-20080-9_4

16. Liu, W., et al.: SSD: single shot multibox detector. In: Leibe, B., Matas, J., Sebe, N., Welling, M. (eds.) ECCV 2016, Part I. LNCS, vol. 9905, pp. 21–37. Springer, Cham (2016). https://doi.org/10.1007/978-3-319-46448-0_2

17. Mersh, A.T., Melesse, D.Y., Chekol, W.B.: A clinical perspective study on the compliance of surgical safety checklist in all surgical procedures done in operation theatres, in a teaching hospital, Ethiopia, 2021: a clinical perspective study. Ann. Med. Surg. **69**, 102702 (2021)

18. Nepogodiev, D., et al.: Global burden of postoperative death. The Lancet **393**(10170), 401 (2019)

19. Nwoye, C.I., et al.: Rendezvous: attention mechanisms for the recognition of surgical action triplets in endoscopic videos. Med. Image Anal. **78**, 102433 (2022)

20. Padoy, N.: Machine and deep learning for workflow recognition during surgery. Minim. Invasive Ther. Allied Technol. **28**(2), 82–90 (2019)

21. Pandey, K., Mukherjee, A., Rai, P., Kumar, A.: DiffuseVAE: efficient, controllable and high-fidelity generation from low-dimensional latents. arXiv preprint arXiv:2201.00308 (2022)

22. Ren, S., He, K., Girshick, R., Sun, J.: Faster R-CNN: towards real-time object detection with region proposal networks. In: Advances in Neural Information Processing Systems, vol. 28 (2015)

23. Ren, S., He, K., Girshick, R., Sun, J.: Faster R-CNN: towards real-time object detection with region proposal networks. IEEE Trans. Pattern Anal. Mach. Intell. **39**(6), 1137–1149 (2016)

24. Rombach, R., Blattmann, A., Lorenz, D., Esser, P., Ommer, B.: High-resolution image synthesis with latent diffusion models. In: 2022 IEEE CVF Conference on Computer Vision and Pattern Recognition (CVPR), pp. 10674–10685 (2022)

25. Sandler, M., Howard, A., Zhu, M., Zhmoginov, A., Chen, L.C.: MobileNetV2: inverted residuals and linear bottlenecks. In: Proceedings of the IEEE Conference on Computer Vision and Pattern Recognition, pp. 4510–4520 (2018)
26. Seenivasan, L., Mitheran, S., Islam, M., Ren, H.: Global-reasoned multi-task learning model for surgical scene understanding. IEEE Robot. Autom. Lett. **7**(2), 3858–3865 (2022). https://doi.org/10.1109/LRA.2022.3146544
27. Twinanda, A.P., Shehata, S., Mutter, D., Marescaux, J., De Mathelin, M., Padoy, N.: EndoNet: a deep architecture for recognition tasks on laparoscopic videos. IEEE Trans. Med. Imaging **36**(1), 86–97 (2016)
28. Vercauteren, T., Unberath, M., Padoy, N., Navab, N.: CAI4CAI: the rise of contextual artificial intelligence in computer-assisted interventions. Proc. IEEE **108**(1), 198–214 (2019)
29. Zhang, H., Chang, H., Ma, B., Wang, N., Chen, X.: Dynamic R-CNN: towards high quality object detection via dynamic training. In: Vedaldi, A., Bischof, H., Brox, T., Frahm, J.-M. (eds.) ECCV 2020, Part XV. LNCS, vol. 12360, pp. 260–275. Springer, Cham (2020). https://doi.org/10.1007/978-3-030-58555-6_16
30. Zhang, J., et al.: Automatic keyframe detection for critical actions from the experience of expert surgeons. In: 2022 IEEE/RSJ International Conference on Intelligent Robots and Systems (IROS), pp. 8049–8056. IEEE (2022)

From Tissue to Sound: Model-Based Sonification of Medical Imaging

Sasan Matinfar[1,2]([✉]), Mehrdad Salehi[1], Shervin Dehghani[1,2], and Nassir Navab[1]

[1] Computer Aided Medical Procedures (CAMP),
Technical University of Munich, 85748 Munich, Germany
sasan.matinfar@tum.de
[2] Nuklearmedizin rechts der Isar, Technical University of Munich,
81675 Munich, Germany

Abstract. We introduce a general design framework for the interactive sonification of multimodal medical imaging data. The proposed approach operates on a physical model that is generated based on the structure of anatomical tissues. The model generates unique acoustic profiles in response to external interactions, enabling the user to learn about how the tissue characteristics differ from rigid to soft, dense to sparse, structured to scattered. The acoustic profiles are attained by leveraging the topological structure of the model with minimal preprocessing, making this approach applicable to a diverse array of applications. Unlike conventional methods that directly transform low-dimensional data into global sound features, this approach utilizes unsupervised mapping of features between an anatomical data model and a sound model, allowing for the processing of high-dimensional data. We verified the feasibility of the proposed method with an abdominal CT volume. The results show that the method can generate perceptually discernible acoustic signals in accordance with the underlying anatomical structure. In addition to improving the directness and richness of interactive sonification models, the proposed framework provides enhanced possibilities for designing multisensory applications for multimodal imaging data.

Keywords: Sonification · Auditory Display · Auditive Augmented Reality · Model-based Sonification · Physical Modeling Sound Synthesis

1 Introduction

Carrying out a surgical procedure requires not only spatial coordination but also the ability to maintain temporal continuity during time-critical situations, underscoring the crucial importance of auditory perception due to the temporal

Supplementary Information The online version contains supplementary material available at https://doi.org/10.1007/978-3-031-43996-4_20.

nature of sound. Perceptual studies have shown that stimuli in different sensory modalities can powerfully interact under certain circumstances, affecting perception or behavior [1]. In multisensory perception, sensory experiences are integrated when they occur simultaneously, i.e., when complementary or redundant signals are received from the same location at the same time. Previous studies have shown that combining independent but causally correlated sources of information facilitates information processing [2]. It has been demonstrated that multisensory integration, particularly auditory cues embedded in complex sensory scenes, improves performance on a wide range of tasks [3]. Biologically, humans have evolved to process spatial dimensions visually [4]. At the same time, their auditory system, which works in an omnidirectional manner, helps to maintain a steady pace in dynamic interaction, facilitating hand-eye coordination. It has been shown that bimanual coordination augmented with auditory feedback, compared to visual, leads to enhanced activation in brain regions involved in motor planning and execution [5].

However, the integration of auditory systems with visual modalities in biomedical research and applications has not been fully achieved. This could be due to the challenges involved in providing perceptually unequivocal and precise sound while reflecting high-resolution and high-dimensional information, as available in medical data. This paper proposes a novel approach, providing an intelligent modeling as a design methodology for interactive medical applications. Considering the limitations of the state-of-the-art, we investigate potential design approaches and the possibility of establishing a new research direction in sonifying high-resolution and multidimensional medical imaging data.

2   State-of-the-Art of Sonification in the Medical Domain

Sonification is the data-dependent generation of sound, if the transformation is systematic, objective and reproducible, so that it can be used as scientific method [6,7]. Researchers have investigated the impact of sound as an integral part of the user experience in interactive systems and interfaces [8]. In this form of interaction design, sound is used to create rich, dynamic, and immersive user experiences and to convey information, provide feedback, and shape user behavior. The use of sonification has expanded in recent years to a wide range of tasks, such as navigation, process monitoring, data exploration, among others [7]. A variety of sonification techniques for time-indexed data such as audification, parameter-mapping sonification (PMSon), and model-based sonification (MBS) translate meaningful data patterns of interaction into perceptual patterns, allowing listeners to gain insight into those patterns and be aware of their changes.

In PMSon, data features are input parameters of a mapping function to determine synthesis parameters. PMSon allows explicit definition of mapping functions in a flexible and adaptive manner, which has made it the most popular approach for medical applications. Medical sonification, as demonstrated by pioneering works such as [9–12], has predominantly used PMSon for sonifying the position or state of surgical instruments with respect to predetermined structures or translating spatial characteristics of medical imaging data into acoustic features. Research in surgical sonification [13, 14] has mainly focused on image-guided navigation, showing significant results in terms of accuracy and facilitating guidance. Fundamental research behind these studies has aimed to expand mapping dimensionality beyond 1D [15] and 2D [16, 17] up to three orthogonal dimensions [18]. A sonification method combining two orthogonal mappings in two alignment phases has been proposed in [19] for the placement of pedicle screws with four degrees of freedom. In addition to expanding the data bandwidth, there have been studies [11, 12, 14, 20, 21] which address the issues of integrability and pleasantness of the resulting sound signal. This is arguably one of the most significant challenges in surgical sonification, which still requires further research.

Medical sonification has shown great potential, despite being a relatively new field of research. However, there are limitations in terms of design and integrability. The tedious and case-specific process of defining a mapping function that can be perceptually resolved, even in low dimensional space, poses a significant challenge, rendering the achievement of a generalized method nearly impossible. Consequently, practitioners endeavor to minimize data complexity and embed data space into low dimensional space, giving rise to sonification models that offer only an abstract and restricted understanding of the data, which is inadequate in the case of complex medical data. Although these methods achieve adequate perceptual resolution and accuracy, they still are not optimized for integration into the surgical workflow and demonstrate low learning rates, which also demand increased cognitive load and task duration. Furthermore, fine-tuning these models involves a considerable amount of artistic creativity in order to avoid undesirable effects like abrasiveness and fatigue.

3   Towards Model-Based Sonic Interaction with Multimodal Medical Imaging Data

The need for an advanced multisensory system becomes more critical in scenarios where the anatomy is accessed through limited means, as in minimally invasive surgery. In such cases, an enriched auditory feedback system that conveys precise information about the tissue or surrounding structures, capturing spatial characteristics of the data (such as density, solidity, softness, or sparsity), can enhance the surgeon's perception and, on occasion, compensate for the lack of haptic feedback in such procedures. Embedding such use cases into a low-dimensional space is not feasible. In the same way that tapping on wood or glass imparts information about the object's construction, an effective sonification design can convey information in a manner that is easily interpreted with minimal cognitive effort. Intuitive embodiment of information in auditory feedback, facilitates the learning process to the extent that subconscious association of auditory cues with events can be achieved. Despite the fact that anatomical structures do not produce sound in human acoustic ranges, and surgeons do not have a direct perceptual experience of them, these structures still adhere to the physical principles of dynamics. The hypothesis of this paper is that sounds based on the principles of physics are more straightforward to learn and utilize due to their association with real-world rules.

MBS is a technique that utilizes mathematical models to represent data and then convert those models into audible sounds. This approach is often used in scientific or data analysis applications such as clustering [22] or data scanning [23], where researchers aim to represent complex data sets or phenomena through sound. MBS employs a mapping approach that associates data features with the features of a sound-capable system, facilitating generalization across different cases. MBS is often coupled with physical modeling synthesis that uses algorithms to simulate the physical properties of real-world instruments and generate sound in a way that mimics the behavior of physical objects [24–26]. This approach is often used to create realistic simulations of acoustic instruments or explore new sonic possibilities that go beyond the limitations of traditional instruments. Sound is a physical phenomenon that arises from the vibration of objects in a medium, such as air, resulting in a complex waveform composed of multiple frequencies at different amplitudes. This vibrational motion can be generated by mechanical impacts, such as striking, plucking, or blowing, applied to a resonant object capable of sustaining vibration. Physical modeling has been previously introduced to the biomedical research community as a potential solution for the issue of unintuitive sounds, as reported in [21]. However, as mentioned earlier, it is limited by the shortcomings of the current state-of-the-art, particularly with respect to reduced data dimensionality.

Physical modeling can be approached using the mass interaction method, which is characterized by its modular design and the capacity to incorporate direct gestural interaction, as demonstrated in [27–29]. This approach offers the advantage of describing highly intricate virtual objects as a construction of elementary physical components. Additionally, such an approach is highly amenable to the iterative and exploratory design of "physically plausible" virtual objects, which are grounded in the principles of Newtonian physics and three laws of motion but are not necessarily limited to the mechanical constraints of the physical world.

Contribution

This paper presents a novel approach for transforming spatial features of multimodal medical imaging data into a physical model that is capable of generating distinctive sound. The physical model captures complex features of the spatial domain of data, including geometric shapes, textures, and complex anatomical structures, and translates them to sound. This approach aims to enhance experts' ability to interact with medical imaging data and improve their mental mapping regarding complex anatomical structures with an unsupervised approach. The unsupervised nature of the proposed approach facilitates generalization, which enables the use of varied input data for the development of versatile sound-capable models.

4   Methods

The proposed method involves multimodal imaging data serving as input, a topology to capture spatial structures, and an interaction module to establish temporal progress. We consider medical imaging data as members of a $\mathbb{R}^{d+m}$ space, where d is the data dimensionality in space domain, and m the dimension of measured features by the imaging systems. A sequence of physics-based sound signals, S, expressed as

$$S = \sum_{i=0}^{n} P_i(act(pos_i, \vec{F}_i), f(A_i), T_i), \tag{1}$$

can be achieved by the sum of n excitations of a physical model P with user interaction act at the position pos with applied force of $\vec{F}$, where f transfers a region of interest (RoI) $A \in \mathbb{R}^{d+m}$ to parameters of the physical model P using the topology matrix T. These components are described in detail in Sect. 4.1. Figure 1 provides an illustration of the method overview and data pipeline.

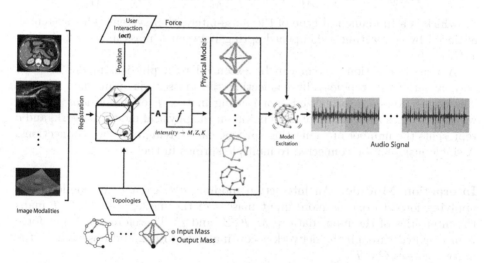

Fig. 1. The diagram shows the transformation of multimodal medical imaging data to sound. The registration method can be arbitrarily chosen depending on the application and data modalities.

4.1   Physical Model

The mass-interaction physics methodology [28] allows the formulation of physical systems, such as the linear harmonic oscillator, which comprise two fundamental constituents: masses, representing material points within a 3D space, with corresponding inertial behaviors, and connecting springs, signifying specific types of physical couplings such as viscoelastic and collision between two mass elements.

Implementation of a Mass-Interaction System in Discrete Time. To represent and compute discretized modular mass-interaction systems, a widely used method involves applying a second-order central difference scheme to Newton's second law, which states that force $\vec{F}$ is equal to mass m times acceleration a, or the second derivative of its position vector x with respect to time t.

The total force exerted by the dampened spring, denoted as $\vec{F}(t_n) = \vec{F}_s(t_n) + \vec{F}_d(t_n)$ at time t_n, where $\vec{F}_s$ represents the elastic force exerted by a linear spring (the interaction) with stiffness K, connecting two masses m_1, m_2 located at positions x_1, x_2, can be expressed using the discrete-time equivalent of Hooke's law. Similarly, the friction force $\vec{F}_d$ applied by a linear damper with damping parameter z can be derived using the Backward Euler difference scheme with the discrete-time inertial parameter $Z = z/\Delta T$. $\vec{F}(t_n)$ is applied symmetrically to each mass in accordance with Newton's third law: $\vec{F}_{2 \to 1}(t_n) = -\vec{F}(t_n)$ and $\vec{F}_{1 \to 2}(t_n) = +\vec{F}(t_n)$. The combination of forces applied to masses and the connecting spring yields a linear harmonic oscillator as described in

$$X(t_{n+1}) = (2 - \frac{K + Z}{M})X(t_n) + (\frac{Z}{M} - 1)X(t_{n-1}) + \frac{\vec{F}(t_n)}{M}, \tag{2}$$

which is a fundamental type of the mass-interaction system. This system is achieved by connecting a dampened spring between a mass and a fixed point $x_1(t_n) = 0$.

A mass-interaction system can be extended to a physical model network with an arbitrary topology by connecting the masses via dampened springs. The connections are formalized as a routing matrix T of dimensions $r \times c$, where r denotes the number of mass points in the physical model network, and c represents the number of connecting springs, each having only two connections. A single mass can be connected to multiple springs in the network.

Interaction Module. An interaction module, *act*, excites the model P by applying force to one or more input masses of the model $I \in T^{r \times c}$. f maps the intensities of the input data to M, K, Z, and $\vec{F}$. Therefore, the input force is propagated through the network according to the Eq. 2 and observed by the output masses $O \in T^{r \times c}$.

To summarize, the output masses are affected by the oscillation of all masses activated in the model with various frequencies and corresponding amplitudes, resulting in a specific sound profile, i.e., tone color. This tone color represents the spatial structure and physical properties of the RoI, which is transformed into the features of the output sound. The wave propagation is significantly influenced by the model's topology and the structure of the inter-mass connections, which have great impact on activating spatial relevant features and sound quality.

4.2   Experiment and Results

The objective is to evaluate the feasibility of the proposed method in creating a model that is capable of generating discernible sound profiles in accordance with the underlying anatomical structures. In particular, we aim to differentiate between the sound of a set of tissue types. Through empirical experimentation on an abdominal CT volume [31], we determined a model configuration that achieves stability and reduces noise to the desired level. The shape of the topology is a 3D cube of size 7 mm³. The inter-mass connections are established at the grid spacing distance of 1 mm between each mass and its adjacent neighbor masses. All the masses located on the surface of the model are set as fixed points. To excite the model, equal forces are applied to the center of the model in a 3D direction and observed at the same position. The RoI is obtained by selecting 3D cubes with the same topology and size in the CT volume. The intensities in the RoI are transformed to define the model parameters. The spring parameters K and Z are derived by averaging the intensities of their adjacent CT voxels, by a linear mapping and M and $\vec{F}$ are set to constant values. We defined a sequence of RoI starting in the heart, passing through lung, liver, bone, muscle, and ending in the air in a 16-steps trajectory, as shown in Fig. 2. For visualizing the trajectory and processing image intensities, we used ImFusion Suite [32]. For generating physical models and generating sound, we used mass-interaction physical modelling library [29] for the Processing sketching environment: https://processing.org/. Visual demonstrations of the models, along with the corresponding sound samples, are provided in the supplementary material, along with additional explanations.

A mel spectrogram is a visual representation of the frequency content of an audio signal, where the frequencies are mapped to a mel scale, which is a perceptual frequency scale based on how humans hear sounds. Therefore, we used this representation to show the frequency content of the resulting sound of the trajectory, presented in Fig. 2.

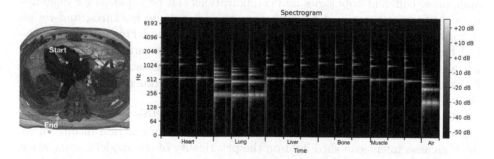

Fig. 2. The spectrogram illustrates the sound profiles of the tissues corresponding to the sequence of RoIs marked as yellow on the abdominal CT. The green spheres highlight the excitation points. (Color figure online)

5 Discussion and Conclusion

This paper introduced a general framework for MBS of multimodal medical imaging data. The preliminary study demonstrates perceptually distinguishable sound profiles between various anatomical tissue types. These profiles are achieved through a minimal preprocessing based on the model topology and a basic mapping definition. This indicates that the model is effective in translating intricate imaging data into a discernible auditory representation, which can conveniently be generalized to a wide range of applications. In contrast to the traditional methods that directly convert low-dimensional data into global sound features, this approach maps features of a data model to the features of a sound model in an unsupervised manner and enables processing of high-dimensional data.

The proposed method presents opportunities for several enhancements and future directions. In the case of CT imaging, it may be feasible to establish modality-specific generalized configurations and tissue-type-specific transfer functions to standardize the approach to medical imaging sonification. Such efforts could lead to the development of an auditory equivalent of 3D visualization for medical imaging, thereby providing medical professionals with a more immersive and intuitive experience. Another opportunity could be in the sonification of intraoperative medical imaging data, such as ultrasound or fluoroscopy, to augment the visual information that a physician would receive by displaying the tissue type or structure which their surgical instruments are approaching. To accommodate various application cases, alternative interaction modes can be designed that simulate different approaches to the anatomy with different tool materials. Additionally, supervised features can be integrated to improve both local and global perception of the model, such as amplifying regions with anomalies. Different topologies and configurations can be explored for magnifying specific structures in data, such as pathologies, bone fractures, and retina deformation. To achieve a realistic configuration of the model regarding the physical behavior of the underlying anatomy, one can use several modalities which correlate with physical parameters of the model. For instance, the masses can be derived by intensities of CT and spring stiffness from magnetic resonance elastography, both registered as a 3D volume.

An evaluation of the model's potential in an interactive setting can be conducted to determine its impact on cognitive load and interaction intuitiveness. Such an assessment can shed light on the practicality of the model's application by considering the user experience, which is influenced by cognitive psychological factors. As with any emerging field, there are constraints associated with the methodology presented in this paper. One of the constraints is the requirement to manually configure the model parameters. Nevertheless, a potential future direction would be to incorporate machine learning techniques and dynamic modeling to automatically determine these parameters from underlying physical structure. For instance, systems such as [30] can provide a reliable reference for such investigations. Considering the surgical auditory scene understanding is vital when incorporating the sonification model into the surgical workflow. Future studies can investigate this by accounting for realistic surgical environments, including existing sound sources, and considering auditory masking effects.

References

1. Ernst, M.O., Di Luca, M.: Multisensory perception: from integration to remapping. In: Sensory Cue Integration, pp. 224–250 (2011)
2. Shams, L., Seitz, A.R.: Benefits of multisensory learning. Trends Cogn. Sci. **12**(11), 411–417 (2008)
3. Van der Burg, E., Olivers, C.N., Bronkhorst, A.W., Theeuwes, J.: Audiovisual events capture attention: evidence from temporal order judgments. J. Vis. **8**(5), 2 (2008)
4. Middlebrooks, J.C., Green, D.M.: Sound localization by human listeners. Annu. Rev. Psychol. **42**(1), 135–159 (1991)
5. Ronsse, R., et al.: Motor learning with augmented feedback: modality-dependent behavioral and neural consequences. Cereb. Cortex **21**(6), 1283–1294 (2011)
6. Hermann, T.: Taxonomy and definitions for sonification and auditory display. In: International Community for Auditory Display (2008)
7. Hermann, T., Hunt, A., Neuhoff, J.G.: The Sonification Handbook, vol. 1. Logos Verlag, Berlin (2011)
8. Franinovic, K., Serafin, S. (eds.): Sonic Interaction Design. MIT Press, Cambridge (2013)
9. Wegner, C.M., Karron, D.B.: Surgical navigation using audio feedback. In: Medicine Meets Virtual Reality, pp. 450–458. IOS Press (1997)
10. Ahmad, A., Adie, S.G., Wang, M., Boppart, S.A.: Sonification of optical coherence tomography data and images. Opt. Express **18**(10), 9934–9944 (2010)
11. Hansen, C., et al.: Auditory support for resection guidance in navigated liver surgery. Int. J. Med. Robot. Comput. Assist. Surg. **9**(1), 36–43 (2013)
12. Matinfar, S., et al.: Surgical soundtracks: towards automatic musical augmentation of surgical procedures. In: Descoteaux, M., Maier-Hein, L., Franz, A., Jannin, P., Collins, D.L., Duchesne, S. (eds.) MICCAI 2017. LNCS, vol. 10434, pp. 673–681. Springer, Cham (2017). https://doi.org/10.1007/978-3-319-66185-8_76
13. Black, D., Hansen, C., Nabavi, A., Kikinis, R., Hahn, H.: A survey of auditory display in image-guided interventions. Int. J. Comput. Assist. Radiol. Surg. **12**, 1665–1676 (2017). https://doi.org/10.1007/s11548-017-1547-z
14. Joeres, F., Black, D., Razavizadeh, S., Hansen, C.: Audiovisual AR concepts for laparoscopic subsurface structure navigation. In: Graphics Interface 2021 (2021)
15. Parseihian, G., Gondre, C., Aramaki, M., Ystad, S., Kronland-Martinet, R.: Comparison and evaluation of sonification strategies for guidance tasks. IEEE Trans. Multimedia **18**(4), 674–686 (2016)
16. Ziemer, T., Black, D., Schultheis, H.: Psychoacoustic sonification design for navigation in surgical interventions. In: Proceedings of Meetings on Acoustics, vol. 30, no. 1, p. 050005. Acoustical Society of America (2017)
17. Ziemer, T., Schultheis, H., Black, D., Kikinis, R.: Psychoacoustical interactive sonification for short range navigation. Acta Acust. Acust. **104**(6), 1075–1093 (2018)
18. Ziemer, T., Schultheis, H.: Psychoacoustical signal processing for three-dimensional sonification. Georgia Institute of Technology (2019)
19. Matinfar, S., et al.: Sonification as a reliable alternative to conventional visual surgical navigation. Sci. Rep. **13**(1), 5930 (2023). https://www.nature.com/articles/s41598-023-32778-z

20. Matinfar, S., Hermann, T., Seibold, M., Fürnstahl, P., Farshad, M., Navab, N.: Sonification for process monitoring in highly sensitive surgical tasks. In: Proceedings of the Nordic Sound and Music Computing Conference 2019 (Nordic SMC 2019) (2019)
21. Roodaki, H., Navab, N., Eslami, A., Stapleton, C., Navab, N.: SonifEye: sonification of visual information using physical modeling sound synthesis. IEEE Trans. Vis. Comput. Graph. **23**(11), 2366–2371 (2017)
22. Hermann, T., Ritter, H.: Listen to your data: model-based sonification for data analysis. In: Advances in Intelligent Computing and Multimedia Systems, vol. 8, pp. 189–194 (1999)
23. Bovermann, T., Hermann, T., Ritter, H.: Tangible data scanning sonification model. Georgia Institute of Technology (2006)
24. Smith, J.O.: Physical modeling using digital waveguides. Comput. Music. J. **16**(4), 74–91 (1992)
25. Cook, P.R.: Physically informed sonic modeling (PhISM): synthesis of percussive sounds. Comput. Music. J. **21**(3), 38–49 (1997)
26. Smith, J.O.: Physical audio signal processing: for virtual musical instruments and audio effects. W3K Publishing (2010)
27. Leonard, J., Cadoz, C.: Physical modelling concepts for a collection of multisensory virtual musical instruments. In: New Interfaces for Musical Expression 2015, pp. 150–155 (2015)
28. Villeneuve, J., Leonard, J.: Mass-interaction physical models for sound and multisensory creation: starting anew. In: Proceedings of the 16th Sound & Music Computing Conference, pp. 187–194 (2019)
29. Mass Interaction Physics in Java/Processing Homepage. https://github.com/mi-creative/miPhysics_Processing. Accessed 4 Mar 2023
30. Illanes, A., et al.: Novel clinical device tracking and tissue event characterization using proximally placed audio signal acquisition and processing. Sci. Rep. **8**(1), 12070 (2018)
31. Luo, X., et al.: WORD: a large scale dataset, benchmark and clinical applicable study for abdominal organ segmentation from CT image. Med. Image Anal. **82**, 102642 (2022)
32. Zettinig, O., Salehi, M., Prevost, R., Wein, W.: Recent advances in point-of-care ultrasound using the *ImFusion Suite* for real-time image analysis. In: Stoyanov, D., et al. (eds.) POCUS/BIVPCS/CuRIOUS/CPM -2018. LNCS, vol. 11042, pp. 47–55. Springer, Cham (2018). https://doi.org/10.1007/978-3-030-01045-4_6

Deep Homography Prediction for Endoscopic Camera Motion Imitation Learning

Martin Huber$^{(\boxtimes)}$ⓘ, Sébastien Ourselinⓘ, Christos Bergelesⓘ,
and Tom Vercauterenⓘ

School of Biomedical Engineering & Image Sciences, King's College London,
London, UK
martin.huber@kcl.ac.uk

Abstract. In this work, we investigate laparoscopic camera motion automation through imitation learning from retrospective videos of laparoscopic interventions. A novel method is introduced that learns to augment a surgeon's behavior in image space through object motion invariant image registration via homographies. Contrary to existing approaches, no geometric assumptions are made and no depth information is necessary, enabling immediate translation to a robotic setup. Deviating from the dominant approach in the literature which consist of following a surgical tool, we do not handcraft the objective and no priors are imposed on the surgical scene, allowing the method to discover unbiased policies. In this new research field, significant improvements are demonstrated over two baselines on the Cholec80 and HeiChole datasets, showcasing an improvement of 47% over camera motion continuation. The method is further shown to indeed predict camera motion correctly on the public motion classification labels of the AutoLaparo dataset. All code is made accessible on GitHub (https://github.com/RViMLab/homography_imitation_learning).

Keywords: Computer vision · Robotic surgery · Imitation learning

1 Introduction

Automation in robot-assisted minimally invasive surgery (RMIS) may reduce human error that is linked to fatigue, lack of attention and cognitive overload [8]. It could help surgeons operate such systems by reducing the learning curve [29]. And in an ageing society with shrinking workforce, it could help to retain accessibility to healthcare. It is therefore expected that parts of RMIS will be ultimately automated [5,30]. On the continuous transition towards different levels of autonomy, camera motion automation is likely to happen first [14].

C. Bergeles and T. Vercauteren—These authors contributed equally to this work.

© The Author(s), under exclusive license to Springer Nature Switzerland AG 2023
H. Greenspan et al. (Eds.): MICCAI 2023, LNCS 14228, pp. 217–226, 2023.
https://doi.org/10.1007/978-3-031-43996-4_21

Initial attempts to automate camera motion in RMIS include rule-based approaches that keep surgical tools in the center of the field of view [4,9,21]. The assumption that surgical tools remain centrally is, however, simplistic, as in many cases the surgeon may want to observe the surrounding anatomy to decide their next course of action.

Contrary to rule-based approaches, data-driven methods are capable to capture more complex control policies. Example data-driven methods suitable for camera motion automation include reinforcement learning (RL) and imitation learning (IL). The sample inefficiency and potential harm to the patient currently restrict RL approaches to simulation [1,22,23], where a domain gap remains. Work to bridge the domain gap and make RL algorithms deployable in real setups have been proposed [3,20], but clinical translation has not yet been achieved. For IL, on the other hand, camera motion automation could be learned from real data, thereby implicitly tackling the domain-gap challenge. The downside is that sufficient data may be difficult to collect. Many works highlight that lack of expert annotated data hinders progress towards camera motion automation in RMIS [7,13,19]. It is thus not surprising that existing literature on IL for camera motion automation utilizes data from mock setups [12,26].

Recent efforts to make vast amounts of laparoscopic intervention videos publicly available [19] drastically change how IL for camera motion automation can be approached. So far, this data is leveraged mainly to solve auxiliary tasks that could contribute to camera motion automation. As reviewed in [18], these tasks include tool and organ segmentation, as well as surgical phase recognition. For camera motion automation specifically, however, there exist no publicly available image-action pairs. Some work, therefore, continues to focus on the tools to infer camera motion [15], or learns on a robotic setup altogether [17] where camera motion is accessible. The realization, however, that camera motion is intrinsic to the videos of laparoscopic interventions and that camera motion could be learned on harvested actions was first realized in [11], and later in [16]. This comes with the additional advantage that no robot is necessary to learn behaviors and that one can directly learn from human demonstrations.

In this work, we build on [11] for computationally efficient image-action pair extraction from publicly available datasets of laparoscopic interventions, which yields more than 20× the amount of data that was used in the closed source data of [16]. Contrary to [16], our camera motion extraction does not rely on image features, which are sparse in surgical videos, and is intrinsically capable to differentiate between camera and object motion. We further propose a novel importance sampling and data augmentation step for achieving camera motion automation IL.

2   Materials and Methods

The proposed approach to learning camera motion prediction is summarized in Fig. 1. The following sections will describe its key components in more detail.

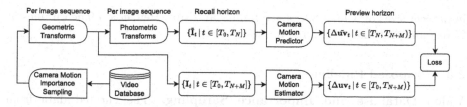

Fig. 1. Training pipeline, refer to Sect. 2.3. From left to right: Image sequences are importance sampled from the video database and random augmentations are applied per sequence online. The lower branch estimates camera motion between subsequent frames, which is taken as pseudo-ground-truth for the upper branch, which learns to predict camera motion on a preview horizon.

2.1 Theoretical Background

Points on a plane, as observed from a moving camera, transform by means of the 3×3 projective homography matrix $\mathbf{G}$ in image space. Thus, predicting future camera motion (up to scale) may be equivalently treated as predicting future projective homographies.

It has been shown in [6] that the four point representation of the projective homography, $i.e.$, taking the difference between four points in homogeneous coordinates $\Delta \mathbf{uv} = \{\mathbf{p}_i - \mathbf{p}_i' \mid i \in [0, 4)\} \in \mathbb{R}^{4 \times 2}$ that are related by $\mathbf{G p}_i \sim \mathbf{p}_i' \ \forall i$, is better behaved for deep learning applications than the 3×3 matrix representation of a homography. Therefore, in this work, we treat camera motion $\mathcal{C}$ as a sequence of four point homographies on a time horizon $[T_0, T_{N+M})$, N being the recall horizon's length, M being the preview horizon's length. Time points lie Δt apart, that is $T_{i+1} = T_i + \Delta t$. For image sequences of length N+M, we work with four point homography sequences $\mathcal{C} = \{\Delta \mathbf{uv}_t \mid t \in [T_0, T_{N+M})\}$.

2.2 Data and Data Preparation

Three datasets are curated to train and evaluate the proposed method: two cholecystectomy datasets (laparoscopic gallbladder removal), namely Cholec80 [25] and HeiChole [27], and one hysterectomy dataset (laparoscopic uterus removal), namely AutoLaparo [28].

To remove status indicator overlays from the laparoscopic videos, which may hinder the camera motion estimator, we identify the bounding circle of the circular field of view using [2]. We crop the view about the center point of the bounding circle to a shape of 240×320, so that no black regions are prominent in the images.

All three datasets are split into training, validation, and testing datasets. We split the videos by frame count into $80 \pm 1\%$ training and $20 \pm 1\%$ testing. Training and testing videos never intersect. We repeat this step to further split the training dataset into (pure) training and validation datasets.

Due to errors during processing the raw data, we exclude videos 19, 21, and 23 from HeiChole, as well as videos 22, 40, 65, and 80 from Cholec80. This results in dataset sizes of: Cholec80 - 4.4e6 frames at 25 fps, HeiChole - 9.5e5 frames at 25 fps, and AutoLaparo - 7.1e4 frames at 25 fps.

2.3 Proposed Pipeline

Video Database and Importance Sampling. The curated data from Sect. 2.2 is accumulated into a video database. Image sequences of length $N + M$ are sampled at a frame increment of Δn between subsequent frames and with Δc frames between the sequence's initial frames. Prior to adding the videos to the database, an initial offline run is performed to estimate camera motion $\Delta \mathbf{uv}$ between the frames. This creates image-motion correspondences of the form $(\mathbf{I}_n, \mathbf{I}_{n+\Delta n}, \Delta \mathbf{uv}_n)$. Image-motion correspondences where $\mathbb{E}(\|\Delta \mathbf{uv}_n\|_2) > \sigma$, with sigma being the standard deviation over all motions in the respective dataset, define anchor indices n. Image sequences are sampled such that the last image in the recall horizon lies at index $n = N - 1$, marking the start of a motion. The importance sampling samples indices from the intersection of all anchor indices, shifted by $-N$, with all possible starting indices for image sequences.

Geometric and Photometric Transforms. The importance sampled image sequences are fed to a data augmentation stage. This stage entails geometric and photometric transforms. The distinction is made because downstream, the pipeline is split into two branches. The upper branch serves as camera motion prediction whereas the lower branch serves as camera motion estimation, also refer to the next section. As it acts as the source of pseudo-ground-truth, it is crucial that the camera motion estimator performs under optimal conditions, hence no photometric transforms, i.e. transforms that change brightness/contrast/fog etc., are applied. Photometrically transformed images shall further be denoted as $\tilde{\mathbf{I}}$. To encourage same behavior under different perspectives, geometric transforms are applied, i.e. transforms that change orientation/up to down/left to right etc. Transforms are always sampled randomly, and applied consistently to the entire image sequence.

Camera Motion Estimator and Predictor. The goal of this work is to have a predictor learn camera motion computed by an estimator. The predictor takes as input a photometrically and geometrically transformed recall horizon $\{\tilde{\mathbf{I}}_t \mid t \in [T_0, T_N)\}$ of length N, and predicts camera motion $\tilde{\mathcal{C}} = \{\Delta \tilde{\mathbf{uv}}_t \mid t \in [T_N, T_{N+M})\}$ on the preview horizon of length M. The estimator takes as input the geometrically transformed preview horizon $\{\mathbf{I}_t \mid t \in [T_M, T_{N+M})\}$ and estimates camera motion $\mathcal{C}$, which serves as a target to the predictor. The estimator is part of the pipeline to facilitate on-the-fly perspective augmentation via the geometric transforms.

3 Experiments and Evaluation Methodology

The following two sections elaborate the experiments we conduct to investigate the proposed pipeline from Fig. 1 in Sect. 2.3. First the camera motion estimator is investigated, followed by the camera motion predictor.

3.1 Camera Motion Estimator

Camera Motion Distribution. To extract the camera motion distribution, we run the camera motion estimator from [11] with a ResNet-34 backbone over all datasets from Sect. 2.2. We map the estimated four point homographies to up/down/left/right/zoom-in/zoom-out for interpretability. Left/right/up/down corresponds to all four point displacements Δ**uv** consistently pointing left/right/up/down respectively. Zoom-in/out corresponds to all four point displacements Δ**uv** consistently pointing inwards/outwards. Rotation left corresponds to all four point displacements pointing up right, bottom right, and so on. Same for rotation right. Camera motion is defined static if it lies below the standard deviation in the dataset. The frame increment is set to 0.25 s, corresponding to $\Delta n = 5$ for the 25 fps videos.

Online Camera Motion Estimation. Since the camera motion estimator is executed online, memory footprint and computational efficiency are of importance. Therefore, we evaluate the estimator from [11] with a ResNet-34 backbone, SURF & RANSAC, and LoFTR [24] & RANSAC. Each estimator is run 1000 times on a single image sequence of length $N+M = 15$ with an NVIDIA GeForce RTX 2070 GPU and an Intel(R) Core(TM) i7-9750H CPU @ 2.60 GHz.

3.2 Camera Motion Predictor

Model Architecture. For all experiments, the camera motion predictor is a ResNet-18/34/50, with the number of input features equal to the recall horizon $N \times 3$ (RGB), where N = 14. We set the preview horizon M = 1. The frame increment is set to 0.25 s, or $\Delta n = 5$ for the 25 fps videos. The number of frames between clips is also set to 0.25 s, or $\Delta c = 5$.

Training Details. The camera motion predictor is trained on each dataset from Sect. 2.2 individually. For training on Cholec80/HeiChole/AutoLaparo, we run 80/50/50 epochs on a batch size of 64 with a learning rate of $2.5e-5/1.e-4/1.e-4$. The learning rates for Cholec80 and HeiChole relate approximately to the dataset's training sizes, see Table 2. For Cholec80, we reduce the learning rate by a factor 0.5 at epochs 50, 75. For Heichole/AutoLaparo we drop the learning rate by a factor 0.5 at epoch 35. The loss in Fig. 1 is set to the mean pairwise distance between estimation and prediction $\mathbb{E}(||\Delta\tilde{\mathbf{uv}}_t - \Delta\mathbf{uv}_t||_2) + \lambda\mathbb{E}(||\Delta\tilde{\mathbf{uv}}_t||_2)$ with a regularizer that discourages the identity $\Delta\tilde{\mathbf{uv}}_t = \mathbf{0}$ (i.e. no motion). We set $\lambda = 0.1$.

Evaluation Metrics. For evaluation we compute the mean pairwise distance between estimated and predicted motion $\mathbb{E}(\|\Delta u\tilde{\mathbf{v}}_t - \Delta u\mathbf{v}_t\|_2)$. All camera motion predictors are benchmarked against a baseline, that is a $\mathcal{O}(1)/\mathcal{O}(2)$-Taylor expansion of the estimated camera motion $\Delta u\mathbf{v}_t$. Furthermore, the model that is found to perform best is evaluated on the multi-class labels (left, right, up, down) that are provided in AutoLaparo.

4 Results

4.1 Camera Motion Estimator

Camera Motion Distribution. The camera motion distributions for all datasets are shown in Fig. 2. It is observed that for a large fraction of the sequences there is no significant camera motion (Cholec80 76.21%, HeiChole 76.2%, AutoLaparo 71.29%). This finding supports the importance sampling that was introduced in Sect. 2.3. It can further be seen that e.g. left/right and up/down motions are equally distributed.

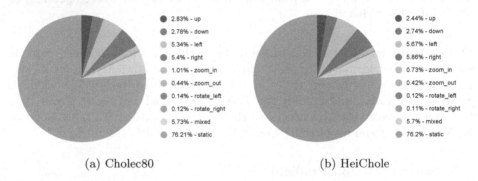

(a) Cholec80   (b) HeiChole

Fig. 2. Camera motion distribution, refer to Sect. 3.1. AutoLaparo: 2.81% - up, 1.88% - down, 4.48% - left, 3.38% - right, 0.45% - zoom_in, 0.2% - zoom_out, 0.3% - rotate_left 0.3%, - rotate_right 14.9% - mixed, 71.29% - static.

Online Camera Motion Estimation. The results of the online camera motion estimation are summarized in Table 1. The deep homography estimation with a Resnet34 backbone executes 11× quicker and has the lowest GPU memory footprint of the GPU accelerated methods. This allows for efficient implementation of the proposed online camera motion estimation in Fig. 1.

4.2 Camera Motion Prediction

The camera motion prediction results for all datasets are highlighted in Table 2. It can be seen that significant improvements over the baseline are achieved on the Cholec80 and HeiChole datasets. Whilst the learned prediction performs better on average than the baseline, no significant improvement is found for the AutoLaparo dataset.

Table 1. Memory footprint and execution time of different camera motion estimators, refer to Sect. 3.1.

Method	Execution time [s]	Speed-up [a.u.]	Model/Batch [Mb]
Resnet34	**0.016 ± 0.048**	**11.1**	**664/457**
LoFTR & RANSAC	0.178 ± 0.06	1.0	669/2412
SURF & RANSAC	0.131 ± 0.024	1.4	NA

The displacement of the image center point under the predicted camera motion for AutoLaparo is plotted against the provided multi-class motion annotations and shown in Fig. 3. It can be seen that the camera motion predictions align well with the ground truth labels.

Table 2. Camera motion predictor performance, refer to Sect. 3.2. Taylor baselines predict based on previous estimated motion, ResNets based on images.

Dataset	Train Size [Frames]	Mean Pairwise Distance [Pixels]				
		Taylor		ResNet (proposed)		
		$\mathcal{O}(1)$	$\mathcal{O}(2)$	18	34	50
Cholec80	3.5e6	27.2 ± 23.1	36.4 ± 31.2	**14.8 ± 11.7**	**14.4 ± 11.4**	**14.4 ± 11.4**
HeiChole	7.6e5	29.7 ± 26.4	39.8 ± 35.9	**15.8 ± 12.5**	**15.8 ± 12.5**	**15.8 ± 12.5**
AutoLaparo	5.9e4	19.4 ± 18.4	25.8 ± 24.7	**11.2 ± 11.0**	**11.3 ± 11.0**	**11.3 ± 11.0**

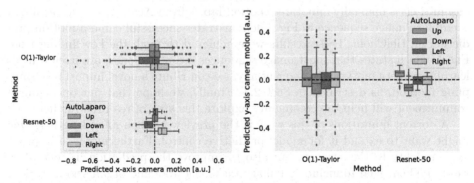

(a) Predicted camera motion along x-axis, scaled by image size to $[-1, 1]$.

(b) Predicted camera motion along y-axis, scaled by image size to $[-1, 1]$.

Fig. 3. Predicted camera motion on AutoLaparo, refer to Sect. 3.2. Camera motion predictor trained on Cholec80 with ResNet-50 backbone, see Table 2. Shown is the motion of the image center under the predicted homography. Clearly, for videos labeled left/right, the center point is predicted to move left/right and for up/down labels, the predicted left/right motion is centered around zero (a). Same is observed for up/down motion in (b), where left/right motion is zero-centered.

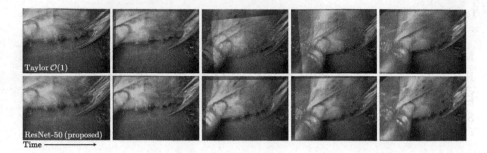

Fig. 4. Exemplary camera motion prediction, refer to Sect. 3.2. In the image sequence, the attention changes from the right to the left tool. We warp the past view (yellow) by the predicted homography and overlay the current view (blue). Good alignment corresponds to good camera motion prediction. Contrary to the baseline, the proposed method predicts the motion well. Data taken from HeiChole test set, ResNet-50 backbone trained on Cholec80, refer Table 2. (Color figure online)

5   Conclusion and Outlook

To the best of our knowledge, this work is the first to demonstrate that camera motion can indeed be learned from retrospective videos of laparoscopic interventions, with no manual annotation. Self-supervision is achieved by harvesting image-motion correspondences using a camera motion estimator, see Fig. 1. The camera motion predictor is shown to generate statistically significant better predictions over a baseline in Table 2 as measured using pseudo-ground-truth and on multi-class manually annotated motion labels from AutoLaparo in Fig. 3. An exemplary image sequence in Fig. 4 demonstrates successful camera motion prediction on HeiChole. These results were achieved through the key finding from Fig. 2, which states that most image sequences, i.e. static ones, are irrelevant to learning camera motion. Consequentially, we contribute a novel importance sampling method, as described in Sect. 2.3. Finally, we hope that our open-source commitment will help the community explore this area of research further.

A current limitations of this work is the preview horizon M of length 1. One might want to extend it for model predictive control. Furthermore, to improve explainability to the surgeon, but also to improve the prediction in general, it would be beneficial to include auxiliary tasks, e.g. tool and organ segmentation, surgical phase recognition, and audio. There also exist limitations for the camera motion estimator. The utilized camera motion estimator is efficient and isolates object motion well from camera motion, but is limited to relatively small camera motions. Improving the camera motion estimator to large camera motions would help increase the preview horizon M.

In future work, we will execute this model in a real setup for investigating transferability. This endeavor is backed by [10], which demonstrates how the learned homography could immediately be deployed on a robotic laparoscope holder. It might proof necessary to fine-tune the presented policy through reinforcement learning with human feedback.

Acknowledgements. This work was supported by core and project funding from the Wellcome/EPSRC [WT203148/Z/16/Z; NS/A000049/1; WT101957; NS/A000027/1]. This project has received funding from the European Union's Horizon 2020 research and innovation programme under grant agreement No 101016985 (FAROS project). TV is supported by a Medtronic/RAEng Research Chair [RCSRF1819\7\34]. SO and TV are co-founders and shareholders of Hypervision Surgical. TV is co-founder and shareholder of Hypervision Surgical. TV holds shares from Mauna Kea Technologies.

References

1. Agrawal, A.S.: Automating endoscopic camera motion for teleoperated minimally invasive surgery using inverse reinforcement learning. Ph.D. thesis, Worcester Polytechnic Institute (2018)
2. Budd, C., Garcia-Peraza Herrera, L.C., Huber, M., Ourselin, S., Vercauteren, T.: Rapid and robust endoscopic content area estimation: a lean GPU-based pipeline and curated benchmark dataset. Comput. Methods Biomech. Biomed. Eng. Imaging Vis. **11**(4), 1215–1224 (2022). https://doi.org/10.1080/21681163.2022.2156393
3. Cartucho, J., Tukra, S., Li, Y., Elson, D.S., Giannarou, S.: VisionBlender: a tool to efficiently generate computer vision datasets for robotic surgery. Comput. Methods Biomech. Biomed. Eng. Imaging Vis. **9**(4), 331–338 (2021)
4. Da Col, T., Mariani, A., Deguet, A., Menciassi, A., Kazanzides, P., De Momi, E.: SCAN: system for camera autonomous navigation in robotic-assisted surgery. In: 2020 IEEE/RSJ International Conference on Intelligent Robots and Systems (IROS), pp. 2996–3002. IEEE (2020)
5. Davenport, T., Kalakota, R.: The potential for artificial intelligence in healthcare. Future Healthc. J. **6**(2), 94 (2019)
6. DeTone, D., Malisiewicz, T., Rabinovich, A.: Deep image homography estimation (2016). http://arxiv.org/abs/1606.03798
7. Esteva, A., et al.: A guide to deep learning in healthcare. Nat. Med. **25**(1), 24–29 (2019)
8. Fiorini, P., Goldberg, K.Y., Liu, Y., Taylor, R.H.: Concepts and trends in autonomy for robot-assisted surgery. Proc. IEEE **110**(7), 993–1011 (2022)
9. Garcia-Peraza-Herrera, L.C., et al.: Robotic endoscope control via autonomous instrument tracking. Front. Robot. AI **9**, 832208 (2022)
10. Huber, M., Mitchell, J.B., Henry, R., Ourselin, S., Vercauteren, T., Bergeles, C.: Homography-based visual servoing with remote center of motion for semi-autonomous robotic endoscope manipulation. In: 2021 International Symposium on Medical Robotics (ISMR), pp. 1–7. IEEE (2021)
11. Huber, M., Ourselin, S., Bergeles, C., Vercauteren, T.: Deep homography estimation in dynamic surgical scenes for laparoscopic camera motion extraction. Comput. Methods Biomech. Biomed. Eng. Imaging Visu. **10**(3), 321–329 (2022)
12. Ji, J.J., Krishnan, S., Patel, V., Fer, D., Goldberg, K.: Learning 2D surgical camera motion from demonstrations. In: 2018 IEEE 14th International Conference on Automation Science and Engineering (CASE), pp. 35–42. IEEE (2018)
13. Kassahun, Y., et al.: Surgical robotics beyond enhanced dexterity instrumentation: a survey of machine learning techniques and their role in intelligent and autonomous surgical actions. Int. J. Comput. Assist. Radiol. Surg. **11**, 553–568 (2016). https://doi.org/10.1007/s11548-015-1305-z

14. Kitaguchi, D., Takeshita, N., Hasegawa, H., Ito, M.: Artificial intelligence-based computer vision in surgery: recent advances and future perspectives. Ann. Gastroenterological Surg. **6**(1), 29–36 (2022)
15. Li, B., Lu, B., Lu, Y., Dou, Q., Liu, Y.H.: Data-driven holistic framework for automated laparoscope optimal view control with learning-based depth perception. In: 2021 IEEE International Conference on Robotics and Automation (ICRA), pp. 12366–12372. IEEE (2021)
16. Li, B., Lu, B., Wang, Z., Zhong, F., Dou, Q., Liu, Y.H.: Learning laparoscope actions via video features for proactive robotic field-of-view control. IEEE Robot. Autom. Lett. **7**(3), 6653–6660 (2022)
17. Li, B., et al.: 3D perception based imitation learning under limited demonstration for laparoscope control in robotic surgery. In: 2022 International Conference on Robotics and Automation (ICRA), pp. 7664–7670. IEEE (2022)
18. Loukas, C.: Video content analysis of surgical procedures. Surg. Endosc. **32**, 553–568 (2018). https://doi.org/10.1007/s00464-017-5878-1
19. Maier-Hein, L., et al.: Surgical data science-from concepts toward clinical translation. Med. Image Anal. **76**, 102306 (2022)
20. Marzullo, A., Moccia, S., Catellani, M., Calimeri, F., De Momi, E.: Towards realistic laparoscopic image generation using image-domain translation. Comput. Methods Programs Biomed. **200**, 105834 (2021)
21. Sandoval, J., Laribi, M.A., Faure, J., Breque, C., Richer, J.P., Zeghloul, S.: Towards an autonomous robot-assistant for laparoscopy using exteroceptive sensors: feasibility study and implementation. IEEE Robot. Autom. Lett. **6**(4), 6473–6480 (2021)
22. Scheikl, P.M., et al.: LapGym-an open source framework for reinforcement learning in robot-assisted laparoscopic surgery. arXiv preprint arXiv:2302.09606 (2023)
23. Su, Y.H., Huang, K., Hannaford, B.: Multicamera 3D viewpoint adjustment for robotic surgery via deep reinforcement learning. J. Med. Robot. Res. **6**(01n02), 2140003 (2021)
24. Sun, J., Shen, Z., Wang, Y., Bao, H., Zhou, X.: LoFTR: detector-free local feature matching with transformers. In: Proceedings of the IEEE/CVF Conference on Computer Vision and Pattern Recognition, pp. 8922–8931 (2021)
25. Twinanda, A.P., Shehata, S., Mutter, D., Marescaux, J., De Mathelin, M., Padoy, N.: EndoNet: a deep architecture for recognition tasks on laparoscopic videos. IEEE Trans. Med. Imaging **36**(1), 86–97 (2016)
26. Wagner, M., et al.: A learning robot for cognitive camera control in minimally invasive surgery. Surg. Endosc. **35**(9), 5365–5374 (2021). https://doi.org/10.1007/s00464-021-08509-8
27. Wagner, M., et al.: Comparative validation of machine learning algorithms for surgical workflow and skill analysis with the heichole benchmark. Med. Image Anal. **86**, 102770 (2023)
28. Wang, Z., et al.: AutoLaparo: a new dataset of integrated multi-tasks for image-guided surgical automation in laparoscopic hysterectomy. In: Wang, L., Dou, Q., Fletcher, P.T., Speidel, S., Li, S. (eds.) MICCAI 2022, Part VII. LNCS, vol. 13437, pp. 486–496. Springer, Cham (2022). https://doi.org/10.1007/978-3-031-16449-1_46
29. van Workum, F., Fransen, L., Luyer, M.D., Rosman, C.: Learning curves in minimally invasive esophagectomy. World J. Gastroenterol. **24**(44), 4974 (2018)
30. Zidane, I.F., Khattab, Y., Rezeka, S., El-Habrouk, M.: Robotics in laparoscopic surgery-a review. Robotica **41**(1), 126–173 (2023)

Learning Expected Appearances for Intraoperative Registration During Neurosurgery

Nazim Haouchine[1]([✉]), Reuben Dorent[1], Parikshit Juvekar[1], Erickson Torio[1], William M. Wells III[1,2], Tina Kapur[1], Alexandra J. Golby[1], and Sarah Frisken[1]

[1] Harvard Medical School, Brigham and Women's Hospital, Boston, MA, USA
nhaouchine@bwh.harvard.edu
[2] Massachusetts Institute of Technology, Cambridge, MA, USA

Abstract. We present a novel method for intraoperative patient-to-image registration by learning Expected Appearances. Our method uses preoperative imaging to synthesize patient-specific expected views through a surgical microscope for a predicted range of transformations. Our method estimates the camera pose by minimizing the dissimilarity between the intraoperative 2D view through the optical microscope and the synthesized expected texture. In contrast to conventional methods, our approach transfers the processing tasks to the preoperative stage, reducing thereby the impact of low-resolution, distorted, and noisy intraoperative images, that often degrade the registration accuracy. We applied our method in the context of neuronavigation during brain surgery. We evaluated our approach on synthetic data and on retrospective data from 6 clinical cases. Our method outperformed state-of-the-art methods and achieved accuracies that met current clinical standards.

Keywords: Intraoperative Registration · Image-guided Neurosurgery · Augmented Reality · Neural Image Analogy · 3D Pose Estimation

1 Introduction

We address the important problem of intraoperative patient-to-image registration in a new way by relying on preoperative data to synthesize plausible transformations and appearances that are *expected* to be found intraoperatively. In particular, we tackle intraoperative 3D/2D registration during neurosurgery, where preoperative MRI scans need to be registered with intraoperative surgical views of the brain surface to guide neurosurgeons towards achieving a maximal safe tumor resection [22]. Indeed, the extent of tumor removal is highly correlated with patients' chances of survival and complete resection must be balanced against the risk of causing new neurological deficits [5] making accurate intraoperative registration a critical component of neuronavigation.

Most existing techniques perform patient-to-image registration using intraoperative MRI [11], CBCT [19] or ultrasound [9,17,20]. For 3D-3D registration,

H. Greenspan et al. (Eds.): MICCAI 2023, LNCS 14228, pp. 227–237, 2023.
https://doi.org/10.1007/978-3-031-43996-4_22

3D shape recovery of brain surfaces can be achieved using near-infrared cameras [15], phase-shift 3D shape measurement [10], pattern projections [17] or stereovision [8]. The 3D shape can subsequently be registered with the preoperative MRI using conventional point-to-point methods such as iterative closest point (ICP) or coherent point drift (CPD). Most of these methods rely on cortical vessels that bring salient information for such tasks. For instance, in [6], cortical vessels are first segmented using a deep neural network (DNN) and then used to constrain a 3D/2D non-rigid registration. The method uses physics-based modeling to resolve depth ambiguities. A manual rigid alignment is however required to initialize the optimization. Alternatively, cortical vessels have been used in [13] where sparse 3D points, manually traced along the vessels, are matched with vessels extracted from the preoperative scans. A model-based inverse minimization problem is solved by estimating the model's parameters from a set of pre-computed transformations. The idea of pre-computing data for registration was introduced by [26], who used an atlas of pre-computed 3D shapes of the brain surface for registration. In [7], a DNN is trained on a set of pre-generated preoperative to intraoperative transformations. The registration uses cortical vessels, segmented using another neural network, to find the best transformation from the pre-generated set.

The main limitation of existing intraoperative registration methods is that they rely heavily on processing intraoperative images to extract image features (eg., 3D surfaces, vessels centerlines, contours, or other landmarks) to drive registration, making them subject to noise and low-resolution images that can occur in the operating room [2,25]. Outside of neurosurgery, the concept of pre-generating data for optimizing DNNs for intraoperative registration has been investigated for CT to x-ray registration in radiotherapy where x-ray images can be efficiently simulated from CTs as digital radiographic reconstructions [12,27]. In more general applications, case-centered training of DNNs is gaining in popularity and demonstrates remarkable results [16].

Contribution: We propose a novel approach for patient-to-image registration that registers the intraoperative 2D view through the surgical microscope to preoperative MRI 3D images by learning *Expected Appearances*. As shown in Fig. 1, we formulate the problem as a camera pose estimation problem that finds the optimal 3D pose minimizing the dissimilarity between the intraoperative 2D image and its pre-generated Expected Appearance. A set of Expected Appearances are synthesized from the preoperative scan and for a set of poses covering the range of plausible 6 Degrees-of-Freedom (DoF) transformations. This set is used to train a patient-specific pose regressor network to obtain a model that is texture-invariant and is cross-modality to bridge the MRI and RGB camera modalities. Similar to other methods, our approach follows a monocular single-shot registration, eliminating cumbersome and tedious calibration of stereo cameras, the laser range finder, or optical trackers. In contrast to previous methods, our approach does not involve processing intraoperative images which have several advantages: it is less prone to intraoperative image acquisition noise; it does not require pose initialization; and is computationally fast thus supporting real-time use. We present results on both synthetic and clinical data and show that our approach outperformed state-of-the-art methods.

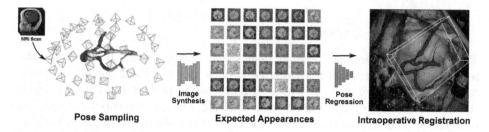

Fig. 1. Our approach estimates the 6-DoF camera pose that aligns a preoperative 3D mesh derived from MRI scans onto an intraoperative RGB image acquired from a surgical camera. We optimize a regressor network $\mathcal{P}_\Omega$ over a set of Expected Appearances that are generated by first sampling multiple poses and appearances from the 3D mesh using neural image analogy through $\mathcal{S}_\Theta$.

2   Method

2.1   Problem Formulation

As illustrated in Fig. 1, given a 3D surface mesh of the cortical vessels $\mathbf{M}$, derived from a 3D preoperative scan, and a 2D monocular single-shot image of the brain surface $\mathbf{I}$, acquired intraoperatively by a surgical camera, we seek to estimate the 6-DoF transformation that aligns the mesh $\mathbf{M}$ to the image $\mathbf{I}$. Assuming a set of 3D points $\mathbf{u} = \{u_j \in \mathbb{R}^3\} \subset \mathbf{M}$ and a set of 2D points in the image $\mathbf{v} = \{v_i \in \mathbb{R}^2\} \subset \mathbf{I}$, solving for this registration problem can be formalized as finding the 6-DoF camera pose that minimizes the reprojection error: $\sum_i^{n_c} \|\mathbf{A}[\mathbf{R}|\mathbf{t}] \begin{bmatrix} u_{c_i}^w \\ 1 \end{bmatrix} - v_i\|_2^2$, where $\mathbf{R} \in SO(3)$ and $\mathbf{t} \in \mathbb{R}^3$ represent a 3D rotation and 3D translation, respectively, and $\mathbf{A}$ is the camera intrinsic matrix composed of the focal length and the principal points (center of the image) while $\{c_i\}_i$ is a correspondence map and is built so that if a 2D point v_i corresponds to a 3D point u_j where $c_i = j$ for each point of the two sets. Note that the set of 3D points $\mathbf{u}$ is expressed in homogenous coordinates in the minimization of the reprojection error.

In practice, finding the correspondences set $\{c_i\}_i$ between $\mathbf{u}$ and $\mathbf{v}$ is non-trivial, in particular when dealing with heterogeneous preoperative and intraoperative modality pairs (MRI, RGB Cameras, ultrasound, etc.) which is often the case in surgical guidance. Existing methods often rely on feature descriptors [14], anatomical landmarks [13], or organ's contours and segmentation [6,18] involving tedious processing of the intraoperative image that is sensitive to the computational image noise. We alleviate these issues by directly minimizing the dissimilarity between the image $\mathbf{I}$ and its Expected Appearance synthesized from $\mathbf{M}$.

By defining a synthesize function $\mathcal{S}_\Theta$ that synthesizes a new image $\widehat{\mathbf{I}}$ given a projection of a 3D surface mesh for different camera poses, i.e. $\widehat{\mathbf{I}} = \mathcal{S}_\Theta(\mathbf{A}[\mathbf{R}|\mathbf{t}], \mathbf{M})$, the optimization problem above can be rewritten as:

$$\underset{\mathbf{A}[\mathbf{R}|\mathbf{t}]}{\mathrm{argmin}} \left\{ \min_\Theta \|\mathbf{I} - \mathcal{S}_\Theta(\mathbf{A}[\mathbf{R}|\mathbf{t}], \mathbf{M})\| \right\} \tag{1}$$

This new formulation is correspondence-free, meaning that it alleviates the requirement of the explicit matching between $\mathbf{u}$ and $\mathbf{v}$. This is one of the major strengths of our approach. It avoids the processing of $\mathbf{I}$ at run-time, which is the main source of registration error. In addition, our method is patient-specific, centered around $\mathbf{M}$, since each model is trained specifically for a given patient. These two aspects allow us to transfer the computational cost from the intraoperative to the preoperative stage thereby optimizing intraoperative performance. The following describes how we build the function $\mathcal{S}_\Theta$ and how to solve Eq. 1.

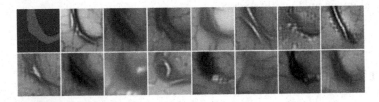

Fig. 2. An example of a set of Expected Appearances showing the cortical brain surface with the parenchyma and vessels. The network $\mathcal{S}_\Theta$ uses the binary image (top-left corner) computed from projecting $\mathbf{M}$ using $[\mathbf{R}|\mathbf{t}]$ to semantically transfer 15 different textures $\{\mathbf{T}\}$ and synthesize the Expected Appearances $\{\widehat{\mathbf{I}}\}$.

2.2   Expected Appearances Synthesis

We define a synthesis network $\mathcal{S}_\Theta : (\mathbf{A}[\mathbf{R}|\mathbf{t}], \mathbf{M}, \mathbf{T}) \rightarrow \widehat{\mathbf{I}}$, that will generate a new image resembling a view of the brain surface from the 2D projection of the input mesh $\mathbf{M}$ following $[\mathbf{R}|\mathbf{t}]$, and a texture $\mathbf{T}$. Several methods can be used to optimize Θ. However, they require a large set of annotated data [3,24] or perform only on modalities with similar sensors [12,27]. Generating RGB images from MRI scans is a challenging task because it requires bridging a significant difference in image modalities. We choose to use a neural image analogy method that combines the texture of a source image with a high-level content representation of a target image without the need for a large dataset [1]. This approach transfers the texture from $\mathbf{T}$ to $\widehat{\mathbf{I}}$ constrained by the projection of $\mathbf{M}$ using $\mathbf{A}[\mathbf{R}|\mathbf{t}]$ by minimizing the following loss function:

$$\mathcal{L}_\Theta = \sum_l \sum_{ij} \left(\mathbf{w}^{l,c}_{\mathbf{T}_{\mathrm{class}}} \mathcal{G}^l_{ij}(\mathbf{T}) - \mathbf{w}^{l,c}_{\mathbf{A}[\mathbf{R}|\mathbf{t}],\mathbf{M}} \mathcal{G}^l_{ij}(\widehat{\mathbf{I}}) \right) \quad \text{for } c \in \{0,1,2\} \tag{2}$$

where $\mathcal{G}_{ij}^{l}(\mathcal{T})$ is the Gram matrix of texture $\mathcal{T}$ at the l-th convolutional layer (pre-trained VGG-19 model), and $\mathbf{w}_{\mathbf{T}_{\mathrm{class}}}^{l,c}$ are the normalization factors for each Gram matrix, normalized by the number of pixels in a label class c of $\mathbf{T}_{\mathrm{class}}$. This allows for the quantification of the differences between the texture image $\mathbf{T}$ and the generated image $\widehat{\mathbf{I}}$ as it is being generated. Importantly, computing the inner-most sum over each label class c allows for texture comparison within each class, for instance: the background, the parenchyma, and the cortical vessels.

In practice, we assume constant camera parameters $\mathbf{A}$ and first sample a set of binary images by randomly varying the location and orientation of a virtual camera $[\mathbf{R}|\mathbf{t}]$ w.r.t. to the 3D mesh $\mathbf{M}$ before populating the binary images with the textures using $\mathcal{S}_\Theta$ (see Fig. 2). We restrict this sampling to the upper hemisphere of the 3D mesh to remain consistent with the plausible camera positions w.r.t. patient's head during neurosurgery.

We use the L-BFGS optimizer and 5 convolutional layers of VGG-19 to generate each image following [1] to find the resulting parameters $\widehat{\Theta}$. The training to synthesize for a single image typically takes around 50 iterations to converge.

2.3 Pose Regression Network

In order to solve Eq. 1, we assume a known focal length that can be obtained through pre-calibration. To obtain a compact representation of the rotation and since poses are restricted to the upper hemisphere of the 3D mesh (No Gimbal lock), the Euler-Rodrigues representation is used. Therefore, there are six parameters to be estimated: rotations r_x, r_y, r_z and translations t_x, t_y, t_z. We estimate our 6-DoF pose with a regression network $P_\Omega : \mathbf{I} \to \mathbf{p}$ and optimize its weights Ω to map each synthetic image $\mathbf{I}$ to its corresponding camera pose $\mathbf{p} = [r_x, r_y, r_z, t_x, t_y, t_z]^\mathsf{T}$.

The network architecture of P_Ω consists of 3 blocks each composed of two convolutional layers and one ReLU activation. To decrease the spatial dimension, an average pooling layer with a stride of 2 follows each block except the last one. At the end of the last hierarchy, we add three fully-connected layers with 128, 64, and 32 neurons and ReLU activation followed by one fully-connected with 6 neurons with a linear activation. We use the set of generated Expected Appearances $T^\mathcal{P} = \{(\mathbf{I}_i; \mathbf{p}_i)\}_i$; and optimize the following loss function over the parameters Ω of the network P_Ω:

$$\mathcal{L}_\Omega = \sum_{i=1}^{|T^\mathcal{P}|} \left(\left\| \mathbf{t}_i - \widehat{\mathbf{t}}_i \right\|_2 + \left\| \mathbf{R}_i^{\mathrm{vec}} - \widehat{\mathbf{R}}_i^{\mathrm{vec}} \right\|_2 \right) \qquad (3)$$

where $\mathbf{t}$ and $\mathbf{R}^{\mathrm{vec}}$ are the translation and rotation vector, respectively. We experimentally noticed that optimizing these entities separately leads to better results. The model is trained for each case (patient) for 200 epochs using mini-batches of size 8 with Adam optimizer and a learning rate of 0.001 and decays exponentially to 0.0001 over the course of the optimization. Finally, at run-time, given an image $\mathbf{I}$ we directly predict the corresponding 3D pose $\mathbf{p}$ so that: $\mathbf{p} \leftarrow \mathcal{P}(\mathbf{I}; \widehat{\Omega})$, where $\widehat{\Omega}$ is the resulting parameters from the training.

3   Results

Dataset. We tested our method retrospectively on 6 clinical datasets from 6 patients (cases) (see Fig. 5). These consisted of preoperative T1 contrast MRI scans and intraoperative images of the brain surface after dura opening. Cortical vessels around the tumors were segmented and triangulated to generate 3D meshes using *3D Slicer*. We generated 100 poses for each 3D mesh (i.e.: each case) and used a total of 15 unique textures from human brain surfaces (different from our 6 clinical datasets) for synthesis using S_Θ. In order to account for potential intraoperative brain deformations [4] we augment the textured projection with elastic deformation [21] resulting in approximately 1500 images per case. The surgical images of the brain (left image of the stereoscopic camera) were acquired with a Carl Zeiss surgical microscope. The ground-truth poses were obtained by manually aligning the 3D meshes on their corresponding images.

We evaluated the pose regressor network on both synthetic and real data. The model training and validation were performed on the synthesized images while the model testing was performed on the real images. Because a conventional train/validation/test split would lead to texture contamination, we created our validation dataset so that at least one texture is excluded from the training set. On the other hand, the test set consisted of the real images of the brain surface acquired using the surgical camera and are never used in the training.

Metrics. We chose the average distance metric (ADD) as proposed in [23] for evaluation. Given a set of mesh's 3D vertices, the ADD computes the mean of the pairwise distance between the 3D model points transformed using the ground truth and estimated transformation. We also adjusted the default 5 cm-5 deg translation and rotation error to our neurosurgical application and set the new threshold to 3 mm-3 deg.

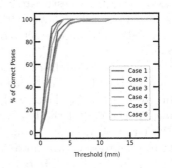

Fig. 3.   Accuracy-threshold curves on the validation set.

Accuracy-Threshold Curves. We calculated the number of 'correct' poses estimated by our model. We varied the distance threshold on the validation sets (excluding 2 textures) in order to reveal how the model performs w.r.t. that threshold. We plotted accuracy-threshold curves showing the percentage of pose accuracy variation with a threshold in a range of 0 mm to 20 mm. We can see in Fig. 3 that a 80.23% pose accuracy was reached within the 3 mm-3 deg threshold for all cases. This accuracy increases to 95.45% with a 5 mm-5 deg threshold.

Validation and Evaluation of Texture Invariance. We chose to follow a Leave-one-*Texture*-out cross-validation strategy to validate our model. This strategy seemed the most adequate to prevent over-fitting on the textures. We measured the ADD errors of our model for each case and report the results in

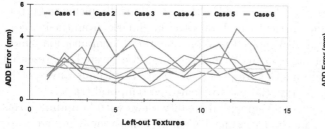
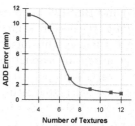

Fig. 4. Evaluation of texture invariance: (left) Leave-one-*Texture*-out cross validation and (right) impact of the number of textures on model accuracy.

Fig. 4-(left) for each left-out texture. We observed a variance in the ADD error that depends on which texture is left out. This supports the need for varying textures to improve the pose estimation. However, the errors remain low, with a 2.01 ± 0.58 mm average ADD error, over all cases. The average ADD error per case (over all left-out textures) is reported in Table 1. We measured the impact of the number of textures on the pose accuracy by progressively adding new textures to the training set, starting from 3 to 12 textures, while leaving 3 textures out for validation. We kept the size of the training set constant to not introduce size biases. Figure 4-(right) shows that increasing the number and variation of textures improved model performances.

Table 1. Validation on synthetic data and comparisons using real data.

Data	Experiment	Metric	Case 1	Case 2	Case 3	Case 4	Case 5	Case 6
Synth.	LOTO CV	Avg. ADD (mm)	1.80	1.56	1.27	2.83	2.58	2.07
	Acc./Thresh.	3mm-3deg (%)	89.33	96.66	98.07	81.81	93.33	80.23
Real	Ours	ADD (mm)	**3.64**	**2.98**	**1.62**	4.83	**3.02**	**3.32**
	ProbSEG	ADD (mm)	4.49	3.82	3.12	**4.69**	4.99	3.67
	BinSEG	ADD (mm)	9.2	4.87	12.12	8.09	11.43	6.29

Test and Comparison on Clinical Images. We compared our method (Ours) with segmentation-based methods (ProbSEG) and (BinSEG) [7]. These methods use learning-based models to extract binary images and probability maps of cortical vessels to drive the registration. We report in Table 1 the distances between the ground truth and estimated poses. Our method outperformed ProbSEG and BinSEG with an average ADM error of 3.26 ± 1.04 mm compared to 4.13 ± 0.70 mm and 8.67 ± 2.84 mm, respectively. Our errors remain below clinically measured neuronavigation errors reported in [4], in which a 5.26 ± 0.75 mm average initial registration error was measured in 15 craniotomy cases using intraoperative ultrasound. Our method outperformed ProbSEG in 5 cases out of 6 and BinSEG in all cases and remained within the clinically measured errors *without* the need to segment cortical vessels or select landmarks from the intraoperative image. Our method also showed fast intraoperative computation times. It required an average of only 45 ms to predict the pose (tested on research code on a laptop with NVidia GeForce GTX 1070 8 GB without any specific optimization), suggesting a potential use for real-time temporal tracking.

Figure 5 shows our results as Augmented Reality views with bounding boxes and overlaid meshes. Our method produced visually consistent alignments for all 6 clinical cases without the need for initial registration. Because our current method does not account for brain-shift deformation, our method produced some misalignment errors. However, in all cases, our predictions are similar to the ground truth.

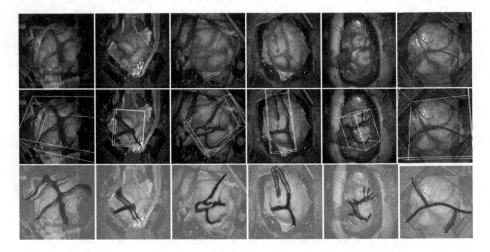

Fig. 5. Qualitative results on 6 patient datasets retrospectively showed in the first row. The second row shows Augmented Reality views with our predicted poses. The third row highlights 3D mesh-to-image projections using the ground-truth poses (green), and our predicted poses (blue). Our predictions are close to the ground truth for all cases. Note: microscope-magnified images with visible brain surface diameter ≈35 mm. (Color figure online)

4   Discussion and Conclusion

Clinical Feasibility. We have shown that our method is clinically viable. Our experiments using clinical data showed that our method provides accurate registration without manual intervention, that it is computationally efficient, and it is invariant to the visual appearance of the cortex. Our method does not require intraoperative 3D imaging such as intraoperative MRI or ultrasound, which require expensive equipment and are disruptive during surgery. Training patient-specific models from preoperative imaging transfers computational tasks to the preoperative stage so that patient-to-image registration can be performed in near real-time from live images acquired from a surgical microscope.

Limitations. The method presented in this paper is limited to 6-DoF pose estimation and does not account for deformation of the brain due to changes in head position, fluid loss, or tumor resection and assumes a known focal length. In the future, we will expand our method to model non-rigid deformations of the 3D mesh and to accommodate expected changes in zoom and focal depth during surgery. We will also explore how texture variability can be controlled and adapted to the observed image to improve model accuracy.

Conclusion. We introduced Expected Appearances, a novel learning-based method for intraoperative patient-to-image registration that uses synthesized expected images of the operative field to register preoperative scans with intraoperative views through the surgical microscope. We demonstrated state-of-the-art, real-time performance on challenging neurosurgical images using our method. Our method could be used to improve accuracy in neuronavigation and in image-guided surgery in general.

Acknowledgement. The authors were partially supported by the following National Institutes of Health grants: R01EB027134, R03EB032050, R01EB032387, and R01EB034223.

References

1. Ulyanov, D., et al.: Texture networks: feed-forward synthesis of textures and stylized images. In: Proceedings of Machine Learning Research, vol. 48, pp. 1349–1357. PMLR (2016)
2. Maier-Hein, L., et al.: Surgical data science - from concepts toward clinical translation. Med. Image Anal. **76**, 102–306 (2022)
3. Fernandez, V., et al.: Can segmentation models be trained with fully synthetically generated data? In: Zhao, C., Svoboda, D., Wolterink, J.M., Escobar, M. (eds.) SASHIMI 2022. LNCS, vol. 13570, pp. 79–90. Springer, Cham (2022). https://doi.org/10.1007/978-3-031-16980-9_8
4. Frisken, S., et al.: A comparison of thin-plate spline deformation and finite element modeling to compensate for brain shift during tumor resection. Int. J. Comput. Assist. Radiol. Surg. **15**, 75–85 (2019). https://doi.org/10.1007/s11548-019-02057-2
5. González-Darder, J.M.: 'State of the art' of the craniotomy in the early twenty-first century and future development. In: González-Darder, J.M. (ed.) Trepanation, Trephining and Craniotomy, pp. 421–427. Springer, Cham (2019). https://doi.org/10.1007/978-3-030-22212-3_34
6. Haouchine, N., Juvekar, P., Wells III, W.M., Cotin, S., Golby, A., Frisken, S.: Deformation aware augmented reality for craniotomy using 3D/2D non-rigid registration of cortical vessels. In: Martel, A.L., et al. (eds.) MICCAI 2020. LNCS, vol. 12264, pp. 735–744. Springer, Cham (2020). https://doi.org/10.1007/978-3-030-59719-1_71

7. Haouchine, N., Juvekar, P., Nercessian, M., Wells III, W.M., Golby, A., Frisken, S.: Pose estimation and non-rigid registration for augmented reality during neurosurgery. IEEE Trans. Biomed. Eng. **69**(4), 1310–1317 (2022)
8. Ji, S., Fan, X., Roberts, D.W., Hartov, A., Paulsen, K.D.: Cortical surface shift estimation using stereovision and optical flow motion tracking via projection image registration. Med. Image Anal. **18**(7), 1169–1183 (2014)
9. Ji, S., Wu, Z., Hartov, A., Roberts, D.W., Paulsen, K.D.: Mutual-information-based image to patient re-registration using intraoperative ultrasound in image-guided neurosurgery. Med. Phys. **35**(10), 4612–4624 (2008)
10. Jiang, J., et al.: Marker-less tracking of brain surface deformations by non-rigid registration integrating surface and vessel/sulci features. Int. J. Comput. Assist. Radiol. Surg. **11**, 1687–1701 (2016). https://doi.org/10.1007/s11548-016-1358-7
11. Kuhnt, D., Bauer, M.H.A., Nimsky, C.: Brain shift compensation and neurosurgical image fusion using intraoperative MRI: current status and future challenges. Crit. Rev. Trade Biomed. Eng. **40**(3), 175–185 (2012)
12. Lecomte, F., Dillenseger, J.L., Cotin, S.: CNN-based real-time 2D–3D deformable registration from a single X-ray projection. CoRR abs/2003.08934 (2022)
13. Luo, M., Larson, P.S., Martin, A.J., Konrad, P.E., Miga, M.I.: An integrated multiphysics finite element modeling framework for deep brain stimulation: preliminary study on impact of brain shift on neuronal pathways. In: Shen, D., et al. (eds.) MICCAI 2019. LNCS, vol. 11768, pp. 682–690. Springer, Cham (2019). https://doi.org/10.1007/978-3-030-32254-0_76
14. Machado, I., et al.: Non-rigid registration of 3d ultrasound for neurosurgery using automatic feature detection and matching. Int. J. Comput. Assist. Radiol. Surg. **13**, 1525–1538 (2018). https://doi.org/10.1007/s11548-018-1786-7
15. Marreiros, F.M.M., Rossitti, S., Wang, C., Smedby, Ö.: Non-rigid deformation pipeline for compensation of superficial brain shift. In: Mori, K., Sakuma, I., Sato, Y., Barillot, C., Navab, N. (eds.) MICCAI 2013. LNCS, vol. 8150, pp. 141–148. Springer, Heidelberg (2013). https://doi.org/10.1007/978-3-642-40763-5_18
16. Mildenhall, B., Srinivasan, P.P., Tancik, M., Barron, J.T., Ramamoorthi, R., Ng, R.: NeRF: representing scenes as neural radiance fields for view synthesis. Commun. ACM **65**(1), 99–106 (2021)
17. Mohammadi, A., Ahmadian, A., Azar, A.D., Sheykh, A.D., Amiri, F., Alirezaie, J.: Estimation of intraoperative brain shift by combination of stereovision and doppler ultrasound: phantom and animal model study. Int. J. Comput. Assist. Radiol. Surg. **10**(11), 1753–1764 (2015). https://doi.org/10.1007/s11548-015-1216-z
18. Nercessian, M., Haouchine, N., Juvekar, P., Frisken, S., Golby, A.: Deep cortical vessel segmentation driven by data augmentation with neural image analogy. In: 2021 IEEE 18th International Symposium on Biomedical Imaging (ISBI), pp. 721–724. IEEE (2021)
19. Pereira, V.M., et al.: Volumetric measurements of brain shift using intraoperative cone-beam computed tomography: preliminary study. Oper. Neurosurg. **12**(1), 4–13 (2015)
20. Rivaz, H., Collins, D.L.: Deformable registration of preoperative MR, pre-resection ultrasound, and post-resection ultrasound images of neurosurgery. Int. J. Comput. Assist. Radiol. Surg. **10**(7), 1017–1028 (2015). https://doi.org/10.1007/s11548-014-1099-4
21. Ronneberger, O., Fischer, P., Brox, T.: U-Net: convolutional networks for biomedical image segmentation. In: Navab, N., Hornegger, J., Wells, W.M., Frangi, A.F. (eds.) MICCAI 2015. LNCS, vol. 9351, pp. 234–241. Springer, Cham (2015). https://doi.org/10.1007/978-3-319-24574-4_28

22. Sanai, N., Polley, M.Y., McDermott, M.W., Parsa, A.T., Berger, M.S.: An extent of resection threshold for newly diagnosed glioblastomas: clinical article. J. Neurosurg. JNS **115**(1), 3–8 (2011)
23. Shotton, J., Glocker, B., Zach, C., Izadi, S., Criminisi, A., Fitzgibbon, A.W.: Scene coordinate regression forests for camera relocalization in RGB-D images. In: 2013 IEEE Conference on Computer Vision and Pattern Recognition, pp. 2930–2937 (2013)
24. Skandarani, Y., Jodoin, P.M., Lalande, A.: GANs for medical image synthesis: an empirical study. J. Imaging **9**(3), 69 (2023)
25. Stoyanov, D.: Surgical vision. Ann. Biomed. Eng. **40**, 332–345 (2012). https://doi.org/10.1007/s10439-011-0441-z
26. Sun, K., Pheiffer, T., Simpson, A., Weis, J., Thompson, R., Miga, M.: Near real-time computer assisted surgery for brain shift correction using biomechanical models. IEEE Transl. Eng. Health Med. **2**, 1–13 (2014)
27. Tian, L., Lee, Y.Z., San José Estépar, R., Niethammer, M.: LiftReg: limited angle 2D/3D deformable registration. In: Wang, L., Dou, Q., Fletcher, P.T., Speidel, S., Li, S. (eds.) MICCAI 2022. LNCS, vol. 13436, pp. 207–216. Springer, Cham (2022). https://doi.org/10.1007/978-3-031-16446-0_20

SENDD: Sparse Efficient Neural Depth and Deformation for Tissue Tracking

Adam Schmidt[1]($\boxtimes$) (iD), Omid Mohareri[2], Simon DiMaio[2],
and Septimiu E. Salcudean[1]

[1] Department of Electrical and Computer Engineering, The University of British
Columbia, Vancouver BC V6T 1Z4, Canada
adamschmidt@ece.ubc.ca

[2] Advanced Research, Intuitive Surgical, Sunnyvale, CA 94086, USA

Abstract. Deformable tracking and real-time estimation of 3D tissue motion is essential to enable automation and image guidance applications in robotically assisted surgery. Our model, Sparse Efficient Neural Depth and Deformation (SENDD), extends prior 2D tracking work to estimate flow in 3D space. SENDD introduces novel contributions of learned detection, and sparse per-point depth and 3D flow estimation, all with less than half a million parameters. SENDD does this by using graph neural networks of sparse keypoint matches to estimate both depth and 3D flow anywhere. We quantify and benchmark SENDD on a comprehensively labelled tissue dataset, and compare it to an equivalent 2D flow model. SENDD performs comparably while enabling applications that 2D flow cannot. SENDD can track points and estimate depth at 10fps on an NVIDIA RTX 4000 for 1280 tracked (query) points and its cost scales linearly with an increasing/decreasing number of points. SENDD enables multiple downstream applications that require estimation of 3D motion in stereo endoscopy.

Keywords: Tissue tracking · Graph neural networks · Scene flow

1 Introduction

Tracking of tissue and organs in surgical stereo endoscopy is essential to enable downstream tasks in image guidance [8], surgical perception [5,13], motion compensation [17], and colonoscopy coverage estimation [28]. Given the difficulty in creating labelled training data, we train an unsupervised model that can estimate motion for anything in the surgical field: tissue, gauze, clips, instruments. Recent models for estimating deformation either use classical features and an underlying model (splines, embedded deformation, etc. [11,21]), or neural networks (eg. CNNs [24] or 2D graph neural networks (GNNs) [19]). The

This work was supported by Intuitive Surgical.

Supplementary Information The online version contains supplementary material available at https://doi.org/10.1007/978-3-031-43996-4_23.

issue with 2D methods is that downstream applications often require depth. For example a correct physical 3D is needed location to enable augmented reality image guidance, motion compensation, and robotic automation (eg. suturing). For our model, SENDD, we extend a sparse neural interpolation paradigm [19] to simultaneously perform depth estimation and 3D flow estimation. This allows us to estimate depth and 3D flow all with one network rather than having to separately estimate dense depth maps. With our approach, SENDD computes motion directly in 3D space, and parameterizes a 3D flow field that estimates the motion of any point in the field of view.

We design SENDD to use few parameters (low memory cost) and to scale with the number of points to be tracked (adaptive to different applications). To avoid having to operate a 3D convolution over an entire volume, we use GNNs instead of CNNs. This allows applications to tune how much computation to use by using more/fewer points. SENDD is trained end-to-end. This includes the detection, description, refinement, depth estimation, and 3D flow steps. SENDD can perform frame-to-frame tracking and 3D scene flow estimation, but it could also be used as a more robust data association term for SLAM. As will be shown in Sect. 2, unlike in prior work, our proposed approach combines feature detection, depth estimation and deformation modeling for scene flow all in one. After providing relevant background for tissue tracking, we will describe SENDD, quantify it with a new IR-labelled tissue dataset, and finally demonstrate SENDD's efficiency. The main novelties are that it is both **3D** and **Efficient** by: estimating scene flow anywhere in 3D space by using a GNN on salient points (**3D**), and reusing salient keypoints to calculate both sparse depth and flow at anywhere (**Efficient**).

2   Background and Related Work

Different components are necessary to enable tissue tracking in surgery: feature detection, depth estimation, deformation modelling, and deformable Simultaneous Localization and Mapping (SLAM). Our model acts as the feature detection, depth estimation, and deformation model for scene flow, all in one.

Recently, SuperPoint [3] features have been applied to endoscopy [1]. These are trained using loss on image pairs that are warped with homographies. In SENDD we use similar detections, but use a photometric reconstruction loss instead. SuperGlue [18] is a GNN method that can be used on top of SuperPoint to filter outliers, but it does not enable estimation of flow at non-keypoint locations, in addition to taking ∼270 ms for 2048 keypoints. Its GNN differs in that we use k-NN connected graph rather than a fully connected one. In stereo depth estimation, BDIS [22] introduces efficient improvements on classical methods, running in ∼70 ms for images of size (1280, 720). For flow estimation, there are the CNN-based RAFT [24], and RAFT3D [25] (45M params ∼386 ms) which downsample by 8x and require computation over full images. KINFlow [19] estimates motion using a GNN, but only in 2D. For SLAM in endoscopy, MIS-SLAM [21] uses classical descriptors and models for deformation. More recent work still does this, mainly due to the high cost of using CNNs when only a

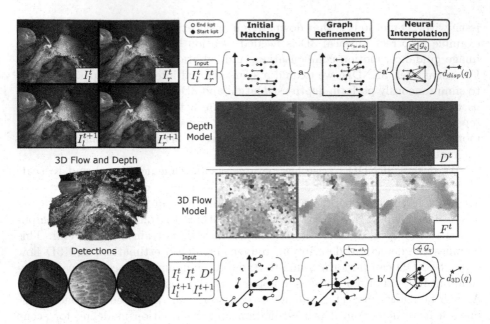

Fig. 1. The SENDD model on a single query point, q (green star). From left to right are different steps in the refinement process for both depth (top) and flow (bottom). They both share feature points, but they use different weights for the graph refinement and interpolation steps. Both the depth and the flow network estimate motion of any point as a function of the nearest detected features using a GNN. The depth of tracked (query) points is passed into the flow model to provide 3D position. $(\mathbf{a}, \mathbf{b})$ are learned (depth, flow) features after nearest-neighbor keypoint matching, and $(\mathbf{a}', \mathbf{b}')$ are node features after graph refinement. Only Neural Interpolation is repeated to track multiple queries. (Color figure online)

few points are actually salient. Specifically, Endo-Depth-and-Motion [16] performs SLAM in rigid scenes. DefSLAM and SD-DefSLAM [6,9] both use meshes along with ORB features or Lucas-Kanade optical flow, respectively, for data association. Lamarca et al. [10] track surfels using photometric error, but they do not have an underlying interpolation model for estimating motion between surfels. SuPer [11], and its extensions [12,13] use classical embedded deformation for motion modelling ($\sim$500 ms/frame). Finally, for full scene and deformation estimation, NERF [15] has recently been applied to endoscopy [26], but requires per-scene training and computationally expensive. With SENDD, we fill the gap between classical and learned models by providing a flexible and efficient deformation model that could be integrated into SuPer [11], SLAM, or used for short-term tracking.

3   Methods

The SENDD model is described in Fig. 1. SENDD improves on prior sparse interpolation methods by designing a solution that: **learns a detector as part**

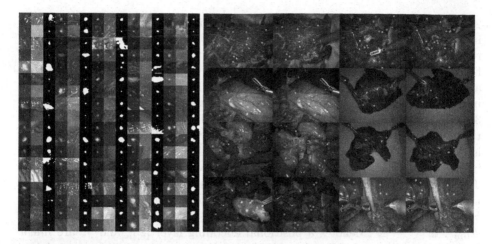

Fig. 2. Test dataset images. Left: patch triplets of IR ims., visible light ims., and ground truth segmentations. Right: IR ims. from the start and end of clips with circles around labelled segment centers. Dataset statistics (*in vivo* 266 min./*ex vivo* 134 min.) with (182/1139) clips and (944/9596) segments.

of the network, uses detected points to evaluate depth in a sparse manner, and uses a sparse interpolation paradigm to evaluate motion in 3D instead of 2D. SENDD consists of two key parts, with the first being the 1D stereo interpolation (Depth), and the second being the 3D flow estimation (Deformation). Both use GNNs to estimate query disparity/motion using a coordinate-based Multi Layer Perceptron (MLP) (aka. implicit functions [23]) of keypoints neighboring it. Before either interpolation step, SENDD detects and matches keypoints. After detailing the detection step, we will explain the 3D flow and disparity models. For the figures in this paper, we densely sample query points on a grid to show dense visualizations, but in practice the query points can be user or application defined. Please see the supplementary video for examples of SENDD tracking tissue points.

Learned Detector: Like SuperPoint [3], we detect points across an image, but we use fewer layers and estimate a single location for each $(32, 32)$ region (instead of $(8, 8)$). Our images (I_l^t, I_r^t) are of size $(1280, 1024)$, so we have $N = 1280$ detected points per frame. Unlike SuperPoint or others in the surgical space [1], we do not rely on data augmentation for training, and instead allow the downstream loss metric to optimize detections to best reduce the final photometric reconstruction error. Thus the detections are directly trained on real-world deformations. Since we have a sparse regime, we use a softmax dot product score in training, and use max in inference. See Fig. 1 for examples of detections.

3D Flow Network: SENDD estimates the 3D flow $d_{3D}(q) \in \mathbb{R}^3$ for each query $q \in \mathbb{R}^2$ using two images (I_l^t, I_l^{t+1}), and their depth maps (D^t, D^{t+1}). For a means to query depth at arbitrary points, see the section on sparse depth estimation. To obtain initial matches, we match the detected points in

2D from frame I_l^t to I_l^{t+1}. We use ReTRo keypoints [20] as descriptors by training the ReTRo network paired with SENDD's learned detector. We do this to
remain as lightweight as possible. Matches further than 256 pixels away are
filtered out. This results in N pairs of points in 2D with positions $p_i^{2D}, p_i'^{2D}$
and feature descriptors $f_i, f_i', \{i \in 1, \ldots, N\}, f_i \in \mathbb{R}^c$. These pairs are nodes
in our graph, $\mathcal{G}$. Given the set of preliminary matches, we can then get their
3D positions by using our sparse depth interpolation and backprojecting using
the camera parameters at each point $p_i^{2D}, p_i'^{2D} \rightarrow p_i^{3D}, p_i'^{3D}$. The graph, $\mathcal{G}$,
is defined with edges connecting each node to its k-nearest neighbors (k-NN)
in 3D, with an optional dilation to enable a wider field of view at low cost.
The positions and features of each correspondence are combined to act as features in this graph. For each node (detected point in image I_l^t), its feature is:
$b_i = \phi_3(p_i^{3D}) + \phi_3(p_i'^{3D}) + \gamma_d(f_i) + \gamma_e(f_i') + \phi_1(\|f_i - f_i'\|_2) + \phi_1(\|p_i - p_i'\|_2)$.
$\phi_{dim} : \mathbb{R}^{dim} \rightarrow \mathbb{R}^c$, denotes a positional encoding layer [23], and γ_* are linear
layers followed by a ReLU, where different subscripts denote different weights.
b_i are used as features in our graph attention network, with the bold version
defined as the set of all node features $\mathbf{b} = \{b_i | i \in N\}$. By using a graph-
attention neural network (GNN), as described in [19], we can refine this graph
in 3D to estimate refined offsets and higher level local-neighborhood features.
$\mathbf{b}' = \text{Ga}(\text{Ga}(\text{Ga}(\text{Ga}(\mathbf{b}, \mathcal{G}))))$, is the final set of node features that are functions
of their neighborhood, $\mathcal{N}$, in the graph. Ga is the graph attention operation. In
practice, for each layer we use dilations of $[1, 8, 8, 1]$, and $k = 4$. The prior steps
only need to run once per image pair. The motion $d_{3D}(q)$ of each query point q is a
function of the nearby nodes in 3D, with $\mathcal{G}_q$ denoting a k-clique graph of the query
point and its $k - 1$ nearest nodes; $d_{3D}(q) \in \mathbb{R}^3 = \text{Lin}_{3D}(\text{Ga}(\{q\} \| \{\mathbf{b}_{\mathcal{N}}'\}, \mathcal{G}_q))$.
Lin_{3D} is a linear layer that converts from c channels to 3.

Sparse Depth Interpolation: Instead of running a depth CNN in parallel, we
estimate disparity sparsely as needed by using the same feature points with
another lightweight GNN. We do this as we found CNNs (eg. GANet [27])
too expensive in terms of training and inference time. We adapt the 3D GNN
interpolation methodology to estimate 1D flow along epipolar lines in the same
way that we modified a 2D sparse flow model to work in 3D. First, matches
are found along epipolar lines between left and right images (I_r^t, I_l^t). Then
matches are refined and query points are interpolated using a GNN. $\mathbf{a}, \mathbf{a}'$ are
the 1D equivalents of $\mathbf{b}, \mathbf{b}'$ from the 3D flow network. For the refined node
features, $\mathbf{a}' \in \mathbb{R}^1 = \text{Ga}(\text{Ga}(\text{Ga}(\text{Ga}(\mathbf{a}, \mathcal{G}))))$, and the final disparity estimate,
$d_{disp}(q) \in \mathbb{R}^1 = \text{Lin}_{1D}(\text{Ga}(\{q\} \| \{\mathbf{a}_{\mathcal{N}}'\}, \mathcal{G}_q))$. Lin_{1D} is a linear layer that converts from c channels to 1. This can be seen like a neural version of the clas-
sic libELAS [4], where libELAS uses sparse support points on a regular grid
along Sobel edge features to match points within a search space defined by the
Delauney triangulation of support point matches.

Loss: We train SENDD using loss on the warped stereo and flow images:

$$\mathcal{L}_p(A, B) = \alpha \frac{1 - SSIM(A, B)}{2} + (1 - \alpha)\|A - B\|_1. \tag{1}$$

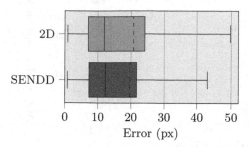

Model	Mean (mm)	Median (mm)
SEND	**7.92** ± 0.28	**6.05**
CSRT	56.66 ± 8.18	17.21
RAFT	33.00 ± 5.01	13.10

Fig. 3. Endpoint error over the full dataset. Left: 2D endpoint error in a boxplot. Mean is the dotted line. Right: 3D endpoint error compared to using RAFT or CSRT to track in both left and right frames. Standard error of the mean is denoted with ±.

$$\mathcal{L}_s(V) = \frac{1}{n} \sum \exp\left(-\frac{\beta}{3}\sum_c \left|\frac{\partial I}{\partial x}\right|\right)\left|\frac{\partial V}{\partial x}\right| + \exp\left(-\frac{\beta}{3}\sum_c \left|\frac{\partial I}{\partial y}\right|\right)\left|\frac{\partial V}{\partial y}\right| \quad (2)$$

$$\mathcal{L}_d = \left|D^{t \to t-1} - D^{t-1}\right| \quad (3)$$

$$\mathcal{L}_{total} = \mathcal{L}_p(I^l, I^{r \to l}) + \mathcal{L}_p(I^t, I^{t+1 \to t}) + \lambda_F \mathcal{L}_s(F) + \lambda_D \mathcal{L}_s(D) + \lambda_d \mathcal{L}_d \quad (4)$$

$\mathcal{L}_p(A, B)$ is a photometric loss function of input images (A, B) that is used for both the warped depth pairs $(\mathcal{L}_p(I^l, I^{r \to l}))$ and warped flow pairs $(\mathcal{L}_p(I^t, I^{t+1 \to t}))$, $\mathcal{L}_s(V)$ is a smoothness loss on a flow field V, and c denotes image color channels [7]. These loss functions are used for both the warped depth $(I^l, I^{r \to l})$ and the flow reconstruction pairs $(I^t, I^{t+1 \to t})$. F and D are densely queried images of flow and depth. We add a depth matching loss, $\mathcal{L}_d$ which encourages the warped depth map to be close to the estimated depth map. We set $\alpha = 0.85, \beta = 150, \lambda_d = 0.001, \lambda_F = 0.01, \lambda_D = 1.0$, using values similar to [7], and weighting depth matching lightly as a guiding term.

4   Experiments

We train SENDD with a set of rectified stereo videos from porcine clinical labs collected with a da Vinci Xi surgical system. We randomly select images and skip between $(1, 45)$ frames for each training pair. We train using PyTorch with a batch size of 2 for 100,000 steps using a one-cycle learning rate schedule with maxlr = 1e−4, minlr = 4e−6. We use 64 channels for all graph attention operations.

Dataset: In order to quantify our results, we generate a new dataset. The primary motivation for this is to have metrics that are not dependent on visible markers or require human labelling (which can bias to salient points). Some other datasets have points that are hand labelled in software [2,11] or use visible markers [12]. Others rely on RMSE of depth maps as a proxy for error, but this does not account for tissue sliding. Our dataset uses tattooed points that flouresce

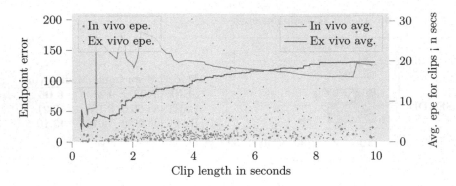

Fig. 4. *In vivo* and *ex vivo* errors of SENDD over the dataset.

under infrared (IR) light by using ICG dye. For each ground truth clip, we capture an IR frame at the start, record a deformation action, and then capture another IR frame. The start and end IR segments are then used as ground truth for quantification. Points that are not visible at both the start and finish are then manually removed. See Fig. 2 for examples of data points at start and end frames. The dataset includes multiple different tissue types in *ex vivo*, including: stomach, kidney, liver, intestine, tongue, heart, chicken breast, pork chop, etc. Additionally the dataset includes recordings from four different *in vivo* porcine labs. This dataset will be publicly released separately before MICCAI. No clips from this labelled dataset are used in training.

Quantification: Instead of using Intersection Over Union (IOU) which would fail on small segments (eg. $IOU = 0$ for a one pixel measurement that is one pixel off), we use endpoint error and chamfer distance between segments to quantify SENDD's performance. We compute endpoint error between the centers of each segmentation region. We use chamfer distance as well because it allows us to see error for regions that are larger or non-circular. Chamfer distance provides a metric that has an lower bound on error, in that if true error is zero then this will also be zero.

Experiments: The primary motivation for creating a 3D model (SENDD) instead of 2D is that it enables applications that require understanding of motion in 3D (automation, image guidance models). That said, we still compare to the 2D version to compare performance in pixel space. To quantify performance on clips, since these models only estimate motion for frame pairs, we run SENDD for each frame pair over n frames in a clip, where n is the clip length. No relocalization or drift prevention is incorporated (the supplementary material shows endpoint error on strides other than one). For all experiments we select clips with length less than 10 s, as we are looking to quantify short term tracking methods. The 2D model has the exact same parameters and model structure as SENDD (3D), except it does not do any depth map calculation, and only runs in image space. First, we compare SENDD to the equivalent 2D model. We do this with endpoint error as seen in Fig. 3, and show that the 3D method outperforms 2D

Table 1. Times in milliseconds for using SENDD with 1280 detected keypoints to track 1280 query points. Times are averaged over 64 runs, with standard error in the second row. Streaming total is calculated by subtracting ($\frac{1}{2}$(col1 + col2) + col4) from the stereo and flow total. This is the time that it would take if reusing features, as is enabled in a streaming manner (10 fps).

Detection & Description (x4)	Auxiliary Buffer (x4)	3. Stereo Refinement	Stereo Total	Flow Refinement	Stereo & Flow Total	Streaming Total
50.1	1.8	12.9	21.4	32.3	145.4	98.1
± 0.4	± 0.02	± 0.6	± 0.8	± 0.9	± 2.1	

over the full dataset. Additionally, we compare to baselines of CSRT [14] which is a high performer in the SurgT challenge [2] and RAFT [24]. To track with CSRT and RAFT, we track in each left and right frame and backproject as done in SurgT [2]. SENDD outperforms these methods used off-the-shelf; see Fig. 3. Performance of the RAFT and CSRT methods could likely be improved with a drift correction, or synchronization between left and right to prevent depth from becoming more erroneous. Then we compare chamfer distance. The 3D method also outperforms the 2D method, on 64 randomly selected clips, with a total average error of 19.18 px vs 20.88 px. The reasons for the SENDD model being close in performance to the equivalent 2D model could be that the 3D method uses the same amount of channels to detect and describe the same features that will be used for both depth and flow. Additionally, lens smudges or specularities can corrupt the depth map, leading to errors in the 3D model that the purely photometric 2D model might not encounter. In 3D all it takes is for one camera artifact to in either image to obscure the depth map, and the resulting flow. The 2D method decreases the likelihood of this happening as it only uses one image. Finally, we compare performance in terms of endpoint error of our 3D model on *in vivo* vs *ex vivo* labelled data. As is shown in Fig. 4, the *in vivo* experiments have a more monotonic performance decrease relative to clip length. Actions in the *ex vivo* dataset were performed solely to evaluate tracking performance, while the *in vivo* data was collected alongside training while performing standard surgical procedures (eg. cholecystectomy). Thus the *in vivo* scenes can have more complicated occlusions or artifacts that SENDD does not account for, even though it is also trained on *in vivo* data.

Benchmarking and Model Size: SENDD has only 366,195 parameters, compared to other models which estimate just flow (RAFT-s [24] with 1.0M params.) or depth (GANet [27] with 0.5M params.) using CNNs. We benchmark SENDD on a NVIDIA Quadro RTX 4000, with a batch size of one, 1280 query points, and 1280 control points (keypoints). As seen in Table 1, the stereo estimation for each frame takes 21.4 ms, and total time for the whole network to estimate stereo (at both t and t + 1) and flow is 145.4 ms. When estimating flow for an image pair, we need to calculate two sets of keypoint features, but for a video, we can reuse the features from the previous frame instead of recalculating the full pair each time. Results with reuse are exactly the same, only timing differs. Subtracting out the time for operations that do not need to be repeated leaves us with a streaming time of 97.3 ms, with a frame rate of 10fps for the entire flow and depth model. This can be further improved by using spatial data structures for nearest neighbor lookup or PyTorch optimizations that have not been enabled (eg. float16). The number of salient (or query) points can also be changed to adjust refinement (or neural interpolation) time.

5   Conclusion

SENDD is a flexible model for estimating deformation in 3D. A limitation of SENDD is that it is unable to cope with occlusion or relocalization, and like all methods is vulnerable to drift over long periods. These could be amended by integrating a SLAM system. We demonstrate that SENDD performs better than the equivalent sparse 2D model while additionally enabling parameterization of deformation in 3D space, learned detection, and depth estimation. SENDD enables real-time applications that can rely on tissue tracking in surgery, such as long term tracking, or deformation estimation.

References

1. Barbed, O.L., Chadebecq, F., Morlana, J., Montiel, J.M.M., Murillo, A.C.: Superpoint features in endoscopy. In: Imaging Systems for GI Endoscopy, and Graphs in Biomedical Image Analysis. LNCS, pp. 45–55. Springer, Cham (2022)
2. Cartucho, J., et al.: SurgT: soft-tissue tracking for robotic surgery, benchmark and challenge (2023). https://doi.org/10.48550/ARXIV.2302.03022
3. DeTone, D., Malisiewicz, T., Rabinovich, A.: SuperPoint: self-supervised interest point detection and description. In: Proceedings of the IEEE Conference on Computer Vision and Pattern Recognition Workshops (2018)
4. Geiger, A., Roser, M., Urtasun, R.: Efficient large-scale stereo matching. In: Kimmel, R., Klette, R., Sugimoto, A. (eds.) ACCV 2010. LNCS, vol. 6492, pp. 25–38. Springer, Heidelberg (2011). https://doi.org/10.1007/978-3-642-19315-6_3
5. Giannarou, S., Ye, M., Gras, G., Leibrandt, K., Marcus, H., Yang, G.: Vision-based deformation recovery for intraoperative force estimation of tool-tissue interaction for neurosurgery. IJCARS 11, 929–936 (2016)
6. Gómez-Rodríguez, J.J., Lamarca, J., Morlana, J., Tardós, J.D., Montiel, J.M.M.: SD-DefSLAM: semi-direct monocular slam for deformable and intracorporeal scenes. In: 2021 IEEE International Conference on Robotics and Automation (ICRA), pp. 5170–5177 (May 2021)
7. Jonschkowski, R., Stone, A., Barron, J.T., Gordon, A., Konolige, K., Angelova, A.: What matters in unsupervised optical flow. In: Vedaldi, A., Bischof, H., Brox, T., Frahm, J.M. (eds.) ECCV 2020. LNCS, pp. 557–572. Springer, Cham (2020). https://doi.org/10.1007/978-3-030-58536-5_33
8. Kalia, M., Mathur, P., Tsang, K., Black, P., Navab, N., Salcudean, S.: Evaluation of a marker-less, intra-operative, augmented reality guidance system for robot-assisted laparoscopic radical prostatectomy. Int. J. CARS 15(7), 1225–1233 (2020)

9. Lamarca, J., Parashar, S., Bartoli, A., Montiel, J.: DefSLAM: tracking and mapping of deforming scenes from monocular sequences. IEEE Trans. Rob. **37**(1), 291–303 (2021). https://doi.org/10.1109/TRO.2020.3020739
10. Lamarca, J., Gómez Rodríguez, J.J., Tardós, J.D., Montiel, J.: Direct and sparse deformable tracking. IEEE Robot. Autom. Lett. **7**(4), 11450–11457 (2022). https://doi.org/10.1109/LRA.2022.3201253
11. Li, Y., et al.: SuPer: a surgical perception framework for endoscopic tissue manipulation with surgical robotics. IEEE Robot. Autom. Lett. **5**(2), 2294–2301 (2020)
12. Lin, S., et al.: Semantic-SuPer: a semantic-aware surgical perception framework for endoscopic tissue identification, reconstruction, and tracking. In: 2023 IEEE International Conference on Robotics and Automation (ICRA), pp. 4739–4746 (2023). https://doi.org/10.1109/ICRA48891.2023.10160746
13. Lu, J., Jayakumari, A., Richter, F., Li, Y., Yip, M.C.: Super deep: a surgical perception framework for robotic tissue manipulation using deep learning for feature extraction. In: ICRA. IEEE (2021)
14. Lukezic, A., Vojir, T., Zajc, L.C., Matas, J., Kristan, M.: Discriminative correlation filter with channel and spatial reliability. In: 2017 IEEE Conference on Computer Vision and Pattern Recognition (CVPR), pp. 4847–4856. IEEE, Honolulu, HI (2017). https://doi.org/10.1109/CVPR.2017.515
15. Mildenhall, B., Srinivasan, P.P., Tancik, M., Barron, J.T., Ramamoorthi, R., Ng, R.: NeRF: representing scenes as neural radiance fields for view synthesis. Commun. ACM **65**(1), 99–106 (2021). https://doi.org/10.1145/3503250
16. Recasens, D., Lamarca, J., Fácil, J.M., Montiel, J.M.M., Civera, J.: Endo-depth-and-motion: reconstruction and tracking in endoscopic videos using depth networks and photometric constraints. IEEE Robot. Autom. Lett. **6**(4), 7225–7232 (2021). https://doi.org/10.1109/LRA.2021.3095528
17. Richa, R., Bó, A.P., Poignet, P.: Towards robust 3D visual tracking for motion compensation in beating heart surgery. Med. Image Anal. **15**, 302–315 (2011)
18. Sarlin, P.E., DeTone, D., Malisiewicz, T., Rabinovich, A.: SuperGlue: learning feature matching with graph neural networks. In: CVPR (2020)
19. Schmidt, A., Mohareri, O., DiMaio, S.P., Salcudean, S.E.: Fast graph refinement and implicit neural representation for tissue tracking. In: ICRA (2022)
20. Schmidt, A., Salcudean, S.E.: Real-time rotated convolutional descriptor for surgical environments. In: de Bruijne, M., et al. (eds.) MICCAI 2021. LNCS, vol. 12904, pp. 279–289. Springer, Cham (2021). https://doi.org/10.1007/978-3-030-87202-1_27
21. Song, J., Wang, J., Zhao, L., Huang, S., Dissanayake, G.: MIS-SLAM: real-time large-scale dense deformable slam system in minimal invasive surgery based on heterogeneous computing. IEEE Robot. Autom. Lett. **3**, 4068–4075 (2018)
22. Song, J., Zhu, Q., Lin, J., Ghaffari, M.: BDIS: Bayesian dense inverse searching method for real-time stereo surgical image matching. IEEE Trans. Rob. **39**(2), 1388–1406 (2023)
23. Tancik, M., et al.: Fourier features let networks learn high frequency functions in low dimensional domains. In: NeurIPS (2020)
24. Teed, Z., Deng, J.: RAFT: recurrent all-pairs field transforms for optical flow. In: Vedaldi, A., Bischof, H., Brox, T., Frahm, J.-M. (eds.) ECCV 2020. LNCS, vol. 12347, pp. 402–419. Springer, Cham (2020). https://doi.org/10.1007/978-3-030-58536-5_24
25. Teed, Z., Deng, J.: RAFT-3D: scene flow using rigid-motion embeddings. In: Proceedings of the IEEE/CVF Conference on Computer Vision and Pattern Recognition, pp. 8375–8384 (2021)

26. Wang, Y., Long, Y., Fan, S.H., Dou, Q.: Neural rendering for stereo 3D reconstruction of deformable tissues in robotic surgery. In: Wang, L., Dou, Q., Fletcher, P.T., Speidel, S., Li, S. (eds.) MICCAI 2022. LNCS, pp. 431–441. Springer, Cham (2022)
27. Zhang, F., Prisacariu, V., Yang, R., Torr, P.H.: GA-Net: guided aggregation net for end-to-end stereo matching. In: 2019 IEEE/CVF Conference on Computer Vision and Pattern Recognition (CVPR), pp. 185–194. IEEE, Long Beach, CA, USA (2019). https://doi.org/10.1109/CVPR.2019.00027
28. Zhang, Y., et al.: ColDE: a depth estimation framework for colonoscopy reconstruction. arXiv:2111.10371 [cs, eess] (2021)

Cochlear Implant Fold Detection in Intra-operative CT Using Weakly Supervised Multi-task Deep Learning

Mohammad M. R. Khan[(✉)], Yubo Fan, Benoit M. Dawant, and Jack H. Noble

Department of Electrical and Computer Engineering, Vanderbilt University, Nashville, TN 37235, USA
mohammad.mahmudur.rahman.khan@vanderbilt.edu

Abstract. In cochlear implant (CI) procedures, an electrode array is surgically inserted into the cochlea. The electrodes are used to stimulate the auditory nerve and restore hearing sensation for the recipient. If the array folds inside the cochlea during the insertion procedure, it can lead to trauma, damage to the residual hearing, and poor hearing restoration. Intraoperative detection of such a case can allow a surgeon to perform reimplantation. However, this intraoperative detection requires experience and electrophysiological tests sometimes fail to detect an array folding. Due to the low incidence of array folding, we generated a dataset of CT images with folded synthetic electrode arrays with realistic metal artifact. The dataset was used to train a multitask custom 3D-UNet model for array fold detection. We tested the trained model on real post-operative CTs (7 with folded arrays and 200 without). Our model could correctly classify all the fold-over cases while misclassifying only 3 non fold-over cases. Therefore, the model is a promising option for array fold detection.

Keywords: Tip fold-over · cochlear implant · synthetic CT

1 Introduction

Cochlear implantations (CIs) are considered to be a standard treatment in case of individuals with severe-to-profound hearing loss [1]. During CI surgical procedure an electrode array (EA) is implanted in the cochlea for stimulating the auditory nerve. Along the cochlear duct length, the neural pathways are arranged in a tonotopic manner by decreasing frequency [2]. Naturally, these pathways get activated according to their characteristic frequencies present in the incoming sound. After implantation, the electrode arrays are used to stimulate the nerve pathways and induce hearing sensation [3].

During the insertion process, one complication surgeons aim to avoid is called "tip fold-over," where the tip of the array curls in an irregular manner inside the cochlear volume resulting in array folding [4]. This can occur when the tip of the electrode array gets stuck within an intracochlear cavity as the surgeon threads the array into the cochlea, largely blind to the intra-cochlear path of the array and with little tactile

H. Greenspan et al. (Eds.): MICCAI 2023, LNCS 14228, pp. 249–259, 2023.
https://doi.org/10.1007/978-3-031-43996-4_24

feedback available to indicate the tip of the array is folding [5]. Therefore, further pushing the base of the EA into the cochlea results in folding of the array (shown in Fig. 1) [5]. Tip fold-over can result in many complications which include trauma, damage to residual hearing and poor positioning of the EA inside cochlea, which ultimately leads to poor hearing restoration for the patient. If intra-operative detection of a tip fold-over is possible, the surgeon can address it through re-implantation of the electrode array [6]. In addition, post-operative detection of minor fold-over can help audiologists to deactivate the affected electrodes to attempt to reach a more satisfactory hearing outcome [4].

Although most CI centers do not currently attempt to detect tip foldovers, the current standard approach among sites that do is visual inspection of intraoperative fluoroscopy. However, these identification methods require experience to align the view optimally and limit the radiation exposure during the fluoroscopy [7]. Various studies (McJunkin et al., Sabban et al., Dirr et al., Timm et al., Sipari et al., Gabrielpillai et al., Jia et al., Garaycochea et al.) have reported on other approaches for detecting tip fold-over through CT imaging, NRT (Neural Response Telemetry) and EcochG (Electrocochleography) [7–9]. In some studies, it is reported that the intraoperative electrophysiological measures, such as NRT or EcochG, sometimes fail to identify the tip fold-over cases [9, 10]. Pile et al. [11] developed a robotic system for tip fold-over detection where the support vector machine classifier is used on the EA insertion force profile. This approach is associated with robot assisted insertion techniques. This broad body of work emphasizes the need for an accurate approach for intra and/or post operative fold-over detection. Therefore, the goal of this study is to develop an approach to detect tip fold-overs in cone beam or conventional CT images using state-of-the-art deep neural network-based image analysis.

As tip fold-over cases are reported to be rare, it would be difficult to acquire a substantial number of cases for fully supervised training of any data-driven method. Zuniga et al. [4] studied CI surgeries in 303 ears and reported 6 tip fold-over cases (less than 2%). Dhanasingh et al. [5] investigated 3177 CI recipients' cases from 13 studies and reported 50 tip fold-over cases (1.57%). Only 0.87% (15 cases) tip fold-over cases were reported among 1722 CI recipients according to Gabrielpillai et al. [12]. Dhanasingh et al. [5] analyzed 38 peer reviewed publications and reported that the rate of tip fold-over with certain types of arrays might be as high as 4.7%. Data scarcity thus makes it difficult to curate a balanced training dataset of tip fold-over cases.

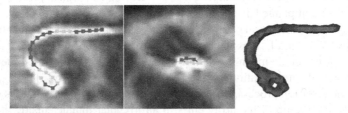

Fig. 1. Coronal view (left) and axial view (middle) of electrode array tip fold-over in a postoperative CT. 3D rendering (right) of a folded over electrode array.

Deep learning methods are the current state-of-the-art in medical image analysis, including image classification tasks. Numerous approaches have been proposed for using 3D networks to solve image classification tasks, e.g. [13–15]. Multi-tasking neural networks have also been used for simultaneous medical image segmentation and classification tasks. These networks have shared layers from the input side which branch into multiple paths for multiple outputs, e.g., [16, 17]. Along with choosing the appropriate network for the CT image classification task one of the typical concerns is the class balance of the training dataset [17]. As tip foldovers are rare, augmentation approaches are needed to reduce the effect of data imbalance.

Therefore, in this work we design a dataset of CT images with folded synthetic arrays with realistic metal artifact to assist with training. We propose a multi-task neural network based on the U-Net [18], train it using the synthetic CT dataset, and then test it on a small dataset of real tip fold-over cases. Our results indicate that the model performs well in detecting fold-over cases in real CT images, and therefore, the trained model could help the intra-operative detection of tip fold-over in CI surgery.

2 Methodology

2.1 Dataset

In this study, we utilize CT images from 312 CI patients acquired under local IRB protocols. This included 192 post-implantation CTs (185 normal and 7 tip-fold), acquired either intra-operatively (cone beam) or post-operatively (conventional or cone beam), and 120 pre-implantation CTs used to create synthetic post-implantation CT. As image acquisition parameters (dimensionality, resolution and voxel intensity) varies among the images, we preprocessed all the CT images to homogenize these parameters. First, the intracochlear structures (e.g., Scala Tympani (ST) and Scala Vestibuli (SV)) were segmented from the CT image using previously developed automatic segmentation techniques [19–21]. Using the ST segmentation, a region-of-interest CT image was cropped from the full-sized CT image keeping the ST at the center of the cropped image. Then the cropped CT resolution was resampled to an isotropic voxel size of 0.3 mm with a $32 \times 32 \times 32$ grid. As the final step of the preprocessing, the voxel intensity of the cropped image was normalized to ensure comparable intensity distribution among all the CT images.

2.2 Synthetic CT Generation

Our synthetic post-operative CT generation approach is inspired by the process of synthetic preoperative CT generation by Khan et al. [22]. First, a random but realistic location for the electrodes is estimated in a real pre-implantation CT image. This was done by considering some constraints, such as the relative location of the ST, electrode spacing, active array length, relative smoothness of the electrode curve, whether a fold exists, and if so, the location of the fold. Randomized variability within plausible margins ensured generating a unique and realistic electrode array shape for each case.

Once we estimated the probable locations of the electrodes, we placed high intensity (around 3–4 times the bone intensity) cubic blocks with a dimensionality of $4 \times 4 \times$

4 voxels in an empty $32 \times 32 \times 32$ grid in locations corresponding to the electrode sites in the preoperative CT. We also added small high intensity cubic blocks (with a dimensionality of $2 \times 2 \times 2$ voxels) between the electrodes to represent the wires that connect to the electrodes. This resulted in an ideal image with a synthetic electrode array (shown in Fig. 2) but lacking realistic reconstruction artifacts. To get a realistic metal artifact, we applied radon transformation on the high intensity blocks to project the data on the detector space. Then, with the resulting sinogram, we applied inverse radon transformation to backproject the sinogram into the world space. The backprojection process is done two times separately: first time with low frequency scaling (to achieve blurry metal edge) and second time with high frequency scaling (to achieve dominant metal artifact). A Hamming filter was used in the backprojection process. Finally, the preoperative CT and the images with back-projected synthetic EA with realistic metal artifact were merged together additively to generate a synthetic postoperative CT image. Wang et al. [23] also presented a method to generate synthetic CT images with metal artifact, however, the aim of the study was to remove metal artifact from post-implant CT images.

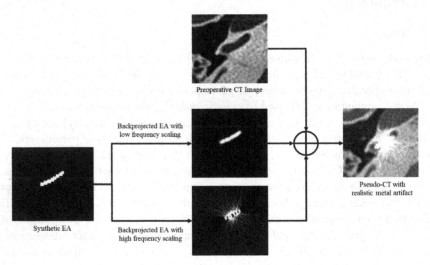

Fig. 2. Synthetic postoperative CT with realistic metal artifact generation process from preoperative CT image.

Table 1. Overall distribution of the dataset.

	Training		Validation		Testing	
	PCT	Real CT	PCT	Real CT	PCT	Real CT
Foldover	135	0	5	0	15	7
Normal	40	160	5	12	15	200
Subtotal	175	160	10	12	30	207
Total	335		22		237	

Using the described synthetic CT generation process, we produced 215 synthetic pseudo CTs (PCTs) with random electrode locations sampled from 100 preoperative real CT images with stratified sampling into the training, validation, and testing datasets. In these PCTs, 155 images have a synthetic EA with tip fold-over and the remaining 60 do not have any fold over. The combined dataset including 379 real CTs and 215 synthetic CTs was divided into training (335), validation (22) and testing (237) subsets. The overall distribution of the dataset is presented in Table 1. As the number of real fold-over cases is very low (about 1.85% in this study), we allotted all of them in the testing dataset.

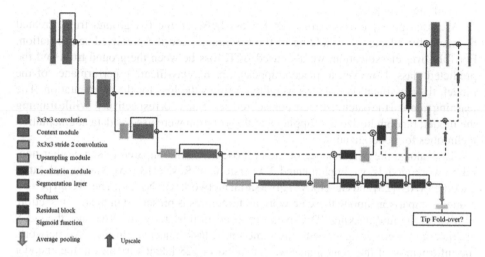

Fig. 3. Multitasking 3D U-Net architecture for EA segmentation and tip fold-over classification.

2.3 Multi-task Deep Learning Network

The neural network model proposed in this study for EA fold-over detection was inspired by the 3D U-Net architectures proposed by Isensee et al. and Ronneberger et al. [18, 24]. The model is a multitasking network where the outputs are the segmentation of the EA and fold-over classification. Our hypothesis is that this multi-task approach helps the network focus attention on the shape of the EA when learning to classify the CT, rather than overfitting to a spurious local minima driven by non-EA features in the training dataset. The architecture is comprised of a context pathway for encoding the increasingly abstract representation of the input as we advance deeper into the neural network. A localization pathway recombines the representation for localizing the EA [20]. In the model, the context modules compute the context pathways. Each of these modules is a pre-activation residual block [25] which has a dropout layer $(p_{dropout} = 0.2)$ in between two convolutional layers of $3 \times 3 \times 3$ dimensionality.

The localization pathways collect features at lower spatial resolution where the contextual information is encoded and transfer it to the higher resolution. This is done by using an upsampling step followed by a convolutional layer. The upsampled features are then concatenated with the corresponding context pathway level. Segmentation layers from different levels of the architecture are convolutional layers that are combined by elementwise summation to build the segmentation in a multi-scale fashion and obtain the final segmentation output. This approach was inspired by Kayalibay et al. [26].

A classification branch is added to the network at the point where the contextual information is encoded at the lowest resolution (shown in Fig. 3). The classification branch consists of 4 residual blocks [27] followed by an average pooling layer and a sigmoid function layer.

We used binary cross entropy (BCE) loss between the EA ground truth, created by manually selected thresholding of the CT image, and the predicted segmentation. For fold-over classification, we also used BCE loss between the ground truth and the predicted class. However, to place emphasis on the classification performance of the model, the classification loss was weighted 5 times the loss for the segmentation. The learning rate and the batch size were considered $3e-5$ and 20, respectively. While training the model, random horizontal flipping and 90° rotation were used as data augmentation techniques for generalization.

To evaluate the performance of the proposed model compared to some other neural network models, we implemented 3D versions of ResNet18 [15], Variational Autoencoder (VAE) [28] and Generative Adversarial Network (GAN) [29]. The overall performance comparison among these network architectures is presented in the result section. Similar to the multitasking 3D U-Net, proposed in this study, the VAE and the GAN architectures were designed with the same multi-task objective. In the VAE network, the information of the input image was encoded in 128 latent variables in the encoder section of the model. With these latent variables the decoder section reconstructs a 3D image with the EA segmentation. A classification branch with the same architectures as proposed above was added to the encoder section for fold-over detection. Similarly, the GAN also had a classification branch in the generator model of the architecture. The ResNet18 performs only classification, and thus represents classification performance achievable without multi-task training.

3   Results

The training and validation loss curves of multitasking 3D U-Net are presented in Fig. 4. The graph at the top presents the overall training and validation loss. A rapid drop is visible in the segmentation loss curves where the validation loss curve swiftly follows the training loss curve. On the other hand, the classification loss demonstrates a gradual drop.

Next, we compare the performance of different models for fold-over classification. In addition to the performance analysis of these networks for the whole testing data, we separately reported the performance analysis for the synthetic as well as the real portions of the testing data. The separate analysis provides insight about the applicability of the trained model for the real CT images.

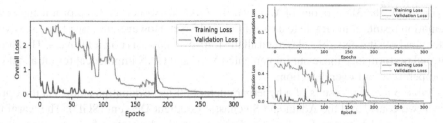

Fig. 4. Training and validation loss curves of 3D U-Net.

As reported in Table 2, the proposed multitask network has a classification accuracy of 98% for all the testing data (237 CT images). Among those 207 (7 tip fold-over and 200 non fold-over cases) are real CT images where the proposed model is 99% accurate regarding the classification by misclassifying 3 non fold-over test cases. In addition, the model misclassified 1 synthetic CT with folded over condition which degraded its accuracy to 97% for synthetic data. Although segmentation is not our primary goal, the model was able to consistently capture the location of the EA (example shown in Fig. 5). Inference time for our network was 0.60 ± 0.10 s.

Table 2. Tip fold-over detection result comparison among different networks.

Network architectures		**3D U-Net**	ResNet18	VAE	GAN
	Overall	0.98	0.93	0.87	0.98
Accuracy	Synthetic	0.97	0.90	1.00	1.00
	Real	**0.99**	0.93	0.83	0.96
	Overall	0.95	0.54	0.88	0.88
Sensitivity	Synthetic	0.93	0.80	1.00	1.00
	Real	**1.00**	0.00	0.00	0.00
	Overall	0.99	0.97	0.87	0.99
Specificity	Synthetic	1.00	1.00	1.00	1.00
	Real	**0.99**	0.97	0.86	0.99

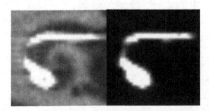

Fig. 5. Real CT with tip fold-over condition (left) and segmentation output of the multitasking 3D U-Net (right).

On the other hand, ResNet18 demonstrated lower classification accuracy (around 87%) while misclassifying several synthetic CTs and the real CTs with tip folded over.

In our study, the 3D versions of VAE and the GAN networks rendered promising segmentation results comparable to those of the 3D U-Net. However, in case of tip fold-over detection both the networks classified all the fold-over cases as non fold-overs. This suboptimal performance made our implemented VAE and GAN impractical for intra and/or postoperative fold-over detection.

Table 3 presents results of a hyperparameter settings evaluation study. As the optimizer, we considered the Adam and Stochastic Gradient Descent (SGD). The learning rate was varied between $1e-3$ to $1e-5$; however, the best classification output was obtained using a learning rate of $3e-4$. As stated in the methodology, we assigned a higher weight with the classification loss for emphasizing on the classification branch during the training process. Better classification accuracy was obtained for real CTs when the weight was either 2 or 5. From the classification accuracy, sensitivity, and specificity analysis in Table 3 it is evident that the Adam optimizer with a learning rate of $3e-5$ outperforms the other hyperparameter combinations for real CT fold-over detection. The classification loss weight and the batch size were considered 5 and 20, respectively. Similar hyperparameter analysis was done to select parameters for the other networks evaluated in this study, but not included here for the sake of brevity.

Using Adam optimizer with learning rate $3e-4$, batch size of 20, and classification loss weight of 5, we repeated the training process 8 times to investigate training stability. In one out of eight cases, the resulting model could again correctly classify all the real CTs (7 with folded EA and 13 without) in the testing dataset. For the remaining 7 of 8 models, the network could correctly classify 19 out of 20 real CTs (misclassifying one fold-over case), which results in a classification accuracy of 95% for real postoperative CT images.

Table 3. Tip fold-over detection result comparison among different hyperparameter settings.

Hyperparameters	Optimizer	Adam	SGD	SGD	Adam	Adam	Adam
	Learning rate	3E-04	3E-04	5E-05	3E-04	3E-04	3E-05
	Classification loss weight	2	2	2	2	5	5
	Batch size	30	30	30	30	20	20
Accuracy	Overall	0.97	0.98	0.99	0.95	0.97	0.98
	Synthetic	0.97	1.00	1.00	1.00	1.00	0.97
	Real	0.96	0.98	0.99	0.95	0.97	0.99
Sensitivity	Overall	0.77	0.82	0.91	0.95	0.95	0.95
	Synthetic	0.93	1.00	1.00	1.00	1.00	0.93
	Real	0.43	0.43	0.71	0.86	0.86	1.00
Specificity	Overall	1.00	0.99	1.00	0.95	0.97	0.99
	Synthetic	1.00	1.00	1.00	1.00	1.00	1.00
	Real	1.00	1.00	1.00	0.95	0.98	0.99

4   Discussion and Conclusion

In a CI surgical procedure, the relative positioning of the EA influences the overall outcome of the surgery [30, 31]. A tip fold-over case results in poor positioning of the apical electrodes and, hence, can lead to severe consequences including trauma, damage to the residual hearing region and poor hearing restoration. Upon intraoperative detection of such a case, the surgeon can extract and reimplant the array to avoid any folding. The conventional detection processes require experience yet are prone to failure. In addition, due to the incidences of fold-over cases being low, training a model to detect these cases with real data is difficult. Therefore, in this study, we generated a dataset of CT images with folded synthetic electrode arrays with realistic metal artifact. A multitask custom network was proposed and trained with the dataset for array fold detection. We tested the trained model on real post-implantation CTs (7 with folded arrays and 200 without). We were able to train a model that could correctly classify all the fold-over cases while misclassifying only 3 non fold-over cases. In future work, clinical deployment of the model will be investigated.

Acknowledgements. This study is conducted under the support of NIH grants R01DC014037, R01DC008408 and T32EB021937. This content is solely the responsibility of the authors and does not necessarily represent the official views of this institute.

References

1. US Department of Health and Human Services, National Institute on Deafness and Other Communication Disorders, Cochlear implants, No. 11–4798 (2014)
2. Yukawa, K., et al.: Effects of insertion depth of cochlear implant electrodes upon speech perception. Audiol. Neurotol. **9**(3), 163–172 (2004)
3. Stakhovskaya, O., et al.: Frequency map for the human cochlear spiral ganglion: implications for cochlear implants. J. Assoc. Res. Otolaryngol. **8**(2), 220 (2007)
4. Zuniga, M.G., et al.: Tip fold-over in cochlear implantation: case series. Otol. Neurotol. Official Publ. Am. Otological Soc. Amer. Neurotol. Soc. Eur. Acad. Otol. Neurotol. **38**(2), 199 (2017)
5. Dhanasingh, A., Jolly, C.: Review on cochlear implant electrode array tip fold-over and scalar deviation. J. Otol. **14**(3), 94–100 (2019)
6. Ishiyama, A., Risi, F., Boyd, P.: Potential insertion complications with cochlear implant electrodes. Cochlear Implants Int. **21**(4), 206–219 (2020)
7. Dirr, F., et al.: Value of routine plain x-ray position checks after cochlear implantation. Otol. Neurotol. **34**(9), 1666–1669 (2013)
8. McJunkin, J.L., Durakovic, N., Herzog, J., Buchman, C.A.: Early outcomes with a slim, modiolar cochlear implant electrode array. Otol. Neurotol. **39**(1), e28–e33 (2018)
9. Garaycochea, O., Manrique-Huarte, R., Manrique, M.: Intra-operative radiological diagnosis of a tip roll-over electrode array displacement using fluoroscopy, when electrophysiological testing is normal: the importance of both techniques in cochlear implant surgery. Braz. J. Otorhinolaryngol. **86**, s38–s40 (2020)
10. Cohen, L.T., Saunders, E., Richardson, L.M.: Spatial spread of neural excitation: comparison of compound action potential and forward-masking data in cochlear implant recipients. Int. J. Audiol. **43**(6), 346–355 (2004)

11. Pile, J., Wanna, G.B., Simaan, N.: Robot-assisted perception augmentation for online detection of insertion failure during cochlear implant surgery. Robotica **35**(7), 1598–1615 (2017)
12. Gabrielpillai, J., Burck, I., Baumann, U., Stöver, T., Helbig, S.: Incidence for tip foldover during cochlear implantation. Otol. Neurotol. **39**(9), 1115–1121 (2018)
13. Ahn, B.B.: The compact 3D convolutional neural network for medical images. Standford University (2017)
14. Jin, T., Cui, H., Zeng, S., Wang, X.: Learning deep spatial lung features by 3d convolutional neural network for early cancer detection. In: 2017 International Conference on Digital Image Computing: Techniques and Applications (DICTA). IEEE, Piscataway (2017)
15. Hara, K., Kataoka, H., Satoh, Y.: Learning spatio-temporal features with 3d residual networks for action recognition. In: Proceedings of the IEEE international Conference on Computer Vision workshops, pp. 3154–3160 (2017)
16. Zhang, D., Wang, J., Noble, J.H., Dawant, B.M.: HeadLocNet: deep convolutional neural networks for accurate classification and multi-landmark localization of head CTs. Med. Image Anal. **61**, 101659 (2020)
17. Jnawali, K., Arbabshirani, M.R., Rao, N., Patel, A.A.: Deep 3D convolution neural network for CT brain hemorrhage classification. In: Medical Imaging 2018: Computer-Aided Diagnosis, vol. 10575, pp. 307–313. SPIE (2018)
18. Ronneberger, O., Fischer, P., Brox, T.: U-net: convolutional networks for biomedical image segmentation. In: Navab, N., Hornegger, J., Wells, W.M., Frangi, A.F. (eds.) MICCAI 2015. LNCS, vol. 9351, pp. 234–241. Springer, Cham (2015). https://doi.org/10.1007/978-3-319-24574-4_28
19. Noble, J.H., Labadie, R.F., Majdani, O., Dawant, B.M.: Automatic segmentation of intra-cochlear anatomy in conventional CT. IEEE Trans. Biomed. Eng **58**(9), 2625–2632 (2011)
20. Noble, J.H., Dawant, B.M., Warren, F.M., Labadie, R.F.: Automatic identification and 3D rendering of temporal bone anatomy. Otol. Neurotol. **30**(4), 436–442 (2009)
21. Noble, J.H., Warren, F.M., Labadie, R.F., Dawant, B.M.: Automatic segmentation of the facial nerve and chorda tympani using image registration and statistical priors. In: Medical Imaging 2008: Image Processing, vol. 6914, p. 69140P. International Society for Optics and Photonics (2008)
22. Khan, M.M., Banalagay, R., Labadie, R.F., Noble, J.H.: Sensitivity of intra-cochlear anatomy segmentation methods to varying image acquisition parameters. In: Medical Imaging 2022: Image-Guided Procedures, Robotic Interventions, and Modeling, vol. 12034, pp. 111–116. SPIE (2022)
23. Wang, Z., et al: Deep learning based metal artifacts reduction in post-operative cochlear implant CT imaging. In: Shen, D., et al. (eds.) MICCAI 2019. LNCS, vol. 11769, pp. 121–129. Springer, Cham (2019). https://doi.org/10.1007/978-3-030-32226-7_14
24. Isensee, F., Kickingereder, P., Wick, W., Bendszus, M., Maier-Hein, K.H.: Brain tumor segmentation and radiomics survival prediction: Contribution to the brats 2017 challenge. In: Crimi, A., Bakas, S., Kuijf, H., Menze, B., Reyes, M. (eds.) BrainLes 2017. LNCS, vol. 10670, pp. 287–297. Springer, Cham (2018). https://doi.org/10.1007/978-3-319-75238-9_25
25. He, K., Zhang, X., Ren, S., Sun, J.: Identity mappings in deep residual networks. In: Leibe, B., Matas, J., Sebe, N., Welling, M. (eds.) ECCV 2016. LNCS, vol. 9908, pp. 630–645. Springer, Cham (2016). https://doi.org/10.1007/978-3-319-46493-0_38
26. Kayalibay, B., Jensen, G., van der Smagt, P.: CNN-based segmentation of medical imaging data (2017). arXiv preprintarXiv:1701.03056
27. He, K., Zhang, X., Ren, S., Sun, J.: Deep residual learning for image recognition. In: Proceedings of the IEEE Conference on Computer Vision and Pattern Recognition, pp. 770–778 (2016)

28. Tan, Q., Gao, L., Lai, Y.K., Xia, S.: Variational autoencoders for deforming 3d mesh models. In: Proceedings of the IEEE Conference on Computer Vision and Pattern Recognition, pp. 5841–5850 (2018)

29. Cirillo, M.D., Abramian, D., Eklund, A.: Vox2Vox: 3D-GAN for brain tumour segmentation. In: Crimi, A., Bakas, S. (eds.) BrainLes 2020. LNCS, vol. 12658, pp. 274–284. Springer, Cham (2021). https://doi.org/10.1007/978-3-030-72084-1_25

30. Rubinstein, J.T.: How cochlear implants encode speech. Curr. Opin. Otolaryngol. Head Neck Surg. **12**(5), 444–448 (2004)

31. Wilson, B.S., Dorman, M.F.: Cochlear implants: current designs and future possibilities. J. Rehabil. Res. Dev. **45**(5), 695–730 (2008)

Detecting the Sensing Area of a Laparoscopic Probe in Minimally Invasive Cancer Surgery

Baoru Huang[1,2(✉)], Yicheng Hu[1,2], Anh Nguyen[3], Stamatia Giannarou[1,2], and Daniel S. Elson[1,2]

[1] The Hamlyn Centre for Robotic Surgery, Imperial College London, London, UK
Baoru.Huang18@imperial.ac.uk
[2] Department of Surgery & Cancer, Imperial College London, London, UK
[3] Department of Computer Science, University of Liverpool, Liverpool, UK

Abstract. In surgical oncology, it is challenging for surgeons to identify lymph nodes and completely resect cancer even with pre-operative imaging systems like PET and CT, because of the lack of reliable intraoperative visualization tools. Endoscopic radio-guided cancer detection and resection has recently been evaluated whereby a novel tethered laparoscopic gamma detector is used to localize a preoperatively injected radiotracer. This can both enhance the endoscopic imaging and complement preoperative nuclear imaging data. However, gamma activity visualization is challenging to present to the operator because the probe is nonimaging and it does not visibly indicate the activity origination on the tissue surface. Initial failed attempts used segmentation or geometric methods, but led to the discovery that it could be resolved by leveraging highdimensional image features and probe position information. To demonstrate the effectiveness of this solution, we designed and implemented a simple regression network that successfully addressed the problem. To further validate the proposed solution, we acquired and publicly released two datasets captured using a custom-designed, portable stereo laparoscope system. Through intensive experimentation, we demonstrated that our method can successfully and effectively detect the sensing area, establishing a new performance benchmark. Code and data are available at https://github.com/br0202/Sensing_area_detection.git.

Keywords: Laparoscopic Image-guided Intervention · Minimally Invasive Surgery · Detection of Sensing Area

Supplementary Information The online version contains supplementary material available at https://doi.org/10.1007/978-3-031-43996-4_25.

1   Introduction

Cancer remains a significant public health challenge worldwide, with a new diagnosis occurring every two minutes in the UK (Cancer Research UK[1]). Surgery is one of the main curative treatment options for cancer. However, despite substantial advances in pre-operative imaging such as CT, MRI, or PET/SPECT to aid diagnosis, surgeons still rely on the sense of touch and naked eye to detect cancerous tissues and disease metastases intra-operatively due to the lack of reliable intraoperative visualization tools. In practice, imprecise intraoperative cancer tissue detection and visualization results in missed cancer or the unnecessary removal of healthy tissues, which leads to increased costs and potential harm to the patient. There is a pressing need for more reliable and accurate intraoperative visualization tools for minimally invasive surgery (MIS) to improve surgical outcomes and enhance patient care.

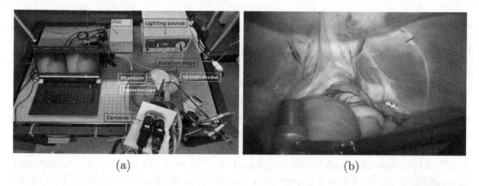

(a)   (b)

Fig. 1. (a) Hardware set-up for experiments, including a customized portable stereo laparoscope system and the 'SENSEI' probe, a rotation stage, a laparoscopic lighting source, and a phantom; (b) An example of the use of the 'SENSEI' probe in MIS.

A recent miniaturized cancer detection probe (i.e., 'SENSEI®' developed by Lightpoint Medical Ltd.) leverages the cancer-targeting ability of nuclear agents typically used in nuclear imaging to more accurately identify cancer intra-operatively from the emitted gamma signal (see Fig. 1b)[6]. However, the use of this probe presents a visualization challenge as the probe is non-imaging and is air-gapped from the tissue, making it challenging for the surgeon to locate the probe-sensing area on the tissue surface.

It is crucial to accurately determine the sensing area, with positive signal potentially indicating cancer or affected lymph nodes. Geometrically, the sensing area is defined as the intersection point between the gamma probe axis and the tissue surface in 3D space, but projected onto the 2D laparoscopic image. However, it is not trivial to determine this using traditional methods due to

---

[1] https://www.cancerresearchuk.org/health-professional/cancer-statistics-for-the-uk.

poor textural definition of tissues and lack of per-pixel ground truth depth data. Similarly, it is also challenging to acquire the probe pose during the surgery.

Problem Redefinition. In this study, in order to provide sensing area visualization ground truth, we modified a non-functional 'SENSEI' probe by adding a miniaturized laser module to clearly optically indicate the sensing area on the laparoscopic images - i.e. the 'probe axis-surface intersection'. Our system consists of four main components: a customized stereo laparoscope system for capturing stereo images, a rotation stage for automatic phantom movement, a shutter for illumination control, and a DAQ-controlled switchable laser module (see Fig. 1a). With this setup, we aim to transform the sensing area localization problem from a geometrical issue to a high-level content inference problem in 2D. It is noteworthy that this remains a challenging task, as ultimately we need to infer the probe axis-surface intersection without the aid of the laser module to realistically simulate the use of the 'SENSEI' probe.

2   Related Work

Laparoscopic images play an important role in computer-assisted surgery and have been used in several problems such as object detection [9], image segmentation [23], depth estimation [20] or 3D reconstruction [13]. Recently, supervised or unsupervised depth estimation methods have been introduced [14]. Ye *et al.* [22] proposed a deep learning framework for surgical scene depth estimation in self-supervised mode and achieved scalable data acquisition by incorporating a differentiable spatial transformer and an autoencoder into their framework. A 3D displacement module was explored in [21] and 3D geometric consistency was utilized in [8] for self-supervised monocular depth estimation. Tao *et al.* [19] presented a spatiotemporal vision transformer-based method and a self-supervised generative adversarial network was introduced in [7] for depth estimation of stereo laparoscopic images. Recently, fully supervised methods were summarized in [1] for depth estimation. However, acquiring per-pixel ground truth depth data is challenging, especially for laparoscopic images, which makes it difficult for large-scale supervised training [8].

Laparoscopic segmentation is another important task in computer-assisted surgery as it allows for accurate and efficient identification of instrument position, anatomical structures, and pathological tissue. For instance, a unified framework for depth estimation and surgical tool segmentation in laparoscopic images was proposed in [5], with simultaneous depth estimation and segmentation map generation. In [12], self-supervised depth estimation was utilized to regularize the semantic segmentation in knee arthroscopy. Marullo *et al.* [16] introduced a multi-task convolutional neural network for event detection and semantic segmentation in laparoscopic surgery. The dual swin transformer U-Net was proposed in [11] to enhance the medical image segmentation performance, which leveraged the hierarchical swin transformer into both the encoder and the decoder of the standard U-shaped architecture, benefiting from the self-attention computation in swin transformer as well as the dual-scale encoding design.

Although the intermediate depth information was not our final aim and can be bypassed, the 3D surface information was necessary in the intersection point inference. ResNet [3] has been commonly used as the encoder to extract the image features and geometric information of the scene. In particular, in [21], concatenated stereo image pairs were used as inputs to achieve better results, and such stereo image types are also typical in robot-assisted minimally invasive surgery with stereo laparoscopes. Hence, stereo image data was also adopted in this paper.

If the problem of inferring the intersection point is treated as a geometric problem, both data collection and intra-operative registration would be difficult, which inspired us to approach this problem differently. In practice, we utilize the laser module to collect the ground truth of the intersection points when the laser is on. We note that the standard illumination image from the laparoscopic probe is also captured with the same setup when the laser module is on. Therefore, we can establish a dataset with an image pair (RGB image and laser image) that shares the same intersection point ground truth with the laser image (see Fig. 2a and Fig. 2b). The assumptions made are that the probe's 3D pose when projected into the two 2D images is the observed 2D pose, and that the intersection point is located on its axis. Hence, we input these axes to the network as another branch and randomly sampled points along them to represent the probe.

3   Dataset

To validate our proposed solution for the newly formulated problem, we acquired and publicly released two new datasets. In this section, we introduce the hardware and software design that was used to achieve our final goal, while Fig. 2 shows a sample from our dataset.

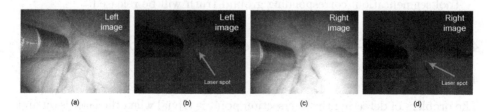

Fig. 2. Example data. (a) Standard illumination left RGB image; (b) left image with laser on and laparoscopic light off; same for (c) and (d) but for right images.

Data Collection. Two miniaturized, high-resolution cameras were coupled onto a stereo laparoscope using a custom-designed connector. The accompanying API allowed for automatic image acquisition, exposure time adjustment, and white balancing. An electrically controllable shutter was incorporated into the standard laparoscopic illumination path. To indicate the probe axis-surface intersection, we incorporated a DAQ controlled cylindrical miniature laser module into a 'SENSEI' probe shell so that the adapted tool was visually identical to the real probe. The laser module emitted a red laser beam (wavelength 650 nm) that was visible as a red spot on the tissue surface.

We acquired the dataset on a silicone tissue phantom which was $30 \times 21 \times 8$ cm and was rendered with tissue color manually by hand to be visually realistic. The phantom was placed on a rotation stage that stepped 10 times per revolution to provide views separated by a 36-degree angle. At each position, stereo RGB images were captured *i)* under normal laparoscopic illumination with the laser off; *ii)* with the laparoscopic light blocked and the laser on; and *iii)* with the laparoscopic light blocked and the laser off. Subtraction of the images with laser on and off readily allowed segmentation of the laser area and calculation of its central point, i.e. the ground truth probe axis-surface intersection.

All data acquisition and devices were controlled by Python and LABVIEW programs, and complete data sets of the above images were collected on visually realistic phantoms for multiple probe and laparoscope positions. This provided 10 tissue surface profiles for a specific camera-probe pose, repeated for 120 different camera-probe poses, mimicking how the probe may be used in practice. Therefore, our first newly acquired dataset, named **Jerry**, contains 1200 sets of images. Since it is important to report errors in 3D and in millimeters, we recorded another dataset similar to **Jerry** but also including ground truth depth map for all frames by using structured-lighting system [8]—namely the **Coffbee** dataset.

These datasets have multiple uses such as:

- Intersection point detection: detecting intersection points is an important problem that can bring accurate surgical cancer visualization. We believe this is an under-investigated problem in surgical vision.
- Depth estimation: corresponding ground truth will be released.
- Tool segmentation: corresponding ground truth will be released.

4   Probe Axis-Surface Intersection Detection

4.1   Overview

The problem of detecting the intersection point is trivial when the laser is on and can be solved by training a deep segmentation network. However, segmentation requires images with a laser spot as input, while the real gamma probe produces no visible mark and therefore this approach produces inferior results.

An alternative approach to detect the intersection point is to reconstruct the 3D tissue surface and estimate the pose of the probe in real time. A tracking and pose estimation method for the gamma probe [6] involved attaching a dual-pattern marker to the probe to improve detection accuracy. This enabled the derivation of a 6D pose, comprising a rotation matrix and translation matrix with respect to the laparoscope camera coordinate. To obtain the intersection point, the authors used the Structure From Motion (SFM) method to compute the

3D tissue surface, combining it with the estimated pose of the probe, all within the laparoscope coordinate system. However, marker-based tracking and pose estimation methods have sterilization implications for the instrument, and the SFM method requires the surgeon to constantly move the laparoscope, reducing the practicality of these methods for surgery.

In this work, we propose a simple, yet effective regression approach to address this problem. Our approach relies solely on the 2D information and works well without the need for the laser module after training. Furthermore, this simple methodology facilitated an average inference time of 50 frames per second, enabling real-time sensing area map generation for intraoperative surgery.

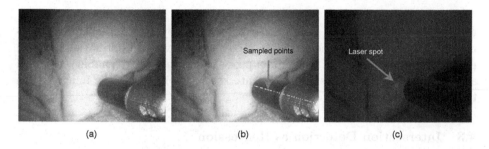

Fig. 3. Sensing area detection. (a) The input RGB image, (b) The estimated line using PCA for obtaining principal points, (c) The image with laser on that we used to detect the intersection ground truth.

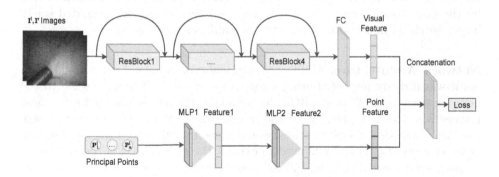

Fig. 4. An overview of our approach using ResNet and MLP.

4.2   Intersection Detection as Segmentation

We utilized different deep segmentation networks as a first attempt to address our problem [10,18]. Please refer to the Supplementary Material for the implementation details of the networks. We observed that when we do not use images with the laser, the network was not able to make any good predictions. This is

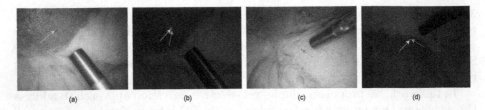

Fig. 5. Qualitative results. (a) and (c) are standard illumination images and (b) and (d) are images with laser on and laparoscopic light off. The predicted intersection point is shown in blue and the green point indicates the ground truth, which are further indicated by arrows for clarity. (Color figure online)

understandable as the red laser spot provides the key information for the segmentation. Therefore the network does not have any visual information to make predictions from images of the gamma probe. We note that to enable real-world applications, we need to estimate the intersection point using the images when the laser module is turned off.

4.3   Intersection Detection as Regression

Problem Formulation. Formally, given a pair of stereo images $\mathbf{I}^l, \mathbf{I}^r$, n points $\{\mathbf{P}_1^l, \mathbf{P}_2^l, ..., \mathbf{P}_n^l\}$ were sampled along the principal axis of the probe, $\mathbf{P}_i^l \in \mathbb{R}^2$ from the left image. The same process was repeated for the right image. The goal was to predict the intersection point $\mathbf{P}_{\text{intersect}}$ on the surface of the tissue. During the training, the ground truth intersection point position was provided by the laser source, while during testing the intersection was estimated solely based on visual information without laser guidance (see Fig. 3).

Network Architecture. Unlike the segmentation approach, the intersection point was directly predicted using a regression network. The images fed to the network were 'laser off' stereo RGB, but crucially, the intersection point for these images was known *a priori* from the paired 'laser on' images. The raw image resolution was 4896×3680 but these were binned to 896×896. Principal Component Analysis (PCA) [15] was used to extract the central axis of the probe and 50 points were sampled along this axis as an extra input dimension. A network was designed with two branches, one branch for extracting visual features from the image and one branch for learning the features from the sequence of principal points using ResNet [3] and Vision Transformer (ViT) [2] as two backbones. The principal points were learned through a multi-layer perception (MLP) or a long short-term memory (LSTM) network [4]. The features from both branches were concatenated and used for regressing the intersection point (see Fig. 4). Finally, the whole network is trained end-to-end using the mean square error loss.

4.4 Implementation

Evaluation Metrics. To evaluate sensing area location errors, Euclidean distance was adopted to measure the error between the predicted intersection points and the ground truth laser points. We reported the mean absolute error, the standard derivation, and the median in pixel units.

Implementation Details. The networks were implemented in PyTorch [17], with an input resolution of 896 × 896 and a batch size of 12. We partitioned the **Jerry** dataset into three subsets, the training, validation, and test set, consisting of 800, 200, and 200 images, respectively, and the same for the **Coffbee** dataset. The learning rate was set to 10^{-5} for the first 300 epochs, then halved until epoch 400, and quartered until the end of the training. The model was trained for 700 epochs using the Adam optimizer on two NVIDIA 2080 Ti GPUs, taking approximately 4 h to complete.

Table 1. Results using ResNet50. Grey color denotes the Jerry dataset and Blue color is for Coffbee dataset (2D errors are in pixels and 3D errors are in mm).

ResNet	✓	✓	✓	✓	✓
MLP		✓			✓
LSTM			✓		
Stereo	✓	✓	✓		
Mono				✓	✓
2D Mean E.	73.5	70.5	73.7	75.6	76.7
2D Std.	65.1	56.8	62.1	62.9	64.4
2D Median	57.5	59.8	56.9	58.8	68.4
2D Mean E.	63.2	52.9	62.0	55.8	60.2
2D Std.	71.4	42.9	63.4	55.3	42.1
2D Median	44.9	44.6	43.4	42.5	52.3
R2 Score	0.55	0.82	0.63	0.73	0.78
3D Mean E.	8.5	7.4	6.5	6.4	11.2
3D Std.	15.7	6.7	6.8	7.1	18.2
3D Median	4.5	4.6	4.0	4.3	5.4

Table 2. Results using ViT. Grey color denotes the Jerry dataset and Blue color is for Coffbee dataset (2D errors are in pixels and 3D errors are in mm).

ViTNet	✓	✓	✓	✓	✓
MLP		✓			✓
LSTM			✓		
Stereo	✓	✓	✓		
Mono				✓	✓
2D Mean E.	77.9	92.3	80.9	87.7	112.1
2D Std.	69.1	71.0	67.4	68.6	84.2
2D Median	59.0	75.0	64.8	74.9	90.0
2D Mean E.	76.3	75.0	88.0	56.5	82.7
2D Std.	69.8	60.6	83.3	75.8	63.9
2D Median	59.9	59.6	68.3	34.5	69.1
R2 Score	0.58	0.66	0.33	0.65	0.60
3D Mean E.	7.9	9.1	11.4	11.6	7.7
3D Std.	6.9	8.2	16.7	21.3	7.0
3D Median	6.0	5.9	7.1	5.3	6.2

5 Results

Quantitative results on the released datasets are shown in Table 1 and Table 2 with different backbones for extracting image features, ResNet and ViT. For the 2D error on two datasets, among the different settings, the combination of ResNet and MLP gave the best performance with a mean error of 70.5 pixels and a standard deviation of 56.8. The median error of this setting was 59.8 pixels while the R2 score was 0.82 (higher is better for R2 score). Comparing the Table 1 and Table 2, we found that the ResNet backbone was better than the ViT backbone in the image processing task, while MLP was better than LSTM in probe pose representation. ResNet processed the input images as a whole, which was better suited for utilizing the global context of a unified scene composed of the tissue and the probe, compared to the ViT scheme, which treated the whole scene as several patches. Similarly, the sampled 50 principal points on the probe axis were better processed using the simple MLP rather than using a recurrent procedure LSTM. It is worth noting that the results from stereo inputs exceeded those from mono inputs, which can be attributed to the essential 3D information included in the stereo image pairs.

For the 3D error, the ResNet backbone still gave generally better performance than the ViT backbone while under the ResNet backbone, LSTM and MLP gave competitive results and they are all in sub-milimeter level. We note that the 3D error subjected to the quality of the acquired ground truth depth maps, which had limited resolution and non-uniformly distributed valid data due to hardware constraints. Hence, we used the median depth value of a square area of 5 pixels around the points where depth value was not available.

Figure 5 shows visualization results of our method using ResNet and MLP. This figure illustrates that our proposed method successfully detected the intersection point using solely standard RGB laparoscopic images as the input. Furthermore, based on the simple design, our method achieved the inference time of 50 frames per second, making it well-suitable for intraoperative surgery.

6   Conclusion

In this work, a new framework for using a laparoscopic drop-in gamma detector in manual or robotic-assisted minimally invasive cancer surgery was presented, where a laser module mock probe was utilized to provide training guidance and the problem of detecting the probe axis-tissue intersection point was transformed to laser point position inference. Both the hardware and software design of the proposed solution were illustrated and two newly acquired datasets were publicly released. Extensive experiments were conducted on various backbones and the best results were achieved using a simple network design, enabling real time inference of the sensing area. We believe that our problem reformulation and dataset release, together with the initial experimental results, will establish a new benchmark for the surgical vision community.

References

1. Allan, M., et al.: Stereo correspondence and reconstruction of endoscopic data challenge. arXiv:2101.01133 (2021)
2. Dosovitskiy, A., et al.: An image is worth 16x16 words: transformers for image recognition at scale. arXiv preprint arXiv:2010.11929 (2020)

3. He, K., Zhang, X., Ren, S., Sun, J.: Deep residual learning for image recognition. In: Proceedings of the IEEE Conference on Computer Vision and Pattern Recognition, pp. 770–778 (2016)
4. Hochreiter, S., Schmidhuber, J.: Long short-term memory. Neural Comput. **9**(8), 1735–1780 (1997)
5. Huang, B., et al.: Simultaneous depth estimation and surgical tool segmentation in laparoscopic images. IEEE Trans. Med. Robot. Bionics **4**(2), 335–338 (2022)
6. Huang, B., et al.: Tracking and visualization of the sensing area for a tethered laparoscopic gamma probe. Int. J. Comput. Assist. Radiol. Surg. **15**(8), 1389–1397 (2020). https://doi.org/10.1007/s11548-020-02205-z
7. Huang, B., et al.: Self-supervised generative adversarial network for depth estimation in laparoscopic images. In: de Bruijne, M., et al. (eds.) MICCAI 2021. LNCS, vol. 12904, pp. 227–237. Springer, Cham (2021). https://doi.org/10.1007/978-3-030-87202-1_22
8. Huang, B., et al.: Self-supervised depth estimation in laparoscopic image using 3d geometric consistency. In: Medical Image Computing and Computer Assisted Intervention (2022)
9. Jo, K., Choi, Y., Choi, J., Chung, J.W.: Robust real-time detection of laparoscopic instruments in robot surgery using convolutional neural networks with motion vector prediction. Appl. Sci. **9**(14), 2865 (2019)
10. Koch, G., Zemel, R., Salakhutdinov, R., et al.: Siamese neural networks for one-shot image recognition. In: ICML Deep Learning Workshop, vol. 2. Lille (2015)
11. Lin, A., Chen, B., Xu, J., Zhang, Z., Lu, G., Zhang, D.: DS-TransUNet: dual Swin transformer u-net for medical image segmentation. IEEE Trans. Instrum. Meas. **71**, 1–15 (2022)
12. Liu, F., Jonmohamadi, Y., Maicas, G., Pandey, A.K., Carneiro, G.: Self-supervised depth estimation to regularise semantic segmentation in knee arthroscopy. In: Martel, A.L., et al. (eds.) MICCAI 2020. LNCS, vol. 12261, pp. 594–603. Springer, Cham (2020). https://doi.org/10.1007/978-3-030-59710-8_58
13. Liu, X., Li, Z., Ishii, M., Hager, G.D., Taylor, R.H., Unberath, M.: Sage: slam with appearance and geometry prior for endoscopy. In: 2022 International Conference on Robotics and Automation (ICRA), pp. 5587–5593. IEEE (2022)
14. Liu, X., et al.: Dense depth estimation in monocular endoscopy with self-supervised learning methods. IEEE Trans. Med. Imaging **39**(5), 1438–1447 (2019)
15. Maćkiewicz, A., Ratajczak, W.: Principal components analysis (PCA). Comput. Geosci. **19**(3), 303–342 (1993)
16. Marullo, G., Tanzi, L., Ulrich, L., Porpiglia, F., Vezzetti, E.: A multi-task convolutional neural network for semantic segmentation and event detection in laparoscopic surgery. J. Personalized Med. **13**(3), 413 (2023)
17. Paszke, A., et al.: Automatic differentiation in pytorch (2017)
18. Ronneberger, O., Fischer, P., Brox, T.: U-Net: convolutional networks for biomedical image segmentation. In: Navab, N., Hornegger, J., Wells, W.M., Frangi, A.F. (eds.) MICCAI 2015. LNCS, vol. 9351, pp. 234–241. Springer, Cham (2015). https://doi.org/10.1007/978-3-319-24574-4_28
19. Tao, R., Huang, B., Zou, X., Zheng, G.: SVT-SDE: spatiotemporal vision transformers-based self-supervised depth estimation in stereoscopic surgical videos. IEEE Trans. Med. Robot. Bionics **5**, 42–53 (2023)
20. Tukra, S., Giannarou, S.: Stereo depth estimation via self-supervised contrastive representation learning. In: Wang, L., Dou, Q., Fletcher, P.T., Speidel, S., Li, S. (eds.) MICCAI 2022. LNCS, vol. 13437, pp. 604–614. Springer, Cham (2022). https://doi.org/10.1007/978-3-031-16449-1_58

21. Xu, C., Huang, B., Elson, D.S.: Self-supervised monocular depth estimation with 3-D displacement module for laparoscopic images. IEEE Trans. Med. Robot. Bionics **4**(2), 331–334 (2022)
22. Ye, M., Johns, E., Handa, A., Zhang, L., Pratt, P., Yang, G.Z.: Self-supervised siamese learning on stereo image pairs for depth estimation in robotic surgery. arXiv preprint arXiv:1705.08260 (2017)
23. Yoon, J., et al.: Surgical scene segmentation using semantic image synthesis with a virtual surgery environment. In: Wang, L., Dou, Q., Fletcher, P.T., Speidel, S., Li, S. (eds.) MICCAI 2022. LNCS, vol. 13437, pp. 551–561. Springer, Cham (2022). https://doi.org/10.1007/978-3-031-16449-1_53

High-Quality Virtual Single-Viewpoint Surgical Video: Geometric Autocalibration of Multiple Cameras in Surgical Lights

Yuna Kato[1]([⊠]), Mariko Isogawa[1], Shohei Mori[1,2], Hideo Saito[1], Hiroki Kajita[1], and Yoshifumi Takatsume[1]

[1] Keio University, Tokyo, Japan
yu01-na10@keio.jp
[2] Graz University of Technology, Graz, Austria

Abstract. Occlusion-free video generation is challenging due to surgeons' obstructions in the camera field of view. Prior work has addressed this issue by installing multiple cameras on a surgical light, hoping some cameras will observe the surgical field with less occlusion. However, this special camera setup poses a new imaging challenge since camera configurations can change every time surgeons move the light, and manual image alignment is required. This paper proposes an algorithm to automate this alignment task. The proposed method detects frames where the lighting system moves, realigns them, and selects the camera with the least occlusion. This algorithm results in a stabilized video with less occlusion. Quantitative results show that our method outperforms conventional approaches. A user study involving medical doctors also confirmed the superiority of our method.

Keywords: Surgical Video Synthesis · Multi-view Camera Calibration · Event Detection

1 Introduction

Surgical videos can provide objective records in addition to medical records. Such videos are used in various applications, including education, research, and information sharing [2,5]. In endoscopic surgery and robotic surgery, the surgical field can be easily captured because the system is designed to place a camera close to it to monitor operations directly within the camera field of view. Conversely, in open surgery, surgeons need to observe the surgical field; therefore, the room for additional cameras can be limited and disturbed [6].

Supplementary Information The online version contains supplementary material available at https://doi.org/10.1007/978-3-031-43996-4_26.

To overcome this issue, Kumar and Pal [7] installed a stationary camera arm to record surgery. However, their camera system had difficulty recording details (i.e., close-up views) since the camera had to be placed far from the surgical field so as not to disturb the surgeons. Instead, Nair et al. [9] used a camera mounted on a surgeon's head, which moved frequently and flexibly.

For solid and stable recordings, previous studies have installed cameras on surgical lights. Byrd et al. [1] mounted a camera on a surgical light, which could easily be blocked by the surgeon's head and body. To address this issue, Shimizu et al. [12] developed a surgical light with multiple cameras to ensure that at least one camera would observe the surgical field (Fig. 1). In such a multi-camera system, automatically switching cameras can ensure that the surgical field is visible in the generated video [4,11,12].

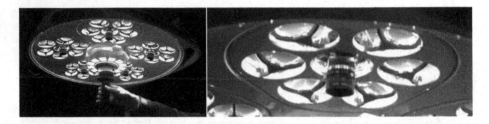

Fig. 1. Surgical light with multiple cameras. The unit consists of five cameras (left), each of which is surrounded by multiple light sources (right).

However, the cameras move every time surgeons move the lighting system, and thus, the image alignment becomes challenging. Obayashi et al. [10] relied on a video player to manually seek and segment a video clip with no camera movement. This unique camera setup and view-switching approaches create a new task to be fulfilled, which we address in this paper: automatic occlusion-free video generation by automated change detection in camera configuration and multi-camera alignment to smoothly switch to the camera with the least occlusion.[1] In summary, our contributions are as follows:

- We are the first to fulfill the task of automatic generation of stable virtual single-view video with reduced occlusion for a multi-camera system installed in a surgical light.
- We propose an algorithm that detects camera movement timing by measuring the degree of misalignment between the cameras.
- We propose an algorithm that finds frames with less occluded surgical fields.
- We present experiments showing greater effectiveness of our algorithm than conventional methods.

---

[1] project page: https://github.com/isogawalab/SingleViewSurgicalVideo.

2   Method

Given a sequence of image frames captured by five cameras installed in a surgical light $x_i = [x_1, x_2, ..., x_T]$ (Fig. 1), our goal is to generate a stable single-view video sequence with less occlusion $z = [z_1, z_2, ..., z_T]$ (Fig. 2). Here, x_i represents a sequence captured with i-th camera c_i, and T indicates the number of frames.

We perform an initial alignment using the method by Obayashi et al. [10] (Fig. 2a), which cumulatively collects point correspondences over none-feature rich frames and calculates homography matrices, M_i, via a common planar scene proxy. Then, we iteratively find a frame ID, t_c, where the cameras started moving (Sect. 2.1, Fig. 2b) and a subsequent frame ID, t_h, to update homography matrices under no moving cameras and the least occlusion (Sect. 2.2, Fig. 2c). Updated homography warping can provide a newly aligned camera view sequence $y_i = [y_1, y_2, ..., y_T]$ after the cameras moved. Finally, using the learning-based object detection method by Shimizu et al. [12], we select camera views with the least occlusion from y_i (Fig. 2d). Collecting such frames results in a stable single-view video sequence with the least occlusion $z = [z_1, z_2, ..., z_T]$.

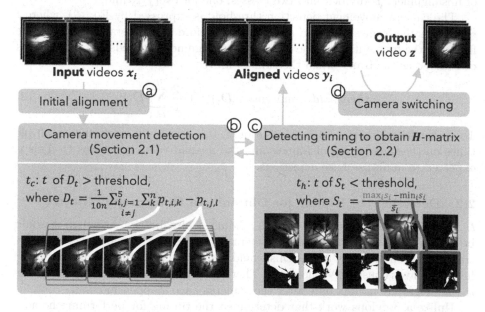

Fig. 2. Overview of the proposed method.

2.1   Camera Movement Detection

To find the t_c, we use the "degree of misalignment" obtained from the sequence y_i of the five aligned cameras. Since the misalignment should be zero if the geometric calibration between the cameras works well and each view overlaps perfectly, it can be used as an indication of camera movement.

First, the proposed method performs feature point detection in each frame of y_i. Specifically, we use the SIFT algorithm [8]. Then, feature point matching is performed for each of the 10 combinations of the five frames. The degree of misalignment D_t at frame t is represented as

$$D_t = \frac{1}{10n} \sum_{\substack{i,j=1 \\ i \neq j}}^{5} \sum_{k}^{n} \boldsymbol{p}_{t,i,k} - \boldsymbol{p}_{t,j,l}, \tag{1}$$

where $\boldsymbol{p}$ and k denote a keypoint position and its index in the i-th camera's coordinates respectively, l represents the corresponding index of k in the j-th camera's coordinates, and n represents the total number of corresponding points.

If D_t exceeds a certain threshold, our method detects camera movement. However, the calculated misalignment is too noisy to be used as is. To eliminate the noise, the outliers are removed, and smoothing is performed by calculating the movement average. Moreover, to determine the threshold, sample clustering is performed according to the degree of misalignment. Assuming that the multi-camera surgical light never moves more than twice in 10 min, the detected degree of misalignment is divided into two classes, one for every 10 min.

The camera movement detection threshold is expressed by Eq. (2), where t represents the frames classified as the frames before the camera movement. The frame at the moment when the degree of misalignment exceeds the threshold is considered as the frame when the camera moved.

$$threshold = \min\left(\max_{t}\left(\boldsymbol{D}_t\right) + 1, \frac{2}{T} \sum_{t=1}^{T} \boldsymbol{D}_t\right) \tag{2}$$

To make the estimation of t_c more robust, this process is performed multiple times on the same group of frames, and the median value is used as t_c. This is expected to minimize false detections.

2.2 Detecting the Timing for Obtaining a Homography Matrix

t_h represents the timing when to obtain homography matrix. Although it would be ideal to generate always-aligned camera sequences by performing homography transformation on every frame, this would incur high computational costs if the homography is constantly calculated. Therefore, the proposed method calculates the homography matrix only after the cameras have stopped moving.

Unlike a previous work that determined the timing for performing homography transformation manually [10], our method automatically detects t_h by using the area of surgical field appearing in surgical videos as an indication. This region indicates the extent of occlusion. Since the five cameras capture the same surgical field, if there is no occluded camera, the area of surgical field will have the same extent in all five camera frames. Eq. (3) is used to calculate the degree to which the area of surgical field is the same in all five camera frames, where $\boldsymbol{s}_i$ is the area of surgical field detected in each camera.

$$S = \{\max_{i}(\boldsymbol{s}_i) - \min_{i}(\boldsymbol{s}_i)\}/\bar{\boldsymbol{s}}_i, \tag{3}$$

where S is calculated every 30 frames, and if it is continuously below a given threshold (0.5 in this method), the corresponding timing is selected as the t_h.

3   Experiments and Results

We conducted two experiments to quantitatively and qualitatively investigate our method's efficacy.

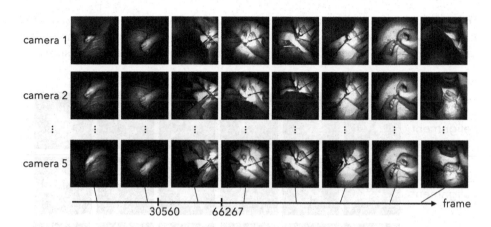

Fig. 3. Video frames after auto-alignment.

Dataset. We used a multi-camera system attached to a surgical light to capture videos of surgical procedures. We captured three types of actual surgical procedures: polysyndactyly, anterior thoracic keloid skin graft, and posttraumatic facial trauma rib cartilage graft. From these videos, we prepared five videos which were trimmed to one minute each. Videos 1 and 2 show the surgery of polysyndactyly, videos 3 and 4 show the anterior thoracic keloid skin graft scene, and video 5 shows the surgery of posttraumatic facial trauma rib cartilage graft.

Implementation Details. We used Ubuntu 20.04 LTS OS, an Intel Core i9-12900 for the CPU, and 62GiB of RAM. We defined the area of surgical field as hue ranging from 0 to 30 or from 150 to 179 in HSV color space.

Virtual Single-View Video Generation. Figure 3 shows a representative frame from the automatic positioning of the surgical video of the polysyndactyly operation using the proposed method. The figure also includes frames with detected camera movement. Once all five viewpoints were aligned, they were fed into the camera-switching algorithm to generate a virtual single-viewpoint video. The method requires about 40 min every time the cameras move. Please note that it is fully automated and needs no human labor, unlike the existing method, i.e., manual-alignment.

Comparison with Conventional Methods. We compared our auto alignment method (auto-alignment) with two conventional methods. In one of these methods, which is used in a hospital camera switching is performed after manual alignment (manual-alignment). The other method switches between camera views with no alignment (no-alignment).

3.1   Qualitative Evaluation

To qualitatively compare our method against baseline methods, we conducted a subjective evaluation. 11 physicians involved in surgical procedures regularly who were expected to actually use the surgical videos were selected as subjects.

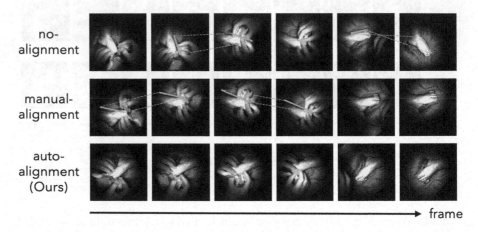

Fig. 4. Video frames obtained using the three methods. The red lines indicate the position and orientation of the instep of patient's foot. (Color figure online)

Figure 4 shows single-view video frames generated by our method and the two conventional methods. The red lines indicate the position and orientation of the instep of patient's foot. In video generated with no-alignment, the position and orientation of the insteps changed every time camera switching was performed, making it difficult to observe the surgical region with comfort. Manual-alignment showed better results than no-alignment. However, it was characterized by greater misalignment than the proposed method. It should also be noted that manual alignment requires time and effort. In contrast, our method effectively reduced misalignment between viewpoints even when camera switching was performed.

To perform a qualitative comparison between the three methods, following a previous work [13], we recruited eleven experienced surgeons who were expected to actually use surgical videos and asked them to conduct a subjective evaluation of the captured videos. The subjects were asked to score five statements, 1 ("disagree") to 5 ("agree").

1. No discomfort when the cameras switched.
2. No fatigue, even after long hours of viewing.
3. Easy to check the operation status.
4. Easy to see important parts of the frame.
5. I would like to use this system.

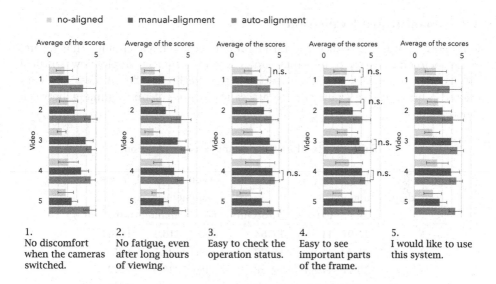

Fig. 5. Results of the subjective evaluation experiment.

The results are shown in Fig. 5. For almost all videos and statements, the Wilcoxon signed-rank test showed that the proposed method (auto-alignment) scored significantly higher than the two conventional methods. Significant differences are indicated by asterisks in the graphs in Fig. 5. The specific p-values are provided in the supplemental material. The results showing that the proposed method outperformed the conventional methods in statements 1 and 2 suggest that our method generates a stable video with little misalignment between camera viewpoints. Additionally, the results indicating that the proposed method outperformed the conventional method in statements 3 and 4 suggest that the video generated by our method makes it easier to confirm the surgical area. Furthermore, as shown in statement 5, the proposed method received the highest score in terms of the subjects' willingness to use the system in actual medical practice. We believe that the proposed method can contribute to improving the quality of medical care by facilitating stable observation of the surgical field.

Although we observed statistically significant differences between the proposed method and the baselines for almost all the test videos, significant differences were not observed only in Video 3, Statement 3 and Video 4, Statements 3 and 4. There may be two reasons for this. One is the small number of participants. Since we limited the participants to experienced surgeons, it was quite difficult to obtain a larger sample size. The second reason is differences in the geometry of the surgical fields. Our method is more effective for scenes with a three-dimensional geometry. If the surgical field is flat with fewer feature points, as in the case of the anterior thoracic keloid skin graft procedure, differences between our method and the manual alignment mehod, which does not take into account three-dimensional structures, are less likely to be observed.

3.2 Quantitative Evaluation

Our method aims to reduce the misalignment between viewpoints that occurs when switching between multiple cameras and generate single-view surgical videos with less occlusion. To investigate the method's effectiveness, we conducted a quantitative evaluation to assess the degree of misalignment between video frames.

Table 1. Results of the quantitative evaluation.

Video ID	ITF [dB] (↑) alignment			AvSpeed [pixel/frame] (↓) alignment		
	no	manual	auto(Ours)	no	manual	auto(Ours)
1	11.97	11.87	**17.54**	406.3	416.1	**166.1**
2	11.30	11.93	**15.77**	339.4	328.7	**195.6**
3	16.17	17.85	**22.26**	448.6	230.9	**92.2**
4	14.43	16.01	**19.26**	379.0	240.7	**77.5**
5	15.19	17.42	**21.66**	551.6	383.2	**169.6**

Following a previous work that calculated degree of misalignment between consecutive time-series frames [3], we used two metrics, the interframe transformation fidelity (ITF) and the average speed (AvSpeed). ITF represents the average peak signal-to-noise ratio (PSNR) between frames as

$$\mathrm{ITF} = \frac{1}{N_f - 1} \sum_{i=1}^{N_f - 1} \mathrm{PSNR}(t), \tag{4}$$

where N_f is the total number of frames. ITF is higher for videos with less motion blur. AvSpeed expresses the average speed of feature points. With the total number of frames N_f and the number of all feature points in a frame N_p, AvSpeed is calculated as

$$\mathrm{AvSpeed} = \frac{1}{N_p(N_f - 1)} \sum_{i=1}^{N_p} \sum_{t=1}^{N_f - 1} \|\dot{z}_i(t)\|, \tag{5}$$

where $z_i(t)$ denotes the image coordinates of the feature point and is calculated as

$$\dot{z}_i(t) = z_i(t + 1) - z_i(t). \tag{6}$$

The results are shown in Table 1. The ITF of the videos generated using the proposed method was 20%–50% higher than that of the videos with manual alignment. The AvSpeed of the videos generated using the proposed method was 40%–70% lower than that of the videos with manual alignment, indicating that the shake was substantially corrected.

4   Conclusion and Discussion

In this work, we propose a method for generating high-quality virtual single-viewpoint surgical videos captured by multiple cameras attached to a surgical light without occlusion or misalignment through automatic geometric calibration. In evaluation experiments, we compared our auto-alignment method with manual-alignment and no-alignment. The results verified the superiority of the proposed method both qualitatively and quantitatively. The ability to easily confirm the surgical field with the automatically generated virtual single-viewpoint surgical video will contribute to medical treatment.

Limitations. Our method relies on visual information to detect the timing of homography calculations (i.e., t_h). However, we may use prior knowledge of a geometric constraint such that cameras are at the pentagon corners (Fig. 1).

We assume that the multi-camera surgical light does not move more than twice in 10 min for a robust calculation of D_t. Although surgeons rarely moved the light more often, fine-tuning the parameter may result in further performance improvement. The current implementation shows misaligned images if the cameras move more frequently.

In the user-involved study, several participants reported noticeable black regions where no camera views were projected. (e.g., Fig. 4). One possible complement is to project pixels from other views.

Acknowledgements. This work was partially supported by JSPS KAKENHI Grant Number 22H03617.

References

1. Byrd, R.J., Ujjin, V.M., Kongchan, S.S., Reed, H.D.: Surgical lighting system with integrated digital video camera. US6633328B1 (2003)
2. Date, I., Morita, A., Kenichiro, K.: NS NOW Updated No.9 Thorough Knowledge and Application of Device and Information Technology (IT) for Neurosurgical Opperation. Medical View Co., Ltd. (2017)
3. Guilluy, W., Beghdadi, A., Oudre, L.: A performance evaluation framework for video stabilization methods. In: 2018 7th European Workshop on Visual Information Processing (EUVIP), pp. 1–6 (2018)
4. Hachiuma, R., Shimizu, T., Saito, H., Kajita, H., Takatsume, Y.: Deep selection: a fully supervised camera selection network for surgery recordings. In: Martel, A.L., et al. (eds.) MICCAI 2020. LNCS, vol. 12263, pp. 419–428. Springer, Cham (2020). https://doi.org/10.1007/978-3-030-59716-0_40

5. Hanada, E.: [special talk] video recording, storing, distributing and editing system for surgical operation. In: ITE Technical Report, pp. 77–80. The Institute of Image Information and Television Engineers (2017). in Japanese
6. Kajita, H.: Surgical video recording and application of deep learning for open surgery. J. Japan Soc. Comput. Aided Surg. **23**(2), 59–64 (2021). in Japanese
7. Kumar, A.S., Pal, H.: Digital video recording of cardiac surgical procedures. Ann. Thorac. Surg. **77**(3), 1063–1065 (2004)
8. Lowe, D.G.: Distinctive image features from scale-invariant keypoints. Int. J. Comput. Vision **60**, 91–110 (2004)
9. Nair, A.G., et al.: Surgeon point-of-view recording: using a high-definition head-mounted video camera in the operating room. Indian J. Ophthalmol. **63**(10), 771–774 (2015)
10. Obayashi, M., Mori, S., Saito, H., Kajita, H., Takatsume, Y.: Multi-view surgical camera calibration with none-feature-rich video frames: toward 3D surgery playback. Appl. Sci. **13**(4), 2447 (2023)
11. Saito, Y., Hachiuma, R., Saito, H., Kajita, H., Takatsume, Y., Hayashida, T.: Camera selection for occlusion-less surgery recording via training with an egocentric camera. IEEE Access **9**, 138307–138322 (2021)
12. Shimizu, T., Oishi, K., Hachiuma, R., Kajita, H., Takatsume, Y., Saito, H.: Surgery recording without occlusions by multi-view surgical videos. In: VISIGRAPP (5: VISAPP), pp. 837–844 (2020)
13. Yoshida, K., et al.: Spatiotemporal video highlight by neural network considering gaze and hands of surgeon in egocentric surgical videos. J. Med. Robot. Res. **7**, 2141001 (2022)

SurgicalGPT: End-to-End Language-Vision GPT for Visual Question Answering in Surgery

Lalithkumar Seenivasan[1]🆔, Mobarakol Islam[2]🆔, Gokul Kannan[3]🆔, and Hongliang Ren[1,4,5(✉)]🆔

[1] Department of Biomedical Engineering, National University of Singapore, Singapore, Singapore
[2] WEISS, University College London, London, UK
[3] Department of Production Engineering, National Institute of Technology, Tiruchirappalli, India
[4] Department of Electronic Engineering, Chinese University of Hong Kong, Shatin, Hong Kong
hlren@ee.cuhk.edu.hk
[5] Shun Hing Institute of Advanced Engineering, Chinese University of Hong Kong, Shatin, Hong Kong

Abstract. Advances in GPT-based large language models (LLMs) are revolutionizing natural language processing, exponentially increasing its use across various domains. Incorporating uni-directional attention, these autoregressive LLMs can generate long and coherent paragraphs. However, for visual question answering (VQA) tasks that require both vision and language processing, models with bi-directional attention or models employing fusion techniques are often employed to capture the context of multiple modalities all at once. As GPT does not natively process vision tokens, to exploit the advancements in GPT models for VQA in robotic surgery, we design an end-to-end trainable Language-Vision GPT (LV-GPT) model that expands the GPT2 model to include vision input (image). The proposed LV-GPT incorporates a feature extractor (vision tokenizer) and vision token embedding (token type and pose). Given the limitations of unidirectional attention in GPT models and their ability to generate coherent long paragraphs, we carefully sequence the word tokens before vision tokens, mimicking the human thought process of understanding the question to infer an answer from an image. Quantitatively, we prove that the LV-GPT model outperforms other state-of-the-art VQA models on two publically available surgical-VQA datasets (based on endoscopic vision challenge robotic scene segmentation 2018 and CholecTriplet2021) and on our newly annotated dataset (based on the holistic surgical scene dataset). We further annotate all three datasets to include question-type annotations to allow sub-type analysis. Furthermore, we extensively study and present the effects of token sequencing, token type and pose embedding for vision tokens in the LV-GPT model.

L. Seenivasan and M. Islam are co-first authors.

Supplementary Information The online version contains supplementary material available at https://doi.org/10.1007/978-3-031-43996-4_27.

1   Introduction

The recent evolution of large language models (LLMs) is revolutionizing natural language processing and their use across various sectors (e.g., academia, healthcare, business, and IT) and daily applications are being widely explored. In medical diagnosis, recent works [23] have also proposed employing the LLM models to generate condensed reports, interactive explanations, and recommendations based on input text descriptions (predicted disease and report). While the current single-modality (language) LLMs can robustly understand the questions, they still require prior text descriptions to generate responses and are unable to directly infer responses based on the medical image. Although language-only models can greatly benefit the medical domain in language processing, there is a need for robust multi-modality models to process both medical vision and language. In the surgical domain, in addition to the scarcity of surgical experts, their daily schedules are often overloaded with clinical and academic work, making it difficult for them to dedicate time to answer inquiries from students and patients on surgical procedures [3]. Although various computer-assisted solutions [1,10,11,16,17] have been proposed and recorded surgical videos have been made available for students to sharpen their skills and learn from observation, they still heavily rely on surgical experts to answer their surgery-specific questions. In such cases, a robust and reliable surgical visual question answering (VQA) model that can respond to questions by inferring from context-enriched surgical scenes could greatly assist medical students, and significantly reduce the medical expert's workload [19].

In the medical domain, MedfuseNet [19], an attention-based model, was proposed for VQA in medical diagnosis. Utilizing the advancements in the transformer models, VisualBert RM [18], a modified version of the VisualBert [12] model was also proposed for VQA in robotic surgery. Compared to most VQA models that require a region proposal network to propose vision patches, the VisualBert RM [18] performed VQA based on features extracted from the whole image, eliminating the need for a region proposal network. However, they were extracted using a non-trainable fixed feature extractor. While VisualBert [12] models and LLMs are transformer models, there are fundamentally different. VisualBert [12] transformers are bidirectional encoder models and are often employed for multi-modality tasks. In contrast, ChatGPT[1] (GPT3.5) and BARD (LaMDA [20]) are language-only uni-directional transformer decoder models employed for language generation. As they are proving to be robust in language generation, exploiting them to process the questions and enabling them to process vision could greatly improve performance in VQA tasks.

In this work, we develop an end-to-end trainable SurgicalGPT model by exploiting a pre-trained LLM and employing a learnable feature extractor to generate vision tokens. In addition to word tokens, vision tokens (embedded with token type and pose embedding) are introduced into the GPT model, resulting in a Language-Vision GPT (LV-GPT) model. Furthermore, we carefully sequence the word and vision tokens to leverage the GPT model's robust language processing ability to process the question and better infer an answer based on the vision

---

[1] chat.openai.com.

tokens. Through extensive experiments, we show that the SurgicalGPT(LV-GPT) outperforms other state-of-the-art (SOTA) models by ~ 3–5% on publically available EndoVis18-VQA [18] and Cholec80-VQA surgical-VQA [18] datasets. Additionally, we introduce a novel PSI-AVA-VQA dataset by adding VQA annotations to the publically available holistic surgical scene dataset(PSI-AVA) and observe similar performance improvement. Furthermore, we study and present the effects of token sequencing, where model performance improved by ~ 2–4% when word tokens are sequenced earlier. Finally, we also study the effects of token type and pose embedding for vision tokens in the LV-GPT model.

2 Proposed Method

2.1 Preliminaries

GPT2 [6], a predecessor to GPT3.5 (ChatGPT), is a transformer decoder model that performs next-word prediction. Auto-regressive in nature, its self-attention blocks attend to earlier word tokens to predict the next word token iteratively, allowing the model to generate complex paragraphs [15]. Although robust in language generation, due to its unidirectional attention [13], in a given iteration, the generated token knows all earlier tokens but does not know any subsequent token (Fig. 1(a)), restricting the model's ability to capture the entire context between all tokens. VisualBert [12], fundamentally different from GPT models, is a non-auto-regressive transformer encoder model. Its bidirectional self-attention blocks attend in both directions (earlier and subsequent tokens) [13], allowing the model to capture the entire context all at once (Fig. 1(b)). Due to this, bi-directional attention models are often preferred for multi-modality tasks.

Vision-Language Processing: Employed mostly for language-only tasks, GPT models do not natively process vision tokens [8]. While it supports robust word embedding, it lacks vision tokenizer and vision embedding layers. This limits exploiting its language processing ability for multi-modality tasks. Alternate to GPT, as the VisualBert model is often preferred for multi-modality tasks, it encompasses dedicated embedding layers for both vision and word tokens.

2.2 LV-GPT: Language-Vision GPT

Overall Network: We design an end-to-end trainable multi-modality (language and vision) LV-GPT model (Fig. 2) for surgical VQA. We integrate a vision tokenizer (feature extractor) module and vision embedding with the GPT model to exploit its language processing ability in performing VQA tasks.

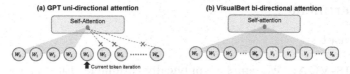

Fig. 1. Uni-directional attention in GPT language model vs bi-direction attention in VisualBert multi-modality model.

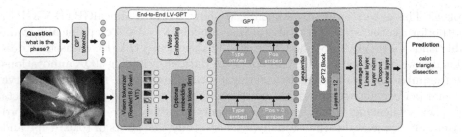

Fig. 2. End-to-End LV-GPT for Surgical VQA: The input question and surgical scene are tokenized, embedded, and sequenced to predict the answer.

Language-Vision Processing: The questions are tokenized using the inherent GPT2 tokenizer. The word tokens are further embedded based on token-id, token type (0) and token position by the inherent GPT2 word embedding layers. To tokenize the input surgical scene (image) into vision tokens, the LV-GPT includes a vision tokenizer (feature extractor): ResNet18 (RN18) [9]/Swin [14]/ViT [7]. Given an image, the tokenizer outputs vision tokens, each holding visual features from an image patch. Additionally, the vision tokens are further embedded based on token type (1) and token position (pos = 0) embeddings. The final embedded word and vision tokens (w_e and v_e) can be formulated as:

$$w_e = T_{t=0}(w_x) + P_{pos}(w_x) + w_x; \quad pos = 0, 1, 2, 3, ..., n.$$

$$v_e = T_{t=1}(v_x) + P_{pos=0}(v_x) + v_x; \quad v_x = \begin{cases} v_t, & dim(v_t^i) = dim(w_x^i) \\ f(v_t), & else \end{cases} \quad (1)$$

where, $T_t()$ is type embedding, $P_{pos}()$ is pose embedding, w_x and v_x are initial word and vision embedding, and v_t are vision tokens. Initial word embeds (w_x) are obtained using word embedding based on word token id. Depending on the size (dim) of each vision token, they undergo additional linear layer embedding ($f()$) to match the size of the word token.

Token Sequencing: LLMs are observed to process long sentences robustly and hold long-term sentence knowledge while generating coherent paragraphs/reports. Considering GPT's superiority in sequentially processing large sentences and its uni-directional attention, the word tokens are sequenced before the vision tokens. This is also aimed at mimicking human behaviour, where the model understands the question before attending to the image to infer an answer.

Classification: Finally, the propagated multi-modality features are then passed through a series of linear layers for answer classification.

3   Experiment

3.1   Dataset

EndoVis18-VQA: We employ publically available EndoVis18-VQA [18] dataset to benchmark the model performance. We use the classification subset that includes classification-based question-and-answer (Q&A) pairs for 14 robotic nephrectomy procedure video sequences of the MICCAI Endoscopic Vision Challenge 2018 [2] dataset. The Q&A pairs are based on the tissue, actions, and locations of 8 surgical tools. The dataset includes 11783 Q&A pairs based on 2007 surgical scenes. The answers consist of 18 classes (1 kidney, 13 tool-tissue interactions, and 4 tool locations). Additionally, we further annotated the validation set (video sequences 1, 5, and 16) on question types to assist in additional analysis. We followed the EndoVis18-VQA [18] dataset's original train/test split.

Cholec80-VQA: The classification subset of the Cholec80-VQA [18] is also employed for model evaluation. It contains Q&A pairs for 40 video sequences of the Cholec80 dataset [21]. The subset consists of 43182 Q&A pairs on the surgical phase and instrument presence for 21591 frames. The answers include 13 classes (2 instrument states, 4 on tool count, and 7 on surgical phase). We additionally annotated the validation set (video sequences: 5, 11, 12, 17, 19, 26, 27 and 31) on the Q&A pairs types for further model analysis. The VQA [18] dataset's original train/test split is followed in this work.

PSI-AVA-VQA: We introduce a novel PSI-AVA-VQA dataset that consists of Q&A pairs for key surgical frames of 8 cases of the holistic surgical scene dataset (PSI-AVA dataset) [22]. The questions and answers are generated in sentence form and single-word (class) response form, respectively. They are generated based on the surgical phase, step, and location annotation provided in the PSI-AVA dataset [22]. The PSI-AVA-VQA consists of 10291 Q&A pairs and with 35 answer classes (4 locations, 11 surgical phases, and 21^1 surgical steps). The Q&A pairs are further annotated into 3 types (location, phase, and step). The fold-1 train/test split of parent PSI-AVA [22] dataset is followed in this work.

3.2 Implementation Details

All variants of our models [2] are trained based on cross-entropy loss and optimized using the Adam optimizer. The models were trained for 80 epoch, with a batch size of 64, except for LV-GPT (ViT) (batch size = 32 due to GPU limitation). learning rates lr = 1×10^{-5}, 1×10^{-5} and 5×10^{-6} are used for EndoVis18-VQA, PSI-AVA-VQA and Cholec80-VQA dataset, respectively. The SOTA VisualBert [12] and VisualBert RM [18] models were implemented using their official code repositories. The Block [5], MUTAN [4], MFB [24] and MFH [25] were implemented using the official codes of Block [5].

4 Results

All our proposed LV-GPT model variants are quantitatively benchmarked (Table 1) against other attention-based/bi-directional encoder-based SOTA models on EndoVis18-VQA, Cholec80-VQA and PSI-AVA-VQA datasets based

---

[1] One class shares a common name with a surgical phase class.
[2] Code available: github.com/lalithjets/SurgicalGPT

Table 1. Quantitaive comparison of our LV-GPT (Swin), LV-GPT (RN18), and (LV-GPT (ViT)) against state-of-the-art models.

MODELS	EndoVis18-VQA [18]			Cholec80-VQA [18]			PSI-AVA-VQA		
	Acc	Recall	FScore	Acc	Recall	FScore	Acc	Recall	FScore
VisualBert [12]	0.6143	0.4282	0.3745	0.9007	0.6294	0.6300	0.5853	0.3307	0.3161
VisualBert RM [18]	0.6190	0.4079	0.3583	0.9001	0.6573	0.6585	0.6016	0.3242	0.3165
Block [5]	0.6088	0.4884	0.4470	0.8948	0.6600	0.6413	0.5990	**0.5136**	**0.4933**
Mutan [4]	0.6303	**0.4969**	<u>0.4565</u>	0.8699	0.6332	0.6106	0.4971	0.3912	0.3322
MFB [24]	0.5238	0.4205	0.3622	0.8410	0.5303	0.4588	0.5712	<u>0.4379</u>	<u>0.4066</u>
MFH [25]	0.5876	0.4835	0.4224	0.8751	0.5903	0.5567	0.4777	0.2995	0.2213
LV-GPT (Swin)	0.6613	0.4460	0.4537	**0.9429**	**0.7339**	**0.7439**	<u>0.6033</u>	0.4137	0.3767
LV-GPT (RN18)	**0.6811**	0.4649	**0.4649**	0.8746	0.5747	0.5794	0.5933	0.3183	0.3168
LV-GPT (ViT)	<u>0.6659</u>	<u>0.4920</u>	0.4336	<u>0.9232</u>	<u>0.6833</u>	<u>0.6963</u>	**0.6549**	0.4132	0.3971

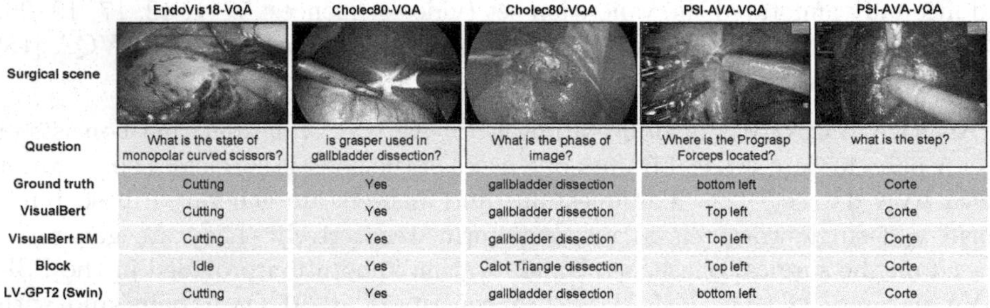

Fig. 3. Qualitative analysis: Comparison of answers predicted by VisualBERT [12], VisualBert RM [18], Block [5], and our LV-GPT (Swin) models against the ground truth based on input surgical scene and question.

on the accuracy (Acc), recall, and Fscore. In most cases, all our variants, LV-GPT (Swin), LV-GPT (RN18) and LV-GPT (ViT), are observed to significantly outperform SOTA models on all three datasets in terms of Acc. Specifically, the LV-GPT (Swin) variant (balanced performance across all datasets) is observed to outperform all SOTA models on all datasets and significantly improve the performance ($\sim$ 3–5% improvement) on EndoVis18-VQA and Cholec80-VQA dataset. Additionally, it should be noted our model variants can be trained end-to-end, whereas, most of the SOTA models requires a region proposal network to process input image into vision tokens. Figure 3 shows the qualitative performance of LV-GPT (Swin) against SOTA models on three datasets. A Comparison of our LV-GPT model performance on the EndoVis18-VQA dataset with default test queries vs rephrased test queries is presented in supplementary materials that highlight the model's robustness in language reasoning.

Early Vision vs Early Word: The performance of LV-GPT based on word and vision token sequencing (Table 2) is also studied. While all three variants of the LV-GPT models processing vision tokens earlier are observed to perform on

Table 2. Comparison of LV-GPT model performance when vision tokens are sequenced earlier vs when word tokens are sequenced earlier.

Token sequencing	Model	EndoVis18-VQA			PSI-AVA-VQA		
		Acc	Recall	FScore	Acc	Recall	FScore
Early vision	LV-GPT (RN18)	0.6338	0.3600	0.3510	0.5542	0.2879	0.2886
	LV-GPT (Swin)	0.6208	0.4059	0.3441	**0.6068**	**0.4195**	**0.3813**
	LV-GPT (ViT)	0.6493	0.4362	0.3701	0.6023	0.2802	0.2628
Early word	LV-GPT (RN18)	**0.6811**	**0.4649**	**0.4649**	0.5933	0.3183	0.3168
	LV-GPT (Swin)	**0.6613**	**0.4460**	**0.4537**	0.6033	0.4137	0.3767
	LV-GPT (ViT)	**0.6659**	**0.4920**	**0.4336**	0.6549	0.4132	0.3971

par with SOTA models reported in Table 1, in most cases, their performances on both datasets further improved by $\sim$ 2–4% when word tokens are processed earlier. This improvement could be attributed to LLM's ability to hold sentence (question) context before processing the vision tokens to infer an answer. This behaviour, in our view, mimics the human thought process, where we first understand the question before searching for an answer from an image.

Pose Embedding for Vision Tokens: The influence of positional embedding of the vision tokens (representing a patch region) in all the LV-GPT variants is studied by either embedded with position information (pos = 1, 2, 3, .., n.) or zero-position (pos = 0). Table 3 shows the difference in the performance of the best-performing LV-GPT variant in each dataset, with its vision tokens

Table 3. Comparison of model performances on EndoVis18-VQA, Cholec80-VQA and PSI-AVA-VQA datasets when vision tokens are embedded with zero-positional embedding vs actual pose embedding.

Dataset	Model	Zero Pose Embedding			Actual Pose Embedding		
	Best LV-GPT	Acc	Recall	FScore	Acc	Recall	FScore
EndoVis18-VQA	LV-GPT (RN18)	**0.6811**	0.4649	0.4649	**0.6811**	**0.4720**	**0.4681**
Cholec80-VQA	LV-GPT (Swin)	**0.9429**	**0.7339**	**0.7439**	0.9414	0.7251	0.7360
PSI-AVA-VQA	LV-GPT (ViT)	**0.6549**	**0.4132**	**0.3971**	0.5905	0.3742	0.3463

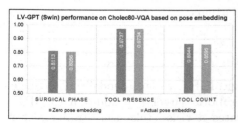 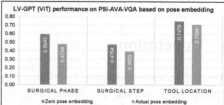

Fig. 4. Sub-type performance analysis of LV-GPT model variants on Cholec80-VQA and PSI-AVA-VQA embedded with zero-pose embedding vs actual-pose embedding.

Table 4. Ablation study on vision token (VT) embedding.

VB-VE	C-VE	VT-TY +VT-PE	VT-TY +VT-ZPE	LV-GPT (RN18)			LV-GPT (ViT)		
				Acc	Recall	FScore	Acc	Recall	FScore
✓				0.6287	0.4061	0.4063	0.6147	0.4199	0.3679
	✓			0.6728	0.4366	0.4455	0.6504	0.4792	0.4323
	✓	✓		**0.6811**	**0.4720**	**0.4681**	0.6259	0.4306	0.3805
	✓		✓	**0.6811**	0.4649	0.4649	**0.6659**	**0.4920**	**0.4336**

embedded with actual-position or zero-position. While we expected the positional embedding to improve the performance (dataset Q&A pairs related to tool location), from the results, we observe that embedding vision tokens with zero-position embedding results in better performance. In-depth analysis shows that our CNN-based LV-GPT (RN18) model improved with positional embedding (Table 3 and Table 4). In the transformer-based LV-GPT (Swin)/LV-GPT (ViT) models, positional embedding is already incorporated at the vision tokenizer (VIT/Swin) layer, and adding positional embedding at the GPT level results in double Position embedding. Thus, "zero-position" can be interpreted as "LV-GPT only requires one layer of positional embedding". A sub-type analysis (Fig. 4) is also performed on the model performance to analyze the effect of positional embedding of the vision tokens. The model in which the vision tokens were embedded with zero-position (at the GPT level), performed marginally better/similar on all sub-types in the Cholec80-VQA dataset. However, its performance improvement was significant in the PSI-AVA-VQA dataset sub-types, including the 'tool location' sub-types that contain questions on tool location.

Ablation Study on Vision Token Embedding: An ablation study on the vision token embedding in the LV-GPT model on the EndoVis18-VQA dataset is also shown in Table 4. VB-VE refers to vision token embedding using VisualBert vision embedding. The C-VE refers to custom embedding, where, in LV-GPT (RN18), the vision token undergoes additional linear layer embedding to match the word-token dimension, and in other variants, vision tokens from the Swin/VIT are directly used. The subsequent VT-TY + VT-PE and VT-TY + VT-ZPE refers to the additional vision token type (TY) and actual-position (PE)/zero-position (ZPE) embedding. We observe that employing C-VE with VT-TY + VT-ZPE results in better performance.

5   Conclusion

We design an end-to-end trainable SurgicalGPT, a multi-modality Language-Vision GPT model, for VQA tasks in robotic surgery. In addition to GPT's inherent word embeddings, it incorporates a vision tokenizer (trainable feature extractor) and vision token embedding (type and pose) to perform multi-modality tasks. Furthermore, by carefully sequencing the word tokens earlier to vision tokens, we exploit GPT's robust language processing ability, allowing the

LV-GPT to significantly perform better VQA. Through extensive quantitative analysis, we show that the LV-GPT outperforms other SOTA models on three surgical-VQA datasets and sequencing word tokens early to vision tokens significantly improves the model performance. Furthermore, we introduce a novel surgical-VQA dataset by adding VQA annotations to the publically available holistic surgical scene dataset. While multi-modality models that process vision and language are often referred to as "vision-language" models, we specifically name our model "language-vision GPT" to highlight the importance of the token sequencing order in GPT models. Integrating vision tokens into GPT also opens up future possibilities of generating reports directly from medical images/videos.

Acknowledgement. This work was supported by Hong Kong Research Grants Council (RGC) Collaborative Research Fund (CRF C4026-21GF and CRF C4063-18G) and Shun Hing Institute of Advanced Engineering (BME-p1-21/8115064) at the Chinese University of Hong Kong. M. Islam was funded by EPSRC grant [EP/W00805X/1].

References

1. Adams, L., et al.: Computer-assisted surgery. IEEE Comput. Graphics Appl. **10**(3), 43–51 (1990)
2. Allan, M., et al.: 2018 robotic scene segmentation challenge. arXiv preprint arXiv:2001.11190 (2020)
3. Bates, D.W., Gawande, A.A.: Error in medicine: what have we learned? (2000)
4. Ben-Younes, H., Cadene, R., Cord, M., Thome, N.: Mutan: Multimodal tucker fusion for visual question answering. In: Proceedings of the IEEE International Conference on Computer Vision, pp. 2612–2620 (2017)
5. Ben-Younes, H., Cadene, R., Thome, N., Cord, M.: Block bilinear superdiagonal fusion for visual question answering and visual relationship detection. In: Proceedings of the AAAI Conference on Artificial Intelligence, vol. 33, pp. 8102–8109 (2019)
6. Brown, T., et al.: Language models are few-shot learners. In: Advance in Neural Information Processing System, vol. 33, pp. 1877–1901 (2020)
7. Dosovitskiy, A., et al.: An image is worth 16x16 words: transformers for image recognition at scale. arXiv preprint arXiv:2010.11929 (2020)
8. Guo, J., et al.: From images to textual prompts: zero-shot VQA with frozen large language models. arXiv preprint arXiv:2212.10846 (2022)
9. He, K., Zhang, X., Ren, S., Sun, J.: Deep residual learning for image recognition. In: Proceedings of the IEEE Conference on Computer Vision and Pattern Recognition, pp. 770–778 (2016)
10. Hong, M., Rozenblit, J.W., Hamilton, A.J.: Simulation-based surgical training systems in laparoscopic surgery: a current review. Virtual Reality **25**, 491–510 (2021)
11. Kneebone, R.: Simulation in surgical training: educational issues and practical implications. Med. Educ. **37**(3), 267–277 (2003)
12. Li, L.H., Yatskar, M., Yin, D., Hsieh, C.J., Chang, K.W.: VisualBERT: a simple and performant baseline for vision and language. arXiv preprint arXiv:1908.03557 (2019)
13. Liu, X., et al.: GPT understands, too. arXiv preprint arXiv:2103.10385 (2021)

14. Liu, Z., et al.: Swin transformer: hierarchical vision transformer using shifted windows. In: Proceedings of the IEEE/CVF International Conference on Computer Vision, pp. 10012–10022 (2021)

15. Peng, B., Li, C., Li, J., Shayandeh, S., Liden, L., Gao, J.: SOLOIST: few-shot task-oriented dialog with a single pretrained auto-regressive model. arXiv preprint arXiv:2005.05298 3 (2020)

16. Rogers, D.A., Yeh, K.A., Howdieshell, T.R.: Computer-assisted learning versus a lecture and feedback seminar for teaching a basic surgical technical skill. Am. J. Surg. **175**(6), 508–510 (1998)

17. Sarker, S., Patel, B.: Simulation and surgical training. Int. J. Clin. Pract. **61**(12), 2120–2125 (2007)

18. Seenivasan, L., Islam, M., Krishna, A.K., Ren, H.: Surgical-VQA: Visual question answering in surgical scenes using transformer. In: Wang, L., Dou, Q., Fletcher, P.T., Speidel, S., Li, S. (eds.) Medical Image Computing and Computer Assisted Intervention-MICCAI 2022. LNCS, vol. 13437, pp. 33–43. Springer, Cham (2022). https://doi.org/10.1007/978-3-031-16449-1_4

19. Sharma, D., Purushotham, S., Reddy, C.K.: MedFuseNet: an attention-based multimodal deep learning model for visual question answering in the medical domain. Sci. Rep. **11**(1), 1–18 (2021)

20. Thoppilan, R., et al.: LAMDA: language models for dialog applications. arXiv preprint arXiv:2201.08239 (2022)

21. Twinanda, A.P., Shehata, S., Mutter, D., Marescaux, J., De Mathelin, M., Padoy, N.: Endonet: a deep architecture for recognition tasks on laparoscopic videos. IEEE Trans. Med. Imaging **36**(1), 86–97 (2016)

22. Valderrama, N., et al.: Towards holistic surgical scene understanding. In: Wang, L., Dou, Q., Fletcher, P.T., Speidel, S., Li, S. (eds.) Medical Image Computing and Computer Assisted Intervention-MICCAI 2022. LNCS, vol. 13437, pp. 442–452. Springer, Cham (2022). https://doi.org/10.1007/978-3-031-16449-1_42

23. Wang, S., Zhao, Z., Ouyang, X., Wang, Q., Shen, D.: ChatCAD: interactive computer-aided diagnosis on medical image using large language models. arXiv preprint arXiv:2302.07257 (2023)

24. Yu, Z., Yu, J., Fan, J., Tao, D.: Multi-modal factorized bilinear pooling with co-attention learning for visual question answering. In: Proceedings of the IEEE International Conference on Computer Vision, pp. 1821–1830 (2017)

25. Yu, Z., Yu, J., Xiang, C., Fan, J., Tao, D.: Beyond bilinear: generalized multimodal factorized high-order pooling for visual question answering. IEEE Trans. Neural Netw. Learn. Syst. **29**(12), 5947–5959 (2018)

Intraoperative CT Augmentation
for Needle-Based Liver Interventions

Sidaty El hadramy[1,2], Juan Verde[3], Nicolas Padoy[2,3], and Stéphane Cotin[1(✉)]

[1] Inria, Strasbourg, France
stephane.cotin@inria.fr
[2] ICube, University of Strasbourg, Strasbourg, CNRS, Strasbourg, France
[3] IHU Strasbourg, Strasbourg, France

Abstract. This paper addresses the need for improved CT-guidance during needle-based liver procedures (i.e., tumor ablation), while reduces the need for contrast agent injection during such interventions. To achieve this objective, we augment the intraoperative CT with the preoperative vascular network deformed to match the current acquisition. First, a neural network learns local image features in a non-contrasted CT image by leveraging the known preoperative vessel tree geometry and topology extracted from a matching contrasted CT image. Then, the augmented CT is generated by fusing the labeled vascular tree and the non-contrasted intraoperative CT. Our method is trained and validated on porcine data, achieving an average dice score of 0.81 on the predicted vessel tree instead of 0.51 when a medical expert segments the non-contrasted CT. In addition, vascular labels can also be transferred to provide additional information. Source code of this work is publicly available at https://github.com/Sidaty1/Intraoperative_CT_augmentation.

Keywords: Liver tumor ablation · Needle-based procedures · Patient-specific interventions · CT-guidance · Medical image augmentation

1 Introduction

Needle-based liver tumor ablation techniques (e.g., radiofrequency, microwave, laser, cryoablation) have a great potential for local curative tumor control [1], with comparable results to surgery in the early stages for both primary and secondary cancers. Furthermore, as it is minimally invasive, it has a low rate of major complications and procedure-specific mortality, and is tissue-sparing, thus, its indications are growing exponentially and extending the limits to more advanced tumors [3]. CT-guidance is a widely used imaging modality for placing the needles, monitoring the treatment, and following up patients. However, it is limited by the exposure to ionizing radiation and the need for intravenous

---

S. El Hadramy and J. Verde—Equal contribution.

© The Author(s), under exclusive license to Springer Nature Switzerland AG 2023
H. Greenspan et al. (Eds.): MICCAI 2023, LNCS 14228, pp. 291–301, 2023.
https://doi.org/10.1007/978-3-031-43996-4_28

injection of contrast agents to visualize the intrahepatic vessels and the target tumor(s).

In standard clinical settings, the insertion of each needle requires multiple check points during its progression, fine-tune maneuvers, and eventual repositioning. This leads to multiple CT acquisitions to control the progression of the needle with respect to the vessels, the target, and other sensible structures [26]. However, intrahepatic vessels (and some tumors) are only visible after contrast-enhancement, which has a short lifespan and dose-related deleterious kidney effects. It makes it impossible to perform each of the control CT acquisitions under contrast injection. A workaround to shortcut these limitations is to perform an image fusion between previous contrasted and intraoperative non-contrasted images. However, such a solution is only available in a limited number of clinical settings, and the registration is only rigid, usually deriving into bad results. In this work, we propose a method for visualizing intrahepatic structures after organ motion and needle-induced deformations, in non-injected images, by exploiting image features that are generally not perceivable by the human eye in common clinical workflows.

To address this challenge, two main strategies could be considered: **image fusion** and **image processing** techniques. Image fusion typically relies on the estimation of rigid or non-rigid transformations between 2 images, to bring into the intraoperative image structures of interest only visible in the preoperative data. This process is often described as an optimization problem [9,10] which can be computationally expensive when dealing with non-linear deformations, making their use in a clinical workflow limited. Recent deep learning approaches [11,12,14] have proved to be a successful alternative to solve image fusion problems, even when a large non-linear mapping is required. When ground-truth displacement fields are not known, state-of-the-art methods use unsupervised techniques, usually an encoder-decoder architecture [7,13], to learn the unknown displacement field between the 2 images. However, such unsupervised methods fail at solving our problem due to lack of similar image features between the contrasted (CCT) and non-contrasted (NCCT) image in the vascular tree region (see Sect. 3.3).

On the other hand, deep learning techniques have proven to be very efficient at solving image processing challenges [15]. For instance, image segmentation [16], image style transfer [17], or contrast-enhancement to cite a few. Yet, segmenting vessels from non-contrasted images remains a challenge for the medical imaging community [16]. Style transfer aims to transfer the style of one image to another while preserving its content [17–19]. However, applying such methods to generate a contrasted intraoperative CT is not a sufficiently accurate solution for the problem that we address. Contrast-enhancement methods could be an alternative. In the method proposed by Seo *et al.* [20], a deep neural network synthesizes contrast-enhanced CT from non contrast-enhanced CT. Nevertheless, results obtained by this method are not sufficiently robust and accurate to provide an augmented intraoperative CT on which needle-based procedures can be guided.

In this paper we propose an alternative approach, where a neural network learns local image features in a NCCT image by leveraging the known preoperative vessel tree geometry and topology extracted from a matching (undeformed) CCT. Then, the augmented CT is generated by fusing the deformed vascular tree with the non-contrasted intraoperative CT. Section 2 presents the method and its integration in the medical workflow. Section 3 presents and discusses the results, and finally we conclude in Sect. 4 and highlight some perspectives.

2   Method

In this section, we present our method and its compatibility with current clinical workflows. A few days or a week before the intervention, a preoperative diagnostic multiphase contrast-enhanced image (MPCECT) is acquired (Fig. 1, yellow box). The day of the intervention, a second MPCECT image is acquired before starting the needle insertion, followed by a series of standard, non-injected acquisitions to guide the needle insertion (Fig. 1, blue box). Using such a non-contrasted intraoperative image as input, **our method performs a combined non-rigid registration and augmentation of the intraoperative CT** by adding anatomical features (mainly intrahepatic vessels and tumors) from the preoperative image to the current image. To achieve this result, our method only requires to process and train on the baseline MPCECT image (Fig. 1, red box). An overview of the method is shown in the Fig 2 and the following sections describe its main steps.

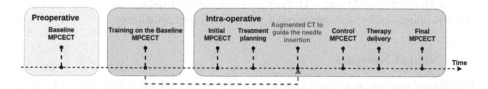

Fig. 1. Integration of our method in the clinical workflow. The neural network trained on preoperative MPCECT avoids contrast agent injections during the intervention.

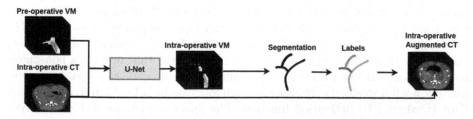

Fig. 2. The neural network takes as input the preoperative vessel map (VM) and the intraoperative NCCT, and outputs the intraoperative vessel map (VM) from which we extract the deformed vascular tree. Finally, the augmented CT is created by fusing the segmented image and labels with the intraoperative NCCT.

2.1 Vessel Map Extraction

We call Vessel Map (VM) the region of interest defining the vascular tree in the NCCT. Since vascular structures are not visible in non-contrasted images, the extraction of this map is done by segmenting the CCT and then using this segmentation as a mask in the NCCT. Mathematical morphology operators, in particular a dilation operation [23], are performed on the segmented region of interest to slightly increase its dimensions. This is needed to compensate for segmentation errors and the slight anatomical motion that may exist between the contrasted and non-contrasted image acquisitions. In practice, the acquisition protocols limit the shift between the NCCT and CCT acquisitions, and only a few sequential dilation operations are needed to ensure we capture the true vessel fingerprint in the NCCT image. Note that the resulting vessel map is not a binary mask, but a subset of the image limited to the volume covered by the vessels.

2.2 Data Augmentation

The preoperative MPCECT provides a couple of registered NCCT and CCT images. This is obviously not sufficient for training purposes, as they do not represent the possible soft tissue deformation that may occur during the procedure. Therefore, we augment the data set by applying multiple random deformations to the original images. Random deformations are created by considering a predefined set of control points for which we define a displacement field with a random normal distribution. The displacement field of the full volume is then obtained by linearly interpolating the control points' displacement field to the rest of the volume. All the deformations are created using the same number of control points and characteristics of the normal distributions.

2.3 Neural Network

Predicting the vascular tree location in the deformed intraoperative NCCT is done using a U-net [5] architecture. The neural network takes as input the preoperative vessel map and the intraoperative NCCT, and outputs the intraoperative vessel map. Our network learns to find the image features (or vessel fingerprint) present in the vessel map, in a given NCCT assuming the knowledge of its geometry, topology, and the distribution of contrast from the preoperative MPCECT. The architecture of our network is illustrated in Fig. 3. It consists of a four layers analysis (left side) and synthesis (right side) paths that provide a non-linear mapping between low resolution input and output images. Both paths include four $3 \times 3 \times 3$ unpadded convolutions, each followed by a Leaky Rectified Linear Unit (LeakyReLU) activation function. The analysis includes a $2 \times 2 \times 2$ max pooling with a stride of 1, while the synthesis follows each convolution by a $2 \times 2 \times 2$ up-convolution with a stride of 1. Shortcut connections from layers of equal resolution in the analysis path provide the essential high-resolution features to the synthesis path. In the last layer, a $1 \times 1 \times 1$ convolution reduces the number of output channels to one, yielding the vessel map in the intraoperative image. Our network is trained by minimizing the mean square error between the predicted and ground truth vessel map. Training details are presented in Sect. 3.

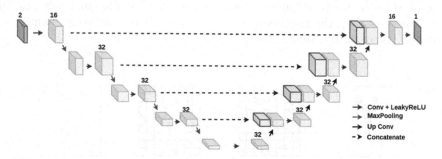

Fig. 3. Our neural network uses a four-path encoder-decoder architecture and takes as input a two-channel image corresponding to the intraoperative NCCT image concatenated with the preoperative vessel map. The output is the intraoperative vessel map.

2.4 Augmented CT

Once the network has been trained on the patient-specific preoperative data, the next step is to augment and visualize the intraoperative NCCT. This is done in 3 steps:

- The dilatation operations introduced in Sect. 2.1 are not reversible (i.e. the segmented vessel tree cannot be recovered from the VM by applying the same number of erosion operations). Also, neighboring branches in the vessel tree could end up being fused, thus changing the topology of the vessel map. Therefore, to retrieve the correct segmented (yet deformed) vascular tree, we compute a displacement field between the pre- and intraoperative VMs. This is done with the Elastix library [21,22]. The resulting displacement field is applied on the preoperative segmentation to retrieve the intraoperative vessel tree segmentation. This is illustrated in Fig. 4.
- The augmented image is obtained by fusing the predicted intraoperative segmentation with the intraoperative NCCT image. The augmented vessels are displayed in green to ensure the clinician is aware this is not a true CCT image (see Fig. 5).
- It is also possible to add anatomical labels to the intraoperative augmented CT to further assist the clinician. To achieve this objective, we compute a graph data structure from the preoperative segmentation. We first extract the vessel centerlines as described in [4]. To define the associated graph structure, we start by selecting all branches with either no parent or no children. The branch with the highest radius is then selected as the root edge. An oriented graph is created using a Breadth First Search algorithm starting from the

root edge. Nodes and edges correspond respectively to vessel tree bifurcations and branches. We use the graph structure to associate each anatomical label (manually defined) with a Strahler [6] graph ordering. The same process is applied to the predicted intraoperative segmentation. This makes it possible to correctly map the preoperative anatomical labels (e.g. vessel name) and display them on the augmented image.

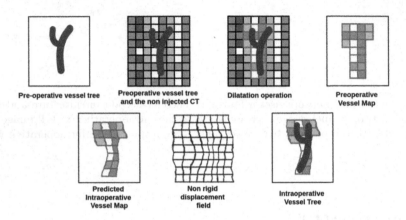

Fig. 4. This figure illustrates the different stages of the pipeline adopted to generate the VM and show how the vessel tree topology is retrieved from the predicted intraoperative VM by computing a displacement field between the preoperative VM and the predicted VM. This field is applied to the preoperative segmentation to get the intraoperative one.

3   Results and Discussion

3.1   Dataset and Implementation Details

To validate our approach, 4 couples of MPCECT abdominal porcine images were acquired from 4 different subjects. For a given subject, each couple corresponds to a preoperative and an intraoperative MPCECT. We recall that an MPCECT contains a set of registered NCCT and CCT images. These images are then cropped and down-sampled to $256 \times 256 \times 256$, and the voxels intensities are scaled between 0 and 255. Finally, we extract the VM from each MPCECT sample and apply 3 dilation operations, which demonstrated the best performance in terms of prediction accuracy and robustness on our data. We note that public data sets such as DeepLesion [24], 3Dircadb-01 [25] and others do not fit our problem since they do not include the NCCT images. Aiming at a patient-specific prediction, we only train on a "subject" at a time. For a given subject, we generate 100 displacement fields using the data augmentation strategy explained above with 50 voxels for the control points spacing in the three spatial directions and a standard deviation of 5 voxels for the normal distributions. The resulting deformation is applied to the **preoperative** MPCECT and its corresponding VM. Thus, we end up with a set of 100 triplets (NCCT, CCT and VM). Two out of the 100 triplets are used for each training batch, where one is considered as the pre-operative MPCECT and the other as the intraoperative one. This makes it possible to generate up to 4950 training and validation samples. The **intraoperative** MPCECT of the same subject is used to test the network. Our method is implemented in Tensorflow 2.4, on a GeForce RTX 3090. We use an Adam optimizer ($\beta_1 = 0.001$, $\beta_2 = 0.999$) with a learning rate of 10^{-4}. The training process converges in about 1,000 epochs with a batch size of 1 and 200 steps per epoch.

3.2 Results

To assess our method, we use a dice score to measure the overlap between our predicted segmentation and the ground truth. Being a commonly used metric for segmentation problems, Dice aligns the nature of our problem as well as the clinical impact of our solution. We ha performed tests on 4 different (porcine) data sets. Results are reported in Table 1. The method achieved a mean dice score of 0.81. An example of a subject intraoperative augmented CT is illustrated in Fig. 5, where the three images correspond respectively to the initial non injected CT, the augmented CT without and with labels. Figure 6 illustrates the results of our method for *Subject 1*. The green vessels correspond to the ground truth intraoperative segmentation, the orange ones to the predicted intraoperative segmentation and finally the gray vessel tree corresponds to the preoperative CCT vessel tree. Such results demonstrate the ability of our method to perform very well even in the presence of large deformations.

Table 1. This table presents our results over 4 subjects in terms of dice score. For each subject the network was trained on the preoperative MPCECT and tested on the intraoperative MPCECT. We achieve a mean dice score of 0.81 vs. 0.51 for the clinical experts.

Dice score	Subject 1	Subject 2	Subject 3	Subject 4	Mean	Std
Ours	0.8	0.77	0.79	0.90	0.81	0.04
Expert clinician	0.52	0.45	0.53	0.52	0.51	0.03

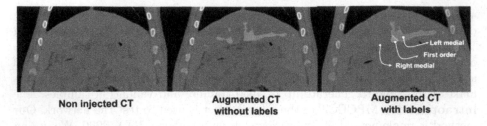

Non injected CT   Augmented CT   Augmented CT
                        without labels   with labels

Fig. 5. In this figure we show the original NCCT on the left. The middle image shows the augmented CT with the predicted vessel tree (in green). The rightmost image shows the augmented image with anatomical labels transferred from the preoperative image segmentation and labelling. (Color figure online)

Fig. 6. Assessment of our method for Subject 1. We show 3 different views of the intraoperative vessel prediction (in orange), the ground truth (in green) and the pre-operative vessels (in grey). (Color figure online)

Qualitative Assessment: To further demonstrate the value of our method, we have asked two clinicians to manually segment the NCCT images in the intraoperative MPCECT data. Their results (mean and standard deviation) are reported in Table 1. Our method outperforms the results of both clinicians, with an average dice score of 0.81 against 0.51 as a mean for the clinical experts.

3.3   Ablation Study and Additional Results

Vessel Map: We have removed the VM from the network input to demonstrate its impact on our results. Using the data of the *Subject 1*, a U-net was trained to segment the vessel tree of the intraoperative NCCT image. The network only managed to segment a small portion of the main portal vein branch. Thus, achieving a dice score of **0.16** vs **0.79** when adding the preoperative VM as additional input. We also studied the influence of the diffusion kernel applied to the initial segmentation. We have seen, on our experimental data, that 3 dilation operations were sufficient to compensate for the possible motion between NCCT and CCT acquisitions.

Comparison with VoxelMorph: The problem that we address can be seen from different angles. In particular, we could attempt to solve it by register-ing the preoperative NCCT to the intraoperative one and then applying the resulting displacement field to the known preoperative segmentation. However, state-of-the-art registration methods such as VoxelMorph [7] and others do not necessarily guarantee a diffeomorphic [8] displacement field that ensures the continuity of the displacement field inside the parenchyma where the intensity is quite homogeneous on the NCCT. To assess this assumption, a VoxelMorph[1] network was trained on the *Subject 1* of our porcine data sets. We trained the network with both MSE and smoothness losses during 100 epochs and given a batch of size 4. Results are illustrated below in Fig. 7. While the VoxelMorph network accurately registers the liver shape, the displacement field is almost null in the region of vessels inside the parenchyma. Therefore, the preoperative vessel segmentation is not correctly transferred into the intraoperative image.

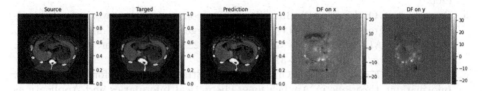

Fig. 7. Illustration of Voxelmorph registration between NCCT preoperative and intraoperative images. The prediction is the output of VoxelMorph method. DF stands for the displacement fields on x and y predicted by VoxelMorph method.

4   Conclusion

In this paper, we proposed a method for augmenting intra-operative NCCT images as a means to improve needle CT-guided techniques while reducing the need for contrast agent injection during tumor ablation procedures, or other needle-based procedures. Our method uses a U-net architecture to learn local vessel tree image features in the NCCT by leveraging the known vessel tree geometry and topology extracted from a matching CCT image. The augmented CT is generated by fusing the predicted vessel tree with the NCCT. Our method is validated on several porcine images, achieving an average dice score of 0.81 on the predicted vessel tree location. In addition, it demonstrates robustness even in the presence of large deformations between the preoperative and intraoperative images. Our future steps will essentially involve applying this method to patient data and perform a small user study to evaluate the usefulness and limitations of our approach.

Aknowledgments. This work was partially supported by French state funds managed by the ANR under reference ANR-10-IAHU-02 (IHU Strasbourg). The authors would like to thank Paul Baksic and Robin Enjalbert for proofreading the manuscript.

---

[1] https://github.com/voxelmorph/voxelmorph.

References

1. Izumi, N., et al.: Risk factors for distant recurrence of hepatocellular carcinoma in the liver after complete coagulation by microwave or radiofrequency ablation. Cancer **91**(5), 949–56 (2001). PMID: 11251946
2. Lencioni, R., et al.: Early-stage hepatocellular carcinoma in patients with cirrhosis: long-term results of percutaneous image-guided radiofrequency ablation. Radiology **234**(3), 961–7 (2005)
3. Schullian, P., Johnston, E.W., Putzer, D., Eberle, G., Laimer, G., Bale, R.: Safety and efficacy of stereotactic radiofrequency ablation for very large (*gt* 8 cm) primary and metastatic liver tumors. Sci. Rep. **10**(1), 1618 (2020)
4. Antiga, L., Ene-Iordache, B.: Centerline computation and geometric analysis of branching tubular surfaces with application to blood vessel modeling. WSCG (2003)
5. Ronneberger, O., Fischer, P., Brox, T.: U-Net: convolutional networks for biomedical image segmentation. In: Navab, N., Hornegger, J., Wells, W.M., Frangi, A.F. (eds.) MICCAI 2015. LNCS, vol. 9351, pp. 234–241. Springer, Cham (2015). https://doi.org/10.1007/978-3-319-24574-4_28
6. Devroye, L., Kruszewski, P.: On the Horton-Strahler number for random tries. Informatique théorique et Applications/Theoretical Informaties and Applications **305**(5), 443–456 (1996)
7. Balakrishnan, G., Zhao, A., Sabuncu, M.R., Guttag, J., Dalca, A.V.: VoxelMorph: a learning framework for deformable medical image registration. IEEE Trans. Med. Imaging **38**, 1788–800 (2019)
8. Cao, Y., Miller, M.I., Winslow, R.L., Younes, L.: Large deformation diffeomorphic metric mapping of vector fields. IEEE Trans. Med. Imaging **24**(9), 1216–1230 (2005)
9. Klein, S., Staring, M., Murphy, K., Viergever, M.A., Pluim, J.P.: Elastix: a toolbox for intensity-based medical image registration. IEEE Trans. Med. Imaging **29**(1), 196–205 (2009)
10. Avants, B.B., Tustison, N.J., Song, G., Cook, P.A., Klein, A., Gee, J.C.: A reproducible evaluation of ants similarity metric performance in brain image registration. Neuroimage **54**(3), 2033–2044 (2011)
11. Cao, X., Yang, J., Wang, L., Xue, Z., Wang, Q., Shen, D.: Deep learning based inter-modality image registration supervised by intra-modality similarity. In: Shi, Y., Suk, H.-I., Liu, M. (eds.) MLMI 2018. LNCS, vol. 11046, pp. 55–63. Springer, Cham (2018). https://doi.org/10.1007/978-3-030-00919-9_7
12. Yang, X., Kwitt, R., Styner, M., Niethammer, M.: Quicksilver: fast predictive image registration-a deep learning approach. Neuroimage **158**, 378–396 (2017)
13. Kim, B., Kim, D.H., Park, S.H., Kim, J., Lee, J.G., Ye, J.C.: CycleMorph: cycle consistent unsupervised deformable image registration. Med. Image Anal. **71**, 102036 (2021)
14. Sokooti, H., de Vos, B., Berendsen, F., Lelieveldt, B.P.F., Išgum, I., Staring, M.: Nonrigid image registration using multi-scale 3D convolutional neural networks. In: Descoteaux, M., Maier-Hein, L., Franz, A., Jannin, P., Collins, D.L., Duchesne, S. (eds.) MICCAI 2017. LNCS, vol. 10433, pp. 232–239. Springer, Cham (2017). https://doi.org/10.1007/978-3-319-66182-7_27
15. Suganyadevi, S., Seethalakshmi, V., Balasamy, K.: A review on deep learning in medical image analysis. Int. J. Multimed. Info. Retr. **11**, 19–38 (2022)

16. Ramesh, K., Kumar, G.K., Swapna, K., Datta, D., Rajest, S.S.: A review of medical image segmentation algorithms. EAI Endorsed Trans Perv Health Tech. 7 (2021)
17. Gatys, L.A., Ecker, A.S., Bethge, M.: A neural algorithm of artistic style. arXiv preprint arXiv:1508.06576 (2015)
18. Gatys, L.A., Ecker, A.S., Bethge, M.: Image style transfer using convolutional neural networks. In: Proceedings of IEEE Conference on Computer Vision and Pattern Recognition (CVPR) (2016)
19. Johnson, J., Alahi, A., Fei-Fei, L.: Perceptual losses for real-time style transfer and super-resolution. In: Leibe, B., Matas, J., Sebe, N., Welling, M. (eds.) ECCV 2016. LNCS, vol. 9906, pp. 694–711. Springer, Cham (2016). https://doi.org/10.1007/978-3-319-46475-6_43
20. Seo, M., et al.: Neural contrast enhancement of CT image. In: 2021 IEEE Winter Conference on Applications of Computer Vision (WACV), Waikoloa, HI, USA (2021)
21. Klein, S., Staring, M., Murphy, K., Viergever, M.A., Pluim, J.P.W.: elastix: a toolbox for intensity based medical image registration. IEEE Trans. Med. Imaging 29(1), 196–205 (2010)
22. Shamonin, D.P., Bron, E.E., Lelieveldt, B.P.F., Smits, M., Klein, S., Staring, M.: Fast parallel image registration on CPU and GPU for diagnostic classification of Alzheimer's disease. Front. Neuroinform. 7(50), 1–15 (2014)
23. Haralick, R.M., Sternberg, S.R., Zhuang, X.: Image analysis using mathematical morphology. IEEE Trans. Pattern Anal. Mach. Intell. PAMI-9(4), 532–550 (1987)
24. Yan, K., Wang, X., Lu, L., Summers, R.M.: DeepLesion: automated mining of large-scale lesion annotations and universal lesion detection with deep learning. J. Med. Imaging (Bellingham), 5(3), 036501 (2018). https://doi.org/10.1117/1.JMI.5.3.036501
25. Soler, L., et al.: 3D image reconstruction for comparison of algorithm database: a patient specific anatomical and medical image database. Technical report IRCAD, Strasbourg, France (2010)
26. Arnolli, M.M., Buijze, M., Franken, M., de Jong, K.P., Brouwer, D.M., Broeders, I.A.M.J.: System for CT-guided needle placement in the thorax and abdomen: a design for clinical acceptability, applicability and usability: system for CT-guided needle placement in the thorax and abdomen. Int. J. Med. Robot. Comput. Assist. Surg. 14(1), e1877 (2018)

LABRAD-OR: Lightweight Memory Scene Graphs for Accurate Bimodal Reasoning in Dynamic Operating Rooms

Ege Özsoy[(✉)], Tobias Czempiel, Felix Holm, Chantal Pellegrini, and Nassir Navab

Computer Aided Medical Procedures, Technische Universität München, Munich, Germany
ege.oezsoy@tum.de

Abstract. Modern surgeries are performed in complex and dynamic settings, including ever-changing interactions between medical staff, patients, and equipment. The holistic modeling of the operating room (OR) is, therefore, a challenging but essential task, with the potential to optimize the performance of surgical teams and aid in developing new surgical technologies to improve patient outcomes. The holistic representation of surgical scenes as semantic scene graphs (SGG), where entities are represented as nodes and relations between them as edges, is a promising direction for fine-grained semantic OR understanding. We propose, for the first time, the use of temporal information for more accurate and consistent holistic OR modeling. Specifically, we introduce memory scene graphs, where the scene graphs of previous time steps act as the temporal representation guiding the current prediction. We design an end-to-end architecture that intelligently fuses the temporal information of our lightweight memory scene graphs with the visual information from point clouds and images. We evaluate our method on the 4D-OR dataset and demonstrate that integrating temporality leads to more accurate and consistent results achieving an +5% increase and a new SOTA of 0.88 in macro F1. This work opens the path for representing the entire surgery history with memory scene graphs and improves the holistic understanding in the OR. Introducing scene graphs as memory representations can offer a valuable tool for many temporal understanding tasks. We will publish our code upon acceptance. The code is publicly available at https://github.com/egeozsoy/LABRAD-OR.

Keywords: Semantic Scene Graphs · Memory Scene Graphs · 3D surgical scene Understanding · Temporal OR Understanding

1 Introduction

Surgical procedures are becoming increasingly complex, requiring intricate coordination between medical staff, patients, and equipment [11,12]. Effective oper-

Supplementary Information The online version contains supplementary material available at https://doi.org/10.1007/978-3-031-43996-4_29.

H. Greenspan et al. (Eds.): MICCAI 2023, LNCS 14228, pp. 302–311, 2023.
https://doi.org/10.1007/978-3-031-43996-4_29

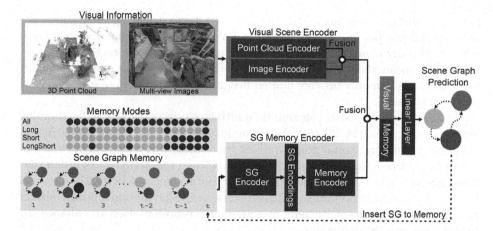

Fig. 1. Overview of our bimodal scene graph generation architecture. We use the visual information extracted from point clouds and images and temporal information represented as memory scene graphs resulting in more accurate and consistent predictions.

ating room (OR) management is critical for improving patient outcomes, optimizing surgical team performance, and developing new surgical technologies [10]. Scene understanding, particularly in dynamic OR environments, is a challenging task that requires holistic and semantic modeling [11,16], where both the coarse and the fine-level activities and interactions are understood. While many recent works addressed different aspects of this understanding, such as surgical phase recognition [2,9,22], action detection [14,15], or tool detection [4,7], these approaches do not focus on holistic OR understanding. Most recently, Özsoy et al. [16] proposed a new dataset, 4D-OR, and an approach for holistic OR modeling. They model the OR using semantic scene graphs, which summarize the entire scene at each timepoint, connecting the different entities with their semantic relationships. However, they did not propose remedies for challenges caused by occlusions and visual similarities of scenes observed at different moments of the intervention. In fact, they rely only on single timepoints for OR understanding, while temporal history is a rich information source that should be utilized for improving holistic OR modeling.

In endoscopic video analysis, using temporality has become standard practice [2,5,19,22]. For surgical workflow recognition in the OR, the use of temporality has been explored in previous studies showcasing their effectiveness [6,13,18]. All of these methods utilize what we refer to as latent temporality, which is a non-interpretable, hidden feature representation. While some works utilize two-stage architectures, where the temporal stage uses precomputed features from a single timepoint neural network, others use 3D(2D + time) methods, directly considering multiple timepoints [3].

Both of these approaches have some downsides. The two-stage approaches are not trained end-to-end, potentially limiting their performance. Additionally, certain design choices must be made regarding which feature from every timepoint should be used as a temporal summary. For scene graph generation, this can be challenging, as most SGG architectures work with representations per relation and not per scene. The end-to-end 3D methods, on the other hand, are computationally expensive both in training and inference and practical hardware limitations mean they can only effectively capture short-term context. Finally, these methods can only provide limited insight into which temporal information is the most useful.

In the computer vision community, multiple studies on scene understanding using scene graphs [24,25] have been conducted. Ji et al. [8] proposes Action Genome, a temporal scene graph dataset containing 10K videos. They demonstrate how scene graphs can be utilized to improve the action recognition performance of their model. While there have been some works [1,21] on using temporal visual information to enhance scene graph predictions, none consider the previous scene graph outputs as a temporal representation to enhance the future scene graph predictions.

In this paper, we propose LABRAD-OR(Lightweight Memory Scene Graphs for Accurate Bimodal ReAsoning in Dynamic Operating Rooms), a novel and lightweight approach for generating accurate and consistent scene graphs using the temporal information available in OR recordings. To this end, we introduce the concept of memory scene graphs, where, for the first time, the scene graphs serve as both the output and the input, integrating temporality into the scene graph generation. Our motivation behind using scene graphs to represent memory is twofold. First, by design, they summarize the most relevant information of a scene, and second, they are lightweight and interpretable, unlike a latent feature-based representation. We design an end-to-end architecture that fuses this temporal information with visual information. This bimodal approach not only leads to significantly higher scene graph generation accuracy than the state-of-the-art but also to better inter-timepoint consistency. Additionally, by choosing lightweight architectures to encode the memory scene graphs, we can integrate the entire temporal information, as human-interpretable scene graphs, with only 40% overhead. We show the effectiveness of our approach through experiments and ablation studies.

2   Methodology

In this section, we introduce our memory scene graph-based temporal modeling approach (LABRAD-OR), a novel bimodal scene graph generation architecture for holistic OR understanding, where both the visual information, as well as the temporal information in the form of memory scene graphs, are utilized. Our architecture is visualized in Fig. 1.

2.1   Single Timepoint Scene Graph Generation

We build up on the 4D-OR [16] method, which uses a single timepoint for generating semantic scene graphs. The 4D-OR method extracts human and object poses and uses them to compute point cloud features for all object pairs. Additionally, image features are incorporated into the embedding to improve the recognition of details. These representations are then further processed to generate object and relation classes and are fused to generate a comprehensive scene graph.

2.2 Scene Graphs as Memory Representations

In this study, we investigate the potential of using scene graphs from previous timepoints, which we refer to as "memory scene graphs", to inform the current prediction. Unlike previous research that treated scene graphs only as the final output, we use them both as input and output. Scene graphs are particularly well-suited for encoding scene information, as they are low-dimensional and interpretable while capturing and summarizing complex semantics. To create a memory representation at a timepoint T, we use the predicted scene graphs from timepoints 0 to T-1 and employ a neural network to compute a feature representation. This generic approach allows us to easily fuse the scene graph memory with other modalities, such as images or point clouds.

Memory Modes: While our efficient memory representation allows us to look at all the previous timesteps, this formulation has two downsides. Surgical duration differs between procedures, and despite the efficiency of scene graphs, prolonged interventions can still be costly. Second, empirically we find that seeing the entire memory leads to prolonged training time and can cause overfitting. To address this, we propose four different memory modes, "All", "Short", "Long", and "LongShort". The entire surgical history is visible only in the "All" mode. In the "Short" mode, only the previous S scene graphs are utilized, while in the "Long" mode, every $S.th$ scene graph is selected using striding. In "LongShort" mode, both "Long" and "Short" modes are combined. The reasoning behind this approach is that short-term context should be observed in detail, while long-term context can be viewed more sparsely. This reduces computational overhead compared to "All" and leads to better results with less overfitting, as observed empirically. The value of S is highly dependent on the dataset and the surgical procedures under analysis (Fig. 2).

2.3 Architecture Design

We extract the visual information from point clouds and, optionally, images using a visual scene encoder, as described in Sect. 2.1. To integrate the temporal information, we convert each memory scene graph into a feature vector using a graph neural network. Then we use a Transformer block [23] to summarize the feature vectors from the $\#T$ memory scene graphs into a single feature vector, which we refer to as the memory representation. Finally, this representation is concatenated with the visual information, forming a bimodal representation. Intuitively, this allows our architecture to consider both the long-term history, such as previous key surgical steps and the short-term history, such as what was just happening.

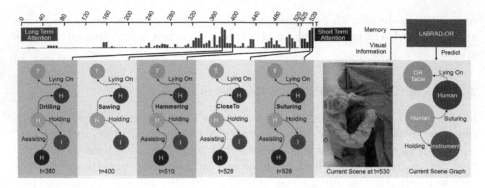

Fig. 2. Visualization of both the "Short" and "Long" memory attention while predicting for the timepoint t = 530. While "Short" attends naturally to the nearest scene graphs, "Long" seems to be concentrating on previous key moments of the surgery, such as "drilling", "sawing", "and hammering". The graphs are simplified for clarity. The shown "current scene graph" is the correct prediction of our model.

Memory Augmentations: While we use the predicted scene graphs from previous timepoints for inference, as these are not available during training, we use the ground truth scene graphs for training. However, training with ground truth scene graphs and evaluating with predicted scene graphs can decrease test performance, as the predicted scene graphs are imperfect. To increase our robustness towards this, we utilize memory augmentations during training. Concretely, we randomly replace part of either the short-term memory (timepoints closest to the timepoint of interest) or the long-term memory (timepoints further away from the timepoint of interest) with a special "UNKNOWN" token. Intuitively, this forces our model to rely on the remaining information and better deal with wrong predictions in the memory during inference.

Timepoint of Interest(ToI) Positional Ids: Transformers [23] employ positional ids to encode the *absolute location* of each feature in a sequence. However, as we are more interested in the relative distance of the features, we introduce Timepoint of Interest(ToI) positional ids to encode the distance of every feature to the current timepoint T. This allows our network to assign meaning to the relative distance to other timepoint features rather than their *absolute locations*.

Multitask Learning: In the scene graph generation task, the visual and temporal information are used together. In practice, we found it valuable to introduce a secondary task, which can be solved only using temporal information. We propose the task of "main action recognition", where instead of the scene graph, only the interaction of the head surgeon to the patient is predicted, such as "sawing" or "drilling". During training, in addition to fusing the memory representation with the visual information, we use a fully connected layer to estimate the main action from the memory representation directly. Learning both scene graph generation and main action recognition tasks simultaneously gives a more direct signal to the memory encoder, resulting in faster training and improved performance.

Table 1. We compare our results to the current SOTA, 4D-OR, on the test set. We experimented with different hyperparameters and found that longer training can improve the 4D-OR results. We report both the original 4D-OR, and the longer trained results, indicated by †, and a latent-based temporality(LBT) baseline, and compare LABRAD-OR to them. All methods use both point clouds and images as visual input.

Method	4D-OR [16]	4D-OR† [16]	LBT	LABRAD-OR
Macro F1	0.75	0.83	0.86	**0.88**

3   Experiments

Dataset: We use the 4D-OR [16] dataset following the official train, validation, and test splits. It comprises ten simulated knee surgeries recorded using six Kinect cameras at 1 fps. Both the 3D point cloud, as well as multiview images are provided for all 6734 scenes. Each scene additionally includes a semantic scene graph label, as well as clinical role labels for staff.

Model Training: Our architecture consists of two components implemented in PyTorch 1.10, a visual model and a memory model. For our visual model, we use the current SOTA model from 4D-OR [16], which uses Pointnet++ [17] as the point cloud encoder, and EfficientNet-B5 [20] as the image encoder. We could improve the original 4D-OR results through longer training than in the original code. As our memory model, we use a combination of Graphormer [26], to extract features from individual scene graphs and Transformers [23], to fuse the features into one memory representation. The visual scene encoder is initialized in all our experiments with the best-performing visual-only model weights. We use the provided human and object pose predictions from 4D-OR and stick to their training setup and evaluation metrics. We use memory augmentations, timepoint of interest positional ids, end-to-end training, and multitask learning. The memory encoders are purposefully designed to be lightweight and fast. Therefore, we use a hidden dimension of 80 and only two layers. We use S, to control both the stride of the "Long" mode and the window size of the "Short" mode and set it to 5. The choice of S ensures we do not miss any phase, as all phases last longer than 5 timepoints while reducing the computational cost. Unless otherwise specified, we use "LongShort" as memory mode and train all models until the validation performance converges.

Evaluation Metrics: We use the official evaluation metrics from 4D-OR for semantic scene graph generation and the role prediction downstream tasks. In both cases, a macro F1 over all the classes is computed. Further, we introduce a consistency metric, where first, for each timepoint, a set of predicates P_t, such as {"assisting", "drilling", "cleaning"} is extracted from the scene graphs. Then, for two timepoints T and T-1, the intersection of union(IoU) between P_t and P_{t-1} is computed. This is repeated for all pairs of adjacent timepoints in a sequence, and the IoU score is averaged over them to calculate the consistency score.

308   E. Özsoy et al.

Table 2. We demonstrate the impact of temporal information on the consistency of our results. We compare only using the point cloud(PC), using images(Img) in addition to point clouds, and temporality(T). We also show the ground truth(GT) consistency score, which should be considered the ceiling for all methods.

Method	PC	PC+Img	PC+T	PC+Img+T	GT
Consistency	0.83	0.84	0.86	**0.87**	0.9

Fig. 3. Qualitative example on the improvement of the scene graph consistency. For clarity, only the "main action" is shown, while only relying on the visual information (V) compared to also using our proposed SG memory (M).

4 Results and Discussion

Scene Graph Generation. In Table 1, we compare our best-performing model LABRAD-OR against the previous SOTA on 4D-OR. We build a latent-based temporal baseline (LBT), which uses a mean of the pairwise relation features as representation per timepoint, which then gets processed analogously with a Transformer architecture. Overall, LABRAD-OR increases the F1 results for all predicates, significantly increasing SOTA from 0.75 F1 to 0.88 F1. We also show that LABRAD-OR improves the F1 score compared to both the longer trained 4D-OR and LBT by 5% and 2%, respectively, demonstrating both the value of temporality for holistic OR modeling as well as the effectiveness of memory scene graphs. Additionally, in Table 2, we show the consistency improvements achieved by using temporal information, from 0.84 to 0.87. Notably, a perfect consistency score is not 1.0, as the scene graph naturally changes over the surgery. Considering the ground truth consistency is at 0.9, LABRADOR-OR (0.87) exhibits superior consistency compared to the baselines without being excessively smooth. A qualitative example of the improvement can be seen in Fig. 3. While the visual-only model confuses "suturing" for "cleaning" or "touching", our LABRAD-OR model with memory scene graphs identifies all correctly "suturing".

Clinical Role Prediction. We also compare LABRAD-OR to 4D-OR on the downstream task of role prediction in Table 5, where the only difference between the two is the improved predicted scene graphs used as input for the downstream task. We improve the results by 4%, showing our improvements in scene graph generation also translate to downstream improvements.

Ablation Studies. We conduct multiple ablation studies to motivate our design choices. In Table 3, we demonstrate the effectiveness of the different components

Table 3. We do an ablation study on the impact of memory augmentations, Timepoint of interest (ToI) positional ids, end-to-end training, and multitask learning by individually disabling them. The experiments use the "All" memory mode and take only point cloud as visual input.

Method	F1
All Techniques	**0.86**
w/o Memory Aug	0.82
w/o ToI Pos Ids	0.83
w/o E2E	0.84
w/o Multitask	0.85

Table 4. Ablation study on using different memory modes, affecting which temporal information is seen. All experiments use memory augmentations, Timepoint of Interest (ToI) positional ids, end-to-end training and multitask learning and only take point clouds as visual input.

All	Short	Long	F1
✓			0.86
	✓		0.85
		✓	**0.87**
	✓	✓	**0.87**

Table 5. Comparison of LABRAD-OR and 4D-OR on the downstream task of role prediction.

Method	4D-OR [16]	LABRAD-OR
Macro F1	0.85	**0.89**

of our method and see that both memory augmentations and ToI positional ids are crucial and significantly contribute to the performance, whereas E2E and multitask have less but still measurable impact. We note that multitask learning also helps in stabilizing the training. In Table 4, we ablate the different memory modes, "Short", "Long", "LongShort", and "All", and find that the strided "Long" mode is the most important. Where the "Short" mode can often lead to insufficient contextual information, "All" can lead to overfitting. Both the "Long" and "LongShort" perform similarly and are more efficient than "All". We use "LongShort" as our default architecture to guarantee no short-term history is overseen. Comparing the "LongShort" F1 result of 0.87 when only using point clouds as visual input to 0.88 when using both point cloud and images, it can be seen that images still provide valuable information, but significantly less than for Özsoy et al. [16].

5 Conclusion

We propose LABRAD-OR, a novel lightweight approach based on human interpretable memory scene graphs. Our approach utilizes both the visual information from the current timepoint and the temporal information from the previous timepoints for accurate bimodal reasoning in dynamic operating rooms. Through experiments, we show that this leads to significantly improved accuracy and consistency in the predicted scene graphs and an increased score in the downstream task of role prediction. We believe LABRAD-OR offers the community an effective and efficient way of using temporal information for a holistic understanding of surgeries.

Acknowledgements. This work has been partially supported by Stryker. The authors would like to thank Carl Zeiss AG for their support of Felix Holm.

References

1. Cong, Y., Liao, W., Ackermann, H., Rosenhahn, B., Yang, M.Y.: Spatial-temporal transformer for dynamic scene graph generation. In: Proceedings of the IEEE/CVF International Conference on Computer Vision, pp. 16372–16382 (2021)
2. Czempiel, T., et al.: TeCNO: surgical phase recognition with multi-stage temporal convolutional networks. In: Martel, A.L., et al. (eds.) MICCAI 2020. LNCS, vol. 12263, pp. 343–352. Springer, Cham (2020). https://doi.org/10.1007/978-3-030-59716-0_33
3. Czempiel, T., Sharghi, A., Paschali, M., Navab, N., Mohareri, O.: Surgical workflow recognition: from analysis of challenges to architectural study. In: Karlinsky, L., Michaeli, T., Nishino, K. (eds.) Computer Vision-ECCV 2022. LNCS, vol. 13803, pp. 556–568. Springer, Cham (2023). https://doi.org/10.1007/978-3-031-25066-8_32
4. Ding, H., Zhang, J., Kazanzides, P., Wu, J.Y., Unberath, M.: Carts: causality-driven robot tool segmentation from vision and kinematics data. In: Wang, L., Dou, Q., Fletcher, P.T., Speidel, S., Li, S. (eds.) Medical Image Computing and Computer Assisted Intervention-MICCAI 2022. LNCS, vol. 13437, pp. 387–398. Springer, Cham (2022). https://doi.org/10.1007/978-3-031-16449-1_37
5. Gao, X., Jin, Y., Long, Y., Dou, Q., Heng, P.-A.: Trans-SVNet: accurate phase recognition from surgical videos via hybrid embedding aggregation transformer. In: de Bruijne, M., et al. (eds.) MICCAI 2021. LNCS, vol. 12904, pp. 593–603. Springer, Cham (2021). https://doi.org/10.1007/978-3-030-87202-1_57
6. Jamal, M.A., Mohareri, O.: Multi-modal unsupervised pre-training for surgical operating room workflow analysis. In: Wang, L., Dou, Q., Fletcher, P.T., Speidel, S., Li, S. (eds.) Medical Image Computing and Computer Assisted Intervention-MICCAI 2022. LNCS, vol. 13437, pp. 453–463. Springer, Cham (2022). https://doi.org/10.1007/978-3-031-16449-1_43
7. Jha, D., et al.: Kvasir-instrument: diagnostic and therapeutic tool segmentation dataset in gastrointestinal endoscopy. In: Lokoč, J., et al. (eds.) MMM 2021. LNCS, vol. 12573, pp. 218–229. Springer, Cham (2021). https://doi.org/10.1007/978-3-030-67835-7_19
8. Ji, J., Krishna, R., Fei-Fei, L., Niebles, J.C.: Action genome: actions as compositions of spatio-temporal scene graphs. In: Proceedings of the IEEE/CVF Conference on Computer Vision and Pattern Recognition, pp. 10236–10247 (2020)
9. Jin, Y., et al.: Multi-task recurrent convolutional network with correlation loss for surgical video analysis. Med. Image Anal. **59**, 101572 (2020)
10. Kennedy-Metz, L.R., et al.: Computer vision in the operating room: opportunities and caveats. IEEE Trans. Med. Robot. Bionics **3**, 2–10 (2020). https://doi.org/10.1109/TMRB.2020.3040002
11. Lalys, F., Jannin, P.: Surgical process modelling: a review. Int. J. Comput. Assist. Radiol. Surg. **9**, 495–511 (2014)

12. Maier-Hein, L., et al.: Surgical data science for next-generation interventions. Nat. Biomed. Eng. **1**(9), 691–696 (2017)
13. Mottaghi, A., Sharghi, A., Yeung, S., Mohareri, O.: Adaptation of surgical activity recognition models across operating rooms. In: Wang, L., Dou, Q., Fletcher, P.T., Speidel, S., Li, S. (eds.) Medical Image Computing and Computer Assisted Intervention-MICCAI 2022. LNCS, vol. 13437, pp. 530–540. Springer, Cham (2022). https://doi.org/10.1007/978-3-031-16449-1_51
14. Nwoye, C.I., et al.: Recognition of instrument-tissue interactions in endoscopic videos via action triplets. In: Martel, A.L., et al. (eds.) MICCAI 2020. LNCS, vol. 12263, pp. 364–374. Springer, Cham (2020). https://doi.org/10.1007/978-3-030-59716-0_35
15. Nwoye, C.I., et al.: Rendezvous: attention mechanisms for the recognition of surgical action triplets in endoscopic videos. Med. Image Anal. **78**, 102433 (2022)
16. Özsoy, E., Örnek, E.P., Eck, U., Czempiel, T., Tombari, F., Navab, N.: 4D-or: semantic scene graphs for or domain modeling. In: Wang, L., Dou, Q., Fletcher, P.T., Speidel, S., Li, S. (eds.) Medical Image Computing and Computer Assisted Intervention-MICCAI. LNCS, vol. 13437, pp. 475–485. Springer, Cham (2022). https://doi.org/10.1007/978-3-031-16449-1_45
17. Qi, C.R., Yi, L., Su, H., Guibas, L.J.: Pointnet++: deep hierarchical feature learning on point sets in a metric space. In: Advances in Neural Information Processing Systems, vol. 30 (2017)
18. Sharghi, A., Haugerud, H., Oh, D., Mohareri, O.: Automatic operating room surgical activity recognition for robot-assisted surgery. In: Martel, A.L., et al. (eds.) MICCAI 2020. LNCS, vol. 12263, pp. 385–395. Springer, Cham (2020). https://doi.org/10.1007/978-3-030-59716-0_37
19. Sharma, S., Nwoye, C.I., Mutter, D., Padoy, N.: Rendezvous in time: an attention-based temporal fusion approach for surgical triplet recognition. arXiv preprint arXiv:2211.16963 (2022)
20. Tan, M., Le, Q.: Efficientnet: rethinking model scaling for convolutional neural networks. In: International Conference on Machine Learning, pp. 6105–6114. PMLR (2019)
21. Teng, Y., Wang, L., Li, Z., Wu, G.: Target adaptive context aggregation for video scene graph generation. In: Proceedings of the IEEE/CVF International Conference on Computer Vision, pp. 13688–13697 (2021)
22. Twinanda, A.P., Shehata, S., Mutter, D., Marescaux, J., De Mathelin, M., Padoy, N.: Endonet: a deep architecture for recognition tasks on laparoscopic videos. IEEE Trans. Med. Imaging **36**(1), 86–97 (2016)
23. Vaswani, A., et al.: Attention is all you need. In: Advances in Neural Information Processing Systems, vol. 30 (2017)
24. Wald, J., Dhamo, H., Navab, N., Tombari, F.: Learning 3D semantic scene graphs from 3D indoor reconstructions. In: Proceedings of the IEEE/CVF Conference on Computer Vision and Pattern Recognition, pp. 3961–3970 (2020)
25. Xu, D., Zhu, Y., Choy, C.B., Fei-Fei, L.: Scene graph generation by iterative message passing. In: Proceedings of the IEEE Conference on Computer Vision and Pattern Recognition, pp. 5410–5419 (2017)
26. Ying, C., et al.: Do transformers really perform badly for graph representation? In: Advances in Neural Information Processing Systems, vol. 34 (2021)

Pelvic Fracture Segmentation Using a Multi-scale Distance-Weighted Neural Network

Yanzhen Liu[1] , Sutuke Yibulayimu[1] , Yudi Sang[3] , Gang Zhu[3],
Yu Wang[1(✉)] , Chunpeng Zhao[2], and Xinbao Wu[2]

[1] Key Laboratory of Biomechanics and Mechanobiology, Ministry of Education,
Beijing Advanced Innovation Center for Biomedical Engineering, School of Biological
Science and Medical Engineering, Beihang University, Beijing 100083, China
{yanzhenliu,wangyu}@buaa.edu.cn
[2] Department of Orthopaedics and Traumatology, Beijing Jishuitan Hospital,
Beijing, China
[3] Beijing Rossum Robot Technology Co., Ltd., Beijing, China

Abstract. Pelvic fracture is a severe type of high-energy injury. Segmentation of pelvic fractures from 3D CT images is important for trauma diagnosis, evaluation, and treatment planning. Manual delineation of the fracture surface can be done in a slice-by-slice fashion but is slow and error-prone. Automatic fracture segmentation is challenged by the complex structure of pelvic bones and the large variations in fracture types and shapes. This study proposes a deep-learning method for automatic pelvic fracture segmentation. Our approach consists of two consecutive networks. The anatomical segmentation network extracts left and right ilia and sacrum from CT scans. Then, the fracture segmentation network further isolates the fragments in each masked bone region. We design and integrate a distance-weighted loss into a 3D U-net to improve accuracy near the fracture site. In addition, multi-scale deep supervision and a smooth transition strategy are used to facilitate training. We built a dataset containing 100 CT scans with fractured pelvis and manually annotated the fractures. A five-fold cross-validation experiment shows that our method outperformed max-flow segmentation and network without distance weighting, achieving a global Dice of 99.38%, a local Dice of 93.79%, and an Hausdorff distance of 17.12 mm. We have made our dataset and source code publicly available and expect them to facilitate further pelvic research, especially reduction planning.

Keywords: CT segmentation · Pelvic fracture · Reduction planning

1 Introduction

Pelvic fracture is a severe type of high-energy injury, with a fatality rate greater than 50%, ranking the first among all complex fractures [8,16]. Surgical planning and reduction tasks are challenged by the complex pelvic structure, as well as the surrounding muscle groups, ligaments, neurovascular and other tissues. Robotic fracture reduction surgery has been studied and put into clinical use in recent

© The Author(s), under exclusive license to Springer Nature Switzerland AG 2023
H. Greenspan et al. (Eds.): MICCAI 2023, LNCS 14228, pp. 312–321, 2023.
https://doi.org/10.1007/978-3-031-43996-4_30

years, and has successfully increased reduction precision and reduced radiation exposure [2]. Accurate segmentation of pelvic fracture is required in both manual and automatic reduction planning, which aim to find the optimal target location to restore the healthy morphology of pelvic bones. Segmenting pelvic fragments from CT is challenging due to the uncertain shape and irregular position of the bone fragments and the complex collision fracture surface. Therefore, surgeons typically annotate the anatomy of pelvic fractures in a semi-automatic way. First, by tuning thresholds and selecting seed points, adaptive thresholding and region-growing methods are used to extract bone regions [1,11,15]. Then, the fracture surfaces are manually delineated by outlining the fragments in 3D view or even modifying the masks in a slice-by-slice fashion. Usually, this tedious process can take more than 30 min, especially when the fractured fragments are collided or not completely separated.

Several studies have been proposed to provide more efficient tools for operators. A semi-automatic graph-cut method based on continuous max-flow has been proposed for pelvic fracture segmentation, but it still requires the manual selection of seed points and trail-and-error [4,22]. Fully automatic max-flow segmentation based on graph cut and boundary enhancement filter is useful when fragments are separated, but it often fails on fragments that are collided or compressed [9,19]. Learning-based bone segmentation has been successfully applied to various anatomy, including the pelvis, rib, skull, etc. [12,13]. Some deep learning methods have been proposed to detect fractures [17,18,21], but the output from these methods cannot provide a fully automated solution for subsequent operations. In FracNet, rib fracture detection was formulated as a segmentation problem, but with a resultant Dice of 71.5%, it merely outlined the fracture site coarsely without delineating the fracture surface [7]. Learning-based methods that directly deal with fracture segmentation have rarely been studied.

Fracture segmentation is still a challenging task for the learning-based method because (1) compared to the more common organ/tumor segmentation tasks where the model can implicitly learn the shape prior of an object, it is difficult to learn the shape information of a bone fragment due to the large variations in fracture types and shapes [10]; (2) the fracture surface itself can take various forms including large space (fragments isolated and moved), small gap (fragments isolated but not moved), crease (fragments not completely isolated), compression (fragments collided), and their combinations, resulting in quite different image intensity profiles around the fracture site; and (3) the variable number of bone fragments in pelvic fracture makes it difficult to prescribe a consistent labeling strategy that applies to every type and case.

This paper proposes a deep learning-based method to segment pelvic fracture fragments from preoperative CT images automatically. Our major contribution includes three aspects. (1) We proposed a complete automatic pipeline for pelvic fractures segmentation, which is the first learning-based pelvic fracture segmentation method to the best of our knowledge. (2) We designed a novel multi-scale distance-weighted loss and integrated it into the deeply supervised training of the fracture segmentation network to boost accuracy near the

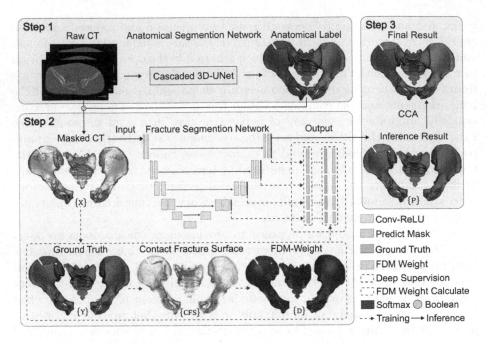

Fig. 1. Overview of the proposed pelvic fracture segmentation framework.

fracture site. (3) We established a comprehensive pelvic fracture CT dataset and provided ground-truth annotations. Our dataset and source code are publicly available at https://github.com/YzzLiu/FracSegNet. We expect them to facilitate further pelvis-related research, including but not limited to fracture identification, segmentation, and subsequent reduction planning.

2   Methods

2.1   The Overall Segmentation Framework

Our study aims to automatically segment the major and minor fragments of target bones (left and right ilia and sacrum) from CT scans. As illustrated in Fig. 1, our method consists of three steps. In the first step, an anatomical segmentation network is used to extract the pelvic bones from the CT scan. With a cascaded 3D nn-Unet architecture, the network is pre-trained on a set of healthy pelvic CT images [5,13] and further refined on our fractured dataset. In the second step, a fracture segmentation network is used to separate the bone fragments from each iliac and sacral region extracted from the first step. To define a consistent labeling rule that is applicable to all fracture types, we prescribe three labels for each bone, namely the background, the main fragment, and other fragments. The main fragment is the largest fragment at the center. In the third step, isolated components are further separated and labeled, and small isolated bone fragments are removed to form the final output.

2.2   Fracture Segmentation Network

The contact fracture surface (CFS) is the part where the bones collide and overlap due to compression and impact, and is the most challenging part for human operators to draw. We are particularly concerned about the segmentation performance in this region. Therefore, we introduce guidance into the network training using fracture distance map (FDM). A 3D UNet is selected as the base model [6]. The model learns a non-linear mapping relationship $M : X \rightarrow Y$, where X and Y are the input and ground truth of a training sample, respectively.

Fracture Distance Map. The FDM is computed on the ground-truth segmentation of each data sample before training. This representation provides information about the boundary, shape, and position of the object to be segmented. First, CFS regions are identified by comparing the labels within each voxel's neighbourhood. Then, we calculate the distance of each foreground voxel to the nearest CFS as its distance value D_v, and divide it by the maximum.

$$D_v = \mathrm{I}(Y_v \geq 1)\min_{u \in CFS}||v - u||_2, \tag{1}$$

$$\hat{D}_v = \frac{D_v}{\max_{v \in V} D_v}, \tag{2}$$

where $v = (h, w, d) \in V$ is the voxel index, Y is the ground-truth segmentation, $\mathrm{I}(Y_v \geq 1)$ is the indicator function for foreground, and $\hat{D}$ is the normalized distance. The distance is then used to calculate the FDM weight $\hat{W}$ using the following formula:

$$W_v = \lambda_{back} + \mathrm{I}\,(Y_v \geq 1)\,\frac{1 - \lambda_{back}}{1 + e^{\lambda_{FDM}\hat{D}_v - 5}}, \tag{3}$$

$$\hat{W}_v = \frac{W_v \cdot |V|}{\sum_{v \in V} W_v}. \tag{4}$$

To ensure the equivalence of the loss among different samples, the weights are normalized so that the average is always 1.

FDM-Weighted Loss. The FDM weight $\hat{W}$ is then used to calculate the weighted Dice and cross-entropy losses to emphasize the performance near CFS by assigning larger weights to those pixels.

$$\mathcal{L}_{dice} = 1 - \frac{2}{|L|}\sum_{l \in L}\frac{\sum_{v \in V} \hat{W}_v P_v^l Y_v^l}{\sum_{v \in V} \hat{W}_v P_v^l + \sum_{v \in V} \hat{W}_v Y_v^l}, \tag{5}$$

$$\mathcal{L}_{ce} = -\frac{1}{|V||L|}\sum_{v \in V}\sum_{l \in L}\hat{W}_v Y_v^l log(P_v^l), \tag{6}$$

where L is the number of labels, P_v^l, Y_v^l are the network output prediction and the one-hot encoding form of the ground truth for the lth label of the vth voxel. The overall loss is their weighted sum:

$$\mathcal{L}_{total} = \mathcal{L}_{dice} + \lambda_{ce}\mathcal{L}_{ce}, \tag{7}$$

where λ_{ce} is a balancing weight.

Multi-scale Deep Supervision. We use a multi-scale deep supervision strategy in model training to learn different features more effectively [20]. The deep layers mainly capture the global features with shape/structural information, whereas the shallow layers focus more on local features that help delineate fracture surfaces. We add auxiliary losses to the decoder at different resolution levels (except the lowest resolution level). The losses are calculated using the corresponding down-sampled FDM $\hat{W}_v^n$, and down-sampled ground truth Y_v^n. We calculate the output of the nth level $\mathcal{L}^n$ by changing λ_{FDM} in Eq. (3). The λ_{FDM} of each layer decreases by a factor of 2 as the depth increases, i.e., $\lambda^{n+1} = \lambda^n/2$. In this way, the local CFS information are assigned more attention in the shallow layers, while the weights become more uniform in the deep layers.

Smooth Transition. To stabilize network training, we use a smooth transition strategy to maintain the model's attention on global features at the early stage of training and gradually shift the attention towards the fracture site as the model evolves [14]. The smooth transition dynamically adjusts the proportion of the FDM in the overall weight matrix based on the number of training iterations. The dynamic weight is calculated using the following formula:

$$W_{st} = \mathrm{I}\,(t < \tau) \left(\frac{1}{1+\delta}\mathcal{J} + \frac{\delta}{1+\delta}\hat{W} \right) + \mathrm{I}\,(t \geq \tau)\,\hat{W}, \tag{8}$$

where $\delta = -ln(1 - \frac{t}{\tau})$, $\mathcal{J}$ is an all-ones matrix with the same size as the input volume, t is the current iteration number, and τ is a hyper-parameter. The dynamic weight W_{st} is adjusted by controlling the relative proportion of $\mathcal{J}$ and $\hat{W}$. The transition terminates when the epoch reaches τ.

2.3   Post-processing

Connected component analysis (CCA) has been widely used in segmentation [3], but is usually unsuitable for fracture segmentation because of the collision between fragments. However, after identifying and removing the main central fragment from the previous step, other fragments are naturally separated. Therefore, in the post-processing step, we further isolate the remaining other fragments by CCA. The isolated components are then assigned different labels. In addition, we remove fragments smaller than a certain threshold, which has no significant impact on planning and robotic surgery.

2.4   Data and Annotation

Although large-scale datasets on pelvic segmentation have been studied in some research [13], to the best of our knowledge, currently there is no well-annotated fractured pelvic dataset publicly available. Therefore, we curated a dataset of 100 preoperative CT scans covering all common types of pelvic fractures. These data is collected from 100 patients (aged 18–74 years, 41 females) who were to undergo pelvic reduction surgery at Beijing Jishuitan Hospital between 2018 and 2022, under IRB approval (202009-04). The CT scans were acquired on a Toshiba Aquilion scanner. The average voxel spacing is $0.82 \times 0.82 \times 0.94$ mm^3. The average image shape is $480 \times 397 \times 310$.

To generate ground-truth labels for bone fragments, a pre-trained segmentation network was used to create initial segmentations for the ilium and sacrum [13]. Then, these labels were further modified and annotated by two annotators and checked by a senior expert.

3   Experiments and Results

3.1   Implementation

We compared the proposed method (FDMSS-UNet) against the network without smooth transition and deep supervision (FDM-UNet) and the network without distance weighting (UNet) in an ablation study. In addition, we also compared the traditional max-flow segmentation method. In the five-fold cross-validation, each model was trained for 2000 epochs per fold. The network input was augmented eight times by mirror flip. The learning rate in ADAM optimizer was set to 0.0001. λ_{back} was set to 0.2. λ_{ce} was set to 1. The initial λ_{FDM} was set to 16. The termination number for smooth transition τ was set to 500 epochs.

The models were implemented in PyTorch 1.12. The experiments were performed on an Intel Xeon CPU with 40 cores, a 256 GB memory, and a Quadro RTX 5000 GPU. The comprehensive code and pertinent details are provided at https://github.com/YzzLiu/FracSegNet.

3.2   Evaluation

We calculate the Dice similarity coefficient (DSC) and the 95th percentile of Hausdorff Distance (HD) to evaluation the performance. In addition, we evaluated the local Dice (LDSC) within the 10 mm range near the CFS to assess the performance in the critical areas. We reported the performance on iliac main fragment (I-main), iliac other fragment(s) (I-other), sacral main fragment (S-main), sacral other fragment(s) (S-other), and all together.

3.3   Results

Figure 2 shows a qualitative comparison among different methods. Max-flow was able to generate reasonable segmentation only when the CFS is clear and mostly

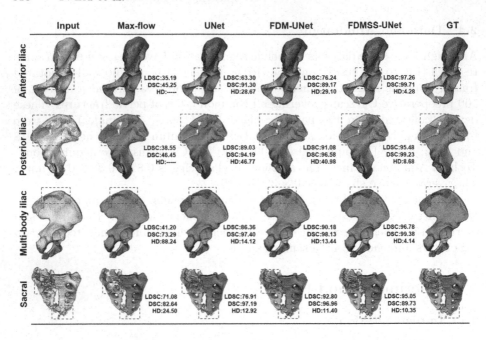

Fig. 2. Pelvic fracture segmentation results from different methods.

non-contact. UNet correctly identified the fracture fragments, but was often confused by the complicated CFS regions, resulting in imprecise fracture lines. The introduction of FDM weighting and deep supervision with smooth transition successfully improved the performance near the CFS, and achieved the overall best result.

Table 1 shows the quantitative results. The results on the main fragments were better than other fragments due to the larger proportion. The deep learning methods had much higher success rates in identifying the fragments, resulting in significantly better results in minor fragments than max-flow, with $p < 0.05$ in paired t-test. Introducing the FDM significantly improved the prediction accuracy in the CFS region ($p < 0.05$). Although FDM-UNet achieved the best LDSC results in several parts, it compromised the global performance of DSC and HD significantly, compared to FDMSS-UNet. The deep supervision and smooth transition strategies stabilized the training, and achieved the overall best results, with balanced local and global performance.

The average inference time for the fracture segmentation network was 12 s. The overall running time for processing a pelvic CT was 0.5 to 2 min, depending on the image size and the number of fractured bones.

Table 1. Quantitative segmentation results using different methods. (L)DSC is reported in percentage value, and HD is in mm. Best values are shown in bold. Statistically significant differences compared to FDMSS-UNet are indicated by * ($p < 0.05$).

		I-main	I-other	S-main	S-other	All
Max-flow	DSC	94.52±0.56*	24.67±3.72*	89.83±1.89*	23.00±3.18*	69.21±4.01*
	LDSC	74.53±0.99*	24.24±3.26*	62.06±1.91*	27.31±3.40*	54.11±3.20*
	HD95	68.56*	77.45*	45.16*	90.76*	73.95*
UNet	DSC	99.38±0.13*	92.88±1.30*	98.66±0.33*	87.91±1.55*	96.60±0.89*
	LDSC	90.68±1.22*	86.58±1.69*	84.44±1.77*	78.50±2.50*	86.86±1.70*
	HD95	43.42*	50.21*	27.14*	76.69*	46.11*
FDM-UNet	DSC	99.37±0.13*	96.15±0.79*	98.73±0.27*	92.20±0.74*	97.88±0.51*
	LDSC	94.43±0.34	**93.94±0.36**	**92.88±0.39***	91.80±0.51	**93.81±0.38**
	HD95	34.75*	52.03*	28.99*	65.26*	45.79*
FDMSS-UNet	DSC	**99.78±0.03**	**99.05±0.10**	**99.57±0.04**	**97.87±0.17**	**99.38±0.09**
	LDSC	**94.79±0.51**	93.75±0.56	91.02±1.62	**93.65±0.70**	93.79±0.78
	HD95	**13.24**	**19.43**	**14.85**	**19.89**	**17.12**

4   Discussion and Conclusion

We have introduced a pelvic fracture CT segmentation method based on deep convolutional networks. A fracture segmentation network was trained with a distance-weighted loss and multi-scale deep supervision to improve fracture surface delineation. We evaluated our method on 100 pelvic fracture CT scans and made our dataset and ground truth publicly available. The experiments demonstrated the method's effectiveness on various types of pelvic fractures. The FDM weighted loss, along with multi-scale deep supervision and smooth transition, improved the segmentation performance significantly, especially in the areas near fracture lines. Our method provides a convenient tool for pelvis-related research and clinical applications, and has the potential to support subsequent automatic fracture reduction planning.

One obstacle for deep learning-based fracture segmentation is the variable number of bone fragments in different cases. The ultimate goal of this study is to perform automatic fracture reduction planning for robotic surgery, where the main bone fragment is held and moved to the planned location by a robotic arm, whereas minor fragments are either moved by the surgeons' hands or simply ignored. In such a scenario, we found isolating the minor fragments usually unnecessary. Therefore, to define a consistent labeling strategy in annotation, we restrict the number of fragments of each bone to three. This rule of labelling applies to all 100 cases we encountered. Minor fragments within each label can be further isolated by CCA or handcrafting when needed by other tasks.

We utilize a multi-scale distance-weighted loss to guide the network to learn features near the fracture site more effectively, boosting the local accuracy without compromising the overall performance. In semi-automatic pipelines where human operators are allowed to modify and refine the network predictions, an accurate initial segmentation near the fracture site is highly desirable because the fracture surface itself is much more complicated, often intertwined and hard to draw by manual operations. Therefore, with the emphasis on fracture surface, even when the prediction from the network is inaccurate, manual operations on 3D view can suffice for most modifications, eliminating the need for the inefficient slice-by-slice handcrafting. In future studies, we plan to integrate the proposed method into an interactive segmentation and reduction planning software and evaluate the overall performance.

Acknowledgements. This work was supported by the Beijing Science and Technology Project (Grants No. Z221100003522007 and Z201100005420033).

References

1. Fornaro, J., Székely, G., Harders, M.: Semi-automatic segmentation of fractured pelvic bones for surgical planning. In: Bello, F., Cotin, S. (eds.) ISBMS 2010. LNCS, vol. 5958, pp. 82–89. Springer, Heidelberg (2010). https://doi.org/10.1007/978-3-642-11615-5_9

2. Ge, Y., Zhao, C., Wang, Y., Wu, X.: Robot-assisted autonomous reduction of a displaced pelvic fracture: a case report and brief literature review. J. Clin. Med. **11**(6), 1598 (2022)

3. Ghimire, K., Chen, Q., Feng, X.: Head and neck tumor segmentation with deeply-supervised 3D UNet and progression-free survival prediction with linear model. In: Andrearczyk, V., Oreiller, V., Hatt, M., Depeursinge, A. (eds.) HECKTOR 2021. LNCS, vol. 13209, pp. 141–149. Springer, Cham (2022). https://doi.org/10.1007/978-3-030-98253-9_13

4. Han, R., et al.: Fracture reduction planning and guidance in orthopaedic trauma surgery via multi-body image registration. Med. Image Anal. **68**, 101917 (2021)

5. Isensee, F., Jaeger, P.F., Kohl, S.A., Petersen, J., Maier-Hein, K.H.: nnU-Net: a self-configuring method for deep learning-based biomedical image segmentation. Nat. Methods **18**(2), 203–211 (2021)

6. Isensee, F., Jäger, P.F., Kohl, S.A., Petersen, J., Maier-Hein, K.H.: Automated design of deep learning methods for biomedical image segmentation. arXiv preprint arXiv:1904.08128 (2019)

7. Jin, L., et al.: Deep-learning-assisted detection and segmentation of rib fractures from CT scans: development and validation of FracNet. EBioMedicine **62**, 103106 (2020)

8. Kowal, J., Langlotz, F., Nolte, L.P.: Basics of computer-assisted orthopaedic surgery. In: Stiehl, J.B., Konermann, W.H., Haaker, R.G., DiGioia, A.M. (eds.) Navigation and MIS in Orthopedic Surgery, pp. 2–8. Springer, Berlin (2007). https://doi.org/10.1007/978-3-540-36691-1_1

9. Krčah, M., Székely, G., Blanc, R.: Fully automatic and fast segmentation of the femur bone from 3D-CT images with no shape prior. In: 2011 IEEE International Symposium on Biomedical Imaging: From Nano to Macro, pp. 2087–2090. IEEE (2011)

10. Kuiper, R.J., et al.: Efficient cascaded V-Net optimization for lower extremity CT segmentation validated using bone morphology assessment. J. Orthop. Res. **40**(12), 2894–2907 (2022)

11. Lai, J.Y., Essomba, T., Lee, P.Y.: Algorithm for segmentation and reduction of fractured bones in computer-aided preoperative surgery. In: Proceedings of the 3rd International Conference on Biomedical and Bioinformatics Engineering, pp. 12–18 (2016)

12. Liu, J., Xing, F., Shaikh, A., Linguraru, M.G., Porras, A.R.: Learning with context encoding for single-stage cranial bone labeling and landmark localization. In: Wang, L., Dou, Q., Fletcher, P.T., Speidel, S., Li, S. (eds.) MICCAI 2022. LNCS, vol. 13438, pp. 286–296. Springer, Cham (2022). https://doi.org/10.1007/978-3-031-16452-1_28

13. Liu, P., et al.: Deep learning to segment pelvic bones: large-scale CT datasets and baseline models. Int. J. Comput. Assist. Radiol. Surg. **16**, 749–756 (2021)

14. Qamar, S., Jin, H., Zheng, R., Ahmad, P., Usama, M.: A variant form of 3D-UNet for infant brain segmentation. Future Gener. Comput. Syst. **108**, 613–623 (2020)

15. Ruikar, D.D., Santosh, K., Hegadi, R.S.: Automated fractured bone segmentation and labeling from CT images. J. Med. Syst. **43**, 1–13 (2019)

16. Sugano, N.: Computer-assisted orthopaedic surgery and robotic surgery in total hip arthroplasty. Clin. Orthop. Surg. **5**(1), 1–9 (2013)

17. Tomita, N., Cheung, Y.Y., Hassanpour, S.: Deep neural networks for automatic detection of osteoporotic vertebral fractures on CT scans. Comput. Biol. Med. **98**, 8–15 (2018)

18. Ukai, K., et al.: Detecting pelvic fracture on 3D-CT using deep convolutional neural networks with multi-orientated slab images. Sci. Rep. **11**(1), 1–11 (2021)

19. Wang, D., Yu, K., Feng, C., Zhao, D., Min, X., Li, W.: Graph cuts and shape constraint based automatic femoral head segmentation in CT images. In: Proceedings of the Third International Symposium on Image Computing and Digital Medicine, pp. 1–6 (2019)

20. Wang, J., Zhang, X., Guo, L., Shi, C., Tamura, S.: Multi-scale attention and deep supervision-based 3D UNet for automatic liver segmentation from CT. Math. Biosci. Eng. MBE **20**(1), 1297–1316 (2023)

21. Yamamoto, N., et al.: An automated fracture detection from pelvic CT images with 3-D convolutional neural networks. In: 2020 International Symposium on Community-Centric Systems (CcS), pp. 1–6. IEEE (2020)

22. Yuan, J., Bae, E., Tai, X.C., Boykov, Y.: A spatially continuous max-flow and min-cut framework for binary labeling problems. Numer. Math. **126**, 559–587 (2014)

Pelvic Fracture Reduction Planning Based on Morphable Models and Structural Constraints

Sutuke Yibulayimu[1], Yanzhen Liu[1], Yudi Sang[3], Gang Zhu[3],
Yu Wang[1](✉), Jixuan Liu[1], Chao Shi[1], Chunpeng Zhao[2], and Xinbao Wu[2]

[1] Key Laboratory of Biomechanics and Mechanobiology, Ministry of Education,
Beijing Advanced Innovation Center for Biomedical Engineering, School of Biological
Science and Medical Engineering, Beihang University, Beijing 100083, China
{sutuk,wangyu}@buaa.edu.cn

[2] Department of Orthopaedics and Traumatology,
Beijing Jishuitan Hospital, Beijing, China

[3] Beijing Rossum Robot Technology Co., Ltd., Beijing, China

Abstract. As one of the most challenging orthopedic injuries, pelvic
fractures typically involve iliac and sacral fractures as well as joint dislocations. Structural repair is the most crucial phase in pelvic fracture
surgery. Due to the absence of data for the intact pelvis before fracture,
reduction planning heavily relies on surgeon's experience. We present a
two-stage method for automatic reduction planning to restore the healthy
morphology for complex pelvic trauma. First, multiple bone fragments
are registered to morphable templates using a novel SSM-based symmetrical complementary (SSC) registration. Then the optimal target reduction pose of dislocated bone is computed using a novel articular surface
(AS) detection and matching method. A leave-one-out experiment was
conducted on 240 simulated samples with six types of pelvic fractures on
a pelvic atlas with 40 members. In addition, our method was tested in
four typical clinical cases corresponding to different categories. The proposed method outperformed traditional SSM, mean shape reference, and
contralateral mirroring methods in the simulation experiment, achieving
a root-mean-square error of 3.4 ± 1.6 mm, with statistically significant
improvement. In the clinical feasibility experiment, the results on various
fracture types satisfied clinical requirements on distance measurements
and were considered acceptable by senior experts. We have demonstrated
the benefit of combining morphable models and structural constraints,
which simultaneously utilizes cohort statistics and patient-specific features.

Keywords: Surgery planning · Pelvic fracture · Morphable models

Supplementary Information The online version contains supplementary material
available at https://doi.org/10.1007/978-3-031-43996-4_31.

1   Introduction

Reduction planning is a crucial phase in pelvic fracture surgery, which aims to restore the anatomical structure and stability of the pelvic ring [21]. Traditional reduction planning relies on surgeons to form a "mental" plan through preoperative CT images, and the results are dependent on the surgeon's skill and experience. Computer-aided diagnosis (CAD) tools have been used for reduction planning, by virtually manipulating 3D bone models [2,23]. For complex fractures, 3D manual planning is not only time-consuming but also error-prone [19]. An automatic reduction planning method that offers improvements in both accuracy and efficiency is in demand.

The most intuitive method for reduction planning is by matching the fracture surface characteristics [16,20]. However, these methods rely on the precise identification and segmentation of fracture lines, which is typically challenging, especially in comminuted fractures where small fragments are "missing". Template-based approaches avoid the fracture line identification process. These methods include the average template statically modeled from pelvic shapes [3] and the template mirrored from the healthy contralateral side [6,22]. The former is not adaptive to individual morphology, usually resulting in poor robustness. Whereas the latter can only handle iliac fractures and is limited by the fact that the two ilia are usually not absolutely symmetric [7]. In addition, the contralateral mirroring (CM) method faces a severe challenge in bilateral injuries, where the mirrored part cannot provide accurate references.

Han et al. [11] used a statistical shape model (SSM) for pelvic fracture reduction, which adaptively matches the target morphology of pelvic bone. However, differences in the shape and integrity of the model bone and target fragments challenge the accurate adaptation of the SSM. Although the SSM method applies to all situations, without fully utilizing the whole pelvic structure, including the symmetricity, it often presents lower robustness. A recent study have pointed out that the CM method achieved better results than the adaptable iliac SSM [14].

While the SSM method addresses the reduction within single bones, it does not apply well to inter-bone matching for joint dislocations [8]. Statistical pose model (SPM) was proposed to jointly model the similarity transformations between bones, and to decouple the morphology of the entire pelvis into single bone shape and inter-bone poses [10]. The SPM models the statistics of bone positions but ignores the details in the joint connections which has rich information useful for aligning the joint surfaces. An under-fitted SPM may result in poor joint alignment and overall pelvic symmetry, which are very important considerations in reduction surgery in clinical practice.

In this study, we present a two-stage method for automatic pelvic reduction planning, addressing bone fractures and joint dislocations sequentially. A novel SSM-based symmetrical complementary (SSC) registration is designed to register multiple bone fragments to adaptive templates to restore the symmetric structure. Then, an articular surface (AS) detection and matching algorithm is designed to search the optimal target pose of dislocated bones with respect to joint alignment and symmetry constraints. The proposed method was evaluated in simulated experiments on a public dataset, and further validated in typical clinical cases. We have made our clinical dataset publicly available at https:// github.com/Sutuk/Clinical-data-on-pelvic-fractures.

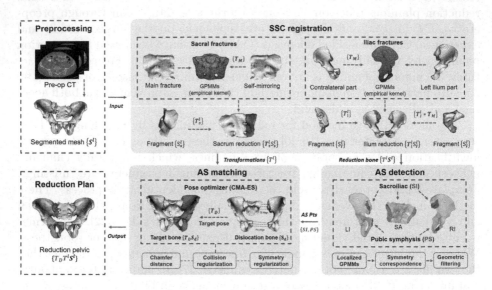

Fig. 1. Overview of the proposed reduction planning method for pelvic fracture surgery.

2 Method

As shown in Fig. 1, 3D models of the bone fragments in ilia and sacrum are obtained by semi-automatic segmentation [15]. Our automatic planning algorithm consists of two stages. The first stage computes the transformation of each bone fragment to restore single-bone morphology. The second stage estimates the target pose of each dislocated bone with respect to the whole pelvic anatomy.

2.1 Gaussian Process Morphable Models for Pelvic Bone

We use Gaussian process morphable models (GPMMs) [17] to model the shape statistics of pelvic bones. A healthy bone template is created for fractured fragments using **empirical kernel**, which enables the model to generate physiologically feasible bone morphology. In addition, **localized kernel** is used to create models with higher flexibility in order to approximate the morphology of the reduced/intact bone and then identify the pelvic AS areas.

Parameterized Model. The GPMMs of the left ilium and sacrum are modeled separately by first aligning bone segmentations to a selected reference Γ_R. A bone shape Γ_B can be obtained by warping the reference Γ_R with a deformation field $u(x)$. The GPMMs models deformation as a Gaussian process $GP(\mu, k)$ with a mean function μ and covariance function k, and is invariant to the choice of reference Γ_R. Parameterized with principal component analysis (PCA) in a finite dimension, the GPMMs can be written as:

$$\Gamma_B = \{x + \mu(x) + Pb \mid x \in \Gamma_R\}, \tag{1}$$

where P and b are the principal components and the weight vector, respectively. GPMMS modeled with the empirical kernel in a finite domain is equivalent to a statistical shape model (SSM), and the parameter is denoted as b_{SM}. The parameters of the localized model are denoted as b_{LC}.

Empirical Kernel for Fracture Reduction. A Gaussian process $GP(\mu_{SM}, k_{SM})$ that models the distinctive deformations is derived from the empirical mean $\mu(x) = \frac{1}{n}\sum_{i=1}^{n} u_i(x)$ and covariance function:

$$k_{SM}(x, y) = \frac{1}{n-1} \sum_{i=1}^{n} (u_i(x) - \mu_{SM}(x))(u_i(y) - \mu_{SM}(y))^T, \tag{2}$$

where n represents the number of training surfaces, and $u_i(x)$ denotes single point deformation on the i-th sample surfaces.

Localized Kernel for as Detection. In order to increase the flexibility of the statistical model, we limit the correlation distance between point clouds. The localized kernel is obtained by multiplying the empirical kernel and a Gaussian kernel $k_g(x, y) = \exp(-\|x - y\|^2/\sigma^2)$:

$$k_{LC}(x, y) = k_{SM}(x, y) \odot I_{3\times3}k_g(x, y), \tag{3}$$

where $\odot$ denotes element-wise multiplication, the identity matrix $I_{3\times3}$ is a 3D vector field, and σ determines the range of correlated deformations. The kernel is then used in AS detection.

2.2 Symmetrical Complementary Registration

A novel SSC registration is designed to register bone fragments to adaptive templates in joint optimization, which alternatingly estimates the desired reduction of fragments and solves the corresponding transformations. A bilateral supplementation strategy is used in model adaptation: mirrored counterparts are aligned to the target point cloud to provide additional guidance. In more detail, both the fragment and its mirrored counterpart are first rigidly aligned to the GPMM with empirical kernel k_{SM}, using the iterative closest point (ICP) algorithm. Then, the GPMM is non-rigidly deformed towards the merged point clouds of the main fragment and its mirrored counterparts. During this step, the model's parameters b_{SM} are optimized using bi-directional vertex correspondences (target to source $v_\rightarrow(x)$ and backward $v_\leftarrow(x)$) based on the nearest neighbour. For each source point x, there is a forward displacement $\delta_\rightarrow(x) = v_\rightarrow(x) - x$, and one or more backward displacement $\delta_\leftarrow(x) = v_\leftarrow(x) - x$. The farthest point within the given threshold ε is selected as the corresponding point displacement $\delta(x)$ of the template:

$$\delta(x) = I\left(|\delta(x)| < \varepsilon\right) \max\left(|\delta_\rightarrow(x)|, |\delta_\leftarrow(x)|\right) + I\left(|\delta(x)| \geq \varepsilon\right) \mathbf{O}, \qquad (4)$$

where I is the indicator function and $\mathbf{O}$ is a zeros vector. The optimal parameter $\hat{b}_{SM}$ of the statistical model can be calculated from Eq. (1), and is constrained within three standard deviations of the eigenvalue λ_i [1]:

$$\hat{b}_{SM} = \mathrm{argmin}\{\delta(x) - \mu(x) - Pb_{SM} \mid x \in \Gamma_R\}, \ |\hat{b}_{SM}| < \pm 3\sqrt{\lambda_i}. \qquad (5)$$

The adaptive shape of the GPMM is regarded as the target morphology of a bone, and the fracture fragments are successively registered to the adaptive template to form the output of the first stage, reduction transformation T^l of each bone fragment S^l, $l = 1, ..., L$. As shown in Fig. 1, in the SSC registration stage, contralateral complementation and self-mirroring complementation are used for ilium and sacrum, respectively.

2.3   Articular Surface Detection and Matching

Articular Surface Detection. The AS of the sacroiliac (SI) joint and pubic symphysis (PS) are detected using the GPMMs for the ilia and sacrum, with the localized kernel. As indicated by the red and blue regions in Fig. 1 - AS detection, surface points in the joint regions are first manually annotated in the mean model template ($b_{LC} = 0$) and then propagated to each instance using non-rigid adaptation of GPMM model. The optimal parameter $\hat{b}_{LC}$ is obtained via the alternative optimization method in Sect. 2.2, and the marked points on the adaptive templates are mapped to the target bones as the AS regions. Due to the symmetric structure of the pelvis, we use a unilateral model to identify bilateral AS through mirror-flipped registration. Each AS landmark (vertex in mesh data) is associated with a normal vector. The average normal direction is computed from these vectors, and any surface landmark pairs deviating from this average direction are removed. The identified surface point sets are denoted as SI and PS.

Articular Surface Matching. For each joint, a ray-tracing collision detection algorithm is performed between opposing surfaces using surface normal [13]. Then, a cost function L_{local} is constructed to measure the degree of alignment between two surfaces, including a distance term and a collision term. The former measures the Chamfer distance between complementary surfaces d_{cd} [4]. The latter uses an exponential function to map the IoU of the joint surface, so that slight collisions can be admitted to compensate for the potential inaccurate segmentation from the preprocessing stage.

$$\mathcal{L}_{local} = d_{cd}\left(S_1, S_2\right) + \gamma \cdot \exp\left(\frac{|S_1 \cap S_2|}{|S_1 \cup S_2|}\right), \tag{6}$$

where S_1 and S_2 are two sets of 3D points, and γ is a balancing weight for the collision term.

In addition to the local cost for each joint, a global cost measuring the degree of symmetry is used to regularize the overall anatomy of the pelvis:

$$\mathcal{L}_{global} = \frac{1}{|D_{\text{left}}|}||D_{\text{left}} - D_{\text{right}}||_2^2 + \langle \vec{V}_{PS}, \vec{V}_{SI}\rangle, \tag{7}$$

where the first term measures the paired difference between the point-wise distance within SI_{left} and SI_{right}. D_{left} and D_{Right} represent the distance between points in SI_{left} and SI_{right}, respectively. The second term measures the angle between the mean vector of the PS, denoted as $\vec{V}_{PS}$, and the mean vector of the bilateral SI joint $\vec{V}_{SI}$. The AS matching problem is formulated as optimizing the transformations with respect to the misalignment of PS and SI joints, plus the pelvic asymmetry:

$$\mathcal{L} = \mathcal{L}_{local}\left(SI_{left}\right) + \mathcal{L}_{local}\left(SI_{right}\right) + \mathcal{L}_{local}\left(PS\right) + \mathcal{L}_{global}, \tag{8}$$

where $\mathcal{L}_{local}\left(SI_{left}\right)$ or $\mathcal{L}_{local}\left(SI_{right}\right)$ can be a constant when the corresponding joint is intact, and can be omitted accordingly.

The cost function in Eq. (8) is optimized with respect to T_D, which determines the pose of the moving bone and its joint surfaces. The covariance matrix adaptation evolution strategy (CMA-ES) is used as the optimizer to avoid local minimum problem [12]. The final output point clouds of the pelvis is obtained by combining the transformation for the dislocated bone and the transformation for each fragment in the first stage $T_D T^l S^l$.

3 Experiments and Results

3.1 Experiments on Simulated Data

We evaluated the proposed fracture reduction planning method in a simulation experiment with leave-one-out cross-validation on the open-source pelvic atlas [9]. Extending from the simulation category in previous research [10], the current study simulated six types of fractures (the first column in Fig. 2), including single-bone fractures, joint dislocation, and combined pelvic trauma. The single-bone fractures include bilateral iliac wing fractures and sacral fractures, and the combined simulation of pelvic trauma includes iliac fracture with dislocation (type A), sacral fracture with dislocation (type B), as well as iliac and sacral fractures with dislocation (type C). In total, 240 instances were simulated (6 fracture categories for each of the 40 samples). On the single-bone fracture data, the proposed SSC registration was tested against the mean shape reference and the SSM reference [11]. On the dislocation-only data, the proposed AS matching method was tested against the pelvic mean shape reference and contralateral mirroring (CM) [22]. On the combined-trauma data, the proposed two-stage method in Fig. 1 was tested against the pelvic mean shape reference and combined SSM and CM (S&C).

The Gaussian kernel parameter σ was set to 30, the collision regularization factor γ was set to 2, and the bidirectional correspondence threshold ε was set to 5. The accuracy of reduction was evaluated by the root-mean-square error (RMSE) on all fragments. Specifically, the average distance between the ground truth and the planned target location of all points was calculated.

As shown in Fig. 2, due to the strong individual variations, the mean reference was insufficient to cope with most fracture reduction problems, resulting in uneven fracture fragments and distorted bone posture. Without the symmetric complement, the SSM planning was more prone to large fragment gaps or overlaps than our method. Solely dependent on sacrum symmetry, the CM method often resulted in bone overlap or tilted reduction poses. The flaws of the cascaded comparison method were further magnified in combined pelvic fracture

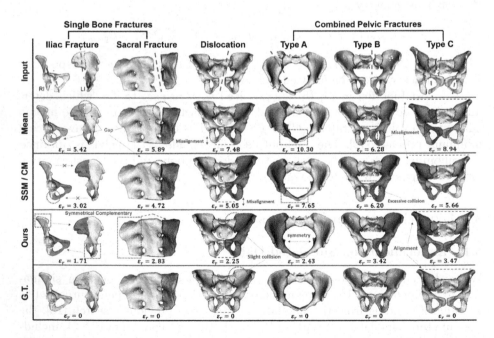

Fig. 2. Planning results on simulated data. Misalignments and good alignments are indicated by red and green dotted lines, respectively. (Color figure online)

reduction. Meanwhile, under the structural constraints, our method payed more attention to the overall symmetry, collision and articular surface anastomosis, and the results were closer to the ground truth.

As shown in Fig. 3, our method achieved the best median RMSE and interquartile range (IQR). Paired t-tests against other methods indicated statistical significance, with $p < 0.001$. For additional results, please refer to the Supplementary Material. The overall running time for the planning algorithm was 5 to 7 min on an Intel i9-12900KS CPU and an NVIDIA RTX 3070Ti GPU.

3.2 Experiments on Clinical Data

To further evaluate the proposed method on real clinical data, we collected CT data from four clinical cases (aged 34–40 years, one female), each representing a distinct type of pelvic fractures corresponding to the simulated categories. The data usage is approved by the Ethics Committee of the Beijing Jishuitan Hospital (202009-04). The proposed method was applied to estimate transformations of bone fragments to obtain a proper reduction in each case. Due to the lack of ground-truth target positions for fractured pelvis in clinical data, we employed

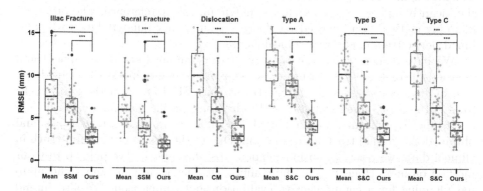

Fig. 3. Boxplots showing the RMSE between the planned target location and the ground truth on the simulated data. "$***$" indicates p-values less than 10^{-3} in t-tests.

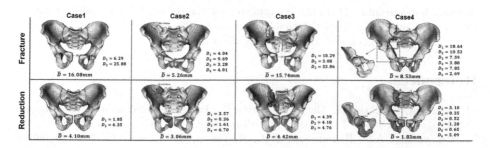

Fig. 4. Planning results on clinical data. D_i measures the distance between fragments in millimeter, and $\overline{D}$ is their average.

geometric measurements to evaluate the planning result. This evaluation method was inspired by Matta's trauma surgery criteria [18]. Surgeons manually measured the distances between all fragments and joints on a 3D slicer platform [5] by identifying a dislocated landmark pair across each fracture line or dislocated joint. The pelvic morphology in the planning result can be appreciated in Fig. 4. The measured distances between fragments or across joints are reduced to 1.8 to 4.4 mm. These planning results were examined by two senior experts and were deemed acceptable.

4 Discussion and Conclusion

We have proposed a two-stage method for pelvic fracture reduction planning based on morphable models and structural constraints. The SSM-based symmetrical complementary registration was successfully applied to both bilateral iliac and sacral fractures, which further extended the unilateral iliac CM reduction. Combining the SSM approach with symmetrical complementation provided more complete morphological guidance for model adaptation, and improved the registration performance significantly. The AS detection and matching method which combines local joint matching and global symmetry constraints achieved significantly higher accuracy than the pelvic mirroring reference. The proposed planning method also achieved satisfactory results on simulated data with combined pelvic trauma and real clinical cases.

We have demonstrated the synergy of the combined statistical model and anatomical constraints. In future work, we plan to further investigate this direction by incorporating feasibility constraints into SPM-based method to benefit from both cohort-based statistics and individual-specific matching criteria.

In the experiments, we simulated the most common fracture types and obtained the ground truth for evaluation. Due to the absence of ground truth in clinical data, we resort to an independent distance metric. We plan to test our method on more clinical data, and combine geometric measurements and manual planning for a comprehensive evaluation and comparison. We also intend to further automate the planning pipeline using a pelvic fracture segmentation network in future research to avoid the tedious manual annotation process.

Acknowledgements. This work is supported by the Beijing Science and Technology Project (Grants No. Z221100003522007 and Z201100005420033). We thank Runze Han et al. for their efforts on the open-source pelvic atlas.

References

1. Albrecht, T., Lüthi, M., Gerig, T., Vetter, T.: Posterior shape models. Med. Image Anal. **17**(8), 959–973 (2013). https://doi.org/10.1016/j.media.2013.05.010
2. Chowdhury, A.S., Bhandarkar, S.M., Robinson, R.W., Yu, J.C.: Virtual multi-fracture craniofacial reconstruction using computer vision and graph matching. Comput. Med. Imaging Graph. **33**(5), 333–342 (2009). https://doi.org/10.1016/j.compmedimag.2009.01.006

3. Ead, M.S., Palızı, M., Jaremko, J.L., Westover, L., Duke, K.K.: Development and application of the average pelvic shape in virtual pelvic fracture reconstruction. Int. J. Med. Robot. **17**(2), e2199 (2021). https://doi.org/10.1002/rcs.2199
4. Fan, H., Hao, S., Guibas, L.: A point set generation network for 3D object reconstruction from a single image. In: 2017 IEEE Conference on Computer Vision and Pattern Recognition (CVPR) (2017)
5. Fedorov, A., et al.: 3D slicer as an image computing platform for the quantitative imaging network. Magn. Reson. Imaging **30**(9), 1323–1341 (2012)
6. Fürnstahl, P., Székely, G., Gerber, C., Hodler, J., Snedeker, J.G., Harders, M.: Computer assisted reconstruction of complex proximal humerus fractures for preoperative planning. Med. Image Anal. **16**(3), 704–720 (2012). https://doi.org/10.1016/j.media.2010.07.012
7. Gnat, R., Saulicz, E., Biały, M., Kłaptocz, P.: Does pelvic asymmetry always mean pathology? Analysis of mechanical factors leading to the asymmetry. J. Hum. Kinet. **21**(2009), 23–32 (2009). https://doi.org/10.2478/v10078-09-0003-8
8. Han, R., et al.: Multi-body 3D-2D registration for image-guided reduction of pelvic dislocation in orthopaedic trauma surgery. Phys. Med. Biol. **65**(13), 135009 (2020). https://doi.org/10.1088/1361-6560/ab843c
9. Han, R., Uneri, A., Silva, T.D., Ketcha, M., Siewerdsen, J.H.: Atlas-based automatic planning and 3D-2D fluoroscopic guidance in pelvic trauma surgery. Phys. Med. Biol. **64**(9) (2019)
10. Han, R., et al.: Fracture reduction planning and guidance in orthopaedic trauma surgery via multi-body image registration. Med. Image Anal. **68**, 101917 (2021). https://doi.org/10.1016/j.media.2020.101917
11. Han, R., et al.: Multi-body registration for fracture reduction in orthopaedic trauma surgery. In: SPIE Medical Imaging, vol. 11315. SPIE (2020). https://doi.org/10.1117/12.2549708
12. Hansen, N., Müller, S.D., Koumoutsakos, P.: Reducing the time complexity of the derandomized evolution strategy with covariance matrix adaptation (CMA-ES). Evol. Comput. **11**(1), 1–18 (2003). https://doi.org/10.1162/106365603321828970
13. Hermann, E., Faure, F., Raffin, B.: Ray-traced collision detection for deformable bodies. In: International Conference on Computer Graphics Theory and Applications (2008)
14. Krishna, P., Robinson, D.L., Bucknill, A., Lee, P.V.S.: Generation of hemipelvis surface geometry based on statistical shape modelling and contralateral mirroring. Biomech. Model. Mechanobiol. **21**(4), 1317–1324 (2022). https://doi.org/10.1007/s10237-022-01594-1
15. Liu, P., et al.: Deep learning to segment pelvic bones: large-scale CT datasets and baseline models. Int. J. Comput. Assist. Radiol. Surg. **16**(5), 749–756 (2021). https://doi.org/10.1007/s11548-021-02363-8
16. Luque-Luque, A., Pérez-Cano, F.D., Jiménez-Delgado, J.J.: Complex fracture reduction by exact identification of the fracture zone. Med. Image Anal. **72**, 102120 (2021). https://doi.org/10.1016/j.media.2021.102120
17. Luthi, M., Gerig, T., Jud, C., Vetter, T.: Gaussian process morphable models. IEEE Trans. Pattern Anal. Mach. Intell. **40**(8), 1860–1873 (2018). https://doi.org/10.1109/tpami.2017.2739743
18. Matta, J.M., Tornetta, P.I.: Internal fixation of unstable pelvic ring injuries. Clin. Orthop. Relat. Res. (1976–2007) **329** (1996)
19. Suero, E.M., Hüfner, T., Stübig, T., Krettek, C., Citak, M.: Use of a virtual 3D software for planning of tibial plateau fracture reconstruction. Injury **41**(6), 589–591 (2010). https://doi.org/10.1016/j.injury.2009.10.053

20. Willis, A.R., Anderson, D.D., Thomas, T.P., Brown, T.D., Marsh, J.L.: 3D reconstruction of highly fragmented bone fractures. In: SPIE Medical Imaging (2007)

21. Yu, Y.H., Liu, C.H., Hsu, Y.H., Chou, Y.C., Chen, I.J., Wu, C.C.: Matta's criteria may be useful for evaluating and predicting the reduction quality of simultaneous acetabular and ipsilateral pelvic ring fractures. BMC Musculoskelet. Disord. **22**(1), 544 (2021). https://doi.org/10.1186/s12891-021-04441-z

22. Zhao, C., et al.: Automatic reduction planning of pelvic fracture based on symmetry. Comput. Methods Biomech. Biomed. Eng. Imaging Vis. **10**(6), 577–584 (2022). https://doi.org/10.1080/21681163.2021.2012830

23. Zhou, B., Willis, A., Sui, Y., Anderson, D.D., Brown, T.D., Thomas, T.P.: Virtual 3D bone fracture reconstruction via inter-fragmentary surface alignment. In: 2009 IEEE 12th International Conference on Computer Vision Workshops, ICCV Workshops, pp. 1809–1816 (2009). https://doi.org/10.1109/ICCVW.2009.5457502

High-Resolution Cranial Defect Reconstruction by Iterative, Low-Resolution, Point Cloud Completion Transformers

Marek Wodzinski[1,2]([✉]) [iD], Mateusz Daniol[1] [iD], Daria Hemmerling[1] [iD], and Miroslaw Socha[1] [iD]

[1] Department of Measurement and Electronics, AGH University of Science and Technology, Krakow, Poland
wodzinski@agh.edu.pl
[2] Information Systems Institute, University of Applied Sciences Western Switzerland (HES-SO Valais), Sierre, Switzerland

Abstract. Each year thousands of people suffer from various types of cranial injuries and require personalized implants whose manual design is expensive and time-consuming. Therefore, an automatic, dedicated system to increase the availability of personalized cranial reconstruction is highly desirable. The problem of the automatic cranial defect reconstruction can be formulated as the shape completion task and solved using dedicated deep networks. Currently, the most common approach is to use the volumetric representation and apply deep networks dedicated to image segmentation. However, this approach has several limitations and does not scale well into high-resolution volumes, nor takes into account the data sparsity. In our work, we reformulate the problem into a point cloud completion task. We propose an iterative, transformer-based method to reconstruct the cranial defect at any resolution while also being fast and resource-efficient during training and inference. We compare the proposed methods to the state-of-the-art volumetric approaches and show superior performance in terms of GPU memory consumption while maintaining high-quality of the reconstructed defects.

Keywords: Cranial Implant Design · Deep Learning · Shape Completion · Point Cloud Completion · SkullBreak · SkullFix · Transformers

1 Introduction

Cranial damage is a common outcome of traffic accidents, neurosurgery, and warfare. Each year, thousands of patients require personalized cranial implants [2]. Nevertheless, the design and production of personalized implants are expensive and time-consuming. Nowadays, it requires trained employees working with

© The Author(s), under exclusive license to Springer Nature Switzerland AG 2023
H. Greenspan et al. (Eds.): MICCAI 2023, LNCS 14228, pp. 333–343, 2023.
https://doi.org/10.1007/978-3-031-43996-4_32

computer-aided design (CAD) software [11]. However, one part of the design pipeline, namely defect reconstruction, can be directly improved by the use of deep learning algorithms [7,8].

The problem can be formulated as a shape completion task and solved by dedicated neural networks. Its importance motivated researchers to organize two editions of the AutoImplant challenge, during which researchers proposed several unique contributions [7,8]. The winning contributions from the first [3] and second editions [18] proposed heavily-augmented U-Net-based networks and treated the problem as segmentation of missing skull fragment. They have shown that data augmentation is crucial to obtain reasonable results [19]. Other researchers proposed similar encoder-decoder approaches, however, without significant augmentation and thus limited performance [10,14]. Another group of contributions attempted to address not only the raw performance but also the computational efficiency and hardware requirements. One contribution proposed an RNN-based approach using 2-D slices taking into account adjacent slices to enforce the continuity of the segmentation mask [21]. The contribution by Li *et al.* has taken into account the data sparsity and proposed a method for voxel rearrangement in coarse representation using the high-resolution templates [6]. The method was able to substantially reduce memory usage while maintaining reasonable results. Another contribution by Kroviakov *et al.* proposed an approach based on sparse convolutional neural networks [5] using Minkowski engine [1]. The method excluded the empty voxels from the input volume and decreased the number of the required convolutions. The work by Yu *et al.* proposed an approach based on principal component analysis with great generalizability, yet limited raw performance [23]. Interestingly, methods addressing the computational efficiency could not compete, in terms of the reconstruction quality, with the resource-inefficient methods using dense volumetric representation [7].

The current state-of-the-art solutions, even though they reconstruct the defects accurately, share some common disadvantages. First, they operate in the volumetric domain and require significant computational resources. The GPU memory consumption scales cubically with the volume size. Second, the most successful solutions do not take into account data sparsity. The segmented skulls are binary and occupy only a limited part of the input volume. Thus, using methods dedicated to 3-D multi-channel volumes is resource-inefficient. Third, the final goal of the defect reconstruction is to propose models ready for printing/manufacturing. Working with volumetric representation requires further postprocessing to transfer the reconstructed defect into a manufacturable model.

Another approach, yet still unexplored, to cranial defect reconstruction is the use of deep networks dedicated to point clouds (PCs) processing. Since the introduction of PointNet [15] and PointNet++ [16], the number of contributions in the area of deep learning for PC processing exploded. Several notable contributions, like PCNet [26], PoinTr [24], AdaPoinTr [25], 3DSGrasp [12], MaS [9], have been proposed directly to the PC completion task. The goal of the PC completion is to predict a missing part of an incomplete PC.

The problem of cranial defect reconstruction can be reformulated into PC completion which has several advantages. First, the representation is sparse, and thus requires significantly less memory than the volumetric one. Second, PCs are unordered collections and can be easily splitted and combined, enabling further optimizations. Nevertheless, the current PCs completion methods focus mostly on data representing object surfaces and do not explore large-scale PCs representing solid objects.

In this work, we reformulate the problem from volumetric segmentation into PC completion. We propose a dedicated method to complete large-scale PCs representing solid objects. We extend the geometric aware transformers [24] and propose an iterative pipeline to maintain low memory consumption. We compare the proposed approach to the state-of-the-art networks for volumetric segmentation and PC completion. Our approach provides high-quality reconstructions while maintaining computational efficiency and good generalizability into previously unseen cases.

2   Methods

2.1   Overview

The input is a 3-D binary volume representing the defective skull. The output is a PC representing the missing skull fragment and (optionally) its meshed and voxelized representation. The processing pipeline consists of: (i) creating the PC from the binary volume, (ii) splitting the PC into a group of coarse PCs, (iii) calculating the missing PC by the geometric aware transformer for each group, (iv) merging the reconstructed coarse PCs, (v) optional voxelization and postprocessing for evaluation. The pipeline is shown in Fig. 1.

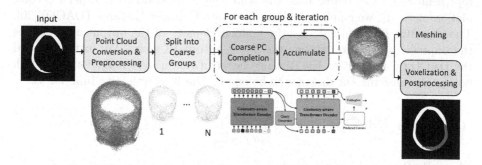

Fig. 1. Visualization of the processing pipeline.

2.2 Preprocessing

The preprocessing starts with converting the binary volume to the PC. The coordinates of the positive voxels are created only from the voxels representing the skull. The PC is normalized to [0–1] range, randomly permuted, and split into N equal groups, where N is calculated based on the number of points in the input PC in a manner that each group contains 32768 randomly sampled input points and outputs 16384 points. Thus, the N depends on the number of positive voxels. The higher the resolution, the more groups are being processed. The number of outputs points is lower than the input points because we assumed that the defect is smaller than the skull.

2.3 Network Architecture - Point Cloud Completion Transformer

We adapt and modify the geometry-aware transformers (PoinTr) [24]. The PoinTr method was proposed and evaluated on coarse PCs representing object surfaces. The full description of the PoinTr architecture is available in [24].

We modify the network by replacing the FoldingNet [22] decoder working on 2-D grids with a folding decoder operating on 3-D representation. The original formulation deforms the 2-D grid into the surface of a 3-D object, while the proposed method focuses on solid 3-D models. Moreover, we modify the original k-NN implementation (with quadratic growth of memory consumption with respect to the input size) to an iterative one, to further decrease and stabilize the GPU memory consumption.

2.4 Objective Function

We train the network using a fully supervised approach where the ground-truth is represented by PCs created from the skull defects. In contrast to other PC completion methods, we employ the Density Aware Chamfer Distance (DACD) [20]. The objective function enforces the uniform density of the output and handles the unpredictable ratio between the input/output PCs size. We further extend the DACD by calculating the distance between the nearest neighbours for each point and enforcing the distance to be equal. The final objective function is:

$$O(P_r, P_{gt}) = DACD(P_r, P_{gt}) + \frac{\alpha}{S} \sum_{i=0}^{S} \sum_{j=0}^{k} \sum_{l=0}^{k} |P_r(i) - P_r(j)| - |P_r(i) - P_r(l)|, \quad (1)$$

where P_r, P_{gt} are the reconstructed and ground-truth PC respectively, S is the number of points in P_{rec}, k is the number of nearest neighbours of point i, α is the weighting parameter. We apply the objective function to all PC ablation studies unless explicitly stated otherwise. The volumetric ablation studies use the soft Dice score.

The traditional objective functions like Chamfer Distance (CD) [20], Extended Chamfer Distance (ECD) [22], or Earth Mover's Distance (EMD) [9] are not well suited for the discussed application. The CD/ECD provide sub-optimal performance for point clouds with uniform density or a substantially different number of samples, tends to collapse, and results in noisy training. The EMD is more stable, however, explicitly assumes bijective mapping (requiring knowledge about the desired number of points) and has high computational complexity.

2.5   Iterative Completion

The coarse PCs are processed by the network separately. Afterwards, the reconstructed PCs are combined into the final reconstruction. To improve the results, the process may be repeated M times with a different initial PC split and a small Gaussian noise added. The procedure improves the method's performance and closes empty holes in the voxelized representation. The optional multi-step completion is performed only during the inference.

The iterative completion allows one to significantly reduce the GPU memory usage and the number of network parameters. The PCs are unordered collections and can be easily split and merged. There is no need to process large PCs in one shot, resulting in the linear growth of inference time and almost constant GPU memory consumption.

2.6   Postprocessing

The reconstructed PCs are converted to mesh and voxelized back to the volumetric representation, mainly for evaluation purposes. The mesh is created by a rolling ball pivoting algorithm using the Open3D library [27]. The voxelization is also performed using Open3D by the PC renormalization and assigning positive values to voxels containing points in their interior. The voxelized representation is further postprocessed by binary closing and connected component analysis to choose only the largest volume. Then, the overlap area between the reconstructed defect and the defective input is subtracted from the reconstructed defect by logical operations.

2.7   Dataset and Experimental Setup

We use the SkullBreak and SkullFix datasets [4] for evaluation. The datasets were used during the AutoImplant I and II challenges and enable comparison to other reconstruction algorithms. The SkullBreak dataset contains 114 high-resolution skulls for training and 20 skulls for testing, each with 5 accompanying defects from various classes, resulting in 570 training and 100 testing cases. All volumes in the SkullBreak dataset are $512 \times 512 \times 512$. The SkullFix dataset is represented by 100 training cases mostly located in the back of the skull with a similar appearance, and additional 110 testing cases. The volumes in the SkullFix dataset are $512 \times 512 \times Z$ where Z is the number of axial slices. The SkullBreak provides more heterogeneity while the SkullFix is better explored and enables direct comparison to other methods.

We perform several ablation studies. We check the influence of the input physical spacing on the reconstruction quality, training time, and GPU memory consumption. Moreover, we check the generalizability by measuring the gap between the results on the training and the external testing set for each method. We compare our method to the methods dedicated to PC completion: (i) PCNet [26], (ii) PoinTr [24], (iii) AdaPoinTr [25], as well as to methods dedicated to volumetric defect reconstruction: (i) 3-D VNet, and (ii) 3-D Residual U-Net. Moreover, we compare the reconstruction quality to results reported by other state-of-the-art methods.

We trained our network separately on the SkullBreak and SkullFix datasets. The results are reported for the external test set containing 100 cases for Skull-Break and 110 cases for the SkullFix datasets, the same as in the methods used for comparison. The models are implemented in PyTorch [13], trained using a single RTX GeForce 3090. We augment the input PCs by random permutation, cropping, rotation, and translation. The volumetric ablation studies use random rotation and translation with the same parameters as for the PCs. All the methods were trained until convergence. The hyperparameters are reported in the associated repository [17].

3   Results

The comparison in terms of the Dice coefficient (DSC), boundary Dice coefficient (BDSC), 95th percentile of Hausdorff distance (HD95), and Chamfer distance (CD), are shown in Table 1. Exemplary visualization, presenting both the PC and volumetric outcomes, is shown in Fig. 2.

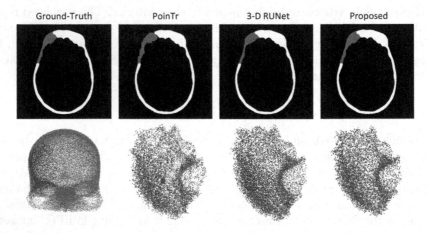

Fig. 2. Exemplary visualization of the reconstructed point clouds/volumes for a case from the SkullBreak dataset. The PCs are shown for the defect only (reconstructed vs ground-truth) for the presentation clarity.

The results of the ablation studies showing the influence of the input size, generalizability, objective function, and the effect of repeating the iterative refinement are presented in Table 2.

The results present that the quantitative outcomes of the proposed method are comparable to the state-of-the-art methods, however, with significantly lower GPU memory consumption that makes it possible to perform the reconstruction at the highest available resolution. The results are slightly worse when compared to the volumetric methods, however, significantly better than other PC-based approaches.

Table 1. Quantitative results on the SkullBreak and SkullFix datasets. The final results are reported for original resolution using the DACD + kNN objective function and 3 iterative refinements (- denotes that results were not reported). The methods used for comparison are reported for the most successful setup (see Table 2).

Method	SkullBreak				SkullFix				GPU Mem ↓[GB]
	DSC ↑	BDSC ↑	HD95 ↓	CD ↓	DSC ↑	BDSC ↑	HD95 ↓	CD ↓	
Point Cloud Completion									
Proposed	0.87	0.85	1.91	0.31	0.90	0.89	1.71	0.29	~2.78
PCNet [26]	0.61	0.58	5.77	1.18	0.77	0.75	3.22	0.41	~2.37
PointTr [24]	0.67	0.66	5.17	0.82	0.82	0.81	3.02	0.36	~3.11
AdaPoinTr [25]	0.66	0.64	5.29	0.84	0.81	0.81	3.05	−0.36	~3.14
Volumetric Segmentation									
3-D VNet	0.87	0.90	1.87	0.21	0.91	0.93	1.66	0.11	21.89
3-D RUNet	0.89	0.91	1.79	0.18	0.91	0.92	1.67	0.09	22.47
State-of-the-art									
Kroviakov et al. [5]	–	–	–	–	0.85	0.94	2.65	–	<6.00
Li et al. [6]	–	–	–	–	0.81	–	–	–	–
Mahdi et al. [10]	–	–	–	–	0.88	0.92	3.59	–	<6.00
Pathak et al. [14]	–	–	–	–	0.90	0.95	2.02	–	–
Wodzinski et al. [18]	0.91	0.95	1.60	–	0.93	0.95	1.48	–	<40.00
Yu et al. [23]	–	–	–	–	0.77	0.77	3.68	–	CPU
Ellis et al. [3]	–	–	–	–	0.94	–	3.60	–	–
Yang et al. [21]	0.85	0.89	3.52	–	–	–	–	–	–

Note: For the "Mahdi et al. [10]" row, the SkullBreak values 0.78, 0.81, 3.42 are shown.

4 Discussion

The reconstruction quality of the method is comparable to the volumetric networks, as shown in Table 1. Meanwhile, the proposed method takes into account the data sparsity, does not require significant computational resources, and scales well with the input size. The proposed method has good generalizability. The gap between the training and testing set is negligible, unlike the volumetric methods that easily overfit and require strong augmentation for practical use. The DACD, as well as the proposed extension, improve the reconstruction quality compared to the CD or ECD by taking into account the uniformity of the expected PC. The original PC completion methods do not scale well with the increase of PC

Table 2. The ablation studies related to the input size, the objective function, and the number of refinements. The results are reported for the SkullBreak dataset at the original scale (except the CD). The Gen. Gap denotes the difference between the training and testing set in terms of the DSC.

Method	DSC ↑	BDSC ↑	HD95 ↓ [mm]	CD ↓ [mm]	GPU Mem ↓ [GB]	Gen. Gap ↓ [% DSC]
Input Size (uniform voxel spacing)						
Proposed: original	0.87	0.85	1.91	0.31	~2.78	4.18
Proposed: 1 mm	0.83	0.77	2.64	0.46	~2.69	4.05
Proposed: 2 mm	0.74	0.71	3.89	0.67	~2.64	4.57
Proposed: 4 mm	0.69	0.64	5.12	0.79	~2.63	3.12
PCNet: original	–	–	–	–	>24	–
PCNet: 1 mm	0.37	0.33	10.57	1.76	~13.22	1.13
PCNet: 2 mm	0.57	0.53	7.18	1.37	~5.37	1.33
PCNet: 4 mm	0.61	0.58	5.77	1.18	~2.37	3.07
PoinTr: original	–	–	–	–	>24	–
PoinTr: 1 mm	0.58	0.55	6.82	1.39	~21.41	1.89
PoinTr: 2 mm	0.65	0.64	5.28	0.94	~6.48	2.19
PoinTr: 4 mm	0.67	0.66	5.17	0.82	~3.11	3.98
3-D RUNet: original	–	–	–	–	>24	–
3-D RUNet: 1 mm	0.89	0.91	1.79	0.18	22.47	10.11
3-D RUNet: 2 mm	0.85	0.85	2.09	0.25	7.84	14.51
3-D RUNet: 4 mm	0.76	0.77	2.89	0.63	3.78	17.48
Objective Function (proposed method, original size, 3 iters)						
DACD + kNN	0.87	0.85	1.91	0.31	~2.78	4.18
DACD	0.85	0.81	2.78	0.42	~2.72	4.22
ECD	0.75	0.71	4.11	0.68	~2.72	3.99
CD	0.83	0.78	2.98	0.28	~2.72	4.58
No. Refinements (proposed method, original size, DACD + kNN)						
1 iters	0.85	0.81	2.51	0.41	~2.78	4.51
2 iters	0.87	0.83	1.98	0.32	~2.78	4.18
3 iters	0.87	0.85	1.91	0.31	~2.78	4.18
4 iters	0.87	0.85	1.90	0.30	~2.78	4.18

size. The possible reason for this is connected with the noisy kNN graph construction when dealing with large PCs and increasing the number of neighbours is unacceptable from the computational point of view. The proposed method has almost constant memory usage, independent of the input shape, in contrast to both the volumetric methods and PC completion methods without the iterative approach. Interestingly, the proposed method outperforms other methods taking into account the data sparsity. The inference speed is slightly lower than for the volumetric methods (several seconds), however, this application does not require real-time processing and anything in the range of seconds is acceptable. However, the required memory consumption is crucial to ensure that the method can be eventually applied in clinical practice.

The disadvantages of the proposed algorithm are connected to long training time, noise at the object boundaries, and holes in the voxelized output. The FoldingNet-based decoder requires a significant number of iterations to converge, thus resulting in training time comparable or even longer than the volumetric methods. Moreover, the voxelization of PCs results in noisy edges and holes that require further morphological postprocessing. What is important, the method has to be extended by another algorithm responsible for converting the reconstruction into implantable implant before being ready to be used in clinical setup.

In future work, we plan to further reformulate the problem and, similarly to Kroviakov *et al.* [5], use only the skull contours. Since the ultimate goal is to propose models ready for 3-D printing, the interior of the skull defect is not required to create the mesh and STL/OBJ file. Another research direction is connected to the PC augmentation to further increase the network generalizability since it was shown that heavy augmentation is crucial to obtain competitive results [3, 18]. Moreover, it is challenging to perform the qualitative evaluation of the context of clinical need and we plan to perform evaluation including clinical experts in the future research.

To conclude, we proposed a method for cranial defect reconstruction by formulating the problem as the PC completion task. The proposed algorithm achieves comparable results to the best-performing volumetric methods while requiring significantly less computational resources. We plan to further optimize the model by working directly at the skull contour and heavily augmenting the PCs.

Acknowledgements. The project was funded by The National Centre for Research and Development, Poland under Lider Grant no: LIDER13/0038/2022 (DeepImplant). We gratefully acknowledge Polish HPC infrastructure PLGrid support within computational grant no. PLG/2023/016239.

References

1. Choy, C., Gwak, J., Savarese, S.: 4D spatio-temporal convnets: Minkowski convolutional neural networks. In: Proceedings of the IEEE Computer Society Conference on Computer Vision and Pattern Recognition, pp. 3070–3079 (2019)
2. Dewan, M.C., et al.: Global neurosurgery: the current capacity and deficit in the provision of essential neurosurgical care. Executive Summary of the Global Neurosurgery Initiative at the Program in Global Surgery and Social Change. J. Neurosurg. **130**(4), 1055–1064 (2019). https://doi.org/10.3171/2017.11.JNS171500
3. Ellis, D.G., Aizenberg, M.R.: Deep learning using augmentation via registration: 1st place solution to the AutoImplant 2020 challenge. In: Li, J., Egger, J. (eds.) AutoImplant 2020. LNCS, vol. 12439, pp. 47–55. Springer, Cham (2020). https://doi.org/10.1007/978-3-030-64327-0_6
4. Kodym, O., et al.: SkullBreak/SkullFix - dataset for automatic cranial implant design and a benchmark for volumetric shape learning tasks. Data Brief **35** (2021)
5. Kroviakov, A., Li, J., Egger, J.: Sparse convolutional neural network for skull reconstruction. In: Li, J., Egger, J. (eds.) AutoImplant 2021. LNCS, vol. 13123, pp. 80–94. Springer, Cham (2021). https://doi.org/10.1007/978-3-030-92652-6_7

6. Li, J., Pepe, A., Gsaxner, C., Jin, Y., Egger, J.: Learning to rearrange voxels in binary segmentation masks for smooth manifold triangulation. In: Li, J., Egger, J. (eds.) AutoImplant 2021. LNCS, vol. 13123, pp. 45–62. Springer, Cham (2021). https://doi.org/10.1007/978-3-030-92652-6_5

7. Li, J., Egger, J. (eds.): AutoImplant 2021. LNCS, vol. 13123. Springer, Cham (2021). https://doi.org/10.1007/978-3-030-92652-6

8. Li, J., et al.: AutoImplant 2020-first MICCAI challenge on automatic cranial implant design. IEEE Trans. Med. Imaging 40(9), 2329–2342 (2021)

9. Liu, M., et al.: Morphing and sampling network for dense point cloud completion. In: Proceedings of the AAAI Conference on Artificial Intelligence, vol. 34, no. 07, pp. 11596–11603 (2020). https://doi.org/10.1609/aaai.v34i07.6827

10. Mahdi, H., et al.: A U-Net based system for cranial implant design with pre-processing and learned implant filtering. In: Li, J., Egger, J. (eds.) AutoImplant 2021. LNCS, vol. 13123, pp. 63–79. Springer, Cham (2021). https://doi.org/10.1007/978-3-030-92652-6_6

11. Marreiros, F., et al.: Custom implant design for large cranial defects. Int. J. Comput. Assist. Radiol. Surg. 11(12), 2217–2230 (2016)

12. Mohammadi, S.S., et al.: 3DSGrasp: 3D shape-completion for robotic grasp (2023). http://arxiv.org/abs/2301.00866

13. Paszke, A., et al.: PyTorch: an imperative style, high-performance deep learning library. In: Advances in Neural Information Processing Systems, vol. 32, pp. 8024–8035. Curran Associates, Inc. (2019)

14. Pathak, S., Sindhura, C., Gorthi, R.K.S.S., Kiran, D.V., Gorthi, S.: Cranial implant design using V-Net based region of interest reconstruction. In: Li, J., Egger, J. (eds.) AutoImplant 2021. LNCS, vol. 13123, pp. 116–128. Springer, Cham (2021). https://doi.org/10.1007/978-3-030-92652-6_10

15. Qi, C.R., Su, H., Mo, K., Guibas, L.J.: PointNet: deep learning on point sets for 3D classification and segmentation (2017). http://arxiv.org/abs/1612.00593

16. Qi, C.R., Yi, L., Su, H., Guibas, L.J.: PointNet++: deep hierarchical feature learning on point sets in a metric space (2017). http://arxiv.org/abs/1706.02413

17. Wodzinski, M.: The associated repository (2023). https://github.com/MWod/DeepImplant_MICCAI_2023. Accessed 13 July 2023

18. Wodzinski, M., Daniol, M., Hemmerling, D.: Improving the automatic cranial implant design in cranioplasty by linking different datasets. In: Li, J., Egger, J. (eds.) AutoImplant 2021. LNCS, vol. 13123, pp. 29–44. Springer, Cham (2021). https://doi.org/10.1007/978-3-030-92652-6_4

19. Wodzinski, M., et al.: Deep learning-based framework for automatic cranial defect reconstruction and implant modeling. Comput. Methods Programs Biomed. 226, 1–13 (2022)

20. Wu, T., et al.: Density-aware chamfer distance as a comprehensive metric for point cloud completion (2021). http://arxiv.org/abs/2111.12702, [cs]

21. Yang, B., Fang, K., Li, X.: Cranial implant prediction by learning an ensemble of slice-based skull completion networks. In: Li, J., Egger, J. (eds.) AutoImplant 2021. LNCS, vol. 13123, pp. 95–104. Springer, Cham (2021). https://doi.org/10.1007/978-3-030-92652-6_8

22. Yang, Y., Feng, C., Shen, Y., Tian, D.: FoldingNet: point cloud auto-encoder via deep grid deformation (2018). http://arxiv.org/abs/1712.07262, [cs]

23. Yu, L., Li, J., Egger, J.: PCA-skull: 3D skull shape modelling using principal component analysis. In: Li, J., Egger, J. (eds.) AutoImplant 2021. LNCS, vol. 13123, pp. 105–115. Springer, Cham (2021). https://doi.org/10.1007/978-3-030-92652-6_9

24. Yu, X., et al.: PoinTr: diverse point cloud completion with geometry-aware transformers (2021). http://arxiv.org/abs/2108.08839
25. Yu, X., et al.: AdaPoinTr: diverse point cloud completion with adaptive geometry-aware transformers (2023). http://arxiv.org/abs/2301.04545
26. Yuan, W., et al.: PCN: point completion network (2019). http://arxiv.org/abs/1808.00671
27. Zhou, Q., Park, J., Koltun, V.: Open3D: a modern library for 3D data processing. CoRR abs/1801.09847 (2018). http://arxiv.org/abs/1801.09847

Regularized Kelvinlet Functions to Model Linear Elasticity for Image-to-Physical Registration of the Breast

Morgan Ringel[1]([✉]), Jon Heiselman[1,2], Winona Richey[1], Ingrid Meszoely[3], and Michael Miga[1]

[1] Department of Biomedical Engineering, Vanderbilt University, Nashville, TN, USA
morgan.j.ringel@Vanderbilt.edu
[2] Department of Surgery, Memorial Sloan-Kettering Cancer Center, New York, NY, USA
[3] Division of Surgical Oncology, Vanderbilt University Medical Center, Nashville, TN, USA

Abstract. Image-guided surgery requires fast and accurate registration to align preoperative imaging and surgical spaces. The breast undergoes large nonrigid deformations during surgery, compromising the use of imaging data for intraoperative tumor localization. Rigid registration fails to account for nonrigid soft tissue deformations, and biomechanical modeling approaches like finite element simulations can be cumbersome in implementation and computation. We introduce regularized Kelvinlet functions, which are closed-form smoothed solutions to the partial differential equations for linear elasticity, to model breast deformations. We derive and present analytical equations to represent nonrigid point-based translation ("grab") and rotation ("twist") deformations embedded within an infinite elastic domain. Computing a displacement field using this method does not require mesh discretization or large matrix assembly and inversion conventionally associated with finite element or mesh-free methods. We solve for the optimal superposition of regularized Kelvinlet functions that achieves registration of the medical image to simulated intraoperative geometric point data of the breast. We present registration performance results using a dataset of supine MR breast imaging from healthy volunteers mimicking surgical deformations with 237 individual targets from 11 breasts. We include analysis on the method's sensitivity to regularized Kelvinlet function hyperparameters. To demonstrate application, we perform registration on a breast cancer patient case with a segmented tumor and compare performance to other image-to-physical and image-to-image registration methods. We show comparable accuracy to a previously proposed image-to-physical registration method with improved computation time, making regularized Kelvinlet functions an attractive approach for image-to-physical registration problems.

Keywords: Deformation · registration · elasticity · image-guidance · breast · finite element · Kelvinlet

© The Author(s), under exclusive license to Springer Nature Switzerland AG 2023
H. Greenspan et al. (Eds.): MICCAI 2023, LNCS 14228, pp. 344–353, 2023.
https://doi.org/10.1007/978-3-031-43996-4_33

1   Introduction

Image-to-physical registration is a necessary process for computer-assisted surgery to align preoperative imaging to the intraoperative physical space of the patient to in-form surgical decision making. Most intraoperatively utilized image-to-physical regis-trations are rigid transformations calculated using fiducial landmarks [1]. However, with better computational resources and more advanced surgical field monitoring sensors, nonrigid registration techniques have been proposed [2, 3]. This has made image-guided surgery more tractable for soft tissue organ systems like the liver, prostate, and breast [4–6]. This work focuses specifically on nonrigid breast registration, although these methods could be adapted for other soft tissue organs. Current guidance technologies for breast conserv-ing surgery localize a single tumor-implanted seed without providing spatial information about the tumor boundary. As a result, resections can have several centimeters of tissue beyond the cancer margin. Despite seed information and large resections, reoperation rates are still high (~17%) emphasizing the need for additional guidance technologies such as computer-assisted surgery systems with nonrigid registration [7].

Intraoperative data available for registration is often sparse and subject to data col-lection noise. Image-to-physical registration methods that accurately model an elastic soft-tissue environment while also complying with intraoperative data constraints is an active field of research. Determining correspondences between imaging space and geo-metric data is required for image-to-physical registration, but it is often an inexact and ill-posed problem. Establishing point cloud correspondences using machine learning has been demonstrated on liver and prostate datasets [8, 9]. Deep learning image registra-tion methods like VoxelMorph have also been used for this purpose [10]. However, these methods require extensive training data and may struggle with generalizability. Other non-learning image-to-physical registration strategies include [11] which utilized a coro-tational linear-elastic finite element method (FEM) combined with an iterative closest point algorithm. Similarly, the registration method introduced in [12] iteratively updated the image-to-physical correspondence between surface point clouds while solving for an optimal deformation state.

In addition to a correspondence algorithm, a technique for modeling a deformation field is required. Both [11] and [12] leverage FEM, which uses a 3D mesh to solve for unique deformation solutions. However, large deformations can cause mesh distortions with the need for remeshing. Mesh-free methods have been introduced to circumvent this limitation. The element-free Galerkin method is a mesh-free method that requires only nodal point data and uses a moving least-squares approximation to solve for a solution [13]. Other mesh-free methods are reviewed in [14]. Although these methods do not require a 3D mesh, solving for a solution can be costly and boundary condi-tion designation is often unintuitive. Having identified these same shortcomings, [15] proposed regularized Kelvinlet functions for volumetric digital sculpting in computer animation applications. This sculpting approach provided de-formations consistent with linear elasticity without large computational overhead.

In this work, we propose an image-to-physical registration method that uses regu-larized Kelvinlet functions as a novel deformation basis for nonrigid registration. Reg-ularized Kelvinlet functions are analytical solutions to the equations for linear elasticity that we superpose to compute a nonrigid deformation field nearly instantaneously [15].

We utilize "grab" and "twist" regularized Kelvinlet functions with a linearized iterative reconstruction approach (adapted from [12]) that is well-suited for sparse data registration problems. Sensitivity to regularized Kelvinlet function hyperparameters is explored on a supine MR breast imaging dataset. Finally, our approach is validated on an exemplar breast cancer case with a segmented tumor by comparing performance to previously proposed registration methods.

2 Methods

In this section, closed-form solutions to linear elastic deformation responses in an infinite medium are derived to obtain regularized Kelvinlet functions. Then, methods for constructing a superposed regularized Kelvinlet function deformation basis for achieving registration within an iterative reconstructive framework are discussed. Equation notation is written such that constants are italicized, vectors are bolded, and matrices are double-struck letters.

2.1 Regularized Kelvinlet Functions

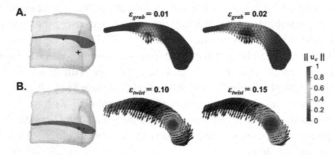

Fig. 1. Visualization of (A) "grab" and (B) "twist" regularized Kelvinlet functions on 2D breast geometry ipsilateral slices at various ε values. (+) denotes nipple location, (·) denotes x_0 location.

Linear elasticity in a homogeneous, isotropic media is governed by the Navier-Cauchy equations in Eq. (1), where E is Young's modulus, v is Poisson's ratio, $\mathbf{u}(x)$ is the displacement vector, and $\mathbf{F}(x)$ is the forcing function. Analytical displacement solutions to Eq. (1) that represent elastostatic states in an infinite solid can be found for specific forcing functions $\mathbf{F}(x)$. Equation (2) represents the forcing function for a point source $\mathbf{F}_\delta(x)$, where f is the point source forcing vector and x_0 is the load location. The closed-form displacement solution for Eq. (1) given the forcing function in Eq. (2) is classically known as the Kelvin state in Eq. (3), rewritten as a function of r where $r = x - x_0$ and $r = \|r\|$. The coefficients are $a = \frac{(1+v)}{2\pi E}$, $b = \frac{a}{[4(1-v)]}$, and $\mathbb{I}$ is the identity matrix.

We note that the deformation response is linear with respect to f, which implies that forcing functions can be linearly superposed. However, practical use of Eq. (3)

becomes numerically problematic in discretized problems because the displacement and displacement gradient become indefinite as x approaches x_0.

$$\frac{E}{2(1+v)}\nabla^2 u(x) + \frac{E}{2(1+v)(1-2v)}\nabla(\nabla \cdot \mathbf{u}(x)) + F(x) = 0 \tag{1}$$

$$\mathbf{F_\delta}(x) = f\delta(x - x_0) \tag{2}$$

$$u(r) = \left[\frac{a-b}{r}\mathbb{I} + \frac{b}{r^3}rr^t\right]f = K(r)f \tag{3}$$

To address numerical singularity, regularization is incorporated with a new forcing function Eq. (4), where $r_\varepsilon = \sqrt{r^2 + \varepsilon^2}$ is the regularized distance, and ε is the regularization radial scale. Solving Eq. (1) using Eq. (4) yields a formula for the first type of regularized Kelvinlet functions used in this work in Eq. (5), which is the closed-form, analytical solution for linear elastic translational ("grab") deformations.

$$\mathbf{F_\varepsilon}(x) = f\left[\frac{15\varepsilon^4}{8\pi}\frac{1}{r_\varepsilon^7}\right] \tag{4}$$

$$\mathbf{u}_{\varepsilon,\text{grab}}(r) = \left[\frac{a-b}{r_\varepsilon}\mathbb{I} + \frac{b}{r_\varepsilon^3}rr^t + \frac{a}{2}\frac{\varepsilon^2}{r_\varepsilon^3}\mathbb{I}\right]f = \mathbb{K}_{\text{grab}}(r)f \tag{5}$$

The second type of regularized Kelvinlet functions represent "twist" deformations which are derived by expanding the previous formulation to accommodate locally affine loads instead of displacement point sources. This is accomplished by associating each component of the forcing function Eq. (4) with the directional derivative of each basis g_i of the affine transformation, leading to the regularized forcing matrix in Eq. (6). An affine loading configuration consisting of pure rotational ("twist") deformation constrains $\mathbb{F}_\varepsilon^{ij}(x)$ to a skew-symmetric matrix that simplifies the forcing function to a cross product about a twisting force vector f in Eq. (7). The pure twist displacement field response $\mathbf{u}_{\varepsilon,\text{twist}}(r)$ to the forcing matrix in Eq. (7) can be represented as the second type of regularized Kelvinlet functions used in this work in Eq. (8).

Superpositions of Eq. (5) and Eq. (8) are used in a registration workflow to model linear elastic deformations in the breast. These deformations are visualized on breast geometry embedded in an infinite medium with varying ε values in Fig. 1.

$$\mathbb{F}_\varepsilon^{ij}(x) = g_i \cdot \nabla f_j\left[\frac{15\varepsilon^4}{8\pi}\frac{1}{r_\varepsilon^7}\right] \tag{6}$$

$$\left[\mathbb{F}_\varepsilon^{ij}\right]_\times(x) = -r \times f\left[\frac{15\varepsilon^4}{8\pi}\frac{1}{r_\varepsilon^7}\right] \tag{7}$$

$$\mathbf{u}_{\varepsilon,\text{twist}}(r) = a\left(\frac{1}{r_\varepsilon^3} + \frac{3\varepsilon^2}{2r_\varepsilon^5}\right)r \times f = [\mathbb{K}_{\text{twist}}(r)]_\times f \tag{8}$$

2.2   Registration Task

For registration, x_0 control point positions for k number of total regularized Kelvinlets "grab" and "twist" functions are distributed in a predetermined configuration. Then, the

f_{grab} and f_{twist} vectors are optimized to solve for a displacement field that minimizes distance error between geometric data inputs.

For a predetermined configuration of regularized Kelvinlet "grab" and "twist" functions centered at different x_0 control point locations, an elastically deformed state can be represented as the summation of all regularized Kelvinlet displacement fields where $\tilde{\mathbf{u}}(x)$ is the superposed displacement vector and $k = k_{grab} + k_{twist}$ in Eq. (9). Equation (9) can be rewritten in matrix form shown in Eq. (10), where α is a concatenated vector of length $3k$ such that $\alpha = [f^1_{grab}, f^2_{grab}, \ldots, f^k_{twist}]$.

$$\tilde{\mathbf{u}}(x) = \sum_{i=1}^{k_{grab}} \mathbf{u}^i_{\varepsilon,\mathbf{grab}}(x) + \sum_{i=1}^{k_{twist}} \mathbf{u}^i_{\varepsilon,\mathbf{twist}}(x) \tag{9}$$

$$\tilde{\mathbf{u}}(x) = \mathbb{K}(x)\alpha \tag{10}$$

This formulation decouples the forcing magnitudes from the Kelvinlet response matrix $\mathbb{K}(x)$, which is composed of column $\mathbf{u}_{\varepsilon,\mathbf{grab}}(x)$ and $\mathbf{u}_{\varepsilon,\mathbf{twist}}(x)$ vectors calculated with unit forcing vectors for each $\mathbb{K}_{grab}(x)$ and $\mathbb{K}_{twist}(x)$ function. This allows for linear scaling of $\tilde{\mathbb{K}}(x)$ using α. By setting x_0 locations, ε_{grab}, and ε_{twist} as hyperparameters, deformation states can be represented by various α vectors with the registration task being to solve for the optimal α vector.

An objective function is formulated to minimize misalignment between the moving space x_{moving} and fixed space x_{fixed} through geometric data constraints. For the breast imaging datasets in this work, we used simulated intraoperative data features that realistically could be collected in a surgical environment visualized in Fig. 2. The first data feature is MR-visible skin fiducial points placed on the breast surface (Fig. 2, red). These fiducials have known point correspondence. The other two data features are an intra-fiducial point cloud of the skin surface (Fig. 2, light blue) and sparse contour samples of the chest wall surface (Fig. 2, yellow). These data features are surfaces that do not have known correspondence. These data feature designations are consistent with implementations in previous work [16, 17].

For a given deformation state, each data feature contributes to the total error measure. For the point data, the error e^i_{point} for each point i is simply the distance magnitude between corresponding points in x_{fixed} and x_{moving} space. For the surface data, the error $e^i_{surface}$ is calculated as the distance from every point i in the x_{fixed} point cloud surface to the closest point in the x_{moving} surface, projected onto the surface unit normal which allows for sliding contact between surfaces.

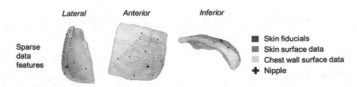

Fig. 2. Sparse data features on breast geometry in the x_{fixed} space.

The optimization using the objective function in Eq. (11) includes two additions to improve the solution. The first is rigid parameters, translation τ and rotation θ, that are optimized simultaneously with the vector α. β represents the deformation state with $\beta = [\alpha, \tau, \theta]$, and this compensates for rigid deformation between x_{fixed} and x_{moving}. The second is a strain energy regularization term e_{SE} which penalizes deformations with large strain energy. e_{SE} is the average strain energy density within the breast geometry, and it is computed for each β at every iteration. It is scaled by weight w_{SE}. The optimal state β is iteratively solved using Levenberg-Marquardt optimization terminating at $|\Delta\Omega(\beta)| < 10^{-12}$.

$$\Omega(\beta) = \frac{1}{n_{point}} \sum_{i=1}^{n_{point}} \left(e_{point}^i\right)^2 + \frac{1}{n_{surface}} \sum_{i=1}^{n_{surface}} \left(e_{surface}^i\right)^2 + w_{SE}(e_{SE})^2 \qquad (11)$$

3 Experiments and Results

In this section, two experiments are conducted. The first explores sensitivity to regularized Kelvinlet function hyperparameters k_{grab}, k_{twist}, ε_{grab}, and ε_{twist} and establishes optimal hyperparameters in a training dataset of 11 breast deformations. The second validates the registration method in a breast cancer patient and compares registration accuracy and computation time to previously proposed methods.

3.1 Hyperparameters Sensitivity Analysis

This dataset consists of supine breast MR images simulating surgical deformations of 11 breasts from 7 healthy volunteers. Volunteers (ages 23–57) were enrolled in a study approved by the Institutional Review Board at Vanderbilt University. Prior to imaging, 26 skin fiducials were distributed on the breast surface. MR images ($0.391 \times 0.391 \times 1$ mm^3 or $0.357 \times 0.357 \times 1$ mm^3) were acquired with the volunteers' arms placed by their sides. This image was used as the x_{moving} space. The volunteers were then instructed to raise one arm above their heads, causing deformation of the ipsilateral breast. A second MR image in the deformed state was acquired to create simulated intraoperative physical data and to use for validation. This second image was used as the x_{fixed} space.

The breast in x_{moving} was segmented at the boundary between the chest wall and breast parenchyma to create a 3D model. The posterior surface was labeled to inform x_0 control point locations. The skin fiducials and intra-fiducial surface point clouds were labeled in both images as data features. Sparse tracked ultrasound data collection patterns were projected on the posterior surface for use as the third data feature. Subsurface anatomical targets were labeled in both images and used to compute target error after registration.

Three configurations were explored to test different distributions of grab and/or twist regularized Kelvinlet functions: grab functions only, twist functions only, and a combination of grab and twist functions. Grab function control points were distributed evenly on the posterior surface of the breast to approximate forces from the chest wall. Twist function control points were distributed evenly within the breast to approximate internal body forces. Three hyperparameter sweeps were used:

- Configuration 1: $k_{grab} = \{10, 40, 70, 100\}$, $\varepsilon_{grab} = \{0.005, 0.05, 0.5\}$
- Configuration 2: $k_{twist} = \{10, 40, 70, 100\}$, $\varepsilon_{twist} = \{0.05, 0.1, 0.2\}$
- Configuration 3: $k_{grab} = 40$, $\varepsilon_{grab} = 0.05$, $k_{twist} = \{1, 5, 10, 20\}$, $\varepsilon_{twist} = \{0.05, 0.1, 0.2\}$

For all registrations, mechanical breast properties were set at $\nu = 0.45$, $E = 2100$ Pa, and $w_E = 10^{-9}$ Pa^{-2} [16, 18]. Accuracy was evaluated by measuring target error (distance magnitude between targets in x_{fixed} and registered x_{moving} spaces) for all targets in 11 breast imaging sets totaling 237 targets per registration.

Target error results from hyperparameter sweep registrations are shown in Fig. 3. The registration with the lowest root mean squared error was from configuration 3 $k_{grab} = 40$, $\varepsilon_{grab} = 0.05$, $k_{twist} = 1$, $\varepsilon_{twist} = 0.1$. These hyperparameters were used on a different dataset for validating and comparing the registration method in Sect. 3.2.

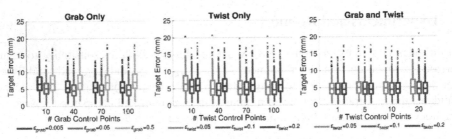

Fig. 3. Target error results from regularized Kelvinlet functions hyperparameter sweeps. Outliers are noted as (x) and are 1.5•IQR.

3.2 Registration Methods Comparison

This dataset consists of supine breast MR images simulating surgical deformations from one breast cancer patient. A 71-year-old patient with invasive mammary carcinoma in the left breast was enrolled in a study approved by the Institutional Review Board at Vanderbilt University. Skin fiducial placement, image acquisition, arm placement, and preprocessing steps followed the same protocol detailed in Sect. 3.1. The tumor was segmented in both images by a subject matter expert, and a 3D tumor model was created to evaluate tumor overlap metrics after registration.

Regularized Kelvinlet function registration was compared to 3 other registration methods: rigid registration, an FEM-based image-to-physical registration method, and an image-to-image registration method. A point-based rigid registration using the skin fiducials provided a baseline comparator for accuracy without deformable correction. The FEM-based image-to-physical registration method, detailed in [12] and implemented in breast in [16], utilizes the same optimization scheme as this method but with an FEM-generated basis. $k = 40$ control points were used for the FEM-based registration. The image-to-image registration method was a symmetric diffeomorphic method with explicit B-spline regularization publicly available in the Advanced Normalization Toolkit (ANTs) repository [19, 20]. Image-to-image registration would not be possible for intraoperative registration in most surgical settings. However, it was included to demonstrate accuracy when volumetric imaging data is available, as opposed to sparse geometric point data as in the surgical application case. The rigid and image-to-physical registrations were performed on a single thread of a 3.6 GHz AMD Ryzen 7 3700X CPU. Image-to-image registration was multithreaded on 2.3 GHz Intel Xeon (E5–4610 v2) CPUs.

Registration results for the 4 methods are shown in Table 1. The regularized Kelvinlet method accuracy was comparable (if not slightly improved) to the FEM-based method for this example case. Runtime for the regularized Kelvinlet method was improved compared to the FEM-based method. As expected, registration without deformable correction was poor, and image-to-image registration had the best accuracy. Registered tumor geometry results are shown in Fig. 4.

Table 1. Registration performance for 4 methods. HD – Hausdorff distance.

		Rigid	Image-to-Physical		Image-to-Image
			FEM	R. Kelvinlets	
Point metrics	Fiducial Error (mm)	7.4 ± 2.0	0.7 ± 0.5	1.4 ± 0.6	2.0 ± 1.7
	Target Error (mm)	6.1 ± 1.4	3.3 ± 1.1	3.0 ± 1.1	2.3 ± 1.5
Tumor overlap metrics	Dice Coefficient	2.3%	32.7%	49.5%	85.8%
	Centroid Distance (mm)	7.3	4.4	3.5	1.3
	Modified HD (mm)	4.1	2.2	1.7	0.6
Runtime (seconds)		< 1	188	14	15,942

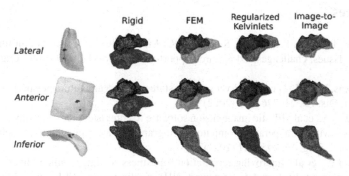

Fig. 4. Tumor overlap after registration. Black – x_{fixed} tumor used for validation. Blue – rigidly registered x_{moving} tumor. Green – FEM-based registered x_{moving} tumor. Pink – regularized Kelvinlet function registered x_{moving} tumor. Orange – image-to-image registered x_{moving} tumor.

4 Limitations and Conclusion

Several limitations should be noted. Regularized Kelvinlet functions describe solutions that assume a physical embedding within an infinite elastic domain, which does not account for organ-specific geometry. This approach may not be well suited for problems where geometry has significant influence. This method is derived from a linear elastic model, and nonlinear models are known to better describe soft tissue mechanics. Additionally, this method assumes homogeneity and isotropy – it does not account for different tissue types and directional structures in the breast. With regards to clinical feasibility, supine MR imaging with skin fiducials is not the standard-of-care. However, using supine MR imaging for surgery is becoming increasingly investigated, and previous work demonstrated the potential of ink-based skin fiducial markings on the breast [21, 22]. Despite these limitations, this method's accuracy and speed may be appropriate for surgical guidance applications.

In this work, we demonstrated the use of regularized Kelvinlet functions for image-to-physical registration of the breast. We achieved near real-time registration with comparable accuracy to previously proposed methods. We believe that this approach is generalizable to other soft-tissue organ systems and is well-suited for improving navigation during image-guided surgeries.

Acknowledgements. This work was supported by the National Institutes of Health through Grant Nos. R01EB027498 and T32EB021937, the National Science Foundation for a Graduate Research Fellowship awarded to M.R., and the Vanderbilt Center for Human Imaging supported by Grant No. 1S10OD021771-01 for the 3T MRI.

References

1. Alam, F., Rahman, S.U., Ullah, S., Gulati, K.: Medical image registration in image guided surgery: Issues, challenges and research opportunities. Biocybern. Biomed. Eng. **38**, 71–89 (2018)
2. Gavriilidis, P., et al.: Navigated liver surgery: state of the art and future perspectives. Hepatob. Pancreat. Dis. Int. **21**, 226–233 (2022)
3. Schmidt, F.A., et al.: Elastic image fusion software to coregister preoperatively planned pedicle screws with intraoperative computed tomography data for image-guided spinal surgery. Int. J. Spine Surg. **15**, 295–301 (2021)
4. Collins, J.A., et al.: Improving registration robustness for image-guided liver surgery in a novel human-to-phantom data framework HHS public access. IEEE Trans. Med. Imaging. **36**, 1502–1510 (2017)
5. Conley, R.H., et al.: Realization of a biomechanical model-assisted image guidance system for breast cancer surgery using supine MRI. Int. J. Comput. Assist. Radiol. Surg. **10**, 1985 (2015)
6. Zettinig, O., et al.: Multimodal image-guided prostate fusion biopsy based on automatic deformable registration. Int. J. Comput. Assist. Radiol. Surg. **10**(12), 1997–2007 (2015). https://doi.org/10.1007/s11548-015-1233-y
7. Kaczmarski, K., et al.: Surgeon re-excision rates after breast-conserving surgery: a measure of low-value care. J. Am. Coll. Surg. **228**, 504-512.e2 (2019)

8. Pfeiffer, M., et al.: Non-rigid volume to surface registration using a data-driven biomechanical model. In: Martel, A.L., et al. (eds.) Medical Image Computing and Computer Assisted Intervention – MICCAI 2020. LNCS, vol. 12264, pp. 724–734. Springer, Cham (2020). https://doi.org/10.1007/978-3-030-59719-1_70

9. Fu, Y., et al.: Biomechanically constrained non-rigid MR-TRUS prostate registration using deep learning based 3D point cloud matching. Med. Image Anal. **67**, 101845 (2021)

10. Balakrishnan, G., Zhao, A., Sabuncu, M.R., Guttag, J., Dalca, A.V.: VoxelMorph: a learning framework for deformable medical image registration. IEEE Trans. Med. Imaging **38**(8), 1788–1800 (2019). https://doi.org/10.1109/TMI.2019.2897538

11. Peterlík, I., et al.: Fast elastic registration of soft tissues under large deformations. Med. Image Anal. **45**, 24–40 (2018)

12. Heiselman, J.S., Jarnagin, W.R., Miga, M.I.: Intraoperative correction of liver deformation using sparse surface and vascular features via linearized iterative boundary reconstruction. IEEE Trans. Med. Imaging. **39**, 2223–2234 (2020)

13. Belytschko, T., Lu, Y.Y., Gu, L.: Element-free Galerkin methods. Int. J. Numer. Methods Eng. **37**, 229–256 (1994)

14. Zhang, L.W., Ademiloye, A.S., Liew, K.M.: Meshfree and particle methods in biomechanics: prospects and challenges. Arch. Comput. Methods Eng. **26**(5), 1547–1576 (2018). https://doi.org/10.1007/s11831-018-9283-2

15. De Goes, F., James, D.L.: Regularized kelvinlets: sculpting brushes based on fundamental solutions of elasticity. ACM Trans. Graph. **36**(4), 1–11 (2017). https://doi.org/10.1145/3072959.3073595

16. Richey, W.L., Heiselman, J.S., Ringel, M.J., Meszoely, I.M., Miga, M.I.: Computational imaging to compensate for soft-tissue deformations in image-guided breast conserving surgery. IEEE Trans. Biomed. Eng. **69**, 3760–3771 (2022)

17. Richey, W.L., Heiselman, J.S., Ringel, M.J., Meszoely, I.M., Miga, M.I.: Tumor deformation correction for an image guidance system in breast conserving surgery. In: SPIE Proceedings, 12034 (2022)

18. Griesenauer, R.H., Weis, J.A., Arlinghaus, L.R., Meszoely, I.M., Miga, M.I.: Breast tissue stiffness estimation for surgical guidance using gravity-induced excitation. Phys. Med. Biol. **62**, 4756–4776 (2017)

19. Tustison, N.J., Avants, B.B.: Explicit B-spline regularization in diffeomorphic image registration. Front. Neuroinform. **7**, 39 (2013)

20. Ringel, M.J., Richey, W.L., Heiselman, J.S., Luo, M., Meszoely, I.M., Miga, M.I.: Supine magnetic resonance image registration for breast surgery: insights on material mechanics. J. Med. Imaging. **9**, 065001 (2022)

21. Gombos, E.C., et al.: Intraoperative supine breast MR imaging to quantify tumor deformation and detection of residual breast cancer: preliminary results. Radiology **281**, 720–729 (2016)

22. Richey, W.L., Heiselman, J.S., Ringel, M.J., Ingrid, M., Meszoely, M.I., Miga, M.I.: Soft tissue monitoring of the surgical field: detection and tracking of breast surface deformations. IEEE Trans. Biomed. Eng. **70**(7), 2002–2012 (2023). https://doi.org/10.1109/TBME.2022.3233909

Synthesising Rare Cataract Surgery Samples with Guided Diffusion Models

Yannik Frisch[1][(✉)], Moritz Fuchs[1], Antoine Sanner[1], Felix Anton Ucar[2],
Marius Frenzel[2], Joana Wasielica-Poslednik[2], Adrian Gericke[2],
Felix Mathias Wagner[2], Thomas Dratsch[3], and Anirban Mukhopadhyay[1]

[1] Technical University Darmstadt, Darmstadt, Germany
yannik.frisch@gris.tu-darmstadt.de
[2] Universitätsmedizin der Johannes Gutenberg-Universität Mainz, Mainz, Germany
[3] Uniklinik Köln, Cologne, Germany

Abstract. Cataract surgery is a frequently performed procedure that demands automation and advanced assistance systems. However, gathering and annotating data for training such systems is resource intensive. The publicly available data also comprises severe imbalances inherent to the surgical process. Motivated by this, we analyse cataract surgery video data for the worst-performing phases of a pre-trained downstream tool classifier. The analysis demonstrates that imbalances deteriorate the classifier's performance on underrepresented cases. To address this challenge, we utilise a conditional generative model based on Denoising Diffusion Implicit Models (DDIM) and Classifier-Free Guidance (CFG). Our model can synthesise diverse, high-quality examples based on complex multi-class multi-label conditions, such as surgical phases and combinations of surgical tools. We affirm that the synthesised samples display tools that the classifier recognises. These samples are hard to differentiate from real images, even for clinical experts with more than five years of experience. Further, our synthetically extended data can improve the data sparsity problem for the downstream task of tool classification. The evaluations demonstrate that the model can generate valuable unseen examples, allowing the tool classifier to improve by up to 10% for rare cases. Overall, our approach can facilitate the development of automated assistance systems for cataract surgery by providing a reliable source of realistic synthetic data, which we make available for everyone.

Keywords: Generative Models · Denoising Diffusion Models · Cataract Surgery

1 Introduction

Cataract surgeries are amongst the most frequently performed treatments, with 4,000 to 10,000 annual operations per million people [27]. The high demand

Supplementary Information The online version contains supplementary material available at https://doi.org/10.1007/978-3-031-43996-4_34.

naturally asks for automation and advanced assistance systems and has seen increasing attention within the CAI community in recent years [1,6,22].

Nevertheless, the publicly available data for training such systems is limited: given the nature of the surgeries, certain surgical *phases* take more time than others. Further, there are variances in their length based on the surgeon's skill and the patient's particular needs. Since surgical *tool* usage is strongly coupled with the surgical phase, certain phases and tools are shown more frequently than others, constituting an inherent imbalance in the data. As displayed in Fig. 1 for the CATARACTS dataset [1], such imbalances impact the performance on downstream tasks, e.g. surgical phase prediction or tool classification.

One cannot simply gather new data showing unusual tools to perform the required actions during a surgical step. Therefore, we must find different ways to represent them in the data and counteract the imbalance. The usual countermeasures in the form of oversampling can increase prediction accuracy [1,6,22]. Still, they only alter the number of times an underrepresented sample is seen during training, resulting in a fragile representation of the tools and phases and hindering generalisation. Generative models [9,18,24,26] can potentially solve this by synthesising unseen examples for underrepresented tool and phase combinations.

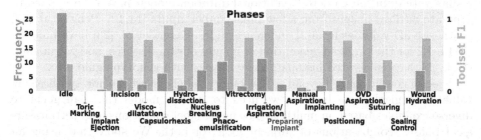

Fig. 1. Distribution of CATARACTS phases. Except for *Idle* - which can appear anytime - all phases are displayed in the usual chronological order. The dataset yields severe class imbalances regarding the available frames per phase (darker blue), which results in performance drops for underrepresented phases (lighter blue). (Color figure online)

Regarding image quality, generative models based on *diffusion models* reached superior performance over alternative methods in the recent past [4,7,16,21,25]. Despite the successes, these models have not yet found application in Surgical Data Science. In the broader medical domain, they have been utilised to generate thorax CT scans [10], brain MRI [5,10] and breast and knee MRI scans [10].

Although these applications have shown promising results, there is a demand for conditional image generation for Surgical Data Science. Since most downstream applications consist of supervised methods, they require training targets, and the likelihood of unconditionally generating diverse samples for unusual cases is very low. For conditional generation with denoising diffusion models, *classifier guidance* has been introduced recently [4,15] but requires computationally extensive parallel training of a separate classifier model. Instead, *Classifier-Free Guidance* (CFG) [8] yields a simple trick to achieve class-constrained generative

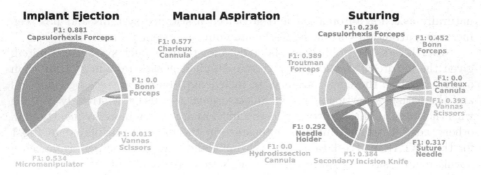

Fig. 2. Ground truth toolset occurrence for the worst performing CATARACTS phases. Some toolsets, e.g. *(Bonn Forceps, Capsulorhexis Forceps)* during *Implant Ejection* (left), are rarely present and poorly detected, deteriorating the overall performance. We focus on such rare toolsets to generate new samples for a phase. E.g. for *Manual Aspiration* (middle), we mainly want additional samples showing the *Hydrodissection Canulla*. The chord diagram for the *Suturing* phase (right) shows the complexity of such occurrences.

results with diffusion models. Conditional diffusion models have been applied to generate medical images from a few binary label inputs [14,19]. Peng et al. [17] synthesise 3D volumes from 2D reference slides. Sagers et al. [23] have built on DALL·E2 for the targeted generation of images of skin diseases, and Moghadam et al. [13] have generated histopathology images with genotype guidance.

Precise conditioning beyond a few binary labels is crucial for synthesising valuable surgical data. Instead, we need to train a model that can generate diverse examples based on multi-class or multi-label conditions, e.g. certain surgical phases, combinations of surgical tools, or both. We show that using an adapted denoising diffusion model together with CFG can yield high-quality samples of cataract surgery data, even for rare cases such as the CATARACTS phase and tool combinations shown in Fig. 2.

To the best of our knowledge, ours is the first work combining CFG with diffusion models to efficiently **generate realistic cataract surgery data** with a complex underlying label structure. Additionally, we examine the cataract video data for the worst-performing phases of a pre-trained tool usage classifier. We then leverage the conditional denoising diffusion model to generate unseen samples for these phases. Our conditioned tools are **recognisable by the tool classifier** and are **hard to differentiate from real images**, even for clinicians with more than five years of experience. Further, we demonstrate how our synthetically extended data can **alleviate the data sparsity problem for the downstream task**. Overall, our evaluations show that the model can generate valuable examples to build the bridge to clinical application.

2 Method

The following section describes how we build our generative diffusion model and integrate CFG to generate cataract surgery frames conditioned on surgical phases and tools. Furthermore, we provide an analysis of the worst-performing surgical steps for a pre-trained tool classifier model. Finally, we demonstrate the sampling procedure using the generative model to improve the classifier.

2.1 Denoising Diffusion Probabilistic Models

The fundamental underlying idea of *Denoising Diffusion Probabilistic Models* (DDPMs) [7] is a *forward diffusion process* that gradually adds Gaussian noise to an image x. This process is defined by $q(x_t|x_{t-1}) = \mathcal{N}(x_t; \sqrt{1 - \beta_t}x_{t-1}, \beta_t\mathbf{I})$, which uses a pre-defined variance schedule $\{\beta_t \in (0,1)\}_{t=1}^{T}$. Eventually, when $T \to \infty$, x_T becomes equivalent to an isotropic Gaussian distribution. When we can learn the *reverse process* $q(x_{t-1}|x_t)$, we can generate samples starting from a simple Gaussian noise $x_T \sim \mathcal{N}(\mathbf{0}, \mathbf{I})$. To achieve this, one can approximate the conditional probabilities by $p_\theta(x_{t-1}|x_t) = \mathcal{N}(x_{t-1}; \mu_\theta(x_t, t), \Sigma_\theta(x_t, t))$. In practice and after some mathematical simplifications, this reduces the reverse process to estimating the noise ϵ_t between x_t and x_{t+1}, as shown by Ho et al. [7]. Usually, the noise is parameterised by a UNet-type architecture ϵ_θ, optimised by

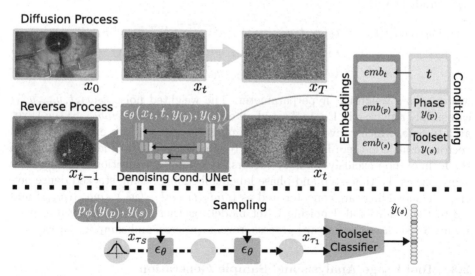

Fig. 3. Illustration of the Conditional Denoising UNet. The model ϵ_θ is trained to reverse the Diffusion Process, mapping a noisy sample x_t to a less noisy x_{t-1}. Condition embeddings based on the surgical phase $y_{(p)}$ and the toolset $y_{(s)}$ are concatenated with the diffusion time step embedding and fed into every level of the UNet to guide targeted sample generation. We utilise the final model to synthesise realistic examples and improve the predictions $\hat{y}_{(s)}$ of a toolset classifier.

minimising the simplified objective

$$\mathcal{L}_t = \mathbb{E}_{t \sim [1,T], x_0, \epsilon_t}[||\epsilon_t - \epsilon_\theta(x_t, t)||^2]$$
$$= \mathbb{E}_{t \sim [1,T], x_0, \epsilon_t}[||\epsilon_t - \epsilon_\theta(\sqrt{\alpha_t}x_0 + \sqrt{1 - \alpha_t}\epsilon_t, t)||^2] \tag{1}$$

where $\alpha_{1:T} \in (0,1]^T, \alpha_t = (1 - \beta_t)\alpha_{t-1}$.

Denoising Diffusion Implicit Models (DDIMs) [25] are a generalisation of these formulations for non-Markovian, more efficient sampling. Instead of the complete Markov chain, they are defined on a reduced set of intermediate latents $\{x_{\tau_1}, ..., x_{\tau_S}\}$, where $[\tau_1, ..., \tau_S] \subseteq [1, ..., T]$. This reduction results in significantly fewer inference steps required to generate samples.

2.2   Classifier-Free Guidance

By utilising *Classifier-Free Guidance* (CFG) [8], we can learn the unconditional model $p_\theta(x)$ and the model $p_\theta(x|y_{(p)}, y_{(s)})$ conditioned on phase $y_{(p)}$ and toolset $y_{(s)}$ using a *single* neural network. The corresponding gradient is given by

$$\nabla_{x_t} \log p(y_{(p)}, y_{(t)}|x_t) = \nabla_{x_t} \log p(x_t|y_{(p)}, y_{(s)}) - \nabla_{x_t} \log p(x_t)$$
$$= -\frac{1}{\sqrt{1 - \bar{\alpha}_t}}(\epsilon_\theta(x_t, t, y_{(p)}, y_{(s)}) - \epsilon_\theta(x_t, t)) \tag{2}$$

This gradient yields

$$\bar{\epsilon}_\theta(x_t, t, y_{(p)}, y_{(s)}) = \epsilon_\theta(x_t, t, y_{(p)}, y_{(s)}) - \sqrt{1 - \bar{\alpha}_t}w\nabla_{x_t} \log p(y_{(p)}, y_{(s)}|x_t)$$
$$= \epsilon_\theta(x_t, t, y_{(p)}, y_{(s)}) + w(\epsilon_\theta(x_t, t, y_{(p)}, y_{(s)}) - \epsilon_\theta(x_t, t)) \tag{3}$$
$$= (w + 1)\epsilon_\theta(x_t, t, y_{(p)}, y_{(s)}) - w\epsilon_\theta(x_t, t)$$

where w is a weighting hyperparameter. The weighted noise $\bar{\epsilon}_\theta(x_t, t, y_{(p)}, y_{(s)})$ can simply replace ϵ in Eq. 1. We add an embedding module $emb_{(p)}$ to the UNet architecture, which converts categorical phase labels $y_{(p)}$ into one-hot encoded vectors to include them into the input. To simultaneously include tool labels in the form of (non-exclusive) binary vectors, we compute projections $emb_{(s)}$ of the same size as the time-step and phase label embeddings using stacked dense layers. All embeddings are concatenated as $\{emb_t(t), emb_{(p)}(y_{(p)}), emb_{(s)}(y_{(s)})\}$ and fed to the conditional denoising UNet model together with the noisy image x_t. Figure 3 visualises the forward process, reverse process, and sampling procedure.

2.3   Tool Usage Analysis and Sample Generation

Following Roychowdhury et al. [1,22], we deploy a ResNet50 architecture to predict tools present in a given frame. An inspection of the phase-wise performance reveals that the model underperforms for underrepresented phases, as shown in Fig. 2. Certain tool combinations are sparsely used in these phases, causing a significant drop in prediction performance. Appendix Fig. 6 displays the distribution of toolset labels for all surgical steps. We synthesise new examples for every phase to smooth out the distribution. We then re-train the classifier model on the original and extended data combined. Adding samples based on toolsets y_s and phase labels y_p requires a throughout pre-selection of query inputs due to the complex underlying latent structure. To automatise this process, we compute the joint probabilities $p_\phi(y_s, y_p)$ from the available CATARACTS annotations. We can then generate rare cases for a given phase by sampling tool labels from the inverse of $p_\phi(y_s, y_p) = p(y_p)p(y_s|y_p)$. This ensures we synthesise examples of underrepresented cases.

3   Experiments and Results

In this section we explain our experimental setup, demonstrate the synthesis of high-quality samples and show how these can improve the downstream model's performance on challenging phases.

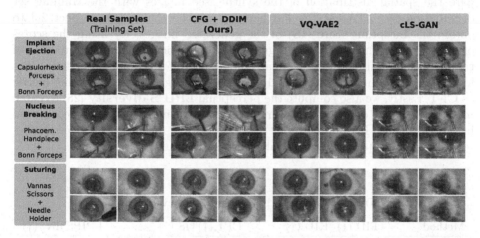

Fig. 4. Qualitative examples for the least common toolsets of the three most challenging phases of CATARACTS. The proposed method produces superior results compared to the baselines, which struggle with the rare combinations of tools and phases shown. The realistically generated tools are especially noteworthy.

3.1   Experimental Setup and Dataset

To evaluate the quality of synthesised images, we generate 30,000 samples with phase and toolset conditions sampled from $p_\phi^{-1}(y_s, y_p) = (1 - p_\phi(y_s, y_p))/\sum(1 - p_\phi(y_s, y_p))$, as explained in Sect. 2.3. The resulting number of examples is close to the test split size of CATARACTS sampled at 3 FPS. We compare the proposed approach to state-of-the-art baselines for conditional generative modelling: A conditional LS-GAN (cLS-GAN) [11,12] and VQ-VAE2 [20]. For the latter, we deploy a PixelSNAIL prior [3] for bottom- and top-level features. Every model is trained on two NVIDIA A40 GPUs for 500 epochs with about 45,000 training examples each. Other hyperparameters vary for each model and can be accessed next to the code to reproduce our results and the generated data at https://github.com/MECLabTUDA/CataSynth. We take $\tau_S = 200$ denoising steps using the DDIM formulation [25] to generate our images, yielding a reasonable inference speed and sample quality trade-off. The displayed and evaluated images are generated with a CFG weight of $\omega = 2.0$. We use a random chance of $p = 0.1$ for the unconditional model during training. All models are trained on images of 128×128 pixels and up-sampled with bilinear interpolation to 270×480 for displaying purposes.

3.2 Quantitative Image Quality

We deploy a variety of quantitative metrics to assess the quality of generated images, for which Table 1 lists the results. Firstly, **FID** [2] and **KID** [2] compare the spatial distribution of the synthesised images with the training set distribution of CATARACTS. Further, we use the classifier from Sect. 2.3 to obtain the Inception Score (**IS**) of the generated images. By using the scores of a classifier trained for tool recognition, this metric yields a measurement of tool realism. Additionally, we evaluate the F1 score for the pre-trained classifier identifying tools in the conditionally generated images. We denote this metric as **CF1**. Lastly, we also compute the perceptual **LPIPS diversity** [28] to catch mode collapses, a common problem with generated models, leading to reduced image variability. In summary, our model generates images of superior quality regarding spatial properties, tool label preservation and diversity.

Table 1. Quantitative image quality evaluation.

Method	FID ($\downarrow$)	KID ($\downarrow$)	CF1 ($\uparrow$)	IS ($\uparrow$)	LPIPS div. ($\uparrow$)
cLS-GAN	284.5	0.319 ± 0.005	0.000	1.376 ± 0.009	0.559 ± 0.119
VQ-VAE2	88.9	0.096 ± 0.002	0.089	2.149 ± 0.035	0.502 ± 0.061
CFG + DDIM	**43.7**	$\mathbf{0.030 \pm 0.002}$	**0.433**	$\mathbf{6.428 \pm 0.115}$	$\mathbf{0.595 \pm 0.070}$

3.3 Qualitative Results and User Study

Figure 4 displays generated samples for the rarest toolsets of CATARACTS' three most challenging phases. Additional examples for the other phases and randomly chosen toolsets are displayed in the Appendix. The qualitative results reflect the quantitative metrics and illustrate our method's superior image quality and tool preservation. Besides qualitative evaluations, we conduct a user study to assess the realism of our generated images. Therefore, we let six clinicians survey 50 generated and 50 real images of size 235×132 pixels in a randomised side-by-side view. In every example, both images show the same phase and tool combination, and participants must distinguish the real image from the synthesised sample. We grouped the participants by domain experience, with domain experts (DE) having more than five years of domain expertise in cataract surgery. On average, the miss rate or false classification rate (FR) was 0.61. This result translates into **clinical experts favouring the generated images 61% of the time** and highlights how realistic they appear. Their answers have an average Matthews Correlation Coefficient (MCC) of -0.216, showing low validity of the subjects' binary decisions. Individual MCC and FR scores are given in Table 2.

Table 2. Results for user study on image realism.

Clinician	NDE1	NDE2	NDE3	DE1	DE2	DE3
MCC	-0.961	-0.288	-0.201	-0.233	0.098	0.288
FR	49/50	32/50	30/50	31/50	23/50	18/50

3.4   Downstream Tool Classification

Finally, we re-train the tool-set classifier on a combined dataset of the original and the synthesised samples. As shown in Table 3, re-training on the combined data (*Extended*) improves the tool-set classifier's prediction performance on the original test data compared to training solely on the original training data (*Original*). For completeness, we also report the performance from fitting the classifier exclusively on the synthetic data and evaluating it on the test split of CATARACTS, denoted *CAS* [20]. Figure 5 displays the individual differences in the classifier's F1 scores for the originally worst-performing phases. Extending the data with synthetic samples yields **performance gains for five of the seven most critical phases.**

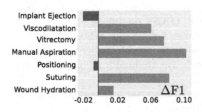

Fig. 5. Phase-wise performance changes for critical phases after re-training including synthetic data.

Table 3. Tool-set prediction performance on the CATARACTS test split for different types of data.

Data	F1 (↑)	AUROC (↑)	Acc. (↑)
Original	0.897	0.986	0.9921
Extended	**0.916**	**0.989**	**0.9924**
Synthetic (CAS)	0.299	0.681	0.9502

4   Conclusion

We present a generative model based on denoising diffusion models and classifier-free guidance, powerful enough to synthesise cataract surgery images that are hard to distinguish for a pre-trained tool classifier and clinical experts. For under-represented phases, state-of-the-art baselines tend to produce frames that show eyes without correct anatomy or barely recognisable tools, resulting in a significant performance gap. Distortions in the dataset further deteriorate their learning. On the contrary, we demonstrate that the proposed approach outperforms these baselines in terms of image quality and tool preservation. As a limitation of our approach, we found that the generalisation capabilities must be strengthened to generate unreasonable samples, e.g. completely wrong tools during a phase. Such samples can happen if they are present in the data. Though, a targeted generation would require a more substantial representation. Additionally, while tool realism is significantly better for the proposed method, the *CF1* and *CAS* scores indicate that it can be further improved. Besides, the underlying class imbalances and lack of available data are even more severe for the downstream task of anatomy and tool segmentation. In future work, we will extend the proposed method to generate segmentation targets, temporally connected data and deploy a tighter structure for conditioning. Overall, we are the first to have shown how conditional diffusion models can successfully be applied to mend data sparsity and generate high-quality cataract surgery images suitable for clinical application. These improvements can bring computer-assisted cataract surgery one step closer to the next level of automation.

References

1. Al Hajj, H., et al.: CATARACTS: challenge on automatic tool annotation for cataRACT surgery. Med. Image Anal. **52**, 24–41 (2019)
2. Bińkowski, M., Sutherland, D.J., Arbel, M., Gretton, A.: Demystifying MMD GANs. arXiv preprint arXiv:1801.01401 (2018)
3. Chen, X., Mishra, N., Rohaninejad, M., Abbeel, P.: PixelSNAIL: an improved autoregressive generative model. In: International Conference on Machine Learning, pp. 864–872. PMLR (2018)
4. Dhariwal, P., Nichol, A.: Diffusion models beat GANs on image synthesis. In: Advances in Neural Information Processing Systems, vol. 34, pp. 8780–8794 (2021)
5. Dorjsembe, Z., Odonchimed, S., Xiao, F.: Three-dimensional medical image synthesis with denoising diffusion probabilistic models. In: Medical Imaging with Deep Learning (2022)
6. Grammatikopoulou, M., et al.: CaDIS: cataract dataset for surgical RGB-image segmentation. Med. Image Anal. **71**, 102053 (2021)

7. Ho, J., Jain, A., Abbeel, P.: Denoising diffusion probabilistic models. In: Advances in Neural Information Processing Systems, vol. 33, pp. 6840–6851 (2020)

8. Ho, J., Salimans, T.: Classifier-free diffusion guidance. arXiv preprint arXiv:2207.12598 (2022)

9. Kalia, M., Aleef, T.A., Navab, N., Black, P., Salcudean, S.E.: Co-generation and segmentation for generalized surgical instrument segmentation on unlabelled data. In: de Bruijne, M., et al. (eds.) MICCAI 2021, Part IV. LNCS, vol. 12904, pp. 403–412. Springer, Cham (2021). https://doi.org/10.1007/978-3-030-87202-1_39

10. Khader, F., et al.: Medical diffusion-denoising diffusion probabilistic models for 3D medical image generation. arXiv preprint arXiv:2211.03364 (2022)

11. Mao, X., Li, Q., Xie, H., Lau, R.Y., Wang, Z., Paul Smolley, S.: Least squares generative adversarial networks. In: Proceedings of the IEEE International Conference on Computer Vision, pp. 2794–2802 (2017)

12. Mirza, M., Osindero, S.: Conditional generative adversarial nets. arXiv preprint arXiv:1411.1784 (2014)

13. Moghadam, P.A., et al.: A morphology focused diffusion probabilistic model for synthesis of histopathology images. In: Proceedings of the IEEE/CVF Winter Conference on Applications of Computer Vision, pp. 2000–2009 (2023)

14. Müller-Franzes, G., et al.: Diffusion probabilistic models beat GANs on medical images. arXiv preprint arXiv:2212.07501 (2022)

15. Nichol, A., et al.: GLIDE: towards photorealistic image generation and editing with text-guided diffusion models. arXiv preprint arXiv:2112.10741 (2021)

16. Nichol, A.Q., Dhariwal, P.: Improved denoising diffusion probabilistic models. In: International Conference on Machine Learning, pp. 8162–8171. PMLR (2021)

17. Peng, W., Adeli, E., Zhao, Q., Pohl, K.M.: Generating realistic 3D brain MRIs using a conditional diffusion probabilistic model. arXiv preprint arXiv:2212.08034 (2022)

18. Pfeiffer, M., et al.: Generating Large Labeled Data Sets for Laparoscopic Image Processing Tasks Using Unpaired Image-to-Image Translation. In: Shen, D., et al. (eds.) MICCAI 2019, Part V. LNCS, vol. 11768, pp. 119–127. Springer, Cham (2019). https://doi.org/10.1007/978-3-030-32254-0_14

19. Pinaya, W.H., et al.: Brain imaging generation with latent diffusion models. In: Mukhopadhyay, A., Oksuz, I., Engelhardt, S., Zhu, D., Yuan, Y. (eds.) DGM4MICCAI 2022. LNCS, vol. 13609, pp. 117–126. Springer, Cham (2022). https://doi.org/10.1007/978-3-031-18576-2_12

20. Razavi, A., Van den Oord, A., Vinyals, O.: Generating diverse high-fidelity images with VQ-VAE-2. In: Advances in Neural Information Processing Systems, vol. 32 (2019)

21. Rombach, R., Blattmann, A., Lorenz, D., Esser, P., Ommer, B.: High-resolution image synthesis with latent diffusion models. In: Proceedings of the IEEE/CVF Conference on Computer Vision and Pattern Recognition, pp. 10684–10695 (2022)

22. Roychowdhury, S., Bian, Z., Vahdat, A., Macready, W.G.: Identification of surgical tools using deep neural networks. Technical report, D-Wave Systems Inc. (2017)

23. Sagers, L.W., Diao, J.A., Groh, M., Rajpurkar, P., Adamson, A.S., Manrai, A.K.: Improving dermatology classifiers across populations using images generated by large diffusion models. arXiv preprint arXiv:2211.13352 (2022)

24. Sommersperger, M., et al.: Surgical scene generation and adversarial networks for physics-based iOCT synthesis. Biomed. Opt. Express **13**(4), 2414–2430 (2022)

25. Song, J., Meng, C., Ermon, S.: Denoising diffusion implicit models. arXiv preprint arXiv:2010.02502 (2020)

26. Uzunova, H., Wilms, M., Forkert, N.D., Handels, H., Ehrhardt, J.: A systematic comparison of generative models for medical images. Int. J. Comput. Assist. Radiol. Surg. **17**(7), 1213–1224 (2022). https://doi.org/10.1007/s11548-022-02567-6
27. Wang, W., et al.: Cataract surgical rate and socioeconomics: a global study. Invest. Ophthalmol. Vis. Sci. **57**(14), 5872–5881 (2016)
28. Zhang, R., Isola, P., Efros, A.A., Shechtman, E., Wang, O.: The unreasonable effectiveness of deep features as a perceptual metric. In: Proceedings of the IEEE Conference on Computer Vision and Pattern Recognition, pp. 586–595 (2018)

A Closed-Form Solution to Electromagnetic Sensor Based Intraoperative Limb Length Measurement in Total Hip Arthroplasty

Tiancheng Li[1(✉)], Yang Song[1], Peter Walker[2], Kai Pan[1],
Victor A. van de Graaf[2], Liang Zhao[1], and Shoudong Huang[1]

[1] Robotics Institute, Faculty of Engineering and Information Technology,
University of Technology Sydney, Ultimo, NSW 2007, Australia
`tiancheng.li-1@student.uts.edu.au`
[2] Concord Repatriation General Hospital, Concord, NSW 2139, Australia

Abstract. Total hip arthroplasty (THA) is an orthopaedic surgery to replace the diseased femoral head and socket of the hip joint with artificial implants. Achieving appropriate leg length and offset in THA is critical to avoid instability, leg length discrepancies, persistent pain, or early implant failure. This paper provides an electromagnetic (EM) sensor based approach for accurately measuring the change in leg length and offset intraoperatively. The proposed approach does not require direct line-of-sight, avoids the need for accurately returning the leg back to the neutral reference position, and has an efficient closed-form solution from least squares optimisation. Validations using simulations, phantom experiments, and cadaver tests demonstrate that the proposed method can provide more accurate results than the conventional method by manual gauge, the standard optical tracking based approach, and the direct use of one EM reading, thus showing significant potential clinical value.

1 Introduction

Hip osteoarthritis, with the top 10% occurrence in all diseases, brought a high demand for total hip arthroplasty (THA) in the past few decades [2]. According to the American Academy of Orthopaedic Surgeons, in the United States, approximately 450,000 THA surgeries are performed each year. One of the main challenges in THA is achieving accurate restoring leg length and femoral offset [3]. Failure in doing so can lead to instability, leg length discrepancies, impingement, persistent pain, and early implant failure [22], thus significantly affecting the clinical outcome and hip durability [19]. Therefore, a reliable intraoperative

Supported by PMSW Research Pty Ltd., Australia.

Supplementary Information The online version contains supplementary material available at https://doi.org/10.1007/978-3-031-43996-4_35.

limb length measurement and restoration method is crucial to optimise patient outcomes and implant survival.

Intraoperatively, leg length and femoral offset can be determined manually using a calliper between two reference points [1,3,11,17], but using a calliper is prone to measurement error [13]. Many computer-assisted methods [6] rely on numerous landmarks, such as condyles, or tibial spines, to determine the limb length [10], which could be inconvenient during surgeries. The optical tracking system is often used in THA for measurement as it has shown higher accuracy and reliability during interventions involving dynamic motion [18]. Sarin et al. [16] fixed optical tracking devices on the pelvis and femur as two references to measure the leg length and offset. Intellijoint HIP [13] is another 3D optical navigation tool, with the camera attached to the pelvis rather than placed next to the patient. Mako combines the preoperative CT 3D reconstruction and intraoperative optical tracking feedback for registration to determine the leg length and offset [4,20]. The main limitation of optical tracking is the requirement for a direct and free line-of-sight between markers and cameras [18].

In contrast, the Electromagnetic (EM) sensor based navigation system can provide fast and accurate tracking without line-of-sight constraints [5,18]. Zhao et al. [23] proposed a real-time robust simultaneous catheter and environment modelling for endovascular navigation, which is based on intravascular ultrasound and EM sensing. Mohammadbagherpoor et al. [12] developed an EM-based inductive proximity sensor system for detecting hip joint implant loosening in the micron range. Intracs$^{®em}$ is an intelligent navigation system based on EM tracking for endoscopic minimally invasive spine surgery [8].

In the commonly used intraoperative leg length equalisation and offset recovery techniques, both traditional and computer-assisted methods require the femur to be held and stored at the preoperative neutral reference position ($0°$ flexion, $0°$ rotation, and $0°$ abduction) prior to hip dislocation, and measure the changes in leg length and femoral offset as the femur is returned to the neutral reference position [1,13,15,16]. Inaccurate repositioning of the femur w.r.t. the pelvis can result in additional measurement errors since only $4°$ of abduction/adduction could cause 5–7 mm error in leg length and 2–4 mm error in offset [9]. In our method, we aim to eliminate the femoral repositioning prior to measurement to avoid the additional errors.

In this work, we propose a robust and accurate intraoperative limb length measurement method for THA based on EM sensing. Using the idea that the femoral movement can be mathematically modelled as a vector rotating around a fixed rotation centre, we develop a closed-form optimisation solution that uses a set of sampled poses from EM readings to calculate the intraoperative limb length change. The experiment results demonstrate that the proposed method can be more accurate. In summary, the key advantages of our method include: (i) the optimisation with a closed-form solution is an active compensation [5,18], which can effectively reduce static errors in the EM tracking itself and significantly improve the accuracy; (ii) different from pivot calibration, only slight movement of the femur is required to obtain accurate limb length measurements;

(iii) no need to return the leg back to the neutral reference position again after replacing the damaged hip joint with artificial implants, which effectively avoids measurement errors due to inaccurate abduction/adduction repositioning; (iv) the proposed method does not require the direct line-of-sight, and can be easily integrated clinically, without interrupting the workflow of THA.

2   Methodology

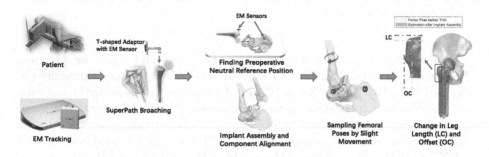

Fig. 1. The proposed EM-based intraoperative limb length measurement framework.

2.1   Problem Formulation

Standard THA Routine. During a standard THA, the surgery is often performed in the lateral position. Two reference points are marked on the pelvis and femur, respectively. The reference can be iliac fixation pins, sensors, or optical trackers. The surgeon then finds the preoperative neutral reference position by experience and records the relative position between the two reference points prior to the femoral head resection. The following routines include damaged femoral head removal, femoral canal broaching, acetabular preparation, and component selection and alignment. The component alignment requires the surgeon to return the femur to the neutral reference position and measure the change in leg length and offset.

Our Setup. In our proposed method (Fig. 1), the EM tracking board is placed under the patient's hip during the THA surgery, and one pin with EM sensor is installed on the pelvis. The supercapsular percutaneously assisted (SuperPath) approach [14] is used to insert a metal stem (or implant) into the hollow centre of the femur, without the need for femoral head resection and removal prior to the femoral canal broaching. A T-shaped adaptor mounted with another EM sensor is rigidly attached to the stem. When measuring changes in leg length and femoral offset, we sample poses of the postoperative femur during a slight femoral movement. *The problem considered in this work is to use the sampling postoperative femoral poses to estimate the leg length and offset change instead of repositioning the femur to the neutral reference position.*

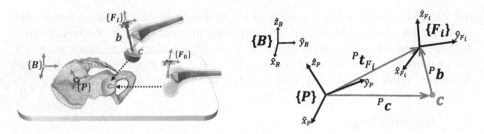

Fig. 2. Left: different coordinate frames involved in our THA setup; Right: relation among different vectors in (1).

Denote the frames of EM sensors on the pelvis and femur as $\{P\}$ and $\{F\}$, respectively. The real-time readings of two EM sensors are represented as $^B\mathbf{X}_P = \{^B\mathbf{R}_P, {}^B\mathbf{t}_P\}$ and $^B\mathbf{X}_F = \{^B\mathbf{R}_F, {}^B\mathbf{t}_F\}$, which are respectively the rotation matrices and translation vectors of frames $\{P\}$ and $\{F\}$ w.r.t. the frame of EM tracking board denoted by $\{B\}$ (Fig. 2). Then, to eliminate the effect of patient motion, the relative pose of frame $\{F\}$ w.r.t. $\{P\}$ is used and denoted by $^P\mathbf{X}_F = \{^P\mathbf{R}_F, {}^P\mathbf{t}_F\}$, where $^P\mathbf{R}_F = {}^B\mathbf{R}_P^\top{}^B\mathbf{R}_F$ and $^P\mathbf{t}_F = {}^B\mathbf{R}_P^\top{}^B\mathbf{t}_F - {}^B\mathbf{t}_P$.

Problem Statement. Suppose $^P\mathbf{X}_{F0}^{pre} = \{^P\mathbf{R}_{F0}^{pre}, {}^P\mathbf{t}_{F0}^{pre}\}$ is the recorded pose of $\{F\}$ w.r.t. $\{P\}$ at the preoperative neutral reference position before femoral head resection, and N sampling postoperative femoral poses in frame $\{P\}$, denoted by $\{^P\mathbf{X}_{Fi} = \{^P\mathbf{R}_{Fi}, {}^P\mathbf{t}_{Fi}\} \mid i \in \{1, \cdots, N\}\}$, are collected through small motion around the neutral position after femoral head replacement. Since the relative rotations of frame $\{F\}$ in $\{P\}$ should be the same at the neutral reference position before and after the femoral head replacement, mathematically, the problem considered in this paper is, *given sampling postoperative femoral poses* $^P\mathbf{X}_{Fi}$ $(i \in \{1, \cdots, N\})$, *accurately estimate the current translation vector* $^P\mathbf{t}_{F0}^{post}$ *when the relative rotation is* $^P\mathbf{R}_{F0}^{pre}$, *and then use it to calculate the change in leg length and offset.*

2.2   Limb Length Measurement Without Femoral Repositioning

Since the joint between the acetabular component and the metal femoral head is a perfect sphere, the femoral movement in the frame $\{P\}$ can be mathematically modelled as a vector $\mathbf{b}$ rotating around a centre $\mathbf{c}$ (Fig. 2: Left), where $\mathbf{c}$ is constant in frame $\{P\}$. Although $\mathbf{b}$ is changing in $\{P\}$, it is a constant in $\{F\}$. Therefore, the relation among the vectors $\mathbf{t}$, $\mathbf{b}$, and $\mathbf{c}$ (refer to Fig. 2: Right) can be described as

$$^P\mathbf{t}_F = {}^P\mathbf{b} + {}^P\mathbf{c} = {}^P\mathbf{R}_F{}^F\mathbf{b} + {}^P\mathbf{c}, \tag{1}$$

where $^P\mathbf{b}$ and $^F\mathbf{b}$ are the rotating vector $\mathbf{b}$ in $\{P\}$ and $\{F\}$, respectively, and $^P\mathbf{c}$ is the rotation centre $\mathbf{c}$ in $\{P\}$.

The relation (1) is valid for all $^P\mathbf{X}_F$, so $^F\mathbf{b}$ and $^P\mathbf{c}$ can be obtained from the N sampled poses $^P\mathbf{X}_{Fi}$ $(i \in \{1, \cdots, N\})$ by solving an optimisation problem.

Then, the postoperative translation vector $^P\mathbf{t}_{F0}^{post}$ can be calculated by the relative rotation $^P\mathbf{R}_{F0}^{pre}$. In contrast to pivot calibration [21] which is commonly used to estimate the tip location of a pointer tool, our method focuses on estimating limb change after femoral head replacement and therefore requires only minor leg movements for sampling, and inaccurate rotation centre estimate due to singularity has almost no effect on limb length estimation. See below for details.

Full Least Squares Solution. Through the iterative Gauss-Newton (GN) method, the solution for $^F\mathbf{b}$ and $^P\mathbf{c}$ can be obtained by solving the following full nonlinear least squares (Full LS) optimisation problem,

$$\underset{^F\mathbf{b},^P\mathbf{c},^P\bar{\mathbf{R}}_{Fi}}{\arg\min} \sum_{i=1}^{N} \|^P\bar{\mathbf{R}}_{Fi}{}^F\mathbf{b} + {}^P\mathbf{c} - {}^P\mathbf{t}_{Fi}\|^2_{\mathbf{\Omega}_{ti}^{-1}} + \|r(^P\bar{\mathbf{R}}_{Fi}) - r(^P\mathbf{R}_{Fi})\|^2_{\mathbf{\Omega}_{Ri}^{-1}}, \quad (2)$$

where $r(^P\mathbf{R}_{Fi})$ and $r(^P\bar{\mathbf{R}}_{Fi})$ are the Euler angles of rotation measurement $^P\mathbf{R}_{Fi}$ and the corresponding rotation variable $^P\bar{\mathbf{R}}_{Fi}$, respectively. $\mathbf{\Omega}_{ti}$ and $\mathbf{\Omega}_{Ri}$ are the covariance matrices of EM measurement noises w.r.t. translation and rotation.

Closed-Form Solution. Since the EM measurements of rotations are accurate enough [7], i.e. $^P\bar{\mathbf{R}}_{Fi} \approx {}^P\mathbf{R}_{Fi}$, the contribution of the second term in (2) is limited. As a result, the optimisation problem can be simplified as a linear least squares problem by letting $^P\bar{\mathbf{R}}_{Fi} = {}^P\mathbf{R}_{Fi}$:

$$\underset{^F\mathbf{b},^P\mathbf{c}}{\arg\min} \sum_{i=1}^{N} \|^P\mathbf{R}_{Fi}{}^F\mathbf{b} + {}^P\mathbf{c} - {}^P\mathbf{t}_{Fi}\|^2_{\mathbf{\Omega}_{ti}^{-1}}, \quad (3)$$

which has an easier and more efficient closed-form solution

$$\begin{bmatrix} ^F\mathbf{b}^* \\ ^P\mathbf{c}^* \end{bmatrix} = \begin{bmatrix} \sum_{i=1}^{N} {}^P\mathbf{R}_{Fi}^\top \mathbf{\Omega}_{ti}^{-1} {}^P\mathbf{R}_{Fi} & \sum_{i=1}^{N} {}^P\mathbf{R}_{Fi}^\top \mathbf{\Omega}_{ti}^{-1} \\ \sum_{i=1}^{N} \mathbf{\Omega}_{ti}^{-1} {}^P\mathbf{R}_{Fi} & \sum_{i=1}^{N} \mathbf{\Omega}_{ti}^{-1} \end{bmatrix}^{-1} \begin{bmatrix} \sum_{i=1}^{N} {}^P\mathbf{R}_{Fi}^\top \mathbf{\Omega}_{ti}^{-1} {}^P\mathbf{t}_{Fi} \\ \sum_{i=1}^{N} \mathbf{\Omega}_{ti}^{-1} {}^P\mathbf{t}_{Fi} \end{bmatrix}.$$
$$(4)$$

The comparison in Sect. 3 will show that the closed-form solution (4) is almost the same as the solution to Full LS (2), but only requires sub-millisecond computational cost which is thousands of times less. Closed-form solution also benefits from the fact that solutions can be obtained in one step, avoiding potential local minima, providing greater robustness, and being easier to implement. So our proposed measurement approach is based on the closed-form solution.

Limb Length Change Measurement. After $^F\mathbf{b}^*$ and $^P\mathbf{c}^*$ are obtained, the postoperative translation vector $^P\mathbf{t}_{F0}^{post}$ at neutral reference position (where the relative rotation is $^P\mathbf{R}_{F0}^{pre}$) can be calculated by (1)

$$^P\mathbf{t}_{F0}^{post} = {}^P\mathbf{R}_{F0}^{pre}{}^F\mathbf{b}^* + {}^P\mathbf{c}^*. \quad (5)$$

The change of relative translation vector at the neutral reference position is $\Delta^P\mathbf{t}_{F0} = {}^P\mathbf{t}_{F0}^{post} - {}^P\mathbf{t}_{F0}^{pre}$. Further, we denote $^P\mathbf{L}$ as the projection of $\Delta^P\mathbf{t}_{F0}$

onto the sagittal plane. Then, the changes in leg length and offset are computed by the norms $\|^P\mathbf{L}\|$ and $\|\Delta^P\mathbf{t}_{F0} - {}^P\mathbf{L}\|$, respectively.

The covariance of $^P\mathbf{t}_{F0}^{post}$ in (5) is calculated by

$$\mathbf{\Omega}_{P\mathbf{t}_{F0}^{post}} = \begin{bmatrix} {}^P\mathbf{R}_{F0}^{pre\top} & \mathbf{I} \end{bmatrix} (\sum_{i=1}^{N} \begin{bmatrix} {}^P\mathbf{R}_{Fi}^{\top} & \mathbf{I} \end{bmatrix}^{\top} \mathbf{\Omega}_{ti}^{-1} \begin{bmatrix} {}^P\mathbf{R}_{Fi}^{\top} & \mathbf{I} \end{bmatrix})^{-1} \begin{bmatrix} {}^P\mathbf{R}_{F0}^{pre\top} & \mathbf{I} \end{bmatrix}^{\top}, \quad (6)$$

where $\mathbf{I}$ is the identity matrix, $\mathbf{\Omega}_{ti}$ is the covariance matrices of EM measurement noises w.r.t. translation. It can be proved that (6) tends to be zero as increasing samples if data are all sampled around the neutral position ($^P\mathbf{R}_{Fi} \approx {}^P\mathbf{R}_{F0}^{pre}$), although the uncertainty of $^F\mathbf{b}^*$ and $^P\mathbf{c}^*$ in (4) is large due to the near singularity in this case. Therefore, a slight movement of the leg (rotating around the neutral position within a few degrees) can guarantee the accuracy of limb length estimate while preventing injury to the patient and the workload of surgeons.

3   Experiments

3.1   Simulation and Robustness Assessment

To compare the closed-form solution (4) and the solution to Full LS (2), five different levels of zero-mean Gaussian noises are added to the rotation angles and translations of sampling femoral poses $^P\mathbf{X}_{Fi}$ ($i \in \{1, \cdots, 100\}$) from EM readings (first row in Table 1). Twenty independent runs are executed for each noise level to test the robustness and accuracy of both two methods. The mean absolute errors compared with the ground truth and standard deviation (STD) of the twenty runs for estimating the neutral femur position are shown in Table 1. It shows that the proposed closed-form solution can achieve similar accuracy compared with the solution to the Full LS problem and the robustness to additionally added sensor noises is high.

Table 1. Estimation error and STD from simulations with five increasing noise levels.

Noises: {Rot, Trans}	{0.1 rad, 2 mm}	{0.2 rad, 4 mm}	{0.3 rad, 6 mm}	{0.4 rad, 8 mm}	{0.5 rad, 10 mm}
Closed-form (mm)	0.2276 (0.0885)	0.7896 (0.3405)	1.4935 (0.8403)	1.9496 (0.7359)	2.7028 (1.1972)
Full LS (mm)	0.2272 (0.0887)	0.7878 (0.3408)	1.4886 (0.8397)	1.9412 (0.7361)	2.6910 (1.1971)

3.2   Phantom Experiments

The phantom experiments (Fig. 3) were performed using two different sawbones models. One experienced surgeon executed a normal surgical routine using standard hip arthroplasty components. Three commonly used standard femoral heads ($\{-4, 0, +4\}$ mm) were used for the alignment (Fig. 3: Right). After placing a

Fig. 3. Left: the first setup for comparison between closed-form solution (4) and solution to Full LS (2); Middle: the second setup for comparison of the proposed method to the other three methods (manual gauge, optical tracking, and one EM reading); Right: standard hip arthroplasty components used in phantom experiments.

metal acetabular shell into the pelvic cavity and inserting a stem into the femur, the surgeon selected one femoral head component and placed it on top of the stem, and the femoral head was placed into the liner. Then the surgeon found and recorded the neutral reference position of the femur model. After that, another femoral head was replaced to change the limb length. Finally, the limb length change before and after the femoral component replacement was calculated by different methods. The ground truth was available from the size changes between femoral head components.

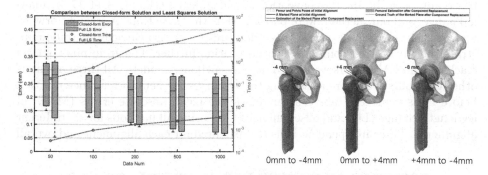

Fig. 4. Results from the first phantom experiment setup (Fig. 3: Left). Left: comparison between closed-form solution (4) and solution to Full LS (2); Right: femur pose estimation with closed-form solution (using 100 data) for three different alignments.

Number of Samples and Comparison with Full LS. The first setup of phantom experiments (Fig. 3: Left) was designed to analyse the effect of the number of sampled poses on the performance of our method and to further compare the closed-form solution (4) and the solution to Full LS (2). Six different alignments of the femoral head components were performed (-4 to 0, -4 to $+4$, 0 to -4, 0 to $+4$, $+4$ to -4, and $+4$ to 0). After installing the replaced femoral head

Table 2. Comparison of the proposed method to the other three methods in the second phantom experiment setup (Fig. 3: Middle).

Setups	−4 to 0		−4 to +4		0 to −4		0 to +4		+4 to −4		+4 to 0		p-value
MAE (mm)	LC	OC	LC	OC	LC	OC	LC	OC	LC	OC	LC	OC	
Closed-form	0.25	0.35	0.21	0.40	0.28	0.22	0.14	0.13	0.17	0.21	0.02	0.15	–
Optical Tracking	1.14	1.99	2.58	2.38	3.04	0.54	1.30	0.41	0.99	0.66	0.39	0.73	3.35e−4
One EM Reading	3.18	1.19	4.45	1.94	2.76	0.75	1.50	0.71	1.87	1.47	0.70	0.80	1.27e−4
Manual Gauge	3.19	2.35	6.23	4.58	3.86	0.82	1.73	1.36	2.30	1.81	1.47	1.86	2.38e−5

components, the surgeon slightly rotated the femur to collect the sampling pose data from EM sensors ({50, 100, 200, 500, 1000} samples for each alignment). As shown in Fig. 4: Left, the closed-form solution performs as good as Full LS, with computational costs ranging from 0.3 ms to 3.3 ms corresponding to data numbers of 50 to 1000, which is thousands of times less than Full LS. Estimation error decreases as the number of samples increases. Three examples of results are shown in Fig. 4: Right. The acquisition frequency of EM tracking is 20 Hz and the improvement in accuracy is limited when the number of data is more than 100. To balance the efficiency and accuracy, the closed-form solution with 100 data is our choice for the experiments in the rest of the paper.

Comparison with Other Methods. We compared our proposed method (Closed-form) with three other different measurement approaches in the second phantom experiment setup (Fig. 3: Middle). The standard optical tracking based approach [13,16], the conventional mechanical method [1,3,11,17] by manual gauge, and our method, were performed at the same time. A straightforward idea of using EM sensor (one EM reading only) was also carried out to demonstrate the advantage of our closed-form solution further. The groin pins, optical trackers, and EM sensors were fixed on the sawbones models. All three methods other than ours require repositioning the femur to the neutral reference position before measurements. Table 2 summarises the mean absolute errors (MAE) of leg length change (LC) and offset change (OC) using all methods for six different alignments. Three independent runs are executed for each alignment and overall

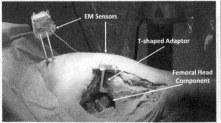

Fig. 5. Left: cadaver experiment; Right: an adjustable trial neck that can align limb length without the need to dislocate the hip to change the head.

the proposed method achieves the highest accuracy (the p-values for the other three methods compared to ours are shown in the last column of Table 2).

Table 3. Cadaver Experiment

Datasets	1		2		3		4		5		6		7		8		MAE (STD)	
Errors (mm)	LC	OC	LC	OC	LC	OC	LC	OC	LC	OC	LC	OC	LC	OC	LC	OC	LC	OC
Closed-form	0.51	0.41	0.42	0.34	0.36	0.28	0.01	0.02	0.35	0.28	0.70	0.56	0.38	0.30	0.60	0.48	0.42 (0.21)	0.33 (0.16)
One EM Reading	1.13	0.98	0.52	1.40	0.74	1.32	1.45	1.55	2.64	1.47	2.99	1.59	4.21	3.73	3.21	2.35	2.11 (1.34)	1.80 (0.87)

3.3 Cadaver Experiments

The proposed method was also tested in cadaver experiments (Fig. 5: Left). The cadaveric body was operated on lateral decubitus. The whole THA routines including SuperPath broaching, hip dislocation, femoral head removal as well as acetabular and femoral preparation were executed in the cadaver experiment. To conduct multiple sets of experiments, a standard adjustable trial neck (Fig. 5: Right) was inserted which allowed limb length changes without dislocating and altering the femoral head. Finally, the cadaver experiment yielded eight sets of alignments by changing the trial neck length. As shown in Table 3, the proposed method can reach a mean absolute error (MAE) of about 0.4 mm, which is much smaller than directly using one EM reading.

4 Conclusion

This paper presents an efficient closed-form solution based on EM sensing to robustly and accurately calculate the intraoperative change in leg length and offset. Simulations, phantom experiments, and cadaver tests demonstrate the efficiency and accuracy of our proposed algorithm compared to the conventional manual gauge method and standard optical tracking based method, showing the potential value in clinical practice. The reasons why the proposed solution can significantly improve the accuracy are: (i) it uses a set of sampled poses from EM readings to optimise the intraoperative limb length change instead of only using one sensor reading; (ii) there is no requirement of repositioning the femur to the neutral position before the measurements. The computational time of our method is only around 0.3 ms mainly due to the closed-form solution.

Some studies assessed that metals in the surgical environment might affect the accuracy of EM tracking [18]. However, the design of EM is not affected by titanium and 300 series stainless steel, which are the main materials of surgical instruments used in THA, and our cadaver experiments in a surgical environment have shown that our method still guarantees measurement accuracy (the mean absolute error is around 0.4 mm). In the future, we aim to further validate our approach through clinical trials and have plans to extend the EM-based intraoperative limb length measurement to total knee arthroplasty surgery.

References

1. Bose, W.J.: Accurate limb-length equalization during total hip arthroplasty. Orthopedics **23**(5), 433 (2000)
2. Cross, M., et al.: The global burden of hip and knee osteoarthritis: estimates from the global burden of disease 2010 study. Ann. Rheum. Dis. **73**(7), 1323–1330 (2014)
3. Desai, A.S., Dramis, A., Board, T.N.: Leg length discrepancy after total hip arthroplasty: a review of literature. Curr. Rev. Musculoskelet. Med. **6**(4), 336–341 (2013). https://doi.org/10.1007/s12178-013-9180-0
4. Fontalis, A., Kayani, B., Thompson, J.W., Plastow, R., Haddad, F.S.: Robotic total hip arthroplasty: past, present and future. Orthop. Trauma **36**(1), 6–13 (2022)
5. Franz, A.M., Haidegger, T., Birkfellner, W., Cleary, K., Peters, T.M., Maier-Hein, L.: Electromagnetic tracking in medicine-a review of technology, validation, and applications. IEEE Trans. Med. Imaging **33**(8), 1702–1725 (2014)
6. Gheewala, R.A., Young, J.R., Villacres Mori, B., Lakra, A., DiCaprio, M.R.: Perioperative management of leg-length discrepancy in total hip arthroplasty: a review. Arch. Orthop. Trauma Surg. **143**, 5417–5423 (2023). https://doi.org/10.1007/s00402-022-04759-w
7. Gomes-Fonseca, J., et al.: Assessment of electromagnetic tracking systems in a surgical environment using ultrasonography and ureteroscopy instruments for percutaneous renal access. Med. Phys. **47**(1), 19–26 (2020)
8. Hagan, M.J., et al.: Navigation techniques in endoscopic spine surgery. BioMed Res. Int. **2022**, 8419739 (2022)
9. Kawamura, H., Watanabe, Y., Nishino, T., Mishima, H.: Effects of lower limb and pelvic pin positions on leg length and offset measurement errors in experimental total hip arthroplasty. J. Orthop. Surg. Res. **16**(1), 1–9 (2021). https://doi.org/10.1186/s13018-021-02347-z
10. Lecoanet, P., Vargas, M., Pallaro, J., Thelen, T., Ribes, C., Fabre, T.: Leg length discrepancy after total hip arthroplasty: can leg length be satisfactorily controlled via anterior approach without a traction table? Evaluation in 56 patients with EOS 3D. Orthop. Traumatol. Surg. Res. **104**(8), 1143–1148 (2018)
11. McGee, H., Scott, J.: A simple method of obtaining equal leg length in total hip arthroplasty. Clin. Orthop. Relat. Res. **194**, 269–270 (1985)
12. Mohammadbagherpoor, H., et al.: An implantable wireless inductive sensor system designed to monitor prosthesis motion in total joint replacement surgery. IEEE Trans. Biomed. Eng. **67**(6), 1718–1726 (2019)
13. Paprosky, W.G., Muir, J.M.: Intellijoint HIP®: a 3D mini-optical navigation tool for improving intraoperative accuracy during total hip arthroplasty. Med. Dev. Evid. Res. **9**, 401–408 (2016)
14. Quitmann, H.: Supercapsular percutaneously assisted (SuperPath) approach in total hip arthroplasty. Oper. Orthop. Traumatol. **31**(6), 536–546 (2019)
15. Renkawitz, T., Schuster, T., Grifka, J., Kalteis, T., Sendtner, E.: Leg length and offset measures with a pinless femoral reference array during THA. Clin. Orthop. Relat. Res.® **468**(7), 1862–1868 (2010). https://doi.org/10.1007/s11999-009-1086-1
16. Sarin, V.K., Pratt, W.R., Bradley, G.W.: Accurate femur repositioning is critical during intraoperative total hip arthroplasty length and offset assessment. J. Arthroplasty **20**(7), 887–891 (2005)
17. Shiramizu, K., Naito, M., Shitama, T., Nakamura, Y., Shitama, H.: L-shaped caliper for limb length measurement during total hip arthroplasty. J. Bone Joint Surg. Br. Vol. **86**(7), 966–969 (2004)

18. Sorriento, A., et al.: Optical and electromagnetic tracking systems for biomedical applications: a critical review on potentialities and limitations. IEEE Rev. Biomed. Eng. **13**, 212–232 (2019)
19. Takamatsu, T., et al.: Radiographic determination of hip rotation center and femoral offset in Japanese adults: a preliminary investigation toward the preoperative implications in total hip arthroplasty. BioMed Res. Int. **2015**, 610763 (2015)
20. Tarwala, R., Dorr, L.D.: Robotic assisted total hip arthroplasty using the MAKO platform. Curr. Rev. Musculoskelet. Med. **4**, 151–156 (2011). https://doi.org/10.1007/s12178-011-9086-7
21. Yaniv, Z.: Which pivot calibration? In: Medical Imaging 2015: Image-Guided Procedures, Robotic Interventions, and Modeling, vol. 9415, pp. 542–550. SPIE (2015)
22. Zahar, A., Rastogi, A., Kendoff, D.: Dislocation after total hip arthroplasty. Curr. Rev. Musculoskelet. Med. **6**(4), 350–356 (2013). https://doi.org/10.1007/s12178-013-9187-6
23. Zhao, L., Giannarou, S., Lee, S.L., Yang, G.Z.: SCEM+: real-time robust simultaneous catheter and environment modeling for endovascular navigation. IEEE Robot. Autom. Lett. **1**(2), 961–968 (2016)

A Unified Deep-Learning-Based Framework for Cochlear Implant Electrode Array Localization

Yubo Fan[1]([✉]), Jianing Wang[2], Yiyuan Zhao[3], Rui Li[4], Han Liu[1], Robert F. Labadie[5], Jack H. Noble[4], and Benoit M. Dawant[4]

[1] Department of Computer Science, Vanderbilt University, Nashville, TN 37235, USA
yubo.fan@vanderbilt.edu
[2] Digital Technology and Innovation, Siemens Healthineers, Princeton, NJ 08540, USA
[3] Digital and Automation, Siemens Healthineers, Malvern, PA 19355, USA
[4] Department of Electrical and Computer Engineering, Vanderbilt University, Nashville, TN 37235, USA
[5] Department of Otolaryngology - Head and Neck Surgery, Medical University of South Carolina, Charleston, SC 29425, USA

Abstract. Cochlear implants (CIs) are neuroprosthetics that can provide a sense of sound to people with severe-to-profound hearing loss. A CI contains an electrode array (EA) that is threaded into the cochlea during surgery. Recent studies have shown that hearing outcomes are correlated with EA placement. An image-guided cochlear implant programming technique is based on this correlation and utilizes the EA location with respect to the intracochlear anatomy to help audiologists adjust the CI settings to improve hearing. Automated methods to localize EA in postoperative CT images are of great interest for large-scale studies and for translation into the clinical workflow. In this work, we propose a unified deep-learning-based framework for automated EA localization. It consists of a multi-task network and a series of postprocessing algorithms to localize various types of EAs. The evaluation on a dataset with 27 cadaveric samples shows that its localization error is slightly smaller than the state-of-the-art method. Another evaluation on a large-scale clinical dataset containing 561 cases across two institutions demonstrates a significant improvement in robustness compared to the state-of-the-art method. This suggests that this technique could be integrated into the clinical workflow and provide audiologists with information that facilitates the programming of the implant leading to improved patient care.

Keywords: Cochlear Implant · Electrode Localization · Object Detection

---

Supplementary Information The online version contains supplementary material available at https://doi.org/10.1007/978-3-031-43996-4_36.

1 Introduction

A cochlear implant (CI) has an electrode array (EA) that is surgically inserted into the cochlea to stimulate the auditory nerves and treat patients with severe-to-profound sensorineural hearing loss. Although CIs have achieved great success, hearing outcomes among recipients vary significantly [1, 2]. Recent studies have shown that factors related to electrode positioning impact on audiological outcomes [3, 4]. These studies require knowing the precise electrode locations, which can be obtained by postoperative CT imaging. Another clinical application that requires precise electrode locations is image-guided cochlear implant programming [5]. After surgical implantation, the CI needs to be programmed such that an optimized frequency mapping [24] can be determined for each patient. Knowledge about the spatial relationship between the electrodes and the intracochlear anatomy permits to generate programming solutions which have been shown to significantly improve the hearing outcomes for both adult and pediatric CI recipients [6, 7].

To facilitate the EA localization process, automated methods have been proposed by several groups. In [8] EAs are categorized as closely-spaced and distantly-spaced. EAs with an interelectrode spacing such that the intensity contrast between electrodes cannot be distinguished are considered closely-spaced, while the opposite applies to distantly-spaced EAs. Note that a given EA model could be categorized as closely-spaced or distantly-spaced, depending on the image resolution. For example, some closely-spaced EAs included in [8] could be considered distantly-spaced in some images acquired at very high resolution (0.08mm isotropic) as in [9]. For algorithms localizing closely-spaced EAs, extracting the centerline of an EA is often an important step [8, 10, 11]. It is achieved by either using intensity-based features only [11] or combining both intensity-based and EA shape-based features [8]. For localizing distantly-spaced EAs, hand-crafted feature extractors are utilized to detect individual blobs [9, 12–14]. To link the electrode candidates in the correct order and remove false positive candidates, graph-based path-finding algorithms [12] or Markov random field models [14] have been proposed. While there has been an emergence of methods based on deep learning (DL) recently [15–17], they cannot be viewed as a complete solution to the automatic EA localization problem because the networks are only trained and validated on one type of EA model [17] or cannot order/link the detected electrodes to form a complete array [15, 16].

In this work, we present a novel DL-based framework that consists of a multi-task network and a set of postprocessing algorithms to localize the cochlear implant electrode array. Our contribution is three-fold: (1) To the best of our knowledge, it is the first unified DL-based framework designed for localizing both distantly- and closely-spaced EAs in CT and cone-beam CT (CBCT) images. (2) We propose four (three detection and one segmentation) tasks for the multi-task network such that this single network can be trained on various kinds of EAs. (3) We extensively evaluate this framework on datasets that significantly exceed the scale of all datasets reported in the literature to date: a heterogeneous clinical _test_ set with CT or CBCT images of 561 implanted ears and a test set with gold standard ground truth for 27 implanted cadaveric ears. Results show that the proposed framework is significantly more robust (generates results that require manual adjustments less often) than the state-of-the-art (SOTA) techniques, while also being slightly more accurate. These findings indicate that the proposed framework could be reliably used to support large-scale quantitative studies and deployed in the clinical workflow to provide clinicians with critical information at the time and point of care.

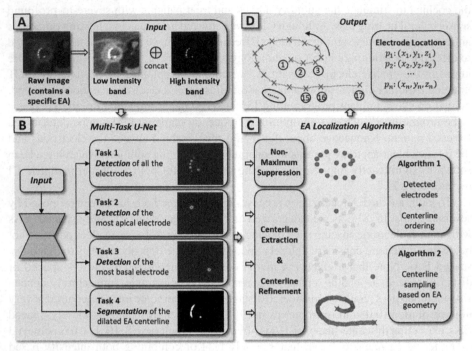

Fig. 1. An overview of the proposed framework. It consists of two major modules: a multi-task U-Net that is generic for various EA types (B) and a set of postprocessing algorithms that utilize the known EA geometry (C). The input to the multi-task U-Net is the two-channel image that contains low and high intensity bands extracted from the raw image (A). The output of the framework is the ordered electrode locations (D).

2   Method

2.1   Data

The data in this study consist of a large-scale clinical dataset from CI recipients (dataset #1) as well as 27 cadaveric samples (dataset #2). Dataset #1 includes 1324 implanted ears from CI recipients treated at two institutions (datasets #1A and #1B). 8 types of distantly- and closely-spaced EA from 3 manufacturers are included in these cases. Dataset #1A has 958 implanted ears of which 97% are scanned with CBCT scanners, and the remaining (3%) are scanned with conventional CT scanners. Dataset #1B includes 366 implanted ears of which most (98%) are scanned with conventional CT scanners and the remaining (2%) are scanned with CBCT scanners. Dataset #2 contains 4 types of distantly-spaced EAs, and these specimens are scanned with conventional CTs.

The training and validation sets are all from dataset #1A and constitute 60% and 20% of that dataset, respectively. The remaining 20% of dataset #1A along with dataset #1B (561 implanted ears in total) are used to test the robustness of the proposed framework. Dataset #2 is used to test its accuracy because the gold standard ground truth can be obtained using their paired micro-CT [16, 18]. Details on the datasets and EA specifications can be found in the supplementary material.

2.2 Multi-task U-Net

The proposed framework is designed to localize various types of EAs in both CT and CBCT. This is challenging because the number of electrodes to be detected and the interelectrode spacing is different among EA models, and the postoperative images can have different intensity characteristics depending on the type of scanner, i.e., CT or CBCT, that is used for their acquisition. To normalize the input image in a way that enhances both the nearby anatomy, which contains contextual information, and the high intensity (usually the electrodes) component, inspired by [16], we separate the raw image into two channels. One is the original image clipped at the 99.9% of its intensity histogram. The other is the original image in the [99.9%-100%] interval of its intensity histogram. Each channel is linearly normalized to [0,1] based on its own min-max values. We refer to these as the low and high intensity bands and an example can be seen in Fig. 1A.

To avoid having to train a network for each EA type, we define four tasks that a single multi-task network can learn simultaneously and whose output can be used to localize and order the electrodes in all arrays. Specifically, as shown in Fig. 1B, the four tasks include (1) detection of all the electrodes on the EA by heatmap regression, (2) detection of the most apical endpoint (tip) of the EA by heatmap regression, (3) detection of the most basal endpoint (the farthest electrode from the tip) of the EA by heatmap regression, and (4) segmentation of the EA centerline that starts and ends with the two endpoints. Although the network is a simple U-Net-like encoder-decoder architecture, the innovation is more focused on the multi-task strategy such that the network can serve as a robust feature extractor that leverages a large heterogeneous dataset.

We train the multi-task network using the electrode positions obtained with the methods described in [8, 12] corrected manually when a large localization error is visually observed. For tasks (#1,2, and 3) aiming at localizing electrodes, the ground truth is a one-channel heatmap for each task with a Gaussian kernel (variance of 2 voxels) at each electrode location. Following [19], we use a penalty-reduced voxel-wise logistic regression with the focal loss [20] as the training objective.

Depending on the image quality and the interelectrode spacing, EAs can appear as a whole bright (i.e., with high Hounsfield Unit (HU) values) tubular structure or distinguishable bright blobs representing individual electrodes. We define the centerline of the EA as a line connecting each electrode in sequential order (from most apical to most basal or the opposite). The motivation for segmenting the EA centerline is two-fold. First, after we extract all the electrodes from the predicted heatmap (Task #1), we need to order them. We will show later in our postprocessing algorithms that these detected electrodes can be linked in the correct order using the segmented centerline and the detected two endpoints (Algorithm 1 in Fig. 1C). The second reason is that it can serve as an alternative EA localization method by sampling the centerline using the known interelectrode spacing specific to a particular EA model (Algorithm 2 in Fig. 1C). It is essential when the electrodes are not discernible due to the low image resolution and/or the small interelectrode spacing. Different from [8], we resort to a DL approach rather than human-crafted feature extractors. To make the training of the centerline segmentation easier, we dilate the ground truth centerline to a tubular-structured mask with a radius of 3 voxels as the learning target. In addition to the Dice loss, we adopt a clDice loss which has shown its superiority in improving the accuracy and preserving the topology of the underlying one-voxel wide centerline [21].

2.3 Postprocessing Algorithms for EA Localization

As said above, although the output of the multi-task network contains essential information to identify the EA, it does not provide the desired final output, i.e., the position (coordinates) of the ordered electrodes in the image. To do so, we have designed a series of postprocessing algorithms that can effectively extract the ordered electrode locations from the four output maps. As shown in Fig. 1C, there are two main algorithms (Algorithm 1 and Algorithm 2) that are suitable for distantly- and closely-spaced EAs, respectively.

Algorithm 1 takes the heatmap of all the electrodes (Task #1) as input and utilizes a non-maximum suppression (NMS) algorithm, which is a common postprocessing step for object detection [22], to obtain the desired number of electrodes. Then, the centerline of the EA is extracted by skeletonizing its segmented mask. Its two endpoints are further refined by merging the detection results of the most apical and basal points. Finally, all the detected electrodes are linked with the guidance of the centerline along the direction from the most apical to the most basal. Algorithm 1 works well if there is an apparent contrast between the electrodes in the heatmap, which is the case for most distantly-spaced EAs.

However, for closely-spaced EAs, it is nearly impossible to differentiate the individual electrodes. Algorithm 2 is designed to localize EAs in such situations. After the extraction and refinement of the EA centerline, the centerline is smoothed with a cubic spline, and the final electrode positions are obtained by resampling along it using known interelectrode spacing for the EA [8]. Note that for distantly-spaced EAs, Algorithm 1 can occasionally lead to abnormal localization results, such as an incorrect number of detected electrodes or the spacing between the detected electrodes being inconsistent with the known interelectrode spacing for the EA. This is most often caused by poor image quality that affects the creation of the subsequent heatmap (Task #1). We have designed simple rules to detect these anomalies. When one is detected, the framework switches to Algorithm 2 for a more reliable sampling-based localization for distantly-spaced EAs.

3   Experiments and Results

3.1   Implementation Details

The multi-task U-Net is trained with PyTorch 1.12 on an NVIDIA RTX 2080 Ti GPU. We use MONAI [23] for data augmentation which contains an additive Gaussian noise and random affine transformations. The images are preprocessed by being rigidly registered to the left ear (mirroring is performed if one case is a right ear) of a template volume (atlas), resampled to a 0.1mm isotropic voxel size, and cropped into a region of interest (ROI) with a dimension of 320 × 320 × 320. Due to GPU memory size limit, we use a patch-based strategy (with a dimension of 256 × 256 × 192) for training. The batch size is set to 1 and we use AdamW optimizer with learning rate of 5e-4. At inference we use a sliding-window approach to merge the results. The network is trained for 250 epochs, and we select the epoch with the lowest validation loss as our final model. Note that a minority (3%) of the images in dataset #1 are reconstructed with a limited HU range, i.e., [-1024,3071]HU. In these images bone and electrodes have similar intensity values. The remaining (97%) images have maximum intensities far larger than 3071HU. In these images electrodes have larger intensity values than bone. To address this issue, we train another DL model with the same training strategy and dataset, but the intensity of all the training images is saturated at 3071HU, i.e., every pixel with an intensity above 3071HU is assigned a value of 3071HU. All the results obtained for limited HU range images presented herein are obtained with this dedicated model. The postprocessing algorithms are implemented in Python 3.9 and NumPy. We use skimage for 3D skeletonization and the csaps package for calculating the cubic spline. The inference time for the proposed framework (from loading the image to outputting the electrode locations) ranges between 5 to 20 s.

3.2 Evaluation and Results

The techniques described in [8] (for closely-spaced EAs) and [12] (for distantly-spaced EAs) are designed and validated with EAs and imaging protocols that are similar to those used in this study. They are also in routine use with over 2000 ears processed at the two institutions in dataset #1. They are considered to be the SOTA methods for comparison.

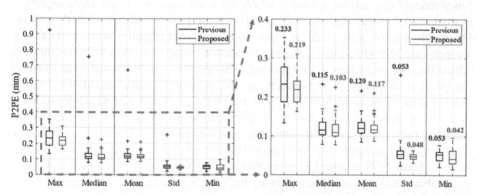

Fig. 2. Box plot of the P2PE in dataset #2 (27 implanted cadaveric ears). The median value for each metric is shown above each box in the right figure.

We first evaluate the accuracy of the proposed framework on dataset #2 (27 implanted cadaveric ears) for which the localization ground truth is known. Since the EAs in dataset #2 are all distantly-spaced, [12] is used as the previous SOTA method for comparison. We define the point-to-point error (P2PE) as the Euclidean distance between the predicted

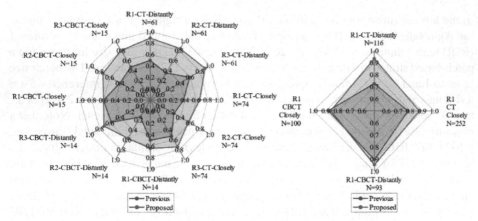

Fig. 3. Acceptance rate (higher means better) of the SOTA and proposed localization results on the large-error subset (left) and the whole clinical test set (right). R1, R2, and R3 are different raters.

electrode location and the ground truth location, and we calculate five P2PE-based metrics (maximum, median, mean, standard deviation (std), and the minimum P2PE) for all the electrodes in each case. The quantitative results are shown in Fig. 2. We can see from the left box plot that the results of the previous method [12] contain an outlier with relatively large P2PE, which is attributed to the low quality of the image. The median values in the right box plot show that the proposed framework has slightly smaller P2PE across the five metrics, but the differences are not statistically significant.

As introduced in Sect. 2.1, the 561 clinical cases in dataset #1 used for testing are highly heterogeneous. They contain CBCT and CT images from dataset #1A and #1B. Since manual annotations are not available for these images, we ask three experts to evaluate and compare the results obtained with the previous SOTA methods ([8] and [12]) and those obtained with the proposed method. Specifically, as shown in Fig. 4, we generate MIP (Maximum Intensity Projection) images and mark the locations at which the electrodes have been localized. This permits a rapid visual assessment of the localization quality. Next, for each case, we present the visualization of the results generated by the previous SOTA methods and by the proposed method side-by-side but in a random order, such that the experts are blind to the methods used to localize the contacts. They are then asked to rate the localization results as acceptable, i.e., no need to adjust the localization result or "Needs Adjustment (NA)", i.e., at least one electrode location needs to be adjusted by more than half a contact size. When both results are acceptable, raters are also asked to decide which one is preferred (i.e., has a more accurate localization result) or there is no preference.

Before performing the expert evaluation, we calculate the maximum P2PE between the results from the previous SOTA and the proposed methods. For cases with maximum P2PE larger than 0.3 mm (we refer to them as the large-error subset), there is a high probability that at least one of the methods generates NA results. For cases with maximum P2PE smaller than 0.3 mm (we refer to them as the small-error subset), we presume that the probability of NA results is relatively small. To limit the demand on the experts'

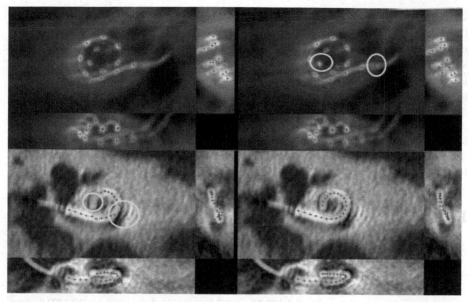

Fig. 4. Two representative cases from the large-error subset are shown with the display used to visually evaluate localization quality. Yellow circles highlight the regions that contain the NA electrode localization. Top row: a distantly-spaced EA. Bottom row: a closely-spaced EA. The Left and right images in each row represent results obtained with one of the methods; the method that is used for each is not disclosed to the evaluator.

time, we ask three experts (R1, R2, and R3) to rate the large error subset (164 cases) and only one expert (R1) to rate the small error subset (397 cases).

The expert evaluation results are shown as radar plots in Fig. 3. The left figure shows the acceptance rate for the large error cases evaluated by raters R1, R2, and R3. Except for the cases rated by R1 that contain closely-spaced EAs in CBCT ("R1-CBCT-Closely"), the localization results generated by the proposed method have a substantially higher acceptance rate than the previous SOTA methods ([8] for closely-spaced and [12] for distantly-spaced). For the cases in which both methods are acceptable, the average preference rates across the three raters are: 20.2% for "SOTA preferred", 30.3% for "Proposed preferred", and 49.5% for "No preference". As can be seen in the right plot of Fig. 3, the overall acceptance rate on the whole clinical test set evaluated by R1 is 80.7% for the previous SOTA methods and 89.7% for the proposed method. It is interesting to note that although the proposed method is trained mostly on CBCT images, it still generalizes well on CT images. Figure 4 shows two representative cases from the large-error subset.

4   Conclusions

In this work, we have proposed a novel DL-based framework for cochlear implant EA localization. To the best of our knowledge, it is the first unified DL-based framework designed for localizing both distantly- and closely-spaced EAs in CT and CBCT images.

Compared to the SOTA methods, the proposed framework is substantially more robust (9% less NA results) when evaluated on a large-scale clinical dataset and achieves slightly more accurate localization results on a dataset containing 27 cadaveric samples with gold standard ground truth. While it may be possible to improve our success rate further, a low percentage of NA results is unavoidable. We are thus developing quality assessment techniques to alert end users when images have poor quality and/or results are unreliable.

Acknowledgments. This research is supported by the National Institutes of Health grant R01DC014037, R01DC008408, and R01DC014462, and T32EB021937. The content is solely the responsibility of the authors and does not necessarily represent the official views of these institutes.

References

1. Sharma, S.D., Cushing, S.L., Papsin, B.C., Gordon, K.A.: Hearing and speech benefits of cochlear implantation in children: a review of the literature. Int. J. Pediatr. Otorhinolaryngol. **133**, 109984 (2020). https://doi.org/10.1016/j.ijporl.2020.109984
2. Boisvert, I., Reis, M., Au, A., Cowan, R., Dowell, R.C.: Cochlear implantation outcomes in adults: a scoping review. PLoS ONE **15**, e0232421 (2020). https://doi.org/10.1371/journal.pone.0232421
3. Holden, L.K., et al.: Factors affecting open-set word recognition in adults with cochlear implants. Ear Hear. **34**, 342 (2013). https://doi.org/10.1097/AUD.0b013e3182741aa7
4. Chakravorti, S., et al.: Further evidence of the relationship between cochlear implant electrode positioning and hearing outcomes. Otol. Neurotol. **40**, 617 (2019). https://doi.org/10.1097/MAO.0000000000002204
5. Noble, J.H., Labadie, R.F., Gifford, R.H., Dawant, B.M.: Image-guidance enables new methods for customizing cochlear implant stimulation strategies. IEEE Trans. Neural Syst. Rehabil. Eng. **21**, 820–829 (2013). https://doi.org/10.1109/TNSRE.2013.2253333
6. Noble, J.H., Gifford, R.H., Hedley-Williams, A.J., Dawant, B.M., Labadie, R.F.: Clinical evaluation of an image-guided cochlear implant programming strategy. Audiol. Neurotol. **19**, 400–411 (2014). https://doi.org/10.1159/000365273
7. Noble, J.H., et al.: Initial results with image-guided cochlear implant programming in children. Otol. Neurotol. **37**, e63 (2016). https://doi.org/10.1097/MAO.0000000000000909
8. Zhao, Y., Dawant, B.M., Labadie, R.F., Noble, J.H.: Automatic localization of closely spaced cochlear implant electrode arrays in clinical CTs. Med. Phys. **45**, 5030–5040 (2018). https://doi.org/10.1002/mp.13185
9. Andersen, S.A.W., et al.: Automated calculation of cochlear implant electrode insertion parameters in clinical cone-beam CT. Otol. Neurotol. **43**, 199–205 (2022). https://doi.org/10.1097/MAO.0000000000003432
10. Zhao, Y., Dawant, B.M., Labadie, R.F., Noble, J.H.: Automatic localization of cochlear implant electrodes in CT. In: Golland, P., Hata, N., Barillot, C., Hornegger, J., Howe, R. (eds.) Medical Image Computing and Computer-Assisted Intervention, pp. 331–338. Springer International Publishing, Cham (2014). https://doi.org/10.1007/978-3-319-10404-1_42
11. Bennink, E., Peters, J.P.M., Wendrich, A.W., Vonken, E., van Zanten, G.A., Viergever, M.A.: Automatic localization of cochlear implant electrode contacts in CT. Ear Hear. **38**, e376 (2017). https://doi.org/10.1097/AUD.0000000000000438
12. Zhao, Y., Chakravorti, S., Labadie, R.F., Dawant, B.M., Noble, J.H.: Automatic graph-based method for localization of cochlear implant electrode arrays in clinical CT with sub-voxel accuracy. Med. Image Anal. **52**, 1–12 (2019). https://doi.org/10.1016/j.media.2018.11.005

13. Braithwaite, B., et al.: Cochlear implant electrode localization in post-operative CT using a spherical measure. In: 2016 IEEE 13th International Symposium on Biomedical Imaging (ISBI), pp. 1329–1333 (2016). https://doi.org/10.1109/ISBI.2016.7493512

14. Hachmann, H., Krüger, B., Rosenhahn, B., Nogueira, W.: Localization of cochlear implant electrodes from cone beam computed tomography using particle belief propagation. In: 2021 IEEE 18th International Symposium on Biomedical Imaging (ISBI), pp. 593–597 (2021). https://doi.org/10.1109/ISBI48211.2021.9433845

15. Chi, Y., Wang, J., Zhao, Y., Noble, J.H., Dawant, B.M.: A deep-learning-based method for the localization of cochlear implant electrodes in CT images. In: 2019 IEEE 16th International Symposium on Biomedical Imaging (ISBI 2019), pp. 1141–1145 (2019). https://doi.org/10.1109/ISBI.2019.8759536

16. Wang, J., Zhao, Y., Noble, J.H., Dawant, B.M.: Metal artifact reduction, intra cochlear anat-omy segmentation, and cochlear implant electrodes localization in CT images with a multi-task 3D network. In: Landman, B.A., Išgum, I. (eds.) Medical Imaging 2021: Image Processing. p. 23. SPIE, Online Only, United States (2021). https://doi.org/10.1117/12.2580931

17. Margeta, J., et al.: A web-based automated image processing research platform for cochlear implantation-related studies. J. Clin. Med. 11, 6640 (2022). https://doi.org/10.3390/jcm11226640

18. Zhao, Y., Labadie, R.F., Dawant, B.M., Noble, J.H.: Validation of automatic cochlear im-plant electrode localization techniques using µCTs. J. Med. Imaging. 5, 035001 (2018). https://doi.org/10.1117/1.JMI.5.3.035001

19. Zhou, X., Wang, D., Krähenbühl, P.: Objects as Points. ArXiv190407850 Cs. (2019)

20. Lin, T.-Y., Goyal, P., Girshick, R., He, K., Dollár, P.: Focal loss for dense object detection. IEEE Trans. Pattern Anal. Mach. Intell. 42, 318–327 (2020). https://doi.org/10.1109/TPAMI.2018.2858826

21. Shit, S., et al.: clDice -- a novel topology-preserving loss function for tubular structure seg-mentation. In: 2021 IEEE/CVF Conference on Computer Vision and Pattern Recognition (CVPR), pp. 16555–16564 (2021). https://doi.org/10.1109/CVPR46437.2021.01629

22. Girshick, R., Donahue, J., Darrell, T., Malik, J.: Rich feature hierarchies for accurate object detection and semantic segmentation. In: Proceedings of the IEEE Conference on Computer Vision and Pattern Recognition (2014)

23. Cardoso, M.J. et al.: MONAI: An open-source framework for deep learning in healthcare, http://arxiv.org/abs/2211.02701 (2022). https://doi.org/10.48550/arXiv.2211.02701

24. Greenwood, D.D.: A cochlear frequency-position function for several species—29 years later. J. Acoust. Soc. Am. 87, 2592–2605 (1990). https://doi.org/10.1121/1.399052

GLSFormer: Gated - Long, Short Sequence Transformer for Step Recognition in Surgical Videos

Nisarg A. Shah[1]([✉]), Shameema Sikder[2,3], S. Swaroop Vedula[3],
and Vishal M. Patel[1]

[1] Johns Hopkins University, Baltimore, MD 21218, USA
snisarg812@gmail.com
[2] Wilmer Eye Institute, Johns Hopkins University School of Medicine,
Baltimore, MD, USA
[3] Malone Center for Engineering in Healthcare,
Johns Hopkins University, Baltimore, USA

Abstract. Automated surgical step recognition is an important task that can significantly improve patient safety and decision-making during surgeries. Existing state-of-the-art methods for surgical step recognition either rely on separate, multi-stage modeling of spatial and temporal information or operate on short-range temporal resolution when learned jointly. However, the benefits of joint modeling of spatio-temporal features and long-range information are not taken in account. In this paper, we propose a vision transformer-based approach to jointly learn spatio-temporal features directly from sequence of frame-level patches. Our method incorporates a gated-temporal attention mechanism that intelligently combines short-term and long-term spatio-temporal feature representations. We extensively evaluate our approach on two cataract surgery video datasets, namely Cataract-101 and D99, and demonstrate superior performance compared to various state-of-the-art methods. These results validate the suitability of our proposed approach for automated surgical step recognition. Our code is released at: https://github.com/nisargshah1999/GLSFormer.

Keywords: Surgical Activity Recognition · Cataract surgery · Phase Recognition · Vision Transformer

1 Introduction

Surgical step recognition is necessary to enable downstream applications such as surgical workflow analysis [4,26], context-aware decision support [21], anomaly detection [14], and record-keeping purposes [5,28]. Some factors that make recognition of steps in surgical videos a challenging problem [20,21] include variability in patient anatomy and surgeon style [11], similarities across steps in a procedure [5,16], online recognition [25] and scene blur [12,21].

Early statistical methods to recognize surgical workflow, such as Conditional Random Fields [19,23], Hidden Markov Models (HMMs) [6,24] and Dynamic

© The Author(s), under exclusive license to Springer Nature Switzerland AG 2023
H. Greenspan et al. (Eds.): MICCAI 2023, LNCS 14228, pp. 386–396, 2023.
https://doi.org/10.1007/978-3-031-43996-4_37

Time Warping [3,18], have limited representation capacity due to pre-defined dependencies. Multiple deep learning based methods have been proposed for surgical step recognition. SV-RCNet [15] jointly trains ResNet [13] and a long short-term memory (LSTM) model and uses a prior knowledge scheme during inference. TMRNet [16] utilizes a memory bank to store long range information on the relationship between the current frame and its previous frames. SV-RCNet and TMR-Net use LSTMs, which are constrained in capturing long-term dependencies in surgical videos due to their limited temporal memory. Furthermore, LSTMs process information in a sequential manner that results in longer inference times. Methods that don't use LSTMs include a 3D-covolutional neural network (3D-CNN) to learn spatio-temporal features [10] and temporal convolution networks (TCNs) [27]. In recent work, a two-stage network called TeCNO included a ResNet to learn spatial features, which are then modeled with a multi-scale TCN to capture long-term dependencies [5]. The previous networks use multi-stage training to exploit spatial and temporal information separately. This approach limits model capacity to learn spatial features using the temporal information. Furthermore, the temporal modeling does not sufficiently benefit from low-dimensional spatial features resulting in low sensitivity to identify transitions between activities [12,15]. [12] attempts to address this issue by adding one more stage to the network using a feature fusion technique that employs a transformer to refine features extracted from the first and second stages of TeCNO [5].

Transformers improve feature representation by effectively modeling long-term dependencies, which is important for recognizing surgical steps in complex videos. In addition, they offer fast processing due to their parallel computing architecture. To exploit these benefits and address the issue of inductive bias (e.g. local connectivity and translation equivariance) in CNNs, recent methods use only transformers as their building blocks, i.e., Vision Transformer. For example, TimesFormer [2], applies self-attention mechanisms to learn spatial and temporal relations in videos, and ViViT [1] utilizes a vision transformer architecture [8] for video recognition. However, these models were proposed for temporal clips and do not specifically focus on capturing long-range dependencies, which is important for surgical step recognition in long-duration untrimmed videos.

In this work, we propose a transformer-based model with the following contributions: (1) Spatio-temporal attention is used as the building blocks to address issues with inductive bias and end-to-end learning of surgical steps; (2) A two-stream model, called **G**ated - **L**ong, **S**hort sequence Trans**former** GLSFormer , is proposed to capture long-range dependencies and a gating module to leverage cross-stream information in its latent space; and (3) The proposed GLSFormer is extensively evaluated on two cataract surgery video datasets to show that it outperforms all compared methods.

2   The **GLSFormer** Model

Given an untrimmed video with $X[1 : T]$ frames, where T represents the total number of frames in the video, the objective is to predict the step $Y[t]$ of a given frame at time t (Fig. 1).

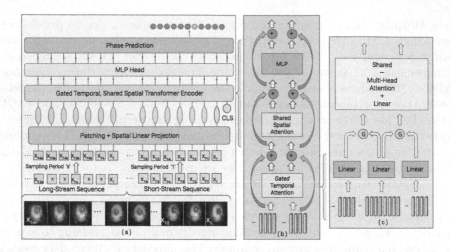

Fig. 1. Overview of the proposed GLSFormer . Specifically, GLSFormer takes two streams (long-stream and short-stream) image sequences as input, sampled with sampling period of s and 1 respectively. Later, each frame is decomposed into non-overlapping patches and each of these patches are linearly mapped to an embedding vector. These embedded features are spatio-temporally encoded into a feature representation using a sequential temporal-spatial attention block as shown in (b). Architecture of gated-temporal attention is showed in detail in (c). The final feature representation is then examined by a multilayer perceptron head and a linear layer to produce a step prediction at every time-point. Our method provides a single-stage, end-to-end trainable model for surgical step recognition.

Long-short Sequence. We propose GLSFormer model that can capture both short-term and long-term dependencies in surgical videos. The input to our GLSFormer are two video sequences, a short-term sequence consisting of the last n^{st} frames from time t, and a long-term sequence composed of n^{lt} frames selected from a sub-sampled set of frames with a sampling period of s. The long-term sequence provides a coarse overview of information distant in time and can aid in accurate predictions of the current frame, overcoming false prediction based on common artifacts in short-term sequences. In addition, the overview of information can address the high variability in surgical scenes [5,9,12,16]. By leveraging both short-term and long-term sequences, our model can accurately capture the complex context present in long surgical videos.

Patch Embedding. We decompose each frame of dimension $H \times W \times 3$ into N non-overlapping patches of size $Q \times Q$ where $N = \frac{HW}{Q^2}$. Each patch is flattened into a vector $x_{p,t} \in \mathbb{R}^{3Q^2}$ for each frame t and spatial location, $p \in (1, N)$. We linearly map the patches of short and long term videos frames into embedding vector of dimension $\mathbb{R}^K$ using a shared learnable matrix $E \in \mathbb{R}^{K \times 3Q^2}$. We concatenate the patch embeddings of the short and long-term streams along the frame dimension to form feature representations $x_{p,t}^{st}$ and $x_{p,t}^{lt}$ of size $N \times n^{st} \times K$

and $N \times n^{lt} \times K$, respectively. along with a learnable positional embedding $e_{p,t}^{st-pos}$ and $e_{p,t}^{lt-pos}$ to encode spatio-temporal position as follows

$$z_{p,t}^{st} = E x_{p,t}^{st} + e_{p,t}^{st-pos}, \quad z_{p,t}^{lt} = E x_{p,t}^{lt} + e_{p,t}^{lt-pos}. \quad (1)$$

Note that a special learnable vector $z_{0,0}^{st} \in \mathbb{R}^K$ representing the step classification token is added in the first position. Our approach is similar to word embeddings in NLP transformer models [1,2,7].

Gated Temporal, Shared Spatial Transformer Encoder. Our Transformer Encoder, consisting of Gated Temporal Attention module and Shared Spatial Attention module takes the sequence of embedding vectors $z_{p,t}^{st}$ and $z_{p,t}^{lt}$ as input. In a self-attention module for spatio-termporal models, computational complexity increases non-linearly $O(T^2 S^2)$ with increase in spatial resolution(S) or temporal frames(T). Thus, to reduce the complexity, we sequentially process our gated temporal cross attention module and spatial attention module [1,2]. Transformer Encoder consists of L Gated-Temporal, Spatial Attention blocks. At each block l, feature representation is computed for both streams from the representation z_{l-1}^{lt} and z_{l-1}^{st} encoded by the preceding block $l-1$. We explain our Gated Temporal attention and shared spatial attention in more detail in the rest of the section.

Gated Temporal Attention. The temporal cross-attention module aligns the long-term(z_{l-1}^{lt}) and short-term features(z_{l-1}^{st}) in the temporal domain, allowing the model to better capture the relationship between the long and short term streams. We concatenate both of these streams to form a strong joint stream that has both fine-grained information from the short-term stream and global context information from the long-term stream. Firstly, a query/key/value vector for each patch in the representations $z_{l-1}^{lt}(p,t)$, $z_{l-1}^{st}(p,t)$ and $z_{l-1}^{lt,st}(p,t)$ using linear transformations with weight matrices U_{qkv}^{lt}, U_{qkv}^{st} and $U_{qkv}^{lt,st}$ respectively and normalization using LayerNorm is computed as follows:

$$(QKV)^{a;st} = z^{st} U_{qkv}^{st}, \qquad\qquad U_{qkv}^{st} \in \mathbb{R}^{K_h \times 3K_h}$$
$$(QKV)^{a;lt} = z^{lt} U_{qkv}^{lt}, \qquad\qquad U_{qkv}^{lt} \in \mathbb{R}^{K_h \times 3K_h} \quad (2)$$
$$(QKV)^{a;lt,st} = [z^{lt} \oplus z^{st}] U_{qkv}^{lt,st}, \qquad U_{qkv}^{lt,st} \in \mathbb{R}^{K_h \times 3K_h}$$

where a ranges from 1 to A representing attention heads. The total number of attention heads is denoted by A, and each has a latent dimensionality of $K_h = \frac{K}{A}$. The computation of these QKV vectors is essential for multi-head attention in transformer.

Now for refining the streams, with most relevant cross-stream information, we gate the individual stream's temporal features(I) $(QKV)^{a;lt/st}$ with the joint stream temporal features(J) $(QKV)^{a;lt,st}$. Gating parameters are calculated by concatenating I and J and passing them through linear and softmax layers which predict Gt^{st} and Gt^{lt} for $(QKV)^{a;st}$ and $(QKV)^{a;lt}$, respectively. By gating the individual stream's temporal features with the joint stream temporal features, the model is able to selectively attend to the most relevant features from both streams, resulting in a more informative representation. This helps in capturing complex relationships between the streams and improves the overall performance of the model. This computation can be described as follows

$$[q_{gt}^{st}, k_{gt}^{st}, v_{gt}^{st}] = (1 - Gt^{st}(0)) * (QKV)^{a;st} + Gt^{st}(0) * (QKV)^{a;lt,st}[lt : lt + st]$$

$$[q_{gt}^{lt}, k_{gt}^{lt}, v_{gt}^{lt}] = (1 - Gt^{lt}(0)) * (QKV)^{a;lt} + Gt^{lt}(0) * (QKV)^{a;lt,st}[: lt].$$

Later, temporal attention is computed by comparing each patch (p, t) with all patches at the same spatial location in other frames of both streams, as follows

$$\alpha_{gt,(p,t)}^{(\bullet)\text{-temporal}(l,a)} = \text{softmax}\left(\frac{q_{gt,(p,t)}^{(\bullet),(l,a)}}{\sqrt{K_h}}\left[k_{gt,(0,0)}^{(l,a)} \ \prod_{t'=1}^{n^{lt}} k_{gt,(p,t')}^{lt,(l,a)} \ \prod_{t'=1}^{n^{st}} k_{gt,(p,t')}^{st,(l,a)}\right]\right)$$

Here, $\alpha_{gt,(p,t)}^{(\bullet)\text{-temporal}(l,a)}$ is separately calculated for the long-stream and short-stream, where $(\bullet) = lt$ or st. Similar to the vision transformer, encoding blocks for each layer (z^{lt} and z^{st}) are computed by taking the weighted sum of value vectors $(\text{SA}_a(z))$ using self-attention coefficients from each attention head as follows

$$\text{SA}_a(z = z^{lt} \oplus z^{st}) = (\alpha_{gt}^{lt,a} \oplus \alpha_{gt}^{st,a}) \cdot (v_{gt}^{lt,a} \oplus v_{gt}^{st,a}). \tag{3}$$

Next, the self-attention block $(\text{SA}_a(z))$ for each attention head is projected along with a residual connection from the previous layer. This multi-head self-attention (MSA) operation can be described as follows

$$\text{MSA}(z) = [\text{SA}_1(z); \text{SA}_2(z); ...; \text{SA}_A(z)] \times Umsa, \qquad Umsa \in \mathbb{R}^{k \cdot D_h \times D} \tag{4}$$

$$(z')^l = \text{MSA}(z^l) + (z)^{l-1} \tag{5}$$

Here, $(z')^l$ is the concatenation of z^{lt} and z^{st}.

Shared Spatial Attention. Next, we apply self-attention to the patches of the same frame to capture spatial relationships and dependencies within the frame. To accomplish this, we calculate new key/query/value using Eq. (2) and use it to perform spatial attention in Eq. (6).

$$\alpha_{gt,(p,t)}^{(\bullet)\text{-spatial}(l,a)} = \text{softmax}\left(\frac{q_{gt,(p,t)}^{(\bullet),(l,a)}}{\sqrt{K_h}}\left[k_{gt,(0,0)}^{(l,a)} \ \prod_{p'=1}^{N} k_{gt,(p',t)}^{(\bullet),(l,a)}\right]\right) \tag{6}$$

The encoding blocks are also calculated using Eq. (4) and (5), and the resulting vector is passed through a multilayer perceptron (MLP) of Eq. (7) to obtain the final encoding $z'(p, t)$ for the patch at block l as follows

$$(z)^l = \text{MLP}(LN((z')^l)) + (z')^l, \quad [z^{lt}, z^{st}] = z^l. \tag{7}$$

The embedding for the entire clip is obtained by taking the output from the final block and passing it through a MLP with one hidden layer. The corresponding computation can be described as $y = \text{LN}(z_{(0,0)}^L) \in \mathbb{R}^D$. The classification token is used as the final input to the MLP for predicting the step class at time t. Our GLSFormer is trained using the cross-entropy loss.

3 Experiments and Results

Datasets. We evaluate our GLS-Former on two video datasets of cataract surgery, namely Cataract-101 [22] and D99 [26]. The Cataract-101 dataset comprises of 101 cataract surgery video recordings, each captured at 25 frames per second and annotated into 10 steps by surgeons. The spatial resolution of these videos is 720 × 540, and temporal resolution of 25 *fps*. In accordance with previous studies [12,24], we use 50 videos for training, 10 for validation and 40 videos for testing. The D99 dataset, which consists of 99 videos with temporal segment annotations of 12 steps by expert physicians, has a frame resolution of 640 × 480 at 59 *fps*. We randomly shuffled videos and select 60, 20 and 19 videos for training, validation and testing respectively. All videos are subsampled to 1 frame per second, as done in previous studies [12,24], and the frames are resized to 250 × 250 resolution.

Evaluation Metrics. To accurately evaluate the results of surgical step prediction models, we use four different metrics, namely Accuracy, Precision, Recall, and Jaccard index [5,12,24].

Implementation Details. We utilized an NVIDIA RTX A5000 GPU to train our GLSFormer model on PyTorch. The batch size was set equal to 64. Data augmentations were applied including 224 × 224 cropping, random mirroring, and color jittering. We employed the Adam optimizer with an initial learning rate of 5e-5 for 50 epochs. Additionally, we initialized the shared parameters of the model from a pre-trained model on Kinetics-400 [2,17]. The model's depth and the number of attention heads were set equal to 12 each. We used 8 frames for both short-stream and long-stream, and sampling rate of 8, unless stated otherwise.

Table 1. Quantitative results of step recognition from different methods on the Cataract-101 and D99 datasets. The average metrics over five repetitions of the experiment with different data partitions for cross-validation are reported (%) along with their respective standard deviation (±).

Method	Cataract-101				D99			
	Accuracy	Precision	Recall	Jaccard	Accuracy	Precision	Recall	Jaccard
ResNet [13]	82.64 ± 1.54	76.68 ± 1.86	74.73 ± 1.27	62.58 ± 1.92	72.06 ± 2.12	54.76 ± 2.77	52.28 ± 2.89	37.98 ± 2.97
SV-RCNet [15]	86.13 ± 0.91	84.96 ± 0.94	76.61 ± 1.18	66.51 ± 1.30	73.39 ± 1.64	58.18 ± 1.67	54.25 ± 1.86	39.15 ± 2.03
OHFM [25]	87.82 ± 0.71	85.37 ± 0.78	78.29 ± 0.81	69.01 ± 0.93	73.82 ± 1.13	59.12 ± 1.33	55.49 ± 1.63	40.01 ± 1.68
TeCNO [5]	88.26 ± 0.92	86.03 ± 0.83	79.52 ± 0.90	70.18 ± 1.15	74.07 ± 1.78	61.56 ± 1.41	55.81 ± 1.58	41.31 ± 1.72
TMRNet [16]	89.68 ± 0.76	85.09 ± 0.72	82.44 ± 0.75	71.83 ± 0.91	75.11 ± 0.91	61.37 ± 1.46	56.02 ± 1.65	41.42 ± 1.76
Trans-SVNet [12]	89.45 ± 0.88	86.72 ± 0.85	81.12 ± 0.93	72.32 ± 1.04	74.89 ± 1.37	60.12 ± 1.55	56.36 ± 1.24	42.06 ± 1.51
ViT [8]	84.56 ± 1.72	78.51 ± 1.42	75.62 ± 1.83	64.77 ± 1.97	72.45 ± 1.91	55.15 ± 2.42	53.60 ± 2.63	38.18 ± 2.79
TimesFormer [2]	90.76 ± 1.05	85.38 ± 0.93	84.47 ± 0.95	75.97 ± 1.26	77.83 ± 0.96	64.24 ± 1.20	55.17 ± 1.26	42.69 ± 1.34
GLSFormer	**92.91 ± 0.67**	**90.04 ± 0.71**	**89.45 ± 0.79**	**81.89 ± 0.92**	**80.24 ± 1.02**	**69.98 ± 1.09**	**56.07 ± 1.12**	**48.35 ± 1.22**

Comparison with State-of-the-art Methods. Table 1 presents a comparison between our proposed approach, GLSFormer , and current state-of-the-art methods for surgical step prediction. The comparison includes six models (1–6) that utilize ResNet [13] as a spatial feature extractor and two models (7–8) that use vision transformer backbones specifically designed for surgical step prediction. While SV-RCNet, OHFM, and TMRNet use ResNet-LSTM to capture short-range dependencies, OHFM uses a multi-step framework and TMRNet uses a multi-stage network to refine predictions using non-trainable long-range memory banks. Our approach achieves a significant improvement of 7%–10% in Jaccard index using a simpler, single-stage training procedure with higher temporal resolution. Although other multi-stage approaches like TeCNO and Trans-SVNet use temporal convolutions to capture long-range dependencies, we achieve a boost of 6%–9% with joint spatiotemporal modeling. In contrast, transformer-based models capture spatial information (ViT) and short-term temporal (TimesFormer) efficiently, but they lack the long-term coarse step information required for complex videos. Our approach combines short-term and long-term spatiotemporal information in a single stage using gated-temporal and shared-spatial attention. This approach outperforms ViT and TimesFormer by a relative improvement of 6% to 11% in Jaccard index across both datasets.

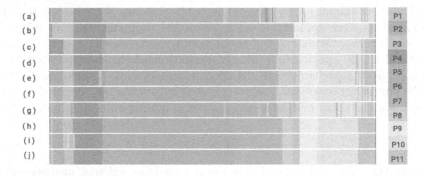

Fig. 2. Qualitative results of step prediction of models on Cataracts-101 dataset in color-coded ribbon format. Predictions from (a) ResNet [13], (b) SV-RCNet [15], (c) OHFM [25], (d) TeCNO [5], (e) TMRNet [16], (f) Trans-SVNet [12], (g) ViT [8], (h) TimesFormer [2], (i) **GLSFormer** , and (j) Ground Truth. P1 to P11 indicates step label.

Qualitative Comparison. In Fig. 2, the color-coded ribbon map of one cataract surgery video from the cataract-101 dataset is shown. Both the ResNet and ViT produce noisy patterns and frequently generate abrupt predictions due to the lack of temporal relations. However, LSTM-based models such as SV-RCNet, OHFM, and TMRNet have comparatively less noisy patterns locally, but suffer from wrong step predictions due to the lack of complete global context. TeCNO and Trans-SVNet achieve smoother results by capturing long-term temporal information in later stages from the spatial embeddings generated by the ResNet. However, these models still suffer from wrong step predictions (such as at P1) and noisy patterns (such as at P10) due to their high reliance on extracted spatial features and error propagation across stages in the model. Specifically, errors in the early stages of spatial modeling ResNet can propagate and accumulate across later stages, and lead to incorrect predictions. The ribbon plot of TimesFormer (i) demonstrates significant improvement compared to previous methods due to joint learning of spatio-temporal features. However, due to the lack of pivotal coarse long-term information, misclassifications are observed in local step transition areas, as seen in the incorrect prediction of P2 at locations of P1 and P11. On the other hand, GLSFormer elegantly aggregates spatio-temporal information from both streams, making the features more reliable. Additionally, our approach uses a single stage to limit the amount of error propagated across stages, contributing to improved accuracy in surgical step recognition.

Ablation Studies. The top part of Table 2 shows the effect of different sampling rates in the long-term stream for step prediction in the Cataract-101 dataset. The results demonstrate that incorporating a coarse long-term stream is crucial for achieving significant performance gains compared to not using any long-term sequence (as in ViT and TimesFormer). Additionally, we observe that gradually increasing the sampling rate from 2 to 8 improves performance across all metrics, except for a slight decline at a sampling rate of 16. This decline may be due to a loss of information and noisy predictions resulting from the high number of frames skipped between each selected frame. Therefore, we chose a sampling rate of 8, as it provided the optimal balance between capturing valuable information and avoiding noise in our long-stream sequence.

Table 2. Ablation testing results for different temporal gating mechanisms and long-term stream sampling rates on the Cataract-101 dataset.

		Accuracy	Precision	Recall	Jaccard
Sampling Rate	2	91.16 ± 0.81	86.47 ± 0.85	87.25 ± 1.00	76.85 ± 1.33
	4	91.43 ± 0.72	87.46 ± 0.91	87.20 ± 0.90	77.48 ± 1.26
	8	**92.91 ± 0.67**	**90.04 ± 0.71**	**89.45 ± 0.79**	**81.89 ± 0.92**
	16	91.35 ± 0.86	88.91 ± 0.92	88.27 ± 0.98	79.24 ± 1.10
Temporal Gating Stream	Only short-term	90.76 ± 1.05	85.38 ± 0.93	84.47 ± 0.95	75.97 ± 1.26
	No Gating	90.24 ± 0.75	86.84 ± 0.88	84.39 ± 0.90	74.82 ± 1.13
	Fix Param. Gating	91.52 ± 0.83	87.13 ± 0.81	88.41 ± 0.94	78.03 ± 1.04
	Feature Gating	**92.91 ± 0.67**	**90.04 ± 0.71**	**89.45 ± 0.79**	**81.89 ± 0.92**

To evaluate our gating mechanism's effectiveness, we conducted ablation experiments with three different settings, as summarized in the bottom part of Table 2. Initially, we passed both short-term and long-term stream features directly in the shared multi-head attention layer, but the model's performance was worse compared to the model trained only with short-term information (75.97 vs 74.82 Jaccard). The reason for this could be the lack of filtering to extract coarse temporal information. For instance, the long-term stream may contain noisy spatial information that is irrelevant to the temporal attention mechanism, which can affect the model's ability to attend to relevant information. Additionally, incorporating a learnable gating parameter to regulate the flow of information between the short-term and long-term streams enhanced our model's performance by 4%, enabling individual stream refinement through cross-stream information. However, we observe that this approach has a limitation as the amount of cross-stream information sharing remains fixed during inference regardless of the quality of the feature representation in both streams at a particular time-frame. To address this limitation, we propose predicting gating parameters directly based on the spatio-temporal representation in both streams for that time frame. This approach allows for dynamic gating parameters, which means that at a particular time-point, short-term temporal feature representation can variably leverage the long-term coarse information as well can prioritize its own representation if it is more reliable. Improvement of 3% Jaccard score is realized by using feature-based gating parameter estimation in GLSFormer compared to a fixed parameter gating mechanism. Our ablation study clearly highlights the significance of our gated temporal mechanism for feature refinement.

4   Conclusion

We propose GLSFormer , a vision transformer-based method for recognizing surgical steps in complex videos. Our approach uses a gated temporal attention mechanism to integrate short and long-term cues, resulting in superior performance compared to recent LSTM and vision transformer-based approaches that only use short-term information. Our end-to-end joint learning captures spatial representations and sequential dynamics more effectively than multi-stage networks. We extensively evaluated GLSFormer and found that it consistently outperformed state-of-the-art models for surgical step recognition.

Acknowledgements. This work was supported by a grant from the National Institutes of Health, USA; R01EY033065. The content is solely the responsibility of the authors and does not necessarily represent the official views of the National Institutes of Health.

References

1. Arnab, A., Dehghani, M., Heigold, G., Sun, C., Lučić, M., Schmid, C.: Vivit: a video vision transformer. In: Proceedings of the IEEE/CVF International Conference on Computer Vision, pp. 6836–6846 (2021)
2. Bertasius, G., Wang, H., Torresani, L.: Is space-time attention all you need for video understanding? In: ICML, vol. 2, p. 4 (2021)
3. Blum, T., Feußner, H., Navab, N.: Modeling and segmentation of surgical workflow from laparoscopic video. In: Jiang, T., Navab, N., Pluim, J.P.W., Viergever, M.A. (eds.) MICCAI 2010. LNCS, vol. 6363, pp. 400–407. Springer, Heidelberg (2010). https://doi.org/10.1007/978-3-642-15711-0_50
4. Bricon-Souf, N., Newman, C.R.: Context awareness in health care: a review. Int. J. Med. Inform. **76**(1), 2–12 (2007)

5. Czempiel, T., et al.: TeCNO: surgical phase recognition with multi-stage temporal convolutional networks. In: Martel, A.L., et al. (eds.) MICCAI 2020. LNCS, vol. 12263, pp. 343–352. Springer, Cham (2020). https://doi.org/10.1007/978-3-030-59716-0_33

6. Dergachyova, O., Bouget, D., Huaulmé, A., Morandi, X., Jannin, P.: Automatic data-driven real-time segmentation and recognition of surgical workflow. Int. J. Comput. Assist. Radiol. Surg. 11(6), 1081–1089 (2016). https://doi.org/10.1007/s11548-016-1371-x

7. Devlin, J., Chang, M.W., Lee, K., Toutanova, K.: Bert: Pre-training of deep bidirectional transformers for language understanding. arXiv preprint arXiv:1810.04805 (2018)

8. Dosovitskiy, A., et al.: An image is worth 16 × 16 words: transformers for image recognition at scale. arXiv preprint arXiv:2010.11929 (2020)

9. Feichtenhofer, C., Fan, H., Malik, J., He, K.: Slowfast networks for video recognition. In: Proceedings of the IEEE/CVF International Conference on Computer Vision, pp. 6202–6211 (2019)

10. Funke, I., Bodenstedt, S., Oehme, F., von Bechtolsheim, F., Weitz, J., Speidel, S.: Using 3D convolutional neural networks to learn spatiotemporal features for automatic surgical gesture recognition in video. In: Shen, D., et al. (eds.) MICCAI 2019. LNCS, vol. 11768, pp. 467–475. Springer, Cham (2019). https://doi.org/10.1007/978-3-030-32254-0_52

11. Funke, I., Mees, S.T., Weitz, J., Speidel, S.: Video-based surgical skill assessment using 3D convolutional neural networks. Int. J. Comput. Assist. Radiol. Surg. 14, 1217–1225 (2019)

12. Gao, X., Jin, Y., Long, Y., Dou, Q., Heng, P.-A.: Trans-SVNet: accurate phase recognition from surgical videos via hybrid embedding aggregation transformer. In: de Bruijne, M., et al. (eds.) MICCAI 2021. LNCS, vol. 12904, pp. 593–603. Springer, Cham (2021). https://doi.org/10.1007/978-3-030-87202-1_57

13. He, K., Zhang, X., Ren, S., Sun, J.: Deep residual learning for image recognition. In: Proceedings of the IEEE Conference on Computer Vision and Pattern Recognition, pp. 770–778 (2016)

14. Huaulmé, A., Jannin, P., Reche, F., Faucheron, J.L., Moreau-Gaudry, A., Voros, S.: Offline identification of surgical deviations in laparoscopic rectopexy. Artif. Intell. Med. 104, 101837 (2020)

15. Jin, Y., et al.: SV-RCNet: workflow recognition from surgical videos using recurrent convolutional network. IEEE Trans. Med. Imaging 37(5), 1114–1126 (2017)

16. Jin, Y., Long, Y., Chen, C., Zhao, Z., Dou, Q., Heng, P.A.: Temporal memory relation network for workflow recognition from surgical video. IEEE Trans. Med. Imaging 40(7), 1911–1923 (2021)

17. Kay, W., et al.: The kinetics human action video dataset. arXiv preprint arXiv:1705.06950 (2017)

18. Lalys, F., Bouget, D., Riffaud, L., Jannin, P.: Automatic knowledge-based recognition of low-level tasks in ophthalmological procedures. Int. J. Comput. Assist. Radiol. Surg. 8, 39–49 (2013)

19. Lea, C., Hager, G.D., Vidal, R.: An improved model for segmentation and recognition of fine-grained activities with application to surgical training tasks. In: 2015 IEEE Winter Conference on Applications of Computer Vision, pp. 1123–1129. IEEE (2015)

20. Lecuyer, G., Ragot, M., Martin, N., Launay, L., Jannin, P.: Assisted phase and step annotation for surgical videos. Int. J. Comput. Assist. Radiol. Surg. 15(4), 673–680 (2020). https://doi.org/10.1007/s11548-019-02108-8

21. Padoy, N.: Machine and deep learning for workflow recognition during surgery. Minim. Invasive Therapy Allied Technol. **28**(2), 82–90 (2019)
22. Schoeffmann, K., Taschwer, M., Sarny, S., Münzer, B., Primus, M.J., Putzgruber, D.: Cataract-101: video dataset of 101 cataract surgeries. In: Proceedings of the 9th ACM Multimedia Systems Conference, pp. 421–425 (2018)
23. Tao, L., Zappella, L., Hager, G.D., Vidal, R.: Surgical gesture segmentation and recognition. In: Mori, K., Sakuma, I., Sato, Y., Barillot, C., Navab, N. (eds.) MICCAI 2013. LNCS, vol. 8151, pp. 339–346. Springer, Heidelberg (2013). https://doi.org/10.1007/978-3-642-40760-4_43
24. Twinanda, A.P., Shehata, S., Mutter, D., Marescaux, J., De Mathelin, M., Padoy, N.: EndoNet: a deep architecture for recognition tasks on laparoscopic videos. IEEE Trans. Med. Imaging **36**(1), 86–97 (2016)
25. Yi, F., Jiang, T.: Hard frame detection and online mapping for surgical phase recognition. In: Shen, D., et al. (eds.) MICCAI 2019. LNCS, vol. 11768, pp. 449–457. Springer, Cham (2019). https://doi.org/10.1007/978-3-030-32254-0_50
26. Yu, F., et al.: Assessment of automated identification of phases in videos of cataract surgery using machine learning and deep learning techniques. JAMA Netw. Open **2**(4), e191860–e191860 (2019)
27. Zhang, J., et al.: Symmetric dilated convolution for surgical gesture recognition. In: Martel, A.L., et al. (eds.) MICCAI 2020. LNCS, vol. 12263, pp. 409–418. Springer, Cham (2020). https://doi.org/10.1007/978-3-030-59716-0_39
28. Zisimopoulos, O., et al.: DeepPhase: surgical phase recognition in CATARACTS videos. In: Frangi, A.F., Schnabel, J.A., Davatzikos, C., Alberola-López, C., Fichtinger, G. (eds.) MICCAI 2018. LNCS, vol. 11073, pp. 265–272. Springer, Cham (2018). https://doi.org/10.1007/978-3-030-00937-3_31

CAT-ViL: Co-attention Gated Vision-Language Embedding for Visual Question Localized-Answering in Robotic Surgery

Long Bai[1], Mobarakol Islam[2], and Hongliang Ren[1,3(✉)]

[1] Department of Electronic Engineering, The Chinese University of Hong Kong
(CUHK), Hong Kong SAR, China
b.long@link.cuhk.edu.hk, hlren@ee.cuhk.edu.hk
[2] Wellcome/EPSRC Centre for Interventional and Surgical Sciences (WEISS),
University College London, London, UK
mobarakol.islam@ucl.ac.uk
[3] Shun Hing Institute of Advanced Engineering, CUHK, Hong Kong SAR, China

Abstract. Medical students and junior surgeons often rely on senior surgeons and specialists to answer their questions when learning surgery. However, experts are often busy with clinical and academic work, and have little time to give guidance. Meanwhile, existing deep learning (DL)-based surgical Visual Question Answering (VQA) systems can only provide simple answers without the location of the answers. In addition, vision-language (ViL) embedding is still a less explored research in these kinds of tasks. Therefore, a surgical Visual Question Localized-Answering (VQLA) system would be helpful for medical students and junior surgeons to learn and understand from recorded surgical videos. We propose an end-to-end Transformer with the **Co-A**ttention ga**T**ed **Vi**sion-**L**anguage (CAT-ViL) embedding for VQLA in surgical scenarios, which does not require feature extraction through detection models. The CAT-ViL embedding module is designed to fuse multimodal features from visual and textual sources. The fused embedding will feed a standard Data-Efficient Image Transformer (DeiT) module, before the parallel classifier and detector for joint prediction. We conduct the experimental validation on public surgical videos from MICCAI EndoVis Challenge 2017 and 2018. The experimental results highlight the superior performance and robustness of our proposed model compared to the state-of-the-art approaches. Ablation studies further prove the outstanding performance of all the proposed components. The proposed method provides a promising solution for surgical scene understanding, and opens up a primary step in the Artificial Intelligence (AI)-based VQLA system for surgical training. Our code is available at github.com/longbai1006/CAT-ViL.

L. Bai and M. Islam—are co-first authors.

Supplementary Information The online version contains supplementary material available at https://doi.org/10.1007/978-3-031-43996-4_38.

1   Introduction

Specific knowledge in the medical domain needs to be acquired through extensive study and training. When faced with a surgical scenario, patients, medical students, and junior doctors usually come up with various questions that need to be answered by surgical experts, and therefore, to better understand complex surgical scenarios. However, the number of expert surgeons is always insufficient, and they are often overwhelmed by academic and clinical workloads. Therefore, it is difficult for experts to find the time to help students individually [22,24]. Automated solutions have been proposed to help students learn surgical knowledge, skills, and procedures, such as pre-recorded videos, surgical simulation and training systems [13,18], etc. Although students may learn knowledge and skills from these materials and practices, their questions still need to be answered by experts. Recently, several approaches [22,24] have demonstrated the feasibility of developing safe and reliable VQA models in the medical field. Specifically, Surgical-VQA [22] made effective answers regarding tools and organs in robotic surgery, but they were still unable to help students make sense of complex surgical scenarios. For example, suppose a student asks a question about the tool-tissue interaction for a specific surgical tool, the VQA model can only simply answer the question, but cannot directly indicate the location of the tool and tissue in the surgical scene. Students will still need help understanding this complex surgical scene. Another problem with Surgical-VQA is that their sentence-based VQA model requires datasets with annotation in the medical domain, and manual annotation is time-consuming and laborious.

Currently, extensive research and progress have been made on VQA tasks in the computer vision domain [17]. Models using long-short term memory modules [28], attention modules [24], and Transformer [17] significantly boost the performance in VQA tasks. Furthermore, FindIt [16] proposed a unified Transformer model for joint object detection and ViL tasks. However, firstly, most of these models acquire the visual features of key targets through object detection models. In this case, the VQA performance strongly depends on the object detection results, which hinders the global understanding of the surgical scene [23], and makes the overall solution not fully end-to-end. Second, many VQA models employ simple additive, averaging, scalar product, or attention mechanisms when fusing heterogeneous visual and textual features. Nevertheless, in heterogeneous feature fusion, each feature represents different meanings, and simple techniques cannot achieve the best intermediate representation from heterogeneous features. Finally, the VQA model cannot highlight specific regions in the image relevant to the question and answer. Supposing the location of the object in the surgical scene can be known along with the answer by VQLA models, students can compare it with the surrounding tissues, different surgical scenes, preoperative scan data, etc., to better understand the surgical scene [4].

In this case, we propose CAT-ViL DeiT for VQLA tasks in surgical scene understanding. Specifically, our contributions are three-fold: (1) We carefully design a Transformer-based VQLA model that can relate the surgical VQA and localization tasks at an instance level, demonstrating the potential of AI-based

VQLA system in surgical training and surgical scene understanding. (2) In our proposed CAT-ViL embedding, the co-attention module allows the text embeddings to have instructive interaction with visual embeddings, and the gated module works to explore the best intermediate representation for heterogeneous embeddings. (3) With extensive experiments, we demonstrate the extraordinary performance and robustness of our CAT-ViL DeiT in localizing and answering questions in surgical scenarios. We compare the performance of detection-based and detection-free feature extractors. We remove the computationally costly and error-prone detection proposals to achieve superior representation learning and end-to-end real-time applications.

2 Methodology

2.1 Preliminaries

VisualBERT [17] generates text embeddings (including token embedding e_t, segment embedding e_s, and position embedding e_p) based on the strategy of natural language model BERT [9], and uses object detection model to extract visual embeddings (consisting of visual features representation f_v, segment embedding f_s and position embedding f_p). Then, it concatenates visual and text embeddings before feeding the subsequent multilayer Transformer module.

Multi-Head Attention [26] can focus limited attention on key and high-value information. In each head $\mathbf{h}_i$, give the certain query $q \in \mathbb{R}^{d_q}$, key matrix $K \in \mathbb{R}^{d_k}$, value matrix $V \in \mathbb{R}^{d_v}$, the attention for each head is calculated as $\mathbf{h}_i = A\left(\mathbf{W}_i^{(q)}\mathbf{q}, \mathbf{W}_i^{(K)}\mathbf{K}, \mathbf{W}_i^{(V)}\mathbf{V}\right)$. $\mathbf{W}_i^{(q)} \in \mathbb{R}^{p_q \times d_q}$, $\mathbf{W}_i^{(k)} \in \mathbb{R}^{p_k \times d_k}$, $\mathbf{W}_i^{(v)} \in \mathbb{R}^{p_v \times d_v}$ are learnable parameters, and A represents the function of single-head attention aggregation. A linear conversion is then applied for the attention aggregation from multiple heads: $\mathbf{h} = MA(\mathbf{W}_o[\mathbf{h}_1\|\ldots\|\mathbf{h}_h])$. $\mathbf{W}_o \in \mathbb{R}^{p_o \times h p_v}$ is the learnable parameters in multiple heads. Each head may focus on a different part of the input to achieve the optimal output.

2.2 CAT-ViL DeiT

We present CAT-ViL DeiT to process the information from different modalities and implement the VQLA task in the surgical scene. DeiT [25] serves as the backbone of our network. As shown in Fig. 1, the network consists of a vision feature extractor, a customized trained tokenizer, a co-attention gated embedding module, a standard DeiT module, and task-specific heads.

Feature Extraction: Taking a given image and the associated question, conventional VQA models usually extract visual features via object proposals [17,28]. Instead, we employ ResNet18 [11] pre-trained on ImageNet [8] as our visual feature extractor. This design enables faster inference speed and global understanding of given surgical scenes. The text embeddings are acquired via a customized pre-trained tokenizer [22]. The CAT-ViL embedding module then processes and fuses the input embeddings from different modalities.

CAT-ViL Embedding: In the following, the extracted features are processed into visual and text embeddings following VisualBERT [17] as described in Sect. 2.1. However, VisualBERT [17] and VisualBERT ResMLP [22] naively concatenate the embeddings from different modalities without optimizing the intermediate representation between heterologous embeddings. In this case, information and statistical representations from different modalities cannot interact perfectly and serve subsequent tasks.

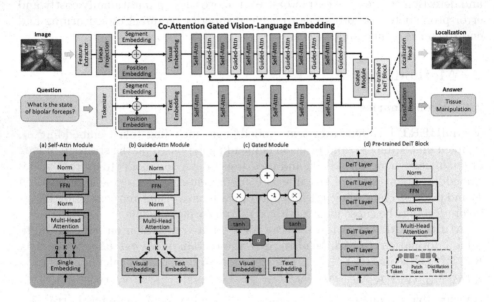

Fig. 1. The proposed network architecture. The network components include a visual feature extractor, tokenizer, CAT-ViL embedding module (embedding setup, co-attention learning, gated module), per-trained DeiT block, and task-specific heads. 'Attn' represents 'Attention'.

Inspired by [3,28], we replace the naive concatenation operation with a co-attention gated ViL module. The gated module can explore the best combination of the two modalities. Co-attention learning enables active information interaction between visual and text embeddings. Specifically, the guided-attention module is applied to infer the correlation between the visual and text embeddings. The normal self-attention module contains the multi-head attention layer, a feed-forward layer, and ReLU activation. The guide-attention module also contains the above components, but its input is from both two modalities, in which the q is from visual embeddings and K,V are from text embeddings:

$$\mathbf{h}_i = \mathrm{A}\left(\mathbf{W}_i^{(q)}\mathbf{q}_{\text{visual}}, \mathbf{W}_i^{(K)}\mathbf{K}_{\text{text}}, \mathbf{W}_i^{(V)}\mathbf{V}_{\text{text}}\right) \tag{1}$$

Therefore, the visual embeddings shall be reconstructed with the original query, and the key and value of the text embeddings, which can realize the text embeddings to have instructive information interaction with the visual embeddings, and help the model to focus on the targeted image context related to the question. Six guided-attention layers are applied in our network. Thus, the correlation between questions and image regions can be gradually constructed. Besides, we also build six self-attention blocks for both visual and text embeddings to boost the internal relationship within each modality. This step can also avoid 'over' guidance and seek a trade-off. Then, the attended text embeddings and text-guided attended visual embedding shall be output from the co-attention module and propagated through the gated module.

Compared to the naive concatenation [17], summation, or the multilayer perceptron (MLP) layer [28], this learnable gated neuron-based model can control the contribution of multimodal input to output through selective activation (set as $tanh$ here). The gate node α is employed to control the weight for selective visual and text embedding aggregation. The equations of the gated module are as follows:

$$\mathbf{E_o} = \mathbf{w} * \tanh\left(\theta_v \cdot \mathbf{E}_v\right) + (1 - \mathbf{w}) * \tanh\left(\theta_t \cdot \mathbf{E}_t\right)$$
$$\mathbf{w} = \alpha\left(\theta_{\mathbf{w}} \cdot [\mathbf{E}_v \| \mathbf{E}_t]\right) \tag{2}$$

$\mathbf{E}_v$ and $\mathbf{E}_t$ denotes visual and text embeddings, respectively. $(\theta_\omega, \theta_v, \theta_t)$ are set as learnable parameters. $[\cdot \| \cdot]$ means the concatenation operation. $\mathbf{E}_o$ is the final output embeddings. The activation function internally encodes the text and visual embeddings separately, and the gate weights are used for embedding fusion. This method is uncomplicated and effective, and can optimize the intermediate aggregation of visual and text embeddings while constraining the model.

Subsequently, the fused embeddings $\mathbf{E}_o$ shall feed the pre-trained DeiT-Base [25] module before the task-specific heads. The pre-trained DeiT-Base module can learn the joint representation, resolve ambiguous groundings from multimodel information, and maximize performance.

Prediction Heads: The classification head, following the normal classification strategy, is a linear layer with Softmax activation. Regarding the localization head, we follow the setup in Detection with Transformers (DETR) [7]. A simple feed-forward network (FFN) with a 3-layer perceptron, ReLU activation, and a linear projection layer is employed to fit the coordinates of the bounding boxes. The entire network is therefore built end-to-end without multi-stage training.

Loss Function: Normally, the cross-entropy loss $\mathcal{L}_{CE}$ serves as our classification loss. The combination of $\mathcal{L}_1$-norm and Generalized Intersection over Union (GIoU) loss [21] is adopted to conduct bounding box regression. GIoU loss [21] further emphasizes both overlapping and non-overlapping regions of bounding boxes. Then, the final loss function is $\mathcal{L} = \mathcal{L}_{CE} + (\mathcal{L}_{GIoU} + \mathcal{L}_1)$.

3   Experiments

3.1   Dataset

EndoVis 2018 Dataset is a public dataset with 14 robotic surgery videos from MICCAI Endoscopic Vision Challenge [1]. The VQLA annotations are publicly accessible by [4], in which the QA pairs are from [23] and the bounding box annotations are from [14]. Specifically, the QA pairs include 18 different single-word answers regarding organs, surgical tools, and tool-organ interactions. When the question is about organ-tool interactions, the bounding box will contain both the organ and the tool. We follow [22] to use video [1, 5, 16] as the test set and the remaining as the training set. Statistically, the training set includes 1560 frames and 9014 QA pairs, and the test set has 447 frames and 2769 QA pairs.

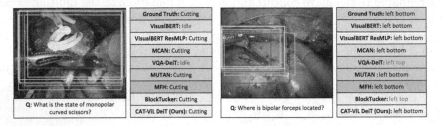

Fig. 2. Qualitative comparison on the VQLA task. Our CAT-ViL DeiT (Yellow) displays state-of-the-art (SOTA) performance on generating the answers and location against VisualBERT (light blue) [17], VisualBERT ResMLP (green) [22], MCAN (orange) [28], VQA-DeiT (purple) [25], MUTAN (gray) [5], MFH (dark blue) [29], and BlockTucker (pink) [6]. The Ground Truth bounding box is red. (Color figure online)

EndoVis 2017 Dataset is also a publicly available dataset from the MIC-CAI Endoscopic Vision Challenge 2017 [2], and the annotations are also available by [4]. We employ this dataset as an external validation dataset to demonstrate the generalization capability of our model in various surgical domains. Specifically, we manually select and annotate frames with common organs, tools, and interactions in EndoVis 2017 Dataset, generating 97 frames with 472 QA pairs. We conduct *no training* but only *testing* on this external validation dataset.

3.2 Implementation Details

We conduct our comparison experiments against VisualBERT [17], VisualBERT ResMLP [22], MCAN [28], VQA-DeiT [25], MUTAN [5], MFH [29], and Block-Tucker [6]. In VQA-DeiT, we use pre-trained DeiT-Base block [25] to replace the multilayer Transformer module in VisualBERT [17]. To keep a fair comparison of VQLA tasks, we use the same prediction heads in and loss function in Sect. 2.2. The evaluation metrics are accuracy, f-score, and mean intersection over union (mIoU) [21]. All models are trained on NVIDIA RTX 3090 GPUs using Adam optimizer [15] with PyTorch. The epoch, batch size, and learning rate are set to 80, 64, and 1×10^{-5}, respectively. The experimental results are the average results with five different random seeds.

3.3   Results

Figure 2 presents the visualization and qualitative comparison of the surgical VQLA system. Quantitative evaluation in Table 1 presents that our proposed model using ResNet18 [11] feature extractor suppresses all SOTA models significantly. Additionally, we compare the performance between using object proposals (Faster RCNN [20]) and using features from the entire image (ResNet18 [11]). The experimental results in EndoVis-18 show that removing the object proposal model improves the performance appreciably on both question-answering and localization tasks, which demonstrates the impact of this approach in correcting potential false detections. Meanwhile, in the external validation set - EndoVis-17, our CAT-ViL DeiT with RCNN feature extractor suffers from domain shift

Table 1. Comparison experiments on EndoVis-18 and EndoVis-17 datasets.

Models	Visual Feature		EndoVis-18			EndoVis-17		
	Detection	Inference Speed	Acc	F-Score	mIoU	Acc	F-Score	mIoU
VisualBERT [17]	FRCNN [20]	55.28 ms	0.5973	0.3223	0.7340	0.4382	**0.3743**	0.6822
VisualBERT R [22]			0.6064	0.3226	0.7305	0.4267	0.3506	0.6947
MCAN [28]			0.6084	0.3428	0.7257	0.4258	0.3035	0.6832
VQA-DeiT [25]			0.6089	0.3217	0.7338	0.4492	0.3213	0.7134
MUTAN [5]			0.6049	0.3238	0.7217	0.4364	0.3206	0.6870
MFH [29]			0.6179	0.3158	0.7227	0.3729	0.2048	**0.7183**
BlockTucker [6]			0.6067	0.3414	0.7313	0.4364	0.3210	0.6825
CAT-ViL DeiT **(Ours)**			**0.6192**	**0.3521**	**0.7482**	**0.4555**	0.3676	0.7049
VisualBERT [17]	✗	6.64 ms	0.6268	0.3329	0.7391	0.4005	0.3381	0.7073
VisualBERT R [22]			0.6301	0.3390	0.7352	0.4190	0.3370	0.7137
MCAN [28]			0.6285	0.3338	0.7526	0.4137	0.2932	0.7029
VQA-DeiT [25]			0.6104	0.3156	0.7341	0.3797	0.2858	0.6909
MUTAN [5]			0.6283	**0.3395**	0.7639	0.4242	0.3482	0.7218
MFH [29]			0.6283	0.3254	0.7592	0.4103	0.3500	0.7216
BlockTucker [6]			0.6201	0.3286	0.7653	0.4221	0.3515	0.7288
CAT-ViL DeiT **(Ours)**			**0.6452**	0.3321	**0.7705**	**0.4491**	0.3622	**0.7322**

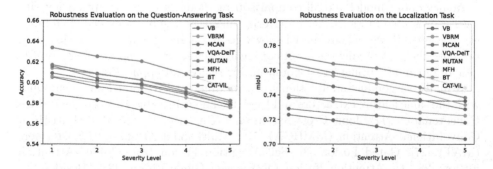

Fig. 3. Robustness experiments on the EndoVis-18 dataset. We process the data with 18 corruption methods at each severity level and average the prediction results.

Table 2. Ablation study on different fusion strategies. All experiments use the same feature extractor, DeiT backbone, and prediction heads. 'Attn' denotes 'Attention'.

Fusion Strategies	EndoVis-18			EndoVis-17		
	Acc	F-Score	mIoU	Acc	F-Score	mIoU
Concatenation [17]	0.6104	0.3156	0.7341	0.3797	0.2858	0.6909
JCA [19]	0.6024	0.3010	0.7527	0.3750	0.2835	0.7145
MMHCA [10]	0.6096	0.3124	0.7449	0.3581	0.3001	0.7077
MAT [27]	0.6186	0.3179	0.7415	0.3369	0.2850	0.6956
Gated Fusion [3]	0.6071	**0.3793**	0.7683	0.4030	0.2824	**0.7388**
Self-Attn [26]	0.5923	0.3095	0.7271	0.3686	0.2673	0.6718
Guided-Attn [28]	0.6194	0.3134	0.7310	0.3517	0.2290	0.7185
Co-Attn (Bi)	0.6056	0.3090	0.7206	0.3644	0.3083	0.7044
Co-Attn (V2T)	0.6392	0.3263	0.7218	0.3453	0.2265	0.7143
Co-Attn (T2V) [28]	0.6136	0.3208	0.7273	0.3805	0.3026	0.6870
Self-Attn Gated (**Ours**)	0.6249	0.3078	0.7314	0.3263	0.2897	0.7086
Guided-Attn Gated (**Ours**)	0.6280	0.3127	0.7651	0.3962	0.3337	0.7145
CAT-ViL (Bi) (**Ours**)	0.6230	0.3121	0.7415	0.4258	0.3593	0.7282
CAT-ViL (V2T) (**Ours**)	0.6352	0.3259	0.7600	0.4301	0.3543	0.7074
CAT-ViL (T2V) (**Ours**)	**0.6452**	0.3321	**0.7705**	**0.4491**	**0.3622**	0.7322

and class imbalance problems, thus achieving poor performance. However, our final model, CAT-ViL DeiT with ResNet18 feature extractor, endows the network with global awareness and outperforms all baselines in terms of accuracy and mIoU, proving the superiority of our method. The inference speed is also enormously accelerated, demonstrating its potential in real-time applications.

Furthermore, a robustness experiment is conducted to observe the model stability when test data is corrupted. We set 18 types of corruption on the test data based on the severity level from 1 to 5 by following [12]. Then, the performance of our model and all comparison methods on each corruption severity level is presented in Fig. 3. As the severity increases, the performance of all models degrades. However, our model shows good stability against corruption, and presents the best prediction results at each severity level. The excellent robustness of our model brings great potential for real-world applications.

Finally, we conduct an ablation study on different ViL embedding techniques with the same feature extractors and DeiT backbone in Table 2. We compare with Concatenation [17], Joint Cross-Attention (JCA) [19], Multimodal Multi-Head Convolutional Attention (MMHCA) [10], Multimodal Attention Transformers (MAT) [27], Gated Fusion [3], Self-Attention Fusion [26], Guided-Attention Fusion [28], Co-Attention Fusion (T2V: Text-Guide-Vision) [28]. Besides, we explore the Co-Attention module with different directions (V2T: Vision-Guide-Text, and Bidirectional). Furthermore, we also incorporate the Gated Fusion with different attention mechanisms (Self-Attention, Guided-Attention, Bidirectional Co-Attention, Co-Attention (V2T), Co-Attention (T2V)) for detailed comparison. They are shown as 'Self-Attn Gated', 'Guided-Attn Gated', 'CAT-ViL (Bi)', 'CAT-ViL (V2T)' and 'CAT-ViL (T2V)' in Table 2. The study proves the superior performance of our ViL embedding strategy against other advanced methods. We also demonstrate that integrating attention feature fusion techniques and the gated module will bring performance improvement.

4   Conclusions

This paper presents a Transformer model with CAT-ViL embedding for the surgical VQLA tasks, which can give the localized answer based on a specific surgical scene and associated question. It brings up a primary step in the study of VQLA systems for surgical training and scene understanding. The proposed CAT-ViL embedding module is proven capable of optimally facilitating the interaction and fusion of multimodal features. Numerous comparative, robustness, and ablation experiments display the leading performance and stability of our proposed model against all SOTA methods in both question-answering and localization tasks, as well as the potential of real-time and real-world applications. Furthermore, our study opens up more potential VQA-related problems in the medical community. Future work can be focused on quantifying and improving the reliability and uncertainty of these safety-critical tasks in the medical domain.

Acknowledgements. This work was funded by Hong Kong RGC CRF C4063-18G, CRF C4026-21GF, RIF R4020-22, GRF 14203323, GRF 14216022, GRF 14211420, NSFC/RGC JRS N_CUHK420/22; Shenzhen-Hong Kong-Macau Technology Research Programme (Type C 202108233000303); Guangdong GBABF #2021B1515120035. M. Islam was funded by EPSRC grant [EP/W00805X/1].

References

1. Allan, M., et al.: 2018 robotic scene segmentation challenge. arXiv preprint arXiv:2001.11190 (2020)
2. Allan, M., et al.: 2017 robotic instrument segmentation challenge. arXiv preprint arXiv:1902.06426 (2019)
3. Arevalo, J., Solorio, T., Montes-y Gómez, M., González, F.A.: Gated multimodal units for information fusion. arXiv preprint arXiv:1702.01992 (2017)
4. Bai, L., Islam, M., Seenivasan, L., Ren, H.: Surgical-VQLA: transformer with gated vision-language embedding for visual question localized-answering in robotic surgery. arXiv preprint arXiv:2305.11692 (2023)
5. Ben-Younes, H., Cadene, R., Cord, M., Thome, N.: MUTAN: multimodal tucker fusion for visual question answering. In: Proceedings of the IEEE International Conference on Computer Vision, pp. 2612–2620 (2017)
6. Ben-Younes, H., Cadene, R., Thome, N., Cord, M.: BLOCK: bilinear superdiagonal fusion for visual question answering and visual relationship detection. In: Proceedings of the AAAI Conference on Artificial Intelligence, vol. 33, pp. 8102–8109 (2019)

7. Carion, N., Massa, F., Synnaeve, G., Usunier, N., Kirillov, A., Zagoruyko, S.: End-to-end object detection with transformers. In: Vedaldi, A., Bischof, H., Brox, T., Frahm, J.-M. (eds.) ECCV 2020. LNCS, vol. 12346, pp. 213–229. Springer, Cham (2020). https://doi.org/10.1007/978-3-030-58452-8_13

8. Deng, J., Dong, W., Socher, R., Li, L.J., Li, K., Fei-Fei, L.: ImageNet: a large-scale hierarchical image database. In: 2009 IEEE Conference on Computer Vision and Pattern Recognition, pp. 248–255. IEEE (2009)

9. Devlin, J., Chang, M.W., Lee, K., Toutanova, K.: BERT: pre-training of deep bidirectional transformers for language understanding. arXiv preprint arXiv:1810.04805 (2018)

10. Georgescu, M.I., et al.: Multimodal multi-head convolutional attention with various kernel sizes for medical image super-resolution. In: Proceedings of the IEEE/CVF Winter Conference on Applications of Computer Vision, pp. 2195–2205 (2023)

11. He, K., Zhang, X., Ren, S., Sun, J.: Deep residual learning for image recognition. In: Proceedings of the IEEE Conference on Computer Vision and Pattern Recognition, pp. 770–778 (2016)

12. Hendrycks, D., Dietterich, T.: Benchmarking neural network robustness to common corruptions and perturbations. arXiv preprint arXiv:1903.12261 (2019)

13. Hsieh, M.C., Lin, Y.H.: VR and AR applications in medical practice and education. Hu Li Za Zhi **64**(6), 12–18 (2017)

14. Islam, M., Seenivasan, L., Ming, L.C., Ren, H.: Learning and reasoning with the graph structure representation in robotic surgery. In: Martel, A.L., et al. (eds.) MICCAI 2020. LNCS, vol. 12263, pp. 627–636. Springer, Cham (2020). https://doi.org/10.1007/978-3-030-59716-0_60

15. Kingma, D.P., Ba, J.: Adam: a method for stochastic optimization. arXiv preprint arXiv:1412.6980 (2014)

16. Kuo, W., Bertsch, F., Li, W., Piergiovanni, A., Saffar, M., Angelova, A.: FindIt: generalized localization with natural language queries. arXiv preprint arXiv:2203.17273 (2022)

17. Li, L.H., Yatskar, M., Yin, D., Hsieh, C.J., Chang, K.W.: VisualBERT: a simple and performant baseline for vision and language. arXiv preprint arXiv:1908.03557 (2019)

18. Lin, H.C., Shafran, I., Yuh, D., Hager, G.D.: Towards automatic skill evaluation: detection and segmentation of robot-assisted surgical motions. Comput. Aided Surg. **11**(5), 220–230 (2006)

19. Praveen, R.G., et al.: A joint cross-attention model for audio-visual fusion in dimensional emotion recognition. In: Proceedings of the IEEE/CVF Conference on Computer Vision and Pattern Recognition, pp. 2486–2495 (2022)

20. Ren, S., He, K., Girshick, R., Sun, J.: Faster R-CNN: towards real-time object detection with region proposal networks. In: Advances in Neural Information Processing Systems, vol. 28 (2015)

21. Rezatofighi, H., Tsoi, N., Gwak, J., Sadeghian, A., Reid, I., Savarese, S.: Generalized intersection over union: a metric and a loss for bounding box regression. In: Proceedings of the IEEE/CVF Conference on Computer Vision and Pattern Recognition, pp. 658–666 (2019)

22. Seenivasan, L., Islam, M., Krishna, A., Ren, H.: Surgical-VQA: visual question answering in surgical scenes using transformer. In: Wang, L., Dou, Q., Fletcher, P.T., Speidel, S., Li, S. (eds.) MICCAI 2022. LNCS, vol. 13437, pp. 33–43. Springer, Cham (2022)

23. Seenivasan, L., Mitheran, S., Islam, M., Ren, H.: Global-reasoned multi-task learning model for surgical scene understanding. IEEE Robot. Autom. Lett. **7**, 3858–3865 (2022)
24. Sharma, D., Purushotham, S., Reddy, C.K.: MedFuseNet: an attention-based multimodal deep learning model for visual question answering in the medical domain. Sci. Rep. **11**(1), 1–18 (2021)
25. Touvron, H., Cord, M., Douze, M., Massa, F., Sablayrolles, A., Jégou, H.: Training data-efficient image transformers & distillation through attention. In: International Conference on Machine Learning, pp. 10347–10357. PMLR (2021)
26. Vaswani, A., et al.: Attention is all you need. In: Advances in Neural Information Processing Systems, vol. 30 (2017)
27. Wu, Z., Liu, L., Zhang, Y., Mao, M., Lin, L., Li, G.: Multimodal crowd counting with mutual attention transformers. In: 2022 IEEE International Conference on Multimedia and Expo (ICME), pp. 1–6. IEEE (2022)
28. Yu, Z., Yu, J., Cui, Y., Tao, D., Tian, Q.: Deep modular co-attention networks for visual question answering. In: Proceedings of the IEEE/CVF Conference on Computer Vision and Pattern Recognition, pp. 6281–6290 (2019)
29. Yu, Z., Yu, J., Xiang, C., Fan, J., Tao, D.: Generalized multimodal factorized high-order pooling for visual question answering. IEEE Trans. Neural Netw. Learn. Syst. **29**(12), 5947–5959 (2018)

Semantic Virtual Shadows (SVS) for Improved Perception in 4D OCT Guided Surgery

Michael Sommersperger[1]([✉]), Shervin Dehghani[1], Philipp Matten[2], Kristina Mach[1], M. Ali Nasseri[3], Hessam Roodaki[4], Ulrich Eck[1], and Nassir Navab[1]

[1] Technical University of Munich, Munich, Germany
michael.sommersperger@tum.de
[2] Medical University of Vienna, Vienna, Austria
[3] Klinikum Rechts der Isar, Augenklinik, Munich, Germany
[4] Carl Zeiss Meditec AG, Munich, Germany

Abstract. Swept-Source Optical Coherence Tomography (SS-OCT) integrated with surgical microscopes has enabled fast, high-resolution, and volumetric visualization of delicate tissue-instrument interactions. However, some visual features, which provide essential perceptual information in microscopic surgery, are not present in 4D OCT. Such a feature is the shadow of the surgical instruments cast onto the retina by the endo-illumination probe, which is among the most important cognitive cues for perceptual distance estimation. In this work, we propose *Semantic Virtual Shadows (SVS)*, a novel concept to artificially generate instrument-specific shadows in OCT volumes, enabling naturally non-existent but important perceptual cues that are present in microscopic surgery. Semantic scene information is leveraged by considering only voxels associated with shadow-casting and shadow-receiving objects, identified using a learning-based approach and efficient volume processing, respectively. Real-time performance is achieved by a precomputed semantic shadow volume texture that assigns a shadowing factor to each voxel associated with a shadow-receiving object. The novelty of the method includes not only instrument-specific shadowing on the surface anatomy but also exclusively on deep-seated subsurface structures, providing advantages for various vitreoretinal procedures. Our user study indicates the benefits of the method for 4D OCT-guided surgery in several cognitive and performance-specific aspects.

Keywords: Optical Coherence Tomography · Real-Time Visualization · Visual Perception

1 Introduction

Navigating surgical instruments in vitreoretinal procedures requires extreme manual dexterity. Surgeons need to manipulate fine structures, such as an

Supplementary Information The online version contains supplementary material available at https://doi.org/10.1007/978-3-031-43996-4_39.

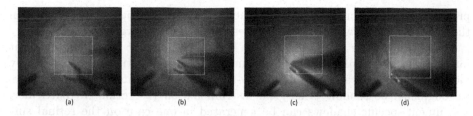

Fig. 1. A surgical forceps is carefully guided towards the retina (a-b), then grasps (c) and peels (d) a thin membrane from the surface. The instrument shadow cast onto the retina acts as a main visual cue for perceptual distance estimation.

Epiretinal Membrane (ERM), which can measure up to only $60\mu m$ [16], or carefully placing a microsurgical cannula within the on average $250\mu m$ [6] thick retina for intra or subretinal injections.

Accurate distance perception regarding the surgical instruments and the retina is of utmost importance for punctuated surgical action. Any unintentional contact with the retina could cause severe damage to important retinal cells. In conventional procedures, surgeons rely only on visual guidance from an operating microscope that only allows a top view. An endo-illumination probe is inserted into the eye to visualize the operating area. When surgical instruments are introduced, they cast a shadow onto the retina, relative to the position of the illumination probe. Such shadows provide one of the most essential cognitive cues to perceive the distance between the instrument and the retina and therefore are important for performing surgical action. Figure 1 shows frames of a video sequence during retinal membrane peeling. While the instrument approaches the retina, the distance to its shadow gradually decreases until the instrument tip coincides with the tip of its shadow at the retina.

Besides performing a procedure through this conventional stereoscopic view, recently, other imaging modalities have also become available. Technical advances have led to the integration of Swept-Source Optical Coherence Tomography (SS-OCT) into surgical microscopes [2,3,7], enabling near video-rate, depth-resolved imaging. Such 4D OCT systems have the potential to generate advanced visualizations of surgical interactions and improved treatments. While previous studies have demonstrated the feasibility of 4D OCT-guided surgery [1,9], some important perceptual cues that are available in the microscopic images are not present in OCT, such as the instrument shadow generated by the illumination probe. The absence of such perceptual cues increases the burden for surgeons to adapt to 4D OCT. While instrument shadows still exist in OCT, they are generated by the OCT laser and occur when the laser beam is blocked by the instrument such that structures below the instrument are fully obscured in the imaging modality. This OCT shadow is fixed in perspective and always occurs directly beneath the instrument. Hence, it does not provide the same intuitive cognitive cues, especially when viewing the volume from a top view, which is the usual view during ophthalmic surgery. To enable perceptually similar features as generated by the illumination probe, advanced visualization techniques need to be explored. Such visualizations could provide cues that improve the general usability of 4D-OCT without deviating from the standard workflow and steepen the learning curve for surgeons.

In this paper, we propose *Semantic Virtual Shadows (SVS)*, a concept that integrates such visual cues for perceptual distance estimation in 4D OCT. By identifying shadow-casting and shadow-receiving objects in the OCT volume, we augment the shadow of surgical instruments on the retina. In particular, instrument-specific shadows can be generated in one case on the retinal surface, and in other cases, exclusively on deep-seated layers below the surface. Precomputing a semantic shadow volume texture prior to direct volume rendering (DVR) enables real-time performance and more flexibility in the rendering approach. We demonstrate the versatility of SVS by proposing two visualization approaches for different scenarios in retinal surgery, including retinal membrane peeling and subretinal injection.

2   Related Work

So far, only a few works have addressed the integration of perceptual cues into OCT volume rendering algorithms for interactive surgical visualizations in 4D OCT. Among these, approaches to color voxels based on distance information have been explored as a means of conveying spatial perceptual understanding. In an early study [1], a color transfer function was applied to each voxel, based on the distance to the center of mass of the volume. In a later work, [14] proposed a visualization that leverages a perceptually linear color map, which is anchored by a retinal layer segmentation, thus encoding distance information to an anatomical reference depth in a color map. They furthermore employ shadow rays [10] to enhance surface structures. However, their method is limited to local shadow rays to achieve real-time performance. There are only a few works that deal with the generation of a shadow for distance estimation in 4D OCT. Only in [5] do the authors propose to define a virtual light direction orthogonal to the OCT A-scans and employ shadow rays to create a shadow projection on an external plane positioned next to the volume. Such visualizations, however, are perceptually similar to 2D lateral volume projections, do not convey correct distance cues in presence of curved structures, and require removing the gaze from the surgical area to perceive instrument distance.

As opposed to previous works, we integrate an instrument-specific shadow augmented directly in the rendered OCT volume. Hence, our method aims to provide surgeons with familiar cognitive cues for perceptual distance estimation, as described in Sect. 1. Additionally, identifying shadow-casting and shadow-receiving objects allows the generation of instrument shadows on a specific retinal layer below the surface, which is not possible with previous approaches due to the self-shadowing of the retina.

3   Methodology

We define a *Semantic Virtual Shadow (SVS)* as a shadow that is only generated by identified shadow-casting objects and is explicitly cast onto identified

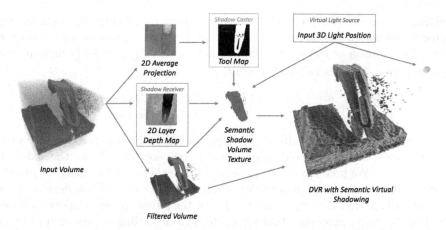

Fig. 2. Method overview: given an OCT volume and a virtual light source position, first shadow-casting and shadow-receiving objects are identified. These allow the generation of a *semantic shadow volume texture*, which is finally consumed by the DVR.

shadow-receiving objects. In our application, the primary use case of SVS is to generate visual cues for spatial distance perception. Taking into consideration shadow-casting and shadow-receiving objects as well as a virtual light source, a semantic shadow volume V_s can be constructed as a 3D texture that assigns a shadowing factor to each voxel associated with a shadow-receiving object in the OCT volume. This precomputed V_s is directly consumed by the direct volume rendering (DVR) to generate the instrument shadow augmentation. An overview of the proposed method is shown in Fig. 2.

While in theory any segmentation method could be employed to identify, we combine efficient volume processing with a learning based approach to approximate the segmentations and achieve the required processing rates for 4D OCT. For our interventional scenarios, we define the instrument as the shadow-casting object. Inspired by [15], to segment the instrument we first generate a 2D average projection image along the OCT A-scan direction (Z-axis) using

$$P_{\text{avg}}(x, y) = \frac{1}{N} \sum_z I(x, y, z) \tag{1}$$

from which A-scans containing parts of the instrument signal can be identified. We train a U-net style [8] architecture with a ResNet34 [4] backbone to generate a 2D binary instrument segmentation map M_{sc}. Since the OCT signal is blocked at the instrument surface, the A-scans corresponding to identified pixels in M_{sc} contain only instrument information. Therefore the instrument can be identified by thresholding the voxel intensities of A-scans obtained by M_{sc}. We further define the retinal layer, on which the SVS is generated as the shadow-receiving object. Due to their terrain-like surface properties, the retinal layer surface can be defined by a 2D depth map M_{sr}, indicating the relative surface position for each A-scan in the volume. Further, a distance parameter d_r is introduced, defining the extent of the shadow-receiving object. To generate V_s, a shadowing factor is calculated for each voxel p associated with a shadow-receiving object. Given the position of the virtual light source p_l and $L(p) = \frac{p_l - p}{|p_l - p|}$ defining the light direction, $V_s(p)$ is calculated with:

$$V_s(p) = \begin{cases} \prod_i (1 - (I(p + i \cdot L(p))^\alpha) \cdot M_{sc}(p + i \cdot L(p)) & \text{if } |M_{sr}(p_x, p_y) - p_z| < d_r \\ 1.0 & \text{otherwise.} \end{cases} \tag{2}$$

sampling volume intensity values $I(p)$ along the light direction through the entire volume. Note that, in our equations, the z axis corresponds to the axial A-scan direction. We precompute V_s prior to volume raymarching using a compute shader program, enabling high update rates and flexibility in the design of the DVR algorithm that directly consumes V_s. In the following, we propose two specific visualization approaches that integrate SVS according to equation (2) with different shadow-receiving objects, demonstrating the versatility and advantages of the method for different vitreoretinal procedures.

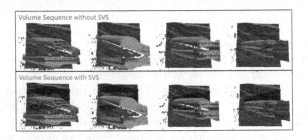

Fig. 3. Comparison of temporal OCT volumes rendered with Phong shading (top) and integrated SVS generating shadows on the retina surface (bottom).

Fig. 4. SVS on the RPE layer. The retina above the RPE does not generate shadow.

3.1   Shadow Augmentation on Surface Structures

During surgical phases, in which the instrument is located above the retinal surface, unintentional contact needs to be avoided. By defining the retinal surface as the shadow-receiving object, the SVS generates cognitive cues for surface distance perception. We use the following 2D projection for M_{sr}:

$$P_{\text{surf}}(x, y) = \min_z I(x, y, z) > t_s \tag{3}$$

where $t_s = 0.17$ is an empirically chosen threshold. We render the OCT volume using classic Phong shading to visualize surface structure details while integrating V_s to augment the instrument-specific shadow $C(p) = C_{Phong}(p) \cdot V_s(p)$. The top row of Fig. 3 shows frames of a volume sequence with Phong shading, while in the bottom the same volumes are rendered with both Phong shading and SVS.

3.2 Shadow Augmentation on Subsurface Structures

In intra- and subretinal injection procedures, once reached the retinal surface, surgeons need to carefully guide the cannula to the injection target without damaging the underlying Retinal Pigment Epithelium (RPE) which is a comparably deep-seated layer in OCT. We propose to augment an instrument shadow explicitly on the RPE and visualize the retinal surface semi-transparent. Compared to the previous section, we define the RPE as the shadow-receiving object. Since in OCT imaging, the RPE typically is a hyper-reflective layer [17], M_{sr} can be efficiently approximated by:

$$P_{\text{subsurf}}(x,y) = \arg\max_z \ I(x,y,z) \tag{4}$$

During volume raymarching, we apply Phong shading models to visualize the instrument and render the retinal surface semi-transparent while preserving surface highlights, as presented in [12]. As previously, Eq. (3) is used to obtain the surface depth. For the RPE and underlying structures, we integrate V_s and apply an RGB color map $C_{RGB}(p)$. During volume raymarching, this leads to the following convention, integrated via a piece-wise linear opacity function:

$$C(p) = \begin{cases} C_{PhongTool}(p) & \text{if } M_{sc}(p) \\ C_{PhongILM}(p) & \text{if } |P_{surf}(p_x,p_y) - p_z| < d_{surf} \\ C_{RGB}(p) \cdot V_s(p) & \text{if } P_{subsurf}(p_x,p_y) - p_z < 0 \\ 0 & \text{otherwise.} \end{cases} \tag{5}$$

Here, $C_{PhongTool}(p)$ and $C_{PhongILM}(p)$ are the Phong shading models for the instrument and the retinal surface, respectively. We refer to Fig. 4 and in particular to our supplementary video demonstrating visualization examples.

4 Experiments and Results

4.1 Implementation and Comparative Time Profiling

Our instrument segmentation network was trained on 2D projection images generated from OCT volumes using Eq. 1. In total 330 OCT volumes of resolution $391 \times 391 \times 720$ (spiral scanning) acquired from a model eye and 160 volumes of resolution $128 \times 512 \times 1024$ (linear scanning) acquired from ex-vivo porcine eyes were used to generate a data set of 3928 images, including data augmentation strategies. The axial projection images were manually labeled by two biomedical engineers. We used Pytorch 1.13 for model training and TensorRT 8.4 for optimization. We implemented our method using C++ and OpenGL 4.6 (Windows 10 with Intel Core i7-8700K @3.7 GHz, Nvidia RTX 3090Ti), employing a compute shader to generate the 2D projection images and V_s, and a fragment shader for DVR.

Table 1 shows the inference times of *SVS* compared to baseline shadow rays (SR), which were previously integrated for OCT visualization [14]. For ablation, we also compare SVS to only using a shadow volume buffer without semantic information (SB). The visualization parameters were adjusted in all methods to achieve a similar visual outcome. The volumes were rendered at a resolution of 3840×2160. To further demonstrate the potential of SVS to support surgical tasks, our method was integrated into the 4D SS-OCT system presented in [2] with a volume acquisition rate of $10Hz$. Surgical actions with a forceps in a phantom eye model are demonstrated in our supplementary material.

Table 1. Average inference time in milliseconds comparing *SVS*, a 3D shadow volume buffer without semantic information (SB) and shadow rays (SR) as baseline.

Method	Filter	Enface	Depth Map	Seg	Shadow	Render	Total
SVS	7.7 ± 0.7	0.7 ± 0.13	0.7 ± 0.1	3.8 ± 0.1	2.3 ± 0.2	13.5 ± 0.2	$\mathbf{28.8 \pm 0.8}$
SB	7.7 ± 0.7	-	-	-	51.9 ± 6.8	14.6 ± 0.3	74.3 ± 6.8
SR	7.7 ± 0.7	-	-	-	-	331.3 ± 4.2	339.1 ± 4.6

4.2   User Study

To evaluate the effect of SVS for distance perception compared to baseline DVR in 4D OCT guided surgery, we conducted a user study with 12 biomedical experts (10 male, 2 female) familiar with OCT. We conducted the study in a virtual environment simulating 4D OCT using the method proposed in [11], which was displayed using stereo rendering in a VR (HTC VIVE Pro) headset. A 3D input device (3D Systems Geomagic Touch) was used to control a calibrated virtual surgical tool, where a motion scaling of $4 : 1$ was applied to not affect the results by the limitations of manual dexterity. Prior to the study, participants were asked to perform vision tests to ensure normal visual acuity, including stereo vision, contrast vision, and normal visual field. The study data was anonymized and performed in accordance with the declaration of Helsinki. Users were asked to familiarize themselves with the interaction in the virtual environment before the study, reducing the impact of a learning curve on the study results. For each trial, we randomly positioned a small target point close to the retina, simulating anatomical structures that could be targeted in a surgical scenario. Participants were asked to move the instrument tip to the target and press a button to confirm the positioning. Each uses performed 10 trials, each in the following three variants: (i) baseline DVR with Phong shading as in [13], (ii) (SVS_{Tool}) treating the instrument as a shadow-casting object and the retinal surface as a shadow-receiving object, and (iii) $SVS_{ToolTarget}$ with both the instrument and the target as shadow-casting objects, as the target represents an anatomical structure that could theoretically be segmented in OCT. We provide examples of the study variants in our supplementary material. During the study, the order of the trial variants and the position of the virtual light source was randomized, while the view onto the retina was fixed at 10°C from the axial

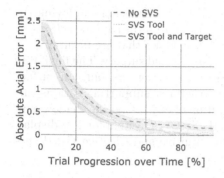
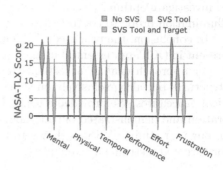

Fig. 5. Depth-specific error over the progression of the trials.

Fig. 6. Task load rating comparing baseline, SVS_{Tool} and $SVS_{ToolTarget}$.

OCT direction. Over the progression of each trial, we extracted the distance error to the target and to the retina. With baseline DVR, a final mean targeting error of 0.29 mm ($\pm$0.18) was achieved, while with SVS_{Tool} and $SVS_{ToolTarget}$, mean errors of 0.19 mm ($\pm$0.15) and 0.12 mm ($\pm$0.07) where achieved respectively. After identifying unequal variances in the distributions, a Kruskal-Wallis test obtained significant differences between the methods ($p < 0.001$). Figure 5 shows the distance error to the target in axial OCT direction measured over the trial progression, along with a 95% confidence interval of the error values. With SVS, users approached with faster convergence to the target and less error variance toward the end of the trial. In addition, Fig. 6 shows the results of the NASA-TLX survey. Statistically significant differences ($p < 0.001$) in all categories could be obtained based on the ANOVA test, while equal variances were found in all categories.

5   Discussion and Conclusion

The results of our user study indicate that the proposed method is able to generate effective perceptual cues for distance estimation in 4D OCT. This is reflected in improved targeting performance, as well as a high acceptance of the generated perceptual cues. Figure 5 suggests more robust targeting in both SVS_{Tool} and $SVS_{ToolTarget}$ variants, implying higher confidence in the approach. Preliminary qualitative feedback from four clinicians confirmed the intuitiveness of SVS and its strong integration potential without changing the existing clinical workflow. An essential component of SVS is the identification of shadow-casting and shadow-receiving objects in the volume. Compared to previous methods, SVS is able to generate perceptual cues exclusively on subsurface structures. Precomputing a semantic shadow volume texture was shown to achieve processing rates suitable for 4D OCT systems with volume acquisition rates of 24.2Hz [7]. In the future, prior knowledge of the instrument's geometrical model could be integrated to enable more accurate shadow representations. We also plan to investigate optimal virtual light source positions and envision employing the illumination probe to control the virtual light for more intuitive interactions.

In conclusion, the proposed SVS augments visual cues that are naturally not present in OCT, but essential in microscopic vitreoretinal surgery. The flexibility of our approach enables object-specific shadow generation not only on surface structures but also exclusively on subsurface structures, supporting various surgical procedures. We provided a general definition of object-specific shadow generation and demonstrated our method on a 4D SS-OCT system for live display. In our user study, SVS was shown to provide intuitive and effective visual cues for targeted instrument maneuvers.

Acknowledgements. This work is partially supported and the data is provided by Carl Zeiss Meditec. The authors wish to thank SynthesEyes (https://syntheseyes.de) for providing the excellent simulation setup for the user study.

References

1. Bleicher, I.D., Jackson-Atogi, M., Viehland, C., Gabr, H., Izatt, J.A., Toth, C.A.: Depth-based, motion-stabilized colorization of microscope-integrated optical coherence tomography volumes for microscope-independent microsurgery. Transl. Vis. Sci. Technol. **7**(6), 1–1 (2018)
2. Britten, A., et al.: Surgical microscope integrated MHz SS-OCT with live volumetric visualization. Biomed. Optics Express **14**(2), 846–865 (2023)
3. Carrasco-Zevallos, O.M., Viehland, C., Keller, B., McNabb, R.P., Kuo, A.N., Izatt, J.A.: Constant linear velocity spiral scanning for near video rate 4D OCT ophthalmic and surgical imaging with isotropic transverse sampling. Biomed. Optics Express **9**(10), 5052 (2018)
4. He, K., Zhang, X., Ren, S., Sun, J.: Deep residual learning for image recognition. In: Proceedings of the IEEE Conference on Computer Vision and Pattern Recognition, pp. 770–778 (2016)
5. Jeong, H., Kim, H.J., Hyeon, M.G., Kim, P., Choi, Y., Kim, B.M.: Shadow extension for ray casting enhances volumetric visualization in real-time 4D-OCT. Optics Commun. **460**, 125237 (2020)
6. Jo, Y.J., Heo, D.W., Shin, Y.I., Kim, J.Y.: Diurnal variation of retina thickness measured with time domain and spectral domain optical coherence tomography in healthy subjects. Invest. Ophthalmol. Vis. Sci. **52**(9), 6497–6500 (2011)
7. Kolb, J.P., Draxinger, W., Klee, J., Pfeiffer, T., Eibl, M., Klein, T., Wieser, W., Huber, R.: Live video rate volumetric OCT imaging of the retina with multi-MHz A-scan rates. PLoS ONE **14**(3), e0213144 (2019)
8. Ronneberger, O., Fischer, P., Brox, T.: U-Net: convolutional networks for biomedical image segmentation. In: Navab, N., Hornegger, J., Wells, W.M., Frangi, A.F. (eds.) MICCAI 2015. LNCS, vol. 9351, pp. 234–241. Springer, Cham (2015). https://doi.org/10.1007/978-3-319-24574-4_28
9. Roodaki, H., di San Filippo, C.A., Zapp, D., Navab, N., Eslami, A.: A surgical guidance system for big-bubble deep anterior lamellar keratoplasty. In: Ourselin, S., Joskowicz, L., Sabuncu, M.R., Unal, G., Wells, W. (eds.) MICCAI 2016. LNCS, vol. 9900, pp. 378–385. Springer, Cham (2016). https://doi.org/10.1007/978-3-319-46720-7_44

10. Ropinski, T., Kasten, J., Hinrichs, K.: Efficient shadows for GPU-based volume raycasting (2008)
11. Sommersperger, M., et al.: Surgical scene generation and adversarial networks for physics-based iOCT synthesis. Biomed. Optics Express **13**(4), 2414–2430 (2022)
12. Trout, R.M., et al.: Methods for real-time feature-guided image fusion of intrasurgical volumetric optical coherence tomography with digital microscopy. Biomed. Optics Express **14**(7), 3308–3326 (2023)
13. Viehland, C., et al.: Enhanced volumetric visualization for real time 4d intraoperative ophthalmic swept-source oct. Biomed. Optics Express **7**(5), 1815–1829 (2016)
14. Weiss, J., Eck, U., Nasseri, M.A., Maier, M., Eslami, A., Navab, N.: Layer-aware iOCT volume rendering for retinal surgery. In: Kozlíková, B., Linsen, L., Vázquez, P.P., Lawonn, K., Raidou, R.G. (eds.) Eurographics Workshop on Visual Computing for Biology and Medicine. The Eurographics Association (2019)
15. Weiss, J., Sommersperger, M., Nasseri, A., Eslami, A., Eck, U., Navab, N.: Processing-aware real-time rendering for optimized tissue visualization in intraoperative 4D OCT. In: Martel, A.L., et al. (eds.) MICCAI 2020. LNCS, vol. 12265, pp. 267–276. Springer, Cham (2020). https://doi.org/10.1007/978-3-030-59722-1_26
16. Wilkins, J.R., et al.: Characterization of epiretinal membranes using optical coherence tomography. Ophthalmology **103**(12), 2142–2151 (1996)
17. Zhang, T., Kho, A.M., Yiu, G., Srinivasan, V.J.: Visible light optical coherence tomography (OCT) quantifies subcellular contributions to outer retinal band 4. Transl. Vis. Sci. Technol. **10**(3), 30–30 (2021)

Intelligent Virtual B-Scan Mirror (IVBM)

Michael Sommersperger[1]([✉]), Shervin Dehghani[1], Philipp Matten[2],
Kristina Mach[1], Hessam Roodaki[3], Ulrich Eck[1], and Nassir Navab[1]

[1] Technical University of Munich, Munich, Germany
michael.sommersperger@tum.de
[2] Medical University of Vienna, Vienna, Austria
[3] Carl Zeiss Meditec AG, Munich, Germany

Abstract. Swept-Source Optical Coherence Tomography (SS-OCT)
allows surgeons to perform certain ophthalmic procedures under the exclusive guidance of real-time volumetric optical coherence tomography (4D
OCT). In such scenarios, surgeons are no longer limited to rigid views
through an operating microscope. Instead, direct volume rendering (DVR)
of 4D OCT enables surgical maneuvers to be performed from arbitrary
viewpoints. While 4D OCT maximizes the use of the depth-resolved OCT
data by displaying it from an oblique perspective, performing complex
instrument maneuvers from such views places a higher mental demand on
the surgeon. In this work, we propose an Intelligent Virtual B-scan Mirror (IVBM), a novel concept for surgical 4D OCT visualization to provide additional guidance for targeted instrument interactions. The IVBM
integrates a virtual mirror into a selected cross-section of the OCT volume. This mirror acts intelligently by only being sensitive to voxels associated with surgical instruments. Furthermore, volume structures aligned
with the IVBM are highlighted, while structures behind the IVBM are preserved through an adaptive opacity transfer function. Unlike previous perceptual OCT visualization concepts, which primarily address depth perception in axial OCT direction, this novel approach aids surgical interactions from arbitrary views. This paper presents the definition and implementation of an IVBM in a 4D OCT integrated microscope. Our user study
in a virtual simulation environment confirms the benefits and provides
insights into the interaction with the concept.

Keywords: Optical Coherence Tomography · Surgical Visualization ·
Volume Raymarching · Virtual Mirrors

1 Introduction

Vitreoretinal surgeries are complex procedures that require extreme manual dexterity. Typically, surgeons operate through a stereoscopic microscope viewing the
surgical area exclusively from an overhead perspective while manipulating delicate anatomical structures with sub-millimeter precision. In an effort to achieve

Supplementary Information The online version contains supplementary material
available at https://doi.org/10.1007/978-3-031-43996-4_40.

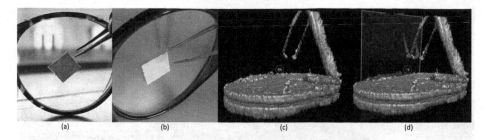

(a)   (b)   (c)   (d)

Fig. 1. (a) In daily life, navigating an instrument to a target is challenging without visual cues for distance perception. (b) Mirrors can provide additional views for improved targeting, see the reflection of the tweezers in the glass. (c) In 4D OCT-guided retinal surgery, navigating to a target is challenging from oblique views. (d) Our IVBM (outlined in green) leverages visual cues inspired by the reflective but transparent behavior illustrated in (b) for 4D OCT visualizations. (Color figure online)

improved surgical visualization, Optical Coherence Tomography (OCT) has been integrated into surgical microscopes, providing high-resolution depth-resolved imaging. Advances in spiral scanning and swept-source OCT [4,5,12] even paved the way for real-time volumetric imaging, enabling 4D visualizations of anatomical structures and surgical instruments. With the validation of intraoperative OCT through clinical studies [7,8], the emergence of 4D OCT systems prompts the anticipation of more precise and efficient microsurgical treatments. A common task in vitreoretinal surgery is to grasp and peel an Epiretinal Membrane (ERM), a 60 μm [24] thin layer that forms on top of the retina, while avoiding to damage the on average 250 μm thick retina [11].

GPU-accelerated direct volume rendering (DVR) was shown to be an effective way of visualizing surgical maneuvers, enabling real-time rendering of 4D OCT data on stereo displays [20]. Instead of viewing the surgical site from the top, as with stereoscopic microscopes, the depth-resolved properties of 4D OCT can be directly and fully utilized when oblique or even more extreme lateral views are provided [6]. This could even lead to more precise tool-tissue interactions since visualization of the surgical area from alternative viewpoints enables improved distance perception. This can be leveraged for instance when approaching small structures located at or above the retina during ERM or Internal Limiting Membrane (ILM) peeling [7,9]. On the other hand, such setups also introduce new complexities since surgical interaction from a lateral perspective is not common in current ophthalmic procedures. In particular, hand-eye coordination is naturally challenged when navigating instruments to a target from an uncommon view, imposing a higher mental demand on surgeons. To exploit the full potential of the 4D depth-resolved imaging modality, advanced visualization techniques could provide additional guidance and support complex maneuvers.

For this reason, we propose an *Intelligent Virtual B-scan Mirror (IVBM)*, a novel visualization concept to improve targeted instrument interactions in 4D OCT-guided surgery. As illustrated in Fig. 1, the concept of an IVBM is inspired

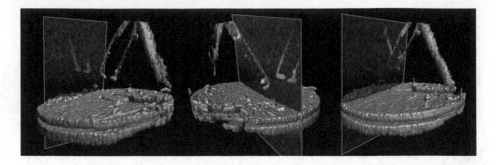

Fig. 2. The advantages of an *IVBM* include: (i) only voxels associated with surgical instruments are mirrored, while a perceptually linear color scheme is applied, which encodes the mirror distance (left), (ii) occluded instrument parts are visualized in the mirror view (middle) and (iii) the mirror is semi-transparent and integrated into a volume cross-section, visualizing tool and tissue structures behind it (right). The examples presented in this Figure show the IVBM outlined in green for better illustration.

by the reflective but at the same time transparent nature of glass. Conceptually, grasping an arbitrarily positioned object in space is challenging in the absence of depth cues. However, the reflections can provide an additional view that aids navigation to the target. Semi-transparency, on the other hand, preserves the background. We transfer this concept to volume raymarching in 4D OCT and integrate a mirror into a selected volume cross-section, which represents a virtual B-scan. While the IVBM highlights volume intensities within this virtual B-scan, it integrates an intelligent mirror, which is only sensitive to voxels that are close to the IVBM and associated with a surgical instrument. Volume structures behind the IVBM are preserved through an adaptive opacity transfer function. In addition, we leverage a perceptually linear color map to encode distance information and enhance the mirrored instrument in the IVBM. Compared to previous works, the proposed method particularly focuses on targeted instrument interactions from alternative viewpoints. The results of our user study provide detailed insights into user perception and interaction with the IVBM and emphasize the benefits for targeted maneuvers in 4D OCT-guided vitreoretinal surgery.

2 Related Work

Virtual mirrors have been initially proposed to support spatial perception and provide secondary views of virtual objects in augmented reality [1,13,19]. Applications have been introduced in angiography or laparoscopic surgery [15,21] and effectiveness of a mirror view in medical scenarios has been shown in user studies with clinical experts [2]. Previous works on OCT volume visualization have so far solely focused on depth perception in axial OCT direction. These include applying a color transfer function based on the distance to a central reference depth [3] or to an identified retinal reference layer [22]. Other ways of providing spatial cues include the augmentation of 2D OCT B-scans [17] or sound feedback to convey distance information [14].

As opposed to previous works on perceptual OCT visualization, the IVBM provides depth cues that aid targeted instrument navigation when viewing the surgical scene from a lateral perspective and targeting a specific cross-section, where axial depth perception is not a primary issue. Augmentation methods proposed in previous works are not designed for such scenarios. Previous works on virtual mirrors have mainly been implemented by rendering the scene from an alternative viewpoint and showing the mirror view next to the virtual objects. In contrast, we augment a selective mirror in-situ into a OCT cross-section.

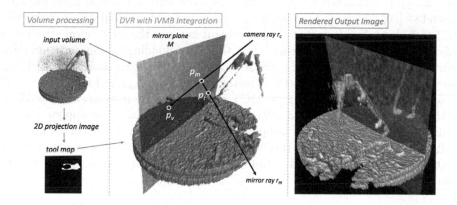

Fig. 3. Method overview: an OCT volume and a 2D instrument segmentation map are provided to the DVR algorithm. During volume raymarching, voxel intensities at the IVBM plane are highlighted and rendered transparent while intersections of the mirror ray with the instrument are integrated using a perceptually linear color map.

3 Methodology

3.1 Definition of an Intelligent Virtual B-Scan Mirror

We define an *Intelligent Virtual B-scan Mirror* (IVBM) as an augmentation of a selected virtual B-scan, which fulfills the following two requirements: (i) voxels integrated into the IVBM are highlighted in the volume rendering, while still visualized semi-transparent to preserve volume structures behind the IVBM. (ii) The selected B-scan cross-section acts as an intelligent mirror by only being sensitive to voxels associated with the surgical instruments in proximity to the IVBM. As illustrated in Fig. 2, the IVBM has the advantage of providing depth cues by visualizing the intersection of the instrument with the cross-section, when the instrument tip meets the tip of the mirrored instrument. It is also capable of visualizing instrument structures that are occluded in the direct view or located behind the IVBM cross-section and amplifies target structures that are aligned with the IVBM. The following sections describe the required components, as well as the composition and direct integration of the IVBM in the volume raymarching algorithm.

3.2   Method Components

Since the IVBM integrates a mirror that is sensitive only to surgical instruments, mirror candidate voxels need to be identified in the volume. To achieve the real-time processing rates necessary for 4D OCT visualizations, we first identify the instrument in a 2D projection image of the volume. Inspired by [23], we create a 2D projection image that encodes the average intensity along each OCT A-scan. This image is forwarded to a Unet-like [16] convolutional neural network with ResNet34 [10] backbone to generate a binary instrument map M_{tool}. Since the OCT signal is fully blocked at the instrument surface, the anatomical structures below the instrument are obscured, and the corresponding A-scans contain only instrument-related voxels. Thus, the 3D position of the voxels associated with the instrument can be obtained by the detected A-scans in M_{tool}, as further described in Sect. 3.3. Before forwarding the OCT volume and the instrument map M_{tool} to the DVR algorithm, we additionally process the volume by applying a 3D median filter to reduce the OCT-typical speckle noise. An overview of the method is illustrated in Fig. 3. With this set of components in place, the composition of the IVBM is fully realized within the DVR algorithm, as described in the following section.

3.3   Composition of an Intelligent Virtual B-Scan Mirror

During volume raymarching, if a camera ray $\vec{r_c}$ intersects with the IVBM, as shown in Fig. 3, three components contribute to the resulting ray color: *(i)* any voxel p_m along $\vec{r_c}$ that is integrated into the IVBM, *(ii)* any voxel p_v along $\vec{r_c}$ behind the IVBM and *(iii)* the specific voxel p_t, at which the mirror ray $\vec{r_m}$ intersects with the surgical instrument. We define the IVBM plane M_{IVBM} as any arbitrary, either manually or automatically selected plane and can thus be defined by the general equation:

$$xn_x + yn_y + zn_z + d = 0 \tag{1}$$

where $\vec{n} = (n_x, n_y, n_z)$ specifies the normal of the plane. We define a distance threshold d_{IVBM} specifying the thickness of the IVBM. During volume raymarching, a voxel p_m with position (p_x, p_y, p_z) is integrated in the IVBM, if the following condition is fulfilled:

$$\left| \frac{|n_x p_x + n_y p_y + n_z p_z + d|}{\sqrt{n_x^2 + n_y^2 + n_z^2}} \right| < d_{IVBM} \tag{2}$$

To visualize the mirror reflections of surgical instruments, a mirror ray $\vec{r_m}$ is cast from each point p_m obtained by (2). The mirror ray direction is determined by:

$$\vec{r_m} = \vec{r_c} - 2 \cdot dot(\vec{r_c}, \vec{n}) \cdot \vec{n} \tag{3}$$

During sampling along $\overrightarrow{r_m}$, we leverage the binary map M_{tool} and an intensity threshold $t_{tool} = 0.25$ (empirically obtained to discard OCT speckle noise), and select the mirrored instrument voxel p_t if the following condition is fulfilled:

$$M_{tool}(p_t) == 1 \wedge I(p_t) > t_{tool} \qquad (4)$$

To mirror the instrument only when close to the IVBM, we limit the number of sampled steps n_{steps} along $\overrightarrow{r_m}$. In case of intersection with an instrument as determined by (4), the raymarching is terminated and the instrument reflection is augmented in the IVBM plane at p_m with color component C_{p_t}. In particular, we encode the distance along $\overrightarrow{r_m}$ between p_m at the IVBM and p_t at the instrument by employing a perceptually linear color map. We employ the $L^*a^*b^*$ color space similar to [22] with the following modifications:

$$C_{L^*a^*b^*}(I, \delta^*) = \gamma(I) \cdot (\delta^* \cdot C_0 + (1 - \delta^*) \cdot C_1) \qquad (5)$$

where $\delta^* = (i_m \cdot 0.7)/n_{steps}$ is a distance predicate that considers the step index $i_m \in [0, n_{steps}]$ along $\overrightarrow{r_m}$ when reaching the instrument at p_t. Further, $\gamma(I)$ is a scaling factor as introduced in [22]. Choosing $C_0 = (I(p_t), -1.5, 1)$ and $C_1 = (I(p_t), -1.0, -1.0)$ achieves a color interpolation between blue hue when the instrument is further from the mirror and green hue when close to the mirror. The RGB component of the instrument reflection is then obtained with:

$$Cp_t = [RGB(C_{L^*a^*b^*}(I(p_t), \delta^*(p)))] \qquad (6)$$

where $RGB(C)$ is an $L^*a^*b^*$ to RGB color space conversion. We additionally decrease the opacity of the reflections with increasing distance of the instrument to the mirror using $\sigma(p_t) = 1.0 - (step_m/n_{steps})^2$.

The overall appearance of a voxel at p_m, integrating instrument reflections while enhancing the natural intensities in the IVBM plane is finally defined by:

$$C(p_m) = [\mu \cdot I(p_m) \cdot (1 - \sigma(p_t)) + Cp_t \cdot \sigma(p_t), \alpha(\mu I(p_m))] \qquad (7)$$

In practice, μ is a dynamically modifiable parameter to amplify the volume intensities of the virtual B-scan and $\alpha(I)$ is a conventional piece-wise linear opacity function. During volume raymarching, we use alpha blending to integrate the IVBM with the remaining volume structures as determined by (2). In general, voxels p_v that are not in the IVBM can be rendered using any convention for direct volume rendering. During our experiments and for the visualizations in Fig. 2 and Fig. 3 we use classic Phong shading as previously employed for OCT volume rendering [20]. In our visualizations, it provides structural surface details while visually highlighting the IVBM in the final rendering.

4   Experimental Setup and User Study

System Integration and Time Profiling. To evaluate if IVBMs can be deployed for the live display of 4D OCT data, we implemented the proposed concept in a C++ visualization framework. The 2D projection image generation and IVBM-integrated volume raymarching were implemented using OpenGL 4.6 and tested on a Windows 10 system with Intel Core i7-8700K @3.7 GHz and NVidia RTX 3090Ti GPU. We train our instrument segmentation network on a custom data set consisting of 3356 2D projection images generated from OCT volumes with a resolution of $391 \times 391 \times 644$ voxels that contain a synthetic eye model and a surgical forceps and include random rotations and flipping for data augmentation. We use PyTorch 1.13 and TensorRT 8.4 for model training and optimization. The data was labeled by two biomedical engineers.

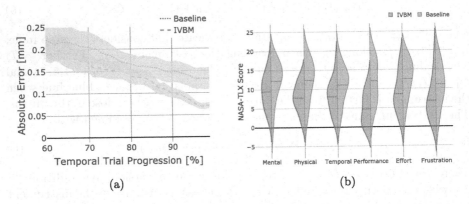

Fig. 4. Distance to the target over the course of the trial progression (a). Results of the subjective task load index based on NASA-TLX (b).

The average overall processing and rendering time, based on 20 test volumes with the same resolution, was 44.3 ($\pm$3.1) ms (filter: 8.1 ($\pm$ 1.2) ms, projection image generation and instrument segmentation: 5.7 ($\pm$2.6) ms, rendering: 30.5 ($\pm$0.6) ms). These benchmarks were achieved with $n_{steps} = 120$ mirror sample steps. To demonstrate the live 4D interactions, our method was integrated into the 4D SS-OCT system presented in [4], acquiring OCT volumes with resolution $391 \times 391 \times 644$ at 10 Hz volume update rates. In the supplementary materials, we include video recordings of instrument maneuvers in 4D OCT with our IVBM visualizations.

User Study. To determine if an IVBM could aid users in performing targeted instrument maneuvers under 4D OCT guidance, we conducted a user study in which we asked participants to move the tip of a surgical instrument to defined target locations. To achieve continuous, accurate, and efficient data collection during the study, we employ a virtual environment (Unity 2021.3) with simulated 4D OCT based on the method proposed in [18]. Additionally, a haptic 3D input device (3D Systems Geomagic Touch) was integrated to navigate the virtual surgical instruments, where motion scaling of 4 : 1 was applied to reduce the influence of the individual manual tremor of the users. Small targets were generated on top of the retina, and the IVBM was automatically positioned at the target locations. To measure the effectiveness of the IVBM when interacting from uncommon perspectives, the virtual scene was rendered from a fixed view approximately orthogonal to the A-scan direction. As stereo vision through a microscope is an inherent part of common ophthalmic procedures, the simulation environment was displayed on an HTC Vive Pro headset leveraging stereo rendering. Users were asked to navigate the instrument tip to the target location and to press a button once satisfied with the positioning. The participants included in total 15 biomedical experts (12 male, 3 female) familiar with ophthalmology and OCT. The study was conducted in accordance with the declaration of Helsinki, the study data were anonymized, and vision tests were performed before the study to ensure healthy vision of all participants. After familiarizing themselves with the interaction in the virtual environment, participants performed 8 trials, with IVBM enabled. For ablation, the same number of trials were performed without IVBM, employing the method proposed in [20] for 4D OCT DVR with Phong shading as a baseline. The accuracy of the positioning and distance to the target was measured over the progression of the trials. Our results show that users reached the target with an average error of $70\,\mu m$ ($\pm 40\,\mu m$) when the IVBM was activated, while in baseline rendering an average error of $135\,\mu m$ ($\pm 128\,\mu m$) was measured, suggesting statistically significant differences between the distributions ($p < 0.002$ based on a Kruskal-Wallis test after detecting unequal variances). Furthermore, we analyzed the distance between the instrument tip and the target with respect to the progress of the trial. Figure 4a shows that when the IVBM was enabled, the deviation of the distance error continuously decreased, especially in the last quarter of the trial progressions, resulting in both more accurate and precise targeting. The outcomes of the NASA-TLX survey (Fig. 4b) conducted after the study showed improvements in all categories when the IVBM was activated. However, statistical significance could only be found in the categories performance ($p < 0.001$) and physical demand ($p < 0.02$) based on an ANOVA test for equal variances.

5   Discussion and Conclusion

Discussion. When users were provided an IVBM, statistically significant improvements regarding the targeting error were found. In a clinical setting, such as during retinal membrane peeling, similar improvements could make a substantial difference and may lead to a safer and more efficient treatment. The results of the subjective task load assessment were also overall improved when the IVBM was enabled, however, statistical significance could not be obtained in categories such as mental demand or frustration. Potential depth conflicts could be investigated in further studies and evaluated with clinical experts. Interestingly, a higher targeting accuracy was achieved with IVBM, even in cases when users reported a higher effort or judged their performance inferior compared to baseline. An advantage of the IVBM is direct in-situ visualization that also highlights target structures, as users do not need to move their gaze from the surgical area to perceive the mirrored view. We envision an automatic identification of anatomical targets to find optimal IVBM positioning in the future.

Conclusion. We presented a novel visualization concept that augments a selected volume cross-section with an intelligent virtual mirror for targeted instrument navigation in 4D OCT. We have provided a definition and implementation of an IVBM and demonstrated its potential to effectively support surgical tasks and to be integrated into a 4D OCT system. We demonstrated the IVBM in simulated vitreoretinal surgery, however, we intend to further apply this concept also to other 4D medical imaging modalities.

Acknowledgements. This work is partially supported and the data is provided by Carl Zeiss Meditec. The authors wish to thank SynthesEyes (https://syntheseyes.de) for providing the excellent simulation setup for the user study.

References

1. Bichlmeier, C., Heining, S.M., Feuerstein, M., Navab, N.: The virtual mirror: a new interaction paradigm for augmented reality environments. IEEE Trans. Med. Imaging **28**(9), 1498–1510 (2009)
2. Bichlmeier, C., Heining, S.M., Rustaee, M., Navab, N.: Laparoscopic virtual mirror for understanding vessel structure evaluation study by twelve surgeons. In: 2007 6th IEEE and ACM International Symposium on Mixed and Augmented Reality, pp. 125–128. IEEE (2007)
3. Bleicher, I.D., Jackson-Atogi, M., Viehland, C., Gabr, H., Izatt, J.A., Toth, C.A.: Depth-based, motion-stabilized colorization of microscope-integrated optical coherence tomography volumes for microscope-independent microsurgery. Transl. Vis. Sci. Technol. **7**(6), 1–1 (2018)
4. Britten, A., et al.: Surgical microscope integrated MHz SS-OCT with live volumetric visualization. Biomed. Optics Express **14**(2), 846–865 (2023)
5. Carrasco-Zevallos, O.M., Viehland, C., Keller, B., McNabb, R.P., Kuo, A.N., Izatt, J.A.: Constant linear velocity spiral scanning for near video rate 4d oct ophthalmic and surgical imaging with isotropic transverse sampling. Biomed. Optics Express **9**(10), 5052 (2018)
6. Draelos, M., Keller, B., Toth, C., Kuo, A., Hauser, K., Izatt, J.: Teleoperating robots from arbitrary viewpoints in surgical contexts. In: 2017 IEEE/RSJ International Conference on Intelligent Robots and Systems (IROS), pp. 2549–2555 (2017)
7. Ehlers, J.P., et al.: Outcomes of intraoperative oct-assisted epiretinal membrane surgery from the pioneer study. Ophthalmol. Retina **2**(4), 263–267 (2018)

8. Ehlers, J.P., et al.: The DISCOVER study 3-year results: feasibility and usefulness of microscope-integrated intraoperative oct during ophthalmic surgery. Ophthalmology **125**(7), 1014–1027 (2018)
9. Falkner-Radler, C.I., Glittenberg, C., Gabriel, M., Binder, S.: Intrasurgical microscope-integrated spectral domain optical coherence tomography-assisted membrane peeling. Retina **35**(10), 2100–2106 (2015)
10. He, K., Zhang, X., Ren, S., Sun, J.: Deep residual learning for image recognition. In: Proceedings of the IEEE Conference on Computer Vision and Pattern Recognition, pp. 770–778 (2016)
11. Jo, Y.J., Heo, D.W., Shin, Y.I., Kim, J.Y.: Diurnal variation of retina thickness measured with time domain and spectral domain optical coherence tomography in healthy subjects. Invest. Ophthalmol. Vis. Sci. **52**(9), 6497–6500 (2011)
12. Kolb, J.P., et al.: Live video rate volumetric OCT imaging of the retina with multi-MHz a-scan rates. PLoS ONE **14**(3), e0213144 (2019)
13. Li, N., Zhang, Z., Liu, C., Yang, Z., Fu, Y., Tian, F., Han, T., Fan, M.: vMirror: enhancing the interaction with occluded or distant objects in VR with virtual mirrors. In: Proceedings of the 2021 CHI Conference on Human Factors in Computing Systems, pp. 1–11 (2021)
14. Matinfar, S., et al.: Surgical Soundtracks: towards automatic musical augmentation of surgical procedures. In: Descoteaux, M., Maier-Hein, L., Franz, A., Jannin, P., Collins, D.L., Duchesne, S. (eds.) MICCAI 2017. LNCS, vol. 10434, pp. 673–681. Springer, Cham (2017). https://doi.org/10.1007/978-3-319-66185-8_76
15. Navab, N., Feuerstein, M., Bichlmeier, C.: Laparoscopic virtual mirror new interaction paradigm for monitor based augmented reality. In: 2007 IEEE Virtual Reality Conference, pp. 43–50. IEEE (2007)
16. Ronneberger, O., Fischer, P., Brox, T.: U-Net: convolutional networks for biomedical image segmentation. In: Navab, N., Hornegger, J., Wells, W.M., Frangi, A.F. (eds.) MICCAI 2015. LNCS, vol. 9351, pp. 234–241. Springer, Cham (2015). https://doi.org/10.1007/978-3-319-24574-4_28
17. Roodaki, H., Filippatos, K., Eslami, A., Navab, N.: Introducing augmented reality to optical coherence tomography in ophthalmic microsurgery. In: 2015 IEEE International Symposium on Mixed and Augmented Reality, pp. 1–6 (2015)
18. Sommersperger, M., et al.: Surgical scene generation and adversarial networks for physics-based iOCT synthesis. Biomed. Optics Express **13**(4), 2414–2430 (2022)
19. Tatzgern, M., Kalkofen, D., Grasset, R., Schmalstieg, D.: Embedded virtual views for augmented reality navigation. In: Proceedings of the International Symposium Mixed Augmented Reality-Workshop Visualization in Mixed Reality Environments, vol. 115, p. 123 (2011)
20. Viehland, C., et al.: Enhanced volumetric visualization for real time 4D intraoperative ophthalmic swept-source OCT. Biomed. Optics Express **7**(5), 1815–1829 (2016)
21. Wang, J., Fallavollita, P., Wang, L., Kreiser, M., Navab, N.: Augmented reality during angiography: integration of a virtual mirror for improved 2D/3D visualization. In: 2012 IEEE International Symposium on Mixed and Augmented Reality (ISMAR), pp. 257–264. IEEE (2012)
22. Weiss, J., Eck, U., Nasseri, M.A., Maier, M., Eslami, A., Navab, N.: Layer-aware iOCT volume rendering for retinal surgery. In: Kozlíková, B., Linsen, L., Vázquez, P.P., Lawonn, K., Raidou, R.G. (eds.) Eurographics Workshop on Visual Computing for Biology and Medicine. The Eurographics Association (2019)

23. Weiss, J., Sommersperger, M., Nasseri, A., Eslami, A., Eck, U., Navab, N.: Processing-aware real-time rendering for optimized tissue visualization in intraoperative 4D OCT. In: Martel, A.L., et al. (eds.) MICCAI 2020. LNCS, vol. 12265, pp. 267–276. Springer, Cham (2020). https://doi.org/10.1007/978-3-030-59722-1_26
24. Wilkins, J.R., et al.: Characterization of epiretinal membranes using optical coherence tomography. Ophthalmology **103**(12), 2142–2151 (1996)

Multi-view Guidance for Self-supervised Monocular Depth Estimation on Laparoscopic Images via Spatio-Temporal Correspondence

Wenda Li[1]([✉])(iD), Yuichiro Hayashi[1], Masahiro Oda[1,2](iD), Takayuki Kitasaka[3](iD), Kazunari Misawa[4](iD), and Kensaku Mori[1,5,6]([✉])(iD)

[1] Graduate School of Informatics, Nagoya University, Aichi, Nagoya 464-8601, Japan
wdli@mori.m.is.nagoya-u.ac.jp
[2] Information and Communications, Nagoya University,
Aichi, Nagoya 464-8601, Japan
[3] Faculty of Information Science, Aichi Institute of Technology,
Yakusacho, Aichi, Toyota 470-0392, Japan
[4] Aichi Cancer Center Hospital, Aichi, Nagoya 464-8681, Japan
[5] Information Technology Center, Nagoya University, Aichi, Nagoya 464-8601, Japan
[6] Research Center of Medical Bigdata, National Institute of Informatics, Tokyo,
Hitotsubashi 101-8430, Japan
kensaku@is.nagoya-u.ac.jp

Abstract. This work proposes an innovative self-supervised approach to monocular depth estimation in laparoscopic scenarios. Previous methods independently predicted depth maps ignoring spatial coherence in local regions and temporal correlation between adjacent images. The proposed approach leverages spatio-temporal coherence to address the challenges of textureless areas and homogeneous colors in such scenes. This approach utilizes a multi-view depth estimation model to guide monocular depth estimation when predicting depth maps. Moreover, the minimum reprojection error is extended to construct a cost volume for the multi-view model using adjacent images. Additionally, a 3D consistency of the point cloud back-projected from predicted depth maps is optimized for the monocular depth estimation model. To benefit from spatial coherence, deformable patch-matching is introduced to the monocular and multi-view models to smooth depth maps in local regions. Finally, a cycled prediction learning for view synthesis and relative poses is designed to exploit the temporal correlation between adjacent images fully. Experimental results show that the proposed method outperforms existing methods in both qualitative and quantitative evaluations. Our code is available at https://github.com/MoriLabNU/MGMDepthL.

Supplementary Information The online version contains supplementary material available at https://doi.org/10.1007/978-3-031-43996-4_41.

Keywords: Monocular depth estimation · Self-supervised learning · Laparoscopic images

1 Introduction

The significance of depth information is undeniable in computer-assisted surgical systems [20]. In robotic-assisted surgery, the depth value is used to accurately map the surgical field and track the movement of surgical instruments [7,22]. Additionally, the depth value is essential to virtual and augmented reality to create 3D models and realize surgical technique training [18].

Learning-based approaches have significantly improved monocular depth estimation (MDE) in recent years. As the pioneering work, Eigen et al. [2] proposed the first end-to-end deep learning framework using multi-scale CNN under supervised learning for MDE. Following this work, ResNet-based and Hourglass-based models were introduced as the variants of CNN-based methods [9,21]. However, supervised learning requires large amounts of annotated data. Collecting depth value from hardware or synthesis scenes is time-consuming and expensive [14]. To address this issue, researchers have explored self-supervised methods for MDE. Zhou et al. [30] proposed a novel depth-pose self-supervised monocular depth estimation from a video sequence. They generate synthesis views through the estimated depth maps and relative poses. Gordon et al. [3] optimized this work by introducing a minimum reprojection error between adjacent images and made a notable baseline named Monodepth2. Subsequently, researchers have proposed a variety of self-supervised methods for MDE, including those based on semantic segmentation [4], adversarial learning [28] and uncertainty [17]. These days, MDE has been applied to laparoscopic images. Ye et al. [27] and Max et al. [1] have provided exceptional laparoscopic scene datasets. Huang et al. [6] used generative adversarial networks to derive depth maps on laparoscopic images. Li et al. [12] combined depth estimation with scene coordinate prediction to improve network performance.

This study presents a novel approach to predict depth values in laparoscopic images using spatio-temporal correspondence. Current self-supervised models for monocular depth estimation face two significant challenges in laparoscopic settings. First, monocular models individually predicted depth maps, ignoring the temporal correlation between adjacent images. Second, accurate point matching is difficult to achieve due to the misleading of large textureless regions caused by the smooth surface of organs. And the homogenous color misleads that the local areas of the edge are regarded with the same depth value. To overcome these obstacles, We introduce multi-view depth estimation (MVDE) with the optimized cost volume to guide the self-supervised monocular depth estimation model. Moreover, we exploit more informative values in a spatio-temporal manner to address the limitation of existing multi-view and monocular models.

Our main contributions are summarized as follows. (i) A novel self-supervised monocular depth estimation guided by a multi-view depth model to leverage adjacent images when estimating depth value. (ii) Cost volume construction for multi-view depth estimation under minimum reprojection error and an optimized point cloud consistency for the monocular depth estimation. (iii) An extended deformable patch matching based on the spatial coherence in local regions and a cycled prediction learning for view synthesis and relative poses to exploit the temporal correlation between adjacent images.

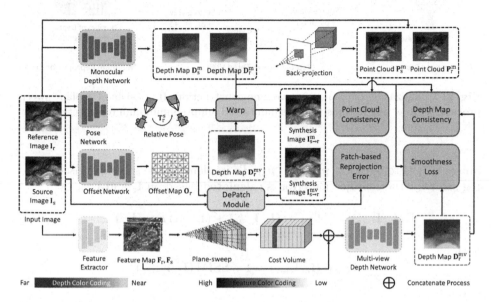

Fig. 1. Overview of our self-supervised monocular depth estimation framework. The proposed method consists of a monocular depth network, a pose network, an offset network, and a multi-view depth network.

2   Method

2.1   Self-supervised Monocular Depth Estimation

Following [3,12], we train a self-supervised monocular depth network using a short clip from a video sequence. The short clip consists of a current frame $\mathbf{I}_t$ as reference image $\mathbf{I}_r$ and adjacent images $\mathbf{I}_s \in \{\mathbf{I}_{t-1}, \mathbf{I}_{t+1}\}$ regarded as source images. As shown in Fig. 1, a monocular depth network $f_D^m(\mathbf{I}_r, \mathbf{I}_s; \theta_D^m)$ respectively predicts pixel-level depth maps $\mathbf{D}_r$ and $\mathbf{D}_s$ corresponding to $\mathbf{I}_r$ and $\mathbf{I}_s$. A pose network $f_T(\mathbf{I}_r, \mathbf{I}_s; \theta_T)$ estimates a transformation matrix $\mathbf{T}_r^s$ as a relative pose of the laparoscope from view $\mathbf{I}_r$ to $\mathbf{I}_s$. We use $\mathbf{D}_r$ and $\mathbf{T}_r^s$ to match the pixels between $\mathbf{I}_r$ and $\mathbf{I}_s$ by

$$p_s = \mathbf{K}\mathbf{T}_r^s \mathbf{D}(p_r)\mathbf{K}^{-1}p_r, \tag{1}$$

where p_r and p_s are the 2D pixel coordinates in $\mathbf{I}_r$ and $\mathbf{I}_s$. $\mathbf{K}$ is the laparoscope's intrinsic parameter matrix. This allows for the generation of a synthetic

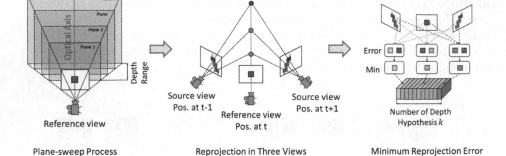

Fig. 2. Visualization depicting cost volume construction. It consists of a plane-sweep process, reprojection in three views as inputs and adopted minimum reprojection error.

image $\mathbf{I}_{s\to r}$ through $\mathbf{I}_{s\to r}(\boldsymbol{p}_r) = \mathbf{I}_s(\boldsymbol{p}_s)$. To implement the self-supervised learning strategy, the reprojection error is calculated based on $\mathbf{I}_r$ and $\mathbf{I}_{s\to r}$ by

$$E(\mathbf{I}_r, \mathbf{I}_{s\to r}) = \frac{\alpha}{2}\mathcal{L}_{SSIM}(\mathbf{I}_r, \mathbf{I}_{s\to r}) + (1-\alpha)\mathcal{L}_{L1}(\mathbf{I}_r, \mathbf{I}_{s\to r}), \tag{2}$$

with

$$\mathcal{L}_{SSIM}(\mathbf{I}_r, \mathbf{I}_{s\to r}) = 1 - \text{SSIM}(\mathbf{I}_r, \mathbf{I}_{s\to r})) \tag{3}$$

and

$$\mathcal{L}_{L1}(\mathbf{I}_r, \mathbf{I}_{s\to r}) = \|\mathbf{I}_r - \mathbf{I}_{s\to r}\|_1, \tag{4}$$

where structured similarity (SSIM) [24] and L1-norm operator both adopt α at 0.85 followed as [3,13]. Instead of adding the auxiliary task proposed in prior work [12], we back-project the 2D pixel coordinates to the 3D coordinates $P(\boldsymbol{p}) = \mathbf{K}^{-1}\mathbf{D}(\boldsymbol{p})\boldsymbol{p}$. All the 3D coordinates from $\mathbf{I}_r$ and $\mathbf{I}_s$ gather as point cloud $\mathbf{S}_r$ and $\mathbf{S}_s$. Then synthesized point cloud is warped as $\mathbf{S}_{s\to r}(\boldsymbol{p}_r) = \mathbf{S}_s(\boldsymbol{p}_s)$. We construct the point cloud consistency by

$$\mathcal{L}_p = \mathcal{L}_{L1}(\mathbf{S}_r, \mathbf{S}_{s\to r}), \tag{5}$$

where $\mathbf{S}_r$ and $\mathbf{S}_{s\to r}$ are based on depth maps from the monocular depth network.

2.2 Improving Depth with Multi-view Guidance

Unlike MDE, MVDE fully leverages the temporal information in the short clip when estimating depth maps. MVDE first samples hypothesized depths value d_k in a depth range from d_{min} to d_{max}. k is the number of sampled depth values. Then, a feature extractor $f_F(\mathbf{I}_r, \mathbf{I}_s; \theta_F)$ obtains the deep feature map $\mathbf{F}_r$ and $\mathbf{F}_s$ from input images $\mathbf{I}_r$ and $\mathbf{I}_s$. Similar to the pixel-coordinate matching between $\mathbf{I}_r$ and $\mathbf{I}_s$ as Eq. 1, 2D pixel coordinates of $\mathbf{F}_s$ is back-projected to the each plane

$\mathbf{Z}^{d_k}$ of hypothesised depth value. $\mathbf{Z}^{d_k}$ shares the same depth value d for each pixels. Then the pixel coordinates are matched by

$$_f\boldsymbol{p}_s^d = \mathbf{K}\mathbf{T}_r^s\mathbf{Z}^{d_k}\left(_f\boldsymbol{p}_r\right)\mathbf{K}^{-1}{}_f\boldsymbol{p}_r, \tag{6}$$

where $_f\boldsymbol{p}_r$ is the pixel coordinates in $\mathbf{F}_r$ and $_f\boldsymbol{p}_s^d$ is the pixel coordinates in $\mathbf{F}_s$ based on the depth value d. Then $\mathbf{F}_s$ is warped to the synthesis feature map by $\mathbf{F}_{s\rightarrow r}^d\left(_f\boldsymbol{p}_r\right) = \mathbf{F}_s\left(_f\boldsymbol{p}_s^d\right)$. Feature volumes $\mathbf{V}_r$ and $\mathbf{V}_{s\rightarrow r}$ are aggregations of feature maps $\mathbf{F}_r$ and $\mathbf{F}_{s\rightarrow r}$. We construct a cost volume $\mathbf{C}$ by

$$\mathbf{C} = \mathcal{L}_{L1}\left(\mathbf{V}_r, \mathbf{V}_{s\rightarrow r}\right) = \|\mathbf{V}_r - \mathbf{V}_{s\rightarrow r}\|_1. \tag{7}$$

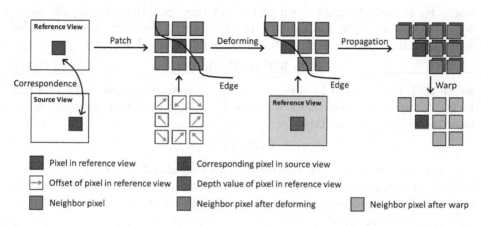

Fig. 3. Illustration of deformable patch matching process with pixel coordinates offset and depth propagation.

Previous approaches average the difference between $\mathbf{V}_r$ and all $\mathbf{V}_{s\rightarrow r}$ from adjacent views to generate cost volumes without considering the occlusion problem and uniform differences between the reference and adjacent feature maps [25,26]. Motivated by these challenges, we introduce the minimum reprojection loss to $\mathbf{C}$, as shown in Fig. 2. We construct the cost volume $\hat{\mathbf{C}}$ via the minor difference value on the corresponding coordinates of the feature volumes as

$$\hat{\mathbf{C}} = \min_s \mathcal{L}_{L1}\left(\mathbf{V}_r, \mathbf{V}_{s\rightarrow r}\right). \tag{8}$$

We construct a consistency between MVDE and MDE by

$$\mathcal{L}_c = \mathcal{L}_{L1}\left(\mathbf{D}_r^m, \mathbf{D}_r^{mv}\right) = \|\mathbf{D}_r^m - \mathbf{D}_r^{mv}\|_1, \tag{9}$$

where $\mathbf{D}_r^m$ and $\mathbf{D}_r^{mv}$ are the depth map predicted by our networks $f_D^m\left(\mathbf{I}_r, \mathbf{I}_s; \theta_D^m\right)$ and $f_D^{mv}\left(\hat{\mathbf{C}}, \mathbf{F}_r, \mathbf{F}_s; \theta_D^{mv}\right)$.

2.3 Deformable Patch Matching and Cycled Prediction Learning

Since large areas of textureless areas and reflective parts will cause brightness-based reprojection errors to become unreliable. Furthermore, the homogeneous color on the edge of organs causes local regions to be regarded in the same depth plane. We introduced deformable patch-matching-based local spatial propagation to MDE. As shown in Fig. 3, an offset map $\mathbf{O}_r(\boldsymbol{p}_r)$ is adopted to obtain a local region for each pixel by transforming the pixel coordinates in the reference image $\mathbf{I}_r$. Inspired by [15,23], to avoid marginal areas affecting the spatial coherence of the local region, an offset network $f_O(\mathbf{I}_r; \theta_O)$ generates a pixel-level add additional offset map $\Delta\mathbf{O}_r(\boldsymbol{p}_r)$. The deformable local region for each pixel can be obtained by

$$\mathbf{R}_r = \boldsymbol{p}_r + \mathbf{O}_r(\boldsymbol{p}_r) + \Delta\mathbf{O}_r(\boldsymbol{p}_r), \tag{10}$$

where $\mathbf{R}_r$ is the deformable local regions for pixel $\boldsymbol{p}_r$. After sharing the same depth value by depth propagation in the local region, we implement the deformable local regions on Eq. 1 to complete patch matching by

$$\mathbf{R}_s = \mathbf{K}\mathbf{T}_r^s\mathbf{D}_r(\mathbf{R}_r)\mathbf{K}^{-1}\mathbf{R}_r, \tag{11}$$

where $\mathbf{R}_s$ is the matched local regions in source views $\mathbf{I}_s$. Based on $\mathbf{R}_r$ and $\mathbf{R}_s$, $\mathbf{I}_s(\mathbf{R}_s)$ is warped to the synthesised regions $\mathbf{I}_{s\to r}(\mathbf{R}_r)$. The patch-matching-based reprojection error is calculated by

$$\mathcal{L}_r = \mathrm{E}(\mathbf{I}_r(\mathbf{R}_r), \mathbf{I}_{s\to r}(\mathbf{R}_r)). \tag{12}$$

To better use the temporal correlation, we considered each image as a reference to construct a cycled prediction learning for depth and pose. The total loss on the final computation is averaged from the error of each combination as

$$\mathcal{L}_{total} = \frac{1}{3}\sum_{i=1}^{3}\mathcal{L}_r^m(i) + \gamma\mathcal{L}_r^{mv}(i) + \mu\mathcal{L}_p(i) + \lambda\mathcal{L}_c(i) + \delta\mathcal{L}_s(i), \tag{13}$$

where $\mathcal{L}_r^m$ is the reprojection error term for MDE, and $\mathcal{L}_r^{mv}$ is for MVDE. i is the index number of views in the short clip. $\mathcal{L}_s$ is the smoothness term [3,12].

3   Experiments

3.1   Datasets and Evaluation Metrics

SCARED [1] datasets were adopted for all the experiments. This dataset contained 35 laparoscopic stereo videos with nine different scenes. And the corresponding depth values obtained through coded structured light images served as ground-truth. We divided the SCARED datasets into a 10:1:1 ratio for each scene based on the video sequence to conduct our experiments. For training, validation, and testing, there were 23,687, 2,405, and 2,405 frames, respectively. Because of limitations in computational resources, we resized the images to 320×256 pixels, a quarter of their original dimensions. Following the previous methods [3,25], we adopted seven classical 2D metrics to evaluate the predicted depth maps. Additionally, we only used the monocular depth model to predict depth values with a single RGB image as input during testing.

Table 1. Depth estimation quantitative results. * denotes the method need multi-frame at test. † denotes that the input images are five frames instead of three during training time.

	Abs Rel↓	Sq Rel↓	RMSE↓	RMSE log↓	$\delta < 1.25$↑	$\delta < 1.25^2$↑	$\delta < 1.25^3$↑
Baseline	0.079	0.999	7.868	0.103	0.918	0.995	**1.000**
Monodepth2 [3]	0.083	0.994	8.167	0.107	0.937	0.995	**1.000**
HR-Depth [13]	0.080	0.938	7.943	0.104	0.940	0.996	**1.000**
Manydepth [25]*	0.075	0.830	7.403	0.099	0.945	0.996	**1.000**
MonoViT [29]	0.074	0.865	7.517	0.097	0.949	0.996	**1.000**
SC-Depth [11]†	0.070	0.744	6.932	0.092	0.951	0.998	0.999
AJ-Depth [10]	0.078	0.896	7.578	0.101	0.937	0.996	**1.000**
GCDepthL [12]	0.071	0.801	7.105	0.094	0.938	0.996	**1.000**
Ours	**0.066**	**0.655**	**6.441**	**0.086**	**0.955**	**0.999**	**1.000**

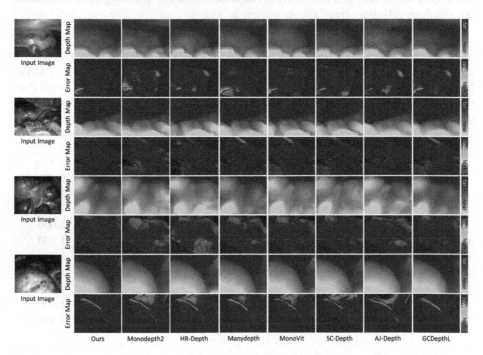

Fig. 4. Depth estimation qualitative results: input images (first column), predicted depth maps and error maps calculated via the absolute relative error metric

3.2 Implementation Details

We utilized PyTorch [16] for model training, employing the Adam optimizer [8] across 25 training epochs. The learning rate started at 1×10^{-4} and dropped by a scale factor of 10 for the final 10 epochs. A batch size is 6, and the total loss function's parameters γ, μ, λ, and δ were set to 0.5, 0.5, 0.2, and 1×10^{-3}, respectively. Additionally, we capped the predicted depth values at 150mm. To construct the cost volume, we adopted the adaptive depth range method [25] and set the number of hypothesized depth values k to 96.

Following [25], we used ResNet-18 [5] with pretrained weights on the ImageNet dataset [19] as encoder module. The feature extractor comprised the first five layers of ResNet-18 [5]. The offset network was two 2D convolution layers.

3.3 Comparison Experiments

We conducted a comprehensive evaluation of our proposed method by comparing it with several classical and state-of-the-art techniques [3, 10–13, 25, 29] retrained on SCARED datasets [1]. Table 1 presents the quantitative results. We also assessed the baseline performance of our proposed method. We compared the depth maps and generated error maps on various laparoscopic scenes based on absolute relative error [25], as shown in Fig. 4.

Table 2. Ablation results for each component contribution in the proposed method. * denotes the method need multi-frame at test. PCC: Point Cloud Consistency; MRE: Minimum Reprojection Error; DPM: Deformable Patch Matching; CPL: Cycled Prediction Learning. M represents monocular depth estimation; MV represents multi-view depth estimation.

	Abs Rel↓	Sq Rel↓	RMSE↓	RMSE log↓	$\delta < 1.25$↑	$\delta < 1.25^2$↑	$\delta < 1.25^3$↑
w/o PCC	0.084	1.364	9.217	0.116	0.925	0.990	0.997
w/o MRE	0.068	0.711	6.576	0.088	0.953	0.998	0.999
w/o DPM	0.069	0.753	6.901	0.091	0.945	0.997	**1.000**
w/o CPL	0.071	0.702	6.868	0.091	0.956	0.998	**1.000**
Baseline (M) [12]	0.071	0.801	7.105	0.094	0.938	0.996	1.000
w/ PCC	0.070	0.763	6.959	0.092	0.948	0.997	1.000
Baseline (MV) [25]*	0.075	0.830	7.403	0.099	0.945	0.997	1.000
w/ MRE*	0.074	0.797	7.241	0.095	0.953	0.997	1.000
Ours (M)	0.068	0.708	6.788	0.089	**0.955**	0.998	1.000
Ours (MV)*	0.070	0.743	6.842	0.091	0.948	0.997	1.000
Ours	**0.066**	**0.655**	**6.441**	**0.086**	**0.955**	**0.999**	1.000

3.4 Ablation Study

We conducted an ablation study to evaluate the influence of different components in our proposed approach. Table 2 shows the results of our method with four different components, namely, point cloud consistency (PCC), minimum reprojection error (MRE), deformable patch matching (DPM), and cycled prediction learning (CPL). We adopted GCDepthL [12] and Manydepth [25] as baselines for our method's monocular and multi-view depth models. We proposed PCC and MRE as two optimized modules for these baseline models and evaluated their impact on each baseline individually. Furthermore, we trained the monocular and multi-view depth models separately in our proposed method without consistency to demonstrate the contribution of combining these two models.

4   Discussion and Conclusions

The laparoscopic scenes typically feature large, smooth regions with organ surfaces and homogeneous colors along the edges of organs. This can cause issues while matching pixels, as the points in the boundary area can be mistaken to be in the same depth plane. Our proposed method's depth maps, as shown in Fig 4, exhibit a smoother performance in the large regions of input images when compared to the existing methods [3, 10–13, 25, 29]. Additionally, the error maps reveal that our proposed method performs better even when the depth maps look similar qualitatively. At the marginal area of organs, our proposed method generates better depth predictions with smoother depth maps and lower errors, despite the color changes being barely perceptible with the depth value changes. Our proposed method outperforms current approaches on the seven metrics we used, as demonstrated in Table 1. The ablation study reveals that the proposed method improves significantly when combining each component, and each component contributes to the proposed method. Specifically, the optimized modules PCC and MRE designed for monocular and multi-view depth models enhance the performance of the baselines [12, 25]. The combination of monocular and multi-view depth models yields better results than the single model trained independently, as seen in the last three rows of Table 2.

In conclusion, we incorporate more temporal information in the monocular depth model by leveraging the guidance of the multi-view depth model when predicting depth values. We introduce the minimum reprojection error to construct the multi-view depth model's cost volume and optimize the monocular depth model's point cloud consistency module. Moreover, we propose a novel method that matches deformable patches in spatially coherent local regions instead of point matching. Finally, cycled prediction learning is designed to exploit temporal information. The outcomes of the experiments indicate an improved depth estimation performance using our approach.

Acknowledgments. We extend our appreciation for the assistance provided through the JSPS Bilateral International Collaboration Grants; JST CREST Grant (JPMJCR20D5); MEXT/JSPS KAKENHI Grants (17H00867, 26108006, 21K19898); and the CIBoG initiative of Nagoya University, part of the MEXT WISE program.

References

1. Allan, M., et al.: Stereo correspondence and reconstruction of endoscopic data challenge. arXiv preprint arXiv:2101.01133 (2021)
2. Eigen, D., Puhrsch, C., Fergus, R.: Depth map prediction from a single image using a multi-scale deep network. In: Advances in Neural Information Processing Systems 27 (2014)

3. Godard, C., Mac Aodha, O., Firman, M., Brostow, G.J.: Digging into self-supervised monocular depth estimation. In: Proceedings of the IEEE/CVF International Conference on Computer Vision, pp. 3828–3838 (2019)

4. Guizilini, V., Hou, R., Li, J., Ambrus, R., Gaidon, A.: Semantically-guided representation learning for self-supervised monocular depth. arXiv preprint arXiv:2002.12319 (2020)

5. He, K., Zhang, X., Ren, S., Sun, J.: Deep residual learning for image recognition. In: Proceedings of the IEEE Conference on Computer Vision and Pattern Recognitionm, pp. 770–778 (2016)

6. Huang, B., et al.: Self-supervised generative adversarial network for depth estimation in laparoscopic images. In: de Bruijne, M., Cattin, P.C., Cotin, S., Padoy, N., Speidel, S., Zheng, Y., Essert, C. (eds.) MICCAI 2021. LNCS, vol. 12904, pp. 227–237. Springer, Cham (2021). https://doi.org/10.1007/978-3-030-87202-1_22

7. Hwang, M., et al.: Applying depth-sensing to automated surgical manipulation with a da Vinci robot. In: 2020 International Symposium on Medical Robotics (ISMR), pp. 22–29. IEEE (2020)

8. Kingma, D.P., Ba, J.: Adam: a method for stochastic optimization. arXiv preprint arXiv:1412.6980 (2014)

9. Laina, I., Rupprecht, C., Belagiannis, V., Tombari, F., Navab, N.: Deeper depth prediction with fully convolutional residual networks. In: 2016 Fourth International Conference on 3D Vision (3DV), pp. 239–248. IEEE (2016)

10. Li, W., Hayashi, Y., Oda, M., Kitasaka, T., Misawa, K., Kensaku, M.: Attention guided self-supervised monocular depth estimation based on joint depth-pose loss for laparoscopic images. Comput. Assist. Radiol. Surg. (2022)

11. Li, W., Hayashi, Y., Oda, M., Kitasaka, T., Misawa, K., Mori, K.: Spatially variant biases considered self-supervised depth estimation based on laparoscopic videos. Comput. Methods Biomech. Biomed. Eng.: Imaging Vis., 1–9 (2021)

12. Li, W., Hayashi, Y., Oda, M., Kitasaka, T., Misawa, K., Mori, K.: Geometric constraints for self-supervised monocular depth estimation on laparoscopic images with dual-task consistency. In: Medical Image Computing and Computer Assisted Intervention-MICCAI 2022: 25th International Conference, Singapore, September 18–22, 2022, Proceedings, LNCS, Part IV, pp. 467–477. Springer (2022). https://doi.org/10.1007/978-3-031-16440-8_45

13. Lyu, X., Liu, L., Wang, M., Kong, X., Liu, L., Liu, Y., Chen, X., Yuan, Y.: HR-Depth: high resolution self-supervised monocular depth estimation. In: Proceedings of the AAAI Conference on Artificial Intelligence, vol. 35, pp. 2294–2301 (2021)

14. Ming, Y., Meng, X., Fan, C., Yu, H.: Deep learning for monocular depth estimation: a review. Neurocomputing **438**, 14–33 (2021)

15. Park, J., Joo, K., Hu, Z., Liu, C.-K., So Kweon, I.: Non-local spatial propagation network for depth completion. In: Vedaldi, A., Bischof, H., Brox, T., Frahm, J.-M. (eds.) ECCV 2020. LNCS, vol. 12358, pp. 120–136. Springer, Cham (2020). https://doi.org/10.1007/978-3-030-58601-0_8

16. Paszke, A., et al.: Automatic differentiation in PyTorch. In: NIPS 2017 Workshop on Autodiff (2017)

17. Poggi, M., Aleotti, F., Tosi, F., Mattoccia, S.: On the uncertainty of self-supervised monocular depth estimation. In: Proceedings of the IEEE/CVF Conference on Computer Vision and Pattern Recognition, pp. 3227–3237 (2020)

18. Qian, L., Zhang, X., Deguet, A., Kazanzides, P.: ARAMIS: augmented reality assistance for minimally invasive surgery using a head-mounted display. In: Shen, D., et al. (eds.) MICCAI 2019. LNCS, vol. 11768, pp. 74–82. Springer, Cham (2019). https://doi.org/10.1007/978-3-030-32254-0_9

19. Russakovsky, O., et al.: ImageNet large scale visual recognition challenge. Int. J. Comput. Vis. **115**(3), 211–252 (2015)
20. Sánchez-González, P., et al.: Laparoscopic video analysis for training and image-guided surgery. Minim. Invasive Therapy Allied Technol. **20**(6), 311–320 (2011)
21. Tosi, F., Aleotti, F., Poggi, M., Mattoccia, S.: Learning monocular depth estimation infusing traditional stereo knowledge. In: Proceedings of the IEEE/CVF Conference on Computer Vision and Pattern Recognition, pp. 9799–9809 (2019)
22. Vecchio, R., MacFayden, B., Palazzo, F.: History of laparoscopic surgery. Panminerva Med. **42**(1), 87–90 (2000)
23. Wang, F., Galliani, S., Vogel, C., Speciale, P., Pollefeys, M.: Patchmatchnet: learned multi-view patchmatch stereo. In: Proceedings of the IEEE/CVF Conference on Computer Vision and Pattern Recognition, pp. 14194–14203 (2021)
24. Wang, Z., Bovik, A.C., Sheikh, H.R., Simoncelli, E.P.: Image quality assessment: from error visibility to structural similarity. IEEE Trans. Image Process. **13**(4), 600–612 (2004)
25. Watson, J., Mac Aodha, O., Prisacariu, V., Brostow, G., Firman, M.: The temporal opportunist: self-supervised multi-frame monocular depth. In: Proceedings of the IEEE/CVF Conference on Computer Vision and Pattern Recognition, pp. 1164–1174 (2021)
26. Yao, Y., Luo, Z., Li, S., Fang, T., Quan, L.: MVSNet: depth inference for unstructured multi-view stereo. In: Proceedings of the European Conference on Computer Vision (ECCV), pp. 767–783 (2018)
27. Ye, M., Johns, E., Handa, A., Zhang, L., Pratt, P., Yang, G.Z.: Self-supervised siamese learning on stereo image pairs for depth estimation in robotic surgery. arXiv preprint arXiv:1705.08260 (2017)
28. Zhao, C., Yen, G.G., Sun, Q., Zhang, C., Tang, Y.: Masked GAN for unsupervised depth and pose prediction with scale consistency. IEEE Trans. Neural Netw. Learn. Syst. **32**(12), 5392–5403 (2020)
29. Zhao, C., et al.: MonoViT: self-supervised monocular depth estimation with a vision transformer. arXiv preprint arXiv:2208.03543 (2022)
30. Zhou, T., Brown, M., Snavely, N., Lowe, D.G.: Unsupervised learning of depth and ego-motion from video. In: Proceedings of the IEEE Conference on Computer Vision and Pattern Recognition, pp. 1851–1858 (2017)

POV-Surgery: A Dataset for Egocentric Hand and Tool Pose Estimation During Surgical Activities

Rui Wang, Sophokles Ktistakis, Siwei Zhang, Mirko Meboldt,
and Quentin Lohmeyer[✉]

ETH Zurich, Zurich, Switzerland
{ruiwang46,ktistaks,meboldtm,qlohmeyer}@ethz.ch, siwei.zhang@inf.ethz.ch

Abstract. The surgical usage of Mixed Reality (MR) has received growing attention in areas such as surgical navigation systems, skill assessment, and robot-assisted surgeries. For such applications, pose estimation for hand and surgical instruments from an egocentric perspective is a fundamental task and has been studied extensively in the computer vision field in recent years. However, the development of this field has been impeded by a lack of datasets, especially in the surgical field, where bloody gloves and reflective metallic tools make it hard to obtain 3D pose annotations for hands and objects using conventional methods. To address this issue, we propose POV-Surgery, a large-scale, synthetic, egocentric dataset focusing on pose estimation for hands with different surgical gloves and three orthopedic surgical instruments, namely scalpel, friem, and diskplacer. Our dataset consists of 53 sequences and 88,329 frames, featuring high-resolution RGB-D video streams with activity annotations, accurate 3D and 2D annotations for hand-object pose, and 2D hand-object segmentation masks. We fine-tune the current SOTA methods on POV-Surgery and further show the generalizability when applying to real-life cases with surgical gloves and tools by extensive evaluations. The code and the dataset are publicly available at http://batfacewayne.github.io/POV_Surgery_io/.

Keywords: Hand Object Pose Estimation · Deep Learning · Dataset · Mixed Reality

1 Introduction

Understanding the movement of surgical instruments and the surgeon's hands is essential in computer-assisted interventions and has various applications, including surgical navigation systems [27], surgical skill assessment [10,15,23] and robot-assisted surgeries [8]. With the rising interest in using head-mounted Mixed Reality (MR) devices for such applications [1,7,21,28], estimating the 3D

R. Wang and S. Ktistakis—Denotes co-first authorship.

© The Author(s), under exclusive license to Springer Nature Switzerland AG 2023
H. Greenspan et al. (Eds.): MICCAI 2023, LNCS 14228, pp. 440–450, 2023.
https://doi.org/10.1007/978-3-031-43996-4_42

pose of hands and objects from the egocentric perspective becomes more important. However, this is more challenging compared to the third-person viewpoint because of the constant self-occlusion of hands and mutual occlusions between hands and objects. While the use of deep neural networks and attention modules has partly addressed this challenge [13,18,19,22], the lack of egocentric datasets to train such models has hindered progress in this field. Most existing datasets that provide 3D hand or hand-object pose annotations focus on the third-person perspective [11,20,29]. FPHA [9] proposed the first egocentric hand-object video dataset by attaching magnetic sensors to hands and objects. However, the attached sensors pollute the RGB frames. More recently, H2O [17] proposed an egocentric video dataset with hand and object pose annotated with a semi-automatic pipeline, based on 2D hand joint detection and object point cloud refinement. However, this pipeline is not applicable to the surgical domain because of the large domain gap between the everyday scenarios in [9,17] and surgical scenarios. For instance, surgeons wear surgical gloves that are often covered with blood during the surgery process, which presents great challenges for vision-based hand keypoint detection methods. Moreover, these datasets focus on large, everyday objects with distinct textures, whereas surgical instruments are often smaller and have featureless, highly reflective metallic surfaces. This results in noisy and incomplete object point clouds when captured with RGB-D cameras. Therefore, in a surgical setting, the annotation approaches proposed in [11,17] are less stable and reliable. Pioneer work in [14] introduces a small synthetic dataset with blue surgical gloves and a surgical drill, following the synthetic data generation approach in [13]. However, being a single-image dataset, it ignores the strong temporal context in surgical tasks, which is crucial for accurate and reliable 3D pose estimation [17,24]. Surgical cases have inherent task-specific information and temporal correlations during surgical instrument usage, such as cutting firmly and steadily with a scalpel. Moreover, it lacks diversity, focusing only on one unbloodied blue surgical glove and one instrument, and only provides low-resolution image patches.

To fill this gap, we propose a novel synthetic data generation pipeline that goes beyond single image cases to synthesize realistic temporal sequences of surgical tool manipulation from an egocentric perspective. It features a body motion capture module to model realistic body movement sequences during artificial surgeries and a hand-object manipulation generation module to model the grasp evolution sequences. With the proposed pipeline, we generate POV-Surgery, a large synthetic egocentric video dataset of surgical activities that features surgical gloves in diverse textures (green, white, and blue) with various bloodstain patterns and three metallic tools that are commonly used in orthopedic surgeries.

In summary, our contributions are:

- A novel, easy-to-use, and generalizable synthetic data generation pipeline to generate temporally realistic hand-object manipulations during surgical activities.
- POV-Surgery: the first large-scale dataset with egocentric sequences for hand and surgical instrument pose estimation, with diverse, realistic surgical glove

textures, and different metallic tools, annotated with accurate 3D/2D hand-object poses and 2D hand-object segmentation masks.
- Extensive evaluations of existing state-of-the-art (SOTA) hand pose estimation methods on POV-Surgery, revealing their shortcomings when dealing with the unique challenges in surgical cases from egocentric view.
- Significantly improved performance for SOTA methods after fine-tuning them on the POV-Surgery training set, on both our synthetic test set and a *real-life test set*.

2   Method

We focus on three tools commonly employed in orthopedic procedures - the scalpel, friem, and diskplacer - each of which requires a unique hand motion. The scalpel requires a side-to-side cutting motion, while the friem uses a quick downward punching motion, similar to using an awl. Finally, the diskplacer requires a screwing motion with the hand. Our pipeline to capture these activities and generate egocentric hand-object manipulation sequences is shown in Fig. 1.

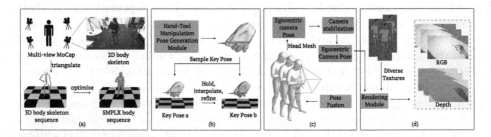

Fig. 1. The proposed pipeline to generate synthetic data sequences. (a) shows the multi-stereo-cameras-based body motion capture module. (b) indicates the optimization-based hand-object manipulation sequence generation pipeline. (c) presents the fused hand-body pose and the egocentric camera pose calculation module. (d) highlights the rendering module with which RGB-D sequences are rendered with diverse textures.

2.1   Multi-view Body Motion Capture

To capture body movements during surgery, we used four temporally synchronized ZED stereo cameras on participants during simulated surgeries. The intrinsic camera parameters were provided by ZED SDK and the extrinsic parameters between the four different cameras were calibrated with a chessboard. We adopt the popular [5] [6] module for SMPLX body reconstruction. OpenPose [2] with hand - and face-detection modules is first used to detect 2D human skeletons with a confidence threshold of 0.3. The 3D keypoints are obtained via triangulation with camera pose, regularized with bone length. The SMPLX body meshed is optimized by minimizing the 2D re-projection and triangulated 3D skeleton errors. Moreover, we enforce a large smoothness constraint, which regularizes the body and hand pose by constraining the between-frame velocities. It vastly reduces the number of unrealistic body poses.

2.2   Hand-Object Manipulation Sequence Generation

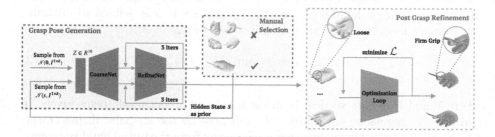

Fig. 2. The Hand manipulation sequence generation pipeline consists of three components: grasp pose generation, pose selection, and pose refinement, highlighted in blue, red, and green, respectively. (Color figure online)

There are two critical differences between the surgical tool and everyday object grasping: surgical tools require to be held in specific poses. Moreover, a surgeon would hold firmly and steadily with a particular pose for some time span during surgeries.

To address this issue, we generate each instrument manipulation sequence by firstly modeling the key poses that are surgically plausible, and then interpolating in between to model pose evolution. The key pose generation pipeline is shown in Fig. 2. The part highlighted in blue is the pose generation component based on GrabNet [25]. We provide the 3D instrument models to GrabNet with arbitrary initial rotation and sample from a Gaussian distribution in latent space to obtain diverse poses. 500 samples are generated for scalpel, diskplacer, and friem, respectively, followed by manual selection to get the best grasping poses as templates. With a pose template as prior, we perform the re-sampling near it to obtain diverse and similar hand-grasping poses as key poses for each sequence. To improve the plausibility of the grasping pose and hand-object interactions, inspired by [16], an optimization module is adopted for post-processing, with the overall loss function defined as:

$$\mathcal{L} = \alpha \cdot L_{penetr} + \beta \cdot L_{contact} + \gamma \cdot L_{keypoint}, \qquad (1)$$

L_{penetr}, $L_{contact}$, and $L_{keypoint}$ denote the penetration loss, contact loss, and keypoint loss, respectively. And α, β, γ are object-specific scaling factors to balance the loss components. For example, the weight for penetration is smaller for the scalpel than the friem and diskplacer to account for the smaller object size. The penetration loss is defined as the overall penetration distance of the object into the hand mesh:

$$L_{penetr} = \frac{1}{|\mathcal{P}_{in}^o|} \sum_{p \in \mathcal{P}_{in}^o} \min_i \|p - \mathcal{V}_i\|_2^2, \tag{2}$$

where $\mathcal{P}_{in}$ denotes the vertices from the object mesh which are inside the hand mesh, and $\mathcal{V}_i$ denotes the hand vertex. The $\mathcal{P}_{in}^o$ is defined as the dot product of the vector from the hand mesh vertices to their nearest neighbors on the object mesh. To encourage hand-object contact, a contact loss is defined to minimize the distance from the hand mesh to the object mesh.

$$L_{contact} = \sum_j \min_i \|\mathcal{V}_i - \mathcal{P}_j\|_2^2, \tag{3}$$

where $\mathcal{V}$ and $\mathcal{P}$ denote vertices from the hand and object mesh, respectively. In addition, we regularize the optimized hand pose by the keypoint displacement, which penalizes hand keypoints that are far away from the initial hand keypoints:

$$L_{keypoint} = \sum_i \|K_i - k_i\|^2, \tag{4}$$

where K is the refined hand keypoint position and k is the source keypoint position. After the grasping pose refinement, a small portion of the generated hand poses are still unrealistic due to the poor initialization. To this end, a post-selection technique similar to [13, 26] is further applied to discard the unrealistic samples with hand-centric interpenetration volume, contact region, and displacement simulation.

For each hand-object manipulation sequence, we select 30 key grasping poses, hold on, and interpolate in between to model pose evolution within the sequence. The number of frames for the transition phase between every two key poses is randomly sampled from 5 to 30. The interpolated hand poses are also optimized via the pose refinement module with the source keypoint in $L_{keypoint}$ defined as the interpolated keypoints between two key poses.

2.3 Body and Hand Pose Fusion and Camera Pose Calculation

In previous sections, we individually obtained the body motion and hand-object manipulation sequences. To merge the hand pose into the body pose to create a whole-body grasping sequence, we established an optimization-based approach based on the SMPLX model. The vertices to vertices loss is defined as:

$$L_{V2V} = \sum_{\hat{v}_i \in P_{hand}} \|v_{M(i)} - (R\hat{v}_i + T)\|_2^2, \tag{5}$$

where $\hat{v}$ is the vertices in the target grasping hand, v is the vertices in the SMPLX body model, with M being the vertices map from MANO's right hand to SMPLX body. R and T are the rotation matrix and translation vector applied to the right hand. The right-hand pose of SMPLX, R, and T are optimized with the Trust Region Newton Conjugate Gradient method (Trust-NCG) for 300 iterations to obtain an accurate and stable whole-body grasping pose. R and T are then applied to the grasped object. The egocentric camera pose for each frame is calculated with head vertices position and head orientation. Afterwards, outlier removal and moving average filter are applied to the camera pose sequence to remove temporal jitterings between frames.

2.4   Rendering and POV-Surgery Dataset Statistics

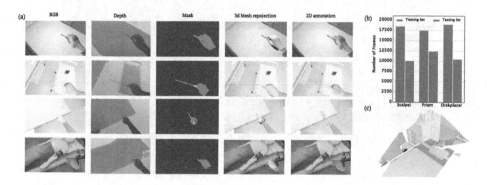

Fig. 3. (a) Dataset samples for RGB-D sequences and annotation. An example of the scalpel, friem, and diskplacer, is shown in the first three rows. The fourth row shows an example of the new scene and blood glove patterns that only appear in the test set. (b) shows the statistics on the number of frames for each surgical instrument in the training and testing sets. (c) shows a point cloud created from an RGB-D frame with simulated Kinect noise.

We use blender [3] and bpycv packages to render the RGB-D sequences and instance segmentation masks. Diverse textures and scenes of high quality are provided in the dataset: it includes 24 SMPLX textures featuring blue, green, and white surgical gloves textures with various blood patterns and a synthetic surgical room scene created by artists. The Cycle rendering engine and de-noising post-processing are adopted to produce high-quality frames. POV-Surgery provides clean depth maps for depth-based methods or point-cloud-based methods, as the artifact of real depth cameras can be efficiently simulated via previous works as [12]. A point cloud example generated from an RGB-D frame with added simulated Kinect depth camera noise is provided in Fig. 3. The POV-Surgery dataset consists of 36 sequences with 55,078 frames in the training set and 17 sequences with 33,161 frames in the testing set, respectively. Three bloodied glove textures and one scene created from a room scanning of a surgery room are only used in the testing set to measure generalizability. Fig. 3 shows the ground truth data samples and the dataset statistics.

3 Experiment

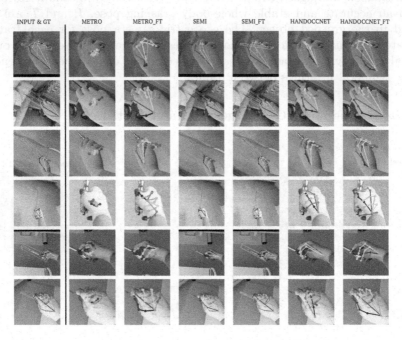

Fig. 4. Qualitative results of METRO [18], SEMI [19], and HANDOCCNET [22] on the test of of POV-Surgery. The FT denotes fine-tuning. We show the 2D re-projection of the predicted 3D hand joints and object control bounding box overlayed on the input image.

We evaluate and fine-tune two state-of-the-art hand pose estimation methods: [18,22], and one hand-object pose estimation [19] method on our dataset with provided checkpoints in their official repositories. 6 out of 36 sequences from the training set are selected as the validation set for model selection. We continue to train their checkpoints on our synthetic training set, with a reduced learning rate (10^{-5}) and various data augmentation methods such as color jittering, scale and center jittering, hue-saturation-contrast value jittering, and motion blur for better generalizability. Afterwards, we evaluate the performance of those methods on our testing set. We set a baseline for object control point error in pixels: 41.56 from fine-tuning [19]. The hand quantitative metrics are shown in Table 1 and qualitative visualizations are shown in Fig. 4, where we highlight the significant performance improvement for existing methods after fine-tuning them on the POV-Surgery dataset.

To further evaluate the generalizability of the methods fine-tuned on our dataset, we collect 6,557 real-life images with multiple surgical gloves, tools, and backgrounds as the *real-life test set*. The data capture setup with four stereo

Table 1. The evaluation result of different methods on the test set of POV Surgery, where the $_{ft}$ denotes fine-tuned on the training set. P_{2d} denotes the 2D hand joint reprojection error (in pixels). MPJPE and PVE denote the 3D Mean Per Joint Position Error and Per Vertex Error, respectively. PA denotes procrustes alignment.

Method	P_{2d} ↓	MPJPE ↓	PVE ↓	PA-MPJPE ↓	PA-PVE ↓
METRO [18]	95.11	77.46	75.06	23.43	22.34
SEMI [19]	77.91	115.67	112.10	12.68	12.76
HandOCCNet [22]	64.70	95.19	90.83	11.71	11.13
METRO$_{ft}$	30.49	14.90	13.80	6.36	4.34
SEMI$_{ft}$	**13.42**	15.14	14.69	**4.29**	**4.23**
HandOCCNet$_{ft}$	13.80	**14.35**	**13.73**	4.49	4.35

cameras is shown in Fig. 5. We adopt a top-down-based method from [4] with manually selected hand bounding boxes for 2D hand joint detection. [5] is used to reconstruct 3D hand poses from different camera observations. We project the hand pose to the egocentric camera view and manually select the frames with accurate hand predictions to obtain reliable 2D hand pose ground truth. We show quantitative examples of the indicated methods and the PCP curve in Fig. 5. After fine-tuning on our synthetic dataset significant performance improvements are achieved for SOTA methods on the *real-life test set*. Particularly, we observe a similar performance improvement for unseen purple-texture gloves, showing the effectiveness of our POV-Surgery dataset towards the challenging egocentric surgical hand-object interaction scenarios in general.

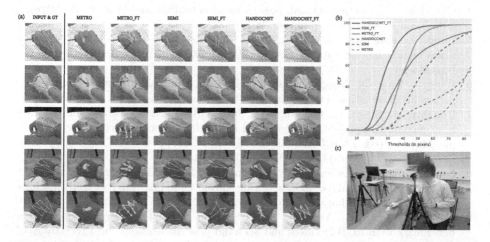

Fig. 5. (a) Ground truth and qualitative results of different methods on the *real-life test set*. (b) Accuracy with different 2D pixel error thresholds, showing large performance improvement after fine-tuning on POV-Surgery (c) Our multi-camera real-life data capturing set-up.

4   Conclusion

This paper proposes a novel synthetic data generation pipeline that generates hand-tool manipulation temporal sequences. Using the data generation pipeline and focusing on three tools used in orthopedic surgeries: scalpel, diskplacer, and friem, we propose a large, synthetic, and temporal dataset on egocentric surgical hand-object pose estimation, with 88,329 RGB-D frames and diverse bloody surgical gloves patterns. We evaluate and fine-tune three current state-of-the-art methods on the POV-Surgery dataset. We prove the effectiveness of the synthetic dataset by showing the significant performance improvement of the SOTA methods in real-life cases with surgical gloves and tools.

Acknowledgement. This work is part of a research project that has been financially supported by Accenture LLP. Siwei Zhang is funded by Microsoft Mixed Reality & AI Zurich Lab PhD scholarship. The authors would like to thank PD Dr. Michaela Kolbe for providing the simulation facilities and the students participating in motion capture.

References

1. Azimi, E., et al.: An interactive mixed reality platform for bedside surgical procedures. In: Martel, A.L., et al. (eds.) MICCAI 2020. LNCS, vol. 12263, pp. 65–75. Springer, Cham (2020). https://doi.org/10.1007/978-3-030-59716-0_7
2. Cao, Z., Hidalgo Martinez, G., Simon, T., Wei, S., Sheikh, Y.A.: OpenPose: real-time multi-person 2D pose estimation using part affinity fields. IEEE Trans. Pattern Anal. Mach. Intell. (2019)
3. Community, B.O.: Blender - a 3D modelling and rendering package. Blender Foundation, Stichting Blender Foundation, Amsterdam (2018). http://www.blender.org
4. Contributors, M.: OpenMMLab pose estimation toolbox and benchmark. https://github.com/open-mmlab/mmpose (2020)
5. Dong, J., Fang, Q., Jiang, W., Yang, Y., Bao, H., Zhou, X.: EasyMocap - make human motion capture easier. Github (2021). https://github.com/zju3dv/EasyMocap
6. Dong, J., Fang, Q., Jiang, W., Yang, Y., Bao, H., Zhou, X.: Fast and robust multi-person 3D pose estimation and tracking from multiple views. In: T-PAMI (2021)
7. Doughty, M., Singh, K., Ghugre, N.R.: SurgeonAssist-Net: towards context-aware head-mounted display-based augmented reality for surgical guidance. In: de Bruijne, M., et al. (eds.) MICCAI 2021. LNCS, vol. 12904, pp. 667–677. Springer, Cham (2021). https://doi.org/10.1007/978-3-030-87202-1_64
8. Fattahi Sani, M., Ascione, R., Dogramadzi, S.: Mapping surgeons hand/finger movements to surgical tool motion during conventional microsurgery using machine learning. J. Med. Robot. Res. **6**(03n04), 2150004 (2021)
9. Garcia-Hernando, G., Yuan, S., Baek, S., Kim, T.K.: First-person hand action benchmark with RGB-D videos and 3D hand pose annotations. In: Proceedings of the IEEE Conference on Computer Vision and Pattern Recognition, pp. 409–419 (2018)
10. Goodman, E.D., et al.: A real-time spatiotemporal AI model analyzes skill in open surgical videos. arXiv preprint arXiv:2112.07219 (2021)

11. Hampali, S., Rad, M., Oberweger, M., Lepetit, V.: HOnnotate: a method for 3D annotation of hand and object poses. In: CVPR (2020)
12. Handa, A., Whelan, T., McDonald, J., Davison, A.J.: A benchmark for RGB-D visual odometry, 3D reconstruction and slam. ICRA (2014)
13. Hasson, Y., et al.: Learning joint reconstruction of hands and manipulated objects. In: Proceedings of the IEEE/CVF Conference on Computer Vision and Pattern Recognition, pp. 11807–11816 (2019)
14. Hein, J., et al.: Towards markerless surgical tool and hand pose estimation. Int. J. Comput. Assist. Radiol. Surgery **16**(5), 799–808 (2021). https://doi.org/10.1007/s11548-021-02369-2
15. Jian, Z., Yue, W., Wu, Q., Li, W., Wang, Z., Lam, V.: Multitask learning for video-based surgical skill assessment. In: 2020 Digital Image Computing: Techniques and Applications (DICTA), pp. 1–8. IEEE (2020)
16. Jiang, H., Liu, S., Wang, J., Wang, X.: Hand-object contact consistency reasoning for human grasps generation. In: Proceedings of the International Conference on Computer Vision (2021)
17. Kwon, T., Tekin, B., Stühmer, J., Bogo, F., Pollefeys, M.: H2O: two hands manipulating objects for first person interaction recognition. In: Proceedings of the IEEE/CVF International Conference on Computer Vision, pp. 10138–10148 (2021)
18. Lin, K., Wang, L., Liu, Z.: End-to-end human pose and mesh reconstruction with transformers. In: CVPR (2021)
19. Liu, S., Jiang, H., Xu, J., Liu, S., Wang, X.: Semi-supervised 3D hand-object poses estimation with interactions in time. In: Proceedings of the IEEE/CVF Conference on Computer Vision and Pattern Recognition, pp. 14687–14697 (2021)
20. Moon, G., Yu, S.I., Wen, H., Shiratori, T., Lee, K.M.: InterHand2. 6M: a dataset and baseline for 3D interacting hand pose estimation from a single RGB image. In: Computer Vision-ECCV 2020: 16th European Conference, Glasgow, UK, August 23–28, 2020, Proceedings, Part XX 16. pp. 548–564. Springer (2020). https://doi.org/10.1007/978-3-030-58565-5_33
21. Palumbo, M.C., et al.: Mixed reality and deep learning for external ventricular drainage placement: a fast and automatic workflow for emergency treatments. In: Medical Image Computing and Computer Assisted Intervention-MICCAI 2022: 25th International Conference, Singapore, September 18–22, 2022, Proceedings, Part VII, pp. 147–156. Springer (2022). https://doi.org/10.1007/978-3-031-16449-1_15
22. Park, J., Oh, Y., Moon, G., Choi, H., Lee, K.M.: HandOccNet: occlusion-robust 3D hand mesh estimation network. In: Conference on Computer Vision and Pattern Recognition (CVPR) (2022)
23. Saggio, G., et al.: Objective surgical skill assessment: an initial experience by means of a sensory glove paving the way to open surgery simulation? J. Surg. Educ. **72**(5), 910–917 (2015)
24. Sener, F., et al.: Assembly101: a large-scale multi-view video dataset for understanding procedural activities. In: Proceedings of the IEEE/CVF Conference on Computer Vision and Pattern Recognition, pp. 21096–21106 (2022)
25. Taheri, O., Ghorbani, N., Black, M.J., Tzionas, D.: GRAB: a dataset of whole-body human grasping of objects. In: European Conference on Computer Vision (ECCV) (2020). https://grab.is.tue.mpg.de
26. Tzionas, D., Ballan, L., Srikantha, A., Aponte, P., Pollefeys, M., Gall, J.: Capturing hands in action using discriminative salient points and physics simulation. Int. J. Comput. Vis. **118**(2), 172–193 (2016)

27. Wesierski, D., Jezierska, A.: Instrument detection and pose estimation with rigid part mixtures model in video-assisted surgeries. Med. Image Anal. **46**, 244–265 (2018)
28. Wolf, J., Luchmann, D., Lohmeyer, Q., Farshad, M., Fürnstahl, P., Meboldt, M.: How different augmented reality visualizations for drilling affect trajectory deviation, visual attention, and user experience. Int. J. Comput. Assist. Radiol. Surgery, 1–9 (2023)
29. Zimmermann, C., Ceylan, D., Yang, J., Russell, B., Argus, M., Brox, T.: FreiHAND: a dataset for markerless capture of hand pose and shape from single RGB images. In: Proceedings of the IEEE/CVF International Conference on Computer Vision, pp. 813–822 (2019)

Surgical Activity Triplet Recognition via Triplet Disentanglement

Yiliang Chen[1], Shengfeng He[2], Yueming Jin[3], and Jing Qin[1(✉)]

[1] Centre for Smart Health, School of Nursing, The Hong Kong Polytechnic University, Hung Hom, Hong Kong
yiliang.chen@connect.polyu.hk, harry.qin@polyu.edu.hk
[2] Singapore Management University, Singapore, Singapore
[3] National University of Singapore, Singapore, Singapore

Abstract. Including context-aware decision support in the operating room has the potential to improve surgical safety and efficiency by utilizing real-time feedback obtained from surgical workflow analysis. In this task, recognizing each surgical activity in the endoscopic video as a triplet *<instrument, verb, target>* is crucial, as it helps to ensure actions occur only after an instrument is present. However, recognizing the states of these three components in one shot poses extra learning ambiguities, as the triplet supervision is highly imbalanced (positive when all components are correct). To remedy this issue, we introduce a triplet disentanglement framework for surgical action triplet recognition, which decomposes the learning objectives to reduce learning difficulties. Particularly, our network decomposes the recognition of triplet into five complementary and simplified sub-networks. While the first sub-network converts the detection into a numerical supplementary task predicting the existence/number of three components only, the second focuses on the association between them, and the other three predict the components individually. In this way, triplet recognition is decoupled in a progressive, easy-to-difficult manner. In addition, we propose a hierarchical training schedule as a way to decompose the difficulty of the task further. Our model first creates several bridges and then progressively identifies the final key task step by step, rather than explicitly identifying surgical activity. Our proposed method has been demonstrated to surpass current state-of-the-art approaches on the CholecT45 endoscopic video dataset.

Keywords: triplet disentanglement · surgical activity recognition · endoscopic videos

1 Introduction

Surgical video activity recognition has become increasingly crucial in surgical data science with the rapid advancement of technology [1–3]. This important task provides comprehensive information for surgical workflow analysis [4–7] and surgical scene understanding [8–10], which supports the implementation of safety

H. Greenspan et al. (Eds.): MICCAI 2023, LNCS 14228, pp. 451–461, 2023.
https://doi.org/10.1007/978-3-031-43996-4_43

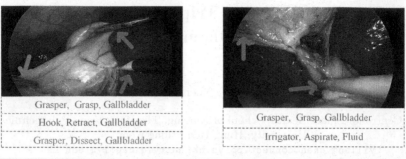

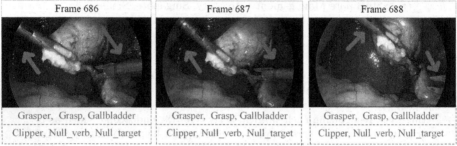

Fig. 1. First Line: Left: Multiple surgical triplets appearing at the same time. **Right:** Instruments located near the boundary. **Second Line:** In consecutive frames of an endoscopic surgery video, an unrelated action (green arrow) is labeled as null. (Color figure online)

warning and computer-assisted systems in the operating room [11]. One of the most popular surgical procedures worldwide is laparoscopic cholecystectomy [12, 13], which is in high demand for the creation of an effective computer-assisted system. Therefore, automated surgical activity triplet recognition is increasingly essential, and learning-based methods are promising solutions to address this need.

Most current works in the field of surgical video analysis primarily focus on surgical phase recognition [6,14–18]. However, only a small portion of the literature is dedicated to surgical activity recognition. The first relevant study [19] dates back to 2020, in which the authors built a relevant dataset and proposed a weakly supervised detection method that uses a 3D interacting space to identify surgical triplets in an end-to-end manner. An updated version Rendezvous (RDV) [20] employs a Transformer [21] inspired semantic attention module in their end-to-end network. Later this method is extended to include the temporal domain, named Rendezvous in Time (RiT) [22] with a Temporal Attention Module (TAM) to better integrates the current and past features of the verb at the frame level. Significantly, benchmark competitions such as the Cholectriplet2021 Challenge [28] have garnered interest in surgical action triplet recognition, with its evaluation centering on 94 valid triplet classes while excluding 6 null classes.

Although significant progress has been made in surgical activity triplet recognition, they suffer from the same limitation of ambiguous supervision from the triplet components. Learning to predict the triplet in one shot is a highly imbalanced process, as only samples with correct predictions across all components are considered positive. This objective is particularly challenging in the following scenarios. Firstly, multiple surgical activities may occur in a single frame (see Fig. 1), with instruments appearing at the edge of the video or being obscured or overlapped, making it difficult to focus on them. Secondly, similar instruments, verbs, and targets that do not need to be recognized can have a detrimental impact on the task. As illustrated in the second row of Fig. 1, an obvious movement of the clipper may be labeled as null in the dataset because it is irrelevant to the recognition task. However, this occurrence is frequent in real situations, and labeling all surgical activities is time-consuming. Furthermore, not all surgical activities are indispensable, and some only appear in rare cases or among surgeons' habits. Moreover, the labels of instruments and triplets for this task are binary, and when multiple duplicate instruments or triplets appear, recognizing them directly only determines whether their categories have appeared or not. To improve recognition results, these factors should also be considered. Lastly, most of the previous methods are end-to-end multi-task learning methods, which means that training may be distracted by other auxiliary tasks and not solely focused on the key task.

To solve the above problems, we propose a triplet disentanglement framework for surgical activity triplet recognition. This approach decomposes the learning objectives, thereby reducing the complexity of the learning process. As stated earlier, surgical activity relies on the presence of tools, making them crucial to our mission. Therefore, our approach concentrates on a simplified numerical representation as a means of mitigating these challenges. Initially, we face challenges such as multiple tools appearing at the same time or irrelevant surgical activities. Therefore, we adopt an intuitive approach to first identify the number/category of tools and whether those activities occur or not, instead of directly recognizing the tool itself. This numerical recognition task helps our network roughly localize the tool's location and differentiate irrelevant surgical activities. Subsequently, we employ a weakly supervised method to detect the tools' locations. However, unlike [20], we extend our architecture to 3D networks to better capture temporal information. Our approach separates different types of instruments using the class activation map (CAM) [23] based on the maximum number of identified instrument categories, allowing our model to slightly minimize the probability of mismatching and reduce the learning difficulty after separation when multiple surgical activities occur simultaneously. Additionally, we propose a hierarchical training schedule that decomposes our tasks into several sub-tasks, starting from easy to hard. This approach improves the efficiency of each individual task and makes training easier.

In summary, this work makes the following contributions: 1) We propose a triplet disentanglement framework for surgical action triplet recognition, which decomposes the learning objectives in endoscopic videos. 2) By further exploit-

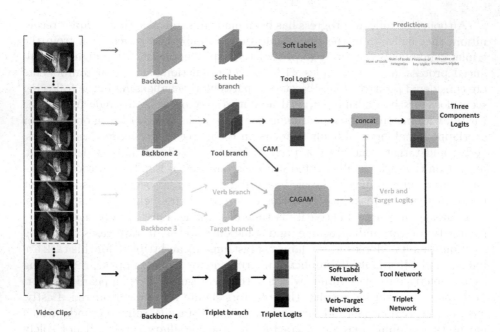

Fig. 2. Overview of Our triplet Disentanglement Network. The parameters of the soft label network are used for the initialization of the other three networks.

ing the knowledge of decomposition, our network is extended to a 3D network to better capture temporal information and make use of temporal class activation maps to alleviate the challenge of multiple surgical activities occurring simultaneously. 3) Our experimental results on the endoscopic video dataset demonstrate that our approaches surpass current state-of-the-art methods.

2   Methods

Our objective is to recognize every triplet at each frame in each video. Let $X = \{X_1, ..., X_n\}$ be the frames of endoscopic videos and $Y = \{Y_1, ..., Y_n\}$ be the set of labels of triplet classes where n is the number of frames and the sets of triplet classes. Moreover, each set of labels of triplet classes can be denoted as $Y = \{Y^I, Y^V, Y^T\}$ where I, V, and T are indicated as Instrument, Verb, and Target. Figure 2 shows the overview of our method.

2.1   Sampling Video Clips

Unlike RDV [20] method which identifies surgical activities in individual frames, the video is cut into different segments to identify them. At the beginning of our process, short video clips are obtained from lengthy and differently-sized videos. Each sample clip drawn from the source videos is represented as $X_s \in {}^{H \times W \times 3 \times M}$, while X_t represents a sample clip from the target. It should be noted that all sample clips have identical specifications, namely a fixed height H, a fixed width W, and a fixed number of frames M.

2.2   Generating Soft Labels

We generate four different soft labels for the number of instruments, the number of instrument categories, the presence of an unrelated surgical activity, and the presence of a critical surgical activity, which are labeled as $<N_I, N_C, U_A, S_A>$ respectively. As for N_I, our goal is to know the number of instruments occurrences because the appearance of an instrument also means the appearance of a surgical activity. For N_C, it is employed to differentiate situations where multiple instruments of the same type are present. In terms of the other two soft labels, they are binary labels, which refer to presence or absence, respectively. In addition, labels with the format *<instrument, null, null>* are marked as irrelevant surgical activity. In contrast, those with the format *<instrument, verb, target>* are marked as critical surgical activity. For example, in the case shown in the second line of Fig. 1, our soft label will be marked as *<2, 2, presence, presence>*.

2.3   Disentanglement Framework

Our encoder backbone is built on the RGB-I3D [26] network (preserving the temporal dimension), which is pre-trained on both ImageNet [25] and Kinetics dataset [26]. Prior to training, each video is segmented into T non-overlapping segments consisting of precisely 5 frames, which are then fed into the soft label network to recognize their corresponding soft labels. This allows our network to more easily identify the location of the tool and to differentiate between crucial and irrelevant actions, which ultimately aids the network's comprehension of triplet associations. Once the network is trained, its parameters are stored and transferred for initialization to the backbones of other networks. In the second stage, we divide our triplet recognition task into three sub-networks: the tool network, verb network, and target branch network. Notably, our backbone operates at the clip level, whereas RDV is frame-based. The features extracted by the backbone from video clips are fed into the three sub-networks. The tool classifier recognizes the tool class and generates its corresponding Class Activation Map (CAM). The features of the verb and target networks, along with the CAMs of the corresponding tool network, are then passed into the Class Activation Guided Attention Mechanism (CAGAM) module [20]. Our CAGAM, a dual-path position attention mechanism, leverages the tool's saliency map to guide the location of the corresponding verb and target. Subsequently, the individual predictions of the three components are generated, and the last objective is to learn their association. Therefore, the CAMs and logits of the three components are aligned into the triplet network to learn their association and generate the final triplet prediction. It is important to note that the tool network, verb-target networks, and triplet network are all initialized by the soft label network and contain branches to predict soft labels.

2.4   Hierarchical Training Schedule

Our framework is a multi-task recognition approach, but training so many tasks simultaneously poses a huge challenge to balancing hyper-parameters. To address this, we propose a hierarchical training schedule method that divides training into different stages. Initially, we train only our soft label network to recognize soft labels, storing its parameters once training is complete. In the next stage, video clips are fed into the tool network to recognize tool categories while simultaneously identifying soft labels. After successful training, the parameters of the tool network are frozen. In the subsequent stage, the verb and target networks identify their respective components and soft labels. At this point, the tool network passes its class activation map to the verb-target networks without updating its parameters. Besides, following previous Tripnet [20], which masks out impossible results, we also mask out improbable outcomes for different components. For instance, predefined masks for the tool are based on possible combinations, while the verb and target masks follow the tool's predictions, excluding scissors and clippers; subsequently, the masking results undergo further refinement for the instrument. Finally, we train the final triplet network using the CAMs and output logits of the three components. Similarly, the parameters of the three components' networks are not updated at this stage. This approach allows us to break down the complexity of the task and improve the accuracy of each individual component at each stage, ultimately leading to higher overall accuracy.

2.5   Separation Processing

Our framework provides a unique approach to address the impact of multiple tool categories that may be present simultaneously. It enables the handling of different surgical instrument categories individually, which reduces the level of complexity for learning. In contrast to RDV [20] and RiT [22], we extend the Grad-CAM [27] approach to 3D-CNN by utilizing the input of 3D tensor data instead of a 2D matrix, resulting in a more precise class activation map. Based on the maximum number of instrument categories (K) predicted by the soft label and the scores obtained by summing each channel along the CAM, we isolate the top-K CAMs. Each isolated CAM is then used to guide the corresponding features of verbs and targets in our CAGAM module to generate individual predictions of the verb and target. Finally, the different predictions of the verb and target are combined. However, differentiating between various instruments, as we do, can help match them with the correct verbs and targets, especially when multiple tool categories appear at the same time, which may be challenging to do without this approach, such as in the RDV method [20].

2.6   Loss Function

For the number of tools and the categories of tools, softmax cross-entropy loss is adopted, while for other soft labels, three components, and triplet, we employ sigmoid cross-entropy losses. Taking sigmoid cross-entropy loss as an example:

$$L = \sum_{c=1}^{C} \frac{-1}{N} (y_c \log_c(\sigma(\hat{y}_c)) + (1 - y_c) \log(1 - \sigma(\hat{y}_c)),$$

where σ is the sigmoid function, while y_c and $\dot{y}_c$ are the ground truth label and the prediction for specific class c. Besides, the balanced weights are also adopted based on previous works [20].

3   Experiments

Dataset. Following previous works [20], an endoscopic video dataset of the laparoscopic cholecystectomy, called CholecT45 [20], is used for all the experiments. The database consists of a total of 45 videos and is stripped of triplet classes of little clinical relevance. There are 6 instruments, 10 verbs, 15 targets, and 100 triplet classes in the CholecT45 dataset. For all the videos, each triplet class is always presented in the format of <*instrument, verb, target*>. In addition, following the data splits in [24], the K-fold cross-validation method was used to divide the 45 videos in the dataset into 5 folds, each containing 9 videos.

Metric. The performance of the method is evaluated based on the mean average precision (mAP) metric to predict the triplet classes. In testing, each triplet class is computed its own average precision (AP) score, and then the AP score of a video will be calculated by averaging the AP scores of all the triplets in this video. Finally, the mean average precision (mAP) of the dataset is measured by averaging the AP scores of all tested videos.

Besides, Top-N recognition performance is also adopted in our evaluation, which means that given a test sample X_i, a model made a correctness if the correct label y_i appears in its top N confident predictions $\hat{Y}_i$. We follow previous works [20] and measure top-5, top-10, and top-20 accuracy in our experiment.

Implementation Details. I3D (Resnet50) [26] is adopted as a backbone network in our framework, which is pre-trained on the ImageNet [25] and the Kinetics dataset [26]. As for the branches of the different networks, they will be slightly modified to fit their corresponding subtasks. We use SGD as an optimizer and apply a step-wise learning rate of $1e-3$ for all sub-tasks, but for the soft-labeled task branches that need to be finetuned, their learning rates start from $1e-6$. Our batch size is 32, and there is no additional database for surgical detection.

3.1   Results and Discussion

Quantitative Evaluation. We demonstrate our experimental results with other state-of-the-art approaches on the CholecT45 [20] dataset.

Table 1 presents the benchmark results on the CholecT45 cross-validation split, and compares our model with current state-of-the-art methods [19,20,22] using the mean Average Precision (mAP) metric. The key mission of surgical activity recognition is to calculate the average precision (AP) score of the triplet, denoted as AP_{IVT}. We present tool recognition, verb recognition, and target recognition as AP_I, AP_V, and AP_T, respectively. Compared to previous methods [19,20], our hierarchical training schedule method enables us to achieve

Table 1. Quantitative Results of the Proposed Method Compared to State-of-the-art.

Method	AP_I	AP_V	AP_T	AP_{IVT}
Tripnet [19]	89.9 ± 1.0	59.9 ± 0.9	37.4 ± 1.5	24.4 ± 4.7
Attention Tripnet [20]	89.1 ± 2.1	61.2 ± 0.6	40.3 ± 1.2	27.2 ± 2.7
RDV [20]	89.3 ± 2.1	62.0 ± 1.3	40.0 ± 1.4	29.4 ± 2.8
RiT [22]	88.6 ± 2.6	64.0 ± 2.5	43.4 ± 1.4	29.7 ± 2.6
Our method	**91.2 ± 1.9**	**65.3 ± 2.8**	**43.7 ± 1.6**	**33.8 ± 2.5**

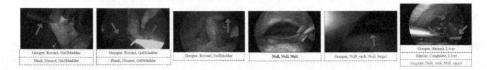

Fig. 3. Qualitative and representative results obtained from our triplet activity recognition model. Color-coded arrows are used to highlight multiple instruments in examples.

better results on individual sub-tasks because we can tune each task individually to achieve the best results. For AP_{IVT}, our framework improves the current SOTA method RiT [22] by 4.1%, demonstrating that decomposing our tasks helps to improve the final result. In addition, we include Fig. 3 in our paper to illustrate the qualitative and representative results obtained from our triplet activity recognition model. Although we are unable to compare our results with other methods as their trained models have not been released, our method can successfully predict many difficult situations, such as the simultaneous appearance of multiple surgical activities, the influence of brightness, and the tool located at the edge.

Following the previous experiments [20], we compare the Top N accuracy of the triplet predictions with different methods. However, the previous method [20] performed this metric only on the CholecT50 [20] dataset, and this dataset has not been published yet. Besides, the source codes of RiT [22] have not been released yet. However, the performance between RDV and RiT is very close according to Table 1. Hence, we try to reproduce the RDV method on this metric. As shown in Table 2, our framework outperforms the RDV method [20] by 8.1%, 5.9% and 2.0% in top-5, top-10, and top-20 respectively.

Ablation Study. In this section, we conduct ablation studies to showcase the effectiveness of each module in our model. As shown in Table 3, the final triplet prediction outcomes experienced a slight decrease of 0.4% and 1.1% in the absence of the separation process or the soft label module, respectively. Additionally, a marginal decrease was observed in the individual sub-tasks. In conclusion, these results demonstrate that the inclusion of these modules can contribute to the overall performance of our model.

Table 2. Top N Accuracy of the Triplet Predictions among Different Methods.

Method	Top-5	Top-10	Top-20
RDV [20]	73.6	84.7	93.2
Our method	**81.7**	**90.6**	**95.2**

Table 3. Comparison of the Different Modules in Our Framework. SC: Separating the category of the instrument in Sect. 2.6; SL: Predicting soft labels in Sect. 2.3.

Method	AP_I	AP_V	AP_T	AP_{IVT}
our method	91.2	65.3	43.7	33.8
our method w/o SC	91.2	64.7	43.1	33.4
our method w/o SL	90.6	63.6	42.8	32.7

4   Conclusion

In this paper, we introduce a novel triplet disentanglement framework for surgical activity recognition. By decomposing the task into smaller steps, our method demonstrates improved accuracy compared to existing approaches. We anticipate that our work will inspire further research in this area and promote the development of more efficient and accurate techniques.

Acknowledgments. The work described in this paper is partly supported by a grant of Hong Kong RGC Theme-based Research Scheme (project no. T45-401/22-N) and a grant of Hong Kong RGC General Research Fund (project no. 15218521).

References

1. Maier-Hein, L., et al.: Surgical data science: enabling next-generation surgery. ArXiv Preprint ArXiv:1701.06482 (2017)
2. Nowitzke, A., Wood, M., Cooney, K.: Improving accuracy and reducing errors in spinal surgery-a new technique for thoracolumbar-level localization using computer-assisted image guidance. Spine J. **8**, 597–604 (2008)
3. Yang, C., Zhao, Z., Hu, S.: Image-based laparoscopic tool detection and tracking using convolutional neural networks: a review of the literature. Comput. Assist. Surg. **25**, 15–28 (2020)
4. Zhang, Y., Bano, S., Page, A., Deprest, J., Stoyanov, D., Vasconcelos, F.: Retrieval of surgical phase transitions using reinforcement learning. In: Wang, L., Dou, Q., Fletcher, P.T., Speidel, S., Li, S. (eds.) MICCAI 2022, Part VII. LNCS, vol. 13437, pp. 497–506. Springer, Cham (2022). https://doi.org/10.1007/978-3-031-16449-1_47
5. Twinanda, A., Shehata, S., Mutter, D., Marescaux, J., De Mathelin, M., Padoy, N.: EndoNet: a deep architecture for recognition tasks on laparoscopic videos. IEEE Trans. Med. Imaging **36**, 86–97 (2016)

6. Zisimopoulos, O., et al.: DeepPhase: surgical phase recognition in CATARACTS videos. In: Frangi, A.F., Schnabel, J.A., Davatzikos, C., Alberola-López, C., Fichtinger, G. (eds.) MICCAI 2018, Part IV. LNCS, vol. 11073, pp. 265–272. Springer, Cham (2018). https://doi.org/10.1007/978-3-030-00937-3_31

7. Nakawala, H., Bianchi, R., Pescatori, L., De Cobelli, O., Ferrigno, G., De Momi, E.: "Deep-Onto" network for surgical workflow and context recognition. Int. J. Comput. Assist. Radiol. Surg. **14**, 685–696 (2019)

8. Valderrama, N., et al.: Towards holistic surgical scene understanding. In: Wang, L., Dou, Q., Fletcher, P.T., Speidel, S., Li, S. (eds.) MICCAI 2022, Part VII. LNCS, vol. 13437, pp. 442–452. Springer, Cham (2022). https://doi.org/10.1007/978-3-031-16449-1_42

9. Lin, W., et al.: Instrument-tissue interaction quintuple detection in surgery videos. In: Wang, L., Dou, Q., Fletcher, P.T., Speidel, S., Li, S. (eds.) MICCAI 2022, Part VII. LNCS, vol. 13437, pp. 399–409. Springer, Cham (2022). https://doi.org/10.1007/978-3-031-16449-1_38

10. Seidlitz, S., et al.: Robust deep learning-based semantic organ segmentation in hyperspectral images. Med. Image Anal. **80**, 102488 (2022)

11. Franke, S., Meixensberger, J., Neumuth, T.: Intervention time prediction from surgical low-level tasks. J. Biomed. Inform. **46**, 152–159 (2013)

12. Pucher, P., et al.: Outcome trends and safety measures after 30 years of laparoscopic cholecystectomy: a systematic review and pooled data analysis. Surg. Endosc. **32**, 2175–2183 (2018)

13. Alli, V., et al.: Nineteen-year trends in incidence and indications for laparoscopic cholecystectomy: the NY State experience. Surg. Endosc. **31**, 1651–1658 (2017)

14. Kassem, H., Alapatt, D., Mascagni, P., AI4SafeChole, C., Karargyris, A., Padoy, N.: Federated cycling (FedCy): semi-supervised federated learning of surgical phases. IEEE Trans. Med. Imaging (2022)

15. Ding, X., Li, X.: Exploring segment-level semantics for online phase recognition from surgical videos. IEEE Trans. Med. Imaging **41**, 3309–3319 (2022)

16. Czempiel, T., Paschali, M., Ostler, D., Kim, S.T., Busam, B., Navab, N.: OperA: attention-regularized transformers for surgical phase recognition. In: de Bruijne, M., et al. (eds.) MICCAI 2021, Part IV. LNCS, vol. 12904, pp. 604–614. Springer, Cham (2021). https://doi.org/10.1007/978-3-030-87202-1_58

17. Jin, Y., et al.: SV-RCNet: workflow recognition from surgical videos using recurrent convolutional network. IEEE Trans. Med. Imaging **37**, 1114–1126 (2017)

18. Sahu, M., Szengel, A., Mukhopadhyay, A., Zachow, S.: Surgical phase recognition by learning phase transitions. Curr. Direct. Biomed. Eng. **6** (2020)

19. Nwoye, C.I., et al.: Recognition of instrument-tissue interactions in endoscopic videos via action triplets. In: Martel, A.L., et al. (eds.) MICCAI 2020, Part III. LNCS, vol. 12263, pp. 364–374. Springer, Cham (2020). https://doi.org/10.1007/978-3-030-59716-0_35

20. Nwoye, C., et al.: Rendezvous: attention mechanisms for the recognition of surgical action triplets in endoscopic videos. Med. Image Anal. **78**, 102433 (2022)

21. Han, K., Xiao, A., Wu, E., Guo, J., Xu, C., Wang, Y.: Transformer in transformer. Adv. Neural. Inf. Process. Syst. **34**, 15908–15919 (2021)

22. Sharma, S., Nwoye, C., Mutter, D., Padoy, N.: Rendezvous in time: an attention-based temporal fusion approach for surgical triplet recognition. Int. J. Comput. Assist. Radiol. Surg. **18**, 1053–1059 (2023)

23. Zhou, B., Khosla, A., Lapedriza, A., Oliva, A., Torralba, A.: Learning deep features for discriminative localization. In: Proceedings of the IEEE Conference on Computer Vision and Pattern Recognition, pp. 2921–2929 (2016)

24. Nwoye, C., Padoy, N.: Data splits and metrics for method benchmarking on surgical action triplet datasets. ArXiv Preprint ArXiv:2204.05235 (2022)
25. He, K., Zhang, X., Ren, S., Sun, J.: Deep residual learning for image recognition. In: Proceedings of the IEEE Conference on Computer Vision and Pattern Recognition, pp. 770–778 (2016)
26. Carreira, J., Zisserman, A.: Quo vadis, action recognition? A new model and the kinetics dataset. In: Proceedings of the IEEE Conference on Computer Vision and Pattern Recognition, pp. 6299–6308 (2017)
27. Selvaraju, R., Cogswell, M., Das, A., Vedantam, R., Parikh, D., Batra, D.: Grad-CAM: visual explanations from deep networks via gradient-based localization. Proceedings of the IEEE International Conference on Computer Vision, pp. 618–626 (2017)
28. Nwoye, C., et al.: CholecTriplet 2021: a benchmark challenge for surgical action triplet recognition. Med. Image Anal. **86**, 102803 (2023)

Automatic Surgical Reconstruction for Orbital Blow-Out Fracture via Symmetric Prior Anatomical Knowledge-Guided Adversarial Generative Network

Jiangchang Xu[1] , Yining Wei[2], Huifang Zhou[2,4], Yinwei Li[2,4(✉)],
and Xiaojun Chen[1,3(✉)]

[1] Institute of Biomedical Manufacturing and Life Quality Engineering,
School of Mechanical Engineering, Shanghai Jiao Tong University, Shanghai, China
xiaojunchen@sjtu.edu.cn
[2] Department of Ophthalmology, Shanghai Ninth People's Hospital,
Shanghai Jiao Tong University School of Medicine, Shanghai, China
dr_yinwei_li@foxmail.com
[3] Institute of Medical Robotics, Shanghai Jiao Tong University, Shanghai, China
[4] Shanghai Key Laboratory of Orbital Diseases and Ocular Oncology,
Shanghai, China

Abstract. Orbital blow-out fracture (OBF) is a complex disease that can cause severe damage to the orbital wall. The ultimate means of treating this disease is orbital reconstruction surgery, where automatic reconstruction of the orbital wall is a crucial step. However, accurately reconstructing the orbital wall is a great challenge due to the collapse, damage, fracture, and deviation in OBF. Manual or semi-automatic reconstruction methods used in clinics also suffer from poor accuracy and low efficiency. Therefore, we propose a symmetric prior anatomical knowledge (SPAK)-guided generative adversarial network (GAN) for automatic reconstruction of the orbital wall in OBF. Above all, a spatial transformation-based SPAK generation method is proposed to generate prior anatomy that guides the reconstruction of the fractured orbital wall. Next, the generated SPAK is introduced into the GAN network, to guide the network towards automatic reconstruction of the fractured orbital wall. Additionally, a multi-function combination supervision strategy is proposed to further improve the network reconstruction performance. Our evaluation on the test set showed that the proposed network achieved a Dice similarity coefficient (DSC) of $92.35 \pm 2.13\%$ and a 95% Hausdorff distance of 0.59 ± 0.23 mm, which is significantly better than other networks. The proposed network is the first AI-based method to implement the automatic reconstruction of OBF, effectively

Supplementary Information The online version contains supplementary material available at https://doi.org/10.1007/978-3-031-43996-4_44.

improving the reconstruction accuracy and efficiency of the fractured orbital wall. In the future, it has a promising application prospect in the surgical planning of OBF.

Keywords: Automatic surgery planning · Orbital blow-out fracture · Surgical reconstruction · Symmetric prior anatomical knowledge

1   Introduction

Orbital fractures represent a frequent occurrence of orbital trauma, with their incidence on the rise primarily attributed to assault, falls, and vehicle collisions [1]. Orbital blow-out fracture (OBF) is a frequent type of fracture where one of the orbital walls fractures due to external force, while the orbital margin remains intact [2]. It is a complex disease that can result in the destruction or collapse of the orbital wall, orbital herniation, invagination of the eyeball, and even visual dysfunction or changes in appearance in severe cases [3]. OBF repair surgery is the ultimate treatment for this disease and involves implanting artificial implants to repair and fill the fractured area. Automatic reconstruction of the orbital wall is a crucial step in this procedure to achieve precise preformed implants and assisted intraoperative navigation.

Orbital wall reconstruction is challenging due to the complex and diverse OBF types, as shown in Fig. 1, including (a) medial wall fracture, (b) floor wall fracture, (c) fractures in both the medial and floor walls, (d) roof wall fracture, (e) and other types. The orbital walls in these cases are often collapsed, damaged, fractured, deviated, or exhibit a large number of defects in severe cases, which makes reconstruction more difficult. Furthermore, the orbital medial and floor walls are thin bones with low CT gradient values and blurred boundaries, which further increases the complexity of reconstruction. Currently, commercial software is typically used for semi-automatic reconstruction in clinical practice. However, this method is inefficient and inaccurate, requiring tedious manual adjustment and correction. As a result, doctors urgently need fast and accurate automated surgical reconstruction methods.

Several automatic segmentation methods for the orbital wall have been explored in previous studies [4], such as Kim et al.'s [5] orbital wall modeling based on paranasal sinus segmentation and Lee et al.'s [6] segmentation algorithm of the orbital cortical bone and thin wall with a double U-Net network structure. However, they did not explore segmentation for OBF. Taghizadeh et al. [7] proposed an orbital wall segmentation method based on a statistical shape model and local template matching, but individualized differences and factors such as orbital wall deviation and collapse greatly affect its performance in fractured orbits. Deep learning-based algorithms for skull defect reconstruction have also been proposed, such as Li et al.'s [8] automatic repair network for skull defects, Xiao et al.'s [9] network model that estimates bone shape using normal facial photos and CT of craniomaxillofacial trauma, and Han et al.'s [10] craniomaxillofacial defect reconstruction algorithm with a statistical shape model and individual features. However, these methods require the removal of the lesion area followed by reconstruction of the defect, which is different from OBF repair that directly performs orbital wall reconstruction without removal of the destroyed bone. The above methods may be less effective in cases of OBF where factors such as deviation, collapse, and fracture greatly impact the reconstruction network, reducing the reconstruction effect. As of now, there are no automated reconstruction methods reported for OBF surgery.

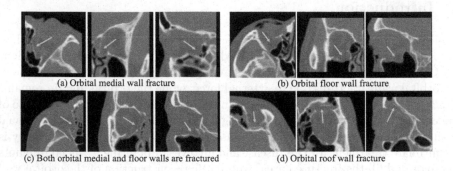

Fig. 1. Common types and reconstruction challenges of orbital blowout fractures.

To address the above challenges, this paper proposes a symmetric prior anatomical knowledge-guided adversarial generative network (GAN) for reconstructing orbital walls in OBF surgery. Firstly, the paper proposes an automatic generation method of symmetric prior anatomical knowledge (SPAK) based on spatial transformation, which considers that the symmetrical normal orbital anatomy can guide the reconstruction of the fractured orbital wall. Secondly, the obtained SPAK is used as a prior anatomical guidance for GAN to achieve more accurate automatic reconstruction of the orbital wall. To further improve the network's reconstruction performance, the paper adopts a supervision strategy of multi-loss function combination. Finally, experimental results demonstrate that the proposed network outperforms some state-of-the-art networks in fracture orbital wall reconstruction.

The main contributions of this paper are summarized as follows: (1) A GAN model guided by SPAK is developed for automatic reconstruction of the orbital wall in OBF surgery, which outperforms the existing methods. (2) The proposed method is the first AI-based automatic reconstruction method of the orbital wall in OBF surgery, which can enhance the repair effectiveness and shorten the surgical planning time.

2   Methods

The structure of the proposed SPAK-guided GAN is illustrated in Fig. 2, and it mainly comprises three components: automatic generation of SPAK, GAN network structure, and multiple loss function supervision. Each component is elaborated below.

2.1 Automatic Generation of SPAK Based on Spatial Transformation

The reconstruction of the orbital wall in OBF is a complex task, made more difficult when severe displacement or defects are present. However, since the left and right orbits are theoretically symmetrical structures, utilizing the normal orbital wall on the symmetrical side as prior anatomical guidance in the reconstruction network can aid in accuracy. Prior knowledge has been demonstrated to be effective in medical image computing [11–13]. Nonetheless, left and right orbits are not perfectly symmetrical, and the mirror plane can be challenging to locate, leading to substantial errors and unwieldy operations. To address these issues, we propose an automatic generation method for SPAK based on spatial transformations, as depicted in Fig. 2. The main steps of this method include: (1) obtaining the orbital ROI containing both orbits and resampling to ensure input image size consistency in the network; (2) using the sagittal plane as the central surface to separate the left and right orbits; (3) employing the 3D V-Net network to automatically segment the normal and fractured orbital walls, respectively, and acquire their respective segmentation results; (4) restoring the left and right orbital segmentation results to their original position to ensure spatial coordinate consistency; (5) mirror the orbital wall normally along the Y-axis plane, ensuring correspondence between the shape of the segmentation result from both the left and right orbital walls; (6) reconstructing the left and right orbital walls

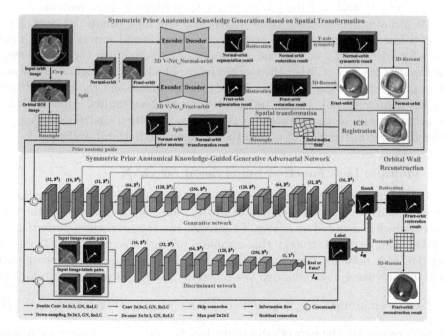

Fig. 2. Structure of the proposed symmetric prior anatomical knowledge-guided generative adversarial network for reconstructing the fractured orbital wall.

in 3D dimensions and obtaining their respective three-dimensional models; (7) using the ICP algorithm to register the normal and fractured orbital walls and acquire their deformation field; (8) based on the previous step, transforming the symmetrical normal orbital wall to the side of the fractured orbital wall using the deformation field; (9) separating the transformed normal orbital wall to obtain a single orbital wall that can serve as a SPAK to guide GAN.

2.2 SPAK-Guided GAN for Orbital Wall Reconstruction

Reconstructing the fractured orbital wall requires generative prediction of the damaged area, which is why we adopted the GAN. Our proposed SPAK-guided GAN consists of a generative network (GN) and a discriminative network (DN), as illustrated in Fig. 2. The GN uses a network structure based on 3D V-Net and consists of five encoded layers and four decoded layers to achieve automatic reconstruction of the fractured orbital wall. The input of the GN is the merged block of SPAK and the original image, which guides the GAN network to make more accurate predictions of the orbital wall. To expand the receptive field, each convolution layer uses two convolution stacks and includes residual connections to reduce gradient dissipation. The activation function is ReLU, and group normalization [14] is added after each convolution to prevent network overfitting. To avoid the loss of shallow features as the number of network layers increases, we added skip connections between the corresponding convolutional layers of the encoded and decoded sections. The DN identifies the authenticity of the reconstructed orbital wall and the ground truth to produce a more realistic orbital wall. It includes five encoded feature layers and one fully connected layer. Each encoded feature layer comprises stacked convolutions and a max-pooling layer with a filter kernel of $2 \times 2 \times 2$ and a step size of 2 to compress the feature map. The input of the DN is either the merged image block of the original image and the reconstructed orbital wall area or the original image and the ground truth. The DN distinguishes between the authenticity of these inputs. After restoring the output result of the network to the original position and resampling, we perform 3D reconstruction to obtain the reconstructed orbital wall model.

2.3 Multiple Loss Function Supervision for GAN

The GAN network's loss function comprises two components: the loss function of the discriminative network, denoted as $Loss_D$, and the loss function of the generative network, denoted as $Loss_G$. To enhance the reconstruction performance, we adopt a supervision strategy that combines multiple loss functions. To ensure that the GAN network can identify the authenticity of samples, we incorporate the commonly used discriminative loss function in GAN, as presented in Eq. (1).

$$Loss_D = \min_G \left[\max_D E_{x,y} \left[\log \left(D \left(x, y \right) \right) \right] + E_x \left[\log \left(1 - D \left(x, G \left(x \right) \right) \right) \right] \right], \quad (1)$$

where D represents the network discriminator, G represents the network generator, $G(x)$ represents reconstruction result, x represents input image, y represents ground truth.

To improve the accuracy of the reconstructed orbital wall, a combination of multiple loss functions is utilized in $Loss_G$. First, to better evaluate the area loss, the dice coefficient loss function $Loss_{dice}$ is adopted, which can calculate the loss between the reconstructed orbital wall region and the ground truth using Eq. (2). Secondly, due to the potential occurrence of boundary fractures and orbital wall holes during GAN reconstruction, the cross-entropy loss function $Loss_{ce}$ is added to $Loss_G$. Its equation is shown in (3) to evaluate the reconstruction of the boundary and holes.

$$Loss_{dice} = 1 - \frac{2\sum_i^N G(x_i)\, y_i}{\sum_i^N G(x_i)^2 + \sum_i^N y_i^2}, \tag{2}$$

$$Loss_{ce} = -\frac{1}{N}\sum_i^N [y_i log(G(x_i)) + (1 - y_i)\, log(1 - G(x_i))], \tag{3}$$

where N represents the total number of voxels, $G(x_i)$ represents the voxels of reconstruction results, y_i represents the voxels of ground truth.

Furthermore, an adversarial loss function L_{adv}, is also incorporated into $Loss_G$. This function evaluates the loss of the generator's output by the discriminator, thus enabling the network to improve its performance in reconstructing the fractured orbital wall. The equation for L_{adv} is shown in (4).

$$L_{adv} = E_x [\log (D(x, G(x)))], \tag{4}$$

Finally, we combine the loss function $Loss_{dice}$, $Loss_{ce}$ and L_{adv} to obtain the final $Loss_G$, whose equation is shown in (5) and λ is the weight parameter.

$$Loss_G = Loss_{dice} + Loss_{ce} + \lambda \bullet L_{adv}. \tag{5}$$

3   Experiments and Results

3.1   Data Set and Settings

The dataset for this study was obtained from Shanghai Ninth People's Hospital Affiliated to Shanghai Jiao Tong University School of Medicine. It included 150 cases of OBF CT data: 100 for training and 50 for testing. Each case had a blowout orbital fracture on one side and a normal orbit on the other. For normal orbit segmentation training, 70 additional cases of normal orbit CT data were used. The images were 512×512 in size, with resolutions ranging from $0.299\,mm \times 0.299\,mm$ to $0.717\,mm \times 0.717\,mm$. The number of slices varied from 91 to 419, with thicknesses ranging from $0.330\,mm$ to $1.0\,mm$. The ground truth was obtained through semi-automatic segmentation and manual repair by experienced clinicians. To focus on the orbital area, CT scans were resampled to 160×160 with a multiple of 32 slices after cutting out both orbits. This resulted in 240 single-orbital CT data for normal orbital segmentation training and 100 for OBF reconstruction training. The proposed networks used patches of size $32 \times 160 \times 160$, with training data augmented using sagittal symmetry.

The proposed and comparison networks all adopted the same input patch size of $32 \times 160 \times 160$, a batch size of 1, a learning rate of 0.001, and were trained for 30,000 iterations using TensorFlow 1.14 on an NVIDIA RTX 8000 GPU. Evaluation of the reconstruction results was based on the dice similarity coefficient (DSC), intersection over union (IOU), precision, sensitivity, average surface distance (ASD), and 95% Hausdorff distance (95HD).

Table 1. Evaluation results from ablation experiments of our method.

Networks	DSC (%) ↑	IOU (%) ↑	Precision (%) ↑	Sensitivity (%) ↑	ASD (mm) ↓	95HD (mm) ↓
SPAK	71.27 ± 4.28	55.53 ± 5.06	74.97 ± 4.98	68.02 ± 4.55	0.50 ± 0.08	1.80 ± 0.30
GN	87.97 ± 4.08	78.75 ± 6.14	89.40 ± 3.23	86.73 ± 5.74	0.20 ± 0.11	1.06 ± 0.59
GN+DN(GAN)	88.98 ± 3.15	80.28 ± 4.95	87.69 ± 3.95	90.41 ± 3.55	0.18 ± 0.09	0.96 ± 0.62
GN+ SPAK	91.87 ± 2.25	85.03 ± 3.75	**92.26 ± 2.65**	91.57 ± 3.30	0.12 ± 0.05	0.64 ± 0.26
Our method	**92.35 ± 2.13**	**85.86 ± 3.53**	92.01 ± 2.58	**92.75 ± 2.66**	**0.11 ± 0.05**	**0.59 ± 0.23**

3.2 Ablation Experiment Results

The proposed network is based on the GN, and an ablation experiment was conducted to verify the effectiveness of the adopted strategy. The experimental results are presented in Table 1. Comparing SPAK with other networks, it is evident that the accuracy of the reconstructed orbital wall is relatively poor, and it cannot be directly employed as a reconstruction network. However, it can be used as a prior guide for GAN. By comparing GN with GN+DN (GAN), it is apparent that the accuracy of GAN, except for precision, is better than GN. This finding shows that DN can promote GN to better reconstruct the orbital wall, indicating the correctness of adopting GAN as a reconstruction network for OBF. Furthermore, the addition of SPAK to GN and GAN significantly improved their reconstruction accuracy. Notably, compared with GAN, the proposed reconstruction network improved the DSC of orbital wall reconstruction accuracy by more than 3.5%, increased IOU by more than 5%, and decreased 95HD by 0.35 mm. These results indicate that SPAK guidance plays a crucial role in orbital wall reconstruction. It further demonstrates that the improved strategy can effectively achieve the precise reconstruction of the orbital wall in OBF.

3.3 Comparative Experiment Results

To demonstrate the superior performance of the proposed reconstruction algorithm, we compared it with several state-of-the-art networks used in medical image processing, including U-Net [15], V-Net [16], Attention U-Net [17], and Attention V-Net. The accuracy of the reconstructed results from these networks is compared in Table 2. The comparison with U-Net, V-Net, and Attention U-Net reveals that the proposed reconstruction network outperforms them significantly in terms of reconstruction accuracy evaluation. Specifically, the DSC is improved

Table 2. Evaluation results from ablation experiments of our method. Att U-Net denotes Attention U-Net, and Att V-Ne denotes Attention V-Net.

Networks	DSC (%) ↑	IOU (%) ↑	Precision (%) ↑	Sensitivity (%) ↑	ASD (mm) ↓	95HD (mm) ↓
U-Net [15]	88.01 ± 3.19	78.73 ± 5.02	88.27 ± 3.39	87.97 ± 5.23	0.20 ± 0.10	1.15 ± 0.10
V-Net [16]	89.42 ± 2.72	80.97 ± 4.39	91.04 ± 3.22	87.97 ± 3.65	0.18 ± 0.12	1.16 ± 1.78
Att U-Net [17]	87.46 ± 3.58	77.88 ± 5.37	91.05 ± 3.98	84.43 ± 5.95	0.22 ± 0.19	1.41 ± 2.50
Att V-Net	89.27 ± 8.81	80.77 ± 12.30	**92.11 ± 12.50**	86.71 ± 5.88	0.19 ± 0.19	0.90 ± 0.80
Our method	**92.35 ± 2.13**	**85.86 ± 3.53**	92.01 ± 2.58	**92.75 ± 2.66**	**0.11 ± 0.05**	**0.59 ± 0.23**

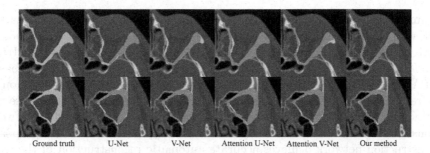

Ground truth   U-Net   V-Net   Attention U-Net   Attention V-Net   Our method

Fig. 3. Orbital wall reconstruction results comparisons between our network and other networks. Green is the ground truth, and red is the network's reconstruction result. (Color figure online)

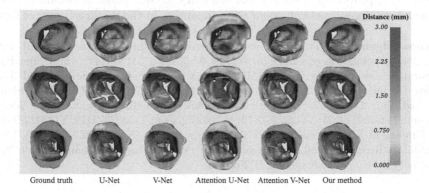

Ground truth   U-Net   V-Net   Attention U-Net   Attention V-Net   Our method

Fig. 4. Surface distance error comparisons between our network and other networks.

by more than 2.5%, the IOU is improved by more than 4.5%, the sensitivity is improved by more than 6.5%, and the 95HD distance error is reduced by more than 0.35 mm. These results indicate that the proposed network has better reconstruction performance for the orbital wall. Comparing with Attention V-Net, it is shown that the proposed reconstruction network has better reconstruction accuracy, except for precision. Although Attention V-Net has higher precision, its standard deviation is relatively large. Figure 3 is a comparison chart of the reconstruction results of each method, showing that other methods are difficult to accurately predict the orbital wall boundary, while the proposed method can generate the orbital wall more accurately. Figure 4 presents a comparison chart of surface distance errors from the reconstruction results, which shows that other methods have relatively large distance errors in the medial and floor walls of the orbit, and even lead to cracks and holes. In contrast, the proposed network significantly addresses these issues, indicating its superior performance.

4   Conclusion

In summary, this paper proposes a SPAK-guided GAN for accurately automating OBF wall reconstruction. Firstly, we propose an automatic generation method of SPAK based on spatial transformation, which maps the segmented symmetrical normal orbital wall to a fractured orbit to form an effective SPAK. On this basis, a SPAK-guided GAN network is developed for the automatic reconstruction of the fractured orbital wall through adversarial learning. Furthermore, we use the strategy of multi-loss function supervision to improve the accuracy of network reconstruction. The final experimental results demonstrate that the proposed reconstruction network achieves accurate automatic reconstruction of the fractured orbital wall, with a DSC of $92.35 \pm 2.13\%$ and a 95% Hausdorff distance of 0.59 ± 0.23 mm, which is significantly better than other networks. This network achieves the automatic reconstruction of the orbital wall in OBF, which effectively improves the accuracy and efficiency of OBF surgical planning. In the future, it will have excellent application prospects in the repair surgery of OBF.

Acknowledgements. This work was supported by grants from the National Natural Science Foundation of China (81971709; M-0019; 82011530141), the China Postdoctoral Science Foundation (2023M732245), the Foundation of Science and Technology Commission of Shanghai Municipality (20490740700; 22Y11 911700), Shanghai Jiao Tong University Foundation on Medical and Technological Joint Science Research (YG2021ZD21; YG2021QN72; YG2022QN056; YG2023ZD19; YG2023ZD15), the Funding of Xiamen Science and Technology Bureau (No. 3502Z20221012), Cross disciplinary Research Fund of Shanghai Ninth People's Hospital, Shanghai Jiao Tong University School of Medicine (JYJC202115).

References

1. Ho, T.Q., Jupiter, D., Tsai, J.H., Czerwinski, M.: The incidence of ocular injuries in isolated orbital fractures. Ann. Plastic Surg. **78**(1), 59–61 (2017)
2. Rossin, E.J., Szypko, C., Giese, I., Hall, N., Gardiner, M.F., Lorch, A.: Factors associated with increased risk of serious ocular injury in the setting of orbital fracture. JAMA Ophthalmol. **139**(1), 77–83 (2021)
3. Ozturker, C., Sari, Y., Ozbilen, K.T., Ceylan, N.A., Tuncer, S.: Surgical repair of orbital blow-out fractures: outcomes and complications. Beyoglu Eye J. **7**(03), 199–206 (2022)
4. Xu, J., Zhang, D., Wang, C., Zhou, H., Li, Y., Chen, X.: Automatic segmentation of orbital wall from CT images via a thin wall region supervision-based multi-scale feature search network. Int. J. Comput. Assist. Radiol. Surg. 1–12 (2023)

5. Kim, H., et al.: Three-dimensional orbital wall modeling using paranasal sinus segmentation. J. Cranio-Maxillofacial Surg. **47**(6), 959–967 (2019)
6. Lee, M.J., Hong, H., Shim, K.W., Park, S.: MGB-net: orbital bone segmentation from head and neck CT images using multi-graylevel-bone convolutional networks. In: 2019 IEEE 16th International Symposium on Biomedical Imaging (ISBI 2019), pp. 692–695. IEEE (2019)
7. Taghizadeh, E., Terrier, A., Becce, F., Farron, A., Büchler, P.: Automated CT bone segmentation using statistical shape modelling and local template matching. Comput. Methods Biomech. Biomed. Eng. **22**(16), 1303–1310 (2019)
8. Li, J., et al.: Automatic skull defect restoration and cranial implant generation for cranioplasty. Med. Image Anal. **73**, 102171 (2021)
9. Xiao, D., et al.: Estimating reference shape model for personalized surgical reconstruction of craniomaxillofacial defects. IEEE Trans. Biomed. Eng. **68**(2), 362–373 (2020)
10. Han, B., et al.: Statistical and individual characteristics-based reconstruction for craniomaxillofacial surgery. Int. J. Comput. Assist. Radiol. Surg. **17**(6), 1155–1165 (2022)
11. Xu, J., et al.: A review on AI-based medical image computing in head and neck surgery. Phys. Med. Biol. (2022)
12. Nijiati, M., et al.: A symmetric prior knowledge based deep learning model for intracerebral hemorrhage lesion segmentation. Front. Physiol. **13**, 2481 (2022)
13. Guo, F., Ng, M., Kuling, G., Wright, G.: Cardiac MRI segmentation with sparse annotations: ensembling deep learning uncertainty and shape priors. Med. Image Anal. **81**, 102532 (2022)
14. Yuxin, W., He, K.: Group normalization. Int. J. Comput. Vision **128**(3), 742–755 (2020)
15. Ronneberger, O., Fischer, P., Brox, T.: U-net: convolutional networks for biomedical image segmentation. In: Navab, N., Hornegger, J., Wells, W.M., Frangi, A.F. (eds.) MICCAI 2015. LNCS, vol. 9351, pp. 234–241. Springer, Cham (2015). https://doi.org/10.1007/978-3-319-24574-4_28
16. Milletari, F., Navab, N., Ahmadi, S.-A., V-net: fully convolutional neural networks for volumetric medical image segmentation. In: 2016 fourth International Conference on 3D Vision (3DV), pp. 565–571. IEEE (2016)
17. Schlemper, J., et al.: Attention gated networks: learning to leverage salient regions in medical images. Med. Image Anal. **53**, 197–207 (2019)

A Multi-task Network for Anatomy Identification in Endoscopic Pituitary Surgery

Adrito Das[1]([envelope]), Danyal Z. Khan[1,2], Simon C. Williams[1,2],
John G. Hanrahan[1,2], Anouk Borg[2], Neil L. Dorward[2], Sophia Bano[1,3],
Hani J. Marcus[1,2], and Danail Stoyanov[1,3]

[1] Wellcome/EPSRC Centre for Interventional and Surgical Sciences,
University College London, London, UK
`adrito.das.20@ucl.ac.uk`
[2] Department of Neurosurgery, National Hospital for Neurology and Neurosurgery,
London, UK
[3] Department of Computer Science, University College London, London, UK

Abstract. Pituitary tumours are in an anatomically dense region of the body, and often distort or encase the surrounding critical structures. This, in combination with anatomical variations and limitations imposed by endoscope technology, makes intra-operative identification and protection of these structures challenging. Advances in machine learning have allowed for the opportunity to automatically identifying these anatomical structures within operative videos. However, to the best of the authors' knowledge, this remains an unaddressed problem in the sellar phase of endoscopic pituitary surgery. In this paper, PAINet (Pituitary Anatomy Identification Network), a multi-task network capable of identifying the ten critical anatomical structures, is proposed. PAINet jointly learns: (1) the semantic segmentation of the two most prominent, largest, and frequently occurring structures (sella and clival recess); and (2) the centroid detection of the remaining eight less prominent, smaller, and less frequently occurring structures. PAINet utilises an EfficientNetB3 encoder and a U-Net++ decoder with a convolution layer for segmentation and pooling layer for detection. A dataset of 64-videos (635 images) were recorded, and annotated for anatomical structures through multi-round expert consensus. Implementing 5-fold cross-validation, PAINet achieved 66.1% and 54.1% IoU for sella and clival recess semantic segmentation respectively, and 53.2% MPCK-20% for centroid detection of the remaining eight structures, improving on single-task performances. This therefore demonstrates automated identification of anatomical critical structures in the sellar phase of endoscopic pituitary surgery is possible.

Keywords: minimally invasive surgery · semantic segmentation · surgical AI · surgical vision

Supplementary Information The online version contains supplementary material available at https://doi.org/10.1007/978-3-031-43996-4_45.

1   Introduction

A difficulty faced by surgeons performing endoscopic pituitary surgery is identifying the areas of the bone which are safe to open. This is of particular importance during the sellar phase as there are several critical anatomical structures within close proximity of each other [9]. The sella, behind which the pituitary tumour is located, is safe to open. However, the smaller structures surrounding the sella, behind which the optic nerves and internal carotid arteries are located, carry greater risk. Failure to appreciate these critical parasellar neurovascular structures can lead to their injury, and adverse outcomes for the patient [9,11].

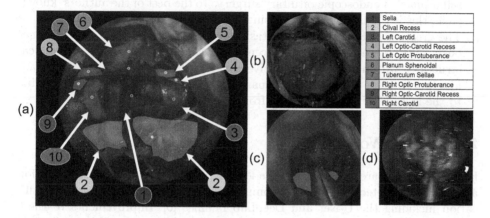

Fig. 1. 10-anatomical-structures semantic segmentation of the sellar phase in endoscopic pituitary surgery. Names and mask colour are given in the legend, with centroids displayed as dots. Example images where each visible critical anatomical structures' mask is annotated by neurosurgeons, are displayed: (a) an image where all structures are clearly seen; (b) an image where only the sella is visible; (c) an image where some structures are occluded by an instrument and biological factors; and (d) an image where none of the structures are identifiable due to blurriness caused by camera movement. (Color figure online)

The human identification of these structures relies on visual clues, inferred from the impressions made on the bone, rather than direct visualisations of the structures [11]. This is especially challenging as the pituitary tumour often compresses; distorts; or encases the surrounding structures [11]. Neurosurgeons utilise identification instruments, such as a stealth pointer or micro-doppler, to aid in this task [9]. However, once an identification instrument is removed, identification is lost upon re-entry with a different instrument, and so the identification can only be used in referenced to the more visible anatomical landmarks. Automatic identification from endoscopic vision may therefore aid surgeons in this effort while minimising disruption to the surgical workflow [11].

This is a challenging computer vision task due to the narrow camera angles enforced by minimally invasive surgery, which lead to: (i) structure occlusions by instruments and biological factors (e.g., blood); and (ii) image blurring caused by rapid camera movements. Additionally, in this specific task there are: (iii) numerous small structures; (iv) visually similar structures; and (v) unclear structure boundaries. Hence, the task can be split into two sub-tasks to account for these difficulties in identification: (1) the semantic segmentation of the two larger, visually distinct, and frequently occurring structures (sella and clival recess); and (2) the centroid detection of the eight smaller structures (Fig. 1).

To solve both tasks simultaneously, PAINet (Pituitary Anatomy Identification Network) is proposed. This paper's contribution is therefore:

1. The automated identification of the ten critical anatomical structures in the sellar phase of endoscopic pituitary surgery. To the best of the authors' knowledge, this is the first work addressing the problem at this granularity.
2. The creation of PAINet, a multi-task neural network capable of simultaneously semantic segmentation and centroid detection of numerous anatomical structures within minimally invasive surgery. PAINet uniquely utilises two loss functions for improved performance over single-task neural networks due to the increased information gain from the complementary task.

2   Related Work

Encoder-decoder architectures are the leading models in semantic segmentation and landmark detection [4], with common architectures for anatomy identification including the U-Net and DeepLab families [6]. Improvements to these models include: adversarial training to limit biologically implausible predictions [14]; spatial-temporal transformers for scene understanding across consecutive frames [5]; transfer learning from similar anatomical structures [3]; and graph neural networks for global image understanding [2]. Multi-task networks improve on the baseline models by leveraging common characteristics between sub-tasks, increasing the total information provided to the network [15], and are effective at instrument segmentation in minimally invasive surgery [10].

The most clinically similar works to this paper are: (1) The semantic segmentation of 3-anatomical-structures in the nasal phase of endoscopic pituitary surgery [12]. Here, U-Net was weakly-supervised on centroids, outputting segmentation masks for each structure. Training on 18-videos (367-images), the model achieved statistically significant results ($P < 0.001$) on the hold-out testing dataset of 5-videos (182-images) when compared to a location prior baseline model [12]. (2) The semantic segmentation of: 2-zones (safe or dangerous); and 3-anatomical-structures in laparoscopic cholecystectomy [7]. Here, two PSPNets were fully-supervised on 290-videos (2627 images) using 10-fold cross-validation, achieving 62% mean intersection over union (MIoU) for the 2-zones; and 74% MIoU for the 3-structures [7]. These works are extended in this paper by increasing the number of anatomical structures and the identification granularity.

3 Methods

PAINet: A multi-task encoder-decoder network is proposed to improve performance by exchanging information between the semantic segmentation and centroid detection tasks. EfficientNetB3, pre-trained on ImageNet, is used as the encoder because of its accuracy, computational efficiency and proven generalisation capabilities [13]. The decoder is based on U-Net++, a state-of-the-art segmentation network widely used in medical applications [16]. The encoder-decoder architecture is modified to output both segmentation and centroid predictions by sending the decoder output into two separate layers: (1) a convolution for segmentation prediction; and (2) an average pooling layer for centroid prediction. Different loss functions were minimised for each sub-task (Fig. 2).

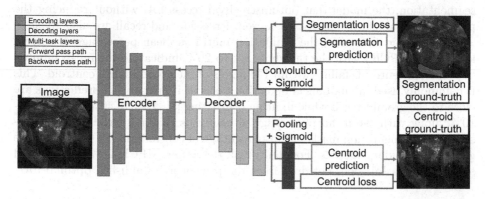

Fig. 2. Multi-task (semantic segmentation and centroid detection) architecture diagram. Notice the two output layers (convolution and pooling) and two loss functions.

Ablation studies and granular details are provided below. The priority was to find the optimal sella segmentation model, as it is required to be opened to access the pituitary tumour behind it, indicating the surgical "safe-zone".

Semantic Segmentation: First, single-class sella segmentation models were trialed. 8-encoders (pre-trained convolution neural networks) and 15-decoders were used, with their selection based off architecture variety. Two loss functions were also used: (1) distribution-based logits cross-entropy; and (2) region-based Jaccard loss. Boundary-based loss functions were not trialed as: (1) the boundary of the segmentation masks are not well-defined; and (2) in the cases of split structures (Fig. 1c), boundary-based loss functions are not appropriate [8]. The decoder output is passed through a convolution layer and sigmoid activation.

For multi-class sella and clival recess segmentation, the optimal single-class model was extended by: (1) sending through each class to the loss function separately (multi-class separate); and (2) sending both classes through together (multi-class together). An extension of logits cross-entropy, logits focal loss, was used instead as it accounts for data imbalance between classes.

Centroid Detection: 5-models were trialed: 3-models consisted of encoders with a convolution layer and linear activation; and 2-models consisted of encoder-decoders with an average pooling layer and sigmoid activation with 0.3 dropout. Two distance-based loss functions were trialed: (1) mean squared error (MSE); and (2) mean absolute error (MAE). Loss was calculated for all structures simultaneously as a 16 dimensional output (8 centroids × 2 coordinates) and set to 0 for a structure if ground-truth centroids of that structure was not present.

4   Experimental Setup

Evaluation Metrics: For sella segmentation the evaluation metric is intersection over union (IoU), as commonly used in the field [4,8]. For multi-class segmentation, the model that optimises clival recess IoU without reducing the previously established sella IoU is chosen. Precision and recall are also given.

For centroid detection, the evaluation metric is mean percentage of correct keypoints (MPCK) with the threshold set to 20%, indicating the mean number of predicted centroids falling within 144 pixels of the ground-truth centroid. This is commonly used in anatomical detection tasks as it ensures the predictions are close to the ground-truth while limiting overfitting [1]. MPCK-40% and MPCK-10%, along with the mean percentage of centroids that fall within their corresponding segmentation mask (mean percentage of centroid masks (MPCM)) are given as secondary metrics. For multi-task detection, MPCK-20% is optimised such that sella IoU does not drop from the previously established optimal IoU.

Network Parameters: 5-fold cross-validation was implemented with no hold-out testing. To account for structure data imbalance, images were randomly split such that the number of structures in each fold is approximately even. Images from a singular video were present in either the training or validation dataset.

Each model was run for with a batch size of 5 for 20 epochs, where the epoch with the best primary evaluation metric on the validation dataset was kept. The optimising method was Adam with varying initial learning rates, with a separate optimiser for each loss function during multi-task training.

All images were scaled to 736 × 1280 pixels for model compatibility, and training images were randomly augmented within the following parameters: shift in any direction by up to 10%; zooming in or out about the image center by up to 10%; rotation about the image center clockwise or anticlockwise by up to $\pi/6$; increasing or decreasing brightness, contrast, saturation, and hue by up to 10%.

The code is written in Python 3.8 using PyTorch 1.8.1, run on a single NVIDIA Tesla V100 Tensor Core 32-GB GPU using CUDA 11.2, and is available at https://github.com/dreets/pitnet-anat-public. For PAINet, a batch size of 5 utilised 29-GB and the runtime was approximately 5-min per epoch. Valuation runtime is under 0.1-s per image and therefore a real-time overlay on-top of the endoscope video feed is feasible intra-operatively.

5   Dataset Description

Images: Images come from 64-videos of endoscopic pituitary surgery where the sellar phase is present [9], recorded between 30 Aug 2018 and 20 Feb 2021 from The National Hospital of Neurology and Neurosurgery, London, United Kingdom. All patients have provided informed consent, and the study was registered with the local governance committee. A high-definition endoscope (Hopkins Telescope, Karl Storz Endoscopy) was used to record the surgeries at 24 frames per second (fps), with at least 720p resolution, and stored as mp4 files. 10-images corresponding to 10-s of the sellar phase immediately preceding sellotomy were extracted from each video at 1 fps, and stored as 720p png files. Video upload and annotation was performed using Touch Surgery™ Enterprise.

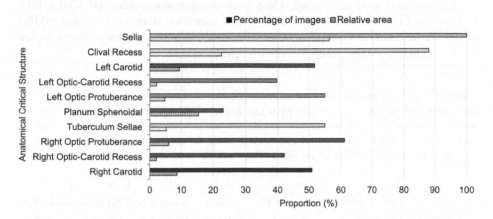

Fig. 3. Distribution of 10 anatomical structures in 635-images: The top bars display structure frequency; and the bottom bars display each structures area relative to the total area covered by all structures in a given image (mean-averaged across all images).

Annotations: Expert neurosurgeons identified 10-anatomical-structures as critical based on the literature (Fig. 1a) [9,11]. 640-images were manually segmented to obtain ground-truth segmentations. A two-stage process was used: (1) two neurosurgeons segmented each image, with any differences settled through discussion; (2) two consultant neurosurgeons independently peer-reviewed the segmentations. Only visible structures were annotated (Fig. 1b); if the structures were occluded, the segmentation boundaries were drawn around these occlusions (Fig. 1c); and if an image is too blurry to see the structures no segmentation boundaries were drawn - this excluded 5 images (Fig. 1d). The center of mass of each segmentation mask was defined as the centroid.

The sella is present in all 635-images (Fig. 3). Other than the clival recess, the remaining 8-structures are found in less than 65% of images, with planum sphenoidal found in less than 25% of images. Moreover, the area covered by these

8-structures are small, with several covering less than 10% of the total area covered by all structures in a given image. Furthermore, most smaller structures boundaries are ambiguous as they are hard to define even by expert neurosurgeons. This emphasizes the challenge of identifying smaller structure in computer vision, and supports the need for detection and multi-task solutions.

6   Results and Discussion

Quantitative evaluation is calculated for: single-class sella segmentation (Table 1); single-class, multi-class, and PAINet 2-structures segmentation (Table 2); multi-class, and PAINet 8-structures centroid detection (Tables 3 and 4).

The optimal model for single-class sella segmentation achieved 65.4% IoU, utilising an EfficientNetB3 encoder; U-Net++ decoder; Jaccard loss; and a 0.001 initial learning rate. Reductions in IoU are seen when alternative parameters are used, highlighting their impact on model performance.

Table 1. Selected models performance on the single-class sella segmentation task (5-fold cross validation). The model with the highest IoU is given in top-most row (in bold), with each row changing one model parameter (in italics), where the highest and lowest IoU for a given model parameter change is shown. Complete results for all models can be found in Supplementary Material Table 1. *Initial learning rate.

Decoder	Encoder	Loss	Rate*	IoU	Recall	Precision
U-Net++	**EfficientNetB3**	**Jaccard**	**0.001**	**65.4±1.6**	**79.3±2.6**	**79.5±3.9**
DeepLabv3+	EfficientNetB3	Jaccard	0.001	63.7 ± 2.9	77.0 ± 4.7	79.5 ± 2.1
PSPNet	EfficientNetB3	Cross-Entropy	0.001	54.5 ± 4.1	74.3 ± 1.9	68.4 ± 3.5
U-Net++	EfficientNetB3	*Cross-Entropy*	0.001	64.8 ± 2.3	79.4 ± 2.5	78.6 ± 2.6
U-Net++	*Xception*	Jaccard	0.001	60.1 ± 4.6	78.2 ± 2.9	77.9 ± 4.3
U-Net++	*ResNet18*	Jaccard	0.001	56.3 ± 3.2	73.3 ± 4.0	74.2 ± 5.2
U-Net++	EfficientNetB3	Jaccard	*0.0001*	63.4 ± 1.3	72.7 ± 4.7	83.6 ± 6.3
U-Net++	EfficientNetB3	Jaccard	*0.01*	34.7 ± 9.5	68.7 ± 8.2	69.6 ± 8.8

Table 2. Selected models performance for sella and clival recess segmentation (5-fold cross-validation). All models use an EfficientNetB3 encoder and U-Net++ decoder. The best performing network, as determined by sella IoU, is displayed in bold.

Training	Loss	Sella IoU	Clival Recess IoU
Single-Class	Jaccard	65.4 ± 1.6	53.4 ± 5.9
Multi-Class Separate	Focal	65.4 ± 2.1	54.2 ± 8.5
Multi-Class Together	Focal	65.6 ± 2.6	49.9 ± 4.8
Multi-task (PAINet)	**Focal**	**66.1 ± 2.3**	**54.1 ± 5.0**

Using the optimal sella model configuration, 53.4% IoU is achieved for single-class clival recess segmentation. Extending this to multi-class and PAINet training improves both sella and clival recess IoU to 66.1% and 54.1% respectively.

The optimal model for centroid detection achieves 51.7% MPCK-20%, with minor deviations during model parameter changes. This model, ResNet18 with MSE loss, outperforms the more sophisticated models, as these models over-learn image features in the training dataset. However, PAINet leverages the additional information from segmentation masks to achieve an improved 53.2%.

The per structure PCK-20% indicate performance is positively correlated with the number of images where the structure is present. This implies the limiting factor is the number of images rather than architectural design.

Table 3. Selected models performance for centroid detection (5-fold cross-validation). The model with the highest MPCK-20% is given in the last row (in bold). Complete results for all models can be found in Supplementary Material Table 2.

Training	Decoder	Encoder	Loss	MPCK-20	MPCK-40	MPCK-10	MPCM
Multi-Class	–	ResNet18	MSE	51.7 ± 9.2	58.0 ± 7.0	28.4 ± 5.9	09.7 ± 2.8
Multi-Class	–	ResNet18	MAE	51.4 ± 9.4	57.9 ± 7.8	34.2 ± 5.3	10.9 ± 2.5
Multi-Class	–	EfficientNetB3	MSE	46.8 ± 8.3	55.5 ± 5.7	19.5 ± 3.2	06.3 ± 3.0
Multi-Class	DeepLabv3+	ResNet18	MSE	50.4 ± 6.2	57.8 ± 5.8	20.1 ± 5.4	06.5 ± 1.8
Multi-Class	U-Net++	ResNet18	MSE	50.1 ± 4.2	58.0 ± 6.8	26.1 ± 2.7	09.7 ± 1.2
Multi-Class	U-Net++	EfficientNetB3	MSE	48.1 ± 3.7	56.2 ± 4.7	26.4 ± 4.8	06.8 ± 1.6
PAINet	**U-Net++**	**EfficientNetB3**	**MSE**	$\mathbf{53.2 \pm 5.9}$	$\mathbf{58.0 \pm 6.9}$	$\mathbf{39.6 \pm 3.2}$	$\mathbf{13.4 \pm 2.9}$

Table 4. PAINet 8-structures centroid detection performance (5-fold cross validation).

Structure	Left Carotid	Left Optic-Carotid Recess	Left Optic Protuberance	Planum Sphenoidale	Tuberculum Sellae	Right Optic Protuberance	Right Optic-Carotid Recess	Right Carotid
PCK-20%	68.3 ± 9.2	37.6 ± 4.8	53.9 ± 7.3	27.6 ± 2.4	76.1 ± 2.8	72.0 ± 9.1	34.8 ± 3.0	55.4 ± 8.5

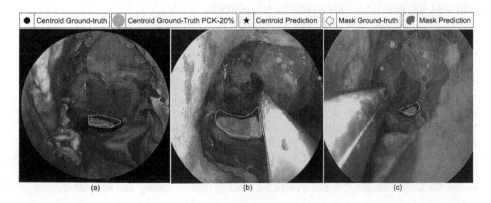

● Centroid Ground-truth ● Centroid Ground-Truth PCK-20% ★ Centroid Prediction ◌ Mask Ground-truth ▨ Mask Prediction

(a)　　　　(b)　　　　(c)

Fig. 4. PAINet predictions for three videos, displaying images with: (a) strong; (b) typical; and (c) poor performances. (The color map is given in Fig. 1.)

Qualitative predictions of the best performing model, PAINet, are displayed in Fig. 4. The segmentation predictions look strong, with small gaps from the ground-truth. However, this is expected as structure boundaries are not well-defined. The centroid predictions are weaker: in (a) the planum sphenoidale (grey) is predicted within the segmentation mask; in (b) three structures are within their segmentation mask, but the left optic-carotid recess (orange) is predicted in a biologically implausible location; and in (c) this is repeated for the right carotid (pink) and no structures are within their segmentation masks.

7   Conclusion

Identification of critical anatomical structures by neurosurgeons during endo-scopic pituitary surgery remains a challenging task. In this paper, the potential of automating anatomical structure identification during surgery was shown. The proposed multi-task network, PAINet, designed to incorporate identification of both large prominent structures and numerous smaller less prominent structures, was trained on images of the sellar phase of endoscopic pituitary surgery. Using 635-images from 64-surgeries annotated by expert neurosurgeons and various model configurations, the robustness of the PAINet was shown over single task networks. PAINet achieved 66.1% (+0.7%) and 54.1% IoU (+0.7%) for sella and clival recess segmentation respectively, a higher performance than other mini-mally invasive surgeries [7]. PAINet also achieved 53.2% MPCK-20% (+1.5%) for detection of the remaining 8-structures. The most important structures to iden-tify and avoid, the carotids and optic protuberances, have high performance, and therefore demonstrate the success of PAINet. This performance is greater than similar studies in endoscopic pituitary surgery for different structures [12] but lower than anatomical detection in other surgeries [1]. Collecting data from more pituitary surgeries will support incorporating anatomy variations and achieving generalisability. Furthermore, introducing modifications to the model architec-ture, such as the use of temporal networks [5], will further boost performance required for real-time video clinical translation.

Acknowledgements. This research was funded in whole, or in part, by the Wellcome/EPSRC Centre for Interventional and Surgical Sciences (WEISS) [203145/Z/16/Z]; the Engineering and Physical Sciences Research Council (EPSRC) [EP/P027938/1, EP/R004080/1, EP/P012841/1, EP/W00805X/1]; and the Royal Academy of Engineering Chair in Emerging Technologies Scheme. AD is supported by EPSRC [EP/S021612/1]. HJM is supported by WEISS [NS/A000050/1] and by the National Institute for Health and Care Research (NIHR) Biomedical Research Cen-tre at University College London (UCL). DZK and JGH are supported by the NIHR Academic Clinical Fellowship. DZK is supported by the Cancer Research UK (CRUK) Predoctoral Fellowship. With thanks to Digital Surgery Ltd, a Medtronic company, for access to Touch Surgery^{TM} Enterprise for both video recording and storage.

References

1. Danks, R.P., et al.: Automating periodontal bone loss measurement via dental landmark localisation. Int. J. Comput. Assist. Radiol. Surg. **16**(7), 1189–1199 (2021). https://doi.org/10.1007/s11548-021-02431-z
2. Gaggion, N., Mansilla, L., Mosquera, C., Milone, D.H., Ferrante, E.: Improving anatomical plausibility in medical image segmentation via hybrid graph neural networks: applications to chest x-ray analysis. IEEE Trans. Med. Imaging **42**(2), 546–556 (2023). https://doi.org/10.1109/tmi.2022.3224660
3. Gu, R., et al.: Contrastive semi-supervised learning for domain adaptive segmentation across similar anatomical structures. IEEE Trans. Med. Imaging **42**(1), 245–256 (2023). https://doi.org/10.1109/tmi.2022.3209798
4. Hao, S., Zhou, Y., Guo, Y.: A brief survey on semantic segmentation with deep learning. Neurocomputing **406**, 302–321 (2020). https://doi.org/10.1016/j.neucom.2019.11.118
5. Jin, Y., Yu, Y., Chen, C., Zhao, Z., Heng, P.A., Stoyanov, D.: Exploring intra- and inter-video relation for surgical semantic scene segmentation. IEEE Trans. Med. Imaging **41**(11), 2991–3002 (2022). https://doi.org/10.1109/tmi.2022.3177077
6. Liu, L., Wolterink, J.M., Brune, C., Veldhuis, R.N.J.: Anatomy-aided deep learning for medical image segmentation: a review. Phys. Med. Biol. **66**(11), 11TR01 (2021). https://doi.org/10.1088/1361-6560/abfbf4
7. Madani, A., et al.: Artificial intelligence for intraoperative guidance using semantic segmentation to identify surgical anatomy during laparoscopic cholecystectomy. Ann. Surg. **276**(2), 363–369 (2020). https://doi.org/10.1097/sla.0000000000004594
8. Maier-Hein, L., Reinke, A., Godau, P., et al.: Metrics reloaded: pitfalls and recommendations for image analysis validation (2022). https://doi.org/10.48550/arxiv.2206.01653
9. Marcus, H.J., et al.: Pituitary society expert Delphi consensus: operative workflow in endoscopic transsphenoidal pituitary adenoma resection. Pituitary **24**(6), 839–853 (2021). https://doi.org/10.1007/s11102-021-01162-3
10. Marullo, G., Tanzi, L., Ulrich, L., Porpiglia, F., Vezzetti, E.: A multi-task convolutional neural network for semantic segmentation and event detection in laparoscopic surgery. J. Personal. Med. **13**(3), 413 (2023). https://doi.org/10.3390/jpm13030413
11. Patel, C.R., Fernandez-Miranda, J.C., Wang, W.H., Wang, E.W.: Skull base anatomy. Otolaryngol. Clin. North Am. **49**(1), 9–20 (2016). https://doi.org/10.1016/j.otc.2015.09.001
12. Staartjes, V.E., Volokitin, A., Regli, L., Konukoglu, E., Serra, C.: Machine vision for real-time intraoperative anatomic guidance: a proof-of-concept study in endoscopic pituitary surgery. Oper. Neurosurg. **21**(4), 242–247 (2021). https://doi.org/10.1093/ons/opab187
13. Tan, M., Le, Q.V.: EfficientNet: rethinking model scaling for convolutional neural networks. arXiv (2019). https://doi.org/10.48550/ARXIV.1905.11946
14. Wang, P., Peng, J., Pedersoli, M., Zhou, Y., Zhang, C., Desrosiers, C.: CAT: constrained adversarial training for anatomically-plausible semi-supervised segmentation. IEEE Trans. Med. Imaging, 1 (2023). https://doi.org/10.1109/tmi.2023.3243069

15. Zhang, Y., Yang, Q.: An overview of multi-task learning. Natl. Sci. Rev. **5**(1), 30–43 (2017). https://doi.org/10.1093/nsr/nwx105
16. Zhou, Z., Rahman Siddiquee, M.M., Tajbakhsh, N., Liang, J.: UNet++: a nested U-net architecture for medical image segmentation. In: Stoyanov, D., et al. (eds.) DLMIA/ML-CDS -2018. LNCS, vol. 11045, pp. 3–11. Springer, Cham (2018). https://doi.org/10.1007/978-3-030-00889-5_1

Robust Vertebra Identification Using Simultaneous Node and Edge Predicting Graph Neural Networks

Vincent Bürgin[1,2](✉), Raphael Prevost[1], and Marijn F. Stollenga[1]

[1] ImFusion GmbH, Munich, Germany
[2] Technical University of Munich, Munich, Germany
vincent.buergin@tum.de

Abstract. Automatic vertebra localization and identification in CT scans is important for numerous clinical applications. Much progress has been made on this topic, but it mostly targets positional localization of vertebrae, ignoring their orientation. Additionally, most methods employ heuristics in their pipeline that can be sensitive in real clinical images which tend to contain abnormalities. We introduce a simple pipeline that employs a standard prediction with a U-Net, followed by a single graph neural network to associate and classify vertebrae with full orientation. To test our method, we introduce a new vertebra dataset that also contains pedicle detections that are associated with vertebra bodies, creating a more challenging landmark prediction, association and classification task. Our method is able to accurately associate the correct body and pedicle landmarks, ignore false positives and classify vertebrae in a simple, fully trainable pipeline avoiding application-specific heuristics. We show our method outperforms traditional approaches such as Hungarian Matching and Hidden Markov Models. We also show competitive performance on the standard VerSe challenge body identification task.

Keywords: graph neural networks · spine localization · spine classification · deep learning

1 Introduction

Vertebra localization and identification from CT scans is an essential step in medical applications, such as pathology diagnosis, surgical planning, and outcome assessment [3,6]. This is however a tedious manual task in a time-sensitive setting that can benefit a lot from automation. However, automatically identifying and determining the location and orientation of each vertebra from a CT

This work is supported by the DAAD program Konrad Zuse Schools of Excellence in Artificial Intelligence, sponsored by the Federal Ministry of Education and Research.

Supplementary Information The online version contains supplementary material available at https://doi.org/10.1007/978-3-031-43996-4_46.

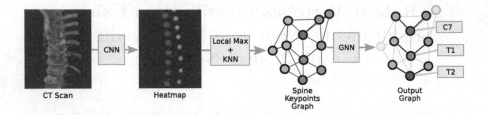

Fig. 1. Summary of the proposed method. A CNN generates a heatmap from which local maxima are connected using k-nearest neighbours to form a graph. Then a single GNN associates the keypoints, performs classification and filters out false postives. The full pipeline is trained and does not require hand-tuned post-processing.

scan can be very challenging: (i) scans vary greatly in intensity and constrast, (ii) metal implants and other materials can affect the scan quality, (iii) vertebrae might be deformed, crushed or merged together due to medical conditions, (iv) vertebrae might be missing due to accidents or previous surgical operations.

Recently, public challenges like the VerSe challenge [24] have offered a common benchmarking platform to evaluate algorithms to automate this task, resulting in a boost in research on this topic. However, these challenges focus on finding the position of vertebrae, ignoring the orientation or direction. Additionally, practically all methods employ manual heuristic methods to identify landmarks and filter out false positives.

In this paper, we introduce a trainable method that performs vertebrae localization, orientation estimation and classification with a single architecture. We replace all hand-crafted rules and post-processing steps with a single trainable Graph Neural Network (GNN) that learns to filter out, associate and classify landmarks. We apply a generalized Message Passing layer that can perform edge and node classification simultaneously. This alleviates the need for sensitive hand-tuned parameter tuning, and increases robustness of the overall pipeline.

The main contributions of our work are: (1) introducing a pipeline that uses a single Graph Neural Network to perform simultaneous vertebra identification, landmark association, and false positive pruning, without the need for any heuristic methods and (2) building and releasing a new spine detection dataset that adds pedicles of vertebrae to create a more complex task that includes orientation estimation of vertebrae, which is relevant for clinical applications.

2 Related Work

Spine Landmark Prediction and Classification. The introduction of standardised spine localization and classification challenges [10,24] resulted in a boost in research on this problem. Convolutional Neural Networks (CNN) became a dominant step in most approaches shortly after their introduction [7]. Most modern methods process the results of a CNN using a heuristic method to create a

1-dimensional sequence of vertebra detections, before applying classification: [27] generate heatmaps for body positions, and refine it into a single sequence graph that uses message passing for classification. [14] generate heatmaps, extract a 1-dimensional sequence and use a recurrent neural network for classification. [18] produce a heatmap using a U-Net [21], but use a simpler approach by taking the local maxima as landmarks, and forming a sequence by accepting the closest vertebra that is within a defined distance range. [16] uses a directional graph and Dynamic Programming to find an optimal classification.

Graph Neural Networks. In recent years, Graph Neural Networks (GNN) have surged in popularity [11], with a wide and growing range of applications [26]. A prominent task in the literature is node-level representation learning and classification. A less prominent task is edge classification, for which early work used a dual representation to turn edge- into node representations [1]. Other approaches model edge embeddings explicitly, such as [4,12]. The most general formulation of GNNs is the message-passing formulation [5] which can be adapted to perform both edge and node classification at the same time. We use this formulation in our method.

Various methods have applied GNNs to keypoint detection, however they all apply to 2-dimensional input data. In [20] GNNs are used to detect cars in images. An edge classification task is used to predict occluded parts of the car. However, the GNN step is ran individually for every car detection and the relation between cars is not taken into account, unlike our task. Also there is no node classification applied. In [15] a GNN is used to group detected keypoints for human-pose estimation on images. Keypoints are grouped using edge prediction where edge-embeddings are used as input the GNN. A separate GNN processes node embeddings to facilitate the final grouping. However, the node and edge embeddings are processed separately from each other.

3   Method

In this paper we tackle vertebra localization and classification, but unlike other methods that only focus on detecting the body of the vertebrae, we also detect the pedicles and associate them with their corresponding body. This allows us to also define the *orientation* of the vertebra defined by the plane passing through the body and pedicles[1]. To this end we introduce a new dataset that includes pedicles for each vertebra, described in Sect. 4.1, creating a challenging keypoint detection and association task.

Our method consists of a two-stage pipeline shown in Fig. 1: we detect keypoints from image data using a *CNN stage*, and form a connected graph from these keypoints that is processed by a *GNN stage* to perform simultaneous node and edge classification, tackling classification, body to pedicle association as well as false positive detection with a single trainable pipeline without heuristics.

---

[1] Another common way to define the orientation is using the end-plates of the vertebra body; however end-plates can be irregular and ill-defined in pathological cases.

CNN Stage. The CNN stage detects candidate body, left pedicle and right pedicle keypoints and provides segment classifications for the body keypoints as either cervical, thoracic, lumbar or sacral. We use a UNet[2] CNN [19] and select all local maxima with an intensity above a certain threshold τ, in this paper we use $\tau = 0.5$. These keypoints are connected to their k nearest neighbours, forming a graph. In rare cases this can result in unconnected cliques, in which case the nearest keypoint of each clique is connected to $\frac{k}{3}$ nearest points in the other cliques, ensuring a fully connected graph. All nodes and edges are associated with information through embeddings, described below.

GNN Stage. The second stage employs a generalized message-passing GNN following [5] to perform several prediction tasks on this graph simultaneously:

1. **keypoint association prediction:** we model association between body keypoints and their corresponding pedicle keypoints as binary edge classification on the over-connected k-NN graph.
2. **body keypoint level prediction:** for body keypoints, we model the spine level prediction as multi-class node classification.
3. **keypoint legitimacy prediction:** to filter out false-positive keypoints, we additionally compute an binary legitimacy prediction for each node.

To perform these task, our message-passing GNN maintains edge and node embeddings which are updated in each layer. A message-passing layer performs a node update and edge update operation. Denoting the feature vector of a node v by x_v, and the feature vector of a directed edge (u, v) by x_{uv}, the node and edge features are updated as follows:

$$\underbrace{x'_u = \bigoplus_{v \in \mathcal{N}_u \cup \{u\}} \psi_{\text{node}}(x_u, x_v, x_{uv}),}_{\text{Node update}} \qquad \underbrace{x'_{uv} = \psi_{\text{edge}}(x_u, x_v, x_{uv})}_{\text{Edge update}} \qquad (1)$$

Here $\bigoplus$ denotes a symmetric pooling operation (in our case max pooling) over the neighborhood $\mathcal{N}_u$. ψ_{node} and $_{\text{edge}}$ are trainable parametric functions: in our case, two distinct two-layer MLPs with ReLU nonlinearities. After N such message-passing layers we obtain an embedding vector for each node and edge. Each node/edge embedding is passed through a linear layer (distinct for nodes and edges) to obtain a vector of node class logits or a single edge prediction logit, respectively. The last entry in the node prediction vector is interpreted as a node legitimacy prediction score: nodes predicted as illegitimate are discarded for the output.

The node input features $x_u \in \mathbb{R}^7$ consist of the one-hot encoded keypoint type (body, left or right pedicle) and the segment input information (a pseudo-probability in $[0, 1]$ for each of the four spine segments of belonging to that segment, computed by applying a sigmoid to the heatmap network's output channels which represent the different spine segments). The edge input features $x_{uv} \in \mathbb{R}^4$ consist of the normalized direction vector of the edge and the distance between the two endpoints.

The output of the GNN contains finer spine-level classification (i.e. C1–C7, T1–T13, L1–L6, S1–S2), keypoint-level legitimacy (legitimate vs. false-positive detection) and body-pedicle association via edge prediction, implicitly defining the orientation of each vertebra. Prediction scores of corresponding directed edges (u, v) and (v, u) are symmetrized by taking the mean.

In our experiments we consider variations to our architecture: weight sharing between consecutive GNN layers, multiple heads with a shared backbone (jointly trained) and dedicated networks (separately trained) for edge/node prediction.

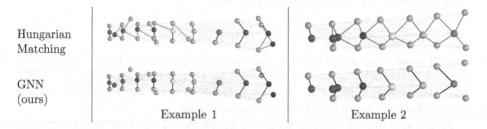

Hungarian Matching

GNN (ours)

Example 1   Example 2

Fig. 2. Examples of challenging cases for the Hungarian Matching edge detection baseline, and corresponding correct detections by our GNN architecture (red = false positive, blue = false negative detection). (Color figure online)

4   Experiments

4.1   Dataset

Our main dataset consists of 2118 scans, of which 1949 form the training dataset, and 169 the validation dataset. This includes 360 scans from the VerSe datasets [24], 1676 scans from the CT Colonography dataset [25] and 82 scans from the CT Pancreas dataset [22]. Of these datasets, only VerSe has labels for spine levels, and none of the datasets have labels for the pedicles of vertebrae. The labeling was done in a two-step process with initial pedicle locations determined through post-processing of ground truth segmentations, then transfered to other datasets via bootstrapping. For each vertebra, the vertebra level is labeled, including the position of the body and the right and left pedicles. This keypoint dataset is publicly available at https://github.com/ImFusionGmbH/VID-vertebra-identification-dataset.

Additionally, we also perform the VerSe body classification task on the original VerSe dataset [24] which contains 160 scans in an 80/40/40 split.

4.2   Training

Heatmap Network. The heatmap is generated by a UNet^2 network [19], with 4 layers and 32 channels. The network is trained on crops of 128^3 voxels with 1.5 mm spacing. The target data consists of Gaussian blobs ($\sigma = 6$ mm) at the landmark positions. We sample 70% of the crops around landmark locations, and 30% randomly from the volume. Additionally we apply 50% mirror augmentations, and 20° rotations around all axes. We use the MADGRAD optimizer [8] with a learning rate of 10^{-3}. We use a binary cross-entropy loss, with an 80% weighting towards positive outputs to counter the data's balancing.

Graph Neural Network. The GNN is implemented in *PyTorch Geometric* 2.0.4 [9]. The three predictions of the graph neural network – edge classification, node classification and node legitimacy prediction – are trained via cross-entropy losses which are weighted to obtain the overall loss:

$$\mathcal{L} = \alpha \mathcal{L}_{\text{edge}}^{\text{BCE}} + \beta \mathcal{L}_{\text{node}_{\text{class}}}^{\text{CE}} + \gamma \mathcal{L}_{\text{node}_{\text{legit}}}^{\text{BCE}} \tag{2}$$

We only make edge predictions on edges that run between a body and a pedicle keypoint – the other edges are only used as propagation edges for the GNN. Similarly, we only make spine level predictions on body keypoints. Solely these subsets of nodes/edges go into the respective losses.

As input data, we use the predicted keypoints of the heatmap network on the training/validation set. The ground-truth keypoints are associated to this graph to create the target of the GNN. We include three synthetic model spines during training (keypoints in a line spaced 30 mm apart) to show the network typical configurations and all potential levels (with all levels/without T13, L6, S2/without T12, T13, L6, S2).

We tune various hyperparameters of our method, such as network depth and weight sharing, the k of the k-NN graph, the loss weighting parameters α, β, γ and the number of hidden channels in the message-passing MLP. We use a short notation for our architectures, such as $(5 \times 1, 4, 1)$ for 5 independent message-passing layers followed by 4 message-passing layers with shared weights and another independent message-passing layer.

Various data augmentations are used to make our network more robust to overfitting: (i) rotation of the spine by small random angles, (ii) mirroring along the saggital axis (relabeling left/right pedicles to keep consistency), (iii) perturbation of keypoints by small random distances, (iv) keypoint duplication and displacement by a small distance (to emulate false-positive duplicate detections of the same keypoint), (v) keypoint duplication and displacement by a large distance (to emulate false-positive detections in unrelated parts of the scan) and (vi) random falsification of spine segment input features. We define four levels of augmentation strength (no/light/default/heavy augmentations) and refer to the supplementary material for precise definitions of these levels.

4.3   Evaluation

We evaluate our method on two tasks. The first one is the full task consisting of node and edge classification for vertebra level and keypoint association detection. We evaluate this on the 2118 scan dataset which comes with pedicle annotations. The second task consists only of vertebra localization, which

Table 1. Results on the 2118 spine dataset (full validation set/hard subset). Comparing best GNN architectures with the baselines. Wilcoxon signed-rank test p-values are given for GNN numbers that outperform the baseline (by the test's construction, p-values agree between the full and hard subset).

Method		identification rate	edge F_1 score	d_{mean}
Edge vs. node classification	Joint node/edge	**97.19/89.88** $(p = 0.080)$	98.81 / 96.22	**1.68/1.95**
	Node only	96.91 / 88.90 $(p = 0.109)$	—	1.68 / 1.96
	Edge only	—	**99.31/97.77** $(p = 0.019)$	—
Baselines	Hidden Markov	94.29 / 79.46	—	1.81 / 2.52
	Hungarian matching	—	98.93 / 96.56	—

we evaluate on the VerSe 2019 dataset [24]. Unless otherwise specified, we use $k = 14$, the (13×1) architecture, batch size 25 and reaugment every 25 epochs for the full task, and $k = 4$, the (9×1) architecture, batch size 1 and reaugment every epoch for the VerSe task. In both cases we use the default augmentation level and $\alpha = \beta = 1$. As evaluation metrics we use the VerSe metrics *identification rate* (ratio of ground-truth body keypoints for which the closest predicted point is correctly classified and within 20mm) and d_{mean} (mean distance of correctly identified body keypoints to their 1-NN predictions) [24]. Furthermore we evaluate the edge and illegitimacy binary predictions by their F_1 scores. Since the identification rate is largely unaffected by false-positive predicted keypoints, we disable legitimacy predictions unless for specific legitimacy prediction experiments to help comparability.

We compare our methods to two baselines: For node prediction, we use a Hidden Markov Model (HMM) [2] that is fitted to the training data using the Baum-Welch algorithm using the *pomegranate* library [23]. The HMM gets the predicted segment labels in sequence. Dealing with false-positive detection outliers is very difficult for this baseline, therefore we filter out non-legitimate detections for the HMM inputs to get a fairer comparison, making the task slightly easier for the HMM. For the VerSe challenge, we also compare our results to the top papers reported in the VerSe challenge [24]. For edge prediction, we compare our method to Hungarian matching on the keypoints from the CNN pipeline.

4.4 Results

The results on our introduced dataset are shown in Table 1. Our method outperforms the baseline methods, both in identification rate and edge accuracy. We also show results on a 'hard' subset, selected as the data samples where either of the methods disagree with the ground truth. This subset contains 47 scans and represents harder cases where the association is not trivial. We give p-values of the Wilcoxon signed-ranked test against the respective baseline for numbers that are better than the baseline. Figure 2 shows a qualitative example where the baseline method fails due to false-positive and false-negative misdetections in the input (these errors in the input were generated by augmentations and are more extreme than usual, to demonstrate several typical baseline failures in one example). The GNN correctly learns to find the correct association and is not derailed by the misdetections.

Table 2. Hyperparameter comparisons on the 2118 spine dataset. Comparing GNN architectures, influence of enabling legitimacy predictions and of different augmentation strengths. Best values within each group highlighted in bold.

Method		identification rate	edge F_1 score	illegitimacy F_1
Single-head architectures	(7×1)	96.74	98.86	—
	(13×1)	**97.19**	98.81	—
	(1,5,1)	96.75	98.91	—
	(1,11,1)	97.10	98.67	—
Multi-head architectures: backbone, edge/node head	$(1 \times 1), (4 \times 1), (12 \times 1)$	96.94	99.16	—
	$(3 \times 1), (2 \times 1), (10 \times 1)$	96.87	99.00	—
	$(5 \times 1), (—), (8 \times 1)$	96.79	98.88	—
Dedicated edge architectures	(3×1)	—	99.28	—
	(5×1)	—	99.28	—
	(1,3,1)	—	**99.35**	—
With legitimacy prediction: different legit. loss weights	$\gamma = 0.1$	95.61	**98.96**	56.00
	$\gamma = 1.0$	**96.50**	98.75	62.41
	$\gamma = 5.0$	96.35	98.18	61.67
	$\gamma = 10.0$	96.42	98.96	**62.78**
Augmentations	None	95.27	96.83	—
	Light	97.18	**98.83**	—
	Default	**97.19**	98.81	—
	Heavy	97.03	98.67	—

Table 3. Results on the VerSe 2019 dataset (validation/test set). We compare our GNN architecture, our Hidden Markov baseline, and the reported numbers of the three top VerSe challenge entries [24]. Best values highlighted in bold.

Method		identification rate	d_{mean}	illegitimacy F_1
Single-head GNN	(9×1)	93.26/93.02 (p=3.1e-6/p = 5.2e-7)	1.28/1.43	—
With legitimacy prediction	$\gamma = 10.0$	90.75/87.51 (p=4.8e-6/p = 9.4e-7)	**1.23/1.32**	77.67/81.69
Baseline	Hidden Markov	48.59/49.06	1.32/1.45	—
VerSe challenge entries [24]	Payer C. [17]	95.65/**94.25**	4.27/4.80	—
	Lessmann N. [13]	89.86/90.42	14.12/7.04	—
	Chen M. [24]	**96.94**/86.73	4.43/7.13	—

The examples where our architecture fails typically have off-by-one errors in the output from the CNN: for example, the last thoracic vertebra is detected as a lumbar vertebra (usually in edge cases where the two types are hard to distinguish). Hence the GNN classifications of the lumbar segment, and possibly the thoracic segment, will be off by one (see Figure S2 in the supplementary).

Table 2 shows the performance differences of various GNN architectures. The single-head architecture with 13 individual layers performs the best on identification rate, although the 13-layer architecture with 11 shared layers performs very similarly. The architectures with fewer layers perform slightly better in edge accuracy since this task is less context dependent. Multi-head architectures perform slightly worse on identification, but retain a good edge accuracy. Training a model solely directed to edge classification does yield the highest edge accuracy, as expected. Enabling legitimacy predictions slightly degrades the performance, likely due to two facts: for one, an extra loss term is added, which makes the model harder to train. Also, the identification rate metric is not majorly affected by having additional false-positive detections, hence there is little to be gained in terms of this metric by filtering out false positives. An optimal legitimacy loss weighting seems to be $\lambda = 1.0$. Finally, augmentations help generalization of the model, but the amount of augmentations seems to have little effect.

Table 3 shows the result on the traditional VerSe body-identification task. Our method yields a competitive performance despite not being optimized on this task (identification rate slightly lower than the leaders but a better average distance to the landmarks).

5 Conclusion

We introduced a simple pipeline consisting of a CNN followed by a single GNN to perform complex vertebra localization, identification and keypoint association. We introduced a new more complex vertebra detection dataset that includes associated pedicles defining the full orientation of each vertebra, to test our method. We show that our method can learn to associate and classify correctly with a single GNN that performs simultaneous edge and node classification.

The method is fully trainable and avoids most heuristics of other methods. We also show competitive performance on the VerSe body-identification dataset, a dataset the method was not optimized for. We believe this method is general enough to be usable for many other detection and association tasks, which we will explore in the future.

References

1. Bandyopadhyay, S., Biswas, A., Murty, M.N., Narayanam, R.: Beyond node embedding: a direct unsupervised edge representation framework for homogeneous networks. arXiv preprint arXiv:1912.05140 (2019)
2. Baum, L.E., Petrie, T.: Statistical inference for probabilistic functions of finite state Markov chains. Ann. Math. Stat. **37**(6), 1554–1563 (1966)
3. Bourgeois, A.C., Faulkner, A.R., Pasciak, A.S., Bradley, Y.C.: The evolution of image-guided lumbosacral spine surgery. Ann. Transl. Med. **3**(5) (2015)
4. Brasó, G., Leal-Taixé, L.: Learning a neural solver for multiple object tracking. In: Proceedings of the IEEE/CVF Conference on Computer Vision and Pattern Recognition, pp. 6247–6257 (2020)
5. Bronstein, M.M., Bruna, J., Cohen, T., Veličković, P.: Geometric deep learning: grids, groups, graphs, geodesics, and gauges. arXiv preprint arXiv:2104.13478 (2021)
6. Burns, J.E., Yao, J., Muñoz, H., Summers, R.M.: Automated detection, localization, and classification of traumatic vertebral body fractures in the thoracic and lumbar spine at ct. Radiology **278**(1), 64 (2016)

7. Chen, H., et al.: Automatic localization and identification of vertebrae in spine CT via a joint learning model with deep neural networks. In: Navab, N., Hornegger, J., Wells, W.M., Frangi, A.F. (eds.) MICCAI 2015. LNCS, vol. 9349, pp. 515–522. Springer, Cham (2015). https://doi.org/10.1007/978-3-319-24553-9_63

8. Defazio, A., Jelassi, S.: Adaptivity without compromise: a momentumized, adaptive, dual averaged gradient method for stochastic optimization. J. Mach. Learn. Res. **23**, 1–34 (2022)

9. Fey, M., Lenssen, J.E.: Fast graph representation learning with PyTorch geometric. arXiv preprint arXiv:1903.02428 (2019)

10. Glocker, B., Zikic, D., Konukoglu, E., Haynor, D.R., Criminisi, A.: Vertebrae localization in pathological spine CT via dense classification from sparse annotations. In: Mori, K., Sakuma, I., Sato, Y., Barillot, C., Navab, N. (eds.) MICCAI 2013. LNCS, vol. 8150, pp. 262–270. Springer, Heidelberg (2013). https://doi.org/10.1007/978-3-642-40763-5_33

11. Gori, M., Monfardini, G., Scarselli, F.: A new model for learning in graph domains. In: Proceedings. 2005 IEEE International Joint Conference on Neural Networks, vol. 2, pp. 729–734 (2005)

12. Kipf, T., Fetaya, E., Wang, K.C., Welling, M., Zemel, R.: Neural relational inference for interacting systems. In: International Conference on Machine Learning, pp. 2688–2697. PMLR (2018)

13. Lessmann, N., Van Ginneken, B., De Jong, P.A., Išgum, I.: Iterative fully convolutional neural networks for automatic vertebra segmentation and identification. Med. Image Anal. **53**, 142–155 (2019)

14. Liao, H., Mesfin, A., Luo, J.: Joint vertebrae identification and localization in spinal CT images by combining short-and long-range contextual information. IEEE Trans. Med. Imaging **37**(5), 1266–1275 (2018)

15. Lin, J.J., Lee, G.H.: Learning spatial context with graph neural network for multi-person pose grouping. In: 2021 IEEE International Conference on Robotics and Automation (ICRA), pp. 4230–4236. IEEE (2021)

16. Meng, D., Mohammed, E., Boyer, E., Pujades, S.: Vertebrae localization, segmentation and identification using a graph optimization and an anatomic consistency cycle. In: Lian, C., Cao, X., Rekik, I., Xu, X., Cui, Z. (eds.) MLMI 2022. LNCS, vol. 13583, pp. 307–317. Springer, Cham (2022). https://doi.org/10.1007/978-3-031-21014-3_32

17. Payer, C., Stern, D., Bischof, H., Urschler, M.: Vertebrae localization and segmentation with SpatialConfiguration-net and U-net. In: Large Scale Vertebrae Segmentation Challenge 2019 (2019)

18. Payer, C., Stern, D., Bischof, H., Urschler, M.: Coarse to fine vertebrae localization and segmentation with SpatialConfiguration-net and U-net. In: VISIGRAPP (5: VISAPP), pp. 124–133 (2020)

19. Qin, X., Zhang, Z., Huang, C., Dehghan, M., Zaiane, O.R., Jagersand, M.: U2-net: going deeper with nested U-structure for salient object detection. Pattern Recogn. **106**, 107404 (2020)

20. Reddy, N.D., Vo, M., Narasimhan, S.G.: Occlusion-net: 2D/3D occluded keypoint localization using graph networks. In: Proceedings of the IEEE/CVF Conference on Computer Vision and Pattern Recognition, pp. 7326–7335 (2019)

21. Ronneberger, O., Fischer, P., Brox, T.: U-net: convolutional networks for biomedical image segmentation. In: Navab, N., Hornegger, J., Wells, W.M., Frangi, A.F. (eds.) MICCAI 2015. LNCS, vol. 9351, pp. 234–241. Springer, Cham (2015). https://doi.org/10.1007/978-3-319-24574-4_28

22. Roth, H.R., Farag, A., Turkbey, E., Lu, L., Liu, J., Summers, R.M.: Data from pancreas-CT. The cancer imaging archive. IEEE Trans. Image Process. (2016)
23. Schreiber, J.: Pomegranate: fast and flexible probabilistic modeling in Python. J. Mach. Learn. Res. **18**(1), 5992–5997 (2017)
24. Sekuboyina, A., et al.: Verse: a vertebrae labelling and segmentation benchmark for multi-detector CT images. Med. Image Anal. **73**, 102166 (2021)
25. Smith, K., et al.: Data from CT_colonography. Cancer Imaging Arch. (2015)
26. Wu, Z., Pan, S., Chen, F., Long, G., Zhang, C., Philip, S.Y.: A comprehensive survey on graph neural networks. IEEE Trans. Neural Netw. Learn. Syst. **32**(1), 4–24 (2020)
27. Yang, D., et al.: Automatic vertebra labeling in large-scale 3D CT using deep image-to-image network with message passing and sparsity regularization. In: Niethammer, M., et al. (eds.) IPMI 2017. LNCS, vol. 10265, pp. 633–644. Springer, Cham (2017). https://doi.org/10.1007/978-3-319-59050-9_50

Imitation Learning from Expert Video Data for Dissection Trajectory Prediction in Endoscopic Surgical Procedure

Jianan Li[1], Yueming Jin[2], Yueyao Chen[1], Hon-Chi Yip[3], Markus Scheppach[4], Philip Wai-Yan Chiu[5], Yeung Yam[6], Helen Mei-Ling Meng[7], and Qi Dou[1(✉)]

[1] Department of Computer Science and Engineering,
The Chinese University of Hong Kong, Hong Kong, China
qidou@cuhk.edu.hk
[2] Department of Biomedical Engineering and Department of Electrical and
Computer Engineering, National University of Singapore, Singapore, Singapore
[3] Department of Surgery, The Chinese University of Hong Kong, Hong Kong, China
[4] Internal Medicine III - Gastroenterology,
University Hospital of Augsburg, Augsburg, Germany
[5] Multi-scale Medical Robotics Center and The Chinese University of Hong Kong,
Hong Kong, China
[6] Department of Mechanical and Automation Engineering,
The Chinese University of Hong Kong, Hong Kong, China
[7] Centre for Perceptual and Interactive Intelligence and The Chinese University of
Hong Kong, Hong Kong, China

Abstract. High-level cognitive assistance, such as predicting dissection trajectories in Endoscopic Submucosal Dissection (ESD), can potentially support and facilitate surgical skills training. However, it has rarely been explored in existing studies. Imitation learning has shown its efficacy in learning skills from expert demonstrations, but it faces challenges in predicting uncertain future movements and generalizing to various surgical scenes. In this paper, we introduce imitation learning to the formulated task of learning how to suggest dissection trajectories from expert video demonstrations. We propose a novel method with implicit diffusion policy imitation learning (iDiff-IL) to address this problem. Specifically, our approach models the expert behaviors using a joint state-action distribution in an implicit way. It can capture the inherent stochasticity of future dissection trajectories, therefore allows robust visual representations for various endoscopic views. By leveraging the diffusion model in policy learning, our implicit policy can be trained and sampled efficiently for accurate predictions and good generalizability. To achieve conditional sampling from the implicit policy, we devise a forward-process guided action inference strategy that corrects the state mismatch. We collected a private ESD video dataset with 1032 short clips to validate our method. Experimental results demonstrate that our solution outperforms SOTA imitation learning methods on our formulated task. To the best of our knowledge, this is the first work applying imitation learning for surgical skill learning with respect to dissection trajectory prediction.

© The Author(s), under exclusive license to Springer Nature Switzerland AG 2023
H. Greenspan et al. (Eds.): MICCAI 2023, LNCS 14228, pp. 494–504, 2023.
https://doi.org/10.1007/978-3-031-43996-4_47

Keywords: Imitation Learning · Surgical Trajectory Prediction · Endoscopic Submucosal Dissection · Surgical Data Science

1 Introduction

Despite that deep learning models have shown success in surgical data science to improve the quality of surgical intervention [20–22], such as intelligent workflow analysis [7,13] and scene understanding [1,28], research on higher-level cognitive assistance for surgery still remains underexplored. One essential task is supporting decision-making on dissection trajectories [9,24,29], which is challenging yet crucial for ensuring surgical safety. Endoscopic Submucosal Dissection (ESD), a surgical procedure for treating early gastrointestinal cancers [2,30], involves multiple dissection actions that require considerable experience to determine the optimal dissection trajectory. Informative suggestions for dissection trajectories can provide helpful cognitive assistance to endoscopists, for mitigation of intraoperative errors, reducing risks of complications [15], and facilitating surgical skill training [17]. However, predicting the desired trajectory for future time frames based on the current endoscopic view is challenging. First, the decision of dissection trajectories is complicated and depends on numerous factors such as safety margins surrounding the tumor. Second, dynamic scenes and poor visual conditions may further hamper scene recognition [27]. To date, there is still no work on data-driven solutions to predict such dissection trajectories, but we argue that it is possible to reasonably learn this skill from expert demonstrations based on video data.

Imitation learning has been widely studied in various domains [11,16,18] with its good ability to learn complex skills, but it still needs adaptation and improvement when being applied to learn dissection trajectory from surgical data. One challenge arises from the inherent uncertainty of future trajectories. Supervised learning such as Behavior Cloning (BC) [3] tends to average all possible prediction paths, which leads to inaccurate predictions. While advanced probabilistic models are employed to capture the complexity and variability of dissection trajectories [14,19,25], how to ensure reliable predictions across various surgical scenes still remains a great challenge. To overcome these issues, implicit models are emerging for policy learning, inspiring us to rely on implicit Behavior Cloning (iBC) [5], which can learn robust representations by capturing the shared features of both visual inputs and trajectory predictions with a unified implicit function, yielding superior expressivity and visual generalizability. However, these methods still bear their limitations. For instance, approaches leveraging energy-based models (EBMs) [4–6,12] suffer from intensive computations due to reliance on the Langevin dynamics, which leads to a slow training process. In addition, the model performance can be sensitive to data distribution and the noise in training data would result in unstable trajectory predictions.

In this paper, we explore an interesting task of predicting dissection trajectories in ESD surgery via imitation learning on expert video data. We propose Implicit Diffusion Policy Imitation Learning (iDiff-IL), a novel imitation learning

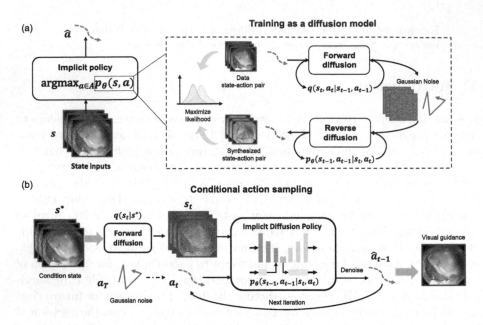

Fig. 1. Overview of our imitation learning method for surgical dissection trajectory prediction. (a) illustrates the modeling of the implicit policy, and its training process. We train a diffusion model to approximate the joint state-action distribution, and (b) depicts the inference loop for trajectory prediction with the forward-diffusion guidance.

approach for dissection trajectory prediction. To effectively model the surgeon's behaviors and handle the large variation of surgical scenes, we leverage implicit modeling to express expert dissection skills. To address the limitations of inefficient training and unstable performance associated with EBM-based implicit policies, we formulate the implicit policy using an unconditional diffusion model, which demonstrates remarkable ability in representing complex high-dimensional data distribution for videos. Subsequently, to obtain predictions from the implicit policy, we devise a conditional action inference strategy with the guidance of forward-diffusion, which further improves the prediction accuracy. For experimental evaluation, we collected a surgical video dataset of ESD procedures, and preprocessed 1032 short clips with dissection trajectories labelled. Results show that our method achieves superior performances in different contexts of surgical scenarios compared with representative popular imitation learning methods.

2   Method

In this section, we describe our approach iDiff-IL, which learns to predict the dissection trajectory from expert video data using the implicit diffusion policy. An overview of our method is shown in Fig. 1. We first present the formulation of the task and the solution with implicit policy for dissection trajectory learning. Next, we present how to train the implicit policy as an unconditional generative diffusion model. Finally, we show the action inference strategy with forward-diffusion guidance which produces accurate trajectory predictions with our implicit diffusion policy.

2.1 Implicit Modeling for Surgical Dissection Decision-Making

In our approach, we formulate the dissection trajectory prediction to an imitation learning from expert demonstrations problem, which defines a Markov Decision Process (MDP) $\mathcal{M} = (\mathcal{S}, \mathcal{A}, \mathcal{T}, \mathcal{D})$, comprising of state space $\mathcal{S}$, action set $\mathcal{A}$, state transition distribution $\mathcal{T}$, and expert demonstrations $\mathcal{D}$. The goal is to learn a prediction policy $\pi^*(a|s)$ from a set of expert demonstrations $\mathcal{D}$. The input state of the policy is a clip of video frames $s = \{I_{t-L+1}, I_{t-L+2}, \ldots, I_t\}$, $I_t \in \mathbb{R}^{H \times W \times 3}$ and the output is an action distribution of a sequence of 2D coordinates $a = \{y_{t+1}, y_{t+2}, \ldots, y_{t+N}\}, y_t \in \mathbb{R}^2$ indicating the future dissection trajectory projected to the image space.

In order to obtain the demonstrated dissection trajectories from the expert video data, we first manually annotate the dissection trajectories on the video frame according to the moving trend of the instruments observed from future frames, then create a dataset $\mathcal{D} = \{(s, a)_i\}_{i=0}^{M}$ containing M pairs of video clip (state) and dissection trajectory (action).

To precisely predict the expert dissection behaviors and effectively learn generalizable features from the expert demonstrations, we use the implicit model as our imitation policy. Extending the formulation in [5], we model the dissection trajectory prediction policy to a maximization of the joint state-action probability density function $\arg\max_{a \in \mathcal{A}} p_\theta(s, a)$ instead of an explicit mapping $F_\theta(s)$. The optimal action is derived from the policy distribution conditioned on the state s, and $p_\theta(s, a)$ represents the joint state-action distribution.

To learn the implicit policy from the demonstrations, we adopt the Behavior Cloning objective which is to essentially minimize the Kullback-Leibler (KL) divergence between the learning policy $\pi_\theta(a|s)$ and the demonstration distribution $\mathcal{D}$, also equivalent to maximize the expected log-likelihood of the joint state-action distribution, as shown:

$$\max_\theta \mathbb{E}_{(s,a) \sim \mathcal{D}}[\log \pi_\theta(a|s)] = \max_\theta \mathbb{E}_{(s,a) \sim \mathcal{D}}[\log p_\theta(s, a)]. \tag{1}$$

In this regard, the imitation of surgical dissection decision-making is converted to a distribution approximation problem.

2.2 Training Implicit Policy as Diffusion Models

Approximating the joint state-action distribution in Eq. 1 from the video demonstration data is challenging for previous EBM-based methods. To address the learning of implicit policy, we rely on recent advances in diffusion models. By representing the data using a continuous thermodynamics diffusion process, which can be discretized into a series of Gaussian transitions, the diffusion model is able to express complex high-dimensional distribution with simple parameterized functions. In addition, the diffusion process also serves as a form of data augmentation by adding a range of levels of noise to the data, which guarantees a better generalization in high-dimensional state space.

As shown in Fig. 1 (a), the diffusion model comprises a predefined forward diffusion process and a learnable reverse denoising process. The forward process gradually diffuses the original data $x_0 = (s, a)$, to a series of noised data $\{x_0, x_1, \cdots, x_T\}$ with a Gaussian kernel $q(x_t|x_{t-1})$, where T denotes the diffusion step. In the reverse process, the data is recovered via a parameterized Gaussian $p_\theta(x_{t-1}|x_t)$ iteratively. With the reverse process, the joint state-action distribution in the implicit policy can be expressed as:

$$p_\theta(x_0) = \sum_{x_{1:T}} p_\theta(x_{0:T}) = \sum_{x_{1:T}} p(x_T) \prod p_\theta(x_{t-1}|x_t) = \mathbb{E}_{p_\theta(x_{1:T})} p_\theta(x_0|x_1). \quad (2)$$

The probability of the noised data x_t in forward diffusion process is a Gaussian distribution expressed as $q(x_t|x_0) = \mathcal{N}(x_t, \sqrt{\alpha_t}x_0, (1 - \alpha_t)\boldsymbol{I})$, where α_t is a scheduled variance parameter, which can be referred from [10], and $\boldsymbol{I}$ is an identity matrix. The trainable reverse transition is a Gaussian distribution as well, whose posterior is $p_\theta(x_{t-1}|x_t) = \mathcal{N}(x_{t-1}; \mu_\theta(x_t, t), \Sigma_\theta(x_t, t))$, in which $\mu_\theta(x_t, t)$ and $\Sigma_\theta(x_t, t)$ are the means and the variances parameterized by a neural network.

To train the implicit diffusion policy, we maximize the log-likelihood of the state-action distribution in Eq. 1. Using the Evidence Lower Bound (ELBO) as the proxy, the likelihood maximization can be simplified to a noise prediction problem, more details can be referred to [10]. Noise prediction errors for the state and the action are combined using a weight $\gamma \in [0, 1]$ as the following:

$$\mathcal{L}_{noise}(\theta) = \mathbb{E}_{\epsilon, t, x_0}[(1 - \gamma)\|\epsilon_\theta^a(x_t, t) - \epsilon^a\| + \gamma\|\epsilon_\theta^s(x_t, t) - \epsilon^s\|], \quad (3)$$

where ϵ^s and ϵ^a are sampled from $\mathcal{N}(0, \boldsymbol{I}^s)$, $\mathcal{N}(0, \boldsymbol{I}^a)$ respectively. To better process features from video frames and trajectories of coordinates, we employ a variant of the UNet as the implicit diffusion policy network, where the trajectory information is fused into feature channels via MLP embedding layers. Then the trajectory noise is predicted by an MLP branch at the bottleneck layer.

2.3   Conditional Sampling with Forward-Diffusion Guidance

Since the training process introduced in Sect. 2.2 is for unconditional generation, the conventional sampling strategy through the reverse process will predict random trajectories in expert data. An intuitive way to introduce the condition into the inference is to input the video clip as the condition state s^* to the implicit diffusion policy directly, then only sample the action part. But there is a mismatch between the distribution of the state s^* and the s_t in the training process, which may lead to inaccurate predictions. Hence, we propose a sampling strategy to correct such distribution mismatch by introducing the forward-process guidance into the reverse sampling procedure.

Considering the reverse process of the diffusion model, the transition probability conditioned by s^* can be decomposed as:

$$p_\theta(x_{t-1}|x_t, s^*) = p_\theta(x_{t-1}|s_t, a_t, s^*) = p_\theta(x_{t-1}|x_t)q(s_t|s^*), \qquad (4)$$

where $x_t = (s_t, a_t)$, $p_\theta(x_{t-1}|x_t)$ denotes the learned denoising function of the implicit diffusion model, and $q(s_t|s^*)$ represents a forward diffusion process from the condition state to the t-th diffused state. Therefore, we can attain conditional sampling via the incorporation of forward-process guidance into the reverse sampling process of the diffusion model.

The schematic illustration of our sampling approach is shown in Fig. 1 (b). At the initial step $t = T$, action a_T is sampled from a pure Gaussian noise, whereas the input state s_T is diffused from the input video clip s^* through a forward-diffusion process. At the t-th step of the denoising loop, the action input a_t comes from the denoised action from the last time step, while the visual inputs s_t are still obtained from s^* via the forward diffusion process. The above forward diffusion process and the denoising step are repeated till $t = 0$. The final action $\hat{a}_0$ is the prediction from the implicit diffusion policy. The deterministic action can be obtained by taking the most probable samples during the reverse process.

3   Experiments

3.1   Experimental Dataset and Evaluation Metrics

Dataset. We evaluated the proposed approach on a dataset assembled from 22 videos of ESD surgery cases, which are collected from the Endoscopy Centre of the Prince of Wales Hospital in Hong Kong. All videos were recorded via Olympus microscopes operated by an expert surgeon with over 15 years of experience in ESD. Considering the inference speed, we downsampled the original videos to 2FPS frames which are resized to 128×128 in resolution. The input state is a 1.5-s length video clip containing 3 consecutive frames, and the expert dissection trajectory is represented by a 6-point polyline indicating the tool's movements in future 3 s. We totally annotated 1032 video clips, which contain 3 frames for each clip. We randomly selected 742 clips from 20 cases for training, consisting of 2226 frames, where 10% of these are for validation. The remaining 290 clips (consisting of 970 frames) were used for testing.

Experiment Setup. First, to study how the model performs on data within the same surgical context as the training data, we define a subset, referred as to the "in-the-context" testing set, which consists of consecutive frames selected from the same cases as included in the training data. Second, to assess the model's ability to generalize to visually distinct scenes, we created an "out-of-the-context" testing set that is composed of video clips sampled from 2 unseen surgical cases. The sizes of these two subsets are 224 and 66 clips, respectively.

Evaluation Metrics. To evaluate the performance of the proposed approach, we adopt several metrics, including commonly used evaluation metrics for trajectory prediction as used in [23, 26], including Average Displacement Error (ADE),

Table 1. Quantitative results on the in-the-context and the out-of-the-context data in metrics of ADE/FDE/FD. Values in parentheses denote video-clip wise standard deviation. The lower is the better.

Method	In-the-context						Out-of-the-context					
	ADE		FDE		FD		ADE		FDE		FD	
BC	10.43	(±5.52)	16.68	(±10.59)	24.92	(±14.59)	13.67	(±6.43)	17.03	(±12.50)	29.23	(±16.56)
iBC[5]	15.54	(±4.79)	22.66	(±8.06)	35.26	(±11.78)	15.81	(±4.66)	19.66	(±7.56)	31.66	(±9.14)
MID[8]	9.90	(±0.66)	15.26	(±1.35)	23.78	(±1.73)	12.42	(±1.89)	16.32	(±3.11)	27.04	(±4.79)
Ours	**9.47**	(±1.66)	**13.85**	(±2.01)	**21.43**	(±3.89)	**10.21**	(±3.17)	**14.14**	(±3.63)	**21.56**	(±5.97)

which respectively reports the overall deviations between the predictions and the ground truths, and Final Displacement Error (FDE) describing the difference from the moving target by computing the L2 distance between the last trajectory points. Besides, we also use the Fréchet Distance (FD) metric, to indicate the geometrical similarity between two temporal sequences. Pixel errors are used as units for all metrics, while the input images are in 128×128 resolution.

3.2   Comparison with State-of-the-Art Methods

To evaluate the proposed approach, we have selected popular baselines and state-of-the-art methods for comparison. We have chosen the fully supervised method, Behavior Cloning, as the baseline, which is implemented using a CNN-MLP network. In addition, we have included iBC [5], an EBM-based implicit policy learning method and MID [8], a diffusion-based trajectory prediction approach, as comparison state-of-the-art approaches.

As shown in Table 1, our method outperforms the comparison approaches in both "in-the-context" and "out-of-the-context" scenarios on all metrics. Compared with the diffusion-based method MID [8], our iDiff-IL is more effective in predicting long-term goals, particularly in the "out-of-the-context" scenes, with the evidence of 2.18 error reduction on FDE. For iBC [5], the performance did not meet our expectations and was even surpassed by the baseline. This exhibits the limitations of EBM-based methods in learning visual representations from complex endoscopic scenes. The superior results achieved by our method demonstrate the effectiveness of the diffusion model in learning the implicit policy from the expert video data. In addition, our method can learn generalizable dissection skills by exhibiting a lower standard deviation of the prediction errors compared to the BC, which severely suffers from over-fitting to the training data. The qualitative results are presented in Fig. 2. We selected three typical scenes in ESD surgery (i.e., submucosa dissection, mucosa dissection and mucosa incision), and showed the predictions of iDiff-IL accompanying the ground truth trajectories. From the results, our method can generate reasonable visual guidance aligning with the expert demonstrations on both evaluation sets.

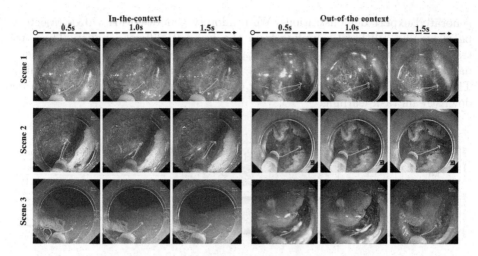

Fig. 2. Typical results of our imitation learning method under settings of in-the-context and out-of-the-context evaluations. Green and yellow arrows respectively denote ground truths and predictions of dissection trajectory. (Color figure online)

3.3 Ablation Study

Implicit Modeling. First, we examined the importance of using implicit modeling as the policy representation. We simulated the explicit form of the imitation policy by training a conditional diffusion model whose conditional input is a video clip. According to the bar charts in Fig. 3, the explicit diffusion policy shows a performance drop for both evaluation sets on ADE compared with the implicit form. The implicit modeling makes a more significant contribution in predicting within the "in-the-context" scenes, suggesting that the implicit model excels at capturing subtle changes in surgical scenes. While our method improves marginally compared with the explicit form on the "out-of-the-context" data, exhibiting a slighter over-fitting with a lower standard deviation.

Forward-Diffusion Guidance. We also investigated the necessity of the forward-diffusion guidance in conditional sampling for prediction accuracy. We remove the forward-diffusion guidance during the action sampling procedure so that the condition state is directly fed into the policy while sampling actions through the reverse process. As shown in Fig. 3, the implicit diffusion policy benefits more from the forward-diffusion guidance in the "in-the-context" scenes, achieving an improvement of 0.33 on ADE. When encountered with the unseen scenarios in "out-of-the-context" data, the performance improvement of such inference strategy is marginal.

Value of Synthetic Data. Since the learned implicit diffusion policy is capable of generating synthetic expert dissection trajectory data, which can potentially reduce the expensive annotation cost. To better explore the value of such synthetic expert data for downstream tasks, we train the baseline model with the generated expert demonstrations. We randomly generated 9K video-trajectory pairs by unconditional sampling from the implicit diffusion policy. Then, we train the BC model with different data, the pure expert data (real), synthetic data only (synt) and the mixed data with the real and the synthetic (mix). The table in Fig. 3 shows the synthetic data is useful as the augmented data for downstream task learning.

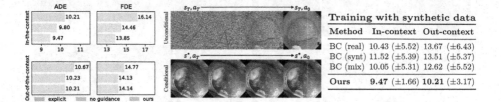

Fig. 3. **Left:** ablation study of key method components; **Middle:** visualization of reverse processes of unconditional/conditional sampling from implicit policy; **Right:** performance of BC trained with synthetic data v.s. our method on ADE.

4   Conclusion

This paper presents a novel approach on imitation learning from expert video data, in order to achieve dissection trajectory prediction in endoscopic surgical procedure. Our iDiff-IL method utilizes a diffusion model to represent the implicit policy, which enhances the expressivity and visual generalizability of the model. Experimental results show that our method outperforms state-of-the-art approaches on the evaluation dataset, demonstrating the effectiveness of our approach for learning dissection skills in various surgical scenarios. We hope that our work can pave the way for introducing the concept of learning from expert demonstrations into surgical skill modelling, and motivate future exploration on higher-level cognitive assistance in computer-assisted intervention.

Acknowledgement. This work was supported in part by Shenzhen Portion of Shenzhen-Hong Kong Science and Technology Innovation Cooperation Zone under HZQB-KCZYB-20200089, in part by Hong Kong Innovation and Technology Commission Project No. ITS/237/21FP, in part by Hong Kong Research Grants Council Project No. T45-401/22-N, in part by Science, Technology and Innovation Commission of Shenzhen Municipality Project No. SGDX20220530111201008.

References

1. Allan, M., et al.: 2018 robotic scene segmentation challenge. arXiv preprint arXiv:2001.11190 (2020)

2. Chiu, P.W.Y., et al.: Endoscopic submucosal dissection (ESD) compared with gastrectomy for treatment of early gastric neoplasia: a retrospective cohort study. Surg. Endosc. **26**, 3584–3591 (2012)
3. Codevilla, F., Santana, E., López, A.M., Gaidon, A.: Exploring the limitations of behavior cloning for autonomous driving. In: Proceedings of the IEEE/CVF International Conference on Computer Vision, pp. 9329–9338 (2019)
4. Du, Y., Mordatch, I.: Implicit generation and modeling with energy based models. In: Advances in Neural Information Processing Systems, vol. 32 (2019)
5. Florence, P., et al.: Implicit behavioral cloning. In: Conference on Robot Learning, pp. 158–168. PMLR (2022)
6. Ganapathi, A., Florence, P., Varley, J., Burns, K., Goldberg, K., Zeng, A.: Implicit kinematic policies: unifying joint and cartesian action spaces in end-to-end robot learning. In: 2022 International Conference on Robotics and Automation (ICRA), pp. 2656–2662. IEEE (2022)
7. Garrow, C.R., et al.: Machine learning for surgical phase recognition: a systematic review. Ann. Surg. **273**(4), 684–693 (2021)
8. Gu, T., et al.: Stochastic trajectory prediction via motion indeterminacy diffusion. In: Proceedings of the IEEE/CVF Conference on Computer Vision and Pattern Recognition, pp. 17113–17122 (2022)
9. Guo, J., Sun, Y., Guo, S.: A novel trajectory predicting method of catheter for the vascular interventional surgical robot. In: IEEE International Conference on Mechatronics and Automation, pp. 1304–1309 (2020)
10. Ho, J., Jain, A., Abbeel, P.: Denoising diffusion probabilistic models. In: Advances in Neural Information Processing Systems, vol. 33, pp. 6840–6851 (2020)
11. Hussein, A., Gaber, M.M., Elyan, E., Jayne, C.: Imitation learning: a survey of learning methods. ACM Comput. Surv. **50**(2), 1–35 (2017)
12. Jarrett, D., Bica, I., van der Schaar, M.: Strictly batch imitation learning by energy-based distribution matching. In: Advances in Neural Information Processing Systems, vol. 33, pp. 7354–7365 (2020)
13. Jin, Y., Long, Y., Gao, X., Stoyanov, D., Dou, Q., Heng, P.A.: Trans-svnet: hybrid embedding aggregation transformer for surgical workflow analysis. Int. J. Comput. Assist. Radiol. Surg. **17**(12), 2193–2202 (2022)
14. Ke, L., Choudhury, S., Barnes, M., Sun, W., Lee, G., Srinivasa, S.: Imitation learning as f-divergence minimization. In: Algorithmic Foundations of Robotics XIV: Proceedings of the Fourteenth Workshop on the Algorithmic Foundations of Robotics, vol. 14. pp. 313–329 (2021)
15. Kim, E., et al.: Factors predictive of perforation during endoscopic submucosal dissection for the treatment of colorectal tumors. Endoscopy **43**(07), 573–578 (2011)
16. Kläser, K., et al.: Imitation learning for improved 3D pet/MR attenuation correction. Med. Image Anal. **71**, 102079 (2021)
17. Laurence, J.M., Tran, P.D., Richardson, A.J., Pleass, H.C., Lam, V.W.: Laparoscopic or open cholecystectomy in cirrhosis: a systematic review of outcomes and meta-analysis of randomized trials. HPB **14**(3), 153–161 (2012)
18. Le Mero, L., Yi, D., Dianati, M., Mouzakitis, A.: A survey on imitation learning techniques for end-to-end autonomous vehicles. IEEE Trans. Intell. Transp. Syst. **23**(9), 14128–14147 (2022)
19. Li, Y., Song, J., Ermon, S.: Infogail: interpretable imitation learning from visual demonstrations. In: Advances in Neural Information Processing Systems, vol. 30 (2017)
20. Loftus, T.J., et al.: Artificial intelligence and surgical decision-making. JAMA Surg. **155**(2), 148–158 (2020)

21. Maier-Hein, L., et al.: Surgical data science-from concepts toward clinical translation. Med. Image Anal. **76**, 102306 (2022)
22. Maier-Hein, L., et al.: Surgical data science for next-generation interventions. Nat. Biomed. Eng. **1**(9), 691–696 (2017)
23. Mohamed, A., Qian, K., Elhoseiny, M., Claudel, C.: Social-STGCNN: a social spatio-temporal graph convolutional neural network for human trajectory prediction. In: Proceedings of the IEEE/CVF Conference on Computer Vision and Pattern Recognition, pp. 14424–14432 (2020)
24. Qin, Y., Feyzabadi, S., Allan, M., Burdick, J.W., Azizian, M.: Davincinet: joint prediction of motion and surgical state in robot-assisted surgery. In: IEEE/RSJ International Conference on Intelligent Robots and Systems, pp. 2921–2928. IEEE (2020)
25. Ren, A., Veer, S., Majumdar, A.: Generalization guarantees for imitation learning. In: Conference on Robot Learning, pp. 1426–1442. PMLR (2021)
26. Sun, J., Jiang, Q., Lu, C.: Recursive social behavior graph for trajectory prediction. In: Proceedings of the IEEE/CVF Conference on Computer Vision and Pattern Recognition, pp. 660–669 (2020)
27. Wang, J., et al.: Real-time landmark detection for precise endoscopic submucosal dissection via shape-aware relation network. Med. Image Anal. **75**, 102291 (2022)
28. Wang, Y., Long, Y., Fan, S.H., Dou, Q.: Neural rendering for stereo 3D reconstruction of deformable tissues in robotic surgery. In: International Conference on Medical Image Computing and Computer-Assisted Intervention, pp. 431–441 (2022)
29. Wang, Z., Yan, Z., Xing, Y., Wang, H.: Real-time trajectory prediction of laparoscopic instrument tip based on long short-term memory neural network in laparoscopic surgery training. Int. J. Med. Robot. Comput. Assist. Surg. **18**(6), e2441 (2022)
30. Zhang, J., et al.: Symmetric dilated convolution for surgical gesture recognition. In: Martel, A.L., et al. (eds.) MICCAI 2020. LNCS, vol. 12263, pp. 409–418. Springer, Cham (2020). https://doi.org/10.1007/978-3-030-59716-0_39

Surgical Action Triplet Detection by Mixed Supervised Learning of Instrument-Tissue Interactions

Saurav Sharma[1]([✉]), Chinedu Innocent Nwoye[1], Didier Mutter[2,3], and Nicolas Padoy[1,2]

[1] ICube, University of Strasbourg, CNRS, Strasbourg, France
{ssharma,nwoye,npadoy}@unistra.fr
[2] IHU Strasbourg, Strasbourg, France
[3] University Hospital of Strasbourg, Strasbourg, France
didier.mutter@ihu-strasbourg.eu

Abstract. Surgical action triplets describe instrument-tissue interactions as ⟨instrument, verb, target⟩ combinations, thereby supporting a detailed analysis of surgical scene activities and workflow. This work focuses on surgical action triplet *detection*, which is challenging but more precise than the traditional triplet *recognition* task as it consists of joint (1) localization of surgical instruments and (2) recognition of the surgical action triplet associated with every localized instrument. Triplet detection is highly complex due to the lack of spatial triplet annotation. We analyze how the amount of instrument spatial annotations affects triplet detection and observe that accurate instrument localization does not guarantee a better triplet detection due to the risk of erroneous associations with the verbs and targets. To solve the two tasks, we propose **MCIT-IG**, a two-stage network, that stands for *Multi-Class Instrument-aware Transformer - Interaction Graph*. The **MCIT** stage of our network models per class embedding of the targets as additional features to reduce the risk of misassociating triplets. Furthermore, the **IG** stage constructs a bipartite dynamic graph to model the interaction between the instruments and targets, cast as the verbs. We utilize a mixed-supervised learning strategy that combines weak target presence labels for **MCIT** and pseudo triplet labels for **IG** to train our network. We observed that complementing minimal instrument spatial annotations with target embeddings results in better triplet detection. We evaluate our model on the CholecT50 dataset and show improved performance on both instrument localization and triplet detection, topping the leaderboard of the CholecTriplet challenge in MICCAI 2022.

Keywords: Activity recognition · surgical action triplets · attention · graph · CholecT50 · instrument detection · triplet detection

Supplementary Information The online version contains supplementary material available at https://doi.org/10.1007/978-3-031-43996-4_48.

1   Introduction

Surgical workflow analysis in endoscopic procedures aims to process large streams of data [8] from the operating room (OR) to build context awareness systems [19]. These systems aim to provide assistance to the surgeon in decision making [9] and planning [6]. Most of these systems focus on coarse-grained recognition tasks such as phase recognition [15], instrument spatial localization and skill assessment [5]. Surgical action triplets [11], defined as ⟨instrument, verb, target⟩, introduce the fine-grained modeling of elements present in an endoscopic scene. In cataract surgery, [7] adopts similar triplet formulation and also provides bounding box details for both instrument and targets. Another related work [1], in prostatectomy, uses bounding box annotations for surgical activities defined as ⟨verb, anatomy⟩. On laparoscopic cholecystectomy surgical data, existing approaches [10,13,16] focus on the challenging triplet recognition task, where the objective is to predict the presence of triplets but ignores their spatial locations in a video frame.

A recently conducted endoscopic vision challenge [14] introduced the triplet detection task that requires instrument localization and its association with the triplet. While most of the contributed methods employ weak supervision to learn the instrument locations by exploiting the model's class activation map (CAM), a few other methods exploit external surgical datasets offering complementary instrument spatial annotations. The significant number of triplet classes left the weakly supervised approaches at subpar performance compared to fully supervised methods. The results of the CholecTriplet2022 challenge [14] have led to two major observations. First, imprecise localization from the CAM-based methods impairs the final triplet detection performance, where the best weakly-supervised method reaches only 1.47% detection mean average precision. Second, the correct association of triplet predictions and their spatial location is difficult to achieve with instrument position information alone. This is mainly due to the possible occurrence of multiple instances of the same instrument and many options of targets/verbs that can be associated with one instrument instance. We set two research questions following these observations: (1) since manual annotation of instrument spatial locations is expensive and tedious, how can we use learned target/verb features to supplement a minimal amount of instrument spatial annotations? (2) since instrument cues are insufficient for better triplet association, how can we generate valid representative features of the targets/verbs that do not require additional spatial labels?

To tackle these research questions, we propose a fully differentiable two-stage pipeline, **MCIT-IG**, that stands for *Multi-Class Instrument-aware Transformer - Interaction Graph*. The MCIT-IG relies on instrument spatial information that we generate with Deformable DETR [22], trained on a subset of Cholec80 [17] annotated with instrument bounding boxes. In the first stage, MCIT, a lightweight transformer, learns class wise embeddings of the target influenced by instrument spatial semantics and high level image features. This allows the embeddings to capture global instrument association features useful in IG. We train MCIT with target binary presence label, providing weak

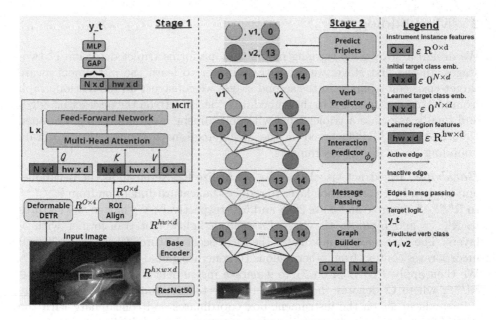

Fig. 1. Model Overview: In Stage 1, **MCIT** learns instrument-aware target class embeddings. In Stage 2, **IG** enforces association between instrument instances and target class embeddings and learns verb on the pairwise edge features. Green edges denotes all interactions and blue edges denotes active instrument-target pair with verb **v**. Red edges denotes no interaction. (Color figure online)

supervision. In the second stage, IG creates an interaction graph that performs dynamic association between the detected instrument instances and the target embeddings, and learns the verb on the interacting edge features, thereby detecting triplets. To train IG, triplet labels for the detected instrument instances are needed, which is unavailable. To circumvent this situation, we generate pseudo triplet labels for the detected instrument instances using the available binary triplet presence labels. In this manner, we provide mixed supervision to train MCIT-IG.

We hypothesize that a precise instrument detector can reveal additional instrument-target associations as more instrument instances are detected. To test this hypothesis, we conduct a study to investigate how the accuracy of the instrument detector affects triplet detection. We train an instrument detector with limited spatial data and evaluate the impact on triplet detection. We find that enhancing instrument localization is strongly linked to improved triplet detection performance. Finally, we evaluate our model on the challenge split of CholecT50 [13,14]. We report both improved instrument localization and triplet detection performance, thanks to the graph-based dynamic association of instrument instances with targets/verbs, that captures the triplet label.

2   Methodology

We design a novel deep-learning model that performs triplet detection in two-stages. In the first stage, we use a transformer to learn the instrument-aware target class embeddings. In the second stage, we construct an interaction graph from instrument instances to the embeddings, learn verb on the interacting edges and finally associate a triplet label with instrument instances. Using a trained Deformable DETR [22] based on the MMDetection [2] framework, we obtain bounding boxes for the instruments.

Backbone: To extract visual features, we utilize ResNet50 [4] as the backbone, and apply a 1×1 convolution layer to reduce the feature dimension from $\mathbb{R}^{h \times w \times c}$ to $\mathbb{R}^{h \times w \times d}$, where c and d are 2048 and 512 respectively. We flatten the features to $\mathbb{R}^{hw \times d}$ and input to the *Base Encoder*, a lightweight transformer with b_l layers. The base encoder modulates the local scene features from ResNet50 and incorporates context from other regions to generate global features $\mathcal{F}_b \in \mathbb{R}^{hw \times d}$. We then apply *ROIAlign* on $\mathcal{F}_b$ to generate instrument instance features $\mathcal{F}_r \in \mathbb{R}^{O \times d}$, where O denotes the number of detected instruments. We also apply a linear layer Φ_b on the instrument box coordinates and concatenate with the embeddings of predicted instrument class category to get d-dimensional features, and finally fuse with $\mathcal{F}_r$ using linear layer Φ_f to produce final d-dimensional instrument features $\mathcal{F}_i$.

Learning Instrument-Aware Target Class Embeddings: To learn target features, we introduce a ***M****ulti-****C****lass* ***I****nstrument-aware* ***T****ransformer (**MCIT**) that generates embeddings for each target class. In the standard transformer [3], a single class-agnostic token models the class distribution, but dilutes crucial class specific details. Inspired from [21], MCIT utilizes N class tokens, $\mathcal{N}_t \in \mathbb{R}^{N \times d}$, to learn class-specific embeddings of the target. However, to make the class embeddings aware of the instruments in the scene, we use the instrument features $\mathcal{F}_i$ along with class tokens and region features. Specifically, MCIT takes input $(\mathcal{F}_b, \mathcal{F}_i)$ and creates learnable queries of dimension $\mathbb{R}^{(hw+N) \times d}$ and keys, values of dimension $\mathbb{R}^{(hw+N+O) \times d}$ to compute attention. Then, MCIT applies t_l layers of attention to generate the output sequence, $\mathcal{P}_t \in \mathbb{R}^{(hw+N) \times d}$. The learned class embeddings are averaged across N and input to a linear layer Φ_t to generate logits $y_t \in \mathbb{R}^N$ following Eq. 1:

$$y_t = \Phi_t \left(\frac{1}{N} \sum_{k=hw}^{hw+N} \mathcal{P}[k, :] \right). \tag{1}$$

MCIT learns meaningful class embeddings of the target enriched with visual and position semantics of the instruments. This instrument-awareness is useful to identify the interacting instrument-target pairs.

Learning Instrument-Target Interactions: To learn the interaction of the instrument and the target, we introduce a graph based framework ***I****nteraction-****G****raph (**IG**) that relies on the discriminative features of instrument instances and target class embeddings. We create an unidirectional complete bipartite graph, $\mathcal{G} = (\mathcal{U}, \mathcal{V}, \mathcal{E})$, where $|\mathcal{U}| = O$ and $|\mathcal{V}| = N$ denotes the source and destination nodes respectively, and edges $\mathcal{E} = \{e_v^u, u \in \mathcal{U} \wedge v \in \mathcal{V}\}$. The node features of $\mathcal{U}$ and $\mathcal{V}$ correspond to the detected instrument instance features $\mathcal{F}_i$ and target class embeddings $\mathcal{N}_t$ respectively. We further project the nodes features to a lower dimensional space d' using a linear layer Φ_p. This setup provides an intuitive way to model instrument-tissue interactions as a set of active edges. Next, we apply message passing using GAT [18], that aggregates instrument features in $\mathcal{U}$ and updates target class embeddings at $\mathcal{V}$.

Learning Verbs: We concatenate the source and destination node features of all the edges $\mathcal{E}$ in $\mathcal{G}$ to construct the edge feature $\mathcal{E}_f = \{e_f, e_f \in \mathbb{R}^{2d'}\}$. Then, we compute the edge confidence score $\mathcal{E}_s = \{e_s, e_s \in \mathbb{R}\}$ for all edges $\mathcal{E}$ in $\mathcal{G}$ by applying a linear layer Φ_e on $\mathcal{E}_f$. As a result, shown in Fig. 1 Stage 2, only active edges (in blue) remain and inactive edges (in red) are dropped based on a threshold. The active edge indicates the presence of interaction between an instrument instance and the target class. To identify the verb, we apply a linear layer Φ_v on $\mathcal{E}_f$ and generate verb logits $y_v \in \mathbb{R}^{V+1}$, where V is the number of verb classes with an additional 1 to denote background class.

Triplet Detection: To perform target and verb association for each instrument instance i, first we select the active edge e_j^i that corresponds to the target class $j = argmax(\alpha(\mathcal{E}_s^i))$, where α denotes softmax function and $\mathcal{E}_s^i = \{e_s^u, \forall e_s^u \in \mathcal{E}_s \wedge u = i\}$. For the selected edge $e' = e_j^i$, we apply softmax on the verb logits to obtain the verb class id, $k = argmax(\alpha(y_v^{e'}))$. The final score for the triplet $\langle i, k, j \rangle$ is given by $p(e_j^i) \times p(y_{v_k}^{e'})$, where p denotes the probability score.

Mixed Supervision: We train our model in two stages. In the first stage, we train MCIT to learn target classwise embeddings with target binary presence label with weighted binary cross entropy on target logits y_t for multi-label classification task following Eq. 2:

$$L_t = \sum_{c=1}^{C} \frac{-1}{N} \left(W_c y_c log\left(\sigma(\hat{y}_c)\right) + (1 - y_c)log\left((1 - \sigma(\hat{y}_c))\right) \right), \qquad (2)$$

where C refers to total number of target classes, y_c and $\hat{y}_c$ denotes correct and predicted labels respectively, σ is the sigmoid function and W_c is the class balancing weight from [13]. For the second stage IG, we generate pseudo triplet labels for each detected instrument instances, where we assign the triplet from the binary triplet presence label if the corresponding instrument class matches. To train IG, we apply categorical cross entropy loss on edge set $\mathcal{E}_s^i$ and verb logits $y_v^{e'}$ for all instrument instances i to obtain losses $L_{\mathcal{G}}^e$ and $L_{\mathcal{G}}^v$ respectively following Eq. 3:

$$L = -\sum_{c=1}^{M} y_c \log(p_c), \qquad (3)$$

where M denotes the number of classes which is N for $L_{\mathcal{G}}^e$ and $V + 1$ for $L_{\mathcal{G}}^v$. The final loss for training follows Eq. 4:

$$L = L_t + \alpha \times L_{\mathcal{G}}^e + \beta \times L_{\mathcal{G}}^v, \tag{4}$$

where α and β denote the weights to balance the loss contribution.

3   Experimental Results and Discussion

3.1   Dataset and Evaluation Metrics

Our experiments are conducted on the publicly available CholecT50 [13] dataset, which includes binary presence labels for 6 instruments, 10 verbs, 15 targets, and 100 triplet classes. We train and validate our models on the official challenge split of the dataset [12]. The test set consists of 5 videos annotated with instrument bounding boxes and matching triplet labels. Since the test set is kept private to date, all our results are obtained by submitting our models to the challenge server for evaluation. The model performance is accessed using video-specific average precision and recall metrics at a threshold ($\theta = 0.5$) using the ivtmetrics library [12]. We also provide box association results in the supplementary material for comparison with other methods on the challenge leaderboard.

3.2   Implementation Details

We first train our instrument detector for 50 epochs using a spatially annotated 12 video subset of Cholec80 and generate instrument bounding boxes and pseudo triplet instance labels for CholecT50 training videos. In stage 1, we set $b_l = 2$, $t_l = 4$, and d to 512. We initialize target class embeddings with zero values. We use 2-layer MLP for Φ_b, Φ_f, and 1-layer MLP for Φ_t. We resize the input frame to 256×448 resolution and apply flipping as data augmentation. For training, we set learning rate $1e^{-3}$ for (backbone, base encoder), and $1e^{-2}$ for MCIT. We use SGD optimizer with weight decay $1e^{-6}$ and train for 30 epochs. To learn the IG, we fine-tune stage 1 and train stage 2. We use learning rate $1e^{-4}$ for (MCIT, base encoder), and $1e^{-5}$ for the backbone. In IG, Φ_p, Φ_e, and Φ_v are 1-layer MLP with learning rate set to $1e^{-3}$, and d' set to 128 in Φ_p to project node features to lower dimensional space. We use Adam optimizer and train both stage 1 and stage 2 for 30 epochs, exponentially decaying the learning rate by 0.99. The loss weights α and β is set to 1 and 0.5 respectively. We set batch size to 32 for both stages. We implement our model in PyTorch and IG graph layers in DGL [20] library. We train the model on Nvidia V100 and A40 GPUs and tune model hyperparameters using random search on 5 validation videos.

3.3   Results

Comparison with the Baseline: We obtain the code and weights of Rendezvous (RDV) [13] model from the public github and generate the triplet predictions on the CholecT50-challenge test set. We then associate these predictions

Table 1. Results on Instrument Localization and Triplet Detection (mAP@0.5 in %).

Method	Instrument Detector	Instrument localization		Triplet detection	
		AP_I	AR_I	AP_{IVT}	AR_{IVT}
RDV (Baseline) [13]	Deformable DETR [22]	**60.1**	**66.6**	6.43	9.50
MCIT (Ours)	Deformable DETR [22]	**60.1**	**66.6**	6.94	9.80
MCIT+IG (Ours)	Deformable DETR [22]	**60.1**	**66.6**	**7.32**	**10.26**

Table 2. Instrument Localization and Triplet Detection (mAP@0.5 in %) vs number of videos/frames used in training the Instrument Detector.

Method	#Videos	#Frames	Instrument localization		Triplet detection	
			AP_I	AR_I	AP_{IVT}	AR_{IVT}
ResNet-CAM-YOLOv5 [14]	51	∼22000	41.9	49.3	4.49	7.87
Distilled-Swin-YOLO [14]	33	∼13000	17.3	30.4	2.74	6.16
MCIT+IG (Ours)	1	4214	33.5	39.6	3.60	4.95
MCIT+IG (Ours)	5	15315	53.1	59.6	5.91	8.73
MCIT+IG (Ours)	12	24536	**60.1**	**66.6**	**7.32**	**10.26**

with the bounding box predictions from the Deformable DETR [22] to generate baseline triplet detections as shown in Table 1. With a stable instrument localization performance (60.1 mAP), our MCIT model leverage the instrument-aware target features to captures better semantics of instrument-target interactions than the baseline. Adding IG further enforces the correct associations, thus improving the triplet detection performance by +0.89 mAP, which is 13.8% increase from the baseline performance. Moreover, the inference time in frame per seconds (FPS) for our MCIT-IG model on Nvidia V100 is ∼29.

Ablation Study on the Spatial Annotation Need: Here, we study the impact of an instrument localization quality on triplet detection and how the target features can supplement fewer spatial annotations of the instruments for better triplet detection. We compare with ResNet-CAM-YOLOv5 [14] and Distilled-Swin-YOLO [14] models which were also trained with bounding box labels. We observed that the triplet detection mAP increases with increasing instrument localization mAP for all the models as shown in Table 2. However, the scale study shows that with lesser bounding box instances, our MCIT-IG model stands tall: outperforming Distilled-Swin-YOLO by +0.86 mAP with ∼9K fewer frames and surpassing ResNet-CAM-YOLOv5 by +1.42 mAP with ∼7K frames to spare. Note that a frame can be annotated with one or more bounding boxes.

Ablation Studies on the Components of MCIT-IG: We analyze the modules used in MCIT-IG and report our results in Table 3. Using both ROI and box features provides a complete representation of the instruments that benefits IG, whereas using just ROI or box features misses out on details about instruments hurting the triplet detection performance. We further test the quality of target

Table 3. Component-wise performance on Triplet Detection (mAP@0.5 in %).

ROI	Box	Graph	ToolAwareness	AP_{IVT}	AR_{IVT}
✓		✓	✓	4.11	5.64
	✓	✓	✓	4.97	6.93
✓	✓	✓		4.98	7.71
✓	✓		✓	6.94	9.80
✓	✓	✓	✓	**7.32**	**10.26**

Table 4. Comparison with top methods from CholecTriplet 2022 Challenge [14], leaderboard results on https://cholectriplet2022.grand-challenge.org/results.

Method	Params (M)	Supervision	Ranking	Instrument localization		Triplet detection	
				AP_I	AR_I	AP_{IVT}	AR_{IVT}
RDV-Det	17.1	Weak	5th	3.0	7.6	0.24	0.79
DualMFFNet	28.3	Weak	4th	4.6	6.6	0.36	0.73
MTTT	181.7	Weak	3rd	11.0	21.1	1.47	3.65
Distilled-Swin-YOLO	88	Full	2nd	17.3	30.4	2.74	6.16
ResNet-CAM-YOLOv5	164	Full	1st	41.9	49.3	4.49	7.87
MCIT+IG (Ours)	100.6	Mixed	–	**60.1**	**66.6**	**7.32**	**10.26**

class embeddings without instrument awareness in MCIT. Results in Table 3 indicates that the lack of instrument context hampers the ability of the target class embeddings to capture full range of associations with the triplets. Also, message passing is key in the IG as it allows instrument semantics to propagate to target class embeddings, which helps distinguish interacting pairs from other non-interacting pairs.

Comparison with the State-of-the-Art (SOTA) Methods: Results in Table 4 show that our proposed model outperforms all the existing methods in the CholecTriplet 2022 challenge [14], obtained the highest score that would have placed our model 1st on the challenge leaderboard in all the accessed metrics. Leveraging our transformer modulated target embeddings and graph-based associations, our method shows superior performance in both instrument localization and triplet detection over methods weakly-supervised on binary presence labels and those fully supervised on external bounding box datasets like ours. More details on the challenge methods are provided in [14].

4   Conclusion

In this work, we propose a fully differentiable two-stage pipeline for triplet detection in laparoscopic cholecystectomy procedures. We introduce a transformer-based method for learning per class embeddings of target anatomical structures in the absence of target instance labels, and an interaction graph that dynamically associates the instrument and target embeddings to detect triplets. We also incorporate a mixed supervision strategy to help train MCIT and IG modules. We show that improving instrument localization has a direct correlation with triplet detection performance. We evaluate our method on the challenge split of the CholecT50 dataset and demonstrate improved performance over the leaderboard.

Acknowledgements. This work was supported by French state funds managed by the ANR within the National AI Chair program under Grant ANR-20-CHIA-0029-01 (Chair AI4ORSafety) and within the Investments for the future program under Grant ANR-10-IAHU-02 (IHU Strasbourg). It was also supported by BPI France under reference DOS0180017/00 (project 5G-OR). It was granted access to the HPC resources of Unistra Mesocentre and GENCI-IDRIS (Grant AD011013710).

References

1. Bawa, V.S., et al.: The saras endoscopic surgeon action detection (ESAD) dataset: challenges and methods. arXiv preprint arXiv:2104.03178 (2021)
2. Chen, K., et al.: MMDetection: open mmlab detection toolbox and benchmark. arXiv preprint arXiv:1906.07155 (2019)
3. Dosovitskiy, A., et al.: An image is worth 16x16 words: transformers for image recognition at scale. In: ICLR (2021)
4. He, K., Zhang, X., Ren, S., Sun, J.: Deep residual learning for image recognition. In: CVPR, pp. 770–778 (2016)
5. Jin, A., et al.: Tool detection and operative skill assessment in surgical videos using region-based convolutional neural networks. In: WACV, pp. 691–699 (2018)
6. Lalys, F., Jannin, P.: Surgical process modelling: a review. IJCARS **9**, 495–511 (2014)
7. Lin, W., et al.: Instrument-tissue interaction quintuple detection in surgery videos. In: Wang, L., Dou, Q., Fletcher, P.T., Speidel, S., Li, S. (eds.) MICCAI 2022. LNCS, vol. 13437, pp. 399–409. Springer, Cham (2022). https://doi.org/10.1007/978-3-031-16449-1_38
8. Maier-Hein, L., et al.: Surgical data science for next-generation interventions. Nat. Biomed. Eng. **1**(9), 691–696 (2017)
9. Mascagni, P., et al.: Artificial intelligence for surgical safety: automatic assessment of the critical view of safety in laparoscopic cholecystectomy using deep learning. Ann. Surg. **275**(5), 955–961 (2022)
10. Nwoye, C.I., et al.: Cholectriplet 2021: a benchmark challenge for surgical action triplet recognition. Med. Image Anal. **86**, 102803 (2023)
11. Nwoye, C.I., et al.: Recognition of instrument-tissue interactions in endoscopic videos via action triplets. In: Martel, A.L., et al. (eds.) MICCAI 2020. LNCS, vol. 12263, pp. 364–374. Springer, Cham (2020). https://doi.org/10.1007/978-3-030-59716-0_35

12. Nwoye, C.I., Padoy, N.: Data splits and metrics for method benchmarking on surgical action triplet datasets. arXiv preprint arXiv:2204.05235 (2022)
13. Nwoye, C.I., et al.: Rendezvous: attention mechanisms for the recognition of surgical action triplets in endoscopic videos. Med. Image Anal. **78**, 102433 (2022)
14. Nwoye, C.I., et al.: Cholectriplet 2022: show me a tool and tell me the triplet - an endoscopic vision challenge for surgical action triplet detection. Med. Image Anal. **89**, 102888 (2023)
15. Padoy, N., Blum, T., Ahmadi, S.A., Feussner, H., Berger, M.O., Navab, N.: Statistical modeling and recognition of surgical workflow. Med. Image Anal. **16**(3), 632–641 (2012)
16. Sharma, S., Nwoye, C.I., Mutter, D., Padoy, N.: Rendezvous in time: an attention-based temporal fusion approach for surgical triplet recognition. IJCARS **18**(6), 1053–1059 (2023)
17. Twinanda, A.P., Shehata, S., Mutter, D., Marescaux, J., de Mathelin, M., Padoy, N.: Endonet: a deep architecture for recognition tasks on laparoscopic videos. IEEE TMI **36**(1), 86–97 (2017)
18. Veličković, P., Cucurull, G., Casanova, A., Romero, A., Lió, P., Bengio, Y.: Graph attention networks. In: ICLR (2018)
19. Vercauteren, T., Unberath, M., Padoy, N., Navab, N.: CAI4CAI: the rise of contextual artificial intelligence in computer-assisted interventions. Proc. IEEE **108**(1), 198–214 (2020)
20. Wang, M., et al.: Deep graph library: a graph-centric, highly-performant package for graph neural networks. arXiv preprint arXiv:1909.01315 (2019)
21. Xu, L., Ouyang, W., Bennamoun, M., Boussaid, F., Xu, D.: Multi-class token transformer for weakly supervised semantic segmentation. In: CVPR, pp. 4310–4319 (2022)
22. Zhu, X., Su, W., Lu, L., Li, B., Wang, X., Dai, J.: DD deformable transformers for end-to-end object detection. In: ICLR (2021)

A Patient-Specific Self-supervised Model for Automatic X-Ray/CT Registration

Baochang Zhang[1,2]([✉]), Shahrooz Faghihroohi[1], Mohammad Farid Azampour[1], Shuting Liu[1], Reza Ghotbi[3], Heribert Schunkert[2,4], and Nassir Navab[1]

[1] Computer Aided Medical Procedures,
Technical University of Munich, Munich, Germany
baochang.zhang@tum.de
[2] German Heart Center Munich, Munich, Germany
[3] HELIOS Hospital west of Munich, Munich, Germany
[4] German Centre for Cardiovascular Research,
Munich Heart Alliance, Munich, Germany

Abstract. The accurate estimation of X-ray source pose in relation to pre-operative images is crucial for minimally invasive procedures. However, existing deep learning-based automatic registration methods often have one or some limitations, including heavy reliance on subsequent conventional refinement steps, requiring manual annotation for training, or ignoring the patient's anatomical specificity. To address these limitations, we propose a patient-specific and self-supervised end-to-end framework. Our approach utilizes patient's preoperative CT to generate simulated X-rays that include patient-specific information. We propose a self-supervised regression neural network trained on the simulated patient-specific X-rays to predict six degrees of freedom pose of the X-ray source. In our proposed network, regularized autoencoder and multi-head self-attention mechanism are employed to encourage the model to automatically capture patient-specific salient information that supports accurate pose estimation, and Incremental Learning strategy is adopted for network training to avoid over-fitting and promote network performance. Meanwhile, an novel refinement model is proposed, which provides a way to obtain gradients with respect to the pose parameters to further refine the pose predicted by the regression network. Our method achieves a mean projection distance of 3.01 mm with a success rate of 100% on simulated X-rays, and a mean projection distance of 1.55 mm on X-rays. The code is available at github.com/BaochangZhang/PSSS_registration.

Keywords: X-ray/CT Registration · Self-supervised Learning · Patient-Specific Model

Supplementary Information The online version contains supplementary material available at https://doi.org/10.1007/978-3-031-43996-4_49.

H. Greenspan et al. (Eds.): MICCAI 2023, LNCS 14228, pp. 515–524, 2023.
https://doi.org/10.1007/978-3-031-43996-4_49

1   Introduction

Augmentation of intra-operative X-ray images using the pre-operative data (e.g., treatment plan) has the potential to reduce procedure time and improve patient outcomes in minimally invasive procedures. However, surgeons must rely on their clinical knowledge to perform a mental mapping between pre- and intra-operative information, since the pre- and intra-operative images are based on different coordinate systems. To utilize pre-operative data efficiently, accurate pose estimation of the X-ray source relative to pre-operative images (or called registration between pre- and intra-operative data) is necessary and beneficial to relieve surgeon's mental load and improve patient outcomes.

Although 2D/3D registration methods for medical images have been widely researched and systematically reviewed [10,17], developing an automatic end-to-end registration method remains an open issue. For conventional intensity-based registration, it is formulated as an iterative optimization problem based on similarity measures. A novel similarity measure called weighted local mutual information is proposed to perform solid vascular 2D-3D registration [11] but has a limited capture range, becoming inefficient and prone to local minima if initial registration error is large. A good approach that directly predicts the spatial mapping relationship between simulated X-rays and real X-rays using a neural network is put forward [12] but requires an initialization. Some supervised learning tasks, such as anatomical landmark detection [1,3], are defined to develop a robust initialization scheme. When used to initialize an optimizer [14] for refinement, they can lead to a fully automatic 2D/3D registration solution [3]. But extensive manual annotation [1,3] or pairwise clinical data [12] is needed for training, which is time- and labor-consuming when expanding to new anatomies, and the robustness of these methods might be challenged due to the neglect of patient's anatomical specificity [1]. A patient-specific landmark refinement scheme is then proposed, which contributes to model's robustness when applied intraoperatively [2]. Nevertheless, the final performance of this automatic registration method still relies on conventional refinement step based on derivative-free optimizer (i.e., BOBYQA), which limits the computational efficiency of deep learning-based registration. Meanwhile, some studies employ regression neural networks to directly predict slice's pose relative to pre-operative 3D image data [4,8,15]. While the simplicity of these approaches is appealing, the applicability of these methods is constrained by their performance and are more suitable as initialization.

In this paper, we propose a purely self-supervised and patient-specific end-to-end framework for fully automatic registration of single-view X-ray to preoperative CT. Our main contributions are as follows: (1) The proposed method eliminates the need for manual annotation, relying instead on self-supervision from simulated patient-specific X-rays and corresponding automatically labeled poses, which makes the registration method easier to extend to new medical applications. (2) Regularized autoencoder and multi-head self-attention mechanism are embedded to encourage the model to capture patient-specific salient information automatically, therefore improving the robustness of registration; and an

novel refinement model is proposed to further improve registration accuracy.
(3) The proposed method has been successfully evaluated on X-rays, achieving
an average run-time of around 2.5 s, which meets the requirements for clinical
applications.

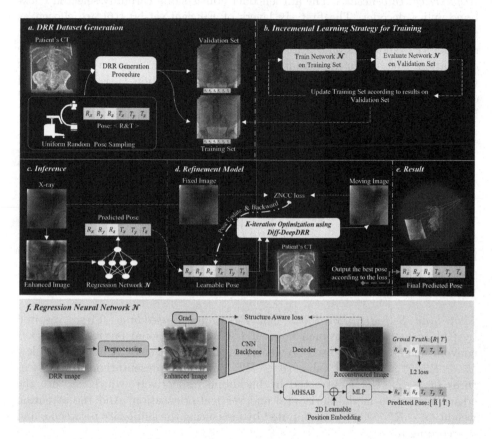

Fig. 1. An overview of our proposed framework (a-e) and the proposed regression
neural network (f). The framework consists of five parts: (a) DRR generation, (b)
Incremental Learning strategy for training, (c) inference phase, (d) refinement model,
and (e) outputting the predicted pose.

2   Method

An overview of the proposed framework is illustrated in Fig. 1, which can be
mainly divided into two parts that are introduced hereafter.

2.1   Pose Regression Model

Network Architecture. A regression neural network architecture is developed to estimate the six degrees of freedom (DOF) pose from an input X-ray image, which consists of a regularized autoencoder, a multi-head self-attention block (MHSAB), a learnable 2D position embedding, and a multilayer perceptron (MLP), as shown in Fig. 1f.

Regularized Autoencoder. The autoencoder consists of a carefully selected backbone and a decoder. The first six layers of EfficientNet-b0 [16] are chosen as the backbone, where the first convolution layer is adapted by setting the input channel to 1 (shown as blue trapezoidal block in Fig. 1f). For image-based pose estimation, edge, as an important structural information, is more advantageous than intensity value. Hence, a structure aware loss function $\mathcal{L}_s$ is defined based on zero-normalized cross-correlation(ZNCC) to constrain the features extracted by encoder to contain necessary structure information, which is formulated as,

$$\mathcal{L}_s(G_I, Y) = \frac{1}{\sigma_{G_I} \sigma_Y} \frac{1}{N} \sum_{u,\,v} (G_I(u,v) - E(G_I))(Y(u,v) - E(Y)) \quad (1)$$

$$G_I = \sqrt{\left(\frac{\partial I}{\partial u}\right)^2 + \left(\frac{\partial I}{\partial v}\right)^2} \quad (2)$$

Here, Y is the output of the decoder, I is the input image, G_I is the normalized gradient of the input image, and N is the number of pixels in the input image. $E()$ is the expectation operator, and σ is standard deviation operator.

Multi-head Self-attention Block and Positional Embedding. First, inspired by [19], a 3×1 convolutional layer, a 1×3 convolutional layer, and a 1×1 convolutional layer are used to generate the query ($q \in \mathcal{R}^{n \times hw \times C/n}$), key ($k \in \mathcal{R}^{n \times hw \times C/n}$), and value ($v \in \mathcal{R}^{n \times hw \times C/n}$) representations in the features ($f \in \mathcal{R}^{h \times w \times C}$) extracted by backbone respectively, which capture the edge information in the horizontal and vertical orientation. And the attention weight ($A \in \mathcal{R}^{n \times hw \times hw}$) is computed by measuring the similarity between q and k according to,

$$A = Softmax\left(\frac{qk^T}{\sqrt{C/n}}\right) \quad (3)$$

where n is the number of heads, C is number of channels, and h, w are the height and width of features. Using the computed attention weights, the output of MHSAB f_a is computed as,

$$f_a = Av + f \quad (4)$$

Second, in order to make the proposed method more sensitive to spatial transformation, 2D learnable positional embedding is employed to explicitly incorporate the object's position information.

Incremental Learning Strategy. Based on Incremental Learning strategy, the network is trained for 100 epochs with a batch size of 32 using the Adam optimizer (learning rate is 0.002, decay of 0.5 per 10 epochs). After training for 40 epochs, the training dataset will be automatically updated if the loss computed on the validation set does not change frequently (i.e., less than 10 percent of the maximum of loss) within 20 epochs, which allows the network to observe a wider range of poses while avoiding over-fitting. Final loss function is a weighted sum of the structure aware loss $\mathcal{L}_s$ for autoencoder and $L2$ Loss between the predicted pose and the ground truth, which is formulated as,

$$loss\,(I,p) = \|P(I) - p\|^2 + \alpha\mathcal{L}_s\,(G_I,\ D(I)) \tag{5}$$

where α is set as 5, I is the input DRR image, p is the ground truth of pose, $P(I)$ is predicted pose, and $D(I)$ is the output of autoencoder.

Pre-processing and Data Augmentation. In order to improve the contrast of the digitally reconstructed radiographs (DRRs) and reduce the gap to real X-rays, the DRR is first normalized by Z-score and then normalized using the sigmoid function, which maps the image into the interval [0, 1] with a mean of 0.5. Then contrast-limited adaptive histogram equalization (CLAHE) is employed to enhance the contrast. For data augmentation, random brightness adjustment, random adding offset, adding Gaussian noise, and adding Poisson noise are adopted.

2.2 Refinement Model

A novel refinement model is proposed to further refine the pose predicted by regression network as shown in Fig. 1(d). Specifically, inspired by DeepDRR [18], a differentiable DeepDRR (Diff-DeepDRR) generator is developed. Our proposed Diff-DeepDRR generator offers two main advantages compared with DeepDRR, including the ability to generate a 256^2 pixel DRR image in just 15.6 ms, making it fast enough for online refinement, and providing a way to obtain gradients with respect to the pose parameters. The refinement model takes the input X-ray image with an unknown pose as the fixed image and has six learnable pose parameters initialized by the prediction of pose regression network. Then the proposed Diff-DeepDRR generator utilizes the six learnable pose parameters to generate DRR image online, which is considered as the moving image. The ZNCC is then used as the loss function to minimize the distance between the fixed image and the moving image. With the powerful PyTorch auto-grad engine, the six pose parameters are learned iteratively. For each refinement process, the refinement model is online optimized for 100 iterations using an Adam optimizer with a learning rate of 5.0 for translational parameters and 0.05 for rotational parameters, and outputs the pose with the minimal loss score.

3   Experiments

3.1   Dataset

Our method is evaluated on six DRR datasets and one X-ray dataset. The six DRR datasets are generated from five common preoperative CTs and one specific CT with previous surgical implants, while the X-ray dataset is obtained from a Pelvis phantom containing a metal bead landmark inside. The X-ray dataset includes a CBCT volume and ten X-ray images with varying poses. For each CT or CBCT, 12800 DRR images with a reduced resolution of 256^2 pixels are generated using the DeepDRR method, which are divided into training (50%), validation (25%), and test (25%) sets. The DRR generation system is configured based on the geometry of a mobile C-arm, with a 432 mm detector, 0.3 mm pixel size, 742.5 mm source-isocenter distance, and 517.15 mm detector-isocenter distance. The center of the CT or CBCT volume is moved to the isocenter of the device. For pose sampling, each 6 DOF pose consists of three rotational and three translational parameters. The angular and orbital rotation are uniformly sampled from $[-40°, 40°]$, and the angle of in-plane rotation in the detector plane is uniformly sampled from $[-10°, 10°]$. Translations are uniformly sampled from $[-70 \text{ mm}, 70 \text{ mm}]$.

3.2   Experimental Design and Evaluation Metrics

To better understand our work, we conduct a series of experiments, for example, 1) A typical pose regression model consists of a CNN backbone and an MLP. To find an efficient backbone for this task, EfficientNet, ResNet [5], and DenseNet [6] were studied. 2) A detailed ablation experiment was performed, where you can learn the evolution process and the superiority of our proposed method. 3) Through the experiment on X-ray data, you can know whether the proposed method trained only on DRRs can be generalized to X-ray applications. To validate the performance of our method on DRR datasets, we employed five measurements including 2D mean projection distance (mPD), 3D mean target registration error (mTRE) [7], Mean Absolute Error (MAE), and success rate (SR) [2], where success is defined as an mTRE of less than 10 mm. Cases with mTRE exceeding 10 mm were excluded from average mPD and mTRE measurements. For the validation on X-rays, projection distance (PD) is used to measure the positional difference of the bead between the X-ray and the final registered DRR. In addition, NCC [13], structural similarity index measure (SSIM), and contrast-structure similarity (CSS) [9] are employed to evaluate the similarity between the input X-ray and final registered DRR.

4   Results

4.1   The Choice of Backbone and Ablation Study

The performance of three models with different backbones were evaluated on the first DRR dataset as reported in Table 1. ***Res-backbone*** [15] means that the first

Table 1. The experimental results of different backbones. Rx, Ry, Rz, Tx, Ty and Tz are the mAE measured on three rotational parameters and three translational parameters.

backbone	mTRE↓ (mm)		mPD↓ (mm)		Rx↓ (degree)		Ry↓ (degree)		Rz↓ (degree)		Tx↓ (mm)		Ty↓ (mm)		Tz↓ (mm)		FLOPs (G)	Paras (M)	SR↑ (%)
	mean	std	mean	std	mean	std	mean	std	mean	std	mean	std	mean	std	mean	std			
Res-backbone [15]	7.68	3.08	7.63	2.05	1.39	1.13	1.31	1.05	0.99	0.88	3.06	2.63	3.03	2.45	4.84	3.70	17.2	9.05	67.13
Dense-backbone	6.63	2.57	6.94	2.10	1.11	0.89	1.28	0.93	0.90	0.72	2.48	2.22	2.46	2.14	4.33	3.28	13.9	4.78	81.94
Ef-backbone	6.71	2.75	6.83	2.05	0.98	0.78	0.93	0.76	0.71	0.56	2.44	2.05	2.57	1.99	3.92	3.05	1.05	0.91	88.66

Table 2. The results of ablation study & our method's results on 6 DRR datasets. # indicates previous surgical implants are included in this dataset.

–	mTRE↓ (mm)		mPD↓ (mm)		Rx↓ (degree)		Ry↓ (degree)		Rz↓ (degree)		Tx↓ (mm)		Ty↓ (mm)		Tz↓ (mm)		SR↑ (%)
	mean	std	mean	std	mean	std	mean	std	mean	std	mean	std	mean	std	mean	std	
A	6.71	2.75	6.83	2.05	0.98	0.78	0.93	0.76	0.71	0.56	2.44	2.05	2.57	1.99	3.92	3.05	88.66
A+B	6.13	2.50	6.39	2.07	0.93	0.78	0.99	0.79	0.62	0.52	2.10	1.82	2.33	1.91	3.49	2.73	90.97
A+C	6.17	2.65	6.12	1.98	1.01	0.80	0.85	0.68	0.70	0.58	1.99	1.75	2.05	1.74	3.72	2.88	91.34
A+B+C	5.92	2.57	6.01	2.02	1.02	0.80	0.83	0.65	0.64	0.54	1.92	1.75	2.05	1.78	3.39	2.71	93.31
[A+B+C]*	5.43	2.18	5.80	2.05	0.90	0.73	0.82	0.65	0.57	0.47	1.80	1.53	1.80	1.50	2.89	2.26	95.25
[A+B+C]*+D proposed	2.85	1.91	2.92	1.93	0.41	0.53	0.31	0.39	0.24	0.35	1.00	0.83	0.90	0.84	1.55	1.35	100.00
dataset1	2.85	1.91	2.92	1.93	0.41	0.53	0.31	0.39	0.24	0.35	1.00	0.83	0.90	0.84	1.55	1.35	100.00
dataset2	2.54	1.28	3.08	1.72	0.38	0.55	0.35	0.38	0.22	0.29	0.65	0.60	0.90	0.51	1.38	0.88	100.00
dataset3	2.71	1.14	3.30	1.45	0.26	0.31	0.33	0.40	0.23	0.26	0.84	0.64	1.04	0.71	1.51	1.09	100.00
dataset4	2.72	1.32	3.12	1.53	0.27	0.48	0.26	0.33	0.23	0.32	0.90	0.86	1.04	0.66	1.58	1.10	100.00
dataset5	2.73	1.03	3.09	1.22	0.35	0.35	0.35	0.33	0.24	0.22	0.74	0.69	0.97	0.58	1.52	1.09	100.00
dataset6 #	2.50	1.48	2.57	1.67	0.36	0.43	0.27	0.39	0.20	0.27	0.75	0.64	0.86	1.00	1.44	1.20	100.00

four layers of ResNet-50 as the backbone; **Dense-backbone** means that the first six layers of DenseNet-121 as the backbone; **Ef-backbone** means that the first six layers of EfficientNet-b0 as the backbone. Compared with the other two networks, **Ef-backbone** achieves a higher SR of 88.66%, and reduces parameter size and FLOPs by an order of magnitude. Therefore, **Ef-backbone** is chosen and regarded as baseline for further studies. Then, a second set of models was trained and evaluated for a detailed ablation study, and the experimental results are shown in Table 2. **A** means the aforementioned baseline; **B** means adding regularized autoencoder; **C** means adding multi-head self-attention block and position embedding; **D** means using refinement model; * means the network is trained via Incremental Learning strategy. It is clear that each module we proposed plays a positive role in this task, and our method makes arresting improvements, e.g., compared with baseline, the SR increased from 88.66% to 100%. In addition, observing the visualized self-attention map shown in Fig. 2(b), we find that the proposed network does automatically capture some salient anatomical regions, which pays more attention to bony structures. Meanwhile, the distribution of attention changes with the view pose, and it seems that the closer to the detector, the bone region gets more attention.

4.2 Experimental Results on DRR Datasets and X-Ray Dataset

The experimental results of our proposed method on six DRR datasets are shown in Table 2. It is worth noticing that the success rate of our method has achieved 100% on all datasets and the average of mTRE on six datasets achieves 2.67 mm.

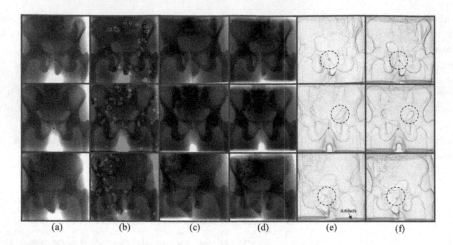

Fig. 2. Results on X-ray cases. (a) X-rays, (b) corresponding attention visualization, (c) initial registered DRRs from pose regression network, (d) final registered DRRs from refinement mode, (e) initial registration error map, (f) final registration error map. Red edges in (e&f) come from (a), green edge in (e) comes from (c), green edge in (f) comes from (d). (Color figure online)

Table 3. The results of our method on X-ray Cases

–	Case1	Case2	Case3	Case4	Case5	Case6	Case7	Case8	Case9	Case10	Mean	Std
PD↓	0.6762	1.8400	1.4251	2.8431	0.6770	1.3244	1.2777	1.4181	1.5259	2.5655	1.5573	0.6687
NCC↑	0.9885	0.9752	0.9867	0.9941	0.9858	0.9870	0.9888	0.9943	0.9913	0.9880	0.9880	0.0051
SSIM↑	0.9395	0.9220	0.9348	0.9736	0.9424	0.9456	0.9436	0.9616	0.9649	0.9412	0.9469	0.0146
CSS↑	0.9427	0.9321	0.9392	0.9750	0.9453	0.9481	0.9463	0.9630	0.9664	0.9448	0.9503	0.0127

For the X-ray dataset, the quantitative evaluation results on 10 X-ray cases are shown in Table 3, achieving a mPD of 1.55 mm, which demonstrates that our method can be successfully generalized to X-ray. More intuitive results are shown in Fig. 2. Observing the dotted circled areas on Fig. 2(e&f), we find that the network only makes coarse predictions for X-rays due to unavoidable artifacts on the CBCT boundaries, and the proposed refinement model facilitates accurate pose estimation, which can be confirmed by the overlapping of edges from the X-rays and the final registered DRRs.

5 Conclusion and Discussion

In this paper, we present a patient-specific and self-supervised end-to-end approach for automatic X-ray/CT rigid registration. Our method effectively addresses the primary limitations of existing methods, such as requirement of manual annotation, dependency on conventional derivative-free optimization, and patient-specific concerns. When field of view of CT is not smaller than that of X-ray, which is often satisfied in clinical routines, our proposed method would perform very well without any additional post-process. The quantitative and qualitative evaluation results of our proposed method illustrates its superiority and its ability to generalize to X-rays even when trained solely on DRRs. For our experiments, the validation on X-ray of phantom cannot fully represent the performance on X-ray of real patients, but it shows that the proposed method has high potential. Meanwhile, domain randomization could reduce the gap between DRR and real X-ray images, which would allow methods validated on phantoms to perform better also on real X-ray. For the runtime aspect, our patient-specific regression network can complete the training phase within one hour using an NVIDIA GPU (Quadro RTX A6000), which meets the requirement of clinical application during pre-operative planning phase. Meanwhile, the proposed network achieves an average inference time of 6ms per image with a size of 256^2, when considering the run-time of the proposed refinement model, the total cost is approximately 2.5 s, which also fully satisfies the requirement for clinical application during intra-operative phase.

Acknowledgements. The project was supported by the Bavarian State Ministry of Science and Arts within the framework of the "Digitaler Herz-OP" project under the grant number 1530/891 02 and the China Scholarship Council (File No.202004910390). We also thank BrainLab AG for their partial support.

References

1. Bier, B., et al.: X-ray-transform invariant anatomical landmark detection for pelvic trauma surgery. In: Frangi, A.F., Schnabel, J.A., Davatzikos, C., Alberola-López, C., Fichtinger, G. (eds.) MICCAI 2018, Part IV. LNCS, vol. 11073, pp. 55–63. Springer, Cham (2018). https://doi.org/10.1007/978-3-030-00937-3_7
2. Grimm, M., Esteban, J., Unberath, M., Navab, N.: Pose-dependent weights and domain randomization for fully automatic X-ray to CT registration. IEEE Trans. Med. Imaging **40**(9), 2221–2232 (2021)
3. Grupp, R.B., et al.: Automatic annotation of hip anatomy in fluoroscopy for robust and efficient 2D/3D registration. Int. J. Comput. Assist. Radiol. Surg. **15**, 759–769 (2020)
4. Guan, S., Meng, C., Sun, K., Wang, T.: Transfer learning for rigid 2D/3D cardiovascular images registration. In: Lin, Z., Wang, L., Yang, J., Shi, G., Tan, T., Zheng, N., Chen, X., Zhang, Y. (eds.) PRCV 2019, Part II. LNCS, vol. 11858, pp. 380–390. Springer, Cham (2019). https://doi.org/10.1007/978-3-030-31723-2_32
5. He, K., Zhang, X., Ren, S., Sun, J.: Deep residual learning for image recognition. In: Proceedings of the IEEE Conference on Computer Vision and Pattern Recognition, pp. 770–778 (2016)
6. Huang, G., Liu, Z., Van Der Maaten, L., Weinberger, K.Q.: Densely connected convolutional networks. In: Proceedings of the IEEE Conference on Computer Vision and Pattern Recognition, pp. 4700–4708 (2017)
7. Van de Kraats, E.B., Penney, G.P., Tomazevic, D., Van Walsum, T., Niessen, W.J.: Standardized evaluation methodology for 2-D-3-D registration. IEEE Trans. Med. Imaging **24**(9), 1177–1189 (2005)

8. Lee, B.C., et al.: Breathing-compensated neural networks for real time C-arm pose estimation in lung CT-fluoroscopy registration. In: 2022 IEEE 19th International Symposium on Biomedical Imaging (ISBI), pp. 1–5. IEEE (2022)

9. Liu, S., et al.: Unpaired stain transfer using pathology-consistent constrained generative adversarial networks. IEEE Trans. Med. Imaging 40(8), 1977–1989 (2021)

10. Markelj, P., Tomaževič, D., Likar, B., Pernuš, F.: A review of 3D/2D registration methods for image-guided interventions. Med. Image Anal. 16(3), 642–661 (2012)

11. Meng, C., Wang, Q., Guan, S., Sun, K., Liu, B.: 2D-3D registration with weighted local mutual information in vascular interventions. IEEE Access 7, 162629–162638 (2019)

12. Miao, S., et al.: Dilated FCN for multi-agent 2D/3D medical image registration. In: Proceedings of the AAAI Conference on Artificial Intelligence, vol. 32 (2018)

13. Penney, G.P., Weese, J., Little, J.A., Desmedt, P., Hill, D.L., et al.: A comparison of similarity measures for use in 2-D-3-D medical image registration. IEEE Trans. Med. Imaging 17(4), 586–595 (1998)

14. Powell, M.J.: The BOBYQA algorithm for bound constrained optimization without derivatives, vol. 26. Cambridge NA Report NA2009/06, University of Cambridge, Cambridge (2009)

15. Salehi, S.S.M., Khan, S., Erdogmus, D., Gholipour, A.: Real-time deep pose estimation with geodesic loss for image-to-template rigid registration. IEEE Trans. Med. Imaging 38(2), 470–481 (2018)

16. Tan, M., Le, Q.: EfficientNet: rethinking model scaling for convolutional neural networks. In: International Conference on Machine Learning, pp. 6105–6114. PMLR (2019)

17. Unberath, M., et al.: The impact of machine learning on 2D/3D registration for image-guided interventions: a systematic review and perspective. Front. Robot. AI 8, 716007 (2021)

18. Unberath, M., et al.: DeepDRR – a catalyst for machine learning in fluoroscopy-guided procedures. In: Frangi, A.F., Schnabel, J.A., Davatzikos, C., Alberola-López, C., Fichtinger, G. (eds.) MICCAI 2018, Part IV. LNCS, vol. 11073, pp. 98–106. Springer, Cham (2018). https://doi.org/10.1007/978-3-030-00937-3_12

19. Vaswani, A., et al.: Attention is all you need. In: Advances in Neural Information Processing Systems, vol. 30 (2017)

Dose Guidance for Radiotherapy-Oriented Deep Learning Segmentation

Elias Rüfenacht[1]([envelope])[iD], Robert Poel[2], Amith Kamath[1][iD], Ekin Ermis[2],
Stefan Scheib[3], Michael K. Fix[2], and Mauricio Reyes[1,2][iD]

[1] ARTORG Center for Biomedical Engineering Research,
University of Bern, Bern, Switzerland
elias.ruefenacht@unibe.ch
[2] Department of Radiation Oncology, Inselspital, Bern University Hospital
and University of Bern, Bern, Switzerland
[3] Varian Medical Systems Imaging Laboratory GmbH, Baden, Switzerland

Abstract. Deep learning-based image segmentation for radiotherapy is intended to speed up the planning process and yield consistent results. However, most of these segmentation methods solely rely on distribution and geometry-associated training objectives without considering tumor control and the sparing of healthy tissues. To incorporate dosimetric effects into segmentation models, we propose a new training loss function that extends current state-of-the-art segmentation model training via a dose-based guidance method. We hypothesized that adding such a dose-guidance mechanism improves the robustness of the segmentation with respect to the dose (i.e., resolves distant outliers and focuses on locations of high dose/dose gradient). We demonstrate the effectiveness of the proposed method on Gross Tumor Volume segmentation for glioblastoma treatment. The obtained dosimetry-based results show reduced dose errors relative to the ground truth dose map using the proposed dosimetry-segmentation guidance, outperforming state-of-the-art distribution and geometry-based segmentation losses.

Keywords: Segmentation · Radiotherapy · Dose Guidance · Deep Learning

1 Introduction

Radiotherapy (RT) has proven effective and efficient in treating cancer patients. However, its application depends on treatment planning involving target lesion and radiosensitive organs-at-risk (OAR) segmentation. This is performed to guide radiation to the target and to spare OAR from inappropriate irradiation. Hence, this manual segmentation step is very time-consuming and must

---

Supplementary Information The online version contains supplementary material available at https://doi.org/10.1007/978-3-031-43996-4_50.

H. Greenspan et al. (Eds.): MICCAI 2023, LNCS 14228, pp. 525–534, 2023.
https://doi.org/10.1007/978-3-031-43996-4_50

be performed accurately and, more importantly, must be patient-safe. Studies have shown that the manual segmentation task accounts for over 40% of the treatment planning duration [7] and, in addition, it is also error-prone due to expert-dependent variations [2,24]. Hence, deep learning-based (DL) segmentation is essential for reducing time-to-treatment, yielding more consistent results, and ensuring resource-efficient clinical workflows.

Nowadays, training of DL segmentation models is predominantly based on loss functions defined by geometry-based (e.g., SoftDice loss [15]), distribution-based objectives (e.g., cross-entropy), or a combination thereof [13]. The general strategy has been to design loss functions that match their evaluation counterpart. Nonetheless, recent studies have reported general pitfalls of these metrics [4,19] as well as a low correlation with end-clinical objectives [11,18,22,23]. Furthermore, from a robustness point of view, models trained with these loss functions have been shown to be more prone to generalization issues. Specifically, the Dice loss, allegedly the most popular segmentation loss function, has been shown to have a tendency to yield overconfident trained models and lack robustness in out-of-distribution scenarios [5,14]. These studies have also reported results favoring distribution-matching losses, such as the cross-entropy being a strictly proper scoring rule [6], providing better-calibrated predictions and uncertainty estimates. In the field of RT planning for brain tumor patients, the recent study of [17] shows that current DL-based segmentation algorithms for target structures carry a significant chance of producing false positive outliers, which can have a considerable negative effect on applied radiation dose, and ultimately, they may impact treatment effectiveness. In RT planning, the final objective is to produce the best possible radiation plan that jointly targets the lesion and spares healthy tissues and OARs. Therefore, we postulate that training DL-based segmentation models for RT planning should consider this clinical objective.

In this paper, we propose an end-to-end training loss function for DL-based segmentation models that considers dosimetric effects as a clinically-driven learning objective. Our contributions are: (i) a dosimetry-aware training loss function for DL segmentation models, which (ii) yields improved model robustness, and (iii) leads to improved and safer dosimetry maps. We present results on a clinical dataset comprising fifty post-operative glioblastoma (GBM) patients. In addition, we report results comparing the proposed loss function, called **Dose-Segmentation Loss** (DOSELO), with models trained with a combination of binary cross-entropy (BCE) and SoftDice loss functions.

2   Methodology

Figure 1 describes the general idea of the proposed DOSELO. A segmentation model (U-Net [20]) is trained to output target segmentation predictions for the Gross Tumor Volume (GTV) based on patient MRI sequences. Predicted segmentations and their corresponding ground-truth (GT) are fed into a dose predictor model, which outputs corresponding dose predictions (denoted as $\widehat{D}_P$ and D_P in Fig. 1). A pixel-wise mean squared error between both dose predictions is then

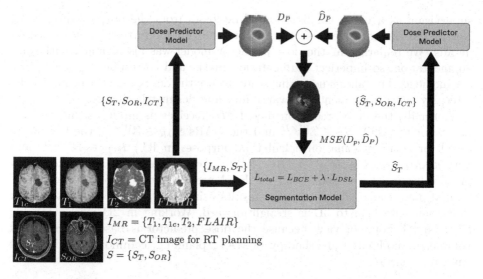

Fig. 1. Schematic overview of the proposed dosimetry-aware training loss function. A segmentation model (U-Net [20]) is trained to output target segmentation predictions ($\widehat{S}_T$) for the Gross Tumor Volume (GTV) based on patient MRI sequences I_{MR}. Predicted ($\widehat{S}_T$) and ground-truth segmentations (S_T) are fed into the dose predictor model along with the CT-image (I_{CT}), and OAR segmentation (S_{OR}). The dose predictor outputs corresponding dose predictions $\widehat{D}_P$ and D_P. A pixel-wise mean squared error between both dose predictions is calculated, and combined with the binary cross-entropy (BCE) loss to form the final loss, $L_{total} = L_{BCE} + \lambda L_{DSL}$.

calculated and combined with the BCE loss to form the final loss. In the next sections we describe the adopted dose prediction model [9,12], and the proposed DOSELO.

2.1   Deep Learning-Based Dose Prediction

Recent DL methods based on cascaded U-Nets have demonstrated the feasibility of generating accurate dose distribution predictions from segmentation masks, approximating analytical dose maps generated by RT treatment planning systems [12]. Originally proposed for head and neck cancer [12], this approach has been recently extended for brain tumor patients [9] with levels of prediction error below 2.5 Gy, which is less than 5% of the prescribed dose. This good level of performance, along with its ability to yield near-instant dose predictions, enables us to create a training pipeline that guides learned features to be dose-aware.

Following [12], the dose predictor model consists of a cascaded U-Net (i.e., the input to the second U-Net is the output of the first concatenated with the input to the first U-Net) trained on segmentation masks, CT images, and reference dose maps. The model's input is a normalized CT volume and segmentation masks for target volume and OARs. As output, it predicts a continuous-valued dose map of the same dimension as the input. The model is trained via deep supervision as a linear combination of L2-losses from the outputs of each U-Net in the cascade. We refer the reader to [9,12] for further implementation details. We remark that the dose predictor model was also trained with data augmentation, so imperfect segmentation masks and corresponding dose plans are included. This allows us in this study to use the dose predictor to model the interplay between segmentation variability and dosimetric changes.

Formally, the dose prediction model M_D receives as inputs: segmentations masks for the GTV $S_T \in \mathbb{Z}^{W \times H}$ and the OARs $S_{OR} \in \mathbb{Z}^{W \times H}$, the CT image (used for tissue attenuation calculation purposes in RT) $I_{CT} \in \mathbb{R}^{W \times H}$, and outputs $M_D(S_T, S_{OR}, I_{CT}) \mapsto D_P \in \mathbb{R}^{W \times H}$, a predicted dose map where each pixel value in D corresponds to the local predicted dose in Gy. Due to the limited data availability, we present results using 2D-based models but remark that their extension to 3D is straightforward. Working in 2D is also feasible from an RT point of view because the dose predictor is based on co-planar volumetric modulated arc therapy (VMAT) planning, commonly used in this clinical scenario.

2.2 Dose Segmentation Loss (DOSELO)

During the training of the segmentation model, we used the dose predictor model to generate pairs of dose predictions for the model-generated segmentations and the GT segmentations. The difference between these two predicted dose maps is used to guide the segmentation model. The intuition behind this is to guide the segmentation model to yield segmentation results being dosimetrically consistent with the dose maps generated via the corresponding GT segmentations.

Formally, given a set of N pairs of labeled training images $\{(I_{MR}, S_P)_i : 1 \leq i \leq N\}$, $I_{MR} \in \mathbb{R}^D$ (with $D : \{T1, T1c, T2, FLAIR\}$ MRI clinical sequences), and corresponding GT segmentations of the GTV $S_T \in \mathbb{Z}^{H \times W}$, a DL segmentation model $M_S(I_{MR}) \mapsto \widehat{S}_T$ is commonly updated by minimizing a standard loss term, such as the BCE loss (L_{BCE}).

To guide the training process with dosimetry information stemming from segmentation variations, we propose to use the mean squared error (MSE) between dose predictions for the GT segmentation (S_T) and the predicted segmentation ($\widehat{S}_T$), and construct the following dose-segmentation loss,

$$L_{DSL} = \frac{1}{H \times W} \sum_{i}^{H \times W} (D_P^i - \widehat{D}_P^i)^2 \tag{1}$$

$$D_P = M_D(S_T, S_{OR}, I_{CT}) \tag{2}$$

$$\widehat{D}_P = M_D(\widehat{S}_T, S_{OR}, I_{CT}), \tag{3}$$

where D_P^i and $\widehat{D}_P^i$ denote pixel-wise dose predictions. The final loss is then,

$$L_{total} = L_{BCE} + \lambda L_{DSL}, \tag{4}$$

where λ is a hyperparameter to weigh the contributions of each loss term. We remark that during training we use standard data augmentations including spatial transformations, which are also subjected to dose predictions, so the model is informed about relevant segmentation variations producing dosimetry changes.

3 Experiments and Results

3.1 Data and Model Training

We divide the descriptions of the two separate datasets used for the dose prediction and segmentation models.

Dose Prediction: The dose prediction model was trained on an in-house dataset comprising a total of 50 subjects diagnosed with post-operative GBM. This includes CT imaging data, segmentation masks of 13 OARs, and the GTV. GTVs were defined according to the ESTRO-ACROP guidelines [16]. The OARs were contoured by one radiotherapist according to [21] and verified by mutual consensus of three experienced radiation oncology experts. Each subject had a reference dose map, calculated using a standardized clinical protocol with Eclipse (Varian Medical Systems Inc., Palo Alto, USA). This reference was generated on basis of a double arc co-planar VMAT plan to deliver 30 times 2 Gy while maximally sparing OARs. We divided the dataset into training (35 cases), validation (5 cases), and testing (10 cases). We refer the reader to [9] for further details.

Segmentation Models: To develop and test the proposed approach, we employed a separate in-house dataset (i.e., different cases than those used to train the dose predictor model) of 50 cases from post-operative GMB patients receiving standard RT treatment. We divided the dataset into training (35 cases), validation (5 cases), and testing (10 cases). All cases comprise a planning CT registered to the standard MRI images (T1-post-contrast (Gd), T1-weighted, T2-weighted, FLAIR), and GT segmentations containing OARs as well as the GTV. We note that for this first study, we decided to keep the dose prediction model fixed during the training of the segmentation model for a simpler presentation of the concept and modular pipeline. Hence, only the parameters of the segmentation model are updated.

Baselines and Implementation Details: We employed the same U-Net [20] architecture for all trained segmentation models, with the same training parameters but two different loss functions, to allow for a fair comparison. As a strong comparison baseline, we used a combo-loss formed by BCE plus SoftDice, which is also used by nnUNet and recommended by its authors [8]. This combo-loss has also been reported as an effective one [13]. For each loss function, we computed a five-fold cross-validation. Our method[1] was implemented in PyTorch 1.13 using Adam optimizer [10] with $\beta_1 = 0.9$, $\beta_2 = 0.999$, batch normalization, dropout set at 0.2, learning rate set at 10^{-4}, $2 \cdot 10^4$ update iterations, and a batch size of 16. The architecture and trained parameters were kept constant across compared models. Training and testing were performed on an NVIDIA Titan X GPU with 12 GB RAM. The input image size is 256×256 pixels with an isotropic spacing of 1 mm.

3.2 Evaluation

To evaluate the proposed DOSELO, we computed dose maps for each test case using a standardized clinical protocol with Eclipse (Varian Medical Systems Inc., Palo Alto, USA). We calculated dose maps for segmentations using the state-of-the-art BCE+SoftDice and the proposed DOSELO. For each obtained dose map, we computed the dose score [12], which is the mean absolute error between the reference dose map (D_{S_T}) and the dose map derived from the corresponding segmentation result ($D_{\widehat{S}_T}$, where $\widehat{S}_T \in \{$BCE+SoftDice, DOSELO$\}$), and set it relative to the reference dose map (D_{S_T}) (see Eq. 5).

$$RMAE = \frac{1}{H \times W} \sum_{i}^{H \times W} \frac{|D_{S_T} - D_{\widehat{S}_T}|}{D_{S_T}} \tag{5}$$

Although it has been shown that geometric-based segmentation metrics poorly correlate with the clinical end-goal in RT [4,11,18,23], we report in supplementary material Dice and Hausdorff summary statistics as well (supplementary Table 3). We nonetheless reemphasize our objective to move away from such proxy metrics for RT purposes and promote the use of more clinically-relevant ones.

3.3 Results

Figure 2 shows results on the test set, sorted by their dosimetric impact. We found an overall reduction of the relative mean absolute error (RMAE) with respect to the reference dose maps, from 0.449 ± 0.545, obtained via the BCE+SoftDice combo-loss, to 0.258 ± 0.201 for the proposed DOSELO (i.e., an effective 42.5% reduction with $\lambda = 1$). This significant dose error reduction shows the ability of the proposed approach to yield segmentation results in better agreement with dose maps obtained using GT segmentations than those obtained using the state-of-the-art BCE+SoftDice combo-loss.

Table 1 shows results for the first and most significant four cases from a RT point of view (due to space limitations, all other cases are shown in supplementary material). We observe the ability of the proposed approach to significantly reduce outliers, generating a negative dosimetry impact on the dose

---

[1] Code available under https://github.com/ruefene/doselo.

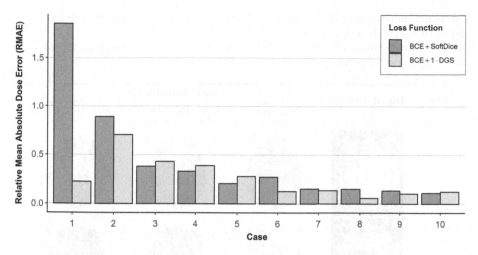

Fig. 2. Relative mean absolute dose errors/differences (RMAE) between the reference dose map and dose maps obtained using the predicted segmentations. Lower is better. Across all tested cases and folds we observe a large RMAE reduction for dose maps using the proposed DOSELO (average RMAE reduction of 42.5%).

maps. We analyzed case number 3, 4, and 5 from Fig. 2 for which the standard BCE+SoftDice was slightly better than the proposed DOSELO. For case no. 3 the tumor presents a non-convex shape alongside the skull's parietal lobe, which was not adequately modeled by the training dataset used to train the segmentation models. Indeed, we remark that both models failed to yield acceptable segmentation quality in this area. In case no. 4, both models failed to segment the diffuse tumor area alongside the skull; however, as shown in Fig. 2-case no. 4, the standard BCE+SoftDice model would yield a centrally located radiation dose, with strong negative clinical impact to the patient. Case no. 5 (shown in supplementary material) is an interesting case called butterfly GBM, which is a rare type of GBM (around 2% of all GBM cases [3]), characterized by bihemispheric involvement and invasion of the corpus callosum. In this case, the training data also lacked characterization for such cases. Despite this limitation, we observed favorable dose distributions with the proposed method.

Although we are aware that classical segmentation metrics poorly correlate with dosimetric effects [18], we report that the proposed method is more robust than the baseline BCE+SoftDice loss function, which yields outliers with Hausdorff distances: 64.06 ± 29.84 mm vs 28.68 ± 22.25 mm (-55.2% reduction) for the proposed approach. As pointed out by [17], segmentation outliers can have a detrimental effect on RT planning. We also remark that the range of HD values is in range with values reported by models trained using much more training data (see [1]), alluding to the possibility that the problem of robustness might not be directly solvable with more data. Dice coefficients did not deviate significantly between the baseline and the DOSELO models (DSC: 0.713 ± 0.203 (baseline) vs. 0.697 ± 0.216 (DOSELO)).

Table 1. Comparison of dose maps and their absolute differences to the reference dose maps (BCE+SoftDice (BCE+SD), and the proposed DOSELO). It can be seen that DOSELO yields improved dose maps, which are in better agreement with the reference dose maps (dose map color scale: 0 (blue) - 70Gy (red)).

Case	Input Image	Dose Simulation		
		Reference	|Ref.-(BCE+SD)|	|Ref. - DOSELO|
1				
2				
3				
4				

4   Discussion and Conclusion

The ultimate goal of DL-based segmentation for RT planning is to provide reliable and patient-safe segmentations for dosimetric planning and optimally targeting tumor lesions and sparing of healthy tissues. However, current loss functions used to train models for RT purposes rely solely on geometric considerations that have been shown to correlate poorly with dosimetric objectives [11,18,22,23]. In this paper, we propose a novel dosimetry-aware training loss function, called DOSELO, to effectively guide the training of segmentation models toward dosimetric-compliant segmentation results for RT purposes. The proposed DOSELO uses a fast-dose map prediction model, enabling model guidance on how dosimetry is affected by segmentation variations. We merge this information into a simple yet effective loss function that can be combined with existing ones. These first results on a dataset of post-operative GBM patients show the ability of the proposed DOSELO to deliver improved dosimetric-compliant segmentation results. Future work includes extending our database of GBM cases and to other anatomies, as well as verifying potential improvements when co-training the segmentation and dose predictor models, and jointly segmenting GTVs and OARs. With this study, we hope to promote more research toward clinically-relevant DL training loss functions.

References

1. Bakas, S., et al.: Identifying the best machine learning algorithms for brain tumor segmentation, progression assessment, and overall survival prediction in the brats challenge. arXiv preprint arXiv:1811.02629 (2018)
2. Cloak, K., et al.: Contour variation is a primary source of error when delivering post prostatectomy radiotherapy: results of the trans-tasman radiation oncology group 08.03 radiotherapy adjuvant versus early salvage (raves) benchmarking exercise. J. Med. Imag. Radiat. Oncol. **63**(3), 390–398 (2019)
3. Dayani, F., et al.: Safety and outcomes of resection of butterfly glioblastoma. Neurosurg. Focus **44**(6), E4 (2018)
4. Fidon, L., et al.: A dempster-shafer approach to trustworthy AI with application to fetal brain MRI segmentation. arXiv preprint arXiv:2204.02779 (2022)
5. Galdran, A., Carneiro, G., Ballester, M.A.G.: On the optimal combination of cross-entropy and soft dice losses for lesion segmentation with out-of-distribution robustness. In: Yap, M.H., Kendrick, C., Cassidy, B. (eds.) Diabetic Foot Ulcers Grand Challenge. DFUC 2022. LNCS, vol. 13797. Springer, Cham (2023). https://doi.org/10.1007/978-3-031-26354-5_4
6. Gneiting, T., Raftery, A.E.: Strictly proper scoring rules, prediction, and estimation. J. Am. Stat. Assoc. **102**(477), 359–378 (2007)
7. Guo, C., Huang, P., Li, Y., Dai, J.: Accurate method for evaluating the duration of the entire radiotherapy process. J. Appl. Clin. Med. Phys. **21**(9), 252–258 (2020)
8. Isensee, F., Jaeger, P.F., Kohl, S.A., Petersen, J., Maier-Hein, K.H.: nnU-Net: a self-configuring method for deep learning-based biomedical image segmentation. Nat. Methods **18**(2), 203–211 (2021)
9. Kamath, A., Poel, R., Willmann, J., Andratschke, N., Reyes, M.: How sensitive are deep learning based radiotherapy dose prediction models to variability in organs at risk segmentation? In: 2023 IEEE 20th International Symposium on Biomedical Imaging (ISBI), pp. 1–4. IEEE (2023)
10. Kingma, D., Ba, J.: Adam: a method for stochastic optimization. arXiv preprint arXiv:1412.6980 (2014)
11. Kofler, F., et al.: Are we using appropriate segmentation metrics? Identifying correlates of human expert perception for CNN training beyond rolling the dice coefficient. arXiv preprint arXiv:2103.06205 (2021)
12. Liu, S., Zhang, J., Li, T., Yan, H., Liu, J.: A cascade 3D U-Net for dose prediction in radiotherapy. Med. Phys. **48**(9), 5574–5582 (2021)
13. Ma, J., et al.: Loss odyssey in medical image segmentation. Med. Image Anal. **71**, 102035 (2021)
14. Mehrtash, A., Wells, W.M., Tempany, C.M., Abolmaesumi, P., Kapur, T.: Confidence calibration and predictive uncertainty estimation for deep medical image segmentation. IEEE Trans. Med. Imaging **39**(12), 3868–3878 (2020). https://doi.org/10.1109/TMI.2020.3006437

15. Milletari, F., Navab, N., Ahmadi, S.A.: V-Net: Fully convolutional neural networks for volumetric medical image segmentation. In: 2016 Fourth International Conference on 3D vision (3DV), pp. 565–571. IEEE (2016)
16. Niyazi, M., et al.: ESTRO-ACROP guideline "target delineation of glioblastomas." Radiotherapy Oncol. **118**(1), 35–42 (2016)
17. Poel, R., et al.: Impact of random outliers in auto-segmented targets on radiotherapy treatment plans for glioblastoma. Radiat. Oncol. **17**(1), 170 (2022)
18. Poel, R., et al.: The predictive value of segmentation metrics on dosimetry in organs at risk of the brain. Med. Image Anal. **73**, 102161 (2021)
19. Reinke, A., et al.: Common limitations of performance metrics in biomedical image analysis. In: Medical Imaging with Deep Learning (2021)
20. Ronneberger, O., Fischer, P., Brox, T.: U-Net: convolutional networks for biomedical image segmentation. In: Navab, N., Hornegger, J., Wells, W.M., Frangi, A.F. (eds.) MICCAI 2015. LNCS, vol. 9351, pp. 234–241. Springer, Cham (2015). https://doi.org/10.1007/978-3-319-24574-4_28
21. Scoccianti, S., et al.: Organs at risk in the brain and their dose-constraints in adults and in children: a radiation oncologist's guide for delineation in everyday practice. Radiother. Oncol. **114**(2), 230–238 (2015)
22. Vaassen, F., et al.: Evaluation of measures for assessing time-saving of automatic organ-at-risk segmentation in radiotherapy. Phys. Imag. Radiat. Oncol. **13**, 1–6 (2020)
23. Vandewinckele, L., et al.: Overview of artificial intelligence-based applications in radiotherapy: recommendations for implementation and quality assurance. Radiother. Oncol. **153**, 55–66 (2020)
24. Vinod, S.K., Jameson, M.G., Min, M., Holloway, L.C.: Uncertainties in volume delineation in radiation oncology: a systematic review and recommendations for future studies. Radiother. Oncol. **121**(2), 169–179 (2016)

Realistic Endoscopic Illumination Modeling for NeRF-Based Data Generation

Dimitrios Psychogyios$^{(\boxtimes)}$ ⓘ, Francisco Vasconcelos ⓘ, and Danail Stoyanov ⓘ

University College London, London, UK
{dimitris.psychogyios.19,f.vasconcelos,danail.stoyanov}@ucl.ac.uk

Abstract. Expanding training and evaluation data is a major step towards building and deploying reliable localization and 3D reconstruction techniques during colonoscopy screenings. However, training and evaluating pose and depth models in colonoscopy is hard as available datasets are limited in size. This paper proposes a method for generating new pose and depth datasets by fitting NeRFs in already available colonoscopy datasets. Given a set of images, their associated depth maps and pose information, we train a novel light source location-conditioned NeRF to encapsulate the 3D and color information of a colon sequence. Then, we leverage the trained networks to render images from previously unobserved camera poses and simulate different camera systems, effectively expanding the source dataset. Our experiments show that our model is able to generate RGB images and depth maps of a colonoscopy sequence from previously unobserved poses with high accuracy. Code and trained networks can be accessed at https://github.com/surgical-vision/REIM-NeRF.

Keywords: Surgical Data Science · Surgical AI · Data generation · Neural Rendering · Colonoscopy

1 Introduction

During colonoscopy screenings, localizing the camera and reconstructing the colon directly from the video feed could improve the detection of polyps and help with navigation. Such tasks can be either treated individually using depth [3,6,15,16] and pose estimation approaches [1,14] or jointly, using structure from motion (SfM) and visual simultaneous localization and mapping (VSLAM) algorithms [5,9,10]. However, the limited data availability present in surgery often makes evaluation and supervision of learning-based approaches difficult.

To address the lack of data in surgery, previous work has explored both synthetic pose and depth data generation [2,15], and real data acquisition [2,4,12].

Supplementary Information The online version contains supplementary material available at https://doi.org/10.1007/978-3-031-43996-4_51.

H. Greenspan et al. (Eds.): MICCAI 2023, LNCS 14228, pp. 535–544, 2023.
https://doi.org/10.1007/978-3-031-43996-4_51

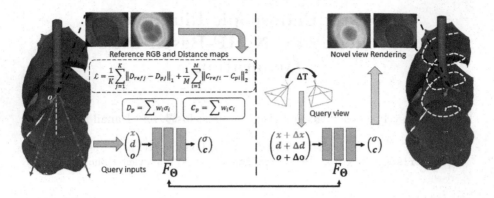

Fig. 1. (Left) The reference depth and RGB images (yellow trajectory) are used to learn an implicit representation of a scene F_θ. (Right) After training, the user can define a new trajectory (white) and extent the original dataset. (Color figure online)

Generating datasets from endoscopic sequences using game engines is scalable and noise-free but often cannot replicate the material properties and lighting conditions of the target environment. In contrast, capturing real data is a laborious process that often introduces sources of error.

Neural Radiance Field (NeRF) [11] networks aim to learn an implicit 3D representation of a 3D scene from a set of images captured from known poses, enabling image synthesis from previously unseen viewpoints. NeRF models render 3D geometry and color, including view-dependent reflections, enabling the rendering of photo-realistic images and geometrically consistent depth maps. EndoNeRF [20] applied NeRF techniques for the first time on surgical video. The method fits a dynamic NeRF [13] on laparoscopic videos, showing tools manipulating tissue from a fixed viewpoint. After training, the video sequences were rendered again without the tools obstructing the tissue. However, directly applying similar techniques in colonoscopy, is challenging because NeRF assumes fixed illumination. As soon as the endoscope moves, changes in tissue illumination result in color ambiguities.

This paper aims to mitigate the depth and pose data scarcity in colonoscopy. Inspired by work in data generation using NeRF [18], we present an extension of NeRF which makes it more suitable for use in endoscopic scenes. Our approach aims to expand colonoscopy VSLAM datasets [4] by rendering views from novel trajectories while allowing simulation of different camera models Fig. 1. Our approach addresses the scalability issues of real data generation techniques while reproducing realistic images. Our main contributions are: 1)The introduction of a depth-supervised NeRF variant conditioned on the location of the endoscope's light source. This extension is important for modeling variation in tissue illumination while the endoscope moves. 2) We evaluate our model design choices on the C3VD dataset and present renditions of the dataset from previously unseen viewpoints in addition to simulating different camera systems.

2   Method

Our method requires a set of images with known intrinsic and extrinsic camera parameters and sparse depth maps of a colonoscopy sequence. This information is already available in high-quality VSLAM [2,4,12] datasets which we wish to expand or can be extracted by running an SfM pipeline such as COLMAP [17] on prerecorded endoscopic sequences. Our pipeline involves first optimizing a special version of NeRF modified to model the unique lighting conditions present in endoscopy. The resulting network is used to render views and their associated dense depth maps from user-defined camera trajectories while allowing to specify of the camera model used during rendering. Images rendered from our models closely resemble the characteristics of the training samples. Similarly, depth maps are geometrically consistent as they share a commonly learned 3D representation. Those properties make our method appealing as it makes realistic data generation easy, configurable, and scalable.

2.1   Neural Radiance Fields

A NeRF [11] implicitly encodes the geometry and radiance of a scene as a continuous volumetric function $F_\Theta : (\mathbf{x}, \mathbf{d}) \rightarrow (\mathbf{c}, \sigma))$. The inputs of F_Θ are the 3D location of a point in space $\mathbf{x} = (x, y, z)$ and the 2D direction from which the point is observed $\mathbf{d} = (\phi, \theta)$. The outputs are the red-green-blue (RGB) color $\mathbf{c} = (r, g, b)$ and opacity σ. F_θ is learned and stored in the weights of two cascaded multi-layer perceptron (MLP) networks. The first, f_σ, is responsible for encoding the opacity σ of a point, based only on $\mathbf{x}$. The second MLP, f_c, is responsible for encoding the point's corresponding $\mathbf{c}$ based on the output of the f_σ and $\mathbf{d}$. NeRFs are learned using differentiable rendering techniques given a set of images of scenes and their associated poses. Optimization is achieved by minimizing the L2 loss between the predicted and reference color of all pixels in the training images. To predict the color of a pixel, a ray $\mathbf{r}(t) = \mathbf{o} + t\mathbf{d}$ is defined in space starting for the origin of the corresponding image o and heading towards $\mathbf{d}$ which is the direction from where light projects to the corresponding pixel. N points $\mathbf{r}(t_i), i \in [n, f]$ are sampled along the ray between a near t_n and a far t_f range to query NeRF for both σ and $\mathbf{c}$. The opacity values σ of all points along the $\mathbf{r}$ can be used to approximate the accumulated transmittance T_i as defined in Eq. (1), which describes the probability of light traveling from $\mathbf{o}$ to $\mathbf{r}(t_i)$. T_i together with the color output of f_σ for every point along the ray, can be used to compute the pixel color C_p using alpha composition as defined in Eq. (2)

$$T_i = \exp\left(-\sum_{j=0}^{i-1} \sigma_j \Delta_j\right), \Delta_k = t_{k+1} - t_k \tag{1}$$

$$C_p = \sum_{i=1}^{N} w_i \mathbf{c}_i, \text{ where } w_i = T_i(1 - \exp(-\Delta_i \sigma_i)) \tag{2}$$

Similarly, the expected ray termination distance D_p can be computed from Eq. (3), which is an estimate of how far a ray travels from the camera until it hits solid geometry. D_p can be converted to z-depth by knowing the uv coordinates of the corresponding pixel and camera model.

$$D_p = \sum_{i=1}^{N} w_i t_i \qquad (3)$$

In practice, NeRF uses two pairs of MLPs. Initially, a coarse NeRF $F_{\Theta c}$ is evaluated on N_c samples along a ray. The opacity output of the coarse network is used to re-sample rays with more dense samples where opacity is higher. The new N_f ray samples are used to query a fine NeRF $F_{\Theta f}$. During both training and inference, both networks are working in parallel. Lastly, to enable NeRF to encapsulate high-frequency geometry and color details, every input of F_Θ is processed by a hand-crafted positional encoding module $\gamma(\cdot)$, using Fourier features [19].

2.2   Extending NeRF for Endoscopy

Light-Source Location Aware MLP. During a colonoscopy, the light source always moves together with the camera. Light source movement results in illumination changes on the tissue surface as a function of both viewing direction (specularities), camera location (exposure changes), and distance between the tissue and the light source (falloff) Fig. 2. NeRF [11], only models radiance as a function of viewing direction as this is enough when the scene is lit uniformly from a fixed light source and the camera exposure is fixed. To model changes in illumination as a function of light source location, we extend the original NeRF formulation by conditioning f_c on both the 2D ray direction $\gamma(\mathbf{d})$ and also the location of the light source $\mathbf{o}$. For simplicity, throughout this work, we assume a single light source co-located with the camera. This parameterization allows the network to learn how light decays as it travels away from the camera and adjusts the scene brightness accordingly.

Depth Supervision. NeRF achieves good 3D reconstruction of scenes using images captured from poses distributed in a hemisphere [11]. This imposes geometric constraints during the optimization because consistent geometry would result in a 3D point projecting in correct pixel locations across different views. Training a NeRF on colonoscopy sequences is hard because the camera moves along a narrow tube-like structure and the colon wall is often texture-less. Supervising depth together with color can guide NeRF to learn a good 3D representation even when pose distribution is sub-optimal [7]. In this work, we compute the distance between the camera and tissue D_{ref} from the reference depth maps and we sample K out of M pixel Eq. (5) to optimize both color and depth as

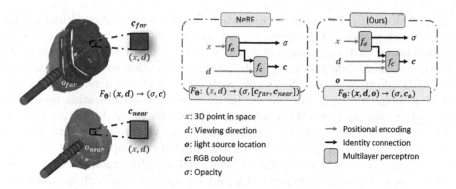

Fig. 2. (left) The color of a point as seen from the colonoscope, from two different distances due to changes in illumination. (center) original NeRF formulation is not able to assign different colors for the same point and viewing direction. (right) ours is conditioned on the location of the light source and the point, modeling light decay.

described in Eq. (4).

$$\mathcal{L} = \frac{1}{K}\sum_{j=1}^{K}\|D_{refj} - D_{pj}\|_1 + \frac{1}{M}\sum_{i=1}^{M}\|C_{refi} - C_{pi}\|_2^2 \tag{4}$$

$$D_{refj} \sim \mathcal{U}[D_{ref1}, D_{refM}], i \in K \leq |M| \tag{5}$$

$\|\cdot\|_1$ is the L1 loss, $\|\cdot\|_2^2$ is the L2 loss, $\mathcal{U}$ denotes uniform sampling.

3   Experiments and Results

3.1   Dataset

We train and evaluate our method on C3VD [4], which provides 22 small video sequences captured from a real wide-angle colonoscopy at 1350×1080 resolution, moving inside 4 different colon phantoms. The videos include sequences from the colon cecum, descending, sigmoid, and transcending. Videos range from 61 to 1142 frames adding to 10.015 in total. We use the per-frame camera poses and set $K/M = 0.03$ in Eq. (5). For each scene, we construct a training set using one out of every 5 frames. We further remove and allocate one out of every 5 poses from the training set for evaluation. Frames not present in either the train or evaluation set are used for testing. We choose to sample both poses and depth information to allow our networks to interpolate more easily between potentially noisy labels and also create a dataset that resembles the sparse output of SfM algorithms.

3.2   Implementation Details

Before training, we spatially re-scale and shift the 3D scene from each video sequence such that every point is enclosed within a cube with a length of two, centered at (0,0,0). Prescaling is important for the positional encoding module $\gamma(\mathbf{x})$ to work properly. We configure positional encoding modules to compute 10 frequencies for each component of $\mathbf{x}$ and 4 frequencies for each component of $\mathbf{d}$ and $\mathbf{o}$. We train models on images of 270×216 resolution to ignore both depth and RGB information outside a circle with a radius of 130 pixels centered at a principal point to avoid noise due to inaccuracies of the calibration model. We used Adam optimizer [8] with a batch size of 1024 for about 140K iterations for all sequences, with an initial learning rate of 5e-4 which we later multiply by 0.5 at 50% and 75% of training. We set the number of samples along the ray to $N_c = 64$ and $N_f = 64$. We configure positional encoding modules to compute 10 frequencies for each component of $\mathbf{x}$ and 4 frequencies for each component of $\mathbf{d}$ and $\mathbf{o}$. Each model from this work is trained for around 30 min on 4 graphics cards from an NVIDIA DGX-A100.

3.3   Model Ablation Study

We ablate our model showing the effects of conditioning NeRF on the light source location and supervising depth. To assess RGB reconstruction, we measure the peak signal-to-noise ratio (PSNR) and structural similarity index measure (SSIM) between reconstructed and reference images at 270×216 resolution. We evaluate depth using the mean squared error (MSE) in the original dataset scale. For each metric, we report the average across all sequences together with the average standard deviation in Table 1. Conditioning NeRF based on the light source location (this work) produces better or equally good results compared to vanilla NeRF for all metrics. Both depth-supervised models, learn practically the same 3D representation but our model achieves better image quality metrics.

Table 1. Mean and standard deviation metrics of every method aggregated across all sequences of C3VD dataset.

Model	PSNR ↑	SSIM ↑	Depth MSE (mm)↓
NeRF	$32.097 \pm (1.173)$	$0.811 \pm (0.021)$	$4.263 \pm (1.178)$
NeRF + ls_loc(Ours)	$32.489 \pm (1.128)$	$0.820 \pm (0.018)$	$1.866 \pm (0.594)$
NeRF + depth	$30.751 \pm (1.163)$	$0.788 \pm (0.022)$	$0.015 \pm (0.016)$
Full model (Ours)	$31.662 \pm (1.082)$	$0.797 \pm (0.020)$	$0.013 \pm (0.018)$

Figure 3 shows renditions of each model of the ablation study for the same frame. Both non-depth-supervised models failed to capture correct geometry but were able to reconstruct accurate RGB information. Since non-depth-supervised networks were optimized only on color with weak geometric constraints, they learn floating artifacts in space which when viewed from a specific viewpoint, closely approximate the training samples. In contrast, depth-supervised networks learned a good representation of (3D) geometry while being able to reconstruct

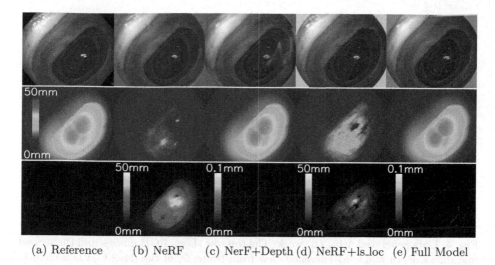

(a) Reference   (b) NeRF   (c) NerF+Depth (d) NeRF+ls_loc (e) Full Model

Fig. 3. Renditions from all model ablations compared to a reference RGB image and depth map. Reference and reconstructed images (top row), depth maps (middle row), and MSE depth difference between the reference and each model prediction in different scales (bottom row).

RGB images accurately. The depth-supervised NeRF model produces flare arti-facts in the RGB image. That is because, during optimization, points are viewed from the same direction but at different distances from the light source. Our Full model is able to cope with illumination changes resulting in artifact-free images and accurate depth. Notably, most of the errors in depth for the depth-supervised approaches are located around sharp depth transitions. Such errors in depth may be a result of inaccuracies in calibration or imperfect camera pose information. Nevertheless, we argue that using RGB images and depth maps produced from our approach can be considered error-free because during inference the learned 3D geometry is fixed and consistent across all rendered views.

3.4   Data Generation

We directly use our proposed model of the d4v2 C3VD scene from the ablation study to render novel views and show results in Fig. 4. In the second column, we show an image rendered from a previously unseen viewpoint, radially off-set from the original camera path. Geometry is consistent and the RGB image exhibits the same photo-realistic properties observed in the training set. In the third column, we render a view by rotating a pose from the training trajectory. In the rotated view, the tissue is illuminated in a realistic way even though the camera never pointed in this direction in the training set. In the fourth column, we show an image rendered using a pinhole camera model whilst only fisheye images were used during training. This is possible because NeRF has captured a good representation of the underlying scene and image formation is done by

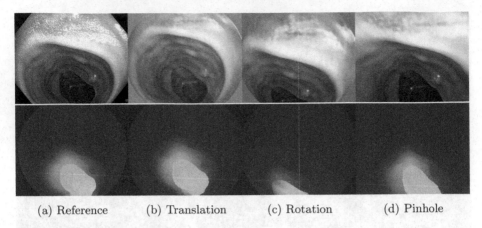

(a) Reference   (b) Translation   (c) Rotation   (d) Pinhole

Fig. 4. Data generated using our work. The top row shows RGB images and the second row shows depth maps. a)Reference view from frame 60 of the d4v2 C3VD sequence. b) Translating (a) along all axis. c) Rotating (a) around both the x and y-axis d) Simulating a pinhole model in (a).

projecting rays in space based on user-defined camera parameters. All the above demonstrate the ability of our method to render images from new, user-defined, trajectories and camera systems similar to synthetic data generation while producing photo-realistic images.

4   Conclusion

We presented an approach for expanding existing VSLAM datasets by rendering RGB images and their associated depth maps from user-defined camera poses and models. To achieve this task, we propose a novel variant of NeRF, conditioned on the location of the light source in 3D space and incorporating sparse depth supervision. We evaluate the effects of our contributions on phantom datasets and show that our work effectively adapts NeRF techniques to the lighting conditions present in endoscopy. We further demonstrate the efficacy of our method by showing RGB images and their associated depth maps rendered from novel views of the target endoscopic scene. 3D information and conditioning NeRF based on the light source location made NeRF suitable for use in Endoscopy. Currently, our method assumes a static environment and requires accurate camera intrinsic and extrinsic information. Subsequent work can incorporate mechanisms to represent deformable scenes [13] and refine camera parameters during training [21]. Further research can investigate adopting the proposed model for data generation in other endoscopic scenes or implementing approaches to perform label propagation for categorical data [22]. We hope this work will mitigate the data scarcity issue currently present in the surgical domain and inspire the community to leverage and improve neural rendering techniques for data generation.

Aknowledgements. This research was funded, in whole, by the Wellcome/EPSRC Centre for Interventional and Surgical Sciences (WEISS) [203145/Z/16/Z]; the Engineering and Physical Sciences Research Council (EPSRC) [EP/P027938/1, EP/R004080/1, EP/P012841/1]; the Royal Academy of Engineering Chair in Emerging Technologies Scheme, and Horizon 2020 FET (863146). For the purpose of open access, the author has applied a CC BY public copyright licence to any author accepted manuscript version arising from this submission.

References

1. Armin, M.A., Barnes, N., Alvarez, J., Li, H., Grimpen, F., Salvado, O.: Learning camera pose from optical colonoscopy frames through deep convolutional neural network (CNN). In: Cardoso, M.J., et al. (eds.) CARE/CLIP -2017. LNCS, vol. 10550, pp. 50–59. Springer, Cham (2017). https://doi.org/10.1007/978-3-319-67543-5_5

2. Azagra, P., et al.: Endomapper dataset of complete calibrated endoscopy procedures. arXiv preprint arXiv:2204.14240 (2022)

3. Batlle, V.M., Montiel, J.M., Tardós, J.D.: Photometric single-view dense 3D reconstruction in endoscopy. In: 2022 IEEE/RSJ International Conference on Intelligent Robots and Systems (IROS), pp. 4904–4910. IEEE (2022)

4. Bobrow, T.L., Golhar, M., Vijayan, R., Akshintala, V.S., Garcia, J.R., Durr, N.J.: Colonoscopy 3D video dataset with paired depth from 2D-3D registration. arXiv preprint arXiv:2206.08903 (2022)

5. Chen, R.J., Bobrow, T.L., Athey, T., Mahmood, F., Durr, N.J.: SLAM endoscopy enhanced by adversarial depth prediction. arXiv preprint arXiv:1907.00283 (2019)

6. Cheng, K., Ma, Y., Sun, B., Li, Y., Chen, X.: Depth estimation for colonoscopy images with self-supervised learning from videos. In: de Bruijne, M., et al. (eds.) MICCAI 2021, Part VI. LNCS, vol. 12906, pp. 119–128. Springer, Cham (2021). https://doi.org/10.1007/978-3-030-87231-1_12

7. Deng, K., Liu, A., Zhu, J.Y., Ramanan, D.: Depth-supervised NeRF: fewer views and faster training for free. In: Proceedings of the IEEE/CVF Conference on Computer Vision and Pattern Recognition, pp. 12882–12891 (2022)

8. Kingma, D.P., Ba, J.: Adam: a method for stochastic optimization. arXiv preprint arXiv:1412.6980 (2014)

9. Lamarca, J., Parashar, S., Bartoli, A., Montiel, J.: DefSLAM: tracking and mapping of deforming scenes from monocular sequences. IEEE Trans. Rob. **37**(1), 291–303 (2020)

10. Ma, R., et al.: RNNSLAM: reconstructing the 3D colon to visualize missing regions during a colonoscopy. Med. Image Anal. **72**, 102100 (2021)

11. Mildenhall, B., Srinivasan, P.P., Tancik, M., Barron, J.T., Ramamoorthi, R., Ng, R.: NeRF: representing scenes as neural radiance fields for view synthesis. In: Vedaldi, A., Bischof, H., Brox, T., Frahm, J.-M. (eds.) ECCV 2020. LNCS, vol. 12346, pp. 405–421. Springer, Cham (2020). https://doi.org/10.1007/978-3-030-58452-8_24

12. Ozyoruk, K.B., et al.: Endoslam dataset and an unsupervised monocular visual odometry and depth estimation approach for endoscopic videos. Med. Image Anal. **71**, 102058 (2021)

13. Pumarola, A., Corona, E., Pons-Moll, G., Moreno-Noguer, F.: D-NeRF: neural radiance fields for dynamic scenes. In: Proceedings of the IEEE/CVF Conference on Computer Vision and Pattern Recognition, pp. 10318–10327 (2021)

14. Rau, A., Bhattarai, B., Agapito, L., Stoyanov, D.: Bimodal camera pose prediction for endoscopy. arXiv preprint arXiv:2204.04968 (2022)
15. Rau, A., et al.: Implicit domain adaptation with conditional generative adversarial networks for depth prediction in endoscopy. Int. J. Comput. Assist. Radiol. Surg. **14**(7), 1167–1176 (2019). https://doi.org/10.1007/s11548-019-01962-w
16. Rodriguez-Puigvert, J., Recasens, D., Civera, J., Martinez-Cantin, R.: On the uncertain single-view depths in colonoscopies. In: Wang, L., Dou, Q., Fletcher, P.T., Speidel, S., Li, S. (eds.) MICCAI 2022, Part III. LNCS, vol. 13433, pp. 130–140. Springer, Cham (2022). https://doi.org/10.1007/978-3-031-16437-8_13
17. Schönberger, J.L., Frahm, J.M.: Structure-from-motion revisited. In: Conference on Computer Vision and Pattern Recognition (CVPR) (2016)
18. Tancik, M., et al.: Block-NeRF: scalable large scene neural view synthesis. In: Proceedings of the IEEE/CVF Conference on Computer Vision and Pattern Recognition, pp. 8248–8258 (2022)
19. Tancik, M., et al.: Fourier features let networks learn high frequency functions in low dimensional domains. Adv. Neural. Inf. Process. Syst. **33**, 7537–7547 (2020)
20. Wang, Y., Long, Y., Fan, S.H., Dou, Q.: Neural rendering for stereo 3D reconstruction of deformable tissues in robotic surgery. In: Wang, L., Dou, Q., Fletcher, P.T., Speidel, S., Li, S. (eds.) MICCAI 2022, Part VII. LNCS, vol. 13437, pp. 431–441. Springer, Cham (2022). https://doi.org/10.1007/978-3-031-16449-1_41
21. Wang, Z., Wu, S., Xie, W., Chen, M., Prisacariu, V.A.: NeRF−−: neural radiance fields without known camera parameters. arXiv preprint arXiv:2102.07064 (2021)
22. Zhi, S., Laidlow, T., Leutenegger, S., Davison, A.J.: In-place scene labelling and understanding with implicit scene representation. In: Proceedings of the IEEE/CVF International Conference on Computer Vision, pp. 15838–15847 (2021)

Multi-task Joint Prediction of Infant Cortical Morphological and Cognitive Development

Xinrui Yuan[1,2], Jiale Cheng[2], Fenqiang Zhao[2], Zhengwang Wu[2], Li Wang[2], Weili Lin[2], Yu Zhang[1], and Gang Li[2(✉)]

[1] School of Biomedical Engineering, Southern Medical University, Guangzhou, Guangdong, China
[2] Department of Radiology and Biomedical Research Imaging Center, University of North Carolina at Chapel Hill, Chapel Hill, NC, USA
gang_li@med.unc.edu

Abstract. During the early postnatal period, the human brain undergoes rapid and dynamic development. Over the past decades, there has been increased attention in studying the cognitive and cortical development of infants. However, accurate prediction of the infant cognitive and cortical development at an individual-level is a significant challenge, due to the huge complexities in highly irregular and incomplete longitudinal data that is commonly seen in current studies. Besides, joint prediction of cognitive scores and cortical morphology is barely investigated, despite some studies revealing the tight relationship between cognitive ability and cortical morphology and suggesting their potential mutual benefits. To tackle this challenge, we develop a flexible multi-task framework for joint prediction of cognitive scores and cortical morphological maps, namely, disentangled intensive triplet spherical adversarial autoencoder (DITSAA). First, we extract the mixed representative latent vector through a triplet spherical ResNet and further disentangles latent vector into identity-related and age-related features with an attention-based module. The identity recognition and age estimation tasks are introduced as supervision for a reliable disentanglement of the two components. Then we formulate the individualized cortical profile at a specific age by combining disentangled identity-related information and corresponding age-related information. Finally, an adversarial learning strategy is integrated to achieve a vivid and realistic prediction of cortical morphology, while a cognitive module is employed to predict cognitive scores. Extensive experiments are conducted on a public dataset, and the results affirm our method's ability to predict cognitive scores and cortical morphology jointly and flexibly using incomplete longitudinal data.

Keywords: Cognitive Prediction · Cortical Morphology Prediction · Disentanglement

1 Introduction

The human brain undergoes rapid and dynamic development during the early postnatal period. Research on infant brain development [1–3] has received significant attention over the last decade. While many infant neuroimaging studies have revealed the brain growth patterns during infancy [4–6] at the population level, knowledge on individual-ized brain development during infancy is still lacking, which is essential for mapping individual brain characteristics to individual phenotypes [7, 8]. Various pioneering meth-ods have been developed using longitudinal scans collected at some pre-defined discrete age time points, such as 3, 6, 9, and 12 months [9–11], which strictly requires each infant has all longitudinal scans at the given time points. However, in practice, longitu-dinal infant images are usually collected at diverse and irregular scan ages due to the inevitable missing data and imaging protocol design, leading to less usable data. As a result, predicting infant development at specific ages from irregular longitudinal data is extremely challenging, yet critical to understanding normal brain development and early diagnosis of neurodevelopmental abnormalities [12–14]. Besides, to the best of our knowledge, joint prediction of cognitive scores and cortical morphology is barely inves-tigated, despite some studies revealing the tight relationship [15–17] between cognitive ability and cortical morphology, suggesting their potential mutual benefits.

To address this issue, we propose a flexible multi-task framework to jointly predict cognitive score and cortical morphological development of infant brains at arbitrary time points with longitudinal data scanned irregularly within 24 months of age. Specifi-cally, the cognitive ability of each infant was estimated using the Mullen Scales of Early Learning (MSEL) [18], including the visual reception scale (VRS), fine motor scale (FMS), receptive language scale (RLS), and expressive language scale (ELS). Our aim is to predict the four cognitive scores and cortical morphological feature maps (e.g., cortical thickness map) flexibly at arbitrary time points given cortical feature maps at a known age. The main contributions of this paper can be summarized as follows: 1) we propose an attention-based feature disentanglement module to separate the identity- and age-related features from the mixed latent features, which not only effectively extracts the discriminative information at individual-level, but also forms the basis for dealing with irregular and incomplete longitudinal data; 2) we introduce a novel identity condi-tional block to fuse identity-related information with designated age-related information, which can model the regression/progression process of brain development flexibly; 3) we propose a unified, multi-task framework to jointly predict the cognitive ability and cortical morphological development and enable flexible prediction at any time points during infancy by concatenating the subject-specific identity information and identity conditional block. We validated our proposed method on the Baby Connectome Project (BCP) [19] dataset, including 416 longitudinal scans from 188 subjects, and achieved superior performance on both cognitive score prediction and cortical morphological development prediction than state-of-the-art methods.

2 Methods

2.1 Network Architecture

The framework of our disentangled intensive triplet spherical adversarial autoencoder (DITSAA) is shown in Fig. 1. Since primary and secondary cortical folds in human brains are well established at term birth and keep relatively stable over time, the identity-related cortical features should be approximately time-invariant and age-related features change over time due to brain development. Based on this prior knowledge, we first disentangle the mixed cortical surface feature maps into age-related and identity-related components. Specifically, in the training stage, we formulate the training samples in triplet units $(S_i^{t_a}, S_i^{t_p}, S_j^{t_n})$, where the first two are surface property maps from the same individual i but at different ages t_a and t_p, and the last one is a surface property map from another individual j at any age t_n. Then we employ the intensive triplet loss [20] to encourage the disentanglement of identity-related features, and age prediction loss to encourage the disentanglement of age-related features.

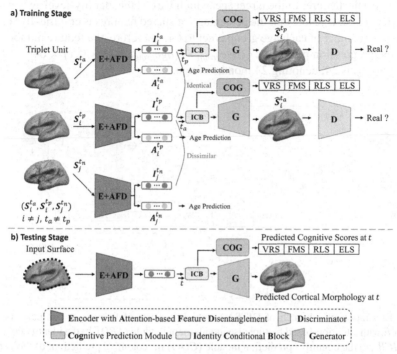

Fig. 1. The framework of the proposed disentangled intensive triplet spherical adversarial autoencoder (DITSAA).

Encoder with Attention-Based Disentanglement. We employ a spherical ResNet, modified based on ResNet [21] and Spherical U-Net [22, 23], denoted as E, as the basic encoder (Fig. 2(a)). The output of the encoder includes multi-level features with different resolutions, and the latent vector z_i^t captures the mixed features of the input cortical surface S_i^t. Then we employ the attention-based feature disentanglement (AFD) module, consisting of a channel attention module and a spatial attention module in parallel, to disentangle the age-related variance A_i^t and the identity-related invariance I_i^t, which is defined as: $z_i^t = \underbrace{z_i^t \odot \phi(z_i^t)}_{A_i^t} + \underbrace{z_i^t \odot (1 - \phi(z_i^t))}_{I_i^t}$. The operation $\odot$ indicates the element-wise multiplication and ϕ denotes an attention module in Fig. 2, which is computed as the average of channel attention [24] and spatial attention [25].

548   X. Yuan et al.

Identity Conditional Block. To preserve the identity-level regression/progression pattern, we extended the identity conditional block (*ICB*) [26] to cortical surface data based on spherical convolution. Specifically, the *ICB* takes the identity-related feature from *AFD* as input to learn an identity-level regression/progression pattern. Then, a weight-sharing strategy is adopted to improve the age smoothness by sharing same spherical convolutional filters across adjacent age groups as shown in Fig. 2(b). The idea of the weight-sharing strategy is that age-related features change gradually over time, and the shared filters can learn common evolving patterns between adjacent age groups. Then, we can select the features of the target age, which we call the identity-level age condition as shown in Fig. 2(b). Note that the percentage of shared features is empirically set to 0.1 and the filter number of each age group is set to 64. Therefore, the feature number of the regression/progression pattern is: $\lfloor (age_{group} - (age_{group} - 1) \times 0.1) \times 64 \rfloor = 1,388$, where age_{group} is 24, denoting 24 months of age.

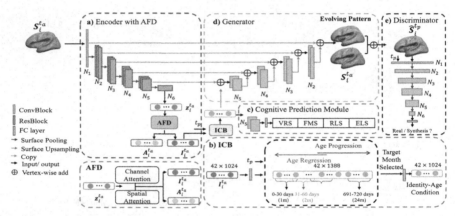

Fig. 2. Illustration of a branch of DITSAA, which consists of five major components: a) the encoder E with attention-based feature disentanglement module (*AFD*), b) the identity conditional block (*ICB*), c) the cognitive prediction module (*COG*) for cognitive prediction, d) the generator G, and e) the discriminator D for cortical morphological map. Each residual block (ResBlock) contains repeated 1-ringConv + BN + ReLU with reslink. Each convolutional block (ConvBlock) contains 1-ringConv + BN + ReLU. Of note, the ConvBlock in generator contains additional spherical transposed convolution layer to deal with the concatenation of multi-level features. N_k denotes the number of vertices in the spherical feature maps at the k-th icosahedron subdivision, where $N_{k+1} = (N_k + 6)/4$, while in our experiments N_1 equals to 40,962. The number of features after each operation block is [64, 128, 256, 512, 1024, 1024] for the k-th icosahedron subdivision ($k \in [1, 6]$).

Cognition Prediction Module. To accurately predict individual-level cognitive scores, we use the identity-age condition features containing subject-specific information and conditional age features to predict the cognitive scores through a spherical convolutional network COG (Fig. 2(c)). The predicted cognitive scores are computed as:
$$\hat{y}_i^{t_p} = COG\left(S_i^{t_a}, t_p\right) = COG\left(ICB\left(I_i^{t_a}, t_p\right)\right).$$

Cortical Morphology Prediction Module. To ensure the quality of the predicted cortical morphological maps, the generator G adopts the architecture of the UNet-based least-square GANs [27]. For the discriminator D, we extend a VGG style classification CNN to spherical surfaces. The generator G takes the selected identity-age conditional features $ICB\left(I_i^{t_a}, t_p\right)$ as inputs and gradually upsamples them and concatenates with the corresponding-level features produced by the encoder as shown in Fig. 2, and aims to predict the target cortical surface map $\widehat{S}_i^{t_p}$. This process can be summarized as:
$$\widehat{S}_i^{t_p} = G\left(S_i^{t_a}, t_p\right) = G\left(\left[E\left(S_i^{t_a}\right); ICB\left(I_i^{t_a}, t_p\right)\right]\right), [\cdot; \cdot] \text{ denotes the concatenation along}$$
the channel dimension.

2.2 Loss Design

Intensive Triplet Loss. As illustrated in Fig. 1, we suppose identity-related invariance I from a pair of same subjects $(S_i^{t_a}, S_i^{t_p})$ should be identical, while that from a pair of different subjects $(S_i^{t_a}, S_j^{t_n})$ should be dissimilar. However, noticing that $(S_i^{t_p}, S_j^{t_n})$ is also a pair of different subjects, to enhance the relative distance between same subject pair and all the different subject pairs, we apply the intensive triplet loss [20] to help AFD extract identity-related features thoroughly from latent features. A simple linear network denoted as ID is designed to extract feature vector of identity-related feature I and the feature similarity is measured by Pearson's correlation coefficient ($Corr$):

$$L_{IT} = -2 * Corr\left(ID\left(I_i^{t_a}\right), ID\left(I_i^{t_p}\right)\right) \\ + \left(Corr\left(ID\left(I_i^{t_a}\right), ID\left(I_j^{t_n}\right)\right) + Corr\left(ID\left(I_i^{t_p}\right), ID\left(I_j^{t_n}\right)\right)\right) \tag{1}$$

Age Estimation Loss. To ensure AFD extracts the age-relation information, a simple network is designed as the regressor $AgeP$ to predict age from age-related variance A. The regressor $AgeP$ has three fully connected layers with 512, 720, and 24 neurons, respectively. It aims to predict the age by computing a softmax expected value [28]:

$$L_{AE} = \|AgeP\left(A_i^{t_a}\right) - t_a\|_1 + \|AgeP\left(A_i^{t_p}\right) - t_p\|_1 + \|AgeP\left(A_j^{t_n}\right) - t_n\|_1. \tag{2}$$

Cognitive Prediction Loss. Given the identity-related age condition derived from ICB, we regress the cognitive scores directly by the cognitive prediction module at a selected age t. The loss function to train the cognitive prediction module is defined as:

$$L_{Cog} = \|y_i^{t_a} - \widehat{y}_i^{t_a}\|_1 + \|y_i^{t_p} - \widehat{y}_i^{t_p}\|_1, \tag{3}$$

where y is the ground truth of cognitive scores.

Adversarial Loss. To improve the quality of the predicted cortical property maps, we used the adversarial loss to train the generator G and discriminator D with respect to least-square GANs [27]. The aim of the discriminator D is to distinguish the generated cortical property maps as fake data, while the original cortical property maps as real data. The generator G aims to generate cortical property maps as close to the real data as possible. Thus, the objective function is expressed as:

$$\min_{D} L_D = 0.5\left[D\left(\left[S_i^{t_a}; C_{t_a}\right]\right) - 1\right]^2 + 0.5\left[D\left(\left[\hat{S}_i^{t_p}; C_{t_p}\right]\right)\right]^2, \qquad (4)$$

$$\min_{G} L_G = 0.5\left[D\left(\left[\hat{S}_i^{t_p}; C_{t_p}\right]\right) - 1\right]^2, \qquad (5)$$

where C_t denotes the one-hot age encoding at different age t.

Reconstruction Loss. To guarantee the high-quality reconstruction and constrain the vertex-wise similarity of the input cortical morphological features and generated cortical morphological features, we adopted the L1 loss to evaluate the reconstruction:

$$L_{rec} = \|\hat{S}_i^{t_a} - S_i^{t_a}\|_1 + \|\hat{S}_i^{t_p} - S_i^{t_p}\|_1. \qquad (6)$$

Full Objective. The objective functions of our model are written as:

$$L_{(E,ID,AgeP)} = \lambda_{IT} L_{IT} + \lambda_{AE} L_{AE}, \qquad (7)$$

$$L_{(ICB,Cog,G,D)} = \lambda_{Cog} L_{Cog} + \lambda_D L_D + \lambda_G L_G + \lambda_{rec} L_{rec}, \qquad (8)$$

where λ_{IT}, λ_{AE}, λ_{Cog}, λ_G, λ_D, and λ_{rec} control the relative importance of the loss terms.

3 Experiments

3.1 Dataset and Experimental Settings

In our experiments, we used the public BCP dataset [19] with 416 scans from 188 subjects (104 females/84 males) within 24 months of age, where 71 subjects have the single-time-point scans, and 117 subjects have multi-time-point scans. Cortical surface maps were generated using an infant-dedicated pipeline (http://www.ibeat.cloud/) [5]. All cortical surface maps were mapped onto the sphere [29] and registered onto the age-specific surface atlases (https://www.nitrc.org/projects/infantsurfatlas/) [6, 30] and further resampled to have 40,962 vertices. We chose 70% of subjects with multi-time-point scans as the train-validation set and conducted a 5-fold cross-validation for tuning the best parameter setting and left 30% for the test set. Of note, subjects with single-time-point scans were unable to evaluate the performance of prediction, hence, they were solely included in the training set to provide additional developmental information.

We trained E, ID, and $AgeP$ using an SGD optimizer under the supervision of L_{AE} and L_{IT} with an initial learning rate of 0.1, momentum of 0.9, and a self-adaption strategy for updating the learning rate, which was reduced by a factor of 0.2 once the training loss stopped decreasing for 5 epochs, and the hyper-parameters in the training loss functions were empirically set as follows: λ_{AE} was 0.1 and λ_{IT} was 0.1. We then fixed E and used an Adam optimizer with an initial learning rate of 0.01, β_1 of 0.5, and β_2 of 0.999 to train ICB, COG, G, and D using Eq. (8) with a self-adaption strategy for updating the learning rate. The hyper-parameters in the loss functions are empirically set as follows: $\lambda_D = 1$, $\lambda_G = 35$, $\lambda_{Cog} = 0.1$, and $\lambda_{rec} = 0.1$.

We used the cortical thickness map at one time point to jointly estimate the four Mullen cognitive scores, i.e., VRS, FMS, RLS, and ELS, and the cortical thickness map at any other time points. To quantitatively evaluate the cognitive prediction performance, the Pearson's correlation coefficient (PCC) between the ground truth and predicted values was calculated. For the cortical morphology prediction, the PCC and mean absolute error (MAE) between the ground truth and predicted values were calculated. In the testing phase, the mean and standard deviation of the 5-fold results were reported.

3.2 Evaluation and Comparison

To comprehensively evaluate the mutual benefits of the proposed modules in cognitive and cortical morphological development prediction, we conducted an ablation study on two modules in Table 1, where *w/Cognitive* and *w/Cortical* respectively denotes the variant for training cognitive prediction module and cortical prediction module only, and *w/o AFD* denotes the variant for training whole framework without *AFD* module. It can be observed that, the performance on cognitive and cortical prediction tasks has been improved by performing two tasks jointly, which highlights the associations and mutual benefits between cortical morphology and cognitive ability during infant brain development.

Table 1. Performance of different variants on both cognitive (first four columns) and cortical morphological property prediction (last column). * indicates statistically significantly better results than other methods with p-value < 0.05.

Variants	ELS	FMS	RLS	VRS	MAE
w/Cognitive	0.745 ± 0.022	0.868 ± 0.005	0.806 ± 0.024	0.769 ± 0.014	/
w/Cortical	/	/	/	/	0.507 ± 0.057
w/o AFD	0.774 ± 0.040	**0.888 ± 0.026**	0.828 ± 0.039	0.790 ± 0.034	0.322 ± 0.044
Proposed	**0.806 ± 0.016***	0.880 ± 0.012	**0.850 ± 0.004***	**0.820 ± 0.014***	**0.312 ± 0.032***

We also compared our DITSAA with various existing methods given different tasks in our multi-task framework. For the cortical morphology and cognitive development prediction, the state-of-the-art spherical networks SUNet [22], UGSCNN [31], and Spherical Transformer [32] were utilized as baselines. However, these networks are unable to flexibly predict cortical morphology at any time point. Therefore, we apply one-hot encoding to specify the age condition and concatenate it with the cortical morphological features along the channel attention as input to predict cognitive scores and cortical thickness at specific time point. Note that these methods with one-hot age condition still lack the generality and exhibit unsatisfactory performance in predicting both cognition and cortical morphological properties with only a few training samples at each month due to the irregularity of the longitudinal dataset [19]. As illustrated in Table 2 and Fig. 3, we can see our proposed DITSAA achieves better performance in both cortical morphological and cognitive development prediction.

Table 2. Comparison with other state-of-the-art methods on both cognitive (first four columns) and cortical morphological property prediction (last two columns) tasks. * indicates statistically significantly better results than other methods with p-value < 0.05.

Methods	ELS	FMS	RLS	VRS	MAE	PCC
SUNet	0.377 ± 0.033	0.339 ± 0.427	0.368 ± 0.055	0.304 ± 0.042	0.524 ± 0.141	0.872 ± 0.182
UGSCNN	0.223 ± 0.117	0.154 ± 0.025	0.185 ± 0.021	0.193 ± 0.089	0.548 ± 0.065	0.849 ± 0.164
Spherical Transformer	0.186 ± 0.045	0.118 ± 0.058	0.198 ± 0.052	0.181 ± 0.035	0.575 ± 0.123	0.861 ± 0.121
Proposed	$\mathbf{0.806 \pm 0.016^*}$	$\mathbf{0.880 \pm 0.012^*}$	$\mathbf{0.850 \pm 0.004^*}$	$\mathbf{0.820 \pm 0.014^*}$	$\mathbf{0.312 \pm 0.032^*}$	$\mathbf{0.893 \pm 0.103^*}$

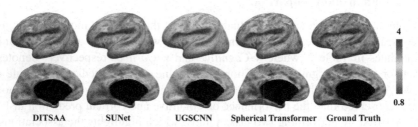

DITSAA   SUNet   UGSCNN   Spherical Transformer   Ground Truth

Fig. 3. The predicted cortical morphology property maps (herein, cortical thickness (mm)) of a randomly selected subject (predicted from 4 to 7 months of age) obtained by proposed DITSAA and other competing methods.

4   Conclusions

In this study, we proposed a novel flexible multi-task joint prediction framework for infant cortical morphological and cognitive development, named disentangled intensive triplet spherical adversarial autoencoder. We effectively and sufficiently leveraged all existing longitudinal infant scans with highly irregular and nonuniform age distribution. Moreover, we leverage the mutual benefits between cortical morphological and cognitive development to improve the performance of both tasks. Our framework enables jointly predicting cortical morphological and cognitive development flexibly at arbitrary ages during infancy, both regression and progression. The promising results on the BCP dataset demonstrate the potential power for individual-level development prediction and modeling.

Acknowledgements. This work was supported in part by NIH grants (MH116225, MH117943, MH127544, and MH123202). This work also utilizes approaches developed by an NIH grant (1U01MH110274) and the efforts of the UNC/UMN Baby Connectome Project Consortium.

References

1. Casey, B., Tottenham, N., Liston, C., Durston, S.: Imaging the developing brain: what have we learned about cognitive development? Trends Cogn. Sci. **9**(3), 104–110 (2005)

2. Dubois, J., Hertz-Pannier, L., Cachia, A., Mangin, J.F., Le Bihan, D., Dehaene-Lambertz, G.: Structural asymmetries in the infant language and sensorimotor networks. Cereb. Cortex **19**(2), 414–423 (2009)
3. Smyser, C.D., Inder, T.E., Shimony, J.S., Hill, J.E., Degnan, A.J., Snyder, A.Z., et al.: Longitudinal analysis of neural network development in preterm infants. Cereb. Cortex **20**(12), 2852–2862 (2010)
4. Gilmore, J.H., Kang, C., Evans, D.D., Wolfe, H.M., Smith, J.K., Lieberman, J.A., et al.: Prenatal and neonatal brain structure and white matter maturation in children at high risk for schizophrenia. Am. J. Psychiatry **167**(9), 1083–1091 (2010)
5. Wang, L., Wu, Z., Chen, L., Sun, Y., Lin, W., Li, G.: iBEAT v2. 0: a multisite-applicable, deep learning-based pipeline for infant cerebral cortical surface reconstruction. Nat. Protocols **18**, 1488–1509 (2023)
6. Li, G., Wang, L., Shi, F., Gilmore, J.H., Lin, W., Shen, D.: Construction of 4D high-definition cortical surface atlases of infants: Methods and applications. Med. Image Anal. **25**(1), 22–36 (2015)
7. Kanai, R., Rees, G.: The structural basis of inter-individual differences in human behaviour and cognition. Nat. Rev. Neurosci. **12**(4), 231–242 (2011)
8. Mueller, S., Wang, D., Fox, M.D., Yeo, B.T., Sepulcre, J., Sabuncu, M.R., et al.: Individual variability in functional connectivity architecture of the human brain. Neuron **77**(3), 586–595 (2013)
9. Fishbaugh, J., Prastawa, M., Gerig, G., Durrleman, S.: Geodesic regression of image and shape data for improved modeling of 4D trajectories. In: 2014 IEEE 11th International Symposium on Biomedical Imaging (ISBI), pp. 385–388. IEEE (2014)
10. Rekik, I., Li, G., Lin, W., Shen, D.: Predicting infant cortical surface development using a 4D varifold-based learning framework and local topography-based shape morphing. Med. Image Anal. **28**, 1–12 (2016)
11. Meng, Y., Li, G., Rekik, I., Zhang, H., Gao, Y., Lin, W., et al.: Can we predict subject-specific dynamic cortical thickness maps during infancy from birth? Hum. Brain Mapp. **38**(6), 2865–2874 (2017)
12. Lin, W., Zhu, Q., Gao, W., Chen, Y., Toh, C.H., Styner, M., et al.: Functional connectivity mr imaging reveals cortical functional connectivity in the developing brain. Am. J. Neuroradiol. **29**(10), 1883–1889 (2008)
13. Ecker, C., Shahidiani, A., Feng, Y., Daly, E., Murphy, C., D'Almeida, V., et al.: The effect of age, diagnosis, and their interaction on vertex-based measures of cortical thickness and surface area in autism spectrum disorder. J. Neural Transm. **121**, 1157–1170 (2014)
14. Querbes, O., et al.: Early diagnosis of alzheimer's disease using cortical thickness: impact of cognitive reserve. Brain **132**(8), 2036–2047 (2009)
15. Girault, J.B., Cornea, E., Goldman, B.D., Jha, S.C., Murphy, V.A., Li, G., et al.: Cortical structure and cognition in infants and toddlers. Cereb. Cortex **30**(2), 786–800 (2020)
16. Kagan, J., Herschkowitz, N.: A Young Mind in a Growing Brain. Psychology Press (2006)
17. Cheng, J., Zhang, X., Ni, H., Li, C., Xu, X., Wu, Z., et al.: Path signature neural network of cortical features for prediction of infant cognitive scores. IEEE Trans. Med. Imaging **41**(7), 1665–1676 (2022)
18. Mullen, E.M., et al.: Mullen Scales of Early Learning. AGS Circle Pines, MN (1995)
19. Howell, B.R., Styner, M.A., Gao, W., Yap, P.T., Wang, L., Baluyot, K., et al.: The UNC/UMN baby connectome project (BCP): an overview of the study design and protocol development. Neuroimage **185**, 891–905 (2019)
20. Hu, D., et al.: Disentangled intensive triplet autoencoder for infant functional connectome fingerprinting. In: Martel, A.L., et al. (eds.) MICCAI 2020. LNCS, vol. 12267, pp. 72–82. Springer, Cham (2020). https://doi.org/10.1007/978-3-030-59728-3_8

21. He, K., Zhang, X., Ren, S., Sun, J.: Deep residual learning for image recognition. In: Proceedings of the IEEE Conference on Computer Vision and Pattern Recognition. pp. 770–778 (2016)
22. Zhao, F., Wu, Z., Wang, L., Lin, W., Gilmore, J.H., Xia, S., et al.: Spherical deformable U-net: application to cortical surface parcellation and development prediction. IEEE Trans. Med. Imaging **40**(4), 1217–1228 (2021)
23. Zhao, F., Xia, S., Wu, Z., Duan, D., Wang, L., Lin, W., et al.: Spherical U-net on cortical surfaces: methods and applications. In: Chung, A., Gee, J., Yushkevich, P., Bao, S. (eds.) IPMI 2019. LNCS, vol. 11492, pp. 855–866. Springer, Cham (2019). https://doi.org/10.1007/978-3-030-20351-1_67
24. Hu, J., Shen, L., Sun, G.: Squeeze-and-excitation networks. In: Proceedings of the IEEE Conference on Computer Vision and Pattern Recognition, pp. 7132–7141 (2018)
25. Woo, S., Park, J., Lee, J.Y., Kweon, I.S.: CBAM: convolutional block attention module. In: Proceedings of the European Conference on Computer Vision (ECCV). pp. 3–19 (2018)
26. Huang, Z., Zhang, J., Shan, H.: When age-invariant face recognition meets face age synthesis: a multi-task learning framework and a new benchmark. IEEE Trans. Pattern Anal. Mach. Intell. **45**(6), 7917–7932 (2023)
27. Mao, X., Li, Q., Xie, H., Lau, R.Y., Wang, Z., Paul Smolley, S.: Least squares generative adversarial networks. In: Proceedings of the IEEE International Conference on Computer Vision, pp. 2794–2802 (2017)
28. Rothe, R., Timofte, R., Van Gool, L.: DEX: deep expectation of apparent age from a single image. In: Proceedings of the IEEE International Conference on Computer Vision Workshops, pp. 10–15 (2015)
29. Fischl, B.: Freesurfer. NeuroImage **62**(2), 774–781 (2012)
30. Wu, Z., Wang, L., Lin, W., Gilmore, J.H., Li, G., Shen, D.: Construction of 4D infant cortical surface atlases with sharp folding patterns via spherical patch-based group-wise sparse representation. Hum. Brain Mapp. **40**(13), 3860–3880 (2019)
31. Jiang, C.M., Huang, J., Kashinath, K., Prabhat, Marcus, P., Nießner, M.: Spherical CNNs on unstructured grids. In: ICLR (Poster) (2019)
32. Cheng, J., et al.: Spherical transformer on cortical surfaces. In: Lian, C., Cao, X., Rekik, I., Xu, X., Cui, Z. (eds.) MLMI 2022. LNCS, vol. 13583, pp. 406–415. Springer, Cham (2022). https://doi.org/10.1007/978-3-031-21014-3_42

From Mesh Completion to AI Designed Crown

Golriz Hosseinimanesh[1]([🖂])(ORCID), Farnoosh Ghadiri[2](ORCID), Francois Guibault[1](ORCID),
Farida Cheriet[1](ORCID), and Julia Keren[3]

[1] Polytechnique Montréal University, Montreal, Canada
{golriz.hosseinimanesh,francois.guibault,farida.cheriet}@polymtl.ca
[2] Centre d'intelligence artificielle appliquée (JACOBB), Montreal, Canada
farnoosh.ghadiri@jacobb.ca
[3] Intellident Dentaire Inc., Montreal, Canada
info@kerenor.ca

Abstract. Designing a dental crown is a time-consuming and labor-intensive process. Our goal is to simplify crown design and minimize the tediousness of making manual adjustments while still ensuring the highest level of accuracy and consistency. To this end, we present a new end-to-end deep learning approach, coined Dental Mesh Completion (DMC), to generate a crown mesh conditioned on a point cloud context. The dental context includes the tooth prepared to receive a crown and its surroundings, namely the two adjacent teeth and the three closest teeth in the opposing jaw. We formulate crown generation in terms of completing this point cloud context. A feature extractor first converts the input point cloud into a set of feature vectors that represent local regions in the point cloud. The set of feature vectors is then fed into a transformer to predict a new set of feature vectors for the missing region (crown). Subsequently, a point reconstruction head, followed by a multi-layer perceptron, is used to predict a dense set of points with normals. Finally, a differentiable point-to-mesh layer serves to reconstruct the crown surface mesh. We compare our DMC method to a graph-based convolutional neural network which learns to deform a crown mesh from a generic crown shape to the target geometry. Extensive experiments on our dataset demonstrate the effectiveness of our method, which attains an average of 0.062 Chamfer Distance. The code is available at: https://github.com/Golriz-code/DMC.git

Keywords: Mesh completion · Transformer · 3D shape generation

1 Introduction

If a tooth is missing, decayed, or fractured, its treatment may require a dental crown. Each crown must be customized to the individual patient in a process, as depicted in Fig. 1. The manual design of these crowns is a time-consuming and labor-intensive task, even with the aid of computer-assisted design software.

H. Greenspan et al. (Eds.): MICCAI 2023, LNCS 14228, pp. 555–565, 2023.
https://doi.org/10.1007/978-3-031-43996-4_53

Designing natural grooves and ensuring proper contact points with the opposing jaw present significant challenges, often requiring technicians to rely on trial and error. As such, an automated approach capable of accelerating this process and generating crowns with comparable morphology and quality to that of a human expert would be a groundbreaking advancement for the dental industry.

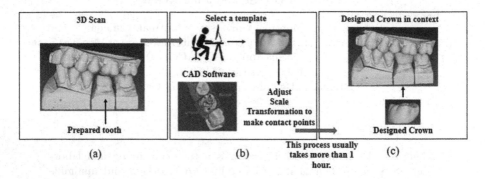

Fig. 1. Dental crown design process: a) Dentist prepares the tooth; b) Technician designs the crown; c) Dentist places the crown on the prepared tooth.

A limited number of studies have focused on how to automate dental crown designs. In [2,3], a conditional generative adversarial based network (GAN) is applied to a 2D depth image obtained from a 3D scan to generate a crown for a prepared tooth. Depth images created from a 3D scan can be used directly with 2D convolutional neural networks (CNNs) such as pix2pix [4], which are well-established in computer vision. However, depth images are limited in their ability to capture fine-grained details and can suffer from noise and occlusion issues. By contrast, point clouds have the advantage of being able to represent arbitrary 3D shapes and can capture fine-grained details such as surface textures and curvatures. [5,8] use point cloud-based networks to generate crowns in the form of 3D point clouds. Input point clouds used by [5] are generated by randomly removing a tooth from a given jaw; then the network estimates the missing tooth by utilizing a feature points-based multi-scale generating network. [8] propose a more realistic setting by generating a crown for a prepared tooth instead of a missing tooth. They also incorporate margin line information extracted from the prepared tooth in their network to have a more accurate prediction in the margin line area. The crowns generated by both approaches are represented as point clouds, so another procedure must convert these point clouds into meshes. Creating a high-quality mesh that accurately represents the underlying point cloud data is a challenging task which is not addressed by these two works. [6] proposed a transformer-based network to generate a surface mesh of the crown for a missing tooth. They use two separate networks, one responsible for generating a point cloud and the other for reconstructing a mesh given the crown generated by the first network. Similar to [5,8], the point completion network used by [6] only uses the Chamfer Distance (CD) loss to learn crown features. This metric's ability to capture shape details in point clouds is limited by the complexity and density of the data.

Although all aforementioned methods are potentially applicable to the task of dental crown design, most of them fail to generate noise-free point clouds, which is critical for surface reconstruction. One way to alleviate this problem is to directly generate a crown mesh. In [28], the authors develop a deep learning-based network that directly generates personalized cardiac meshes from sparse contours by iteratively deforming a template mesh, mimicking the traditional 3D shape reconstruction method. To our knowledge, however, the approach in [28] has not been applied to 3D dental scans.

In this paper, we introduce Dental Mesh Completion (DMC), a novel end-to-end network for directly generating dental crowns without using generic templates. The network employs a transformer-based architecture with self-attention to predict features from a 3D scan of dental preparation and surrounding teeth. These features deform a 2D fixed grid into a 3D point cloud, and normals are computed using a simple MLP. A differentiable point-to-mesh module reconstructs the 3D surface. The process is supervised using an indicator grid function and Chamfer loss from the target crown mesh and point cloud. Extensive experiments validate the effectiveness of our approach, showing superior performance compared to existing methods as measured by the CD metric. In summary, our main contributions include proposing the first end-to-end network capable of generating crown meshes for all tooth positions, employing a non-template-based method for mesh deformation (unlike previous works), and showcasing the advantages of using a differentiable point-to-mesh component to achieve high-quality surface meshes.

2   Related Work

In the field of 3D computer vision, completing missing regions of point clouds or meshes is a crucial task for many applications. Various methods have been proposed to tackle this problem. Since the introduction of PointNet [13,14], numerous methods have been developed for point cloud completion [12]. The recent works PoinTr [11] and SnowflakeNet [16] leverage a transformer-based architecture with geometry-aware blocks to generate point clouds. It is hypothesized that using transformers preserves detailed information for point cloud completion. Nonetheless, the predicted point clouds lack connections between points, which complicates the creation of a smooth surface for mesh reconstruction.

Mesh completion methods are usually useful when there are small missing regions or large occlusions in the original mesh data. Common approaches based on geometric priors, self-similarity, or patch encoding can be used to fill small missing regions, as demonstrated in previous studies [26,27], but are not suitable for large occlusions. [18] propose a model-based approach that can capture the variability of a particular shape category and enable the completion of large missing regions. However, the resulting meshes cannot achieve the necessary precision required by applications such as dental crown generation. Having a mesh prior template can also be a solution to generate a complete mesh given a sparse point cloud or a mesh with missing parts. In [28], cardiac meshes are reconstructed from sparse point clouds using several mesh deformation blocks. Their network can directly generate 3D meshes by deforming a template mesh under the guidance of learned features.

We combine the advantages of point cloud completion techniques with a differentiable surface reconstruction method to generate a dental mesh. Moreover, we used the approach in [28] to directly produce meshes from 3D dental points and compared those results with our proposed method.

3 Method

3.1 Network Architecture

Our method is an end-to-end supervised framework to generate a crown mesh conditioned on a point cloud context. The overview of our network is illustrated in Fig. 2. The network is characterized by two main components: a transformer encoder-decoder architecture and a mesh completion layer. The following sections explain each part of the network.

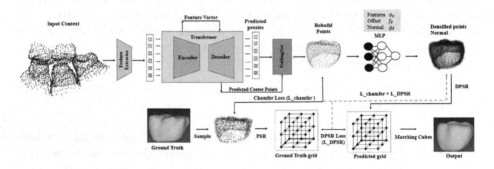

Fig. 2. Pipeline of our proposed network.

Transformer Encoder-Decoder. We adapt the transformer encoder-decoder architecture from [11] to extract global and local 3D features from our input (context) using the encoder and generate crown points via the decoder. A dynamic graph convolution network (DGCNN) [29] is used to group the input points into a smaller set of feature vectors that represent local regions in the context. The generated feature vectors are then fed into the encoder with a geometry-aware block. This block is used to model the local geometric relationships explicitly. The self-attention layer in the encoder updates the feature vectors using both long-range and short-range information. The feature vectors are further updated by a multi-layer perceptron. The decoder's role is to reason about the crown based on the learnable pairwise interactions between features of the input context and the encoder output. The decoder incorporates a series of transformer layers with a self-attention and cross-attention mechanisms to learn structural knowledge. The output of the transformer decoder is fed into a folding-based decoder [15] to deform a canonical 2D grid onto the underlying 3D surface of the crown points.

Mesh Completion Layer. In this stage, to directly reconstruct the mesh from the crown points, we use a differentiable Poisson surface reconstruction (DPSR) method introduced by [17]. We reconstruct a 3D surface as the zero level set of an indicator function. The latter consists in a regular 3D point grid associated with values indicating whether a point is inside the underlying shape or not. To compute this function, We first densify the input unoriented crown points. This is done by predicting additional points and normals for each input point by means of an MLP network. After upsampling the point cloud and predicting normals, the network solves a Poisson partial differential equation (PDE) to recover the indicator function from the densified oriented point cloud. We represent the indicator function as a discrete Fourier basis on a dense grid (of resolution 128^3) and solve the Poisson equation (PDE) with the spectral solver method in [17].

During training, we obtain the estimated indicator grid from the predicted point cloud by using the differentiable Poisson solver. We similarly acquire the ground truth indicator grid on a dense point cloud sampled from the ground truth mesh, together with the corresponding normals. The entire pipeline is differentiable, which enables the updating of various elements such as point offsets, oriented normals, and network parameters during the training process. At inference time, we leverage our trained model to predict normals and offsets using Differentiable Poisson Surface Reconstruction (DPSR) [17], solve for the indicator grid, and finally apply the Marching Cubes algorithm [22] to extract the final mesh.

Loss Function. We use the mean Chamfer Distance (CD) [20] to constrain point locations. The CD measures the mean squared distance between two point clouds S_1 and S_2. Individual distances are measured between each point and its closest point in the other point set, as described in Eq. (1). In addition, we minimize the L_2 distance between the predicted indicator function x and a ground truth indicator function x_{gt}, each obtained by solving a Poisson PDE [17] on a dense set of points and normals. We can express the Mean Square Error (MSE) loss as Eq. (2), where $f_\theta(X)$ represents a neural network (MLP) with parameters θ conditioned on the input point cloud X, D is the training data distribution, along with indicator functions x_i and point samples X_i on the surface of shapes. The sum of the CD and MSE losses is used to train the overall network.

$$CD(S_1, S_2) = \frac{1}{|S_1|} \sum_{x \in S_1} \min_{y \in S_2} |x - y|^2 + \frac{1}{|S_2|} \sum_{y \in S_2} \min_{x \in S_1} |y - x|^2 \qquad (1)$$

$$L_{DPSR}(\theta) = E_{X_i, x_i \sim D} ||Poisson(f_\theta(X_i)) - x_i||_2^2 \qquad (2)$$

4 Experimental Results

4.1 Dataset and Preprocessing

Our dataset consisted of 388 training, 97 validation, and 71 test cases, which included teeth in various positions in the jaw such as molars, canines, and incisors.

The first step in the preprocessing was to generate a context from a given 3D scan. To determine the context for a specific prepared tooth, we employed a pre-trained semantic segmentation model [21] to separate the 3D scan into 14 classes representing the tooth positions in each jaw. From the segmentations, we extracted the two adjacent and three opposing teeth of a given prepared tooth, as well as the surrounding gum tissue. To enhance the training data, we conducted data augmentation on the entire dental context, which included the master arch, opposing arch, and shell, treated as a single entity. Data augmentation involved applying 3D translation, scaling, and rotation, thereby increasing the training set by a factor of 10. To enrich our training set, we randomly sampled 10,240 cells from each context to form the input during training. We provide two types of ground truth: mesh and point cloud crowns. To supervise network training using the ground truth meshes, we calculate the gradient from a loss on an intermediate indicator grid. We use the spectral method from [17] to compute the indicator grid for our ground truth mesh.

4.2 Implementation Details

We adapted the architecture of [11] for our transformer encoder-decoder module. For mesh reconstruction, we used Differentiable Poisson Surface Reconstruction (DPSR) from [17]. All models were implemented in PyTorch with the AdamW optimizer [23], using a learning rate of 5e-4 and a batch size of 16. Training the model took 400 epochs and 22 h on an NVIDIA A100 GPU.

4.3 Evaluation and Metrics

To evaluate the performance of our network and compare it with point cloud-based approaches, we used the Chamfer distance to measure the dissimilarity between the predicted and ground truth points. We employed two versions of CD: CD_{L_1} uses the L_1-norm, while CD_{L_2} uses the L_2-norm to calculate the distance between two sets of points. Additionally, we used the F-score [24] with a distance threshold of 0.3, chosen based on the distance between the predicted and ground truth point clouds. We also used the Mean Square Error (MSE) loss [17] to calculate the similarity between the predicted and ground truth indicator grids or meshes.

We conducted experiments to compare our approach with two distinct approaches from the literature, as shown in Table 1. The first such approach, PoinTr+margin line [8], uses the PoinTr [11] point completion method as a baseline and introduces margin line information to their network. To compare our work to [8], we used its official implementation provided by the author. In the second experiment, PoinTr+graph, we modified the work of [28] to generate a dental crown mesh. To this end, we use deformation blocks in [28] to deform a generic template mesh to output a crown mesh under the guidance of the learned features from PoinTr. The deformation module included three Graph Convolutional Networks (GCNs) as in [9].

Table 1. Comparison of proposed method (DMC) with two alternate methods. Evaluation metrics: CD_{L_1}, CD_{L_2} ; MSE on output meshes; F-Score@0.3 .

Method	CD-L1 ($\downarrow$)	CD-L2 ($\downarrow$)	MSE ($\downarrow$)	$F1^{0.3}$ ($\uparrow$)
PoinTr + margin line [8]	0.065	0.018		0.54
PoinTr + graph	1.99	1.51		0.08
DMC	**0.0623**	**0.011**	0.0028	**0.70**

All experiments used the same dataset, which included all tooth positions, and were trained using the same methodology. To compare the results of the different experiments, we extracted points from the predicted meshes of our proposed network (DMC), as illustrated in Fig. 3. Table 1 shows that DMC outperforms the two other networks in terms of both CD and F-score. PoinTr+graph achieves poor CD and F-score results compared to the other methods. While the idea of using graph convolutions seems interesting, features extracted from the point cloud completion network don't carry enough information to deform the template into an adequate final crown. Therefore, these methods are highly biased toward the template shape and need extensive pre-processing steps to scale and localize the template. In the initial two experiments, the MSE metric was not applicable as it was calculated on the output meshes. Figure 4 showcases the visual results obtained from our proposed network (DMC). Furthermore, Fig. 5 presents a visual comparison of mesh surfaces generated by various methods for a sample molar tooth.

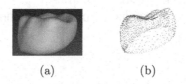

(a)   (b)

Fig. 3. Crown mesh predicted by DMC (a) and extracted point set (b).

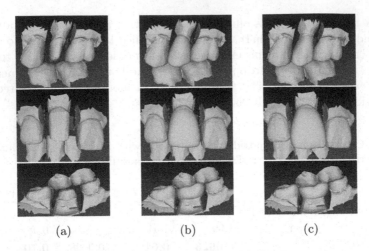

(a)   (b)   (c)

Fig. 4. Examples of mesh completions by the proposed architecture (DMC). a) Input context containing master arch, prepped tooth and opposing arch; b) Generated mesh in its context; c) Ground truth mesh in its context.

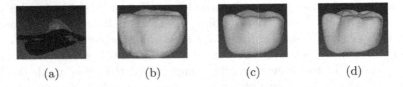

(a)   (b)   (c)   (d)

Fig. 5. Qualitative comparison of crown mesh generation approaches: a) Standard Poisson surface reconstruction; b) PoinTr [11] for point cloud and Shape as points [17] for mesh; c) Proposed method (DMC); d) Ground truth shape.

Table 2. Results of ablation study. Metrics are the same as in Table 1.

Method	CD-L1 ($\downarrow$)	CD-L2 ($\downarrow$)	MSE ($\downarrow$)	$F1^{0.3}$ ($\uparrow$)
PoinTr [11]	0.070	0.023		0.24
PoinTr + SAP	0.067	0.021	0.031	0.50
DMC without MSE	0.0641	0.015		0.65
DMC (Full Model)	**0.0623**	**0.011**	**0.0028**	**0.70**

4.4   Ablation Study

We conducted an ablation study to evaluate the components of our architecture. We started with the baseline PoinTr [8], a point completion method. To enhance it, we integrated Shape as Points (SAP) [17] as a separate network for mesh reconstruction from the PoinTr-generated point cloud. Next, we tested our proposed method (DMC) by excluding the Mean Square Error (MSE) loss function. Finally, we assessed DMC's performance, including the MSE loss function. The results, shown in Table 2, demonstrate the consistent improvements achieved by our full model across all evaluation metrics.

5   Conclusion

Existing deep learning-based dental crown design solutions require additional steps to reconstruct a surface mesh from the generated point cloud. In this study, we propose a new end-to-end approach that directly generates high-quality crown meshes for all tooth positions. By utilizing transformers and a differentiable Poisson surface reconstruction solver, we effectively reason about the crown points and convert them into mesh surfaces using Marching Cubes. Our experimental results demonstrate that our approach produces accurately fitting crown meshes with superior performance. In the future, incorporating statistical features into our deep learning method for chewing functionality, such as surface contacts with adjacent and opposing teeth, could be an interesting avenue to explore.

Acknowledgments. This work was funded by Kerenor Dental Studio, Intellident Dentaire Inc..

References

1. https://www.healthdirect.gov.au/dental-crown-procedure (2021)
2. Hwang, J.-J., Azernikov , S., Efros, A.A., Yu, S.X.: Learning beyond human expertise with generative models for dental restorations (2018)
3. Yuan, F., et al.: Personalized design technique for the dental occlusal surface based on conditional generative adversarial networks. Int. J. Numer. Methods Biomed. Eng. **36**(5), e3321 (2020)
4. Isola, P., Zhu, J.-Y., Zhou, T., Efros, A.A. : Image-to-image translation with conditional adversarial networks. In: CVPR (2017)
5. Lessard, O., Guibault, F., Keren, J., Cheriet, F.: Dental restoration using a multi-resolution deep learning approach. In: IEEE 19th International Symposium on Biomedical Imaging, (ISBI) (2022)
6. Zhu, H., Jia, X., Zhang, C., Liu, T.: ToothCR: a two-stage completion and reconstruction approach on 3D dental model. In: Gama, J., Li, T., Yu, Y., Chen, E., Zheng, Y., Teng, F. (eds.) Advances in Knowledge Discovery and Data Mining. PAKDD 2022. LNCS, vol. 13282. Springer, Cham (2022). https://doi.org/10.1007/978-3-031-05981-0_13
7. Ping, Y., Wei, G., Yang, L., Cui, Z., Wang, W.: Self-attention implicit function networks for 3D dental data completion. Comput. Aided Geometr. Design **90**, 102026 (2021)
8. Hosseinimanesh, G., et al. : Improving the quality of dental crown using a transformer-based method. Medical Imaging 2023: Physics of Medical Imaging, vol. 12463. SPIE (2023)
9. Wang, N., Zhang, Y., Li, Z., Fu, Y., Liu, W., Jiang, Y.-G.: Pixel2Mesh: generating 3D mesh models from single RGB images. In: Ferrari, V., Hebert, M., Sminchisescu, C., Weiss, Y. (eds.) ECCV 2018. LNCS, vol. 11215, pp. 55–71. Springer, Cham (2018). https://doi.org/10.1007/978-3-030-01252-6_4
10. Huang, Z., Yu, Y., Xu, J., Ni, F., Le, X.: PF-Net: point fractal network for 3D point cloud completion. In: Proceedings of the IEEE Conference on Computer Vision and Pattern Recognition, pp. 7662–7670 (2020)

11. Yu, X., Rao, Y., Wang, Z., Liu, Z., Lu, J., Zhou, J.: PoinTr: diverse point cloud completion with geometry-aware transformers. In Proceedings of the IEEE/CVF International Conference on Computer Vision, pp. 12498–12507

12. Fei, B., Yang, W., Chen, W., Li, Z., et al.: Comprehensive review of deep learning-based 3D point clouds completion processing and analysis. arXiv Prepr. arXiv2203.03311 (2022)

13. Qi, C.R., Su, H., Mo, K., Guibas, L.J.: PointNet: deep learning on point sets for 3D classification and segmentation. In: Proceedings of the IEEE Conference on Computer Vision and Pattern Recognition, pp. 652–660 (2017)

14. Qi, C.R., Yi, L., Su, H., Guibas, L.J.: PointNet++: deep hierarchical feature learning on point sets in a metric space. Deep Hierarchical Feature Learning on Point Sets in a Metric Space (2017)

15. Yang, Y., Feng, C., Shen, Y., Tian, D.: FoldingNet: point cloud auto-encoder via deep grid deformation. In: Proceedings of the IEEE Conference on Computer Vision and Pattern Recognition, pp. 206–215 (2018)

16. Xiang, P., et al.: SnowflakeNet: point cloud completion by snowflake point deconvolution with skip-transformer. In: Proceedings of the IEEE/CVF International Conference on Computer Vision, pp. 5499–5509 (2021). https://doi.org/10.48550/arxiv.2108.04444

17. Peng, S., Jiang, C., Liao, Y., Niemeyer, M., Pollefeys, M., Geiger, A.: Shape as points: a differentiable poisson solver. Journal, Advances in Neural Information Processing Systems (2021)

18. Litany, O., Bronstein, A., Bronstein, M., Makadia, A.: Deformable shape completion with graph convolutional autoencoders. In: Proceedings of the IEEE Computer Society Conference on Computer Vision and Pattern Recognition, pp. 1886–1895 (2017)

19. Hui, K.-H., Li, R., Hu, J., Fu, C.-W.: Neural template: topology-aware reconstruction and disentangled generation of 3D meshes. In: Proceedings of the IEEE/CVF Conference on Computer Vision and Pattern Recognition, pp. 18572–18582 (2022)

20. Yuan, W., Khot, T., Held, D., Mertz, C., Hebert, M.: PCN: point completion network. In: 3DV, pp. 728–737 (2018)

21. Alsheghri, A., et al. : Semi-supervised segmentation of tooth from 3D scanned dental arches. In: Medical Imaging, SPIE (2022)

22. Lorensen, W.E., Cline, H.E.: Marching cubes: a high-resolution 3D surface construction algorithm. In: Proceedings of the 14th Annual Conference on Computer Graphics and Interactive Techniques, SIGGRAPH 87, pp. 163–169, New York, NY, USA, 1987. Association for Computing Machinery

23. Loshchilov, I., Hutter, F.: Decoupled weight decay regularization. Journal, arXiv preprint arXiv:1711.05101 (2017)

24. Tatarchenko, M., Richter, S.R., Ranftl, R., Li, Z., Koltun, V., Brox, T.: What do single-view 3D reconstruction networks learn? CoRR, abs/1905.03678 (2019)

25. Foti, S., et al.: Clarkson: intraoperative liver surface completion with graph convolutional VAE. In: Uncertainty for Safe Utilization of Machine Learning in Medical Imaging, and Graphs in Biomedical Image Analysis, pp. 198–207 (2020)

26. Sarkar, K., Varanasi, K., Stricker, D.: Learning quadrangulated patches for 3D shape parameterization and completion. In: International Conference on 3D Vision (3DV), pp. 383–392 (2017)

27. Kazhdan, M., Hoppe, H.: Johns Hopkins University: screened poisson surface reconstruction. ACM Trans. Graph. **32**(3), 1–13 (2013)

28. Chen, X., et al.: Shape registration with learned deformations for 3D shape reconstruction from sparse and incomplete point clouds. J. Med. Image Anal. **74**, 102228 (2021). Elsevier
29. Wang, Y., et al.: Dynamic Graph CNN for Learning on Point Clouds. ACM Trans. Graph. (TOG) **38**, 1–12 (2019)

Spatiotemporal Incremental Mechanics Modeling of Facial Tissue Change

Nathan Lampen[1], Daeseung Kim[2(✉)], Xuanang Xu[1], Xi Fang[1],
Jungwook Lee[1], Tianshu Kuang[2], Hannah H. Deng[2],
Michael A. K. Liebschner[3], James J. Xia[2], Jaime Gateno[2],
and Pingkun Yan[1(✉)]

[1] Department of Biomedical Engineering and Center for Biotechnology and
Interdisciplinary Studies, Rensselaer Polytechnic Institute, Troy, NY 12180, USA
yanp2@rpi.edu
[2] Department of Oral and Maxillofacial Surgery, Houston Methodist Research
Institute, Houston, TX 77030, USA
dkim@houstonmethodist.org
[3] Department of Neurosurgery, Baylor College of Medicine, Houston, TX 77030, USA

Abstract. Accurate surgical planning for orthognathic surgical procedures requires biomechanical simulation of facial soft tissue changes. Simulations must be performed quickly and accurately to be useful in a clinical pipeline, and surgeons may try several iterations before arriving at an optimal surgical plan. The finite element method (FEM) is commonly used to perform biomechanical simulations. Previous studies divided FEM simulations into incremental steps to improve convergence and model accuracy. While incremental simulations are more realistic, they greatly increase FEM simulation time, preventing integration into clinical use. In an attempt to make simulations faster, deep learning (DL) models have been developed to replace FEM for biomechanical simulations. However, previous DL models are not designed to utilize temporal information in incremental simulations. In this study, we propose Spatiotemporal Incremental Mechanics Modeling (SIMM), a deep learning method that performs spatiotemporally-aware incremental simulations for mechanical modeling of soft tissues. Our method uses both spatial and temporal information by combining a spatial feature extractor with a temporal aggregation mechanism. We trained our network using incremental FEM simulations of 18 subjects from our repository. We compared SIMM to spatial-only incremental and single-step simulation approaches. Our results suggest that adding spatiotemporal information may improve the accuracy of incremental simulations compared to methods that use only spatial information.

Keywords: Surgical Planning · Deep Learning · Spatiotemporal · Biomechanics

---

Supplementary Information The online version contains supplementary material available at https://doi.org/10.1007/978-3-031-43996-4_54.

H. Greenspan et al. (Eds.): MICCAI 2023, LNCS 14228, pp. 566–575, 2023.
https://doi.org/10.1007/978-3-031-43996-4_54

1   Introduction

Orthognathic surgery corrects facial skeletal deformities that may cause functional and aesthetic impairments. In orthognathic surgical procedures, the jaws are cut into several bony segments and repositioned to desired locations to achieve an ideal alignment. Surgeons plan the movement of osteotomized bony segments before surgery to obtain the best surgical outcome. While surgeons do not operate on the soft tissue of the face during the procedure, it is passively moved by the underlying bone, causing a change in facial appearance. To visualize the proposed outcome, simulations of the planned procedure may be performed to predict the final facial tissue appearance. Current simulation techniques use the finite element method (FEM) to estimate the change of the facial tissue caused by the movement of the bony segments [2,5,6,8,9]. Subject-specific FEM models of the facial tissue are created from preoperative Computed Tomography (CT) imaging. The planned bony displacement is fed to the model as an input boundary condition. Previous studies divided the input bony displacement into smaller, incremental steps to improve model accuracy [7]. While incremental FEM simulations are very accurate, they require significant computational time, sometimes approaching 30 min to perform a single simulation [7]. This restricts the utility of FEM since surgeons may try several iterations before arriving at an optimal surgical plan, and simulations must be performed quickly to be useful in a clinical pipeline.

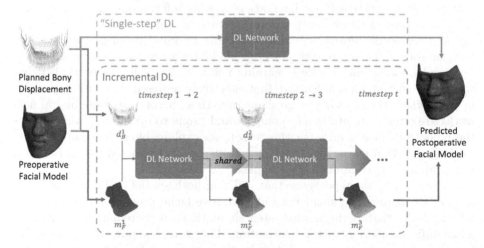

Fig. 1. An overview of the difference between the "single-step" DL method and the incremental DL method, which replicates incremental FEM simulations.

Previous studies have implemented deep learning (DL) methods to shorten soft tissue simulations. Physics-informed neural networks (PINNs) learn the constitutive model of soft tissues, however, these models do not generalize well to new settings [1,11,15]. Other studies learned a general biomechanical model by training a UNet-style network on FEM simulations, but these methods are limited to grid-like structures and cannot represent irregular meshes used in FEM [12,14]. PointNet++ has also been used to learn biomechanical deformation based on point cloud data representing nodes within a mesh [10]. While PointNet++ can represent nodes in an irregular mesh, it does not capture the connections (edges) between nodes. PhysGNN utilizes edge information by learning biomechanical deformation based on graphs created from FE meshes [16]. However, the main limitation of all the aforementioned DL methods is that they perform "single-step" simulations (Fig. 1), which are not ideal for modeling non-linear materials such as facial soft tissue, especially when large deformation (> 1mm) is involved [7]. Incremental simulations help deal with the non-linearity of soft tissue by simulating several transitional states, providing a more realistic and traceable final postoperative prediction.

A simplistic approach to performing incremental simulations using DL is to break a large simulation into smaller steps and perform them in sequential order, but independently. However, this does not allow the model to correct errors, causing them to accumulate over incremental steps. One previous work implemented a data augmentation strategy to minimize network error in incremental simulations of soft tissue [17]. However, the network used was not designed to utilize temporal information or capture temporal trends across incremental steps, such as deformation that follows a general trajectory, which can be utilized to improve incremental predictions. While another previous work incorporated long short-term memory into a convolutional neural network (CNN) to learn across multiple timesteps, the use of a CNN means the network is limited to representing regular grids without integration of edge information [4]. Therefore, a method is needed which can perform incremental simulations using both spatial and temporal information while accepting irregular mesh geometry in the form of graphs. There are several technical challenges to combining spatial and temporal information in a deep learning network. First, it can be challenging to capture spatiotemporal features that effectively represent the data, especially in cases where the objects change shape over time. Second, spatiotemporal networks are often computationally complex and prone to overfitting, which limits the model's clinical utility. For this reason, an explainable spatiotemporal network, i.e. one that learns temporal trends from already extracted spatial features, while minimizing computational complexity is needed.

In this study, we hypothesize that utilizing both spatial and temporal information for incremental simulations can improve facial prediction accuracy over DL networks that perform single-step simulations or incremental simulations while only considering spatial information. Therefore, the purpose of this study is to introduce a spatiotemporal deep learning approach for incremental biomechanics simulation of soft tissue deformation. We designed a network that learns spatial features from incremental simulations and aggregates them across multiple incremental simulations to establish sequential continuity. The contributions of this work are (1) a method for combining spatial and temporal learning in an incremental manner (2) a method for performing incremental simulations while preserving knowledge from previous increments. Our proposed method successfully implements spatiotemporal learning by observing temporal trends from spatial features while minimizing computational complexity.

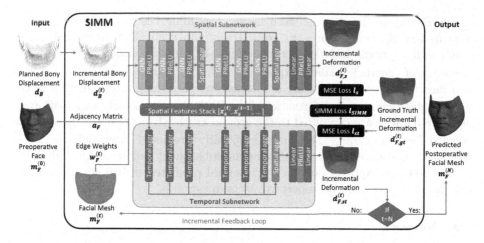

Fig. 2. The proposed framework of Spatiotemporal Incremental Mechanics Modeling (SIMM) for incremental mechanical modeling of facial tissue deformation.

2 Method

The Spatiotemporal Incremental Mechanics Modeling (SIMM) method predicts the incremental deformation of a three-dimensional (3D) soft tissue mesh based on incrementally planned input displacement. The SIMM network consists of two main operations: 1) spatial feature encoding and 2) temporal aggregation of spatial features (Fig. 2). The SIMM method is designed to first encode spatial features using a graph neural network optimized for spatial tasks, then observe how these features change over time using a temporal aggregation mechanism. SIMM is designed to be explainable and efficient by using the already extracted spatial features to observe temporal trends.

2.1 Data Representation

The SIMM method uses an incremental simulation scheme, shown in Fig. 2. SIMM is designed to accept input data in graph format that represents volumetric meshes used in FEM. In the input graphs, vertices correspond to nodes in an FE mesh and the edges correspond to the connections between nodes within an element [13]. For this reason, we will refer to our input data as "meshes" consisting of "nodes" and "edges" in the rest of this manuscript. In the SIMM method, the planned bony displacement d_B for a surgical procedure is broken into smaller steps and applied incrementally. The incremental planned bony displacement $d_B^{(t)}$ and geometric information from an input facial mesh $m_F^{(t)}$ are fed to the spatial sub-network to predict the incremental facial deformation $d_{F,s}^{(t)}$. The geometric information from the input facial mesh consists of an adjacency matrix a_F and a set of edge weights $w_F^{(t)}$. The adjacency matrix is a binary matrix describing the connections between nodes. The edge weights are calculated as the inverse of the Euclidean distance between nodes, providing the network with spatial information. Spatial features x_s are extracted in each GNN layer and aggregated using jumping-knowledge connections [18]. In the temporal sub-network, the stack of encoded spatial features $[x_s^{(t)}, x_s^{(t-1)}, ...]$ are aggregated across several incremental steps, resulting in a separate spatiotemproal prediction of the incremental facial tissue deformation $d_{F,st}^{(t)}$. Both the spatial-only and spatiotemproal predictions are compared to the ground truth FEM predicted incremental facial deformation $d_{F,gt}^{(t)}$. The spatiotemporally predicted facial deformation is then used to update the facial mesh for the subsequent timestep. After all incremental steps are applied (at $t = N$), the final predicted deformation is used to calculate the predicted postoperative facial mesh m_F^N.

2.2   Network

Spatial sub-network: We used a PhysGNN network to learn the spatial features [16]. PhysGNN consists of six GNN layers that extract features in an increasing neighborhood around a given node. PhysGNN also uses jumping-knowledge connections which aggregate information from the first three and last three GNN layers. The jumping-knowledge connections use an aggregation function to fuse features across the GNN layers using the equation:

$$y_s = f_{aggr}(x^{(0)}, ..., x^{(l)}) \tag{1}$$

where y_s is the spatially aggregated features, and $x^{(l)}$ are the features from layers 0 to l. f_{aggr} can be any form of aggregation function, such as concatenation. In this work, we use a long-short-term memory (LSTM) attention aggregation mechanism to achieve optimal results, which has been demonstrated in our ablation study (Sect. 3.4) [3,18].

Temporal sub-network: To capture temporal trends, we implemented additional aggregation layers to process features across a sequence of incremental timepoints. Instead of aggregating features across several spatial "neighborhoods" as seen in Eq. (1), these additional aggregation layers aggregate features from a given spatial neighborhood across several sequential timepoints (t), as seen in Eq. (2):

$$y_t = f_{aggr}(x^{(t)}, ..., x^{(t-N)}) \tag{2}$$

where y_t is the temporally aggregated features and $x^{(t)}$ are the features to be aggregated across timepoints (t) to $(t - N)$. We hypothesized that including more than one previous timepoint in the aggregation layer may improve the network performance (see Sect. 3.4). We used the same LSTM-attention aggregation mechanism as used in the PhysGNN spatial sub-network.

Training Strategy: The SIMM network produces two predictions of the incremental tissue deformation, one from the spatial sub-network and the other after spatiotemporal aggregation (Fig. 2). We used incremental FEM simulations as ground truth while training our network. The loss was calculated as the mean-squared error between the predicted deformation and the ground truth FEM deformation. We use the following equation to calculate the loss of the network:

$$l_{SIMM} = l_s + l_{st} \tag{3}$$

where l_{SIMM} is the total loss of the SIMM network, l_s is the loss of the spatial sub-network predicted deformation $d^{(t)}_{F,s}$, and l_{st} is the loss of the spatiotemporal predicted deformation $d^{(t)}_{F,st}$. The combined loss was used to train the network to ensure the spatial sub-network still learns to adequately encode the spatial features.

3 Experiments and Results

3.1 Dataset

The SIMM method was evaluated on a dataset of incremental FEM simulations from 18 subjects using a leave-one-out cross-validation (LOOCV). The subjects were chosen from our digital archive of patients who underwent double-jaw orthognathic surgery (IRB# PRO00008890). FEM meshes were generated from CT scans and FE simulations were performed using an existing clinical pipeline [7]. Boundary conditions including bony movements and sliding effect were applied to the FEM mesh. FEM simulation was performed to simulate the facial change with at least 200 incremental steps to have enough training data. The total number of incremental steps varied, depending on the stability of each incremental simulation step. Meshes for all subjects consisted of 3960 nodes and 2784 elements. The incremental bony displacement information was fed to the network as an input feature vector for each node. For facial nodes assumed to move together with the underlying moving bony surface, the input displacement vector was equal to the bony surface displacement $d^{(t)}_B$, which describes the boundary condition between the moving bony structures and the facial soft tissue. For facial nodes not on the surface of a moving bony segment, or that are on the surface of a fixed bony segment, the input displacement vector was initialized as all zeros. To increase the number of simulations for training, we further split the simulation for a given subject into several "sub-simulations" consisting of a subset of the incremental timesteps. The timesteps for each sub-simulation were determined by using a maximum deformation threshold, d_{max}. For example, starting from timepoint 0, the first incremental timepoint with a maximum deformation exceeding the threshold would be included in the sub-simulation (Fig. S1). The starting point of each sub-simulation was unique, however, all sub-simulations ended on the postoperative face as the last incremental step, allowing for a direct comparison between all sub-simulations on the postoperative mesh for a given subject. We used a maximum deformation threshold of 0.5mm when splitting each subject's simulation into sub-simulations, which was chosen based on an ablation study (see Sect. 3.4). The number of sub-simulations for each subject can be found in Tab. S1.

Table 1. Mean Euclidean distance error, mean simulation time, and model size comparison between SIMM, Incremental, and Single-step deep learning methods.

Method	Single-Step	Incremental	SIMM
Mean Error ± StdDev [mm]	0.44 ± 0.14^a	0.47 ± 0.19^a	0.42 ± 0.13^b
Mean Simulation Time [s]	0.08	1.14	1.36
Model Size [MB]	2052	2774	3322
# of Parameters [M]	0.9	0.9	3.3

* Values with different superscript letters are significantly different ($p < 0.05$)

3.2   Implementation Details and Evaluation Metrics

All networks were trained in pytorch using the Adam optimizer with an initial learning rate of 5e-3 for 100 epochs on an NVIDIA Tesla V100. We trained all networks in leave-one-out cross-validation across all 18 subjects. The loss was calculated as the mean squared error between the network-predicted and FEM-predicted deformation vectors for each node within the facial mesh. The final accuracy was calculated as the Euclidean distance between the network-predicted and FEM-predicted node coordinates on the final postoperative face. The distribution of the Euclidean distances was not found to be normal using the Kolmogorov-Smirnov test, so statistical significance between methods was tested using the Wilcoxon signed-rank test and a p-value of 0.05.

3.3   Comparison with Baselines

We compared our SIMM method with two baseline methods: 1) a single-step method, and 2) an incremental method. For the single-step method, we trained a PhysGNN network to predict the total facial deformation in a single step. The same sub-simulations used in the incremental and SIMM methods (see Sect. 3.1) were used to train the single-step network, however, all intermediate timepoints between the first and final timepoints were removed. For the incremental method, we used PhysGNN to perform incremental simulations. The incremental method used the prediction from timepoint $t-1$ to update the edge weights in timepoint t, similar to the feedback loop in SIMM (Fig. 2). However, no temporal aggregation or separate spatiotemporal prediction was used.

Our SIMM method achieved a mean error of 0.42 mm on all subjects (Table 1) with subject-specific mean error between 0.23 and 0.77 mm (Tab. S1). In comparison, the single-step method achieved a mean error of 0.44 mm on all subjects (Table 1) with subject-specific mean error between 0.30 and 0.91 mm (Tab. S1). The incremental method achieved a mean error of 0.47 mm on all subjects (Table 1) with subject-specific mean error between 0.25 and 1.00 mm (Tab. S1). Statistical analysis showed SIMM performed significantly better than the single-step and incremental methods, while the single-step and incremental methods were not significantly different.

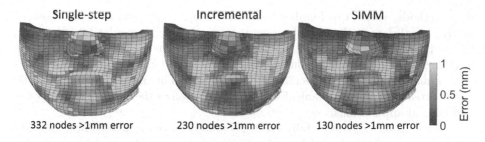

Fig. 3. Error of a) the single-step method, b) incremental method, and c) SIMM method for subject 9

3.4   Ablation Studies

We also performed ablation studies to investigate the effects of several key components in the SIMM method. First, when splitting subject simulations into sub-simulations, we varied the maximum deformation threshold d_{max} to 1.0, 0.5, and 0.1 mm. We found a d_{max} of 0.5 achieved the best performance in the incremental method, although the best d_{max} may change for different cases (Tab. S2). Second, we investigated the effect of the type of aggregation mechanism used in the PhysGNN network of the incremental method by replacing the LSTM-attention aggregation mechanism with a concatenation aggregation mechanism. The mean error increased to 1.25 mm when using a concatenation aggregation. Third, we tried increasing the number of previous timepoints in the spatial features stack to be used in the temporal aggregation layers from 1 previous timepoint to 5 previous timepoints. We hypothesized that including multiple previous timepoint in the aggregation layer may improve performance. The mean error increased to 0.44 mm when using 5 previous timepoints.

4   Discussions and Conclusions

The SIMM method achieved a lower mean error than the single-step and incremental methods, as seen in the quantitative results (Table 1). The results of the incremental method suggest the network accumulates errors across incremental steps, as seen in a plot of the predicted deformation vectors over time (Fig. S2). Figure 3 shows an example of the improvement of SIMM over the single-step and incremental methods. The results of the ablation study showed concatenation aggregation did not perform as well as LSTM-attention aggregation, which is consistent with other studies that investigated aggregation methods in GNNs [3,18]. Our ablation study also demonstrated that increasing the number of previous timepoints in the temporal feature aggregation did not improve network performance. We think this result is likely caused by over-smoothing the temporal information when aggregating features from many previous timepoints.

One limitation of our SIMM method is that it requires incremental FEM simulations to be trained. This greatly increases training requirements over single-step methods, which can feasibly be trained only on preoperative and postoperative facial data without performing incremental FEM simulations. However, this is true of any incremental simulation method. Once SIMM is trained, the inference simulation time is considerably faster than FEM, which can take several minutes to perform an incremental simulation. The SIMM method could be easily extended to biomechanical simulations of many other types of soft tissues and clinical applications.

In conclusion, we successfully created a spatiotemporal incremental mechanics modeling (SIMM) method for simulating incremental facial deformation for surgical planning. The SIMM method shows potential to improve simulation accuracy over methods based on spatial information alone, suggesting the importance of spatiotemporal learning for incremental simulations.

Acknowledgements. This work was partially supported by NIH awards R01 DE022676, R01 DE027251, and R01 DE021863.

References

1. Buoso, S., Joyce, T., Kozerke, S.: Personalising left-ventricular biophysical models of the heart using parametric physics-informed neural networks. Med. Image Anal. 71 (2021)
2. Chabanas, M., Luboz, V., Payan, Y.: Patient specific finite element model of the face soft tissues for computer-assisted maxillofacial surgery. Med. Image Anal. **7**(2), 131–151 (2003)
3. Hamilton, W.L., Ying, R., Leskovec, J.: Inductive representation learning on large graphs. In: Advances in Neural Information Processing Systems 2017-December, pp. 1025–1035 (2017)
4. Karami, M., Lombaert, H., Rivest-Hénault, D.: Real-time simulation of viscoelastic tissue behavior with physics-guided deep learning. Comput. Med. Imaging Graph. **104**, 102165 (2023)
5. Kim, D., et al.: A clinically validated prediction method for facial soft-tissue changes following double-jaw surgery. Med. Phys. **44**(8), 4252–4261 (2017)
6. Kim, D., et al.: A new approach of predicting facial changes following orthognathic surgery using realistic lip sliding effect. In: Medical Image Computing and Computer-Assisted Intervention: MICCAI ... International Conference on Medical Image Computing and Computer-Assisted Intervention, vol. 11768, pp. 336–344 (2019)
7. Kim, D., et al.: A novel incremental simulation of facial changes following orthognathic surgery using FEM with realistic lip sliding effect. Med. Image Anal. **72**, 102095 (2021)
8. Knoops, P.G., et al.: Three-dimensional soft tissue prediction in orthognathic surgery: a clinical comparison of Dolphin, ProPlan CMF, and probabilistic finite element modelling. Int. J. Oral Maxillofac. Surgery **48**(4), 511–518 (2019)
9. Knoops, P.G., et al.: A novel soft tissue prediction methodology for orthognathic surgery based on probabilistic finite element modelling. PLOS ONE **13**(5), e0197209 (2018)
10. Lampen, N., et al.: Deep learning for biomechanical modeling of facial tissue deformation in orthognathic surgical planning. Int. J. Comput. Assist. Radiol. Surgery **17**(5), 945–952 (2022)

11. Liu, M., Liang, L., Sun, W.: A generic physics-informed neural network-based constitutive model for soft biological tissues. Comput. Methods Appl. Mech. Eng. **372**, 113402 (2020)
12. Mendizabal, A., Márquez-Neila, P., Cotin, S.: Simulation of hyperelastic materials in real-time using deep learning. Med. Image Anal. **59**, 101569 (2020)
13. Pfaff, T., Fortunato, M., Sanchez-Gonzalez, A., Battaglia, P.: Learning mesh-based simulation with graph networks. In: International Conference on Learning Representations (2021)
14. Pfeiffer, M., Riediger, C., Weitz, J., Speidel, S.: Learning soft tissue behavior of organs for surgical navigation with convolutional neural networks. Int. J. Comput. Assist. Radiol. Surgery **14**(7), 1147–1155 (2019)
15. Raissi, M., Perdikaris, P., Karniadakis, G.E.: Physics-informed neural networks: a deep learning framework for solving forward and inverse problems involving nonlinear partial differential equations. J. Comput. Phys. **378**, 686–707 (2019)
16. Salehi, Y., Giannacopoulos, D.: PhysGNN: a physics-driven graph neural network based model for predicting soft tissue deformation in image-guided neurosurgery. arXiv preprint arXiv:2109.04352 (2021)
17. Wu, J.Y., Munawar, A., Unberath, M., Kazanzides, P.: Learning Soft-Tissue Simulation from Models and Observation. In: 2021 International Symposium on Medical Robotics, ISMR 2021 (2021)
18. Xu, K., Li, C., Tian, Y., Sonobe, T., Kawarabayashi, K.I., Jegelka, S.: representation learning on graphs with jumping knowledge networks. In: 35th International Conference on Machine Learning, ICML 2018, vol. 12, pp. 8676–8685 (2018)

Analysis of Suture Force Simulations for Optimal Orientation of Rhomboid Skin Flaps

Wenzhangzhi Guo[1,2,4]([✉])[iD], Ty Trusty[2], Joel C. Davies[3], Vito Forte[4,5], Eitan Grinspun[2], and Lueder A. Kahrs[1,2,6,7][iD]

[1] Medical Computer Vision and Robotics Lab,
University of Toronto, Toronto, ON, Canada
wenzhi.guo@mail.utoronto.ca

[2] Department of Computer Science, University of Toronto, Toronto, ON, Canada

[3] Department of Otolaryngology - Head and Neck Surgery, Sinai Health System,
University of Toronto, Toronto, ON, Canada

[4] The Wilfred and Joyce Posluns CIGITI,
The Hospital for Sick Children, Toronto, ON, Canada

[5] Department of Otolaryngology – Head and Neck Surgery,
University of Toronto, Toronto, ON, Canada

[6] Department of Mathematical and Computational Sciences,
University of Toronto Mississauga, Mississauga, ON, Canada

[7] Institute of Biomedical Engineering, University of Toronto, Toronto, ON, Canada

Abstract. Skin flap is a common technique used by surgeons to close the wound after the resection of a lesion. Careful planning of a skin flap procedure is essential for the most optimal functional and aesthetic outcome. However, currently surgical planning is mostly done based on surgeons' experience and preferences. In this paper, we introduce a finite element method (FEM) simulation that is used to make objective recommendations of the most optimal flap orientation. Rhomboid flap is chosen as the main flap type of interest because it is a very versatile flap. We focus on evaluating suture forces required to close a wound as large tension around it could lead to complications. We model the skin as an anisotropic material where we use a single direction to represent the course of relaxed skin tension lines (RSTLs). We conduct a thorough search by rotating the rhomboid flap in small increments (1°–10°) and find the orientation that minimizes the suture force. We repeat the setup with different material properties and the recommendation is compared with textbook knowledge. Our simulation is validated with minimal error in comparison with other existing simulations. Our simulation shows to minimize suture force, the existing textbook knowledge recommendation needs to be further rotated by 15°–20°.

Supplementary Information The online version contains supplementary material available at https://doi.org/10.1007/978-3-031-43996-4_55.

Keywords: FEM simulation · Rhomboid (Limberg) flap · Flap orientation recommendation

1 Introduction

Skin flap is a widely used technique to close the wound after the resection of a lesion. In a skin flap procedure, a healthy piece of tissue is harvested from a nearby site to cover the defect [1]. Careful planning of such procedure is often needed, especially in the facial region where aesthetic of the final outcome is crucial. The ideal wound closure is dependent on many factors including incision path, anisotropic skin tension, location with respect to aesthetic sub-units, and adjacent tissue types [1]. For each patient, there often exists many valid reconstructive plans and a surgeon needs to take the above factors into account to strive for the most optimal aesthetic result while making sure facial sub-units functionalities are not affected. While there are basic guidelines in medical textbooks, there is often no single recipe that a surgeon can follow to determine the most optimal surgical plan for all patients. Surgeons require a lot of training and experience before they can start to make such decisions. This poses a challenge as each surgeon has their own opinion on the best flap design for each case (so it is hard to reach a consensus) and training a new surgeon to perform this task remains difficult.

In this paper, we build a system to help surgeons pick the most optimal flap orientation. We focus on the rhomboid flap as it is a very versatile flap. According to medical textbooks [1,2], the course of Relaxed Skin Tension Lines (RSTLs) is important information for surgeons when deciding on the orientation of rhomboid flaps. We validate this claim with our simulation and provide new insights into designing rhomboid flaps. The main contributions of the paper are:

- We created a skin flap FEM simulation and validated the simulation outputs of rhomboid flaps against other simulators;
- We performed quantitative evaluation of the suture forces for rhomboid flaps and compared the result against textbook knowledge;
- We provided an objective and methodical way to make recommendations for aligning rhomboid flaps relative to RSTLs;
- We generated a database of rhomboid flap simulations under various material properties and relative angles to RSTLs.

2 Related Works

There exists various virtual surgical simulators for facial skin flap procedures. In a recent work, Wang and Sifakis et al. built an advanced skin flap simulator that allows users to construct free-form skin flaps on a 3D face model interactively [3] (We further reference this work as "UWG simulator"). Mitchell and Sifakis et al. also worked on earlier versions of interactive simulators where a novel computational pipeline was presented in GRIDiron [4] and an embedding framework was introduced [5]. Additionally, there are also many other interactive tutorials created for educational purposes [6–8]. While the above simulators are all very valuable for surgical education, they do not provide recommendations for flap type or alignment.

To gain insights into deformations in a skin flap procedure, there are various FEM simulations with stress/strain analysis. Most of those simulations are built based on a commercial FEM software such as Abaqus [9–16], ANSYS [17–22], or MSC. Marc/Mentat [23,24]. In those simulations, one or more skin flap types are analyzed and visualizations of different stress/strain measures are provided. Stowers et al. [9] constructed a surrogate model for creating simulated result with different material properties efficiently. With this model, they ran simulations with various material properties of three different flaps (including rhomboid flap) and provided visualizations of strain distribution and optimal flap alignment with various material properties. However, they did not come up with a consistent conclusion for the most optimal alignment for rhomboid flap. Spagnoli et al. [10] and Rajabi et al. [20] both evaluated different internal angles of the rhomboid flap and provided recommendations for the most optimal design. Rajabi et al. [20] also provided a modified design for the rhomboid flap based on stress optimization. Rhomboid flaps were also analyzed for comparison between different age groups [13], for comparison between different skin models [17] and for analysis on 3D face models [19]. Although the above works all offer valuable insights, they do not provide recommendations for the most optimal orientation for rhomboid flap nor evaluation/comparison for their results.

3   Methods

We built our skin flap FEM simulation based on the open-source library Bartels [25] and different components of the simulation are described in this section.

3.1   Constitutive Model of Skin

The skin patch is modelled as a thin membrane that resides in two dimensional space (planar stress formulation), as it is a common setup in the existing literature [9–12]. We model the skin using the adapted artery strain energy model proposed by Gasser-Ogden-Holzapfel (GOH) [26]. The strain energy density function is the sum of a volumetric component (Ψ^{vol}), an isochoric component ($\bar{\Psi}^{iso}$) and an anisotropic component ($\bar{\Psi}^{f}$):

$$\Psi = \Psi^{vol} + \bar{\Psi}^{iso} + \bar{\Psi}^{f}. \tag{1}$$

The isochoric component is proportional to μ, the shear modulus and the anisotropic component is parameterized by k_1 and k_2, which are stiffness parameters. More details of the skin constitutive model can be found in Stowers et al. [9].

3.2   Cutting and Suturing

Cutting is accomplished by duplicating vertices along the cutting line and re-meshing the local neighborhood to setup the correct connections with the duplicated vertices. This mechanism can run very efficiently and is sufficient for our setup since we always triangulate the mesh area based on the specific flap type so cutting can only happen on edges. An illustration of such mechanism is shown in the Supplementary Material.

The suture process is implemented by adding zero rest-length springs between vertices. Different stages of a rhomboid flap procedure is shown in Fig. 1b to d.

3.3   Triangulation

In our setup, we assume the lesion is within a circular area at the center of the patch. Similar to [9], for each flap type, a predefined template is used to build the geometric shape of the flap and the corresponding suture points are manually specified. To ensure the symmetry of the patch boundary, the vertices on the outer bounds of the circular patch are generated so that there is one vertex every 90° (i.e. at 0°, 90°, 180°, and 270°). Then the vertices in each quadrant are evenly spaced, and the number of vertices in each quadrant is the same (the exact number depends on the triangulation density and size of the patch). The triangulation is generated with the publicly available library Triangle [27,28]. An example triangulation for rhomboid flap can be seen in Fig. 1.

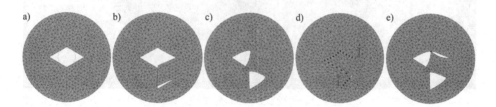

Fig. 1. FEM simulation of a rhomboid flap procedure. a) Initial setup for the rhomboid flap (flap design and cutting lines are shown in magenta and lesion is shown in cyan). b) Start of simulation. c) Completed suturing for one side (with green line representing the completed suture line). d) Completed procedure (colored dots indicate suture points). f) Example single-suture setup (the suture point is shown in blue). (Color figure online)

3.4   Validation

To validate our FEM simulation, we compared it with UWG simulator [3,29]. We first performed a skin flap procedure in UWG simulator and took a screenshot of the head model with the flap design ("pre-operative" image) and another one with the completed flap procedure ("post-operative" image). Then we aligned the flap design in our simulation with the "pre-operative" image taken from UWG simulator. Next, we ran our simulation and obtained the final suture line.

Lastly, we compared the suture line of our simulation with the "post-operative" image. We used normalized Hausdorff distance [30] (nHD) to quantify similarities between two suture lines, where Hausdorff distance is divided by the larger of the height and width of the bounding box of the suture line generated by UWG simulator (an illustration of the pipeline is shown in Fig. 2).

3.5   Suture Force Analysis

Multi-sutures. As excess force along the suture line can cause wound dehiscence and complications [9], the flap rotation angle that requires the lowest suture force should be the optimal orientation. For this experiment, we set the direction of RSTLs to be the horizontal direction. The initial pose of the flap was oriented with its first segment of the tail (the one that is connected to the rhomboid) perpendicular to RSTLs, as seen in Fig. 1a. To reduce the effect of fixed boundary, we set the radius of the patch to be 1 unit and the radius of the lesion to be 0.1 unit. To find the optimal alignment, we rotated the flap design counter-clockwisely in increment of 10°. After we found the rotation angle with minimum suture force, we fine-tuned the result by searching between −10° and +10° of that minimum rotation angle at increments of 1°. For each rotation, after reaching steady state, we took the maximum suture force among all the sutures shown as colored dots in Fig. 1d (see Algorithm 1 in Fig. 3).

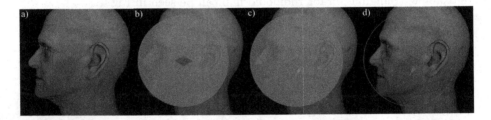

Fig. 2. Steps for comparing with UWG simulator output [3,29]. a) Flap design in UWG simulator. b) Mapping our simulation mesh to UWG simulator. c) Overlaying suture line generated by our simulation. d) Comparing our suture line with UWG simulator (our generated suture line is in green), $nHD = 0.0715$.

Algorithm 1 Multi-sutures
Require: All sutures are completed
$max\_force = 0$
$index = 1$
while $index <= j$ **do**
if $max\_force < suture\_force(index)$ **then**
$max\_force = suture\_force(index)$
end if
Increment $index$
end while
return max_force

Algorithm 2 Single-suture
$max\_force = 0$
$index = 1$
while $index <= j$ **do**
Complete suture at the current index
Set $current\_force$
if $max\_force < current\_force$ **then**
$max\_force = current\_force$
end if
Remove suture at the current index
Increment $index$
end while
return max_force

Fig. 3. Pseudocode of two different suture force analyses

To investigate the effect of material properties, we repeated the same procedure using the material properties listed in Table 1 (same range as in [9]). We changed one material property at a time while keeping other properties at the average (default) value. The material properties used in each trial can be seen in Table 1.

Table 1. Material properties of different trials (rows represent material parameters and columns represent trial setups). The ranges of valid material properties of the skin were taken from [9], where μ is the shear modulus, k_1 and k_2 are stiffness parameters.

	$default$	$small\_\mu$	$large\_\mu$	$small\_k_1$	$large\_k_1$	$small\_k_2$	$large\_k_2$
$\mu[kPa]$	5.789	4.774	6.804	5.789	5.789	5.789	5.789
$k_1[kPa]$	106.550	106.550	106.550	3.800	209.300	106.550	106.550
$k_2[-]$	107.195	107.195	107.195	107.195	107.195	52.530	161.860

Single-suture. Instead of taking the maximum suture force after completing all sutures, we also experimented with making one suture at a time (see Algorithm 2 in Fig. 3) and picking the maximum suture force for any single suture (example of a single-suture step is shown in Fig. 1e).

4   Results

The simulation results were generated using a Macbook Pro and an iMac. It required around 30 s - 2 min to compute suture force for each rotation.

4.1   Validation with UWG Simulator

Further comparisons of skin flap placed at various locations of the face model are shown in Fig. 4. The normalized Hausdorff distances shown in Fig. 2 and 4 are all very small (below 0.1), indicating a good match between the two simulators.

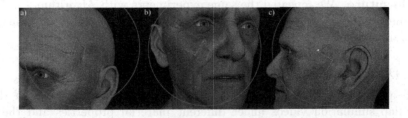

Fig. 4. Comparisons between our simulation and UWG simulator [3,29] (our generated suture line is in green). a) Forehead ($nHD = 0.0893$). b) Cheek ($nHD = 0.0926$). c) Temple ($nHD = 0.0629$). (Color figure online)

4.2 Suture Force Analysis

Multi-sutures. The suture forces of various flap rotation angles under different material properties are shown in Table 2 and Fig. 5. Further analysis shows that the minimum suture force occurred at 99° and 281° (median value). As seen in Fig. 5b, the maximum force occurred at vertex 6 for all material properties (yellow dot in Fig. 1d), which is consistent with what is reported in literature [31].

Table 2. Optimal flap rotation angles (rows represent optimal configurations and columns represent trials of material properties presets).

	$default$	$small\_\mu$	$large\_\mu$	$small\_k_1$	$large\_k_1$	$small\_k_2$	$large\_k_2$	**avg**	**med**
multi1 (deg)	99	103	99	127	100	99	99	104	99
multi2 (deg)	281	281	281	304	282	281	281	284	281
single1 (deg)	106	108	105	115	104	101	99	105	105
single2 (deg)	284	287	284	299	284	278	280	285	284

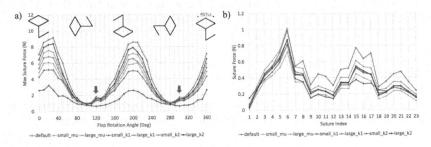

Fig. 5. Multi-sutures force visualization. a) Forces for various rotations (red arrows represent textbook recommendations). b) Suture force when rotation angle is at 99° (with default material properties); the suture index goes from i to j as shown in Fig. 1d. (Color figure online)

Single-suture. We repeated the same experiments with Algorithm 2 in Fig. 3 and the results are shown in Table 2 and Fig. 6. For our single-suture analysis, the minimum suture force occurred at 105° and 284° (median value).

4.3 Database Creation

While performing different experiments, we also created a dataset of rhomboid flap simulation videos under different material properties and the corresponding mesh at the end of each simulation. There are in total over 500 video clips of around 30 s. Overall, there are more than 10,000 frames available. A sample video can be found in the Supplementary Material. The dataset can be accessed through the project website: https://medcvr.utm.utoronto.ca/miccai2023-rhomboidflap.html.

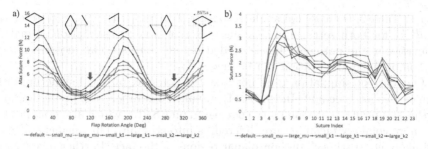

Fig. 6. Single-suture force visualization. a) Forces for various rotations (red arrows represent textbook recommendations). b) Suture force when rotation angle is at 89° (with default material properties); the suture index goes from i to j as shown in Fig. 1d. (Color figure online)

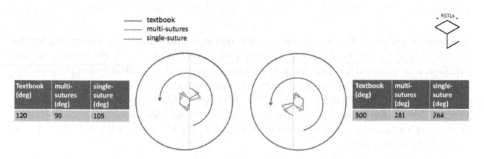

Fig. 7. Comparison between our suture force minimization result and textbook knowledge (yellow dotted line represents LMEs). (Color figure online)

5   Discussion and Conclusion

We built a FEM simulation to find the most optimal flap orientation for rhomboid flap based on suture force analysis. Through minimizing suture forces, we found that the optimal orientation occurred at 99°/281° and 105°/284° for multi-sutures and single-suture settings, respectively. The range of forces we obtained in our simulation is similar to what was reported in literature [22,24,32]. For Limberg flap (rhomboid flap with equal edge lengths and a top angle of 120°), there is a textbook description that aligns it based on lines of maximum skin extensibility (LMEs), which are perpendicular to RSTLs (see Supplementary Material). For our setting, this textbook alignment occurs at 120° and 300°. A comparison between our simulation and textbook knowledge is shown in Fig. 7.

Our experiment suggests that the minimal suture force occurs at a configuration that is close to textbook knowledge recommendation, but we found that rhomboid flaps have to be rotated further 15° to 20° away from the LMEs as seen in Fig. 7. This could be because that the textbook knowledge is trying to optimize a different goal or has taken other factors into account. Additionally, as seen in Table 2, the optimal angle is consistent among all trials, except for trial $small\_k_1$ for both settings, which suggests that a good estimation of material properties of the patient's skin is needed before more specific recommendation of rhomboid flap orientation can be made.

Our current simulation does not take the complex physical structure of the human face into account. In this study, we aim to compare the prediction made by this setup with textbook knowledge based on RSTLs, which also does not take the more complex facial feature differences into account. This means our model will likely not work well on non-planar regions of the face. Further investigation and comparison with a physical setup (with either synthetic model or clinical setting) is also needed to show clinical significance of our result (for both suture forces comparison and flap orientation design). In the future, it would be interesting to continue investigating this topic by comparing behaviors of different flap types.

Acknowledgment. We would like to acknowledge and thank the Harry Barberian Research fund in the Department of Otolaryngology-Head and Neck Surgery at the University of Toronto for providing the financial support to initiate this project. We also acknowledge the support of the Natural Sciences and Engineering Research Council of Canada (NSERC), RGPIN- 2020-05833. WG would like to thank the support from NSERC Alexander Graham Bell Canada Graduate Scholarship-Doctoral (CGS D) and VinBrain Graduate Student Fellowship. We also would like to thank David I.W. Levin for his helpful suggestions.

References

1. Baker, S.R.: Local Flaps in Facial Reconstruction, 2nd edn. Mosby, St. Louis (2007)
2. Weerda, H.: Reconstructive Facial Plastic Surgery: A Problem-solving Manual, 1st edn. Thieme, New York (2001)
3. Wang, Q., Tao, Y., Cutting, C., et al.: A computer based facial flaps simulator using projective dynamics. Comput. Methods Program. Biomed. **218**, 106730 (2022)
4. Mitchell, N., Cutting, C., Sifakis, E.: GRIDiron: an interactive authoring and cognitive training foundation for reconstructive plastic surgery procedures. ACM Trans. Graph. **34**(4), 1–12 (2015)
5. Sifakis, E., Hellrung, J., Teran, J., et al.: Local flaps: a real-time finite element based solution to the plastic surgery defect puzzle. Stud. Health Technol. Inform. **142**, 313–318 (2009)
6. Naveed, H., Hudson, R., Khatib, M., et al.: Basic skin surgery interactive simulation: system description and randomised educational trial. Adv. Simul. **3**(14), 1–9 (2018)
7. Shewaga, R., Knox, A., Ng, G., et al.: Z-DOC: a serious game for Z-plasty procedure training. Stud. Health Technol. Inform. **184**, 404–406 (2013)
8. Khatib, M., Hald, N., Brenton, H., et al.: Validation of open inguinal hernia repair simulation model: a randomized controlled educational trial. Am. J. Surg. **208**(2), 295–301 (2014)
9. Stowers, C., Lee, T., Bilionis, I., et al.: Improving reconstructive surgery design using Gaussian process surrogates to capture material behavior uncertainty. J. Mech. Behav. Biomed. Mater. **118**, 104340 (2021)

10. Spagnoli, A., Alberini, R., Raposio, E., et al.: Simulation and optimization of reconstructive surgery procedures on human skin. J. Mech. Behav. Biomed. Mater. **131**, 105215 (2022)
11. Alberini, R., Spagnoli, A., Terzano, M., et al.: Computational mechanical modeling of human skin for the simulation of reconstructive surgery procedures. Procedia Struct. Integr. **33**, 556–563 (2021)
12. Lee, T., Turin, S.Y., Stowers, C., et al.: Personalized computational models of tissue-rearrangement in the scalp predict the mechanical stress signature of rotation flaps. Cleft Palate Craniofac. J. **58**(4), 438–445 (2021)
13. Lee, T., Gosain, A.K., Bilionis, I., et al.: Predicting the effect of aging and defect size on the stress profiles of skin from advancement, rotation and transposition flap surgeries. J. Mech. Phys. Solids **125**, 572–590 (2019)
14. Lee, T., Turin, S.Y., Gosain, A.K., Bilionis, I., Buganza Tepole, A.: Propagation of material behavior uncertainty in a nonlinear finite element model of reconstructive surgery. Biomech. Model. Mechanobiol. **17**(6), 1857–1873 (2018). https://doi.org/10.1007/s10237-018-1061-4
15. Tepole, A.B., Gosain, A.K., Kuhl, E.: Computational modeling of skin: using stress profiles as predictor for tissue necrosis in reconstructive surgery. Comput. Struct. **143**, 32–39 (2014)
16. Flynn, C.: Finite element models of wound closure. J. Tissue Viab. **19**(4), 137–149 (2010)
17. Retel, V., Vescovo, P., Jacquet, E., et al.: Nonlinear model of skin mechanical behaviour analysis with finite element method. Skin Res. Technol. **7**(3), 152–158 (2001)
18. Pauchot, J., Remache, D., Chambert, J., et al.: Finite element analysis to determine stress fields at the apex of V-Y flaps. Eur. J. Plast. Surg. **36**, 185–190 (2012)
19. Kwan, Z., Khairu Najhan, N.N., Yau, Y.H., et al.: Anticipating local flaps closed-form solution on 3D face models using finite element method. Int. J. Numer. Method. Biomed. Eng. **36**(11), e3390 (2020)
20. Rajabi, A., Dolovich, A.T., Johnston, J.D.: From the rhombic transposition flap toward Z-plasty: an optimized design using the finite element method. J. Biomech. **48**(13), 3672–3678 (2015)
21. Remache, D., Chambert, J., Pauchot, J., et al.: Numerical analysis of the V-Y shaped advancement flap. Med. Eng. Phys. **37**(10), 987–994 (2015)
22. Topp, S.G., Lovald, S., Khraishi, T., et al.: Biomechanics of the rhombic transposition flap. Otolaryngol. Head Neck Surg. **151**(6), 952–959 (2014)
23. Yang, Z.L., Peng, Y.H., Yang, C. et al.: Preoperative evaluation of V-Y flap design based on computer-aided analysis. Comput. Math. Methods Med. 8723571 (2020)
24. Capek, L., Jacquet, E., Dzan, L., et al.: The analysis of forces needed for the suturing of elliptical skin wounds. Med. Biol. Eng. Comput. **50**(2), 193–198 (2012)
25. Bartels. https://github.com/dilevin/Bartels. Accessed 20 Feb 2023
26. Gasser, T.C., Ogden, R.W., Holzapfel, G.A.: Hyperelastic modelling of arterial layers with distributed collagen fibre orientations. J. R. Soc. Interface **3**(6), 15–35 (2006)
27. Shewchuk, J.R.: Triangle: engineering a 2D quality mesh generator and Delaunay triangulator. In: Lin, M.C., Manocha, D. (eds.) WACG 1996. LNCS, vol. 1148, pp. 203–222. Springer, Heidelberg (1996). https://doi.org/10.1007/BFb0014497
28. Jacobson, A., et al.: gptoolbox: geometry processing toolbox. http://github.com/alecjacobson/gptoolbox. Accessed 20 Feb 2023
29. SkinFlaps - a soft tissue surgical simulator using projective dynamics. https://github.com/uwgraphics/SkinFlaps. Accessed 10 Jan 2023

30. Huttenlocher, D.P., Klanderman, G.A., Rucklidge, W.J.: Comparing images using the Hausdorff distance. PAMI **15**(9), 850–863 (1993)
31. Borges, A.F.: Choosing the correct Limberg flap. Plast. Reconstr. Surg. **62**(4), 542–545 (1978)
32. Lear, W., Roybal, L.L., Kruzic, J.J.: Forces on sutures when closing excisional wounds using the rule of halve. Clin. Biomech. (Bristol Avon) **72**, 161–163 (2020)

FLIm-Based in Vivo Classification of Residual Cancer in the Surgical Cavity During Transoral Robotic Surgery

Mohamed A. Hassan[1], Brent Weyers[1], Julien Bec[1], Jinyi Qi[1], Dorina Gui[2],
Arnaud Bewley[3], Marianne Abouyared[3], Gregory Farwell[4], Andrew Birkeland[3],
and Laura Marcu[1(✉)]

[1] Department of Biomedical Engineering, University of California, Davis, USA
lmarcu@ucdavis.edu
[2] Department of Pathology and Laboratory Medicine, University of California, Davis, USA
[3] Department of Otolaryngology – Head and Neck Surgery, University of California, Davis, USA
[4] Department of Otorhinolaryngology – Head and Neck Surgery, University of Pennsylvania, Philadelphia, USA

Abstract. Incomplete surgical resection with residual cancer left in the surgical cavity is a potential sequelae of Transoral Robotic Surgery (TORS). To minimize such risk, surgeons rely on intraoperative frozen sections analysis (IFSA) to locate and remove the remaining tumor. This process, may lead to false negatives and is time-consuming. Mesoscopic fluorescence lifetime imaging (FLIm) of tissue fluorophores (i.e., collagen and metabolic co-factors NADH and FAD) emission has demonstrated the potential to demarcate the extent of head and neck cancer in patients undergoing surgical procedures of the oral cavity and the oropharynx. Here, we demonstrate the first label-free FLIm-based classification using a novelty detection model to identify residual cancer in the surgical cavity of the oropharynx. Due to highly imbalanced label representation in the surgical cavity, the model employed solely FLIm data from healthy surgical cavity tissue for training and classified the residual tumors as an anomaly. FLIm data from $N = 22$ patients undergoing upper aerodigestive oncologic surgery were used to train and validate the classification model using leave-one-patient-out cross-validation. Our approach identified all patients with positive surgical margins ($N = 3$) confirmed by pathology. Furthermore, the proposed method reported a point-level sensitivity of 0.75 and a specificity of 0.78 across optically interrogated tissue surface for all $N = 22$ patients. The results indicate that the FLIm-based classification model can identify residual cancer by directly imaging the surgical cavity, potentially enabling intraoperative surgical guidance for TORS.

Keywords: TORS · Positive Surgical Margin · FLIm · Head and Neck Oncology

Supplementary Information The online version contains supplementary material available at https://doi.org/10.1007/978-3-031-43996-4_56.

1 Introduction

Residual tumor in the cavity after head and neck cancer (HNC) surgery is a significant concern as it increases the risk of cancer recurrence and can negatively impact the patient's prognosis [1]. HNC comprises the third highest positive surgical margins (PSM) rate across all oncology fields [2]. Achieving clear margins can be challenging in some cases, particularly in tumors with involved deep margins [3, 4].

During transoral robotic surgery (TORS), surgeons may assess the surgical margin via visual inspection, palpation of the excised specimen and intraoperative frozen sections analysis (IFSA) [5]. In the surgical cavity, surgeons visually inspect for residual tumors and use specimen driven or defect-driven frozen section analysis to identify any residual tumor [6, 7]. The latter involves slicing a small portion of the tissue at the edge of the cavity and performing a frozen section analysis. These approaches are error-prone and can result in PSMs and a higher risk of cancer recurrence [7]. In an effort to improve these results, recent studies reported the use of exogenous fluorescent markers [8] and wide-field optical coherence tomography [9] to inspect PSMs in the excised specimen. While promising, each modality presents certain limitations (e.g., time-consuming analysis, administration of a contrast agent, controlled lighting environment), which has limited their clinical adoption [10, 11].

Label-free mesoscopic fluorescence lifetime imaging (FLIm) has been demonstrated as an intraoperative imaging guidance technique with high classification performance (AUC = 0.94) in identifying in vivo tumor margins at the epithelial surface prior to tumor excision [12]. FLIm can generate optical contrast using autofluorescence derived from tissue fluorophores such as collagen, NADH, and FAD. Due to the sensitivity of these fluorophores to their microenvironment, the presence of tumor changes their emission properties (i.e., intensity and lifetime characteristics) relative to healthy tissue, thereby enabling the optical detection of cancer [13].

However, ability of label-free FLIm to identify residual tumors in vivo in the surgical cavity (deep margins) has not been reported. One significant challenge in developing a FLIm-based classifier to detect tumor in the surgical cavity is the presence of highly imbalanced labels.

Surgeons aim to perform an en bloc resection, removing the entire tumor and a margin of healthy tissue around it to ensure complete excision. Therefore, in most cases, only healthy tissue in left in the cavity. To address the technical challenge of highly imbalanced label distribution and the need for intraoperative real-time cavity imaging, we developed an intraoperative FLIm guidance model to identify residual tumors by classifying residual cancer as anomalies. Our proposed approach identified all patients with PSM. In contrast, the IFSA reporting a sensitivity of 0.5 [6, 7].

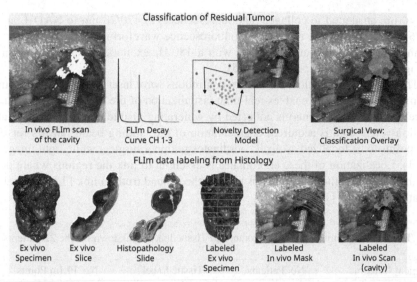

Fig. 1. Overview methodology of the label-free FLIm-based intraoperative surgical guidance, data collection, histopathology registration, and data processing. *Upper panel*: overview of developing the classification method to detect residual HNC in the surgical cavity during the TORS procedure. *Lower panel*: describes the workflow involving generating labels for classifier training and testing. The labels were derived directly from histopathology, evaluated, and annotated by a clinical pathologist (DG). Each annotated H&E section was registered with the ex vivo and in vivo FLIm scan images. The red annotations correspond to cancer labels, and the green annotations correspond to healthy.

2 Method

As illustrated in Fig. 1, the proposed method uses a clinically-compatible FLIm system coupled to the da Vinci SP transoral robotic surgical platform to scan the surgical cavity in vivo and acquire FLIm data. We used the cumulative distribution transform (CDT) of the fluorescence decay curves extracted from the FLIm data as the input feature. The novelty detection model classified FLIm points closer to the healthy distribution as healthy and further from the healthy distribution as a residual tumor. We implemented the image guidance by augmenting the classification map to the surgical view using the predictor output and point locations of the scan.

2.1 FLIm Hardware and Data Acquisition

This study used a multispectral fluorescence lifetime imaging (FLIm) device to acquire data [14]. The FLIm device features a 355 nm UV laser for fluorescence excitation, which is pulsed at a 480 Hz repetition rate. A 365 μm multimode optical fiber (0.22 NA) delivers excitation light to tissue and relays the corresponding fluorescence signal to a set of dichroic mirrors and bandpass filters to spectrally resolve the autofluorescence. Three variable gain UV enhanced Si APD modules with integrated trans-impedance amplifiers receive the autofluorescence, which is spectrally resolved as follows: (1)

390/40 nm attributed to collagen autofluorescence, (2) 470/28 nm to NADH, and (3) 542/50 nm to FAD. The resulting autofluorescence waveform measurements for each channel are averaged four times, thus with a 480 Hz excitation rate, resulting in 120 averaged measurements per second [15].

The FLIm device includes a 440 nm continuous wave laser that serves as an aiming beam; this aiming beam enables real-time visualization of the locations where fluorescence (point measurements) is collected by generating visible blue illumination at the location where data is acquired. Segmentation of the 'aiming beam' allows for FLIm data points to be localized as pixel coordinates within a surgical white light image (see Fig. 1). Localization of these coordinates is essential to link the regions where data is obtained to histopathology, which is used as the ground truth to link FLIm optical data to pathology status [16].

Table 1. Anatomy, Surgical outcome, and Tissue label breakdown for the 22 patients

Surgical Outcome	No. Patients	Tissue Label	No. FLIm Points
Clear margin	19	Healthy	170,535
PSM	3	Cancer	2,451

FLIm data was acquired using the da Vinci SP robotic surgical platform. As part of the approved protocol for this study, the surgeon performed in vivo FLIm scan on the tumor epithelial surface and the surrounding uninvolved benign tissue. Upon completing the scan, the surgeon proceeded with en bloc excision of the tissue suspected of cancer. An ex vivo FLIm scan was then performed on the surgically excised specimen. Finally, the patient's surgical cavity was scanned to check for residual tumor.

2.2 Patient Cohort and FLIm Data Labeling

The research was performed under the approval of the UC Davis Institutional Review Board (IRB) and with the patient's informed consent. All Patients were anesthetized, intubated, and prepared for surgery as part of the standard of care. N = 22 patients are represented in this study, comprising HNC in the palatine tonsil (N = 15) and the base of the tongue (N = 7). For each patient, the operating surgeon conducted an en bloc surgical tumor resection procedure (achieved by TORS-electrocautery instruments), and the resulting excised specimen was sent to a surgical pathology room for grossing. The tissue specimen was serially sectioned to generate tissue slices, which were then formalin-fixed, paraffin-embedded, sectioned, and stained to create Hematoxylin & Eosin (H&E) slides for pathologist interpretation (see Fig. 1).

After the surgical excision of the tumor, an in vivo FLIm scan of approximately 90 s was conducted within the patient's surgical cavity, where the tumor was excised. To validate optical measurements to pathology labels (e.g., benign tissue vs. residual tumor), pathology labels from the excision margins were digitally annotated by a pathologist on each H&E section. The aggregate of H&E sections was correspondingly labeled on the ex vivo specimen at the cut lines where the tissue specimen was serially sectioned.

Thereafter, the labels were spatially registered in vivo within the surgical cavity. This process enables the direct validation of FLIm measurements to the pathology status of the electrocauterized surgical margins (see Table 1).

2.3 FLIm Preprocessing

The raw FLIm waveform contains background noise, instrument artifacts, and other types of interference, which need to be carefully processed and analyzed to extract meaningful information (i.e., the fluorescence signal decay characteristics). To account for background noise, the background signal acquired at the beginning of each clinical case was subtracted from the measured raw FLIm waveform. To retrieve the fluorescence function, we used a non-parametric model based on a Laguerre expansion polynomials and a constrained least-square deconvolution with the instrument impulse response function as previously described [17]. In addition, an SNR threshold of ≥ 50 dB was applied as a filtering criterion to select FLIm points with good signal quality.

2.4 Novelty Detection Model

The state-of-the-art novelty detection models were comprehensively reviewed in the literature [18, 19]. Due to its robust performance, we chose the Generalized One-class Discriminative Subspaces (GODS) classification model [20] to classify healthy FLIm points from the residual tumor. The model trained only on the healthy FLIm points and use a semi-supervised technique to classify residual tumor from healthy. The GODS is a pairwise complimentary classifier defined by two separating hyperplanes to minimize the distance between the two classifiers, limiting the healthy FLIm data within the smallest volume and maximizing the margin between the hyperplanes and the data, thereby avoiding overfitting while improving classification robustness. The first hyperplane (w_1, b_1) projects most of the healthy FLIm points to the positive half of the space, whereas the second hyperplane (w_2, b_2) projects most of the FLIm points in the negative half.

$$\min_{W \in S_d^K, b} F = \frac{1}{2n} \sum_{i=1}^{n} \sum_{j=1}^{2} \| W_j^T x_i + b_j \|_2^2$$

$$+ \frac{\nu}{2\pi} \sum_i \left[\eta - \min\left(W_1^T x_i + b_1 \right) \right]_+^2 + \left[\eta - \max\left(W_2^T x_i + b_2 \right) \right]_+^2 \quad (1)$$

where W_1, W_2 are the orthonormal frames, $\min_{W \in S_d^K, b}$ is the Stiefel manifold, η is the sensitivity margin, and was set $\eta = 0.4$ for our experiments. ν denote a penalty factor on these soft constraints, and b is the biases. x_i denotes the training set containing CDT of the concatenated FLIm decay curve across channels 1–3 along the time axis. The CDT of the concatenated decay curves is computed as follows: Normalize the decay curves to 0–1. Compute and normalize the cumulative distribution function (CDF). Transforming the normalized CDF into the cumulative distribution transform by taking the inverse cumulative distribution function of the normalized CDF [21].

2.5 Classifier Training and Evaluation

The novelty detection model used for detecting residual cancer is evaluated at the point-measurement level to assess the diagnostic capability of the method over an entire tissue surface. The evaluation followed a leave-one-patient-out cross-validation approach. The study further compared GODS with two other novelty detection models: robust covariance and, one-class support vector machine (OC-SVM) [22]. Novelty detection model solely used healthy labels from the in vivo cavity scan for training. The testing data contained both healthy and residual cancer labels. We used grid search to optimize the hyper-parameters and features used in each model and are tabulated in the supplementary section Table S1. The sensitivity, specificity, and accuracy were used as evaluation metrics to assess the performance of classification models in the context of the study. Results of a binary classification model using SVM are also shown in the supplementary section Table S2.

2.6 Classifier Augmented Display

The classifier augmentation depends on three independent processing steps: aiming beam localization, motion correction, and interpolation of the point measurements. A detailed description of implementing the augmentation process is discussed in [23]. The interpolation consists of fitting a disk to the segmented aiming beam pixel location for each point measurement and applying a color map (e.g., green: healthy and red: cancer) for each point prediction. Individual pixels from overlapping disks are averaged to produce the overall classification map and augmented to the surgical field as a transparent overlay.

3 Results

Table 2 tabulates the classification performance comparison of novelty detection models for classifying residual cancer vs. healthy on in vivo FLIm scans in the cavity. Three novelty detection models were evaluated, and all three models could identify the presence of residual tumors in the cavity for the three patients. However, the extent of the tumor classification over the entire tissue surface varied among the models. The GODS reported the best classification performance with an average sensitivity of 0.75 ± 0.02 (see Fig. 2). The lower standard deviation indicates that the model generalizes well. The OC-SVM and robust covariance reported a high standard deviation, indicating that the performance of the classification model is inconsistent across different patients.

The model's ability to correctly identify negative instances is essential to its reliability. The GODS model reported the highest mean specificity of 0.78 ± 0.14 and the lowest standard deviation. The Robust Covariance model reported the lowest specificity, classifying larger portions of healthy tissue in the cavity as a residual tumor; indicating that the model did not generalize well to the healthy labels. We also observed that changing the hyper-parameter, such as the anomaly factor, biased the model toward a single class indicating overfitting (see supplementary section Fig. S1).

The GODS uses two separating hyperplanes to minimize the distance between the two classifiers by learning a low-dimensional subspace containing FLIm data properties

Table 2. Classification performance comparison of novelty detection models for classifying residual cancer vs. healthy on in vivo FLIm scans in the cavity, mean (sd). **Bold font:** Best-performing model.

Novelty Detection Model	Sensitivity (N = 3)	Specificity (N = 22)	Accuracy (N = 3)
OC-SVM	0.68 (0.13)	0.72 (0.18)	0.58 (0.14)
Robust Covariance	0.67 (0.15)	0.63 (0.17)	0.49 (0.11)
GODS	**0.75 (0.02)**	**0.78 (0.14)**	**0.76 (0.02)**

of healthy labels. Residual tumor labels are detected by calculating the distance between the projected data points and the learned subspace. Points that are far from the subspace are classified as residual tumors. We observed that the GODS with the FLIm decay curves in the CDT space achieve the best classification performance compared to other novelty detection models with a mean accuracy of 0.76 ± 0.02. This is mainly due to the robustness of the model, the ability to handle high-dimensional data, and the contrast in the FLIm decay curves. The contrast in the FLIm decay curves was further improved in the CDT space by transforming the FLIm decay curves to a normalized scale and improving linear separability.

4   Discussion

Curent study demonstrates that label-free FLIm parameters-based classification model, using a novelty detection aproach, enables identification of residual tumors in the surgical cavity. The proposed model can resolve residual tumor at the point-measurement level over a tissue surface. The model reported low point-level false negatives and positives. Moreover, the current approach correctly identified all patients with PSMs (see Fig. 2). This enhances surgical precision for TORS procedures otherwise limited to visual inspection of the cavity, palpation of the excised specimen, and IFSA. The FLIm-based classification model could help guide the surgical team in real-time, providing information on the location and extent of cancerous tissue.

In context to the standard of care, the proposed residual tumor detection model exhibits high patient-level sensitivity (sensitivity $= 1$) in detecting patients with PSMs. In contrast, defect-driven IFSA reports a patient-level sensitivity of 0.5 [6, 7]. Our approach exhibits a low patient-level specificity compared to IFSA. Surgeons aim to achieve negative margins, meaning the absence of cancer cells at the edges of the tissue removed during surgery. The finding of positive margins from final histology would result in additional surgical resection, potentially impacting the quality of life. Combining the proposed approach and IFSA could lead to an image-guided frozen section analysis to help surgeons achieve negative margins in a more precise manner. Therefore, completely resecting cancerous tissue and improving patient outcomes.

The false positive predictions from the classification model presented two trends: false positives in an isolated region and false positives spreading across a larger region. Isolated false positives are often caused by the noise of the FLIm system and are

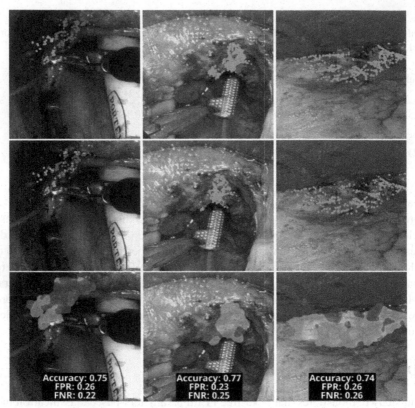

Fig. 2. GODS classification overlay of in vivo cavity scans of three patients presenting with residual tumor. The columns represent each patient, and the rows depict the ground truth labels, the point-prediction overlay, and the augmented surgical view. FPR-False Positive Rate, FNR-False Negative Rate.

accounted for by the interpolation approach used for the classifier augmentation (refer to supplementary section Fig. S2). On the other hand, false positives spreading across a larger region are much more complex to interpret. One insight is that the electrocautery effects on the tissues in the cavity may have influenced them [24]. According to Jackson's burn wound model, the thermal effects caused by electrocautery vary with the different burnt zones. We observed a correlation between a larger spread of false positive predictions associated with a zone of coagulation to a zone of hyperemia.

The novelty detection model generalizes to the healthy labels and considers data falling off the healthy distribution as residual cancer. The FLIm properties associated with the healthy labels in the cavity are heterogeneous due to the electrocautery effects. Electrocautery effects are mainly thermal and can be observed by the levels of charring in the tissue. Refining the training labels based on the levels of charring could lead to a more homogeneous representation of the training set and result in an improved classification model with better generalization.

5 Conclusion

This study demonstrates a novel FLIm-based classification method to identify residual cancer in the surgical cavity of the oropharynx. The preliminary results underscore the significance of the proposed method in detecting PSMs. The model will be validated on a larger patient cohort in future work and address the limitations of the point-level false positive and negative predictions. This work may enhance surgical precision for TORS procedures as an adjunctive technique in combination with IFSA.

Acknowledgment. This work was supported by the National Institutes of Health under Grant 2R01CA187427 in collaboration with Intuitive Surgical, Inc; and P41-EB032840-01. Authors would like to acknowledge Dr. Jonathan Sorger (Intuitive Surgical, Sunnyvale CA) for his support for our ongoing industry collaboration; key areas of his industry support include FLIm visualization aspects and integration of FLIm fiber optic probes into the da Vinci SP TORS platform.

References

1. Gorphe, P., Simon, C.: A systematic review and meta-analysis of margins in transoral surgery for oropharyngeal carcinoma. Oral Oncol. **98**, 69–77 (2019)
2. Orosco, R.K., et al.: Positive surgical margins in the 10 most common solid cancers. Sci. Rep. **8**(1), 1–9 (2018)
3. Li, M.M., Puram, S.V., Silverman, D.A., Old, M.O., Rocco, J.W., Kang, S.Y.: Margin analysis in head and neck cancer: state of the art and future directions. Ann. Surg. Oncol. **26**(12), 4070–4080 (2019)
4. Williams, M.D.: Determining adequate margins in head and neck cancers: practice and continued challenges. Curr. Oncol. Rep. **18**(9), 1–7 (2016). https://doi.org/10.1007/s11912-016-0540-y
5. Poupore, N.S., Chen, T., Nguyen, S.A., Nathan, C.-A.O., Newman, J.G.: Transoral robotic surgery for oropharyngeal squamous cell carcinoma of the tonsil versus base of tongue: a systematic review and meta-analysis. Cancers (Basel) **14**(15), 3837 (2022)
6. Nentwig, K., Unterhuber, T., Wolff, K.-D., Ritschl, L.M., Nieberler, M.: The impact of intraoperative frozen section analysis on final resection margin status, recurrence, and patient outcome with oral squamous cell carcinoma. Clin. Oral Investig. **25**, 6769–6777 (2021)
7. Horwich, P., et al.: Specimen oriented intraoperative margin assessment in oral cavity and oropharyngeal squamous cell carcinoma. J. Otolaryngol. - Head Neck Surg. **50**(1), 1–12 (2021)
8. van Keulen, S., et al.: Rapid, non-invasive fluorescence margin assessment: optical specimen mapping in oral squamous cell carcinoma. Oral Oncol. **88**, 58–65 (2019)
9. Badhey, A.K., et al.: Intraoperative use of wide-field optical coherence tomography to evaluate tissue microstructure in the oral cavity and oropharynx. JAMA Otolaryngol. Head Neck Surg. **149**(1), 71–78 (2023)
10. Zhang, R.R., et al.: Beyond the margins: Real-time detection of cancer using targeted fluorophores. Nat. Rev. Clin. Oncol. **14**(6), 347–364 (2017)
11. Wu, C., Gleysteen, J., Teraphongphom, N.T., Li, Y., Rosenthal, E.: In-vivo optical imaging in head and neck oncology: Basic principles, clinical applications and future directions review-Article. Int. J. Oral Sci. **10**(2), 10 (2018)

12. Hassan, M.A., et al.: Anatomy-specific classification model using label-free FLIm to aid intraoperative surgical guidance of head and neck cancer. IEEE Trans. Biomed. Eng. 1–11 (2023)
13. Marcu, L., French, P.M.W., Elson, D.S.: Fluorescence Lifetime Spectroscopy and Imaging : Principles and Applications in Biomedical Diagnostics. CRC Press, Boca Raton (2014)
14. Gorpas, D., et al.: Autofluorescence lifetime augmented reality as a means for real-time robotic surgery guidance in human patients. Sci. Rep. 9(1), 1187 (2019)
15. Zhou, X., Bec, J., Yankelevich, D., Marcu, L.: Multispectral fluorescence lifetime imaging device with a silicon avalanche photodetector. Opt. Express 29(13), 20105 (2021)
16. Weyers, B.W., et al.: Fluorescence lifetime imaging for intraoperative cancer delineation in transoral robotic surgery. Transl. Biophotonics 1(1–2), e201900017 (2019)
17. Liu, J., Sun, Y., Qi, J., Marcu, L.: A novel method for fast and robust estimation of fluorescence decay dynamics using constrained least-squares deconvolution with Laguerre expansion. Phys. Med. Biol. 57(4), 843–865 (2012)
18. Perera, P., Oza, P., Patel, V.M.: One-class classification: a survey (2021)
19. Seliya, N., Abdollah Zadeh, A., Khoshgoftaar, T.M.: A literature review on one-class classification and its potential applications in big data. J. Big Data 8(1), 1–31 (2021)
20. Cherian, A., Wang, J.: Generalized one-class learning using pairs of complementary classifiers. IEEE Trans. Pattern Anal. Mach. Intell. 44, 6993–7009 (2022)
21. Rubaiyat, A.H.M., Hallam, K.M., Nichols, J.M., Hutchinson, M.N., Li, S., Rohde, G.K.: Parametric signal estimation using the cumulative distribution transform. IEEE Trans. Signal Process. 68, 3312–3324 (2020)
22. Pedregosa, F., et al.: Scikit-learn: machine learning in python (2011)
23. Gorpas, D., Ma, D., Bec, J., Yankelevich, D.R., Marcu, L.: Real-time visualization of tissue surface biochemical features derived from fluorescence lifetime measurements. IEEE Trans. Med. Imaging 35(8), 1802–1811 (2016)
24. Lagarto, J.L., et al.: Electrocautery effects on fluorescence lifetime measurements: an in vivo study in the oral cavity. J. Photochem. Photobiol. B Biol. 185, 90–99 (2018)

Spinal Nerve Segmentation Method and Dataset Construction in Endoscopic Surgical Scenarios

Shaowu Peng[1(✉)], Pengcheng Zhao[1], Yongyu Ye[2], Junying Chen[1(✉)], Yunbing Chang[2], and Xiaoqing Zheng[2]

[1] School of Software Engineering, South China University of Technology, Guangzhou, China
{swpeng,jychense}@scut.edu.cn
[2] Department of Spine Surgery, Guangdong Provincial People's Hospital (Guangdong Academy of Medical Sciences), Southern Medical University, Guangzhou, China

Abstract. Endoscopic surgery is currently an important treatment method in the field of spinal surgery and avoiding damage to the spinal nerves through video guidance is a key challenge. This paper presents the first real-time segmentation method for spinal nerves in endoscopic surgery, which provides crucial navigational information for surgeons. A finely annotated segmentation dataset of approximately 10,000 consecutive frames recorded during surgery is constructed for the first time for this field, addressing the problem of semantic segmentation. Based on this dataset, we propose FUnet (Frame-Unet), which achieves state-of-the-art performance by utilizing inter-frame information and self-attention mechanisms. We also conduct extended experiments on a similar polyp endoscopy video dataset and show that the model has good generalization ability with advantageous performance. The dataset and code of this work are presented at: https://github.com/zzzzzzpc/FUnet.

Keywords: Video endoscopic spinal nerve segmentation · Self-attention · Inter-frame information

1 Introduction

Spinal nerves play a crucial role in the body's sensory, motor, autonomic, and other physiological functions. Injuries to these nerves carry an extremely high risk and may even lead to paralysis. Minimally invasive endoscopic surgery is a common treatment option for spinal conditions, with great care taken to avoid damage to spinal nerves. However, the safety of such surgeries still heavily relies on the experience of the doctors, and there is an urgent need for computers to provide effective auxiliary information, such as real-time neural labeling and guidance in videos. Ongoing studies using deep learning methods to locate spinal nerves in endoscopic videos can be classified into two categories based on the level of visual granularity: coarse-grained and fine-grained tasks.

Supplementary Information The online version contains supplementary material available at https://doi.org/10.1007/978-3-031-43996-4_57.

Coarse-grained vision task focuses on object detection of spinal nerve locations. In this task, object detection models applied to natural images are widely transferred to endoscopic spinal nerve images. Peng Cui et al. [1] used the Yolov3 [2] model to transfer training to the recognition of spinal nerves under endoscopy, which has attracted widespread attention. Sue Min Cho et al. [3] referred to Kaiming's work and achieved certain results using RetinaNet [4] for instrument recognition under spinal neuro-endoscopic images.

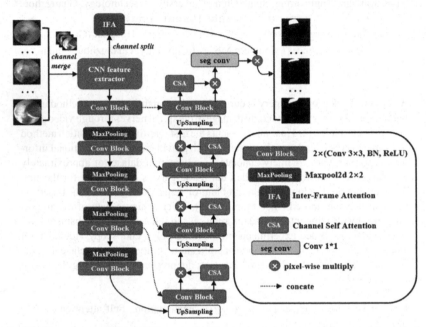

Fig. 1. Pipeline of the proposed FUnet, including the inter-frame attention module and channel self-attention module with global information.

Fine-grained tasks require semantic segmentation of spinal nerves, which provides finer contours and better visualization of the position and morphology of nerves in endoscopic view, leading to greater clinical significance and surgical auxiliary value. However, there are still very few deep learning-based studies on fine-grained segmentation of nerves, one important reason is a lack of semantic segmentation models suitable for spinal nerve endoscopy scenarios. The endoscopic image has the characteristics of blur, blisters, and the lens movement angle is not large, which is quite different from the natural scene image. Therefore, a segmentation model that performs well under natural images may not still be applicable under endoscopic images. Another reason is medical data involves ethical issues such as medical privacy, and the labeling of pixel-level data also relies on professional doctors. Furthermore, endoscopic images of the inside of the spine are often only available during surgery, which is much more difficult than obtaining image datasets from ordinary medical examinations. These lead to the scarcity of labeled data, and ultimately it is difficult to drive the training of neural network models.

In response to the above two problems, our contribution is as follows:

- We innovatively propose inter-frame and channel attention modules for the spinal neural segmentation problem. These two modules can be readily inserted into popular traditional segmentation networks such as Unet [5], resulting in a segmentation network proposed in this paper, called Frame-Unet (FUnet, Fig. 1). The purpose of the inter-frame attention module is to capture the highly similar context between certain adjacent frames, which are characterized by the slower movement of the lens and high similarity of information such as background, elastic motion, and texture between frames. Moreover, we devised a channel self-attention module with global information to overcome the loss of long-distance dependent information in traditional convolutional neural networks. FUnet achieved leading results in many indicators of the dataset we created. Furthermore, FUnet was verified on similar endoscopic video datasets (such as polyps), and the results demonstrate that it outperforms others, confirming our model's strong generalization performance instead of overfitting to a single dataset.
- We propose the first dataset on endoscopic spinal nerve segmentation from endoscopic surgeries, and each frame is finely labeled by professional labelers. The annotated results are also rechecked by professional surgeons.

2 Method

2.1 Details of Dataset

The dataset was taken by the professional SPINENDOS, SP081375.030 machines, and we collected nearly 10000 consecutive frames of video images, each with a resolution of up to 1080*1080 (Fig. 2). The dataset aims to address the specific task of avoiding nerves during spinal endoscopic surgery. To this end, we selected typical scenes from the authors' surgical videos that not only have neural tissue as the target but also reflect the visual and motion characteristics of such scenes. Unlike other similar datasets with interval frame labeling [6, 7], we labeled each image frame by frame, whether it contains spinal nerves or not. Each frame containing spinal nerves was outlined in detail as a semantic segmentation task. Additionally, we used images containing spinal nerves (approximately 4–5 k images) to split the dataset into training, validation, and test sets for model construction (65%: 17.5%: 17.5%).

2.2 Network Architecture

The overall network architecture is based on a classical Unet network [5] (Fig. 1). In the input part of the network, we integrate T frames ($T > 1$) in a batch to fully utilize and cooperate with the **inter-frame attention module** (**IFA**). The input features dimensions are (B, T, C, H, W), where B is batchsize, C denotes number of channels, H and W are the height and width of the image. Since we use a convolutional neural network for feature extraction, the 5-dimensional feature map is first merged into the convolutional network with B and T channels. Afterwards, the convolutional features are fed into the **IFA** for inter-frame information integration.

Unet's skip-connection method is a good complement to the information lost in the down-sampling reduction process, but it is difficult to capture the global contextual information due to the local dependency of the convolutional network. However, this global information is also crucial to the spinal nerve segmentation problem in the endoscopic scenario, so we inserted a **channel self-attention module (*CSA*)** in each up-sampling section with the global capability of the self-attention mechanism.

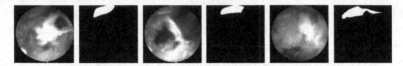

Fig. 2. Illustration of the original and labeled images.

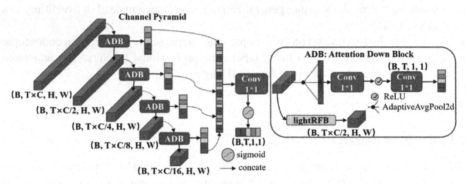

Fig. 3. Pipeline of inter-frame attention Module (IFA)

2.3 *IFA* Module

To exploit the rich semantic information (such as blur, blisters, and other background semantics) between endoscopic frames, we designed the *IFA* to correlate the weight assignment between T frames. (Fig. 3). If the features extracted by convolutional feature extraction are $(B \times T, C, H, W)$, we first need to split the features to obtain $(B, C \times T, H, W)$, which expands the channel dimension and allows subsequent convolutional operations to share the information between frames. After that, the feature matrix is down-sampled by four *Attention Down Block*'s (*ADB*) channel pyramid architecture to obtain the $(B, T, 1, 1)$ vector, and each $(1, 1)$ weight value of this vector in T dimension will be assigned to the attention weight of the segmentation result of T frames.

Channel Pyramid. Although the multiscale operation in traditional convolutional neural networks can improve the generalization ability of the model to targets of different sizes [8, 9], it is difficult to capture information across frames, and down-sampling losses spatial precision at each cross-frame scale. Meanwhile, in many cases of endoscopic video, only the keyframes are clear enough to localize and segment the target, which can guide other blurred frames.

Hence, we propose a channel pyramid architecture to compress the channel dimension for capturing the cross-frame multiscale information, as well as to keep the feature map size unchanged in dimensions of height and width for preserving spatial precision. Such channel down-sampling obtains multi-scale information and semantic information in different cross-frame ranges. The result can adjust the frame weight on keyframe segmenting guidance. For detail, the feature matrix obtained by each ADB is compressed in the channel dimension, which avoids the loss of image size information. Like the perceptual field in the multi-scale approach, the number of inter-frame channels in the channel pyramid at different scales represents the magnitude of the scale across frames, and this inter-frame information at different scales is contacted for further fusion calculations, which is used to generate attention weights.

Attention Down Block (ADB). This module (Fig. 3) is responsible for downsampling the channel dimension and generating the weight information between frames at one scale. We use the Light-RFB [10, 11] module for channel down-sampling without changing the size in the height and width dimensions. In terms of the generation of attention weights, the feature vector after the adaptive global average pooling operation will be scaled to the T dimension by two 1×1 convolutions.

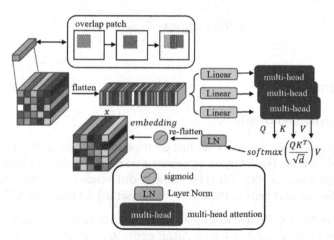

Fig. 4. Pipeline of Channel Self-Attention Module

2.4 *CSA* Module

Inspired by the related work of vision transformer [12, 13], we propose *CSA* mechanism (Fig. 4) to address the problem of the lack of global information of traditional convolutional networks under spinal nerve endoscopy videos. Different from the classical vision transformer work, firstly, in the image patching stage, we use the convolution operation to obtain a feature matrix with a smaller dimension (such as $B \times C \times 32 \times 32$), and the length corresponding to each patch is the length of each pixel channel (32×32), which reduces the amount of computation while sharing of information between different patches. This is because, in the process of convolution down-sampling, the convolution kernels will naturally cause overlapping calculations on the feature map, which will lead to the increase of the receptive field and the overlap of information between different patches. We use three fully connected layers Lq, Lk, Lv to generate Query, Key and Value feature vectors, this can be expressed as follows:

$$Query = Lq(X), Key = Lk(X), Value = Lv(X) \tag{1}$$

$X \in \mathbb{R}^{B \times (H \times W) \times C}$ is the original feature matrix expanded by the flatten operation. At the same time, to supplement the loss of spatial position information, we supplement the pos embedding vector by addition operation.

The multi-headed attention mechanism is implemented by a split operation. For calculation of self-attention, we use the common method [14], the Query matrix and the transpose matrix of Key are multiplied and then divided by the length d of the embedding vector, and this part of the result will be multiplied with Value after soft-max operation. Finally, this part of the features operated by self-attention will enter the LN layer, and after the sigmoid operation, the dimensions of the final attention weight matrix are consistent with the input vector. The formula for this part is as follows:

$$\sigma \left\{ LN \left[softmax \left(\frac{QK^T}{\sqrt{d}} \right) V \right] \right\} \tag{2}$$

3 Experiments

3.1 Implementation Details

Dataset. The self-built dataset is our first contribution. We set up the training set, test set, and validation set. For extended experiments on the polyp dataset, we used the same dataset configuration as PNS-Net [11]. The only difference is that in the testing phase, we used *Valid* and *Test* parts of the CVC-612 dataset [6] for testing (CVC-612-VT).

Training. On the self-built dataset and the CVC-612 dataset, we both use learning rate and weight decay of 1e−4 with Adam optimizer. On the self-built dataset, our model converges after about 30 epochs, while on the CVC-616 dataset, we use the same pre-training and fine-tuning approach as PNS-Net. Methods involved in the data augmentation phase include flipping and cropping. A single TITAN RTX 24 GB graphics card is used for training.

Testing. Five images ($T = 5$) are input to the *IFA* and use the same input resolution of 256*448 as PNSNet to ensure consistency in subsequent tests. Our FUnet is capable of inference at 75 fps on a single TITAN RTX 24 GB, which means that real-time endoscopic spinal nerve segmentation is possible in surgery.

3.2 Experimental Results

On our spinal nerve dataset, we tested the classical and leading medical segmentation networks Unet [5], Unet++ [15], TransUnet (ViT-Base) [16], SwinUnet (ViT-Base) [17] and PNSNet [11] respectively. For comparison, we used the default hyperparameter settings of the networks and employed a CNN feature extractor consistent with that of PNSNet.

Four metrics that are widely used in the field of medical image segmentation are chosen, maxIOU, maxDice, meanSpe/maxSpe and MAE. The quantitative result is in Table 1. Our FUnet achieves state of the art performance on different medical segmentation metrics. The Our-VT dataset means that we use the validation and test datasets for the testing phase (neither of which is involved in training).

The qualitative comparison is in Fig. 5, which shows our FUnet can more accurately segment the contour of the model and the texture of the edges.

Table 1. The quantitative results on our spinal nerve datasets. Our-VT means we use the valid and test part for testing. (Validation part is unseen during testing).

Dataset	Model	Metrics			
		maxDice↑	maxIOU↑	MAE↓	meanSpe↑
Our-VT	Unet	0.778	0.702	0.044	0.881
	Unet + +	0.775	0.705	0.026	0.878
	TransUnet	0.825	0.758	0.082	0.881
	SwinUnet	0.823	0.752	0.026	0.877
	PNSNet	0.882	0.823	0.016	0.885
	FUnet	**0.890**	**0.833**	**0.016**	**0.885**

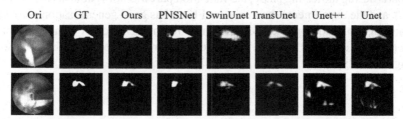

Fig. 5. The qualitative results on our dataset, for more results please refer to the supplementary material.

3.3 Ablation Experiments

The baseline model uses Res2Net [18] as the basic feature extractor with the Unet model, and in the first output feature layer, we adopt a feature fusion strategy consistent with

PNSNet. We have gradually verified the performance of the *IFA* and *CSA* modules on the baseline model (Table 2), and experiments have proved that our two modules can stably improve the segmentation performance of spinal nerves in endoscopic scene. A comparison of qualitative results is available in Fig. 6.

Table 2. Ablation studies. B* for baseline. I* for *IFA* module. C* for *CSA* module. Our-VT means we use the valid and test part for testing.

Dataset	Model	Metrics			
		maxDice↑	maxIOU↑	MAE↓	meanSpe↑
Our-VT	B*	0.862	0.798	0.016	0.885
	B* + I*	0.884	0.826	0.015	0.884
	B* + I* + C*	0.890	0.833	0.016	0.885

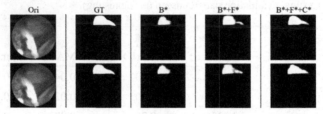

Fig. 6. The qualitative results on ablation study, for more results please refer to the supplementary material.

In addition, more extended experiments are carried out under a similar endoscopic polyp dataset CVC-612 [6] (Table 3, CVC-612-TV means that we used both test and valid parts during the testing phase, the validation part was not visible during the training phase.), and the experiments show that our FUnet has good generalization performance and can adapt to endoscopic segmentation in different scenarios. A comparison of qualitative results is available in Fig. 7, our FUnet still performs well in its ability to segment the edges of polyps.

Table 3. The quantitative results on polyp datasets. CVC-612-TV means that we used both test and valid parts during the testing phase.

Dataset	Model	Metrics			
		maxDice↑	maxIOU↑	MAE↓	maxSpe↑
CVC-612-VT	Unet	0.727	0.623	0.041	0.971
	Unet + +	0.713	0.603	0.042	0.963
	PNSNet	0.866	0.797	0.025	**0.991**
	FUnet	**0.873**	**0.805**	**0.025**	0.989

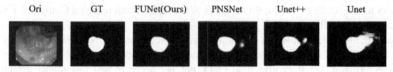

Ori   GT   FUNet(Ours)   PNSNet   Unet++   Unet

Fig. 7. The qualitative results on CVC-612-Valid. For more results please refer to the supplementary material.

4 Conclusion

In this paper, we propose the industry's first semantic segmentation dataset of spinal nerves from endoscopic surgery to date and design the FUnet segmentation network based on inter-frame information and self-attention mechanism. FUnet has achieved state of the art performance on our dataset and shows strong generalization performance on polyp dataset with similar scenes. The *IFA* and *CSA* modules of FUnet can be easily incorporated into other networks. We plan to expand the dataset in the future with the help of self-supervised methods, to improve the performance of the model to provide better computer-assisted surgery capabilities for spinal endoscopy.

Acknowledgement. This work is supported by Guangdong Basic and Applied Basic Research Foundation (2021A1515011349, 2021A1515012651, 2022A1515111091, 2022A1515012557), the National Natural Science Foundation of China Incubation Project (KY0120220040), and the Guangzhou Key R&D Project (202103000053).

References

1. Cui, P., et al.: Tissue recognition in spinal endoscopic surgery using deep learning. In: 2019 IEEE 10th International Conference on Awareness Science and Technology (iCAST), pp. 1–5. IEEE (2019)
2. Redmon, J., Farhadi, A.: YOLOv3: an incremental improvement. arXiv preprint arXiv:1804. 02767 (2018)
3. Cho, S.M., Kim, Y.G., Jeong, J., Kim, I., Lee, H.J., Kim, N.: Automatic tip detection of surgical instruments in biportal endoscopic spine surgery. Comput. Biol. Med. **133**, 104384 (2021)

4. Lin, T.Y., Goyal, P., Girshick, R., He, K., Dollár, P.: Focal loss for dense object detection. In: Proceedings of the IEEE International Conference on Computer Vision, pp. 2980–2988 (2017)
5. Ronneberger, O., Fischer, P., Brox, T.: U-net: Convolutional networks for biomedical image segmentation. In: Navab, N., Hornegger, J., Wells, W.M., Frangi, A.F. (eds.) MICCAI 2015. LNCS, vol. 9351, pp. 234–241. Springer, Cham (2015). https://doi.org/10.1007/978-3-319-24574-4_28
6. Bernal, J., Sánchez, F.J., Fernández -Esparrach, G., Gil, D., Rodríguez, C., Vilariño, F.: WM-DOVA maps for accurate polyp highlighting in colonoscopy: validation vs. saliency maps from physicians. Comput. Med. Imaging Graph. **43**, 99–111 (2015)
7. Bernal, J., Sánchez, J., Vilarino, F.: Towards automatic polyp detection with a polyp appearance model. PR **45**(9), 3166–3182 (2012)
8. Lin, T.Y., Dollár, P., Girshick, R., He, K., Hariharan, B., Belongie, S.: Feature pyramid networks for object detection. In: Proceedings of the IEEE Conference on Computer Vision and Pattern Recognition, pp. 2117–2125 (2017)
9. Zhao, H., Shi, J., Qi, X., Wang, X., Jia, J.: Pyramid scene parsing network. In: Proceedings of the IEEE Conference on Computer Vision and Pattern Recognition, pp. 2881–2890 (2017)
10. Liu, S., Huang, D.: Receptive field block net for accurate and fast object detection. In: Proceedings of the European Conference on Computer Vision (ECCV), pp. 385–400 (2018)
11. Ji, G.P., et al.: Progressively normalized self-attention network for video polyp segmentation. In: de Bruijne, M., et al. (eds.) Medical Image Computing and Computer Assisted Intervention – MICCAI 2021. MICCAI 2021. LNCS, vol. 12901, pp. 142–152. Springer, Cham (2021). https://doi.org/10.1007/978-3-030-87193-2_14
12. Dosovitskiy, A., et al.: An image is worth 16x16 words: Transformers for image recognition at scale. arXiv preprint arXiv:2010.11929 (2020)
13. Liu, Z., et al.: Swin transformer: hierarchical vision transformer using shifted windows. In: Proceedings of the IEEE/CVF International Conference on Computer Vision, pp. 10012–10022 (2021)
14. Vaswani, A., et al.: Attention is all you need. In: Advances in Neural Information Processing Systems, vol. 30 (2017)
15. Zhou, Z., Siddiquee, M.M.R., Tajbakhsh, N., Liang, J.: UNet++: a nested U-Net architecture for medical image segmentation. In: IEEE TMI, pp. 3–11 (2019)
16. Chen, J., et al.: TransUNet: transformers make strong encoders for medical image segmentation. arXiv preprint arXiv:2102.04306 (2021)
17. Cao, H., et al.: Swin-UNet: UNet-like pure transformer for medical image segmentation. In: Karlinsky, L., Michaeli, T., Nishino, K. (eds.) Computer Vision – ECCV 2022 Workshops. ECCV 2022. LNCS, vol. 13803, pp. 205–218. Springer, Cham (2023). https://doi.org/10.1007/978-3-031-25066-8_9
18. Gao, S.H., Cheng, M.M., Zhao, K., Zhang, X.Y., Yang, M.H., Torr, P.: Res2Net: a new multi-scale backbone architecture. IEEE Trans. Pattern Anal. Mach. Intell. **43**(2), 652–662 (2019)

Optical Coherence Elastography Needle for Biomechanical Characterization of Deep Tissue

Robin Mieling[(✉)] [ID], Sarah Latus, Martin Fischer, Finn Behrendt
and Alexander Schlaefer

Institute of Medical Technology and Intelligent Systems,
Hamburg University of Technology, Hamburg, Germany
robin.mieling@tuhh.com

Abstract. Compression-based optical coherence elastography (OCE) enables characterization of soft tissue by estimating elastic properties. However, previous probe designs have been limited to surface applications. We propose a bevel tip OCE needle probe for percutaneous insertions, where biomechanical characterization of deep tissue could enable precise needle placement, e.g., in prostate biopsy. We consider a dual-fiber OCE needle probe that provides estimates of local strain and load at the tip. Using a novel setup, we simulate deep tissue indentations where frictional forces and bulk sample displacement can affect biomechanical characterization. Performing surface and deep tissue indentation experiments, we compare our approach with external force and needle position measurements at the needle shaft. We consider two tissue mimicking materials simulating healthy and cancerous tissue and demonstrate that our probe can be inserted into deep tissue layers. Compared to surface indentations, external force-position measurements are strongly affected by frictional forces and bulk displacement and show a relative error of 49.2% and 42.4% for soft and stiff phantoms, respectively. In contrast, quantitative OCE measurements show a reduced relative error of 26.4% and 4.9% for deep indentations of soft and stiff phantoms, respectively. Finally, we demonstrate that the OCE measurements can be used to effectively discriminate the tissue mimicking phantoms.

Keywords: Optical Coherence Tomography · Tissue Elasticity · Prostate Biopsy

1 Introduction

Healthy and cancerous soft tissue display different elastic properties, e.g. for breast [19], colorectal [7] and prostate cancer [4]. Different imaging modalities

Supplementary Information The online version contains supplementary material available at https://doi.org/10.1007/978-3-031-43996-4_58.

can be used to detect the biomechanical response to an external load for the characterization of cancerous tissue, e.g., ultrasound, magnetic resonance and optical coherence elastography (OCE). The latter is based on optical coherence tomography (OCT), which provides excellent visualization of microstructures and superior spatial and temporal resolution in comparison to ultrasound or magnetic resonance elastography [8]. One common approach for quantitative OCE is to determine the elastic properties from the deformation of the sample and the magnitude of a quasi-static, compressive load [10]. However, due to the attenuation and scattering of the near-infrared light, imaging depth is generally limited to approximately 1 mm in soft tissue. Therefore, OCE is well suited for sampling surface tissue and commonly involves bench-top imaging systems [26], e.g. in ophthalmology [21,22] or as an alternative to histopathological slice examination [1,16]. Handheld OCE systems for intraoperative assessment [2,23] have also been proposed. While conventional OCE probes have been demonstrated at the surface, regions of interest often lie deep within the soft tissue, e.g., cancerous tissue in percutaneous biopsy.

Taking prostate cancer as an example, biomechanical characterization could guide needle placement for improved cancer detection rates while reducing complications associated with increased core counts, e.g. pain and erectile dysfunction [14,18]. However, the measurement of both the applied load and the local sample compression is challenging. Friction forces superimpose with tip forces as the needle passes through tissue, e.g., the perineum. Furthermore, the prostate is known to display large bulk displacement caused by patient movement and needle insertions [20,24] in addition to actual sample compression (Fig. 1, left). Tip force sensing for estimating elastic properties has been proposed [5] but bulk tissue displacement of deep tissue was not considered. In principle, compression and tip force could be estimated by OCT. Yet, conventional OCE probes typically feature flat tip geometry [13,17].

To perform OCE in deep tissue structures, we propose a novel bevel tip OCE needle design for the biomechanical characterization during needle insertions. We consider a dual-fiber setup with temporal multiplexing for the combined load and compression sensing at the needle tip. We design an experimental setup that can simulate friction forces and bulk displacement occurring during needle biopsy (Fig. 1). We consider tissue-mimicking phantoms for surface and deep tissue indentation experiments and compare our results with force-position curves externally measured at the needle shaft. Finally, we consider how the obtained elasticity estimates can be used for the classification of both materials.

2   Methods

In the following, we first present our OCE needle probe and outline data processing for elasticity estimates. We then present an experimental setup for simulating friction and bulk displacement and describe the conducted surface and deep tissue indentation experiments.

2.1 OCE Needle for Deep Tissue Indentation

Our OCE needle approach is illustrated in Fig. 2. It consists of an OCT imaging system, a time-division multiplexer and our OCE needle probe. The needle features two single-mode glass fibers (SMF-28, Thorlabs GmbH, GER) embedded into a bevel tip needle. The forward viewing fiber (Fiber 1) images sample compression while the load sensing fiber (Fiber 2) visualizes the displacement of a reference epoxy layer that is deformed under load. We cleave the distal ends of both fibers to enable common path interference imaging. The outer diameter of the OCE needle prototype is 2.0 mm. We use a spectral domain OCT imaging system (Telesto I, Thorlabs GmbH, GER) with a center wavelength λ_0 of 1325 nm to acquire axial scans (A-scans) at a sampling rate of 91.3 kHz. A solid state optical switch (NSSW 1x2 NanoSpeed^{TM}, Agiltron, USA), a 100 kHz switch driver (SWDR DC-100KHz NS Driver, Agiltron, USA) and a microcontroller (Arduino^{TM} Mega 2560, Arduino, USA) alternate between the two fibers every second A-scan. Compared to spatial multiplexing [17], our temporal multiplexing maximizes the field-of-view and signal strength while effectively halving the acquisition frequency.

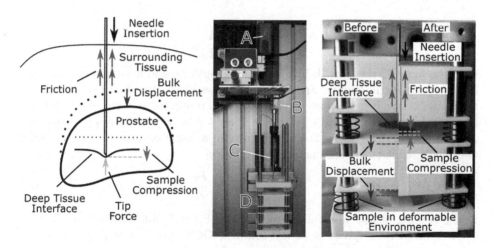

Fig. 1. Left: Schematic of deep tissue indentation during needle insertion. Friction forces (red) and tip forces (grey) are superimposed and the forward motion of the needle (black) only partially results in sample compression (green) due to bulk displacement (blue). Middle: Experimental setup used for indentation experiments, with a linear actuator (A), an axial force sensor (B), the OCE needle probe (C) and the sample layers (D). Right: Simulated deep tissue indentation before and after needle motion. Friction can be added by puncturing multiple layers and bulk displacement is simulated by placing the sample on springs. (Color figure online)

2.2   OCE Measurement

In unconfined compression, the elasticity of the sample can be determined by the relation between stress σ and bulk strain ϵ denoted by the Young's modulus

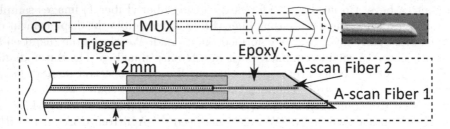

Fig. 2. Illustration of our OCE needle probe for deep tissue indentation. Axial scans (blue) are alternately recorded from fiber 1 and 2 with the OCT and multiplexer (MUX) setup. We use optical fiber 1 to measure sample compression and fiber 2 for the displacement of a reference epoxy layer (green) that is deformed under tip forces. (Color figure online)

$$E = \frac{\sigma}{\epsilon} = \frac{F/A}{\Delta L/L_0}, \tag{1}$$

with the force F, the area A, initial sample length L_0 and assuming incompressibility, quasi-static loading and neglecting viscoelasticity. However, the indentation with our bevel tipped needle will not result in uniform stress and we hypothesize instead that the elasticity is only relative to the applied tip force F_T and the resulting local strain ϵ_l. To obtain a single parameter for comparing two measurements, we assume a linear relation

$$E_{OCE}(F_T, \epsilon_l) \approx \frac{F_T}{\epsilon_l} \tag{2}$$

in the context of this work. To detect strain (Fiber 1) and applied force (Fiber 2), we consider the phase ϕ of the complex OCT signals for fiber i at time t and depth z. The phase shift between two A-scans is proportional to the depth dependent displacement $\delta u_i(z,t)$

$$\delta\phi_i(z,t) = \frac{4\,\pi\,n\,\delta u_i(z,t)}{\lambda_0}, \tag{3}$$

assuming a refractive index n of 1.45 and 1.5 for tissue (Fiber 1) and epoxy (Fiber 2), respectively. We obtain the deformation $u_i(z,t)$ from the unwrapped phase and perform spatial averaging to reduce noise. For fiber 1, we employ a moving average with a window size of 0.1 mm. We estimate local strain based on the finite difference along the spatial dimension over an axial depth Δz of 1 mm.

$$\epsilon_l(t) = \frac{u_1(z_0 + \Delta z, t) - u_1(z_0, t)}{\Delta z} \tag{4}$$

For fiber 2, we calculate the mean $u_2(t)$ over the entire depth of the epoxy. We assume a linear coefficient a_F to model the relation between the applied tip force F_T and the mean deformation $\bar{u}_2$ of the reference epoxy layer.

$$F_T(t) = a_F * \bar{u}_2(t). \tag{5}$$

2.3   Experimental Setup

We build an experimental setup for surface and deep tissue indentations with simulated force and bulk displacement (Fig. 1). For deep tissue indentations, different tissue phantoms are stacked on a sample holder with springs in between. For surface measurements, we position the tissue phantoms separately without additional springs or tissue around the needle shaft. We use a motorized linear stage (ZFS25B, Thorlabs GmbH, GER) to drive the needle while simultaneously logging motor positions. An external force sensor (KD24s 20N, ME-Meßsysteme GmbH, GER) measures combined axial forces. We consider two gelatin gels as tissue mimicking materials for healthy and cancerous tissue. The two materials (Mat. A and Mat. B) display a Young's modulus of 53.4 kPa and 112.3 kPa, respectively. Reference elasticity is determined by unconfined compression experiments of three cylindrical samples for each material according to Eq. 1, using force and position sensor data (See supplementary material). The Young's modulus is obtained by linear regression for the combined measurements of each material. We calibrate tip force estimation (Fiber 2) by indentation of silicone samples with higher tear resistance to ensure that no partial rupture has taken place. We

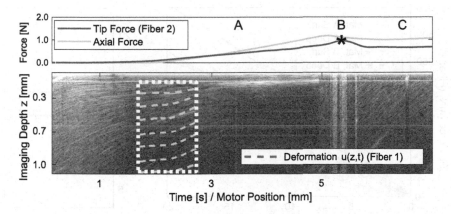

Fig. 3. Demonstration of the dual-fiber needle in insertion with puncture event (B) and post-rupture cutting phase (C). The needle is inserted with a velocity of $1\,\mathrm{mm\,s^{-1}}$. Visualization of the magnitude of the complex OCT signal from fiber 1 (bottom) displayed over needle motion/time. Estimated tip forces from fiber 2 and axial forces are displayed at the top. Acquisition window during pre-deformation phase (A) considered for OCE measurements is indicated by dashed white line. Local strain is calculated based on the tracked deformation from the OCT phase difference as visualized in red.

then determine a linear fit according to Eq. 5 and obtain $a_F = 174.4\,\text{mN}\,\text{mm}^{-1}$ from external force sensor and motor position measurements (See supplementary material).

2.4   Indentation Experiments

In total, we conduct ten OCE indentation measurements for each material. Three surface measurements with fixed samples and seven deep tissue indentations with simulated friction and bulk displacement. For each indentation, we place the needle in front of the surface or deep tissue interface and acquire OCT data while driving the needle for 3 mm (Fig. 1). As the beginning of the needle movement might not directly correspond to the beginning of sample indentation, we evaluate OCE measurements only if the estimated tip force is larger than 50 mN. To further ensure that measurements occur within the pre-rupture deformation phase [6,15], only samples below 20 % local strain are considered. A visualization of the OCE acquisition window from an example insertion with surface rupture and post-rupture cutting phase [6,15] is shown in Fig. 3. We evaluate external needle shaft measurements of relative axial force and relative motor position with the same endpoint obtained from local strain estimates. We perform linear regression to determine the slopes $E_{OCE}[\text{mN}\,\%^{-1}]$ and $E_{EXT}[\text{mN}\,\text{mm}\,\%^{-1}]$ from tip-force-strain and axial-force-position curves, respectively. As we can consider surface measurements as equivalents to the known elasticity, we regard the relative error (RE) of the mean value obtained for deep indentations, with respect to the average estimate during surface indentations. We report the RE

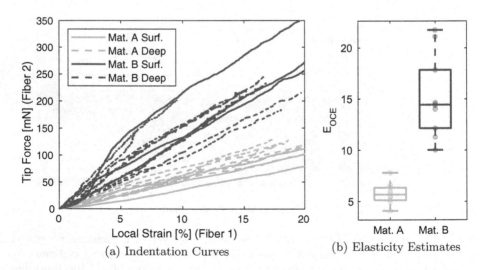

(a) Indentation Curves   (b) Elasticity Estimates

Fig. 4. (a) OCE needle measurements for surface and deep tissue indentations based on the estimated tip force (fiber 2) and the detected local strain (fiber 1). (b) Resulting OCE elasticity estimates show good separation between the two materials, enabling quantitative biomechanical characterization during deep tissue indentations.

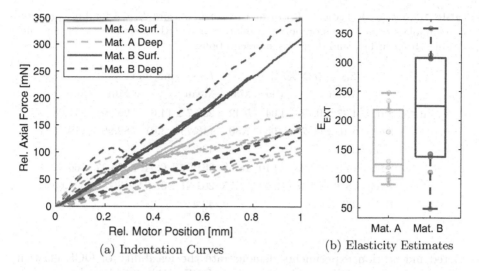

(a) Indentation Curves (b) Elasticity Estimates

Fig. 5. (a) Axial force-position-curves for surface and deep tissue indentations from external force sensor and motor encoder. Friction increases the measured axial force, while bulk displacement decreases the observed slope. Bulk displacement can occur suddenly due to stick-slip, as seen in two cases of material B. (b) Resulting elasticity estimates show overlap between the two materials, hampering quantitative biomechanical characterization.

for both OCE and external measurements and material A and B, respectively. Finally, we consider the measured elasticities for the biomechanical classification of the material. We report the area under the receiver operating characteristic (AUROC) and area under the precision recall curve (AUPRC) for both external and OCE sensing.

3 Results

The OCE measurements for surface and deep tissue indentations are displayed in Fig. 4a. In comparison, external force-position curves are shown in Fig. 5a. The resulting estimates E_{OCE} and E_{EXT} are shown in Fig. 4b and Fig. 5b, respectively. It can be seen that OCE measurements result in separation of both materials while an overlap is visible for external sensors. The sample elasticities and relative error are also accumulated in Table 1. Biomechanical characterization based on the OCE estimates allows complete separation between materials, with AUROC and AUPRC scores of 1.00 (See supplementary material). External measurements do not enable robust discrimination of materials and yielded AUROC and AUPRC scores of only 0.85 and 0.861, respectively.

4 Discussion and Conclusion

We demonstrate our approach on two tissue mimicking materials that have similar elastic properties as healthy and cancerous prostate tissue [5,11]. The con-

Table 1. Measured elasticity E_{OCE} and external sensors E_{EXT} for both materials with sample size n. We also report the relative error (RE) between the mean values obtained from surface and deep tissue indentations.

		E_{OCE} [mN %$^{-1}$]			E_{EXT} [mN mm^{-1}]			n
		Mean	Min	Max	Mean	Min	Max	
Mat. A	Surf	4.87 ± 0.72	4.04	5.42	232.7 ± 14.6	217.97	247.08	3
	Deep	6.18 ± 0.87	5.05	7.80	118.26 ± 3.01	89.99	179.75	7
	RE	26.4 %			49.2 %			
Mat. B	Surf	14.92 ± 3.28	12.10	18.53	361.02 ± 6.94	353.37	366.92	3
	Deep	15.39 ± 4.48	10.02	21.75	201.31 ± 89.04	136.56	354.60	7
	RE	4.9 %			42.4 %			

ducted indentation experiments demonstrate the feasibility of OCE elasticity estimates for deep tissue needle insertions. OCE estimates show better agreement between surface and deep tissue indentations compared to external measurements, as displayed by reduced relative errors of 26.4% and 4.9% for both phantoms, respectively. Bulk displacement causes considerate underestimation of elasticity estimates when only needle position and axial forces are considered, shown by relative errors of 49.2% and 42.4% for material A and B, respectively. Additionally, quantitative OCE estimates allow the robust discrimination between the two materials as shown by Fig. 4b and the AUROC and AUPRC scores of 1. Note that the high errors for external measurements at the needle shaft are systematic, as friction and bulk displacement are unknown. In contrast, our probe does not suffer from these systematic errors. Moreover, considering the standard deviation for OCE estimates, improved calibration of our dual-fiber needle probe is expected to further improve performance. Deep learning-based approaches for tip force estimation could provide increased accuracy and sensitivity compared to the assumed linear model [3]. Weighted strain estimation based on OCT signal intensity [26] could address the underestimation of local strain during segments of low signal-to-noise-ratio (See supplementary material). We are also currently only considering the loading cycle and linear elastic models for our approach. However, soft-tissue displays strong non-linearity in contrast to the mostly linear behavior of gelatin gels. Compression OCE theoretically enables the analysis of non-linear elastic behavior [26] and future experiments will consider non-linear models and unloading cycles better befitting needle-tissue-interaction [15,25].

Interestingly, our needle works with a beveled tip geometry that allows insertion into deep tissue structures. During insertion, tip force estimation can be used to detect interfaces and select the pre-rupture deformation phase for OCE estimates (Fig. 3). This was previously not possible with flat tip needle probes [9,13,17]. While the cylindrical tip is advantageous for calculating the

Young's modulus, it has been shown that the calculation of an equivalent Young's modulus is rarely comparable across different techniques and samples [4,12]. Instead, it is important to provide high contrast and high reproducibility to reliably distinguish samples with different elastic properties. We show that our dual-fiber OCE needle probe enables biomechanical characterization by deriving quantitative biomechanical parameters as demonstrated on tissue mimicking phantoms. Further experiments need to include biological soft tissue to validate the approach for clinical application, as our evaluation is currently limited to homogeneous gelatin. This needle probe could also be very useful when considering robotic needle insertions, e.g., to implement feedback control based on elasticity estimates.

Acknowledgements. This work was partially funded by Deutsche Forschungsgemeinschaft under Grant SCHL 1844/6-1, the i^3 initiative of Hamburg University of Technology, and the Interdisciplinary Competence Center for Interface Research (ICCIR) on behalf of the University Medical Center Hamburg-Eppendorf and the Hamburg University of Technology.

References

1. Allen, W.M., et al.: Wide-field quantitative micro-elastography of human breast tissue. Biomed. Opt. Express **9**(3), 1082–1096 (2018). https://doi.org/10.1364/BOE.9.001082
2. Fang, Q., et al.: Handheld probe for quantitative micro-elastography. Biomed. Opt. Express **10**(8), 4034–4049 (2019). https://doi.org/10.1364/BOE.10.004034
3. Gessert, N., et al.: Needle tip force estimation using an OCT fiber and a fused convGRU-CNN architecture. In: Frangi, A.F., Schnabel, J.A., Davatzikos, C., Alberola-López, C., Fichtinger, G. (eds.) MICCAI 2018. LNCS, vol. 11073, pp. 222–229. Springer, Cham (2018). https://doi.org/10.1007/978-3-030-00937-3_26
4. Good, D.W., et al.: Elasticity as a biomarker for prostate cancer: a systematic review. BJU Int. **113**(4), 523–534 (2014). https://doi.org/10.1111/bju.12236
5. Iele, A., et al.: Miniaturized optical fiber probe for prostate cancer screening. Biomed. Opt. Express **12**(9), 5691–5703 (2021). https://doi.org/10.1364/BOE.430408
6. Jiang, S., Li, P., Yu, Y., Liu, J., Yang, Z.: Experimental study of needle-tissue interaction forces: effect of needle geometries, insertion methods and tissue characteristics. J. Biomech. **47**(13), 3344–3353 (2014). https://doi.org/10.1016/j.jbiomech.2014.08.007
7. Kawano, S., et al.: Assessment of elasticity of colorectal cancer tissue, clinical utility, pathological and phenotypical relevance. Cancer Sci. **106**(9), 1232–1239 (2015). https://doi.org/10.1111/cas.12720
8. Kennedy, B.F., Kennedy, K.M., Sampson, D.D.: A review of optical coherence elastography: fundamentals, techniques and prospects (2014). https://doi.org/10.1109/JSTQE.2013.2291445
9. Kennedy, K.M., et al.: Needle optical coherence elastography for the measurement of microscale mechanical contrast deep within human breast tissues. J. Biomed. Opt. **18**(12), 121510 (2013). https://doi.org/10.1117/1.JBO.18.12.121510

10. Kennedy, K.M., et al.: Quantitative micro-elastography: imaging of tissue elasticity using compression optical coherence elastography. Sci. Rep. **5**(Apr), 1–12 (2015). https://doi.org/10.1038/srep15538

11. Krouskop, T.A., Wheeler, T.M., Kallel, F., Garra, B.S., Hall, T.: Elastic moduli of breast and prostate tissues under compression. Ultrason. Imaging **20**(4), 260–274 (1998). https://doi.org/10.1177/016173469802000403

12. McKee, C.T., Last, J.A., Russell, P., Murphy, C.J.: Indentation versus tensile measurements of young's modulus for soft biological tissues. Tissue Eng. Part B Rev. **17**(3), 155–164 (2011). https://doi.org/10.1089/ten.TEB.2010.0520

13. Mieling, R., Sprenger, J., Latus, S., Bargsten, L., Schlaefer, A.: A novel optical needle probe for deep learning-based tissue elasticity characterization. Curr. Dir. Biomed. Eng. **7**(1), 21–25 (2021). https://doi.org/10.1515/cdbme-2021-1005

14. Oderda, M., et al.: Accuracy of elastic fusion biopsy in daily practice: results of a multicenter study of 2115 patients. Int. J. Urol. **25**(12), 990–997 (2018). https://doi.org/10.1111/IJU.13796

15. Okamura, A.M., Simone, C., O'Leary, M.D.: Force modeling for needle insertion into soft tissue. IEEE Trans. Biomed. Eng. **51**(10), 1707–1716 (2004). https://doi.org/10.1109/TBME.2004.831542

16. Plekhanov, A.A., et al.: Histological validation of in vivo assessment of cancer tissue inhomogeneity and automated morphological segmentation enabled by optical coherence elastography. Sci. Rep. **10**(1), 11781 (2020). https://doi.org/10.1038/s41598-020-68631-w

17. Qiu, Y., et al.: Quantitative optical coherence elastography based on fiber-optic probe for in situ measurement of tissue mechanical properties. Biomed. Opt. Express **7**(2), 688 (2016). https://doi.org/10.1364/boe.7.000688

18. Rosenkrantz, A.B., et al.: Prostate magnetic resonance imaging and magnetic resonance imaging targeted biopsy in patients with a prior negative biopsy: a consensus statement by AUA and SAR. J. Urol. **196**(6), 1613–1618 (2016). https://doi.org/10.1016/j.juro.2016.06.079

19. Samani, A., Zubovits, J., Plewes, D.: Elastic moduli of normal and pathological human breast tissues: an inversion-technique-based investigation of 169 samples. Phys. Med. Biol. **52**(6), 1565 (2007)

20. Schouten, M.G., et al.: Evaluation of a robotic technique for transrectal MRI-guided prostate biopsies. Eur. Radiol. **22**(2), 476–483 (2012). https://doi.org/10.1007/s00330-011-2259-3

21. Singh, M., Nair, A., Aglyamov, S.R., Larin, K.V.: Compressional optical coherence elastography of the cornea. Photonics **8**(4), 111 (2021). https://doi.org/10.3390/photonics8040111

22. de Stefano, V.S., Ford, M.R., Seven, I., Dupps, W.J.: Live human assessment of depth-dependent corneal displacements with swept-source optical coherence elastography. PLoS ONE **13**(12), e0209480 (2018). https://doi.org/10.1371/journal.pone.0209480

23. Wang, X., Wu, Q., Chen, J., Mo, J.: Development of a handheld compression optical coherence elastography probe with a disposable stress sensor. Opt. Lett. **46**(15), 3669 (2021). https://doi.org/10.1364/ol.429955

24. Xu, H., et al.: MRI-guided robotic prostate biopsy: a clinical accuracy validation. In: Jiang, T., Navab, N., Pluim, J.P.W., Viergever, M.A. (eds.) MICCAI 2010. LNCS, vol. 6363, pp. 383–391. Springer, Heidelberg (2010). https://doi.org/10.1007/978-3-642-15711-0_48

25. Yang, C., Xie, Y., Liu, S., Sun, D.: Force modeling, identification, and feedback control of robot-assisted needle insertion: a survey of the literature. Sensors **18**(2) (2018). https://doi.org/10.3390/S18020561

26. Zaitsev, V.Y., et al.: Strain and elasticity imaging in compression optical coherence elastography: the two-decade perspective and recent advances (2021). https://doi.org/10.1002/jbio.202000257

Semantic Segmentation of Surgical Hyperspectral Images Under Geometric Domain Shifts

Jan Sellner[1,2,3,4(✉)], Silvia Seidlitz[1,2,3,4], Alexander Studier-Fischer[4,5,6], Alessandro Motta[1], Berkin Özdemir[5,6], Beat Peter Müller-Stich[5,6], Felix Nickel[2,5,6], and Lena Maier-Hein[1,2,3,4,6]

[1] Division of Intelligent Medical Systems (IMSY), German Cancer Research Center (DKFZ), Heidelberg, Germany
`j.sellner@dkfz-heidelberg.de`
[2] Helmholtz Information and Data Science School for Health, Karlsruhe/Heidelberg, Germany
[3] Faculty of Mathematics and Computer Science, Heidelberg University, Heidelberg, Germany
[4] National Center for Tumor Diseases (NCT), NCT Heidelberg, a Partnership Between DKFZ and University Medical Center Heidelberg, Heidelberg, Germany
[5] Department of General, Visceral, and Transplantation Surgery, Heidelberg University Hospital, Heidelberg, Germany
[6] Medical Faculty, Heidelberg University, Heidelberg, Germany

Abstract. Robust semantic segmentation of intraoperative image data could pave the way for automatic surgical scene understanding and autonomous robotic surgery. Geometric domain shifts, however – although common in real-world open surgeries due to variations in surgical procedures or situs occlusions – remain a topic largely unaddressed in the field. To address this gap in the literature, we (1) present the first analysis of state-of-the-art (SOA) semantic segmentation networks in the presence of geometric out-of-distribution (OOD) data, and (2) address generalizability with a dedicated augmentation technique termed 'Organ Transplantation' that we adapted from the general computer vision community. According to a comprehensive validation on six different OOD data sets comprising 600 RGB and yperspectral imaging (HSI) cubes from 33 pigs semantically annotated with 19 classes, we demonstrate a large performance drop of SOA organ segmentation networks applied to geometric OOD data. Surprisingly, this holds true not only for conventional RGB data (drop of Dice similarity coefficient (DSC) by 46 %) but also for HSI data (drop by 45 %), despite the latter's rich information content per pixel. Using our augmentation scheme improves on the SOA

J. Sellner and S. Seidlitz—Equal contribution.

The original version of this chapter was previously published non-open access. A correction to this chapter is available at https://doi.org/10.1007/978-3-031-43996-4_71

Supplementary Information The online version contains supplementary material available at https://doi.org/10.1007/978-3-031-43996-4_59.

DSC by up to 67% (RGB) and 90% (HSI)) and renders performance on par with in-distribution performance on real OOD test data. The simplicity and effectiveness of our augmentation scheme makes it a valuable network-independent tool for addressing geometric domain shifts in semantic scene segmentation of intraoperative data. Our code and pretrained models are available at https://github.com/IMSY-DKFZ/htc.

Keywords: deep learning · domain generalization · geometrical domain shifts · semantic organ segmentation · hyperspectral imaging · surgical data science

1   Introduction

Automated surgical scene segmentation is an important prerequisite for context-aware assistance and autonomous robotic surgery. Recent work showed that deep learning-based surgical scene segmentation can be achieved with high accuracy [7,14] and even reach human performance levels if using hyperspectral imaging (HSI) instead of RGB data, with the additional benefit of providing functional tissue information [15]. However, to our knowledge, the important topic of geometric domain shifts commonly present in real-world surgical scenes (e.g., situs occlusions, cf. Fig. 1) so far remains unaddressed in literature. It is questionable whether the state-of-the-art (SOA) image-based segmentation networks in [15] are able to generalize towards an out-of-distribution (OOD) context. The only related work by Kitaguchi et al. [10] showed that surgical instrument segmentation algorithms fail to generalize towards unseen surgery types that involve known instruments in an unknown context. We are not aware of any investigation or methodological contribution on geometric domain shifts in the context of surgical scene segmentation.

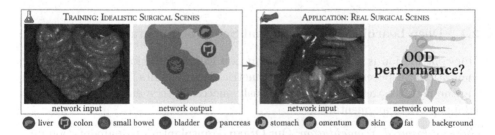

Fig. 1. State-of-the-art (SOA) surgical scene segmentation networks show promising results on idealistic datasets. However, in real-world surgeries, geometric domain shifts such as occlusions of the situs by operating staff are common. The generalizability of SOA algorithms towards geometric out-of-distribution (OOD) has not yet been addressed.

Generalizability in the presence of domain shifts is being intensively studied by the general machine learning community. Here, data augmentation evolved as a simple, yet powerful technique [1,16]. In deep learning-based semantic image segmentation, geometric transformations are most common [8]. This holds particularly true for surgical applications. Our analysis of the SOA (35 publications on tissue or instrument segmentation) exclusively found geometric (e.g., rotating), photometric (e.g., color jittering) and kernel (e.g., Gaussian blur) transformations and only in a single case elastic transformations and Random Erasing (within an image, a rectangular area is blacked out) [22] being applied. Similarly, augmentations in HSI-based tissue classification are so far limited to geometric transformations. To our knowledge, the potential benefit of complementary transformations proposed for image classification and object detection, such as Hide-and-Seek (an image is divided into a grid of patches that are randomly blacked out) [17], Jigsaw (images are divided into a grid of patches and patches are randomly exchanged between images) [2], CutMix (a rectangular area is copied from one image onto another image) [21] and CutPas (an object is placed onto a random background scene) [4] (cf. Fig. 2), remains unexplored.

Given these gaps in the literature, the contribution of this paper is twofold:

1. We show that geometric domain shifts have disastrous effects on SOA surgical scene segmentation networks for both conventional RGB and HSI data.
2. We demonstrate that topology-altering augmentation techniques adapted from the general computer vision community are capable of addressing these domain shifts.

2   Materials and Methods

The following sections describe the network architecture, training setup and augmentation methods (Sect. 2.1), and our experimental design, including an overview of our acquired datasets and validation pipeline (Sect. 2.2).

2.1   Deep Learning-Based Surgical Scene Segmentation

Our contribution is based on the assumption that application-specific data augmentation can potentially address geometric domain shifts. Rather than changing the network architecture of previously successful segmentation methods, we adapt the data augmentation.

Surgery-Inspired Augmentation: Our Organ Transplantation augmentation illustrated in Fig. 2 has been inspired by the image-mixing augmentation CutPas that was originally proposed for object detection [4] and recently adapted for instance segmentation [5] and low-cost dataset generation via image synthesis from few real-world images in surgical instrument segmentation [19]. It is based on placing an organ into an unusual context while keeping shape and texture consistent. This is achieved by transplanting all pixels belonging to one object class (e.g., an organ class or background) into a different surgical scene. Our selection of further computer vision augmentation methods that could potentially improve

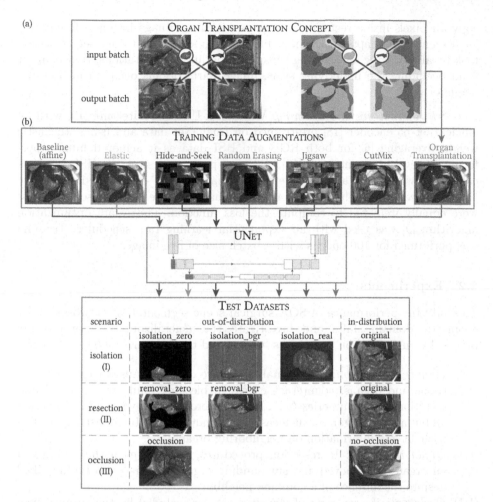

Fig. 2. (a) *Organ Transplantation* augmentation concept inspired from [4]. Image features and corresponding segmentations of randomly selected organs are transferred between images in one batch (in the example, the stomach is transferred from the left to the right and the spleen from the right to the left image). (b) Illustration of our validation experiments. We assess the generalizability under geometric domain shifts of seven different data augmentation techniques in deep learning-based organ segmentation. We validate the model performance on a range of out-of-distribution (OOD) scenarios, namely (1) organs in isolation (*isolation_zero*, *isolation_bgr* and *isolation_real*), (2) organ resections (*removal_zero* and *removal_bgr*), and (3) situs occlusions (*occlusion*), in addition to in-distribution data (*original* and *no-occlusion* (subset of *original* without occlusions)).

geometric OOD performance (cf. Fig. 2) was motivated by the specific conditions encountered in surgical procedures (cf. Sect. 2.2 for an overview). The noise augmentations Hide-and-Seek and Random Erasing black out all pixels inside rectangular regions within an image, thereby generating artificial situs occlusions. Instead of blacking out, the image-mixing techniques Jigsaw and CutMix copy all pixels inside rectangular regions within an image into a different surgical scene. We adapted the image-mixing augmentations to our segmentation task by also copying and pasting the corresponding segmentations. Hence, apart from occluding the underlying situs, image parts/organs occur in an unusual neighborhood.

Network Architecture and Training: We used a U-Net architecture [13] with an efficientnet-b5 encoder [18] pre-trained on ImageNet data and using stochastic weight averaging [6] for both RGB and HSI data as it achieved human performance level in recent work [15]. As a pre-processing step, the HSI data was calibrated with white and dark reference images and ℓ^1-normalized to remove the influence of multiplicative illumination changes. Dice and cross-entropy loss were equally weighted to compute the loss function. The Adam optimization algorithm [9] was used with an exponential learning rate scheduler. Training was performed for 100 epochs with a batch size of five images.

2.2   Experiments

To study the performance of SOA surgical scene segmentation networks under geometric domain shifts and investigate the generalizability improvements offered by augmentation techniques, we covered the following OOD scenarios:

(I)   *Organs in isolation*: Abdominal linens are commonly used to protect soft tissue and organs, counteract excessive bleeding, and absorb blood and secretion. Some surgeries (e.g., enteroenterostomy), even require covering all but a single organ. In such cases, an organ needs to be robustly identified without any information on neighboring organs.

(II)   *Organ resections*: In resection procedures, parts or even the entirety of an organ are removed and surrounding organs thus need to be identified despite the absence of a common neighbor.

(III)   *Occlusions*: Large parts of the situs can be occluded by the surgical procedure itself, introducing OOD neighbors (e.g., gloved hands). The non-occluded parts of the situs need to be correctly identified.

Real-World Datasets: In total, we acquired 600 intraoperative HSI cubes from 33 pigs using the HSI system Tivita® Tissue (Diaspective Vision GmbH, Am Salzhaff, Germany). These were semantically annotated with background and 18 tissue classes, namely heart, lung, stomach, small intestine, colon, liver, gallbladder, pancreas, kidney with and without Gerota's fascia, spleen, bladder, subcutaneous fat, skin, muscle, omentum, peritoneum, and major veins. Each HSI cube captures 100 spectral channels in the range between 500nm and 1000nm at an image resolution of 640×480 pixels. RGB images were reconstructed by aggregating spectral channels in the blue, green, and red ranges. To study organs in isolation, we acquired 94 images from 25 pigs in which all but a specific organ were covered by abdominal linen for all 18 different organ classes (dataset *isolation_real*). To study the effect of occlusions, we acquired 142 images of 20 pigs with real-world situs occlusions (dataset *occlusion*), and 364 occlusion-free images (dataset *no-occlusion*). Example images are shown in Fig. 2.

Manipulated Data: We complemented our real-world datasets with four manipulated datasets. To simulate organs in isolation, we replaced every pixel in an image I that does not belong to the target label l either with zeros or spectra copied from a background image. We applied this transformation to all images in the dataset *original* and all target labels l, yielding the datasets *isolation_zero* and *isolation_bgr*. Similarly, we simulated organ resections by replacing all pixels belonging to the target label l either with zeros or background spectra, yielding the datasets *removal_zero* and *removal_bgr*. Example images are shown in Fig. 2.

Train-Test Split and Hyperparameter Tuning: The SOA surgical scene segmentation algorithms are based on a union of the datasets *occlusion* and *no-occlusion*, termed dataset *original*, which was split into a hold-out test set (166 images from 5 pigs) and a training set (340 images from 15 pigs). To enable a fair comparison, the same train-test split on pig level was used across all networks and scenarios. This also holds for the occlusion scenario, in which the dataset *no-occlusion* was used instead of *original* for training. All networks used the geometric transformations shift, scale, rotate, and flip from the SOA prior to applying the augmentation under examination. All hyperparameters were set according to the SOA. Only hyperparameters related to the augmentation under examination, namely the probability p of applying the augmentation, were optimized through a grid search with $p \in \{0.2, 0.4, 0.6, 0.8, 1\}$. We used five-fold-cross-validation on the datasets *original*, *isolation_zero*, and *isolation_bgr* to tune p such that good segmentation performance was achieved on both in-distribution and OOD data.

Validation Strategy: Following the recommendations of the Metrics Reloaded framework [11], we combined the Dice similarity coefficient (DSC) [3] as an overlap-based metric with the boundary-based metric ormalized surface distance (NSD) [12] for validation for each class l. To respect the hierarchical test set structure, metric aggregation was performed by first macro-averaging the class-level metric value M_l ($M \in \{DSC, NSD\}$) across all images of one pig and subsequently across pigs. The organ removal experiment required special attention in this context, as multiple M_l values per image could be generated corresponding to all the possible neighbour organs that could be removed. In this case, we selected for each l the minimum of all M_l values, which corresponds to the segmentation performance obtained after removing the most important neighbour of l. The same class-specific NSD thresholds as in the SOA were used.

3   Results

Effects of Geometric Domain Shifts: When applying a SOA segmentation network to geometric OOD data, the performance drops radically (cf. Fig. 3). Starting from a high DSC for in-distribution data (RBG: 0.83 (standard deviation (SD) 0.10); HSI: 0.86 (SD 0.10)), the performance drops by 10%-46% for RGB

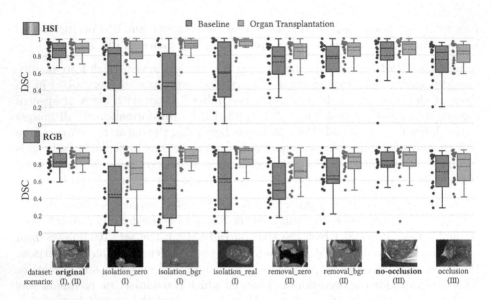

Fig. 3. Segmentation performance of the hyperspectral imaging (HSI) and RGB modality for all eight test datasets (six out-of-distribution (OOD) and two in-distribution datasets (bold)) comparing the baseline network with the Organ Transplantation network. Each point denotes one out of 19 class-level Dice similarity coefficient (DSC) values after hierarchical aggregation across images and subjects. The boxplots show the quartiles of the class-level DSC. The whiskers extend up to 1.5 times the interquartile range and the median and mean are represented as a solid and dashed line, respectively.

and by 5 %-45% for HSI, depending on the experiment. In the organ resection scenario, the largest drop in performance of 63% occurs for the gallbladder upon liver removal (cf. Suppl. Fig. 1). Similar trends can be observed for the boundary-based metric NSD, as shown in Suppl. Fig. 2.

Performance of Our Method: Figure 3 and Suppl. Fig. 2 show that the Organ Transplantation augmentation (gold) can address geometric domain shifts for both the RGB and HSI modality. The latter yields consistently better results, indicating that the spectral information is crucial in situations with limited context. The performance improvement compared to the baseline ranges from 9 %-67% (DSC) and 15 %-79% (NSD) for RGB, and from 9%-90% (DSC) and 16 %-96% (NSD) for HSI, with the benefit on OOD data being largest for organs in isolation and smallest for situs occlusions. The Organ Transplantation augmentation even slightly improves performance on in-distribution data (*original* and *no-occlusion*). Upon encountering situs occlusions, the largest DSC improvement is obtained for the organ classes pancreas (283 %) and stomach (69 %). For organs in isolation, the performance improvement on manipulated data (DSC increased by 57% (HSI) and 61% (RGB) on average) is comparable to that on real data (DSC increased by 50% (HSI) and 46% (RGB)).

Comparison to SOA Augmentations: There is no consistent ranking across all six OOD datasets except for Organ Transplantation always ranking first and baseline usually ranking last (cf. Fig. 4 for DSC- and Suppl. Fig. 3 for NSD-based ranking). Overall, image-mixing augmentations outperform noise augmentations. Augmentations that randomly sample rectangles usually rank better than comparable augmentations using a grid structure (e.g., CutMix vs. Jigsaw).

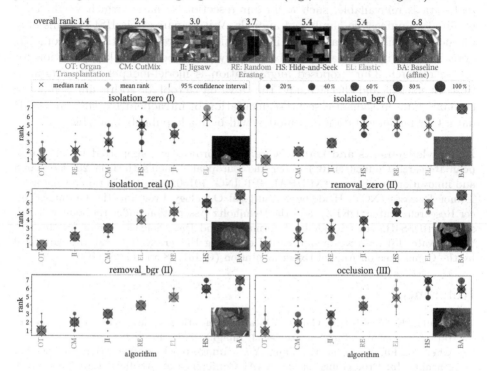

Fig. 4. Uncertainty-aware ranking of the seven augmentation methods for all six geometric out-of-distribution (OOD) test datasets. Organ Transplantation consistently ranks first and baseline last. The area of each blob for one rank and algorithm is proportional to the relative frequency of that algorithm achieving the respective rank across 1000 bootstrap samples consisting of 19 hierarchically aggregated class-level Dice similarity coefficient (DSC) values each (concept from [20]). The numbers above the example images denote the overall ranking across datasets (mean of all mean ranks).

4 Discussion

To our knowledge, we are the first to show that SOA surgical scene segmentation networks fail under geometric domain shifts. We were particularly surprised by the large performance drop for HSI data, rich in spectral information. Our results clearly indicate that SOA segmentation models rely on context information.

Aiming to address the lack of robustness to geometric variations, we adapted so far unexplored topology-altering data augmentation schemes to our target application and analyzed their generalizability on a range of six geometric OOD datasets specifically designed for this study. The Organ Transplantation augmentation outperformed all other augmentations and resulted in similar performance to in-distribution performance on real OOD data. Besides its effectiveness and computational efficiency, we see a key advantage in its potential to reduce the amount of real OOD data required in network training. Our augmentation networks were optimized on simulated OOD data, indicating that image manipulations are a powerful tool for judging geometric OOD performance if real data is unavailable, such as in our resection scenario, which would have required an unfeasible number of animals. With laparoscopic HSI systems only recently becoming available, the investigation and compensation of geometric domain shifts in minimally-invasive surgery could become a key direction for future research. Our proposed augmentation is model-independent, computationally efficient and effective, and thus a valuable tool for addressing geometric domain shifts in semantic scene segmentation of intraoperative HSI and RGB data. Our implementation and models will be made publicly available.

Acknowledgements and Data Usage. This project was supported by the European Research Council (ERC) under the European Union's Horizon 2020 research and innovation programme (NEURAL SPICING, 101002198), the National Center for Tumor Diseases (NCT) Heidelberg's Surgical Oncology Program, the German Cancer Research Center (DKFZ), and the Helmholtz Association under the joint research school HIDSS4Health (Helmholtz Information and Data Science School for Health). The private HSI data was acquired at Heidelberg University Hospital after approval by the Committee on Animal Experimentation (G-161/18 and G-262/19).

References

1. Alomar, K., Aysel, H.I., Cai, X.: Data augmentation in classification and segmentation: a survey and new strategies. J. Imaging **9**(2), 46 (2023)
2. Chen, Z., Fu, Y., Chen, K., Jiang, Y.G.: Image block augmentation for one-shot learning. In: Proceedings of the AAAI Conference on Artificial Intelligence, vol. 33, no. 01, pp. 3379–3386 (2019)
3. Dice, L.R.: Measures of the amount of ecologic association between species. Ecology **26**(3), 297–302 (1945)
4. Dwibedi, D., Misra, I., Hebert, M.: Cut, Paste and Learn: Surprisingly Easy Synthesis for Instance Detection (2017)
5. Ghiasi, G., et al.: Simple copy-paste is a strong data augmentation method for instance segmentation. In: 2021 IEEE/CVF Conference on Computer Vision and Pattern Recognition (CVPR), Nashville, TN, USA, pp. 2917–2927. IEEE (2021)
6. Izmailov, P., Podoprikhin, D., Garipov, T., Vetrov, D., Wilson, A.G.: Averaging weights leads to wider optima and better generalization. In: Proceedings of the International Conference on Uncertainty in Artificial Intelligence (2018)
7. Kadkhodamohammadi, A., Luengo, I., Barbarisi, S., Taleb, H., Flouty, E., Stoyanov, D.: Feature aggregation decoder for segmenting laparoscopic scenes. In: Zhou, L., et al. (eds.) OR 2.0/MLCN -2019. LNCS, vol. 11796, pp. 3–11. Springer, Cham (2019). https://doi.org/10.1007/978-3-030-32695-1_1
8. Kar, M.K., Nath, M.K., Neog, D.R.: A review on progress in semantic image segmentation and its application to medical images. SN Comput. Sci. **2**(5), 397 (2021)
9. Kingma, D.P., Ba, J.: Adam: A Method for Stochastic Optimization. arXiv:1412.6980 (2017)
10. Kitaguchi, D., Fujino, T., Takeshita, N., Hasegawa, H., Mori, K., Ito, M.: Limited generalizability of single deep neural network for surgical instrument segmentation in different surgical environments. Sci. Rep. **12**(1), 12575 (2022)

11. Maier-Hein, L., et al.: Metrics reloaded: pitfalls and recommendations for image analysis validation (2023)
12. Nikolov, S., et al.: Clinically applicable segmentation of head and neck anatomy for radiotherapy: deep learning algorithm development and validation study. J. Med. Internet Res. **23**(7) (2021)
13. Ronneberger, O., Fischer, P., Brox, T.: U-Net: convolutional networks for biomedical image segmentation. In: Navab, N., Hornegger, J., Wells, W.M., Frangi, A.F. (eds.) MICCAI 2015. LNCS, vol. 9351, pp. 234–241. Springer, Cham (2015). https://doi.org/10.1007/978-3-319-24574-4_28
14. Scheikl, P., et al.: Deep learning for semantic segmentation of organs and tissues in laparoscopic surgery. Curr. Dir. Biomed. Eng. **6**, 20200016 (2020)
15. Seidlitz, S., et al.: Robust deep learning-based semantic organ segmentation in hyperspectral images. Med. Image Anal. **80**, 102488 (2022)
16. Shorten, C., Khoshgoftaar, T.M.: A survey on image data augmentation for deep learning. J. Big Data **6**(1), 60 (2019)
17. Singh, K.K., Lee, Y.J.: Hide-and-seek: forcing a network to be meticulous for weakly-supervised object and action localization. In: 2017 IEEE International Conference on Computer Vision (ICCV), pp. 3544–3553 (2017)
18. Tan, M., Le, Q.V.: EfficientNet: rethinking model scaling for convolutional neural networks. In: International Conference on Machine Learning, pp. 6105–6114 (2019)
19. Wang, A., Islam, M., Xu, M., Ren, H.: Rethinking surgical instrument segmentation: a background image can be all you need. In: Wang, L., Dou, Q., Fletcher, P.T., Speidel, S., Li, S. (eds.) MICCAI 2022. LNCS, vol. 13437, pp. 355–364. Springer, Cham (2022). https://doi.org/10.1007/978-3-031-16449-1_34
20. Wiesenfarth, M., et al.: Methods and open-source toolkit for analyzing and visualizing challenge results. Sci. Rep. **11**(1), 2369 (2021)
21. Yun, S., Han, D., Chun, S., Oh, S.J., Yoo, Y., Choe, J.: CutMix: regularization strategy to train strong classifiers with localizable features. In: 2019 IEEE/CVF International Conference on Computer Vision (ICCV), Seoul, Korea (South), pp. 6022–6031. IEEE (2019)
22. Zhong, Z., Zheng, L., Kang, G., Li, S., Yang, Y.: Random erasing data augmentation. In: Proceedings of the AAAI Conference on Artificial Intelligence, vol. 34, no. 07, pp. 13001–13008 (2020)

Ultrasonic Tracking of a Rapid-Exchange Microcatheter with Simultaneous Pressure Sensing for Cardiovascular Interventions

Sunish Mathews[1]([✉]) [iD], Richard Caulfield[2] [iD], Callum Little[1] [iD], Malcolm Finlay[3] [iD], and Adrien Desjardins[1] [iD]

[1] University College London, London WC1E 6BT, UK
sunish.mathews@ucl.ac.uk
[2] Echopoint Medical, London W1T 3JL, UK
[3] Barts Heart Centre, London EC1A 7BE, UK

Abstract. Ultrasound imaging is widely used for guiding minimally invasive cardiovascular procedures such as structural heart repair and renal denervation. Visualization of medical devices such as catheters is critically important and it remains challenging in many clinical contexts. When 2D ultrasound imaging is used, the catheter can readily stray from the imaging plane; with 3D imaging, there can be a loss of visibility at steep angles of insonification. When the catheter tip is not accurately identified, there can be damage to critical structures and procedural inefficiencies. In this paper, we present a tracking system to directly visualize a custom fiber optic ultrasound sensor integrated into a rapid-exchange microcatheter, in the coordinate system of an external ultrasound imaging probe. Pairs of co-registered images were acquired in rapid succession: a tracking image obtained from the ultrasonic sensor signals that were time-synchronized to the ultrasound imaging probe transmissions, and a conventional B-mode ultrasound image. The custom fiber-optic sensor comprised a free-standing membrane originally developed for blood pressure sensing, which was optically interrogated with a wavelength-tunable laser for ultrasound reception. The measured axial and lateral tracking accuracies in water were both within the range of 0.2 to 1 mm. To obtain a preliminary indication of the clinical potential of this ultrasonic tracking system, the microcatheter was delivered over a guidewire into the femoral and renal arteries in an *in vivo* porcine model and intravascular blood pressure waveforms were obtained concurrently. The results demonstrate that ultrasonic catheter tracking using optically-interrogated fiber optic blood pressure sensors is viable, and that it could be useful to guide minimally invasive cardiovascular procedures by providing accurate, real-time position measurements.

Keywords: Ultrasound tracking · minimally invasive cardiovascular interventions · fiber optic sensing

H. Greenspan et al. (Eds.): MICCAI 2023, LNCS 14228, pp. 628–636, 2023.
https://doi.org/10.1007/978-3-031-43996-4_60

1   Introduction

In cardiology, endovascular microcatheters are widely used to provide sensing or interventional imaging for diagnostic and therapeutic applications. One example is a pressure-sensing microcatheter, which measures blood pressure waveforms within coronary arteries to assess the severity of a stenosis and thereby to guide decisions about stent deployment. A "rapid exchange" microcatheter has a lumen in its distal section that allows it to be delivered over a guidewire positioned within the patient's vasculature. Rapid-exchange microcatheters are typically guided to their target destination with fluoroscopic (X-ray) imaging. The use of fluoroscopic guidance has several disadvantages, including exposure of the patient and clinician to X-rays, back pain experienced by practitioners from wearing heavy X-ray protective aprons, and the need for X-ray imaging systems that are not always available in resource-constrained environments. Across a wide range of cardiovascular applications, it is of significant interest to explore alternatives to fluoroscopic guidance of microcatheters.

Ultrasound (US) tracking is an emerging method for localizing medical devices within the body that involves ultrasonic communication between the device and an external imaging system. This method can be performed in "receive-mode" with an ultrasound sensor in the device that receives transmissions from the imaging probe; the time delays between transmission and reception are processed to obtain estimates of the sensor position in 2D [1, 2] and 3D [3, 4]. In reciprocal "transmit-mode" US tracking, the imaging probe receives transmissions from the device [5]. Fiber optic receivers are well suited to receive-mode US tracking: they have broad bandwidth for compatibility with different imaging probes and for high tracking resolution, they are largely omnidirectional, and their small lateral dimensions and flexibility are well suited to integration into minimally invasive cardiovascular devices such as microcatheters. Previous studies with fiber optic receivers have been focused on tracking needles, for instance in the contexts of peripheral nerve blocks and fetal medicine [2, 6].

US tracking of microcatheters would potentially enable ultrasound imaging to be used in place of X-ray imaging for guidance, particularly in applications where there are unobstructed ultrasonic views of the vasculature; these devices typically have very low echogenicity due to their small dimensions and polymeric construction. To the authors' knowledge, US microcatheter tracking has not previously been performed. This endeavor leads to several questions that are addressed in this study: first, how can a fiber optic sensor that was originally developed for blood pressure sensing be adapted to obtain concurrent ultrasound signals; second, how does spatial resolution depend on the angular orientation and spatial position of microcatheter; third, how does this combination perform within a clinically realistic environment? In this study, we address these questions, with validation in a porcine model *in vivo*.

2   Materials and Methods

2.1   Ultrasonic Tracking and Concurrent Pressure Sensing System

The ultrasonic tracking system comprised three components: a clinical US imaging system (SonixMDP, Ultrasonix, Canada) with an external 2-D linear array probe (L14–5; 128 elements), a coronary microcatheter with the integrated pressure/ultrasound sensor (as described in Sect. 2.2), and an US tracking console. The tracking console interrogated the fiber optic sensor in the microcatheter to receive transmissions from the array probe, and obtained processed B-mode US images and two triggers (start of each B-mode frame; start of each A-line) from the US imaging system.

The US signals received by the fiber optic sensor were parsed according to the onsets of the A-lines using the acquired triggers, and concatenated to create a 2-D tracking image of the sensor that was inherently co-registered to the corresponding B-mode US images. Envelope detection of the US signals was performed with a Hilbert transform. To localize the sensor, a region of interest (9 mm × 9 mm) was selected from the tracking image, which was centered on the maximum value. After zeroing values less than 70% of this maximum value (an empirically-obtained threshold value), the sensor position was calculated as the location of the center of mass within this region. The coordinates of the catheter tip were then superimposed on the acquired US images frame-by-frame to show the tracked position of the catheter tip.

Automatic electronic focusing of the B-mode images was performed in order to improve the lateral tracking accuracy [7]. With this method, the estimated axial (depth) coordinate of the estimated sensor position was relayed to the US imaging system. In this way, the sensor was maintained in the electronic focus of transmissions from the imaging probe without operator intervention.

2.2 Sensor and Microcatheter

The sensor comprised a cylindrical capillary structure (diameter: 250 μm; length: 1 mm) with an inorganic membrane at the distal end. This membrane was separated by an air gap from the distal end of a single mode fiber, thereby creating a low-finesse Fabry-Pérot (F-P) cavity. As the blood pressure increases, the membrane is deflected inward, and vice-versa; this deflection is measured using phase-sensitive low-coherence interferometry with a broadband light source [8]. The sensor was integrated within a rapid-exchange coronary microcatheter (minimum/maximum diameters: 0.6 mm/0.9 mm), designed for deployment over a coronary guidewire (diameter: $0.014'' = 0.36$ mm). The sensor was positioned on one side of the microcatheter in the rapid-exchange region, which allowed for a guidewire to pass through the central lumen (Fig. 1). Reference pressure data were obtained from the aorta with a fluid-line and an external pressure sensor, acquired by the console synchronously with the sensor data. The sensor, microcatheter, and pressure sensing console were provided by Echopoint Medical (London, UK).

For US tracking, the sensor was interrogated concurrently with a wavelength that was continuously tuned so that the change of reflectivity with membrane deflection was maximized [9]. Light from this source was distinct from that of the broadband source used for pressure sensing, thereby avoiding optical cross-talk. The two light sources were combined using a fiber optic coupler (50/50; Thorlabs, UK). With this arrangement, invasive pressure measurements and ultrasound reception were obtained concurrently.

2.3 Relative Tracking Accuracy

The relative tracking accuracy of the system was evaluated on the benchtop with the microcatheter immersed in a water tank. With the sensor and the surrounding region of

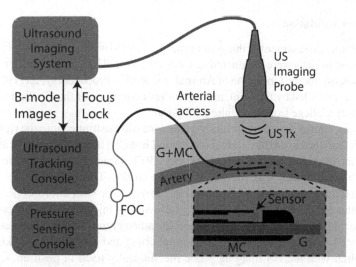

Fig. 1. Schematic of the ultrasound (US) tracking and intravascular pressure sensing system. The rapid exchange microcatheter was progressed over a guidewire within the artery; it contained a fiber optic sensor for receiving ultrasound transmissions (Tx) from the US imaging probe that was also used for concurrent intravascular pressure sensing via a fiber optic coupler (FOC). G: guidewire; MC: microcatheter.

the microcatheter held stationary within the imaging plane, the US imaging probe was translated in steps in the lateral and axial positions with two motorized linear translation stages (MTS50/M-Z-8, Thorlabs, UK) arranged orthogonally. At each step, 100 tracking images were acquired; a digital frequency filter (low-pass, 4th-order Butterworth; 4 MHz cut-off) was applied for noise rejection. The corresponding estimated changes in sensor position relative to the starting position were averaged. These changes were subtracted from the actual changes effected by the linear translation stages to measure the relative tracking accuracy.

2.4 Impact of Microcatheter Orientation

To determine the extent to which the hyperechoic guidewire within the central lumen of the microcatheter shadows ultrasound transmissions by the imaging probe, the signal-to-noise ratio (SNR) of the tracking signals was measured for different axial orientations. These orientations included 0°, where the sensor has a smaller depth than the guidewire and one could expect an absence of shadowing, and 180°, where the sensor has a larger depth so that the guidewire is directly above it and one could expect maximal shadowing. The catheter tip was positioned on a mount at different orientations, with the sensor in the imaging plane at a depth of the 4.5 cm. At each orientation angle, the tracking signals were recorded and the mean SNR was estimated over 100 US tracking frames. For the SNR calculation, the signal was taken as the maximum within the tracking frame; the noise was estimated as the standard deviation from deeper regions of the frame for which signals were visually absent.

2.5 *In Vivo* Validation

An initial clinical assessment of the system was performed with a swine model *in vivo*. All procedures on animals were conducted in accordance with U.K. Home Office regulations and the Guidance for the Operation of Animals (Scientific Procedures) Act (1986). Ethics approval was provided by the joint animal studies committee of the Griffin Institute and the University College London, United Kingdom.

Following arterial access, a coronary guidewire was positioned into the right femoral artery, with guidance from the external ultrasound imaging probe. The microcatheter was subsequently inserted over the guidewire and 250 US tracking frames were acquired, while the microcatheter was moved inside the artery. The microcatheter was then removed and the guidewire was positioned in a renal artery with fluoroscopic guidance. A guide catheter delivered over the guidewire allowed for injections of radio-opaque contrast for locating the renal artery. With the guide catheter removed, the microcatheter was then advanced over the guidewire into the renal artery and another 250 tracking frames were acquired with the US imaging probe mechanically fixed in position. Concurrent blood pressure data were obtained from the renal artery.

3   Results and Discussion

The relative spatial locations of the tracked and actual sensor positions in water were in good agreement for all sensor positions (Fig. 2a, b). To measure the relative axial accuracy, the depth of the sensor relative to the imaging probe (Z) was varied from 18 to 68 mm, corresponding to relative positions of 0 to 50 mm, whilst its lateral position (Y) was held in the center of the imaging plane (Y = 0). Conversely, to measure lateral accuracy, Y was varied from 7 to 33 mm (full range: 0 to 38 mm), corresponding to relative positions of 0 to 26 mm, whilst Z was held constant at 68 mm. The accuracy, defined here as the absolute difference between estimated and actual relative positions, was finer in the Y dimension (<0.6 mm) than in the Z dimension (<0.9 mm).

The SNR of the US tracking signals varied with the orientation of the microcatheter. The axial orientation was changed from 0° to 270° in steps of 90° (Fig. 2c). The SNR was a maximum with the sensor head at 0° (SNR: 202) and a minimum at 180° (SNR: 107) (Fig. 2d). These differences in SNR can be attributed to partial US shadowing of the metallic guidewire at an orientation of 180°. Despite these shadowing effects, the SNR was sufficiently high to permit tracking at all orientations.

When the microcatheter was in the femoral artery of the swine (depth range: 10 to 20 mm) and also in the imaging plane, SNR values as high as 72 were obtained. As the microcatheter was advanced along the guidewire inside the artery, the tracking images each had singular locations from which strong signals were obtained, which corresponded to received transmissions from the US imaging probe (Fig. 3a, b, c). During the experiment, strong visual correspondences between the estimated positions and the B-mode ultrasound images were observed.

When the microcatheter was in the renal artery (depth range: 30 to 35 mm), SNR values as high as 19 were obtained (Fig. 3d, e). This lower maximum SNR value relative to the one obtained from the femoral artery can be attributed in part to the greater depth of the microcatheter and corresponding larger ultrasound attenuation, although slight

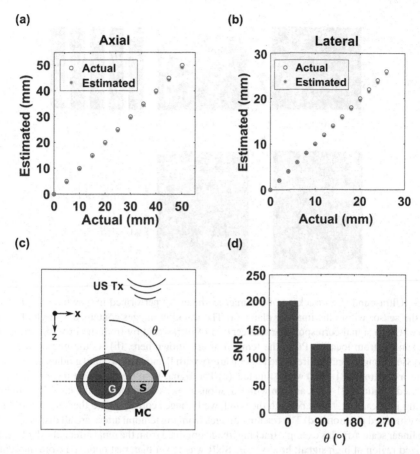

Fig. 2. Measurements of relative tracking accuracy in the axial (a) and lateral (b) directions, performed with the microcatheter and the ultrasound imaging probe in water. With a motorized translation, the axial position was increased from 18 to 68 mm, corresponding to the plotted relative axial positions of 0 to 50 mm. Likewise, the lateral position was increased from 7 to 33 mm, corresponding to the plotted relative lateral positions of 0 to 26 mm. The signal-to-noise ratio (SNR) of the tracking images was measured as a function of the angular orientation of the microcatheter (c). The region of the microcatheter (MC) containing the sensor (S) was within the imaging plane; in this cross-sectional view, the out-of-plane dimension is denoted as "x". US Tx: ultrasound transmission; G: guidewire. The SNR varied as a function of the angular orientation, assuming a maximum when the sensor was facing the imaging probe (0°) and a minimum at 180°.

out-of-plane deviations from the imaging plane could also be responsible. Whilst there was axial spread in the tracking images (Fig. 3d), its impact on tracking accuracy was mitigated with the use of center of mass for sensor position estimation. The pressure trace signal recorded by the microcatheter from the renal artery was in good agreement with the reference fluid column measurement from the aorta (Fig. 3f).

To the authors' knowledge, this is the first study in which US tracking of a rapid-exchange microcatheter was performed, and also the first in which concurrent ultrasound

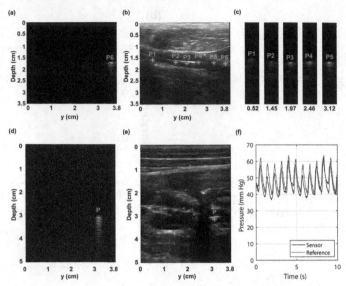

Fig. 3. Ultrasound (US) tracking of the microcatheter tip performed in a swine model *in vivo*, with the sensor within the imaging plane. (a) The tracking images contained a single, localized region of high signal corresponding to reception of ultrasound by from the imaging probe, with one example from location P6 in the femoral artery shown here. (b) As the microcatheter was progressed along the guidewire in the femoral artery with B-mode ultrasound guidance, a sequence of sensor positions (P1—P6) was estimated. (c) The signal-to-noise (SNR) ratio varied across the estimated positions P1—P5, as seen with variations in peak brightness in the laterally windowed tracking images. The values beneath the windowed images correspond to the lateral coordinates of the estimated positions. (d) The tracking images from the femoral artery are all displayed on the same linear scale. (d) This example tracking image obtained from the renal artery also had a single, localized region of high signal; however, its SNR was lower than that obtained from the femoral artery. It is plotted on a smaller linear scale than that of the tracking images from the femoral artery shown above, so that background noise is apparent. (e) The microcatheter intersected the ultrasound imaging plane only in the vicinity of the sensor. (f) Blood pressure measurements acquired concurrently with ultrasound reception for tracking showed good agreement with the reference fluid line measurement.

and invasive blood pressure measurements were obtained with a single fiber optic sensor. The multimodal ultrasound/pressure sensing capability that was achieved with one optical fiber could be critically important for vessels with small lumens, to minimize the complexity, size, and flexibility of the device. With its diminutive size, this sensor could be readily incorporated into a wide range of cardiovascular devices and could find widespread utility in cardiovascular medicine. In addition to intracoronary sensing, it could be used for US tracking during endovascular repair of the tricuspid valve, where visualizing therapeutic devices with a transesophageal probe is very challenging, and during renal denervation procedures. This tracking technology is compatible with other types of US transducers than the one used here, such as curvilinear and phased array probes. Future optimizations to the sensor could include reflective surfaces on the distal end of the fiber and the pressure-sensing membrane to increase the ultrasound sensitivity via an increase in optical finesse. Upstream, diffuse delivery of light and temperature measurements from the sensor will enable invasive flow measurements [10], and optically-absorbing nanocomposite coatings applied to the distal end of the membrane could be used for concurrent optical ultrasound imaging [11]. Reproducibility of the sensor and device will be an important area of focus to broaden the range of applications for clinical translation.

The system presented here has the advantage of providing tracking images in which signals derive solely from the sensor, which are inherently co-registered with the B-mode US images. The sub-mm tracking accuracy of the system was similar to those previously obtained with receive- and transmit-mode US tracking of medical needles [1, 2]. A key observation made in this study was that high sensitivity to ultrasound transmissions could be achieved even when the microcatheter was oriented with the sensor on the opposite side of the guidewire from the imaging probe. In future studies, significant signal attenuation from adipose tissue and sound speed heterogeneities could play a confounding role. Additionally, artifacts might arise from reflections of ultrasound transmissions from strongly echoic structures such as bones or implanted medical devices, which could give rise to multiple objects in tracking images. These artifacts could be potentially mitigated using deep learning approaches developed for ultrasonic tracking [12, 13] and photoacoustic imaging [14]. Mechanical resonance of the sensor membrane, which can lead to axial spread in the tracking images, could be mitigated with frequency filtering. Ultimately, the advantages of US tracking relative to other tracking methods, including automatic image-based device detection, stereo camera tracking, electromagnetic tracking and magnetic sensing depend on a multitude of factors such as size and cost requirements. As demonstrated with an *in vivo* model in this study, ultrasonic tracking is a promising method for guiding endovascular interventions and many other minimally invasive procedures.

Acknowledgements. This work was supported by the Wellcome/EPSRC Centre for Interventional and Surgical Sciences, which is funded by the Wellcome (Grant number 203145/Z/16/Z) and the Engineering and Physical Sciences Research Council (ESPRC; NS/A000050/1). Additional funding was provided by an Innovative Engineering for Health award by the Wellcome (WT101957) and the EPSRC (NS/A000027/1), by a Healthcare Technologies Challenge Award by the EPSRC (EP/N021177/1), and by a Starting Grant by the European Research Council (Proposal 310970 MOPHIM).

References

1. Baker, C., et al.: Intraoperative needle tip tracking with an integrated fibre-optic ultrasound sensor. Sensors **22**, 9035 (2022)
2. Xia, W., et al.: In-plane ultrasonic needle tracking using a fiber-optic hydrophone. Med. Phys. **42**, 5983–5991 (2015). https://doi.org/10.1118/1.4931418
3. Mung, J., Vignon, F., Jain, A.: A non-disruptive technology for robust 3D tool tracking for ultrasound-guided interventions. In: Fichtinger, G., Martel, A., Peters, T. (eds.) MICCAI 2011. LNCS, vol. 6891, pp. 153–160. Springer, Heidelberg (2011). https://doi.org/10.1007/978-3-642-23623-5_20

4. Xia, W., West, S.J., Mari, J.-M., Ourselin, S., David, A.L., Desjardins, A.E.: 3D ultrasonic needle tracking with a 1.5D transducer array for guidance of fetal interventions. In: Ourselin, S., Joskowicz, L., Sabuncu, M.R., Unal, G., Wells, W. (eds.) MICCAI 2016. LNCS, vol. 9900, pp. 353–361. Springer, Cham (2016). https://doi.org/10.1007/978-3-319-46720-7_41

5. Xia, W., et al.: Ultrasonic needle tracking with a fibre-optic ultrasound transmitter for guidance of minimally invasive fetal surgery. In: Descoteaux, M., Maier-Hein, L., Franz, A., Jannin, P., Collins, D.L., Duchesne, S. (eds.) MICCAI 2017. LNCS, vol. 10434, pp. 637–645. Springer, Cham (2017). https://doi.org/10.1007/978-3-319-66185-8_72

6. Xia, W., et al.: Coded excitation ultrasonic needle tracking: an *in vivo* study. Med. Phys. **43**, 4065–4073 (2016)

7. Mathews, S.J., et al.: Ultrasonic needle tracking with dynamic electronic focusing. Ultrasound Med. Biol. **48**, 520–529 (2022)

8. Coote, J.M., et al.: Dynamic physiological temperature and pressure sensing with phase-resolved low-coherence interferometry. Opt. Express **27**, 5641–5654 (2019)

9. Guggenheim, J.A., et al.: Ultrasensitive plano-concave optical microresonators for ultrasound sensing. Nat. Photonics **11**, 714–719 (2017)

10. Carr, E., et al.: Optical interferometric temperature sensors for intravascular blood flow measurements. In: European Conference on Biomedical Optics, p. 11075_1 (2019)

11. Colchester, R.J., Alles, E.J., Desjardins, A.E.: A directional fibre optic ultrasound transmitter based on a reduced graphene oxide and polydimethylsiloxane composite. Appl. Phys. Lett. **114**, 113505 (2019)

12. Maneas, E., et al.: Enhancement of instrumented ultrasonic tracking images using deep learning. Int. J. Comput. Assist. Radiol. Surg. **18**, 395–399 (2023)

13. Maneas, E., et al.: Deep learning for instrumented ultrasonic tracking: from synthetic training data to in vivo application. IEEE Trans. Ultrason. Ferroelectr. Freq. Control **69**, 543–552 (2021)

14. Allman, D., Reiter, A., Bell, M.A.L.: Photoacoustic source detection and reflection artifact removal enabled by deep learning. IEEE Trans. Med. Imaging **37**, 1464–1477 (2018)

Self-distillation for Surgical Action Recognition

Amine Yamlahi[1,2(✉)], Thuy Nuong Tran[1,5], Patrick Godau[1,2,3,5],
Melanie Schellenberg[1,2,3,5], Dominik Michael[1,2], Finn-Henri Smidt[1],
Jan-Hinrich Nölke[1,5], Tim J. Adler[1,2,5], Minu Dietlinde Tizabi[1],
Chinedu Innocent Nwoye[4], Nicolas Padoy[4], and Lena Maier-Hein[1,2,5,6]

[1] Division of Intelligent Medical Systems, German Cancer Research Center (DKFZ),
Heidelberg, Germany
[2] National Center for Tumor Diseases (NCT), NCT Heidelberg a Partnership
between DKFZ and University Medical Center Heidelberg, Heidelberg, Germany
m.elyamlahi@dkfz-heidelberg.de
[3] HIDSS4Health - Helmholtz Information and Data Science School for Health,
Karlsruhe/Heidelberg, Germany
[4] ICube Laboratory, University of Strasbourg, Strasbourg, France
[5] Faculty of Mathematics and Computer Science,
Heidelberg University, Heidelberg , Germany
[6] Medical Faculty, Heidelberg University, Heidelberg , Germany

Abstract. Surgical scene understanding is a key prerequisite for
context-aware decision support in the operating room. While deep
learning-based approaches have already reached or even surpassed human
performance in various fields, the task of surgical action recognition
remains a major challenge. With this contribution, we are the first to
investigate the concept of self-distillation as a means of addressing class
imbalance and potential label ambiguity in surgical video analysis. Our
proposed method is a heterogeneous ensemble of three models that use
Swin Transformers as backbone and the concepts of self-distillation and
multi-task learning as core design choices. According to ablation studies
performed with the CholecT45 challenge data via cross-validation, the
biggest performance boost is achieved by the usage of soft labels obtained
by self-distillation. External validation of our method on an independent test set was achieved by providing a Docker container of our inference model to the challenge organizers. According to their analysis, our
method outperforms all other solutions submitted to the latest challenge
in the field. Our approach thus shows the potential of self-distillation
for becoming an important tool in medical image analysis applications.
Code available at https://github.com/IMSY-DKFZ/self-distilled-swin.

Keywords: Surgical action recognition · Self-distillation ·
Laparoscopic surgery · Surgical workflow

Supplementary Information The online version contains supplementary material
available at https://doi.org/10.1007/978-3-031-43996-4_61.

H. Greenspan et al. (Eds.): MICCAI 2023, LNCS 14228, pp. 637–646, 2023.
https://doi.org/10.1007/978-3-031-43996-4_61

1   Introduction

Surgical scene understanding is an important prerequisite for artificial intelligence (AI)-empowered surgery [12], underlying a range of application areas such as context-aware decision support, autonomous robotics, and workflow optimization. One of its key components is the fully-automatic recognition of the surgical action performed at a given point in time - a task not yet solved by state-of-the-art-methods [4,20]. To advance the field, the CholecTriplet challenge was organized in the scope of the Medical Image Computing and Computer Assisted Interventions (MICCAI) conferences 2021 and 2022. However, according to the organizers analysis [15,18,20], the task still remains unsolved. The guiding hypothesis of our work was that self-distillation could address some of the challenges in surgical action recognition, namely the high number of classes (100 in the case of CholecTriplet); high class imbalance, and label ambiguity. Self-distillation builds upon the widespread concept of knowledge distillation (KD) [6], in which the knowledge is transferred from one deep model (i.e., a teacher) to another shallow model (i.e., a student). Self-distillation diverges from traditional KD by distilling knowledge within the network itself. While KD is already used in various communities, the purpose of this work was to pioneer the concept of self-distillation in the context of surgical data science. Based on the CholecTriplet training data set, we developed a method for surgical action recognition (Fig. 2) that leverages self-distillation, Swin Transformers [11], multi-task learning, and ensembling. The following Sect. 2 presents our methodological contribution in detail. Sect. 3 presents ablation studies on the challenge data set that reveal the most important design choices as well as an external validation of our solution on an independent surgical video data set. We conclude with a brief discussion of the most relevant aspects of our work in Sect. 4.

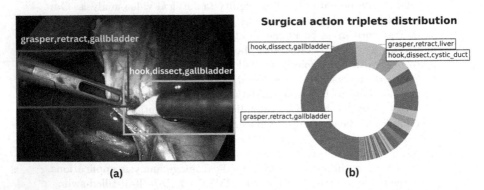

Fig. 1. Task of surgical action recognition. (a) Each action is represented by a triplet comprising instrument, verb and target. Multiple triplets can be present in one image, as shown in the example. (b) CholecTriplet training data set illustrating the heavy class imbalance. Of 100 possible triplet classes, the prevalence ranges from 0.01% to 44.6%.

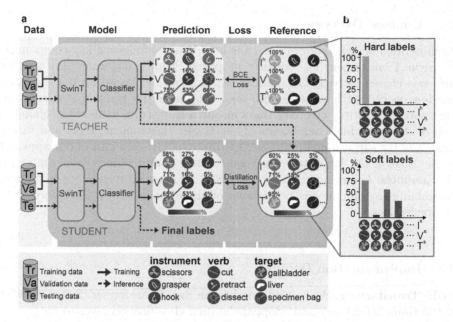

Fig. 2. Approach to surgical action triplet recognition. (a) Our architecture leverages Swin Transformer (SwinT) as a backbone and the concepts of self-distillation, multi-task learning, and ensembling as core strategies. The teacher model is trained on hard labels using binary cross-entropy (BCE) loss. Inferencing the training data, the sigmoid probabilities are used as input in the next step. The student model is trained with the noisy soft labels in a multi-task fashion to minimize the BCE loss, commonly referred to as distillation loss, between the teacher and the student's predictions. (b) Visualization of label distribution in hard and soft labels.

2 Methods

2.1 Task Description and Dataset

Our study is based on the CholecTriplet Challenge 2022 [18], which was conducted under the umbrella of the Endoscopic Vision Challenges (EndoVis) in conjunction with MICCAI. The Surgical Action Recognition task required participants to submit solutions that recognize surgical action triplets in laparoscopic videos, as illustrated in Fig. 1. The challenge granted access to the CholecT45 [19] dataset which consists of 45 video recordings of laparoscopic cholecystectomy with a total of 90,489 frames. CholecT45 is annotated with 100 action triplet classes, with one instrument, verb, and target forming a triplet. The annotations include six different instrument classes, ten verbs (denoting the action performed), and 15 targets such as organs, tissues, or foreign bodies (clip, specimen bag, etc.). The theoretical maximum of $6 \cdot 10 \cdot 15$ classes was reduced to the above-mentioned 100 based on medical relevance and prevalence. An example image from the CholecT45 dataset containing two triplet annotations can be seen in Fig. 1 (a). A chart depicting the highly imbalanced class distribution is shown in Fig. 1 (b).

2.2   Concept Overview

As illustrated in Fig. 2, our approach is based on the following key components: (1) Swin Transformer: The recently proposed Swin Transformer [11] architecture was chosen as backbone. (2) Multi-task learning: Based on the success of previous work that leveraged multi-task learning as its training paradigm, we incorporated multiple auxiliary tasks in our architecture, namely the classification of the individual components of the triplet (instrument, verb, and target) as well as the surgical phase. (3) Self-distillation: The core idea of our approach is the usage of soft labels to reduce overconfidence and address label ambiguity. (4) Ensemble: Following common successful training strategies, we implement ensembling to combine the predictions of three trained Swin Transformers of different scales.

2.3   Implementation Details

Swin Transformer. We base our method on Swin Transformer (SwinT) models of the timm [24] library and adopted the final classification layer to output the 100 triplet predictions, as well as the individual instruments, verbs, targets and surgical phase as auxiliary tasks to leverage the interconnection between them in a multi-task fashion (+Multi).

Self-distillation. The concept of self-distillation was achieved by training a teacher Swin transformer on one-hot encoded hard labels for 20 epochs, with a batch size of 64, an Adam [10] optimizer, a learning rate of 2×10^{-4}, a cosine annealing scheduler decreasing to a minimum learning rate of 2×10^{-6}, and a binary cross-entropy loss function. The model was trained with light augmentations that comprise resizing the images to 224×224 pixels, horizontal and vertical flips, rotation, brightness and saturation perturbations with a probability of 0.5. We trained five teacher and five student models; one for each fold of the official five-fold cross validation splitting introduced by the challenge. The teacher was trained on four of the five splits of its fold. After convergence, the soft labels (i.e., the sigmoid probabilites) for the same four splits were computed and the student was trained using these soft labels. The validation was performed on the fifth split, using hard (i.e., the original) labels for both the teacher and the student. During inference, the sigmoid probabilities of the five student models were averaged to yield the final result.

The five teacher models shared a common weight intialization seed. The five student models shared a separate weight initialization seed. The student models were trained for 40 epochs with the same augmentations as the teacher models. We saved the weights on the epoch with the best mean Average Precision (mAP) score based on the validation split for the current fold.

Ensemble. We combined three trained Swin Transformers (SwinT) of different scales (SwinT base/SwinT large) and configurations for our final ensemble (Ens) model: First, we employed a SwinT base model with multi-task learning of instrument, verb, and target and trained it using self-distillation. Second, we used a SwinT large model using the same approach, and added label smoothing to the soft labels. Third, we included phase annotations as an additional task for the multi-task training of a SwinT base model still employing self-distillation. Please note that every single model mentioned here corresponds to the five aggregated student models of the previous paragraph. All the models were trained using Nvidia GPUs Geforce RTX 3090 and Tesla V100 32GB.

3   Experiments and Results

The purpose of the experiments was to validate the performance of our method and to quantify the (potential) benefit of each individual component. To this end, we conducted (1) comprehensive ablation studies using the CholecT45 official 5-Fold cross-validation split [17], (2) an analysis of the specific benefit of soft labels, and (3) an external validation based on a Docker container submitted to the CholecTriplet 2022 challenge organizers. The Rendezvous Net [18], provided by the challenge organizers, served as the benchmark. In line with the challenge design [13], we validated the performance using mAP (following the aggregation scheme in [15]) and the top K=5 Accuracy as metrics. All scores were computed using the ivtmetrics library [17].

Ablation Studies. We designed the ablation studies as follows: We first calculated the performance of our Swin Transformer backbone as a stand-alone triplet classifier (SwinT). Next, we added multiple auxiliary targets (instruments, verbs, targets, and phases) for multi-task classification (+MultiT). As a third component, we implemented self-distillation by training a student model on soft labels, acquired by training the teacher model (+SelfD). The fourth step was the ensembling of three student model SwinT (+Ens). The results are shown in Tab. 1. A single SwinT as model backbone yields a higher mAP for triplet classification (mAP=32.3%) than the benchmark (mAP=28.8%), which corresponds to a relative improvement by 10.3%. The biggest boost was achieved by including self-distillation, which improved the Triplet mAP and top-5 accuracy by 3.8% points (pp) and 2.4pp, respectively, compared to our baseline. The final model yielded a mAP of 38.5% and a top-5 accuracy of 86.5%, which corresponds to a boost of 6.2pp in mAP and 2.7pp in top-5 accuracy compared to our own baseline, and a relative improvement by 33.7% for mAP compared to the state-of-the-art method. For transparency, we also provide per-video results, depicted in Fig. 3. With a few exceptions, our final model consistently provides the best results.

Analysis of Soft Labels. The addition of self-distillation resulted in the highest boost in performance. This holds true despite the fact that the mAP of the teacher model, trained on hard labels, was about 88% on the training set, which is sub-optimal. The question is thus why the poorer soft labels still yielded a performance improvement. While part of the answer is provided in the literature on soft/noisy labels [9,14,23,26], we also speculated that the soft labels

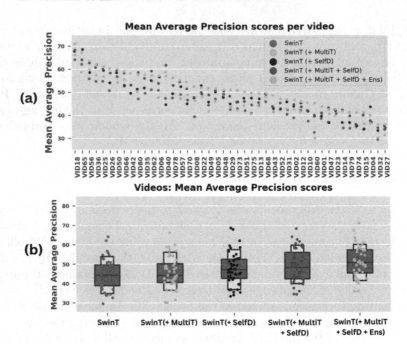

Fig. 3. Quantitative results on the validation data. (a) mean Average Precision (mAP) plotted separately for each video for five configurations of our method with increasing complexity. A Swin Transformer (SwinT) was gradually complemented by multi-task learning (MultiT), self-distillation (SelfD), and ensembling (Ens). Videos were sorted by mAP score of final ensemble performance from highest to lowest. (b) Corresponding dot- and boxplots of mAP scores, aggregated over all videos

may address the issue of ambiguous/erroneous labels in our particular use case. More specifically, we assumed that if the teacher model increases the probability of semantically close triplets in the soft labels, it could lead to an enhanced level of confidence in the student model's prediction of the ground truth, potentially accounting for the observed performance improvement. To investigate whether this is the case, we first defined a pragmatic proxy metric for semantic similarity: the number of identical triplet items (max: two for different triplets). We then selected all frames with only one unique triplet label and retrieved the top five triplets (excluding the reference) with the highest soft label score. Figure 4, depicts an example of such a comparison. The reference triplet "bipolar, dissect, cystic_plate" is shown with five soft label triplets ranked by probability. In the example, the top five triplets share an average of 1.6 components with the reference, indicating that they contain similar semantic information. We found that over all samples, the average number of component matches between reference and top five triplets is 1.0±0.002. In contrast, when comparing the reference with five triplets randomly drawn (while respecting prevalence), the agreement is 0.5±0.002.

Table 1. Main quantitative results Starting from our backbone model - a Swin Transformer (Swin T) - we gradually added individual components, namely multi-task learning (MultiT), self-distillation (SelfD), and ensembling (Ens). Each component addition leads to an increase in mAP and top-5 accuracy, in both cross-validation (left) and independent external validation (right).

	Cross Validation		External validation (CholecTriplet 2022 test set)	
	Triplet mAP [%]	Top-5 accuracy [%]	Triplet mAP [%]	Top-5 accuracy [%]
Rendezvous	29.4	79.3	32.7	69.4
Ours: SwinT	32.3	83.8	32.9	70.7
Ours: SwinT + MultiT	33.1	84.6	33.8	71.6
Ours: SwinT + SelfD	35.0	85.2	36.1	72.9
Ours: SwinT + MultiT + SelfD	36.1	86.2	37.3	73.3
Ours: SwinT + MultiT + SelfD + Ens	**38.5**	**86.5**	**37.4**	**74.0**

This shows that self-distillation specifically leads to increased scores for semantically related classes.

Independent External Validation. External validation was conducted on the CholecTriplet challenge test set. The results, shown in Table 1, confirm the results from cross-validation experiments, with the final ensemble scoring 37.4% in mAP. This equals an absolute improvement of 4.7pp and a relative improvement of 14.4% compared to the Rendezvous benchmark (mAP= 32.7%). With a previous version of the method presented in this paper (scoring slightly lower) we won the challenge in 2022 as the only team that explored the concept of self-distillation.

4   Discussion

This paper pioneers the concept of self-distillation in the medical image analysis domain. Specifically, we are the first to tackle key challenges in surgical action recognition, namely the high number of classes and class imbalance, with self-distillation. Comprehensive ablation studies combined with external validation yielded the following findings:

1. Swin Transformers, as recently introduced by the computer vision community, can serve as a strong backbone in endoscopic vision tasks. This is suggested by the fact that even our most ablated model, consisting of a single SwinT, surpasses the state-of-the-art surgical action recognition method *Rendezvouz*.
2. Multi-task learning, here using the classification of instrument, verb, and target as well as of the surgical phase as auxiliary tasks, yielded a notable increase in performance.
3. Self-distillation yielded the biggest boost in performance, suggesting that soft labels are better suited for surgical action recognition.
4. Ensembling increased performance further, as also suggested by various publications in a wide range of fields.

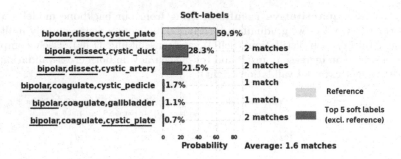

Fig. 4. Example of generated soft labels with reference (in green) and top 5 soft label triplets ranked by probability (in red). Average number of component matches between reference "bipolar, dissect, cystic_plate" and top five triplets is 1.6.

Overall, the addition of self-distillation (in combination with the SwinT as a backbone) resulted in the highest performance boost. While label noise has been shown to be beneficial in various work [9,14,23,26], the concept of self-distillation may not necessarily be intuitive; although the mAP achieved by the teacher model, trained on hard labels, is sub-optimal (32%), the teacher's noisy labels lead to an overall improvement in performance when compared to the (presumably better-quality) hard labels. In the general machine learning literature, the knowledge encoded in noisy labels is referred to as "dark knowledge" because it is not yet well-understood. Aiming to shed light on this topic, our experiments on semantic similarity suggest that soft labels may actually address the issue of ambiguous/erroneous labels. Further analyses with more sophisticated metrics for semantic similarity are, however, needed to support this finding.

Related work has so far tackled the challenge of surgical action recognition with various strategies including multi-task learning [16,21], and different attention mechanisms [3,19] incorporated into diverse architectures based on temporal convolutional networks [2,21], transformers [3,5,19], or combinations of convolutional neural networks (CNN) with recurrent neural networks (RNN) [7,8,16,25] or hidden Markov models (HMM) [22]. While our approach was particularly successful according to the challenge analysis, the overall performance is still not optimal. Advancing the methods will require more data that features a sufficient number of samples for each triplet and captures the full variability of scenes that might be encountered in practice. From a methodological perspective, future work should be directed to efficiently taking temporal context into account and addressing potential domain shifts [1].

In conclusion, our study is the first to demonstrate the benefit of self-distillation for surgical vision tasks. Based on the substantial performance boost obtained, the usage of soft labels could become a valuable tool in the endoscopic vision community.

Acknowledgements. This project was supported by a Twinning Grant of the German Cancer Research Center (DKFZ) and the Robert Bosch Center for Tumor Diseases

(RBCT). Part of this work was funded by HELMHOLTZ IMAGING, a platform of the Helmholtz Information & Data Science Incubator and the Helmholtz Association under the joint research school "HIDSS4Health - Helmholtz Information and Data Science School for Health" and by French state funds managed within the Plan Investissements d'Avenir by the ANR under references: National AI Chair AI4ORSafety [ANR-20-CHIA-0029-01], Labex CAMI [ANR-11-LABX-0004], DeepSurg [ANR-16-CE33-0009], IHU Strasbourg [ANR-10-IAHU-02] and by BPI France under references: project CONDOR, project 5G-OR. Model Docker evaluation were performed with servers/HPC resources managed by CAMMA, IHU Strasbourg, Unistra Mésocentre, and GENCI-IDRIS [Grant 2021-AD011011638R1, 2021-AD011011638R2, 2021-AD011011638R3].

References

1. Castro, D.C., Walker, I., Glocker, B.: Causality matters in medical imaging. Nat. Commun. **11**(1), 3673 (2020)
2. Czempiel, T., et al.: TeCNO: surgical phase recognition with multi-stage temporal convolutional networks. In: Martel, A.L., et al. (eds.) MICCAI 2020. LNCS, vol. 12263, pp. 343–352. Springer, Cham (2020). https://doi.org/10.1007/978-3-030-59716-0_33
3. Czempiel, T., Paschali, M., Ostler, D., Kim, S.T., Busam, B., Navab, N.: OperA: attention-regularized transformers for surgical phase recognition. In: de Bruijne, M., et al. (eds.) MICCAI 2021. LNCS, vol. 12904, pp. 604–614. Springer, Cham (2021). https://doi.org/10.1007/978-3-030-87202-1_58
4. Eisenmann, M., et al.: Biomedical image analysis competitions: The state of current participation practice. arXiv preprint arXiv:2212.08568 (2022)
5. Gao, X., Jin, Y., Long, Y., Dou, Q., Heng, P.-A.: Trans-SVNet: accurate phase recognition from surgical videos via hybrid embedding aggregation transformer. In: de Bruijne, M., et al. (eds.) MICCAI 2021. LNCS, vol. 12904, pp. 593–603. Springer, Cham (2021). https://doi.org/10.1007/978-3-030-87202-1_57
6. Hinton, G., Vinyals, O., Dean, J.: Distilling the knowledge in a neural network. arXiv preprint arXiv:1503.02531 (2015)
7. Jin, Y., et al.: Sv-rcnet: workflow recognition from surgical videos using recurrent convolutional network. IEEE Trans. Med. Imaging **37**(5), 1114–1126 (2017)
8. Jin, Y., Long, Y., Chen, C., Zhao, Z., Dou, Q., Heng, P.A.: Temporal memory relation network for workflow recognition from surgical video. IEEE Trans. Med. Imaging **40**(7), 1911–1923 (2021)
9. Kim, K., Ji, B., Yoon, D., Hwang, S.: Self-knowledge distillation with progressive refinement of targets. In: Proceedings of the IEEE/CVF International Conference on Computer Vision, pp. 6567–6576 (2021)
10. Kingma, D.P., Ba, J.: Adam: a method for stochastic optimization. arXiv preprint arXiv:1412.6980 (2014)
11. Liu, Z., et al.: Swin transformer: hierarchical vision transformer using shifted windows. In: Proceedings of the IEEE/CVF International Conference on Computer Vision (ICCV), pp. 10012–10022 (October 2021)
12. Maier-Hein, L., et al.: Surgical data science-from concepts toward clinical translation. Med. Image Anal. **76**, 102306 (2022)
13. MICCAI SIG for Challenges: MICCAI registered challenges (2022). https://www.miccai.org/special-interest-groups/challenges/miccai-registered-challenges/
14. Mobahi, H., Farajtabar, M., Bartlett, P.: Self-distillation amplifies regularization in hilbert space. Adv. Neural. Inf. Process. Syst. **33**, 3351–3361 (2020)

15. Nwoye, C.I., et al.: Cholectriplet 2021: a benchmark challenge for surgical action triplet recognition. arXiv preprint arXiv:2204.04746 (2022)
16. Nwoye, C.I., et al.: Recognition of instrument-tissue interactions in endoscopic videos via action triplets. In: Martel, A.L., et al. (eds.) MICCAI 2020. LNCS, vol. 12263, pp. 364–374. Springer, Cham (2020). https://doi.org/10.1007/978-3-030-59716-0_35
17. Nwoye, C.I., Padoy, N.: Data splits and metrics for benchmarking methods on surgical action triplet datasets. arXiv preprint arXiv:2204.05235 (2022)
18. Nwoye, C.I., Padoy, N.: Surgical action triplet detection 2022 (2022). https://cholectriplet2022.grand-challenge.org/
19. Nwoye, C.I., et al.: Rendezvous: attention mechanisms for the recognition of surgical action triplets in endoscopic videos. Med. Image Anal. **78**, 102433 (2022)
20. Nwoye, C.I., , et al.: Cholectriplet 2022: show me a tool and tell me the triplet-an endoscopic vision challenge for surgical action triplet detection. arXiv preprint arXiv:2302.06294 (2023)
21. Ramesh, S., et al.: Multi-task temporal convolutional networks for joint recognition of surgical phases and steps in gastric bypass procedures. Int. J. Comput. Assist. Radiol. Surg. **16**(7), 1111–1119 (2021). https://doi.org/10.1007/s11548-021-02388-z
22. Twinanda, A.P., Shehata, S., Mutter, D., Marescaux, J., De Mathelin, M., Padoy, N.: Endonet: a deep architecture for recognition tasks on laparoscopic videos. IEEE Trans. Med. Imaging **36**(1), 86–97 (2016)
23. Vu, D.Q., Le, N., Wang, J.C.: Teaching yourself: a self-knowledge distillation approach to action recognition. IEEE Access **9**, 105711–105723 (2021)
24. Wightman, R.: Pytorch image models. https://github.com/rwightman/pytorch-image-models (2019). https://doi.org/10.5281/zenodo.4414861
25. Yu, T., Mutter, D., Marescaux, J., Padoy, N.: Learning from a tiny dataset of manual annotations: a teacher/student approach for surgical phase recognition. arXiv preprint arXiv:1812.00033 (2018)
26. Yun, S., Park, J., Lee, K., Shin, J.: Regularizing class-wise predictions via self-knowledge distillation. In: Proceedings of the IEEE/CVF Conference on Computer Vision and Pattern Recognition, pp. 13876–13885 (2020)

Encoding Surgical Videos as Latent Spatiotemporal Graphs for Object and Anatomy-Driven Reasoning

Aditya Murali[1]([✉]), Deepak Alapatt[1], Pietro Mascagni[2,3],
Armine Vardazaryan[2], Alain Garcia[2], Nariaki Okamoto[4], Didier Mutter[2],
and Nicolas Padoy[1,2]

[1] ICube, University of Strasbourg, CNRS, Strasbourg, France
murali@unistra.fr
[2] IHU-Strasbourg, Institute of Image-Guided Surgery, Strasbourg, France
[3] Fondazione Policlinico Universitario Agostino Gemelli IRCCS, Rome, Italy
[4] Institute for Research Against Digestive Cancer (IRCAD), Strasbourg, France

Abstract. Recently, spatiotemporal graphs have emerged as a concise and elegant manner of representing video clips in an object-centric fashion, and have shown to be useful for downstream tasks such as action recognition. In this work, we investigate the use of latent spatiotemporal graphs to represent a surgical video in terms of the constituent anatomical structures and tools and their evolving properties over time. To build the graphs, we first predict frame-wise graphs using a pre-trained model, then add temporal edges between nodes based on spatial coherence and visual and semantic similarity. Unlike previous approaches, we incorporate long-term temporal edges in our graphs to better model the evolution of the surgical scene and increase robustness to temporary occlusions. We also introduce a novel graph-editing module that incorporates prior knowledge and temporal coherence to correct errors in the graph, enabling improved downstream task performance. Using our graph representations, we evaluate two downstream tasks, critical view of safety prediction and surgical phase recognition, obtaining strong results that demonstrate the quality and flexibility of the learned representations. Code is available at github.com/CAMMA-public/SurgLatentGraph.

Keywords: Scene Graphs · Surgical Scene Understanding · Representation Learning

1 Introduction

Surgical video analysis is a rapidly growing field that aims to improve and gain insights into surgical practice by leveraging increasingly available surgical

Supplementary Information The online version contains supplementary material available at https://doi.org/10.1007/978-3-031-43996-4_62.

H. Greenspan et al. (Eds.): MICCAI 2023, LNCS 14228, pp. 647–657, 2023.
https://doi.org/10.1007/978-3-031-43996-4_62

video footage [12,25]. Several key applications have been well explored, ranging from surgical skill assessment to workflow analysis to intraoperative safety enhancement [4,5,11,27]. Yet, effectively learning and reasoning based on surgical anatomy remains a challenging problem, as evidenced by lagging performance in fine-grained tasks such as surgical action triplet detection and critical view of safety prediction [15,16].

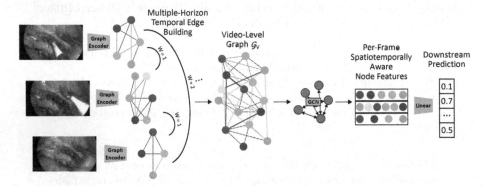

Fig. 1. Overview of our proposed approach. We begin by computing graphs for each frame using [15], then add temporal edges (shown with solid lines) between graphs at different horizons to obtain the video-level graph $\mathcal{G}_V$. We process $\mathcal{G}_V$ with a GNN to yield spatiotemporally-aware node features, which we use for downstream prediction. Each node color corresponds to an object class, and each edge color to a relation class. We retain spatial edges (shown with dotted lines) from the graph encoder in $\mathcal{G}_V$. (Color figure online)

Such anatomy-based reasoning can be accomplished through *object-centric modeling*, which is gaining popularity in general computer vision [6,14,19,26,28]. Object-centric models represent images or clips according to their constituent objects by running an object detector and then using the detections to factorize the visual feature space into per-object features. By retaining implicit visual features, these approaches maintain differentiability, allowing them to be fine-tuned for various downstream tasks. Meanwhile, they can also be extended to include object attributes such as class, location, and temporal order for tasks that rely heavily on scene semantics. Recent works have explored object-centric representations in the surgical domain, but they are characterized by one of several limitations: they often include only surgical tools [9,22], preventing anatomy-driven reasoning; they are limited to single-frames or short clips [9,15,18,22], preventing video-level understanding; or they formulate the object-centric representation as a final output (e.g. scene graph prediction) and only include scene semantics, which limits their effectiveness for downstream tasks [8,17,23].

In this work, we tackle these challenges by proposing to build latent spatiotemporal graph representations of *entire surgical videos*, with each node representing a surgical tool or anatomical structure and edges representing relationships between nodes across space and time. To build our graphs, rather than use an off-the-shelf object detector, we employ the latent graph encoder of [15] to generate per-frame graphs that additionally encode object semantics, segmentation details, and inter-object relations, all of which are important for downstream anatomy-driven reasoning. We then add edges between nodes in different graphs, resulting in a spatiotemporal graph representation of the entire video. We encounter two main challenges when building these graphs for surgical videos: (1) surgical scenes evolve slowly over time, calling for long-term modeling, and (2) object detection is often error-prone due to annotated data scarcity. To address the first challenge, we introduce a framework to add temporal edges at multiple horizons, enabling reasoning about the short-term and long-term evolution of the underlying video. Then, to address the error-prone object detection, we propose a Graph Editing Module that leverages the spatiotemporal graph structure and predicted object semantics to efficiently correct errors in object detection.

We evaluate our method on two downstream tasks: critical view of safety (CVS) clip classification and surgical phase recognition. CVS clip classification is a fine-grained task that requires accurate identification and reasoning about anatomy, and is thus an ideal target application for our object-centric approach. On the other hand, phase recognition is a coarse-grained task that requires holistic understanding of longer video segments, which can demonstrate the effectiveness of our temporal edge building framework. We achieve competitive performance in both of these tasks and show that our graph representations can be used with or without task-specific finetuning, thereby demonstrating their value as general-purpose video representations.

In summary, we contribute the following:

1. A method to encode surgical videos as latent spatiotemporal graphs that can then be used without modification for two diverse downstream tasks.
2. A framework for effectively modeling long-range relationships in surgical videos via multiple-horizon temporal edges.
3. A Graph Editing Module that can correct errors in the predicted graph based on temporal coherence cues and prior knowledge.

2 Methods

In this section, we describe our approach to encode a T-frame video $V = \{I_t \mid 1 \leq t \leq T\}$ as a latent spatiotemporal graph G_V (illustrated in Fig. 1). Our method consists of a frame-wise object detection step followed by a temporal graph building step and a graph editing module to correct errors in the predicted graph representation. We also describe our graph neural network decoder to process the resulting representation G_V for downstream tasks.

2.1 Graph Construction

Object Detection. To construct a latent spatiotemporal graph representation, we must first detect the objects in each frame, along with any additional properties. We do so by employing the graph encoder ϕ_{SG} proposed in [15], yielding a graph $\mathcal{G}_t$ for each frame $I_t \in V$. The resulting $\mathcal{G}_t$ is composed of nodes $\mathcal{N}_t$ and edges $\mathcal{E}_t$; $\mathcal{N}_t$ and $\mathcal{E}_t$ are in turn composed of features h_i, $h_{i,j}$, bounding boxes b_i, $b_{i,j}$, and object/relation class r_i, $r_{i,j}$.

Spatiotemporal Graph Building. Once we have computed the graphs $\mathcal{G}_t$, we add temporal edges to construct a single graph $\mathcal{G}_V$ that describes the entire video. $\mathcal{G}_V$ retains the spatial edges from the various $\mathcal{G}_t$ to describe geometric relations between objects in the same frame (i.e. to the left of, above), while also including temporal edges between spatially and visually similar nodes. It can then be processed with a graph neural network during downstream evaluation to efficiently propagate object-level information across space and time.

We add temporal edges to $\mathcal{G}_V$ based on object bounding box overlap and visual feature similarity, inspired by [26]; however, we extend their approach to construct edges at multiple temporal horizons rather than between adjacent frames alone. Specifically, we design an operator ϕ_{TE} that takes a pair of graphs G_t, G_{t+w} and outputs a list of edges, which are defined by their connectivity $\mathcal{C}_{t,t+w}$ containing pairs of adjacent nodes, and their relation class $\mathcal{R}_{t,t+w}$ containing relation class ids. To compute the edges, ϕ_{TE} calculates pairwise similarities between nodes in G_t and nodes in G_{t+w} using two separate kernels: K_B, which computes the generalized IoU between node bounding boxes, and K_F, which computes the cosine similarity between node features. This yields similarity matrices M_B and M_F, each of size $\mathcal{N}_t \times \mathcal{N}_{t+w}$. Using each matrix, we select the most similar node $n_{j,t+w} \in \mathcal{N}_{t+w}$ for each $n_{i,t} \in \mathcal{N}_t$ and vice-versa. Altogether, this yields $4 * (|\mathcal{N}_t| + |\mathcal{N}_{t+w}|)$ edges consisting of connectivity tuples $c_{m,n} = ((i,t),(j,t+w))$ and relation classes $r_{m,n}$, which we store in $\mathcal{C}_{t,t+w}$ and $\mathcal{R}_{t,t+w}$ respectively. We apply ϕ_{TE} to all pairs of graphs $\mathcal{G}_t, \mathcal{G}_{t+w}$ for various temporal horizons $w \in \mathcal{W}$, then combine the resulting $\mathcal{C}_{t,t+w}$ and $\mathcal{R}_{t,t+w}$ to obtain temporal edge connectivities $\mathcal{C}_{ST}$ and relation classes $\mathcal{R}_{ST}$. Finally, we augment each temporal edge with features $h_{m,n}$ and bounding boxes $b_{m,n}$, yielding the video-level graph $\mathcal{G}_V$:

$$h_{m,n} = h_{i,t_x} + h_{j,t_y}, \ b_{m,n} = \cup(b_{i,t_x}, b_{j,t_y}), \ \text{where } (i,t_x),(j,t_y) = c_{m,n}$$
$$\mathcal{E}_{ST} = \{b_{m,n}, r_{m,n}, h_{m,n} \mid (m,n) \in \mathcal{C}_{ST}\}; \ \mathcal{E}_V = \{\mathcal{E}_t \mid 1 \le t \le T\} \cup \mathcal{E}_{ST} \tag{1}$$
$$\mathcal{G}_V = \{\mathcal{N}_V, \mathcal{E}_V\}, \ \text{where } \mathcal{N}_V = \bigcup_{1 \le t \le T} \mathcal{N}_t.$$

Edge Horizon Selection. While ϕ_{TE} is designed to construct edges between arbitrarily distant graphs, effective selection of temporal horizons $\mathcal{W}$ is nontrivial. We could naively include every possible temporal horizon, setting $\mathcal{W} = \{1, 2, ..., T-1\}$ to maximize temporal information flow; however, making $\mathcal{W}$ too dense results in redundancies in the resulting graph, which can have an oversmoothing effect during downstream processing with a graph neural network (GNN) [29]. To avoid this issue, we take inspiration from temporal convolutional networks (TCN) [10], which propagate information over long input sequences using a series of convolutions with exponentially increasing dilation. We similarly use exponentially increasing temporal horizons, setting $\mathcal{W} = \{1, 2, 4, ..., 2^l\}$ to enable efficient information flow at each GNN layer and long-horizon message passing via a stack of GNN layers.

2.2 Graph Editing Module

One limitation of object-centric approaches is a reliance on high quality object detection, which is particularly steep in surgical videos. These difficulties in object detection could be tackled by incorporating prior knowledge such as anatomical scene geometry, but incorporating these constraints into the learning process often requires complex constraint formulations and methodologies. We posit that our spatiotemporal graph structure represents a simpler framework to incorporate such constraints; to demonstrate this, we introduce a module to filter detections of anatomical structures, which are particularly difficult to detect, incorporating the constraint that there is only one of each structure in each frame. Specifically, after building the spatiotemporal graph, we compute a dropout probability $p_{i,t} = \frac{1}{deg(n_{i,t})}$ for each node, where deg is the degree operator. Then, for each frame t, for each object class r_j, we select the highest scoring node n_t from $\{n_{i,t}|r_{i,t}\}$. During training, we apply graph editing with probability p_{edit}, providing robustness to a wide range of input graphs.

2.3 Downstream Task Decoder

For downstream prediction from $\mathcal{G}_V$, we first apply a GNN using the architecture proposed in [3], yielding spatiotemporally-aware node features. Then, we pool the node features within each frame and apply a linear layer to yield frame-wise predictions (see Fig. 1). We process these predictions differently depending on the task: for clip classification, we output only the prediction for the last frame, while for temporal video segmentation, we output the frame-wise predictions.

2.4 Training

We adopt a two-stage training approach, starting by training ϕ_{SG} as proposed in [15] and then extracting graphs for all images. Then, in the second stage, we process a sequence of graphs with our model to predict frame-wise outputs. We supervise each prediction with the corresponding frame label, propagating the clip label to each frame when per-frame labels are unavailable.

3 Experiments and Results

In this section, we describe our evaluation tasks and datasets, describe baseline methods and our model, then present results for each task. We conclude with an ablation study that illustrates the impact of our various model components.

3.1 Evaluation Tasks and Datasets

Critical View of Safety (CVS) Prediction. The CVS consists of three independent criteria, and can be viewed as a multi-label classification problem [13]. For our experiments, we use the Endoscapes+ dataset introduced in [15], which contains 11090 images annotated with CVS evenly sampled from the dissection phase of 201 cholecystectomies at 0.2 fps; it also includes a subset of 1933 images with segmentation masks and bounding box annotations. We model CVS prediction as a clip classification problem, constructing clips of length 10 at 1 fps, and use the label of the last frame as the clip label. As in [15], we investigate CVS prediction performance in two experimental settings to study the label-efficiency of various methods: (1) using only the bounding box labels and CVS labels and (2) additionally using the segmentation labels. We report mean average precision (mAP) across the three criteria for all methods.

Surgical Phase Recognition. For surgical phase recognition, we use the publically available Cholec80 dataset [24], which includes 80 videos with frame-wise phase annotations ($\{1, 2, ..., 7\}$). We use the first 40 videos for training, the next 8 for validation, and the remaining 32 for testing, as in [2,20]. In addition, to enable object-centric approaches, we use the publically available CholecSeg8k dataset [7] as it represents multiple surgical phases unlike Endoscapes+. As CholecSeg8k is a subset of Cholec80, we split the images into training, validation, and testing following the aforementioned video splits. We model phase recognition as a temporal video segmentation problem, and process the entire video at once. Again, we explore two experimental settings: (1) temporal phase recognition without single-frame finetuning to evaluate the surgical video representations learned by each method and (2) temporal phase recognition with single-frame finetuning, the classical setting [2,5,20]. We report mean F1 score across videos for all methods, as suggested in [20].

3.2 Baseline Methods

Single-Frame Methods. As CVS clip classification is unexplored, we compare to two recent single-frame methods for reference: LG-CVS [15], a graph-based approach, and DeepCVS [13], a non-object-centric approach. We quote results from [15], which improves DeepCVS and enables training with bounding boxes.

DeepCVS-Temporal. We also extend DeepCVS for clip classification by replacing the last linear layer of the dilated ResNet-18 with a Transformer decoder, referring to this model as DeepCVS-Temporal.

STRG. Space-Time Region Graphs (STRG) [26] is a spatiotemporal graph-based approach for action recognition that builds a latent graph by predicting region proposals and extracting per-region visual features using an I3D backbone; we repurpose STRG for CVS clip classification and phase recognition. Because STRG is trained end-to-end, it can only process clips of limited length; consequently, we train STRG on clips of 15 frames for phase recognition rather the entire video as in other methods. We also only consider phase recognition with finetuning. For CVS clip classification, we additionally pre-train the I3D feature extractor in STRG on bounding box/segmentation labels using a Faster-RCNN box head [21] or DeepLabV3+ decoder [1].

TeCNO. TeCNO [2] is a temporal model for surgical phase recognition consisting of frame-wise feature extraction followed by temporal decoding with a causal TCN [10] to classify phases. For phase recognition without single-frame finetuning, we use a ResNet-50 pre-trained on CholecSeg8k using a DeepLabV3+ head to extract features, enabling fair comparisons with our method. For the other setting, we report performance from [20].

Table 1. CVS Clip Classification Performance. Standard deviation is across three runs of each method. Single frame methods from prior works are also reported for reference.

Method		CVS mAP	
		Box	Seg
Single Frame	DeepCVS-R18 [13,15]	54.1 ± 1.3	60.2 ± 1.6
	LG-CVS [15]	63.6 ± 0.8	67.3 ± 1.4
Temporal	DeepCVS-Temporal	57.8 ± 3.2	63.8 ± 2.1
	STRG [26]	60.5 ± 0.7	61.7 ± 1.5
	Ours	$\mathbf{66.3 \pm 0.9}$	$\mathbf{69.7 \pm 1.3}$

Ours. We train our model in two stages, starting by training the graph encoder ϕ_{SG} as described in [15] on the subset of Endoscapes+ annotated with segmentation masks or bounding boxes for CVS clip classification, or on CholecSeg8k for phase recognition. We then extract frame-wise graphs for the entire dataset and apply our spatiotemporal graph approach to predict CVS or phase. In the second experimental setting for phase recognition, we additionally finetune ϕ_{SG} on all training images with the frame-wise phase labels before extracting the graphs. Finally, we evaluate a version of our method that additionally applies a TCN to the un-factorized image features and adds the TCN-processed features to the pooled temporally-aware node features prior to linear classification. We set $l = 3$, $p_{edit} = 0.5$, and use a 5-layer GNN for CVS prediction and an 8-layer GNN for phase recognition.

3.3 Main Experiments

CVS. Our first takeaway from Table 1 is that temporal models provide a method-agnostic boost of 3% mAP for CVS prediction. Furthermore, our approach outperforms both non-object-centric and object-centric temporal baselines, achieving a substantial performance boost in the label-efficient bounding box setting while remaining competitive when also trained with segmentation masks. In the

Table 2. Surgical Phase Recognition Performance.

Method		Phase F1
	TeCNO [2]	64.3
No Single-Frame Finetuning	Ours	70.3
	Ours + TCN	**74.1**
	STRG [26]	77.1
	TeCNO [2], reported from [20]	80.3
With Single-Frame Finetuning	Ours	79.9
	Ours + TCN	**81.4**

Table 3. Ablation Study of Model Components.

Ablated Feature	Performance Drop (↑ is worse)	
	CVS mAP (Seg)	Phase F1 (No FT)
No Long Term Edges ($\mathcal{W} = \{1\}$)	1.6	7.2
Naive Edge Horizon Selection ($\mathcal{W} = \{1, 2, ..., T - 1\}$)	1.4	4.1
No Graph Editing Module	1.1	0.2

box setting, we observe that the non-object-centric DeepCVS approaches perform rather poorly due to an over-reliance on predicted semantics rather than effective visual encodings [15]. Object-centric modeling addresses some of these limitations, as evidenced by STRG outperforming DeepCVS-Temporal. Nevertheless, our method achieves a much stronger performance boost, owing to multiple factors: (1) our model is based on the underlying LG-CVS, which constructs its frame-wise object-centric representation by using the final bounding box predictions rather than just region proposals like STRG, and also encodes semantic information, and (2) our proposed improvements (multiple-horizon edges, graph editing) are critical to improving model performance. Meanwhile, in the segmentation setting, the object-centric STRG is ineffective, performing worse than DeepCVS-Temporal; this discrepancy arises because, as previously mentioned, STRG relies on region proposals rather than object-specific bounding boxes in its graph representation, and as a result, cannot fully take advantage of the additional information provided by the segmentation masks. Our approach translates the ideas of STRG but importantly builds on top of the already effective representations learnt by LG-CVS to achieve universal effectiveness for spatiotemporal modeling of CVS.

Phase. Table 2 shows the phase recognition results for various methods with (bottom) and without (top) finetuning the underlying single-frame model. Our model is already highly effective for phase recognition without any finetuning, outperforming the corresponding TeCNO model by 6.1% F1 in its original form and by nearly 10% F1 when also using a TCN. This shows that the graph representations contain general-purpose information about the surgical scenes and their evolution. Finally, by finetuning the underlying single-frame graph encoder, we match the existing state-of-the-art, highlighting our method's flexibility.

3.4   Ablation Studies

Table 3 illustrates the impact of each model component on CVS clip classification and phase recognition performance. The first two rows illustrate the importance of using exponential edge horizons. Without any long-term edges (as in STRG), we observe a staggering 7.20% drop in Phase F1; naively building edges between all the graphs improves performance but is still 4.14% worse than our proposed method. We observe similar trends for the CVS mAP but with lower magnitude, as CVS prediction is not as reliant on long-term video understanding. Meanwhile, we observe the opposite effect for the graph editing module, which is quite effective for CVS clip classification but does not considerably impact phase F1. This is again consistent with the nature of the tasks, as CVS requires fine-grained understanding of the surgical anatomy, and performance can suffer greatly from errors in object detection, while phase recognition is more coarse-grained and is thus less impacted by errors at this fine-grained level.

4   Conclusion

We introduce a method to encode surgical videos in their entirety as latent spatiotemporal graph representations. Our graph representations enable fine-grained anatomy-driven reasoning as well as coarse long-range video understanding due to the inclusion of edges at multiple-temporal horizons, robustness against errors in object detection provided by a graph editing module, and memory- and computational-efficiency afforded by a two-stage training pipeline. We believe that the resulting graphs are powerful general-purpose representations of surgical videos that can fuel numerous future downstream applications.

Acknowledgement. This work was supported by French state funds managed by the ANR within the National AI Chair program under Grant ANR-20-CHIA-0029-01 (Chair AI4ORSafety) and within the Investments for the future program under Grants ANR-10-IDEX-0002-02 (IdEx Unistra) and ANR-10-IAHU-02 (IHU Strasbourg). This work was granted access to the HPC resources of IDRIS under the allocation 2021-AD011011640R1 made by GENCI.

References

1. Chen, L.-C., Zhu, Y., Papandreou, G., Schroff, F., Adam, H.: Encoder-decoder with atrous separable convolution for semantic image segmentation. In: Ferrari, V., Hebert, M., Sminchisescu, C., Weiss, Y. (eds.) ECCV 2018. LNCS, vol. 11211, pp. 833–851. Springer, Cham (2018). https://doi.org/10.1007/978-3-030-01234-2_49
2. Czempiel, T.: TeCNO: surgical phase recognition with multi-stage temporal convolutional networks. In: Martel, A.L., et al. (eds.) MICCAI 2020. LNCS, vol. 12263, pp. 343–352. Springer, Cham (2020). https://doi.org/10.1007/978-3-030-59716-0_33

3. Dhamo, H., et al.: Semantic image manipulation using scene graphs. In: CVPR, pp. 5213–5222 (2020)
4. Funke, I., Bodenstedt, S., Oehme, F., von Bechtolsheim, F., Weitz, J., Speidel, S.: Using 3D convolutional neural networks to learn spatiotemporal features for automatic surgical gesture recognition in video. In: Medical Image Computing and Computer Assisted Intervention (2019)
5. Gao, X., Jin, Y., Long, Y., Dou, Q., Heng, P.-A.: Trans-SVNet: accurate phase recognition from surgical videos via hybrid embedding aggregation transformer. In: de Bruijne, M., et al. (eds.) MICCAI 2021. LNCS, vol. 12904, pp. 593–603. Springer, Cham (2021). https://doi.org/10.1007/978-3-030-87202-1_57
6. Herzig, R., et al.: Object-region video transformers. In: Proceedings of the IEEE/CVF Conference on Computer Vision and Pattern Recognition (CVPR), pp. 3148–3159 (June 2022)
7. Hong, W.Y., Kao, C.L., Kuo, Y.H., Wang, J.R., Chang, W.L., Shih, C.S.: Cholecseg8k: a semantic segmentation dataset for laparoscopic cholecystectomy based on cholec80. arXiv preprint arXiv:2012.12453 (2020)
8. Islam, M., Seenivasan, L., Ming, L.C., Ren, H.: Learning and reasoning with the graph structure representation in robotic surgery. In: Martel, A.L., et al. (eds.) MICCAI 2020. LNCS, vol. 12263, pp. 627–636. Springer, Cham (2020). https://doi.org/10.1007/978-3-030-59716-0_60
9. Khan, S., Cuzzolin, F.: Spatiotemporal deformable scene graphs for complex activity detection. In: BMVC (2021)
10. Lea, C., Flynn, M.D., Vidal, R., Reiter, A., Hager, G.D.: Temporal convolutional networks for action segmentation and detection. In: proceedings of the IEEE Conference on Computer Vision and Pattern Recognition, pp. 156–165 (2017)
11. Madani, A., et al.: Artificial intelligence for intraoperative guidance: using semantic segmentation to identify surgical anatomy during laparoscopic cholecystectomy. Annals Surgery (2022)
12. Maier-Hein, L., et al.: Surgical data science for next-generation interventions. Nat. Biomed. Eng. **1**(9), 691–696 (2017)
13. Mascagni, P., et al.: Artificial intelligence for surgical safety: automatic assessment of the critical view of safety in laparoscopic cholecystectomy using deep learning. Annals Surgery (2021)
14. Materzynska, J., Xiao, T., Herzig, R., Xu, H., Wang, X., Darrell, T.: Somethingelse: compositional action recognition with spatial-temporal interaction networks. In: CVPR, pp. 1049–1059 (2020)
15. Murali, A., et al.: Latent graph representations for critical view of safety assessment. arXiv preprint arXiv:2212.04155 (2022)
16. Nwoye, C.I., et al.: Cholectriplet 2021: a benchmark challenge for surgical action triplet recognition. arXiv preprint arXiv:2204.04746 (2022)
17. Özsoy, E., Örnek, E.P., Eck, U., Czempiel, T., Tombari, F., Navab, N.: 4d-or: semantic scene graphs for or domain modeling. arXiv preprint arXiv:2203.11937 (2022)
18. Pang, W., Islam, M., Mitheran, S., Seenivasan, L., Xu, M., Ren, H.: Rethinking feature extraction: gradient-based localized feature extraction for end-to-end surgical downstream tasks. IEEE Robot. Autom. Lett. **7**(4), 12623–12630 (2022)
19. Raboh, M., Herzig, R., Berant, J., Chechik, G., Globerson, A.: Differentiable scene graphs. In: Proceedings of the IEEE/CVF Winter Conference on Applications of Computer Vision, pp. 1488–1497 (2020)
20. Ramesh, S., et al.: Dissecting self-supervised learning methods for surgical computer vision. arXiv preprint arXiv:2207.00449 (2022)

21. Ren, S., He, K., Girshick, R., Sun, J.: Faster r-cnn: towards real-time object detection with region proposal networks. In: Advances in Neural Information Processing Systems 28 (2015)
22. Sarikaya, D., Jannin, P.: Towards generalizable surgical activity recognition using spatial temporal graph convolutional networks. arXiv preprint arXiv:2001.03728 (2020)
23. Seenivasan, L., Mitheran, S., Islam, M., Ren, H.: Global-reasoned multi-task learning model for surgical scene understanding. IEEE Robot. Autom. Lett. **7**(2), 3858–3865 (2022)
24. Twinanda, A.P., Shehata, S., Mutter, D., Marescaux, J., De Mathelin, M., Padoy, N.: Endonet: a deep architecture for recognition tasks on laparoscopic videos. IEEE Trans. Med. Imaging **36**(1), 86–97 (2016)
25. Vercauteren, T., Unberath, M., Padoy, N., Navab, N.: Cai4cai: the rise of contextual artificial intelligence in computer-assisted interventions. Proc. IEEE **108**(1), 198–214 (2019)
26. Wang, X., Gupta, A.: Videos as space-time region graphs. In: Ferrari, V., Hebert, M., Sminchisescu, C., Weiss, Y. (eds.) ECCV 2018. LNCS, vol. 11209, pp. 413–431. Springer, Cham (2018). https://doi.org/10.1007/978-3-030-01228-1_25
27. Wu, J.Y., Tamhane, A., Kazanzides, P., Unberath, M.: Cross-modal self-supervised representation learning for gesture and skill recognition in robotic surgery. IJCARS **16**(5), 779–787 (2021)
28. Zhang, C., Gupta, A., Zisserman, A.: is an object-centric video representation beneficial for transfer? In: Proceedings of the Asian Conference on Computer Vision (ACCV), pp. 1976–1994 (December 2022)
29. Zhang, S., Tong, H., Xu, J., Maciejewski, R.: Graph convolutional networks: a comprehensive review. Comput. Soc. Netw. **6**(1), 1–23 (2019)

Deep Reinforcement Learning Based System for Intraoperative Hyperspectral Video Autofocusing

Charlie Budd[1]([✉]), Jianrong Qiu[2], Oscar MacCormac[1,3], Martin Huber[1],
Christopher Mower[1], Mirek Janatka[1,4], Théo Trotouin[1,4], Jonathan Shapey[1,4],
Mads S. Bergholt[2], and Tom Vercauteren[1,4]

[1] Biomedical Engineering and Imaging Science, King's College London, London, UK
charles.budd@kcl.ac.uk
[2] School of Craniofacial and Regenerative Biology, King's College London,
London, UK
[3] Department of Neurosurgery, King's College Hospitals, Denmark Hill, London, UK
[4] Hypervision Surgical Limited, 1st Floor 85 Great Portland Street, London, UK

Abstract. Hyperspectral imaging (HSI) captures a greater level of spectral detail than traditional optical imaging, making it a potentially valuable intraoperative tool when precise tissue differentiation is essential. Hardware limitations of current optical systems used for handheld real-time video HSI result in a limited focal depth, thereby posing usability issues for integration of the technology into the operating room. This work integrates a focus-tunable liquid lens into a video HSI exoscope, and proposes novel video autofocusing methods based on deep reinforcement learning. A first-of-its-kind robotic focal-time scan was performed to create a realistic and reproducible testing dataset. We benchmarked our proposed autofocus algorithm against traditional policies, and found our novel approach to perform significantly ($p < 0.05$) better than traditional techniques ($0.070 \pm .098$ mean absolute focal error compared to $0.146 \pm .148$). In addition, we performed a blinded usability trial by having two neurosurgeons compare the system with different autofocus policies, and found our novel approach to be the most favourable, making our system a desirable addition for intraoperative HSI.

Keywords: Autofocus · Deep Reinforcement Learning · Hyperspectral Imaging · Computer Assisted Intervention

1 Introduction

1.1 Background

Traditional optical imaging samples the visual spectrum in three diffuse spectral bands (RGB), while hyperspectral imaging (HSI) provides much more detailed

Supplementary Information The online version contains supplementary material available at https://doi.org/10.1007/978-3-031-43996-4_63.

spectral information. This information is potentially valuable for making intraoperative decisions, particularly in cases where tissue differentiation is critical but challenging to perform using traditional visualisation techniques. In the case of brain tumour excision, fluorescence-guided resection is commonly used to minimize damage to healthy tissue [2] but is limited to high-grade gliomas, and results in added cost and workflow disruptions. Thanks to a more detailed definition between tissue types [5], HSI is seen as a promising alternative with wider applicability and smoother integration into the workflow.

While HSI has been integrated into surgical microscope systems [11], it is suggested that handheld systems are better suited to translational research [4]. Such handheld systems consist of an exoscope coupled to a draped optical stack, as shown in Fig. 1. The optics in the exoscope typically result in a short focal depth, making manual focusing tricky, particularly as the tuning must be performed through the drape. As such, these systems are commonly left at a fixed focal power and the surgeon must keep the working distance fixed to keep the subject in focus. Furthermore, the narrow spectral bands of HSI sensors reduce the amount of light collected [12]. To avoid increasing exposure time, a large aperture size is needed, at a cost of further reducing focal depth. This exacerbates the focusing issues, making current real-time handheld HSI imaging systems particularly challenging to focus, posing significant usability issues. Figure 1 highlights the limited focal depth of our system, and shows a typical target that the surgeon must manually bring into focus during surgery.

The issue of reduced focal depth in real-time HSI systems could be mitigated by the introduction of a video autofocus system. Autofocus methods are divided into active methods, which use transmission to probe the scene, and passive methods, which rely only on incoming light. Passive methods are further split into phase-based, which require specialised hardware, and contrast-based, which compare images captured at different focal powers. Our investigation focuses on contrast-based methods, which require minimal hardware development.

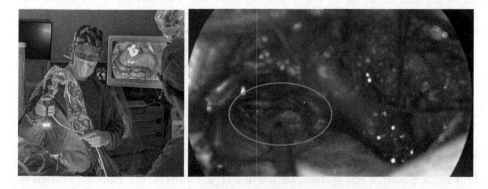

Fig. 1. Left) Existing fixed-focus HSI system being used during neurosurgery in an ethically approved study. Right) RGB reconstruction of an image taken with the fixed-focus HSI system following a craniotomy. The focus has been manually adjusted for the cavity visible through the craniotomy (circled).

1.2 Related Autofocusing Works

While autofocusing systems are prevalent in consumer device, the scientific literature is sparse, especially for dynamic video autofocusing. Many publications in the field are concerned with benchtop microscope autofocus systems [7,8,15]. This environment is conducive to autofocus as the scene is typically static with a single focal plane across the whole image. Additionally, the focus can be adjusted easily by moving the stage vertically. [8] take a traditional approach, making use of a Laplacian focal metric combined with a modified hill-climber optimisation scheme. [15] input a stack of sequential images to a 3d convolutional neural network (CNN) trained as a deep reinforcement agent trained to output changes in stage height. [7] train a CNN to regress the optimal focal power from just two images taken at different focal powers. Beyond benchtop microscopy, [6] also use a CNN to directly regress optimal focal powers, this time from varying number of samples from the full focal stacks. [1] take a novel approach by using pre-trained object detection models to generate latent vector representations of images and using these as inputs to a deep reinforcement agent. [14] train two CNNs, one to regress focal steps from a single image, the other to determine if the current image is in focus.

1.3 Contributions

This work aims to improve intraoperative handheld HSI systems by alleviating one of their main usability drawbacks, that of shortened focal depth. We introduce an autofocus system to an existing handheld intraoperative real-time HSI system [4]. The focus adjustments are handled by a focus tunable liquid lens which is integrated into the setup. We propose autofocusing policies based on deep reinforcement learning and compare these to traditional heuristic approaches. Our final model is similar to that presented in [15] but differs in its use of a weight shared image encoder, software simulated defocusing for training data, and small input patch size. In addition, our method is designed and trained to handle dynamic environments, something entirely missing in the literature. We performed a robotic focal-time scan to create a reproducible testing benchmark and allow quantitative comparison of autofocus policies. Finally, we demonstrate the utility of our approach in a blinded user study involving two neurosurgeons.

2 Materials and Methods

2.1 Optical System

Our intraoperative HSI system, shown in Fig. 2, builds on our existing system [4] by integrating an Optotune EL-10-30-Ci focus-tunable liquid lens to allow electrical control of the focal length. The hyperspectral camera is based on an IMEC 2/3" snapshot mosaic CMV2K-SSM4X4-VIS sensor, which acquires 16 spectral bands in a 4×4 mosaic between the spectral range of 460 nm and 600 nm. With

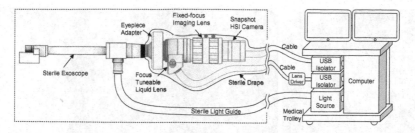

Fig. 2. Schematic diagram of our intraoperative video HSI system with focus-tunable liquid lens, allowing electrically controllable focal length. The handheld portion of the system is shown in the dashed line box.

a sensor resolution of 2048×1088 pixels, hyperspectral data is acquired with a spatial resolution of 512×272 pixels per spectral band. Video-rate imaging of snapshot data is achieved with a speed of up to 50 FPS depending on acquisition parameters.

2.2 Datasets

Software Simulated Focal-Time Scans. We define a focal-time scan as a time series of focal stacks, with a focal stack being a single image captured at multiple focal lengths. In order to assemble a large and diverse focal-time scan dataset, we choose to simulate focal-time scans using existing in-focus video data. To ensure the resulting focal-time scan features diverse camera motion, we implement a smooth random walk to step a cropping rectangle across the video after each frame. This also allows for the construction of plausible focal-time scans from single images, although features such as dynamic subjects or imaging noise will be missing. In order to simulate defocus, we implement another random walk to simulate a dynamic optimal focal power. When an agent is interacting with the simulated scan, a Gaussian filter is used to approximate focal blurring with $\sigma = \sigma_0 |f^* - f|$ where f and f^* are the current and optimal focal powers and σ_0 is chosen randomly from the range 2–8 for each scan. We use this technique to create a training and testing dataset consisting of 1000 and 200 simulated focal-time scans based on 200 10-second video clips sampled from Cholec80 [13], a popular endoscopic dataset. In addition, we created simulated focal-time scans from 200 in focus images taken of a brain phantom with our HSI system. These act as a validation dataset to help prevent over fitting and aid generalisation. While Gaussian blur is a reasonable approximation, we note that more rigours methods exist to simulate defocus blur that may produce better simulated data [9].

Robotic Focal-Time Scan. As a testing dataset similar to our intended use case, we chose to approximate a real focal-time scan by controlling conditions during capture of the individual focal stacks. Our optical system was fixed to

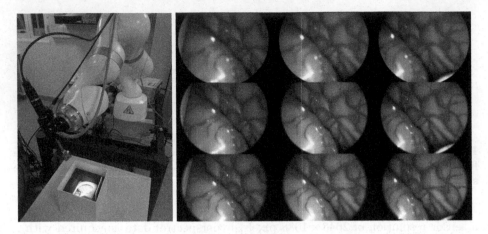

Fig. 3. Left) Robotic arm holding our optical system imaging a brain phantom. Right) Sample from our robotic focal-time scan, with the columns representing sequential focal stacks sampled at focal powers of 0.2, 0.5 and 0.8 (top to bottom). For low focal powers, the focal plane is behind the phantom (upper row). As the focal power increases, the focal plane intersects with the fissure (middle row), and then with the area surrounding the fissure (bottom row).

a robotic arm, which was then used in a compliant control mode to record a natural hand-guided trajectory whilst imaging a brain phantom. The motion was performed to try to emulate typical usage during a surgery, whilst also trying to cover the range of plausible working distances. The focal range of the liquid lens is discretised into a set of focal powers, and the recorded trajectory is discretised into a sequence of 1184 poses. For each discrete pose, the robotic arm is fixed, and an image captured for each focal power. We randomise the order of the focal powers to reduce systematic bias caused by the response of the liquid lens. Auto-exposure was implemented in order to ensure good exposure across all working distances. To ensure consistency within a given focal stack, auto-exposure was only stepped in-between discrete poses. The robotic arm holding our optical system and a sample of the resulting focal-time scan can be seen in Fig. 3. The optimal focus for all focal stacks was computed via global search of a traditional focal metric (mean gradient magnitude) as detailed below. This was then validated visually and corrected where appropriate.

Integration and Usability Trial. To ensure the validity of our quantitative evaluation, and to get feedback on the system in general, a blinded trial was set up with two practising neurosurgeons. A set was made containing two repeats of three selected autofocus policies. This set was then shuffled, and the surgeons remained blinded to the autofocus policy until after the trial. Each surgeon used our optical system to inspect a brain phantom with each policy in the set. The surgeon was made aware when the policy was changed and prompted to make comments throughout the trial, which were recorded.

2.3 Autofocus Policies

As seen in Fig. 1, the area of surgical interest can make up a rather small amount of the overall image, as such, we limit ourselves to a patch size of just 32×32 pixels. The positioning of the patch could be dictated by a second algorithm or user input, but this is outside the scope of this work. Here, we simply position the patch at the centre of the circular content area, which is detected using the method presented in [3]. All of our autofocus policies deal with the grayscale reconstruction of the HSI images. Throughout this work, we further deal with a normalised focal power range (0–1).

Traditional Approach. We implement two traditional autofocus policy based on different focal metrics combined with a simple hill-climber optimisation policy. We choose mean gradient magnitude (MGM) and mean local ratio (MLR). Two focal metrics which are conceptually simple but competitive [6] and implemented in quite different ways. They are defined as

$$\phi^{\text{MGM}}(I) = \frac{1}{n} \sum_p \sqrt{I_x^2(p) + I_y^2(p)} \tag{1}$$

$$\phi^{\text{MLR}}(I) = \frac{1}{n} \sum_p \max \left(\frac{G_\sigma(I)(p) + 1}{I(p) + 1}, \frac{I(p) + 1}{G_\sigma(I)(p) + 1} \right) \tag{2}$$

where p is the set of all pixels in the image, I_x and I_y are defined as the x and y responses of a Sobel filter, and G_σ is a Gaussian blur. The kernel size is chosen as $\sigma = 4$ for all our experiments. Our hill-climber optimisation policy O^{HC} sets the focal power f at time $t + 1$ based on information at time t and is defined as

$$f_{t+1} = O^{\text{HC}}(\phi_t, f_t, \phi_{t-1}) = \begin{cases} f_t + d_{prev}h, & \text{if } 0 < f_t < 1 \text{ and } \phi_t > \phi_{t-1} \\ f_t - d_{prev}h, & \text{otherwise} \end{cases} \tag{3}$$

where $d_{prev} = \text{sign}(f_t - f_{t-1})$ is the direction of the previous step and h is a step size which we set to $h = 0.05$ for all our experiments. We note that our definition is different from standard hill-climber. A normal hill-climber will repeat a step while the focal metric is increasing, and either stop or change direction with a smaller step size when the focal metric decreases, but this does not translate to a continuous and dynamic environment.

Learned Optimisation Policy. Due to our dynamic environment, it seems likely that considering a sequence of the N last focal metrics, rather than the last two, would help to build a strong optimisation policy. However, as N increases, it quickly becomes unclear how to incorporate this information effectively. It is likely that a learning based solution would uncover a better strategy than heuristic approaches. While regression based approcahes may work, reinforcement learning provides a natural framework for this problem by allowing the policy to model the trade-off between maximisation and exploration. By modelling the autofocus task as a Markov process, we can define a Q-function $Q(s, a)$ which maps state-action pairs to expected future rewards. We define our state, actions, and reward function as

$$s_t = \{\phi_t, f_t, ..., \phi_{t-(N-1)}, f_{t-(N-1)}\}$$
$$A = \{-h, 0, +h\}$$
$$r_t = -|f_t^* - f_t|$$

where f_t^* is the optimal focal power at t which can only be known in controlled environment. As before, we take $h = 0.05$. Our learned optimisation policy O_{RL} can then be defined as

$$f_{t+1} = O^{\mathrm{RL}}(f_t, s_t) = f_t + \max_a Q(s_t, a) \tag{4}$$

To model $Q(s, a)$, we use an MLP consisting of 2 hidden layers of 256 ReLUs each and a third layer with 3 outputs corresponding to the 3 possible actions. The MLP takes as input the state vector s containing the N most recent focal metrics and focal powers, we take $N = 8$ for all our experiments. To train the model, we use Deep Q Learning following the recommendations set out by the DQN method [10] to improve training stability. We use an experience memory with size 2.5×10^6, and an ϵ-greedy exploration policy where ϵ exponentially decays from 1.0 to 0.1 over the first 2×10^6 experiences. Our target model is updated with exponential moving average (EMA) weight updates with a $\beta = 0.005$, and we use $\gamma = 0.99$ in our Bellman equation. Finally, we use a smoothed L1 loss function and optimise with RMSProp with learning rate 1×10^{-5} and momentum 0.95. We trained on our software simulated focal-time scans created from real endoscopy videos and validated against our simulated focal-time scans created with HSI images taken with our optical system mounted on a robotic arm.

End-to-End Model. In addition to learning the optimisation policy, we can also learn the focal metric. By learning the two together, we are no longer constrained to a scalar metric and can instead learn a latent vector encoding of the image patches. To do this, we construct a CNN consisting of 4 convolutions with 8 filters each and a stride of 2, outputting a vector of 8 logits for our patch size of 32×32. The CNN is run on each of the N most recent image patches as a batch during training, but only the most recent during inference, with the previous encodings stored between steps. The encodings are concatenated with the N most recent focal powers and fed into an MLP. The MLP and training procedure are the same as before.

3   Results

We evaluated each autofocus policies on both our simulated focal-time scan test set, and the robotically recorded focal-time scan. The mean focal errors are shown in Table 1. The scores show an improvement in almost all cases by the introduction of a learned optimiser. The paths taken and the focal error over time for the robotic focal-time scan for a selection of policies are plotted in Fig. 4.

During the usability trial, the surgeon participants were positive about all presented policies. In line with our quantitative results, the participants both showed preference for the CNN-based policy. It was thought by both to be smoother and more deliberate in its adjustments, and felt more robust to minor accidental motions inherent to hand-operated system. One commented that it felt slower to focus but more stable, going on to state that this was desirable behaviour. All algorithms handled the brain fissure well, this is likely due to the small patch size used, allowing for precise targeting. Overall, the surgeons were very positive about the integration of autofocus into optical imaging systems.

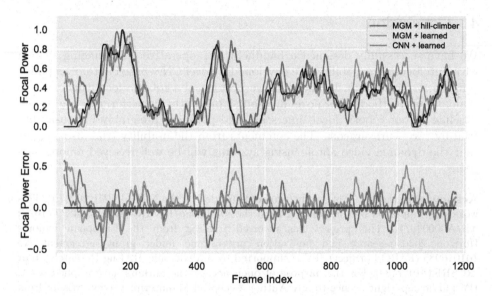

Fig. 4. Focal path (top) and error in focal power (bottom) for three autofocus policies on the robotic focal-time scan. The optimal focal power is shown in black. All paths have been smoothed with a moving average with a window of 5 frames for visualisation purposes.

Table 1. Mean absolute focal power error (0 to 1), and the percentage of in focus frames (focal power error < 0.1), for different autofocus policies on both the simulated and robotic focal-time scan testing sets.

Autofocus policy		Focal power error (MAE)		In focus frames	
Metric	Optimiser	Simulated	Robotic	Simulated	Robotic
n/a	fixed	0.236 ± .152	0.262 ± .161	19.0%	26.4%
MGM	hill-climber	0.102 ± .138	0.146 ± .148	67.9%	46.4%
MLR	hill-climber	0.092 ± .118	0.163 ± .168	68.2%	44.1%
MGM	learned	0.085 ± .115	0.126 ± .118	70.4%	50.8%
MLR	learned	0.098 ± .120	0.156 ± .131	66.4%	39.5%
CNN	learned	**0.049 ± .072**	**0.070 ± .099**	**84.9%**	**79.1%**

4   Conclusion

We have successfully designed a handheld intraoperative HSI imaging system with autofocusing capability. We developed a novel CNN-based autofocus policy suitable for video data. In addition, we performed a robotic focal-time scan to evaluate our methods. Our novel method significantly outperforms a traditional baseline on our robotic focal-time scan, and performs preferably in a usability trial by two neurosurgeons. The comments from the usability trial also suggest that the dynamic video autofocusing systems will be well received among surgeons.

Acknowledgements. This study/project is funded by the NIHR [NIHR202114]. This work was supported by core funding from the Wellcome/EPSRC [WT203148/Z/16/Z; NS/A000049/1]. This project has received funding from the European Union's Horizon 2020 research and innovation programme under grant agreement No 101016985 (FAROS project). TV is supported by a Medtronic/RAEng Research Chair [RCSRF1819\7\34]. For the purpose of open access, the authors have applied a CC BY public copyright licence to any Author Accepted Manuscript version arising from this submission. TV is a co-founder and shareholder of Hypervision Surgical.

References

1. Anikina, A., Rogov, O.Y., Dylov, D.V.: Dasha: decentralized autofocusing system with hierarchical agents. arXiv preprint arXiv:2108.12842 (2021)
2. Bogaards, A., et al.: Increased brain tumor resection using fluorescence image guidance in a preclinical model. Lasers Surg. Med. **35**(3), 181–190 (2004)
3. Budd, C., Herrera, L.C.G.P., Huber, M., Ourselin, S., Vercauteren, T.: Rapid and robust endoscopic content area estimation: a lean gpu-based pipeline and curated benchmark dataset. In: Computer Methods in Biomechanics and Biomedical Engineering: Imaging & Visualization, pp. 1–10 (2023)

4. Ebner, M., et al.: Intraoperative hyperspectral label-free imaging: from system design to first-in-patient translation. J. Phys. D Appl. Phys. **54**(29) (2021). https://doi.org/10.1088/1361-6463/abfbf6. https://www.scopus.com/inward/record.url?scp=85107008535&partnerID=8YFLogxK

5. Halicek, M., Fabelo, H., Ortega, S., Callico, G.M., Fei, B.: In-vivo and ex-vivo tissue analysis through hyperspectral imaging techniques: revealing the invisible features of cancer. Cancers **11**(6) (2019). https://doi.org/10.3390/cancers11060756. https://www.mdpi.com/2072-6694/11/6/756

6. Herrmann, C., et al.: Learning to autofocus. In: Proceedings of the IEEE/CVF Conference on Computer Vision and Pattern Recognition (CVPR) (2020)

7. Ivanov, T., Kumar, A., Sharoukhov, D., Ortega, F., Putman, M.: DeepFocus: a deep learning model for focusing microscope systems. In: Zelinski, M.E., Taha, T.M., Howe, J., Awwal, A.A.S., Iftekharuddin, K.M. (eds.) Applications of Machine Learning 2020, vol. 11511, p. 1151103. International Society for Optics and Photonics, SPIE (2020). https://doi.org/10.1117/12.2568990

8. Jia, D., Zhang, C., Wu, N., Zhou, J., Guo, Z.: Autofocus algorithm using optimized laplace evaluation function and enhanced mountain climbing search algorithm. Multimedia Tools Appl. **81**(7), 10299–10311 (2022)

9. Liu, Y.Q., Du, X., Shen, H.L., Chen, S.J.: Estimating generalized gaussian blur kernels for out-of-focus image deblurring. IEEE Trans. Circ. Syst. Video Technol. **31**(3), 829–843 (2021). https://doi.org/10.1109/TCSVT.2020.2990623

10. Mnih, V., et al.: Playing atari with deep reinforcement learning. arXiv preprint arXiv:1312.5602 (2013)

11. Pichette, J.: Intraoperative video-rate hemodynamic response assessment in human cortex using snapshot hyperspectral optical imaging. Neurophotonics **3**(4), 045003 (2016)

12. Shapey, J., et al.: Intraoperative multispectral and hyperspectral label-free imaging: a systematic review of in vivo clinical studies. J. Biophoton. **12**(9), e201800455 (2019)

13. Twinanda, A.P., Shehata, S., Mutter, D., Marescaux, J., de Mathelin, M., Padoy, N.: Endonet: a deep architecture for recognition tasks on laparoscopic videos. CoRR abs/1602.03012 (2016). https://arxiv.org/abs/1602.03012

14. Wang, C., Huang, Q., Cheng, M., Ma, Z., Brady, D.J.: Deep learning for camera autofocus. IEEE Trans. Comput. Imaging **7**, 258–271 (2021). https://doi.org/10.1109/TCI.2021.3059497

15. Yu, X., Yu, R., Yang, J., Duan, X.: A robotic auto-focus system based on deep reinforcement learning. In: 2018 15th International Conference on Control, Automation, Robotics and Vision (ICARCV), pp. 204–209. IEEE (2018)

Towards Multi-modal Anatomical Landmark Detection for Ultrasound-Guided Brain Tumor Resection with Contrastive Learning

Soorena Salari[1]([✉]), Amirhossein Rasoulian[1], Hassan Rivaz[2], and Yiming Xiao[1]

[1] Department of Computer Science and Software Engineering,
Concordia University, Montreal, Canada
{soorena.salari,ah.rasoulian,yiming.xiao}@concordia.ca
[2] Department of Electrical and Computer Engineering,
Concordia University, Montreal, Canada
hassan.rivaz@concordia.ca

Abstract. Homologous anatomical landmarks between medical scans are instrumental in quantitative assessment of image registration quality in various clinical applications, such as MRI-ultrasound registration for tissue shift correction in ultrasound-guided brain tumor resection. While manually identified landmark pairs between MRI and ultrasound (US) have greatly facilitated the validation of different registration algorithms for the task, the procedure requires significant expertise, labor, and time, and can be prone to inter- and intra-rater inconsistency. So far, many traditional and machine learning approaches have been presented for anatomical landmark detection, but they primarily focus on mono-modal applications. Unfortunately, despite the clinical needs, inter-modal/contrast landmark detection has very rarely been attempted. Therefore, we propose a novel contrastive learning framework to detect corresponding landmarks between MRI and intra-operative US scans in neurosurgery. Specifically, two convolutional neural networks were trained jointly to encode image features in MRI and US scans to help match the US image patch that contain the corresponding landmarks in the MRI. We developed and validated the technique using the public RESECT database. With a mean landmark detection accuracy of 5.88 ± 4.79 mm against 18.78 ± 4.77 mm with SIFT features, the proposed method offers promising results for MRI-US landmark detection in neurosurgical applications for the first time.

Keywords: Deep learning · Anatomical landmark · Contrastive learning · Inter-modality · Neurosurgery · Intraoperative ultrasound

1 Introduction

Gliomas are the most common central nervous system (CNS) tumors in adults, accounting for 80% of primary malignant brain tumors [1]. Early surgical treatment to remove the maximum amount of cancerous tissues while preserving the eloquent brain regions can improve the patient's survival rate and functional outcomes of the procedure [2]. Although the latest multi-modal medical imaging (e.g., PET, diffusion/functional MRI) allows more precise pre-surigcal planning, during surgery, brain tissues can deform under multiple factors, such as gravity, intracranial pressure change, and drug administration. The phenomenon is referred to as brain shift, and often invalidates the pre-surgical plan by displacing surgical targets and other vital anatomies. With high flexibility, portability, and cost-effectiveness, intra-operative ultrasound (US) is a popular choice to track and monitor brain shift. In conjunction with effective MRI-US registration algorithms, the tool can help update the pre-surgical plan during surgery to ensure the accuracy and safety of the intervention.

As the true underlying deformation from brain shift is impossible to obtain and the differences of image features between MRI and US are large, quantitative validation of automatic MRI-US registration algorithms often rely on homologous anatomical landmarks that are manually labeled between corresponding MRI and intra-operative US scans [3]. However, manual landmark identification requires strong expertise in anatomy and is costly in labor and time. Moreover, inter- and intra-rater variability still exists. These factors make quality assessment of brain shift correction for US-guided brain tumor resection challenging. In addition, due to the time constraints, similar evaluation of inter-modal registration quality during surgery is nearly impossible, but still highly desirable. To address these needs, deep learning (DL) holds the promise to perform efficient and automatic inter-modal anatomical landmark detection.

Previously, many groups have proposed algorithms to label landmarks in anatomical scans [4–9]. However, almost all earlier techniques were designed for mono-modal applications, and inter-modal landmark detection, such as for US-guided brain tumor resection, has rarely been attempted. In addition, unlike other applications, where the full anatomy is visible in the scan and all landmarks have consistent spatial arrangements across subjects, intra-operative US of brain tumor resection only contains local regions of the pathology with non-canonical orientations. This results in anatomical landmarks with different spatial distributions across cases. To address these unique challenges, we proposed a new contrastive learning (CL) framework to detect matching landmarks in intra-operative US with those from MRI as references. Specifically, the technique leverages two convolutional neural networks (CNNs) to learn features between MRI and US that distinguish the inter-modal image patches which are centered at the matching landmarks from those that are not. Our approach has two major novel contributions to the field. First, we proposed a multi-modal landmark detection algorithm for US-guided brain tumor resection for the first time. Second, CL is employed for the first time in inter-modal anatomical landmark detection. We developed and validated the proposed technique with the public RESECT database [10] and compared its landmark detection accuracy against the popular scale-invariant feature transformation (SIFT) algorithm in 3D [11].

2   Related Work

Contrastive learning has recently shown great results in a wide range of medical image analysis tasks [12–18]. In short, it seeks to boost the similarity of feature representations between counterpart samples and decrease those between mismatched pairs. Often, these similarities are calculated based on deep feature representations obtained from DL models in the feature embedding space. This self-supervised learning set-up allows robust feature learning and embedding without explicit guidance from fine-grained image annotations, and the encoded features can be adopted in various downstream tasks, such as segmentation. A few recent works [19–21] explored the potential of CL in anatomical landmark annotation in head X-ray images for 2D skull landmarks. Quan et al. [19,20] attempted to leverage CL for more efficient and robust learning. Yao et al. [21] used multiscale pixel-wise contrastive proxy tasks for skull landmark detection in X-ray images. With a consistent protocol for landmark identification, they trained the network to learn signature features within local patches centered at the landmarks. These prior works with CL focus on single-modal 2D landmark identification with systematic landmark localization protocols and sharp image contrast (i.e., skull in X-ray). In contrast, our described application is more challenging due to the 3D nature, difficulty in inter-modal feature learning, weaker anatomical contrast (i.e., MRI vs US), and variable landmark locations. In CL, many works have employed the InfoNCE loss function [22,23] in attaining good outcomes. Inspired by Yao *et al.* [21], we aimed to use InfoNCE as our loss function with a patch-based approach. To date, CL has not been explored in multi-modal landmark detection, a unique problem in clinical applications. In this paper, to bridge this knowledge gap, we proposed a novel CL-based framework for MRI-US anatomical landmark detection.

3   Methods and Materials

3.1   Data and Landmark Annotation

We employed the publicly available EASY-RESECT (REtroSpective Evaluation of Cerebral Tumors) dataset [10] (https://archive.sigma2.no/pages/public/dataset Detail.jsf?id=10.11582/2020.00025) to train and evaluate our proposed method. This dataset is a deep-learning-ready version of the original RESECT database, and was released as part of the 2020 Learn2Reg Challenge [24]. Specifically, EASY-RESECT contains MRI and intra-operative US scans (before resection) of 22 subjects who have undergone low-grade glioma resection surgeries. All images were resampled to a unified dimension of 256 × 256 × 288 voxels, with an isotropic resolution of ∼0.5mm. Between MRI and the corresponding US images, matching anatomical landmarks were manually labeled by experts and 15∼16 landmarks were available per case. A sample illustration of corresponding inter-modal scans and landmarks is shown in Fig. 1. For the target application, we employed the T2FLAIR MRI to pair with intra-operative US since low-grade gliomas are usually more discernible in T2FLAIR than in T1-weighted MRI[10].

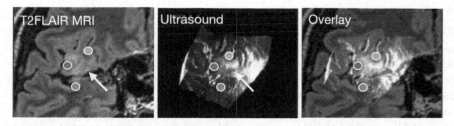

Fig. 1: Sample corresponding landmarks on co-registered T2FLAIR MRI and US. The arrows point to the brain tumor region.

3.2   Contrastive Learning Framework

We used two CNNs with identical architectures in parallel to extract robust image features from MRI and US scans. Specifically, these CNNs are designed to acquire relevant features from MRI and US patches, and maximize the similarity between features of corresponding patches while minimizing those between mismatched patches. Each CNN network contains six successive blocks, and each block consists of one convolution layer and one group norm, with Leaky ReLU as the activation function. Also, the convolution layer of the first and last three blocks of the network has 64 and 32 convolutional filters, respectively, and a kernel size of 3 is used across all blocks. After the convolution layers, the proposed network has two multi-layer perceptron (MLP) layers with 64 and 32 neurons and Leaky ReLU as the activation function. These MLP layers compress the extracted features from convolutional layers and produce the final feature vectors. The resulting CNN network is depicted in Fig. 2.

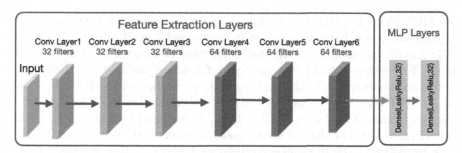

Fig. 2: The proposed CNN for feature encoding from MRI and US scans.

3.3   Landmark Matching with a 2.5D Approach

Working with 3D images is computationally expensive and can make the model training unstable and prone to overfitting, especially when the size of the database is limited. Therefore, instead of a full 3D processing, we decided to implement a 2.5D approach [25] to leverage the efficiency of 2D CNN in the

672   S. Salari et al.

CL framework for the task. In this case, we extracted a series of three adjacent 2D image patches in one canonical direction (x-, y-, or z-direction), with the middle slice centred at the true or candidate landmarks in a 3D scan to provide slight spatial context for the middle slice of interest. To construct the full 2.5D formulation, we performed the same image patch series extraction in all x-, y-, and z-directions for a landmark, and this 2.5D patch forms the basis to compute the similarity between the queried US and reference MRI patches. Note that the setup of CL requires three types of samples, anchor, positive sample pairs, and negative sample pairs. Specifically, the anchor is defined as the 2.5D MRI patch centred at a predefined landmark, a positive pair is represented by an anchor and a 2.5D US patch at the corresponding landmark, and finally, a negative pair means an anchor and a mismatched 2.5D US patch. Note that during network training, instead of 2.5D patches, we compared the 2D image patch series in one canonical direction between MRI and US, and 2D patch series in all three directions were used. During the inference stage, the similarity between MRI and US 2.5D patches was obtained by summing the similarities of corresponding 2D image patch series in each direction, and a match was determined with the highest similarity from all queried US patches. With the assumption that the brain shift moves the anatomy within a limited range, during the inference time, we searched within a range of [-5,5] mm in each direction in the US around the reference MRI landmark location to find the best match. Note that this search range is an adjustable parameter by the user (e.g., surgeons/clinicians), and when no match is found in the search range, an extended search range can be used. The general overview of the utilized framework for 2D image patch extraction is shown in Fig. 3.

3.4   Landmark Matching with 3D SIFT

The SIFT algorithm [11] is a well-known tool for keypoint detection and image registration. It has been widely used in multi-modal medical registration, such as landmark matching for brain shift correction in image-guided neurosurgery [8,26]. To further validate the proposed CL-based method for multi-modal anatomical landmark detection in US scans, we replicated the procedure using the 3D SIFT algorithm as follows. First, we calculated the SIFT features at the reference landmark's location in MRI. Then, we acquired a set of candidate SIFT points in the corresponding US scan. Finally, we identified the matching US landmark by selecting the top ranking candidate based on SIFT feature similarity measured with cosine similarity. Note that, for SIFT-based landmark matching, we have attempted to impose a similar spatial constraint like in the CL-based approach. However, as the SIFT algorithm pre-selects keypoint candidates based on their feature strengths, with this in consideration, we saw no major benefits by imposing the spatial constraint.

4   Experimental Setup

4.1   Data Preprocessing

For CL training, both positive and negative sample pairs need to be created. All 2D patch series were extracted according to Sect. 3.3 with a size of $42 \times 42 \times 3$ voxels. These sample pairs were used to train two CNNs to extract relevant image features across MRI and US leveraging the InfoNCE loss.

4.2   Loss Function

We used the InfoNCE loss [23] for our CL framework. The loss function has been widely used and demonstrated great performance in many vision tasks. Like other contrastive loss functions, InfoNCE requires a similarity function, and we chose commonly used cosine similarity. The formulas for InfoNCE ($L_{InfoNCE}$) and cosine similarity ($CosSim$) are as follows:

$$
\begin{aligned}
\mathcal{L}_{\text{InfoNCE}} &= -\mathbb{E}\left[\log \frac{\exp(\alpha)}{\exp(\alpha) + \sum \exp(\alpha')}\right]; \\
\alpha &= s[F_\theta \circ X_i^A, G_\beta \circ X_i^P]; \\
\alpha' &= s[F_\theta \circ X_i^A, G_\beta \circ X_j^N],
\end{aligned}
\tag{1}
$$

$$
s\,[v, w] = \text{Cos Sim}\,(v, w) = \frac{\langle v \cdot w \rangle}{\|v\| \cdot \|w\|}
\tag{2}
$$

where F_θ and G_β are the CNN feature extractors for MR and US patches. X_i^A and X_i^P are the cropped image patches around the corresponding landmarks in MR and US scans, respectively, and X_i^N is a mismatched patch in the US image to that cropped around the MRI reference landmark. Here, $F_\theta \circ X_i^A$, $G_\beta \circ X_i^P$, and $G_\beta \circ X_j^N$ give the extracted feature vectors for MR and US patches.

4.3   Implementation Details and Evaluation

To train our DL model, we made subject-wise division of the entire dataset into 70%:15%:15% as the training, validation, and testing sets, respectively. Also, to improve the robustness of the network, we used data augmentation for the training data by random rotation, random horizontal flip, and random vertical flip. Furthermore, an AdamW optimizer with a learning rate of 0.00001 was used, and we trained our model for 50 epochs with a batch size of 256.

In order to evaluate the performance of our technique, we used the provided ground truth landmarks from the database and calculated the Euclidean distance between the ground truths and predictions. The utilized metric is as follows:

$$
\text{Mean landmark identification error} = \frac{1}{N}\sum_{i=1}^{N}\|x_i - x_i'\|
\tag{3}
$$

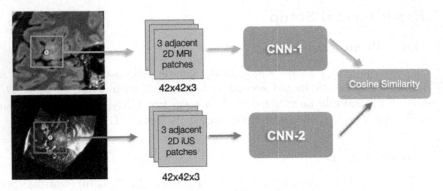

Fig. 3: An overview of the framework for image feature learning.

where x_i and x'_i, and N are the ground truth landmark location, model prediction, and the total number of landmarks per subject, respectively.

5   Results

Table 1 lists the mean and standard deviation of landmark identification errors (in mm) between the predicted position and the ground truth in intra-operative US for each patient of the RESECT dataset. In the table, we also provide the severity of brain shift for each patient. Here, tissue deformation measured as mean target registration errors (mTREs) with the ground truth anatomical landmarks is classified as small (mTRE below 3 mm), median (3-6 mm), or large (above 6 mm). The results show that our CL-based landmark selection technique can locate the corresponding US landmarks with a mean landmark identification error of 5.88±4.79 mm across all cases while the SIFT algorithm has an error 18.78±4.77 mm. With a two-sided paired-samples t-test, our method outperformed the SIFT approach with statistical significance (p <1e-4). When reviewing the mean landmark identification error using our proposed technique, we also found that the magnitude is associated with the level of brain shift. However, no such trend is observed when using SIFT features for landmark identification. When inspecting landmark identification errors across all subjects between the CL and SIFT techniques, we also noticed that our CL framework has significantly lower standard deviations (p <1e-4), implying that our technique has a better performance consistency.

6   Discussion

Inter-modal anatomical landmark localization is still a difficult task, especially for the described application, where landmarks have no consistent spatial arrangement across different cases and image features in US are rough. We tackled the challenge with the CL framework for the first time. As the first step

Table 1: Landmark identification errors (mean±std) per case in mm. Our proposed CL-based algorithm achieved a mean landmark identification error of **5.88±4.79** mm across all cases while the SIFT algorithm obtained an error of **18.78±4.77** mm. The level of brain shift is listed beside the patient ID.

Patient ID	Proposed CL	SIFT Algorithm	Patient ID	Proposed CL	SIFT Algorithm
1 (Small)	1.80 ± 0.78	12.16 ± 4.75	15 (Medium)	3.07 ± 1.39	19.33 ± 4.49
2 (Medium)	6.16 ± 1.40	17.84 ± 8.50	16 (Medium)	6.42 ± 0.92	12.48 ± 4.84
3 (Large)	7.79 ± 0.55	13.37 ± 6.08	17 (Large)	8.13 ± 0.72	18.57 ± 6.27
4 (Small)	3.59 ± 0.73	13.97 ± 5.50	18 (Medium)	4.19 ± 0.87	13.71 ± 5.815
5 (Large)	10.65 ± 1.13	20.71 ± 6.12	19 (Medium)	3.97 ± 0.93	27.70 ± 16.49
6 (Medium)	2.20 ± 0.93	28.11 ± 11.90	21 (Medium)	6.01 ± 0.77	24.40 ± 13.73
7 (Small)	1.96 ± 0.96	24.07 ± 8.02	23 (Large)	6.97 ± 1.03	22.50 ± 6.72
8 (Small)	2.56 ± 0.79	19.30 ± 6.09	24 (Small)	1.33 ± 0.49	14.91 ± 6.09
12 (Large)	23.77 ± 0.96	22.01 ± 6.64	25 (Large)	9.94 ± 2.43	15.37 ± 5.42
13 (Medium)	6.18 ± 1.43	13.86 ± 6.99	26 (Small)	2.95 ± 0.88	17.93 ± 10.15
14 (Medium)	3.39 ± 0.69	21.67 ± 6.46	27 (Medium)	6.42 ± 0.76	19.20 ± 8.65

towards more accurate inter-modal landmark localization, there are still aspects to be improved. First, while the 2.5D approach is memory efficient and quick, 3D approaches may better capture the full corresponding image features. This is partially reflected by the observation that the quality of landmark localization is associated with the level of tissue shift. However, due to limited clinical data, 3D approaches caused overfitting in our network training. Second, in the current setup, we employed landmarks in pre-operative MRIs as references since its contrast is easier to understand and it allows sufficient time for clinicians to annotate the landmarks before surgery. Future exploration will also seek techniques to automatically tag MRI reference landmarks. Finally, we only employed US scans before resection since tissue removal can further complicate feature matching between MRI and US, and requires more elaborate strategies, such as those involving segmentation of resected regions [27]. We will explore suitable solutions to extend the application scenarios of our proposed framework as part of the future investigation. As a baseline comparison, we employed the SIFT algorithm, which has demonstrated excellent performance in a large variety of computer vision problems for keypoint matching. However, in the described inter-modal landmark identification for US-guided brain tumor resection, the SIFT algorithm didn't offer satisfactory results. This could be due to the coarse image features and textures of intra-operative US and the differences in the physical resolution between MRI and US. One major critique for using the SIFT algorithm is that it intends to find geometrically interesting keypoints, which may not have good anatomical significance. In the RESECT dataset, eligible anatomical landmarks were defined as deep grooves and corners of sulci, convex points of gyri, and vanishing points of sulci. The relevant local features may be hard to capture with the SIFT algorithm. In this sense, DL-based approaches may be a better choice for the task. With the CL framework, our method learns the common features between two different modalities via the training process. Besides better landmark identification accuracy, the tighter standard deviations also imply that our DL approach serves a better role in grasping the local image features within the image patches.

7   Conclusions

In this project, we proposed a CL framework for MRI-US landmark detection for neurosurgery for the first time by leveraging real clinical data, and achieved state-of-the-art results. The algorithm represents the first step towards efficient and accurate inter-modal landmark identification that has the potential to allow intra-operative assessment of registration quality. Future extension of the method in other inter-modal applications can further confirm its robustness and accuracy.

Acknowledgment. We acknowledge the support of the Natural Sciences and Engineering Research Council of Canada (NSERC) and Fonds de Recherche du Québec Nature et technologies (FRQNT).

References

1. Holland, E.C.: Progenitor cells and glioma formation. Curr. Opin. Neurol. **14**(6), 683–688 (2001)
2. Dolecek, T.A., Propp, J.M., Stroup, N.E., Kruchko, C.: CBTRUS statistical report: primary brain and central nervous system tumors diagnosed in the united states in 2005–2009. Neuro-oncology **14**(suppl_5), v1–v49 (2012)
3. Xiao, Y., et al.: Evaluation of MRI to ultrasound registration methods for brain shift correction: the CuRIOUS2018 challenge. IEEE Trans. Med. Imaging **39**(3), 777–786 (2019)
4. Yao, Q., Xiao, L., Liu, P., Zhou, S.K.: Label-free segmentation of COVID-19 lesions in lung CT. IEEE Trans. Med. Imaging **40**(10), 2808–2819 (2021)
5. Ghesu, F.C., Georgescu, B., Mansi, T., Neumann, D., Hornegger, J., Comaniciu, D.: An artificial agent for anatomical landmark detection in medical images. In: Ourselin, S., Joskowicz, L., Sabuncu, M.R., Unal, G., Wells, W. (eds.) MICCAI 2016. LNCS, vol. 9902, pp. 229–237. Springer, Cham (2016). https://doi.org/10.1007/978-3-319-46726-9_27
6. Zhu, H., Yao, Q., Xiao, L., Zhou, S.K.: You only learn once: universal anatomical landmark detection. In: de Bruijne, M., et al. (eds.) MICCAI 2021. LNCS, vol. 12905, pp. 85–95. Springer, Cham (2021). https://doi.org/10.1007/978-3-030-87240-3_9
7. Tripathi, A., et al.: Unsupervised landmark detection and classification of lung infection using transporter neural networks. Comput. Biol. Med. **152**, 106345 (2023)
8. Toews, M., Wells, W.M., III.: Efficient and robust model-to-image alignment using 3d scale-invariant features. Med. Image Anal. **17**(3), 271–282 (2013)

9. Salari, S., Rasoulian, A., Battie, M., Fortin, M., Rivaz, H., Xiao, Y.: Uncertainty-aware transformer model for anatomical landmark detection in paraspinal muscle MRIs. In: Medical Imaging,: Image Processing, vol. 12464, pp. 238–244. SPIE (2023)

10. Xiao, Y., Fortin, M., Unsgård, G., Rivaz, H., Reinertsen, I.: Re trospective evaluation of cerebral tumors (RESECT): a clinical database of pre-operative MRI and intra-operative ultrasound in low-grade glioma surgeries. Med. Phys. **44**(7), 3875–3882 (2017)

11. Rister, B., Horowitz, M.A., Rubin, D.L.: Volumetric image registration from invariant keypoints. IEEE Trans. Image Process. **26**(10), 4900–4910 (2017)

12. You, K., Lee, S., Jo, K., Park, E., Kooi, T., Nam, H.: Intra-class contrastive learning improves computer aided diagnosis of breast cancer in mammography. In: Medical Image Computing and Computer Assisted Intervention-MICCAI: 25th International Conference, Singapore, 18–22 September 2022, Proceedings, Part III, pp. 55–64. Springer (2022). https://doi.org/10.1007/978-3-031-16437-8_6

13. Cheng, L.-H., Sun, X., van der Geest, R.J.: Contrastive learning for echocardiographic view integration. In: Medical Image Computing and Computer Assisted Intervention-MICCAI,: 25th International Conference, Singapore, v September 2022, Proceedings, Part IV, pp. 340–349. Springer (2022). https://doi.org/10.1007/978-3-031-16440-8_33

14. Bhattacharya, D., et al.: Supervised contrastive learning to classify paranasal anomalies in the maxillary sinus. In: Medical Image Computing and Computer Assisted Intervention-MICCAI, et al.: 25th International Conference, Singapore, 18–22 September 2022, Proceedings, Part III, pp. 429–438. Springer (2022). https://doi.org/10.1007/978-3-031-16437-8_41

15. Emre, T., Chakravarty, A., Rivail, A., Riedl, S., Schmidt-Erfurth, U., Bogunović, H.: Tinc: temporally informed non-contrastive learning for disease progression modeling in retinal OCT volumes. In: Medical Image Computing and Computer Assisted Intervention-MICCAI,: 25th International Conference, Singapore, 18–22 September 2022, Proceedings, Part II, pp. 625–634. Springer (2022). https://doi.org/10.1007/978-3-031-16434-7_60

16. Liu, T., Liu, W., Yu, L., Wan, L., Han, T., Zhu, L.: Joint prediction of meningioma grade and brain invasion via task-aware contrastive learning. In: Medical Image Computing and Computer Assisted Intervention-MICCAI,: 25th International Conference, Singapore, 18–22 September 2022, Proceedings, Part III, pp. 355–365. Springer (2022). https://doi.org/10.1007/978-3-031-16437-8_34

17. Pan, Y., Gernand, A.D., Goldstein, J.A., Mithal, L., Mwinyelle, D., Wang, J.Z.: Vision-language contrastive learning approach to robust automatic placenta analysis using photographic images. In: Medical Image Computing and Computer Assisted Intervention-MICCAI: 25th International Conference, Singapore, September 18–22, 2022, Proceedings, Part III, pp. 707–716. Springer (2022). https://doi.org/10.1007/978-3-031-16437-8_68

18. Hang, W., Huang, Y., Liang, S., Lei, B., Choi, K.-S., Qin, J.: Reliability-aware contrastive self-ensembling for semi-supervised medical image classification. In: Medical Image Computing and Computer Assisted Intervention-MICCAI,: 25th International Conference, Singapore, 18–22 September 2022, Proceedings, Part I, pp. 754–763. Springer (2022). https://doi.org/10.1007/978-3-031-16431-6_71

19. Quan, Q., Yao, Q., Li, J., Zhou, S.K.: Which images to label for few-shot medical landmark detection?. In: Proceedings of the IEEE/CVF Conference on Computer Vision and Pattern Recognition, pp. 20606–20616 (2022)

20. Quan, Q., Yao, Q., Li, J., et al.: Information-guided pixel augmentation for pixel-wise contrastive learning, arXiv preprint arXiv:2211.07118 (2022)
21. Yao, Q., Quan, Q., Xiao, L., Kevin Zhou, S.: One-shot medical landmark detection. In: de Bruijne, M., et al. (eds.) MICCAI 2021. LNCS, vol. 12902, pp. 177–188. Springer, Cham (2021). https://doi.org/10.1007/978-3-030-87196-3_17
22. Gutmann, M., Hyvärinen, A.: Noise-contrastive estimation: a new estimation principle for unnormalized statistical models. In: Proceedings of the Thirteenth International Conference on Artificial Intelligence and Statistics. JMLR Workshop and Conference Proceedings, pp. 297–304 (2010)
23. Oord, A.v.d., Li, Y., Vinyals, O.: Representation learning with contrastive predictive coding, arXiv preprint arXiv:1807.03748 (2018)
24. Hering, A., et al.: Learn2reg: comprehensive multi-task medical image registration challenge, dataset and evaluation in the era of deep learning. IEEE Trans. Med. Imaging (2022)
25. Pirhadi, A., Salari, S., Ahmad, M.O., Rivaz, H., Xiao, Y.: Robust landmark-based brain shift correction with a Siamese neural network in ultrasound-guided brain tumor resection. Inter. J. Comput. Assisted Radiol. Surgery, 1–8 (2022)
26. Luo, J., et al.: A feature-driven active framework for ultrasound-based brain shift compensation. In: Frangi, A.F., Schnabel, J.A., Davatzikos, C., Alberola-López, C., Fichtinger, G. (eds.) MICCAI 2018. LNCS, vol. 11073, pp. 30–38. Springer, Cham (2018). https://doi.org/10.1007/978-3-030-00937-3_4
27. Canalini, L., Klein, J., Miller, D., Kikinis, R.: Segmentation-based registration of ultrasound volumes for glioma resection in image-guided neurosurgery. Int. J. Comput. Assist. Radiol. Surg. **14**, 1697–1713 (2019)

ConTrack: Contextual Transformer for Device Tracking in X-Ray

Marc Demoustier, Yue Zhang, Venkatesh Narasimha Murthy[✉],
Florin C. Ghesu, and Dorin Comaniciu

Digital Technology and Innovation, Siemens Healthineers, Princeton, NJ, USA
venkatesh.n.murthy@siemens-healthineers.com

Abstract. Device tracking is an important prerequisite for guidance
during endovascular procedures. Especially during cardiac interventions,
detection and tracking of guiding the catheter tip in 2D fluoroscopic
images is important for applications such as mapping vessels from angiography (high dose with contrast) to fluoroscopy (low dose without contrast). Tracking the catheter tip poses different challenges: the tip can
be occluded by contrast during angiography or interventional devices;
and it is always in continuous movement due to the cardiac and respiratory motions. To overcome these challenges, we propose ConTrack,
a transformer-based network that uses both spatial and temporal contextual information for accurate device detection and tracking in both
X-ray fluoroscopy and angiography. The spatial information comes from
the template frames and the segmentation module: the template frames
define the surroundings of the device, whereas the segmentation module
detects the entire device to bring more context for the tip prediction.
Using multiple templates makes the model more robust to the change in
appearance of the device when it is occluded by the contrast agent. The
flow information computed on the segmented catheter mask between the
current and the previous frame helps in further refining the prediction by
compensating for the respiratory and cardiac motions. The experiments
show that our method achieves 45% or higher accuracy in detection and
tracking when compared to state-of-the-art tracking models.

Keywords: Device tracking · Transformer network · X-Ray navigation

1 Introduction

Tracking of interventional devices plays an important role in aiding surgeons
during catheterized interventions such as percutaneous coronary interventions
(PCI), cardiac electrophysiology (EP), or trans arterial chemoembolization
(TACE). In cardiac image-guided interventions, surgeons can benefit from visual

Supplementary Information The online version contains supplementary material
available at https://doi.org/10.1007/978-3-031-43996-4_65.

guidance provided by mapping vessel information from angiography (Fig. 1b) to fluoroscopy (Fig. 1a) [6, 10] for which the catheter tip is used as an anchor point representing the root of the vessel tree structure. This visual feedback helps in reducing the contrast usage [7] for visualizing the vascular structures and it can also aid in effective placements of stents or balloons.

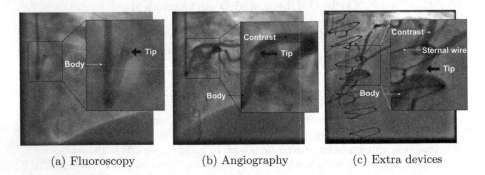

(a) Fluoroscopy   (b) Angiography   (c) Extra devices

Fig. 1. Example frames from X-ray sequences showing the catheter tip: (a) Fluoroscopy image; (b) Angiographic image with injected contrast medium; (c) Angiographic image with sternum wires. Tracking the tip in angiography is challenging due to occlusion from surrounding vessels and interferring devices.

Recently, deep learning-based siamese networks have been proposed for medical device tracking [1, 4, 5]. These networks achieve high frame rate tracking, but are limited by their online adaptability to changes in target's appearance as they only use spatial information. Cycle Ynet [5] uses the cycle consistency of a sequence and relies on a semi-supervised learning approach by doing a forward and a backward tracking. In practice, this method suffers from drifting for long sequences and cannot recover from misdetections because of the single template usage. The closest work related ours is [6], they use a convolutional neural network (CNN) followed by particle filtering as a post processing step. The drawback of this method is that, it does not compensate for the cardiac and respiratory motions as there is no explicit motion model for capturing temporal information. A similar method adds a graph convolutional neural network for aggregating both spatial information and appearance features [3] to provide a more accurate tracking but its effectiveness is limited by its vulnerability to appearance changes and occlusion resulting from detection techniques. Optical flow based network architectures [8] utilize keypoint tracking throughout the entire sequence to estimate the motion of the whole image. However, such approaches are not adapted for tracking a single point, such as a catheter tip.

For general computer vision applications, transformer [9] based-trackers have achieved state-of- the-art performance [2, 11, 12]. Initially proposed for natural language processing (NLP), Transformers learn the dependencies between elements in a sequence, making it intrinsically well suited at capturing global information. Thus, our proposed model consists of a transformer encoder that helps in capturing the underlying relationship between template and search image using self and cross attentions, followed by multiple transformer decoders to accurately track the catheter tip.

To overcome the limitations of existing works, we propose a generic, end-to-end model for target object tracking with both spatial and temporal context. Multiple template images (containing the target) and a search image (where we would identify the target location, usually the current frame) are input to the system. The system first passes them through a feature encoding network to encode them into the same feature space. Next, the features of template and search are fused together by a fusion network, i.e., a vision transformer. The fusion model builds complete associations between the template feature and search feature and identifies the features of the highest association. The fused features are then used for target (catheter tip) and context prediction (catheter body). While this module learns to perform these two tasks together, spatial context information is offered implicitly to provide guidance to the target detection. In addition to the spatial context, the proposed framework also leverages the temporal context information which is generated using a motion flow network. This temporal information helps in further refining the target location.

Our main contributions are as follows: 1) Proposed network consists of segmentation branch that provides spatial context for accurate tip prediction; 2) Temporal information is provided by computing the optical flow between adjacent frames that helps in refining the prediction; 3) We incorporate dynamic templates to make the model robust to appearance changes along with the initial template frame that helps in recovery in case of any misdetection; 4) To the best of our knowledge, this is the first transformer-based tracker for real-time

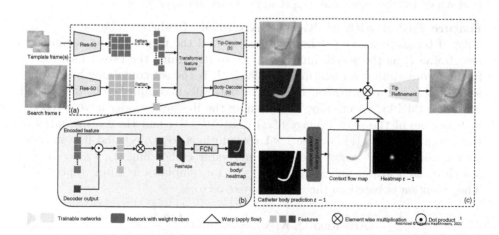

Fig. 2. Proposed ConTrack architecture: (a) Transformer feature fusion backbone: ResNet-50 for feature extraction followed by a transformer encoder/decoder; (b) Prediction head for catheter tip (heatmap) and catheter body segmentation (mask segmentation); (c) Flow refinement: use prediction on previous frame to refine the tip detection.

device tracking in medical applications; 5) We conduct numerical experiments and demonstrate the effectiveness of the proposed model in comparison to other state-of-the-art tracking models.

2   Methodology

Given a sequence of consecutive X-ray images $\{I_t\}_{t=0}^n$ and an initial location of the target catheter tip $x_0 = (u_0, v_0)$, our goal is to track the location of the target $x_t = (u_t, v_t)$ at any time t, $t > 0$. The proposed model framework is summarized in Fig. 2. It consists of two stages, target localization stage and motion refinement stage. First, given a selective set of template image patches and the search image, we leverage the CNN-transformer architecture to jointly localize the target and segment the neighboring context, i.e., body of the catheter. Next, we estimate the context motion via optical flow on the catheter body segmentation between neighboring frames and use this to refine the detected target location. We detail these two stages in the following subsections.

2.1   Target Localization with Multi-template Feature Fusion

To identify the target in the search frame, existing approaches build a correlation map between the template and search features. Limited by definition, the template is a single image, either static or from the last frame tracked result. A transformer naturally extends the bipartite relation between template and search images to complete feature associations which allow us to use multiple templates. This improves model robustness against suboptimal template selection which can be caused by target appearance changes or occlusion.

Feature Fusion with Multi-head Attention. In the encoding stage, given a set of template image patches centered around the target $\{T_{ti}\}_{ti \in \mathcal{H}}$ and current frame I_s as the search image, we aim to determine the target location by fusing information from multiple templates. $\mathcal{H}$ is the set containing historically selected frames for templates. This can be naturally accomplished by multi-head attention (MHA). Specifically, let us denote the ResNet encoder by θ, given the feature map of the search image $\theta(I_s) \in \mathbb{R}^{C \times H_s \times W_s}$, and the feature maps of the templates $\{\theta(T_{ti})\}\}$, we use 1×1 convolutions to project and flatten them into d−dimensional vector query, key and value embedding, q_s, k_s, v_s for the search image features and $\{q_{ti}\}$, $\{k_{ti}\}$, $\{v_{ti}\}$ for templates features respectively. The attention is based on the concatenated vectors,

$$\text{Attention(Q, K, V)} := \text{softmax}(\frac{QK^T}{\sqrt{d}})V, \tag{1}$$

where $Q = \text{Concat}(q_s, q_{t1}, q_{t2}, ..., q_{tn})$, $K = \text{Concat}(k_s, k_{t1}, k_{t2}, ..., k_{tn})$, $V = \text{Concat}(v_s, v_{t1}, v_{t2}, ..., v_{tn})$. The definition of MHA then follows [9].

Joint Target Localization and Context Segmentation. In the decoding stage, we follow [12] and adjust the transformer decoder to a multi-task setting.

As the catheter tip represents a sparse object in the image, solely detecting it suffers from class imbalance issue. To guide the catheter tip tracking with spatial information, we incorporate additional contextual information by simultaneously segmenting the catheter body in the same frame. Specifically, two object queries (e_1, e_2) are employed in the decoder, where e_1 defines the position of the catheter tip, and e_2 defines the mask of the catheter body. As is illustrated in Fig. 2 (b), we first calculate similarity scores between decoder and the encoder output via dot product. We then use element-wise product between the similarity scores and the encoder features to promote regions with high similarity. After reshaping the processed features to $d \times H_s \times W_s$, an encoder-decoder structured 6-layer FCN is attached to process the features to probability maps with the same size as the search image. A combination of the binary cross-entropy and the dice loss is then used,

$$L = \lambda_{bce}^x L_{bce}(G(x_i; \mu, \sigma), \hat{x}_i^s) + \lambda_{dice}^x L_{dice}(G(x_i; \mu, \sigma), \hat{x}_i^s) + \\ \lambda_{bce}^m L_{bce}(m_i, \hat{m}_i) + \lambda_{dice}^m L_{dice}(m_i, \hat{m}_i), \tag{2}$$

where x_i, m_i represent the ground truth annotation of the catheter tip and mask, $\hat{x}_i^s$, $\hat{m}_i^s$ are predictions respectively. Here we use sup-script "s" to denote the predictions from this spatial stage. $G(x_i; \mu, \sigma) := \exp(-\|x_i - \mu\|^2/\sigma^2)$ is the smoothing function that transfers dot location of x_i to probability map. $\lambda_{bce}^*, \lambda_{dice}^* \in \mathbb{R}$ are hyperparameters that are empirically optimized.

2.2   Localization Refinement with Context Flow

In interventional procedures, one common challenge for visual tracking comes from occlusion. This can be caused by injected contrast medium (in the angiographic image) or interferring devices such as sternal wires, stent and additional guiding catheters. If the target is occluded in the search image, using only spatial information for localization is inadequate. To address this challenge, we impose a motion prior of the target to further refine the tracked location. As the target is a sparse object, this is done via optical flow estimation of the context.

Context Flow Estimation. Obtaining ground truth optical flow in real world data is a challenging task and may require additional hardware such as motion sensors. As such, training a model for optical flow estimation directly in the image space is difficult. Instead, we propose to estimate the flow in the segmentation space, i.e., on the predicted heatmaps of the catheter body between neighboring frames. We use the RAFT [8] model for this task. Specifically, given the predicted segmentation maps m_{t-1} and m_t, we first use a 6-block ResNet encoder g_θ to extract the features $g_\theta(m_{t-1}), g_\theta(m_t) \in R^{H_f \times W_f \times D_f}$. Then we construct the correlation volume pyramid $\{C_i\}_{i=0}^3$, where

$$C_i = \text{AvgPool}(corr(g_\theta(m_{t-1}), g_\theta(m_t)), \text{stride} = 2^i). \tag{3}$$

Here $corr(g_\theta(m_{t-1}), g_\theta(m_t)) \in \mathbb{R}^{H_f \times W_f \times H_f \times W_f}$ stands for correlation evaluation:

$$corr(g_\theta(m_{t-1}), g_\theta(m_t))_{ijkl} = \sum_{h=1}^{D_f} g_\theta(m_{t-1})_{ijh} \cdot g_\theta(m_t)_{klh}, \qquad (4)$$

which can be computed via matrix multiplication. Starting with an initial flow $f_0 = 0$, we follow the same model setup as [8] to recurrently refine the flow estimates to $f_k = f_{k-1} + \triangle f$ with a gated recurrent unit (GRU) and a delta flow prediction head of 2 convolutional layers. Given the tracked tip result from the previous frame $\hat{x}_{t-1}$, we can then predict the new tip location at time t by warping with the context flow $\hat{x}_t^f = f^k(\hat{x}_{t-1})$. Here we use sup-script "f" to denote the prediction by flow warping.

We note here that since the segmentations of the catheter body are sparse objects compared to the entire image, computation of the correlation volume and subsequent updates can be restricted to a cropped sub-image which reduces computation cost and flow inference time. As the flow estimation is performed on the segmentation map, one can simply generate synthetic flows and warp them with the existing catheter body annotation to generate data for model training.

Refinement with Combined Spatial-temporal Prediction. Finally, we generate a score map with combined information from the spatial localization stage and the temporal prediction by context flow,

$$S_t(u, v) = \begin{cases} (\alpha + \hat{m}_t^s(u, v))(\hat{x}_t^s(u, v) + \hat{x}_t^f(u, v)) & \hat{m}_t^s(u, v) > 0, \\ \hat{x}_t^s(u, v) + \hat{x}_t^f(u, v) & \text{otherwise.} \end{cases} \qquad (5)$$

Here α is a positive scalar. It helps the score map to promote coordinates that are activated jointly on both the spatial prediction $\hat{x}_t^s$ and the temporal prediction $\hat{x}_t^f$. Finally, we forward the score map through a refinement module to finalize the prediction. The refinement module consists of a stack of 3 convolutional layers. Similar to the spatial localization stage, a combination of the binary cross-entropy and the dice loss is used as the final loss.

3   Experiments and Results

Dataset. Our study uses an internal dataset of X-ray sequences captured during percutaneous coronary intervention procedures, featuring a field of view displaying the catheter within the patient's heart. The test dataset is divided into two primary categories: fluoroscopic and angiographic sequences. Fluoroscopic sequences are real-time videos of internal movements captured by low-dose X-rays without radiopaque substances, while angiographic sequences display blood vessels in real-time after the introduction of radiopaque substances.

We further separate the test dataset into a third category, "devices", presenting a unique challenge for both fluoroscopic and angiographic sequences. In these cases, devices such as wires can obscure the catheter tip and have a similar appearance to the catheter, making tracking more challenging.

The dataset includes frames annotated with the coordinates of the catheter tip and, in some cases, a catheter body mask annotation. For training and validation, we use 2,314 sequences consisting of 198,993 frames, of which 44,957 are annotated. As the model training only requires image pairs, *i.e.*templates and search images, in order to reduce annotation effort, a nonadjacent subset of frames in each sequence is annotated. Their neighboring unannotated frames are also used to provide flow estimation, as is shown in Fig. 2(c). For testing, we use 219 sequences consisting of 17,988 frames, all annotated. The test dataset split is as follows: Fluoro (i.e., fluoroscopy), consisting of 94 sequences, 8,494 frames, from 82 patients; Angio (i.e., angiography), consisting of 101 sequences, 6,904 frames, from 81 patients; and devices, consisting of 24 sequences, 2,593 frames, from 10 patients. All frames undergo a preprocessing pipeline with resampling and padding to size of 512×512 with 0.308 mm isotropic pixel spacing.

Training. The template frame is of size 64×64. The search frame is of size 160×160. With this, the inference speed reaches 12 fps. We train our model for 300 epochs using a learning rate of 0.0001.

Comparison Study. We compare the proposed approach with existing arts and summarize the results in Table 1. The proposed approach achieves best performance in all testing dataset. In contrast to our method, SiameseRPN [4], STARK [12] and MixFormer [2] focus on spatial localization of the target. Temporal information is being incorporated only with the setting of multi-templates thus target motion modeling is limited. While such approaches can achieve good performance with low median errors ($\sim$2 mm), the high 5–7 mm standard deviations indicate the stability issues, especially in data with devices where occlusions are present. Cycle Ynet [5] uses cycle-consistency loss for motion learning directly on the target. As catheter tip is a sparse object, our approach leverages the motion information of the neighboring context which provide more robust guidance for target location refinement.

Overall, ConTrack outperforms all other methods, with a median tracking error of less than 1.08 mm. Our model is particularly effective at tracking the catheter tip when other devices are in the field of view, where all other methods

Table 1. Comparison study on different testing set. The results are the average distance in mm. Best numbers are marked in bold. Accuracy improvement is statistically significant (p-value $<$ 0.005) over the second best in the table.

Models	Fluoroscopy		Angiography		Devices		All		
	median	mean	median	mean	median	mean	median	mean	std
SiameseRPN [4]	6.93	8.19	7.74	9.42	7.89	10.51	7.13	9.01	6.81
STARK [12]	2.38	3.02	2.82	4.49	4.35	7.01	2.65	4.14	4.93
MixFormer [2]	2.02	4.42	2.76	4.86	5.00	9.20	2.68	5.15	7.10
Cycle Ynet [5]	2.05	2.92	1.69	2.09	4.39	4.23	1.96	2.68	2.40
ConTrack (Ours)	**0.73**	**1.04**	**1.27**	**1.91**	**1.61**	**2.73**	**1.08**	**1.63**	**1.70**

tend to underperform. Compared to Cycle Ynet on all test datasets, our model is 45% more accurate, with an average distance of less than 1mm between the prediction and ground truth. Further, we show the accuracy distributions in Fig. 3. It can be seen that the proposed approach shows superior performance to all other approaches in various percentiles.

Ablation Study. We conduct an ablation study to investigate the effectiveness of different model components. Results are summarized in Table 2. Our ablation study revealed three key findings: 1) The addition of the mask segmentation branch improved tracking performance on Fluoro, where the device appearance remains consistent and there is no occlusion. However, when there are distractors, the results are less accurate; 2) The inclusion of a mask segmentation enabled the estimation of motion. The resulting flow helped to stabilize tracking in the presence of distractors; and 3) Multiple templates were employed to better handle changes in appearance. The combined model showed the best performance in dataset of angiography and data with devices, while yielding similar results in dataset of fluoroscopy.

Despite our framework's incorporation of various temporal and spatial contexts, catheter tracking remains a challenging task, particularly in cases where other devices or contrast agents obscure the catheter tip and create visual similarities with the catheter itself. Nonetheless, our results demonstrate the promise of ConTrack as a valuable tool for enhancing catheter tracking accuracy.

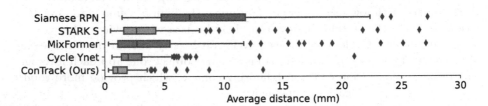

Fig. 3. Benchmark study on average distance distribution over all test datasets.

Table 2. Ablation study on model components. ✗ denotes the component is removed, while ✓ represents the component is used. Performance is evaluated on the same test cases as before and the results are the average distance in mm.

Multitask	Flow	Multi-templates	Fluoro		Angio		Devices		All		
			median	mean	median	mean	median	mean	median	mean	std
✗	✗	✗	0.81	1.40	1.29	1.94	2.99	6.20	1.13	2.17	3.75
✓	✗	✗	0.67	1.03	1.53	2.15	3.97	10.49	1.11	2.58	6.10
✓	✓	✗	**0.65**	**0.96**	1.49	1.95	1.93	4.52	**0.99**	1.81	2.30
✓	✓	✓	0.73	1.05	**1.27**	**1.91**	**1.61**	**2.73**	1.08	**1.63**	**1.70**

4   Conclusion

Device tracking is an important task in interventional procedures. In this paper, we propose a generic model framework, ConTrack, that leverages both spatial and temporal information of the surrounding context for accurate target localization and tracking in X-ray. Through extensive experimentation on large datasets, our approach demonstrated superior tracking performance, outperforming other state-of-the-art tracking models, especially in challenging scenarios where occlusions and distractors are present. Current approach has its limitations. Motion estimation is learned from neighboring two frames and thus target historical trajectory information is missing. Further, transformer-based model training require large amount of annotated data, which is challenging to collect in interventional applications. Finally, throughout the paper we follow established setups and focus on the development on the tracking model with manual initialization. In general, long-term visual tracking with automatic (re-)initialization is a challenging problem and require a system of approaches. A safe and automatic system of device and anatomy tracking is of great clinical relevance and will be an important future work for us.

Disclaimer. The concepts and information presented in this paper/presentation are based on research results that are not commercially available. Future commercial availability cannot be guaranteed.

References

1. Bromley, J., Guyon, I., LeCun, Y., Säckinger, E., Shah, R.: Signature verification using a Siamese time delay neural network. In: Cowan, J., Tesauro, G., Alspector, J. (eds.) Advances in Neural Information Processing Systems, vol. 6. Morgan-Kaufmann (1993)
2. Chen, Q., et al.: Mixformer: mixing features across windows and dimensions. In: Proceedings of the IEEE/CVF Conference on Computer Vision and Pattern Recognition (CVPR), pp. 5249–5259 (2022)
3. Huang, L., Liu, Y., Chen, L., Chen, E.Z., Chen, X., Sun, S.: Robust landmark-based stent tracking in X-ray fluoroscopy. In: Avidan, S., Brostow, G., Cisse, M., Farinella, G.M., Hassner, T. (eds.) Computer Vision – ECCV 2022. ECCV 2022. LNCS, vol. 13682, pp. 201–216. Springer, Cham (2022). https://doi.org/10.1007/978-3-031-20047-2_12
4. Li, B., Yan, J., Wu, W., Zhu, Z., Hu, X.: High performance visual tracking with Siamese region proposal network. In: 2018 IEEE/CVF Conference on Computer Vision and Pattern Recognition, pp. 8971–8980 (2018)
5. Lin, J., Zhang, Y., Amadou, A., Voigt, I., Mansi, T., Liao, R.: Cycle ynet: semi-supervised tracking of 3D anatomical landmarks. In: Liu, M., Yan, P., Lian, C., Cao, X. (eds.) MLMI 2020. LNCS, vol. 12436, pp. 593–602. Springer, Cham (2020). https://doi.org/10.1007/978-3-030-59861-7_60
6. Ma, H., Smal, I., Daemen, J., van Walsum, T.: Dynamic coronary roadmapping via catheter tip tracking in x-ray fluoroscopy with deep learning based Bayesian filtering, vol. 61, p. 101634 (2020)

7. Piayda, K., et al.: Dynamic coronary roadmapping during percutaneous coronary intervention: a feasibility study, vol. 23, p. 36 (2018)
8. Teed, Z., Deng, J.: RAFT: recurrent all-pairs field transforms for optical flow. In: Vedaldi, A., Bischof, H., Brox, T., Frahm, J.-M. (eds.) ECCV 2020. LNCS, vol. 12347, pp. 402–419. Springer, Cham (2020). https://doi.org/10.1007/978-3-030-58536-5_24
9. Vaswani, A., et al.: Attention is all you need. In: Guyon, I., et al. (eds.) Advances in Neural Information Processing Systems, vol. 30. Curran Associates, Inc. (2017)
10. Wang, P., Chen, T., Ecabert, O., Prummer, S., Ostermeier, M., Comaniciu, D.: Image-based device tracking for the co-registration of angiography and intravascular ultrasound images. In: Fichtinger, G., Martel, A., Peters, T. (eds.) MICCAI 2011. LNCS, vol. 6891, pp. 161–168. Springer, Heidelberg (2011). https://doi.org/10.1007/978-3-642-23623-5_21
11. Yan, B., et al.: Towards grand unification of object tracking. In: Avidan, S., Brostow, G., Cisse, M., Farinella, G.M., Hassner, T. (eds.) Computer Vision – ECCV 2022. ECCV 2022. LNCS, vol. 13681, pp. 733–751. Springer, Cham (2022). https://doi.org/10.1007/978-3-031-19803-8_43
12. Yan, B., Peng, H., Fu, J., Wang, D., Lu, H.: Learning spatio-temporal transformer for visual tracking. In: 2021 IEEE/CVF International Conference on Computer Vision (ICCV), pp. 10428–10437 (2021)

FocalErrorNet: Uncertainty-Aware Focal Modulation Network for Inter-modal Registration Error Estimation in Ultrasound-Guided Neurosurgery

Soorena Salari[1]([✉]), Amirhossein Rasoulian[1], Hassan Rivaz[2], and Yiming Xiao[1]

[1] Department of Computer Science and Software Engineering,
Concordia University, Montreal, Canada
{soorena.salari,ah.rasoulian,yiming.xiao}@concordia.ca
[2] Department of Electrical and Computer Engineering,
Concordia University, Montreal, Canada
hassan.rivaz@concordia.ca

Abstract. In brain tumor resection, accurate removal of cancerous tissues while preserving eloquent regions is crucial to the safety and outcomes of the treatment. However, intra-operative tissue deformation (called brain shift) can move the surgical target and render the pre-surgical plan invalid. Intra-operative ultrasound (iUS) has been adopted to provide real-time images to track brain shift, and inter-modal (i.e., MRI-iUS) registration is often required to update the pre-surgical plan. Quality control for the registration results during surgery is important to avoid adverse outcomes, but manual verification faces great challenges due to difficult 3D visualization and the low contrast of iUS. Automatic algorithms are urgently needed to address this issue, but the problem was rarely attempted. Therefore, we propose a novel deep learning technique based on 3D focal modulation in conjunction with uncertainty estimation to accurately assess MRI-iUS registration errors for brain tumor surgery. Developed and validated with the public RESECT clinical database, the resulting algorithm can achieve an estimation error of 0.59 ± 0.57 mm.

Keywords: Registration · Inter-modal · Error estimation · Deep learning

1 Introduction

Resection of early-stage brain tumors can greatly reduce the mortality rate of patients. During the surgery, brain tissue deformation (called brain shift) can occur due to various causes, such as gravity, drug administration, and pressure change after craniotomy. While modern magnetic resonance imaging (MRI) techniques can provide rich anatomical and physiological information with various contrasts (e.g., fMRI) for more elaborate pre-surgical planning, intra-operative MRI that can track brain shift requires a complex setup and is costly. In contrast,

H. Greenspan et al. (Eds.): MICCAI 2023, LNCS 14228, pp. 689–698, 2023.
https://doi.org/10.1007/978-3-031-43996-4_66

intra-operative ultrasound (iUS) has gained popularity for real-time imaging during surgery to monitor tissue deformation and surgical tools because of its lower cost, portability, and flexibility [1]. Accurate and robust MRI-iUS registration techniques [2] can greatly enhance the value of iUS for updating pre-surgical plans and guiding the interpretation of iUS, which has an unintuitive contrast and non-standard orientations. This can greatly enhance the safety and outcomes of the surgical procedure by allowing maximum brain tumor removal while avoiding eloquent regions [3]. However, as the true underlying tissue deformation is unknown due to the 3D nature of the surgical data and the time constraint, real-time manual inspection of MRI-iUS registration results is challenging and error-prone, especially for precision-sensitive neurosurgery. Therefore, algorithms that can detect and quantify unreliable inter-modal medical image registration results are highly beneficial.

Recently, automatic quality assessment for medical image registration has attracted increasing attention [4] from the domains of big medical data analysis and surgical interventions. With high efficiency, machine, and deep learning techniques have been proposed to allow automatic grading and dense estimation of medical image registration errors. Early endeavors on this topic primarily relied on hand-crafted features, including information theory-based metrics [5–10]. More recently, deep learning (DL) techniques that learn task-specific features have also been adopted in automatic evaluation of medical image registration, with a primary focus on intra-contrast/modal applications, including CT [9,10] and MRI [11]. Unfortunately, so far, error grading and estimation in inter-contrast/modal registration have rarely been explored, despite the particular demand in surgical applications. In this direction, Bierbrier *et al.* [12] made the first attempt using simulated iUS from MRI to train 3D convolutional neural networks (CNNs) to perform dense error regression for MRI-iUS registration in brain tumor resection. Although their algorithm performed well in simulated cases, the results on real clinical scans still required improvements. In this paper, we propose a novel 3D CNN to perform patch-wise error estimation for MRI-iUS registration in neurosurgery, by using focal modulation [13], a recent alternative DL technique to self-attention [14] for encoding contextual information, and uncertainty estimation. We call our method FocalErrorNet, which has three main novelties. **First**, we adapted the focal modulation network [13] from 2D to 3D and employed the technique in registration error assessment for the first time. **Second**, we incorporated uncertainty estimation using Monte Carlo (MC) dropouts [15] to offer assurance for error regression. **Lastly**, we developed and thoroughly evaluated our technique against a recent baseline model [12] using real clinical data and showed excellent results.

2   Methods and Materials

2.1   Dataset and Preprocessing

For methodological development and assessment, we used the RESECT (REtro-Spective Evaluation of Cerebral Tumors) dataset [16], which has pre-operative

MRI, and iUS scans at different surgical stages from 23 subjects who underwent low-grade glioma resection surgeries. As it is still challenging to model iUS scans with tissue resection, we took 22 cases with T2FLAIR MRI that better depicts tumor boundaries and iUS acquired before resection. An example of an MRI-iUS pair from a patient is shown in Fig. 1. We hypothesized that directly leveraging clinical iUS could help learn more realistic image features with potentially better outcomes in clinical applications than with simulated contrasts [9,12]. However, since the true brain shift model is impossible to obtain, we followed the strategy of creating silver ground truths for image alignment [9,12], upon which simulated misalignment is augmented in the iUS to build and test our DL model. To create the silver registration ground truths, we used the homologous landmarks between MRI and iUS in the RESECT dataset to perform landmark-based 3D B-Spline nonlinear registration to register iUS to the corresponding MRI for all 22 cases. To tackle the limited field of view (FOV) in iUS, we cropped the T2FLAIR MRI to the same FOV of the iUS, which was resampled to a $0.5 \times 0.5 \times 0.5$ mm^3 resolution.

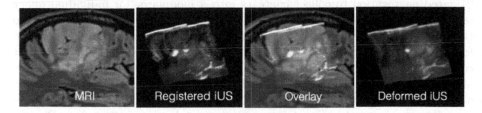

Fig. 1. Left to right: demonstration of sample pre-operative MRI, perfectly registered, and deformed iUS with a mean registration error of 1.4 mm.

To perform spatial misalignment augmentation, we continued to leverage 3D B-Spline transformation, similar to earlier reports on the same topic [10,12,17]. In short, B-Spline transformation can be modeled by a grid of regularly spaced control points and the associated parameters to allow various levels of nonlinear deformation. While the spacing of the control points determines the levels of details in local deformation fields, the displacement parameters control the magnitude of the deformation. To ensure that simulated registration errors are of different varieties and sizes, we randomly selected the number of control points and the associated displacements (in each 3D axis) with a maximum of 20 points and 30 mm, respectively. Note that the control point grid is isotropic, and the density is arbitrarily determined per deformation in our case. Each co-registered iUS scan was deformed ten times. After misalignment augmentation on the previously co-registered iUS, matching pairs of 3D image patches of size $33 \times 33 \times 33$ voxels were taken from both the iUS volume and the corresponding MRI. As iUS has limited FOV and may contain no anatomical features, to ensure that the patches we extracted contain useful information (e.g. to avoid the dark background) in iUS, we focused on acquiring patches centered around the anatomical landmark locations available through the RESECT database. Since B-spline transformation offers a displacement vector at each voxel of the iUS volume, we directly considered the norm of the vector as the simulated registration error at the associated voxel. In our design, we determined the registration error of the image patch pair as the mean of all voxel-wise errors within the iUS patch. Finally, the image patch pairs, along with corresponding registration errors were then fed to the proposed DL algorithm for training and validation.

2.2   Network Architecture

We proposed a novel 3D neural network, named FocalErrorNet, based on the recent focal modulation networks [13] that was originally proposed for 2D vision tasks to estimate the registration error between MRI and iUS patches. With a similar goal as the Vision Transformer (ViT), the focal modulation network was designed to model contextual information in images. It incorporates three main elements to achieve the goal: 1) focal contextualization that comprises a stack of depth-wise convolutional layers to account for long- to short-range dependencies, 2) gated aggregation to collect contexts into a modulator for individual query tokens, and 3) element-wise affine transformation to inject the modulator into the query. In the architecture of FocalErrorNet (see Fig. 2), all layers contain two focal modulator blocks, where two depth-wise convolutional layers focally extract contexts around each voxel, selectively aggregate and inject them into the query, and pass the information to the next block. We designed the FocalErrorNet as a ResNet-like variant of the focal modulation network to better encode relevant features across the input image and ensure a better gradient flow. Finally, the information from the backbone was propagated to a multi-layer perceptron (MLP) to regress registration errors, and two MC dropout layers were added to the MLP to allow uncertainty quantification for the results.

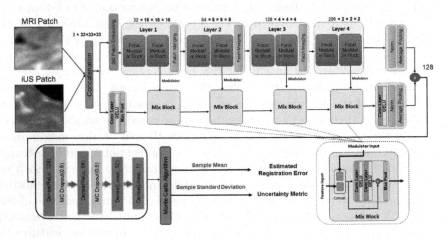

Fig. 2. The proposed FocalErrorNet for registration error and uncertainty estimation.

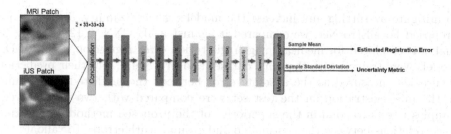

Fig. 3. The baseline 3D CNN [9,12] with the added MC dropout layer.

2.3   Uncertainty Quantification

For registration error regression in surgical applications, knowledge regarding the reliability of the automated results is instrumental for the safety and well-being of the patients. Uncertainty estimation has gained popularity in probing the trustworthiness and credence of DL algorithms. Although the concept has been widely applied in image segmentation and classification, it has not been employed for registration error estimation, especially in the case of multi-modal situations, such as MRI-iUS alignment. Therefore, we incorporated uncertainty estimation in our proposed FocalErrorNet. For each MRI-iUS patch pair, 200 regression samples were collected by random sampling from MC dropouts [15] at test time. While the final patch registration error was obtained as the mean of all the samples, the sample standard deviation was used as the uncertainty metric.

Table 1. Accuracy comparison of different models for registration error estimation.

Model	Absolute error (mm)	Correlation
3D CNN [9,12]	1.69 ± 1.37	0.61
FocalErrorNet	**0.59 ± 0.57**	**0.82**

2.4   Experimental Setup and Implementation

From the transformation augmentation, we acquired 3380 samples of MRI-iUS pairs. For our experiments, we arbitrarily split the subjects into training, validation, and test sets with the proportion of 60%, 20%, and 20%, respectively. To prevent information leakage, we ensured that each patient was included in only one of the split sets. For model training, we adopted the Adam optimization with a learning rate of 5×10^{-5} and a batch size of 64. For the loss function, we used mean squared error (MSE) to minimize the difference between the predicted MRI-iUS registration error and the ground truths. Furthermore, in addition to the transformation augmentation, we also included additional data augmentation, including random noise addition and random image flipping on training sets to mitigate overfitting and increase the model's generalizability. To assess our proposed FocalErrorNet, we compared it against a 3D CNN [9,12] (see Fig. 3) that was employed for medical image registration error regression. The two DL models were trained with the same dataset and procedure, and their prediction accuracies, measured as the absolute error between the predicted and ground truths mis-registration on the test set were compared with two-sided paired-samples t-tests to confirm the superiority of the proposed method, in addition to correlations between their estimated and ground truth errors. To validate the proposed uncertainty estimation method, we calculated the correlation between the uncertainty measure and absolute error of FocalErrorNet, and the correlation between the uncertainty and mutual information between MRI and iUS, which is often used to measure the information overlap in multi-modal registration. Finally, to test the robustness of the FocalErrorNet, we acquired additional MRI-iUS patch pairs from the test subjects, by introducing random linear shifts (the max displacement from landmark locations is 10 voxels) from the selected locations in the original set, and evaluated the DL model performance.

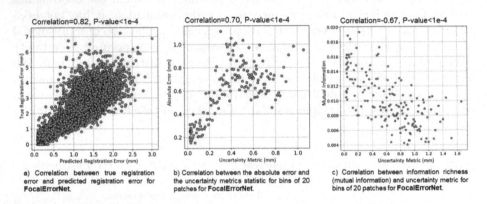

Fig. 4. Left to right: scatter plots and correlations between true registration error vs. predicted registration error, uncertainty metric vs. absolute error, and mutual information vs. uncertainty metric for the proposed FocalErrorNet.

3   Results

3.1   Error Regression Accuracy

The accuracy comparison between the proposed FocalErrorNet and the baseline 3D CNN [9,12] is shown in Table 1. Across all samples in the testing data, we achieved an accuracy of 0.59 ± 0.57 mm, while the counterpart obtained a prediction error of 1.69 ± 1.37 mm. With the t-test, our FocalErrorNet outperformed the 3D CNN [9,12] (p < 1e-4). In addition, the correlations between the predicted and ground truths errors are 0.82 (p < 1e-4) and 0.61 (p < 1e-3) for FocalErrorNet and 3D CNN, respectively, further confirming the advantage of the proposed technique. To allow a qualitative comparison, scatter plots for predicted vs. ground truth errors of the two models are depicted in Fig. 4a and 5a. At larger error levels, it is evident that the point clouds exhibit a wider shape.

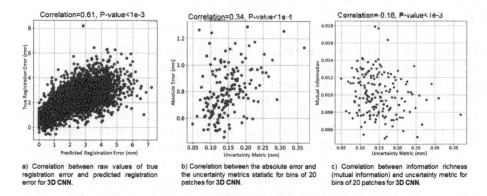

a) Correlation between raw values of true registration error and predicted registration error for 3D CNN.

b) Correlation between the absolute error and the uncertainty metrics statistic for bins of 20 patches for 3D CNN.

c) Correlation between information richness (mutual information) and uncertainty metric for bins of 20 patches for 3D CNN.

Fig. 5. Left to right: scatter plots and correlations between true registration error vs. predicted registration error, uncertainty metric vs. absolute error, and mutual information vs. uncertainty metric for the 3D CNN [9,12].

3.2 Validation of the Uncertainty Evaluation

We obtained correlations of 0.70 (p < 1e-4) and 0.34 (p < 1e-4) between estimated uncertainty and prediction error for FocalErrorNet and the baseline 3D CNN, respectively. Additionally, the uncertainty vs. mutual information uncertainties was assessed at −0.67 (p < 1e-4) for our proposed method and −0.18 (p < 1e-3) for the baseline. To allow better visual comparisons, the associated scatter plots are illustrated in Fig. 4 and 5. These metrics proved the validity of our uncertainty measure and further confirmed the performance of FocalErrorNet. Note the scatter plots for uncertainty measure validation were performed using value binning (with 20 values per bin) for each axis to better reveal the trends of the metrics.

3.3 Robustness of the Proposed Model

To examine the performance of our proposed method for image regions that contain fewer potent anatomical features, we acquired additional image pairs from test subjects, according to Sect. 2.4. With the new test set, the prediction errors for our method and the baseline model were 1.28 ± 0.99 mm and 2.49 ± 1.87 mm, respectively. Furthermore, the correlations between estimated and true error were calculated at 0.41 for FocalErrorNet and 0.20 for the baseline. These results supported the benefits of focal modulation in registration error estimation. In this test, patches can contain large areas of zeros (image content out of the scanning FOV of the iUS). The main reason for the observed performance decline is due to the reduction in sufficient image features in iUS. However, despite these challenges, we saw an acceptable outcome from FocalErrorNet (absolute error = 1.28 mm or ∼1 voxel in clinical MRIs).

4   Discussion

In image-guided interventions, there is an urgent need for automatic assessment of image registration quality. Multi-modal registration quality evaluation poses major challenges due to three main factors. First, dissimilar contrasts between images require more elaborate strategies to derive relevant features for error assessment. Second, unlike segmentation or classification, the ground truths of registration errors are difficult to obtain. Finally, compared with classification, regression tasks tend to be more error-prone for deep learning algorithms. To tackle these challenges, we employed 3D focal modulation with depth-wise convolution to encode contextual information for the image pair. Compared with the ViT and its variants, focal modulation allows a more lightweight setup, which could be desirable for 3D data. Although we admit that residual errors still remain after landmark-based B-Spline nonlinear alignment, this approach has been adopted in different prior studies, considering the residual landmark registration error is fairly low (mTRE of 0.0008 ± 0.0010mm). Although simulated ultrasound has been used to provide a perfect alignment with MRIs, the fidelity of the simulated results is still suboptimal, and this may explain the underperformance of the previous technique in real clinical data [12]. To ensure the performance of our FocalErrorNet, we opted to regress the mean registration error of image patches than simplistic error grades or voxel-wise error maps. We believe that this design choice offers a more stable performance, which is supported by our validation. We adopted uncertainty estimation in inter-modal registration error assessment for the first time. While other techniques exist to provide model uncertainty [18], MC dropout is more flexible for various DL models. Furthermore, the use of standard deviation as an uncertainty measurement maintains the same unit as the regressed errors, thus making the interpretation more intuitive. From quantitative and qualitative evaluations using correlation coefficients and scatter plots to assess the association of uncertainty measures with the prediction errors and image entropy, we confirmed the validity of the proposed uncertainty estimation approach. For our FocalErrorNet, we achieved a prediction error of 0.59 ± 0.57 mm, which is on par with the image resolution (0.5 mm). Additionally, the standard deviation of our results is lower than the baseline model [12]. These signify a robust performance of the FocalErrorNet. One limitation of our work lies in the limited patient data, as public iUS datasets are scarce, while the settings and properties of US scanners can vary, potentially affecting the DL model designs. Therefore, we created random deformations for patch-wise error estimation, and will further explore data-efficient approaches for registration error assessment.

5   Conclusion

We proposed FocalErrorNet, a novel DL model for uncertainty-aware inter-modal registration error estimation in iUS-guided neurosurgery, leveraging the latest focal modulation technique and MC dropout. With thorough assessments of the accuracy and uncertainty measures, we have confirmed the performance of the proposed method against a baseline model previously adopted for the same task. As the first to introduce uncertainty measures and 3D focal modulation in registration error evaluation, our work provides the first step for fast and reliable feedback in inter-modal medical image registration to guide clinical decisions in surgery. We plan to adapt the presented framework for other inter-modal/contrast image registration applications in the future.

Acknowledgment. We acknowledge the support of the Natural Sciences and Engineering Research Council of Canada (NSERC) and Fonds de Recherche du Québec Nature et Technologies (FRQNT).

References

1. Rivaz, H., Collins, D.L.: Near real-time robust non-rigid registration of volumetric ultrasound images for neurosurgery. Ultrasound Med. Biol. **41**(2), 574–587 (2015)
2. Xiao, Y., et al.: Evaluation of MRI to ultrasound registration methods for brain shift correction: the CuRIOUS2018 challenge. IEEE Trans. Med. Imaging **39**(3), 777–786 (2019)
3. Marko, N.F., Weil, R.J., Schroeder, J.L., Lang, F.F., Suki, D., Sawaya, R.E.: Extent of resection of glioblastoma revisited: personalized survival modeling facilitates more accurate survival prediction and supports a maximum-safe-resection approach to surgery. J. Clin. Oncol. **32**(8), 774 (2014)
4. Bierbrier, J., Gueziri, H.-E., Collins, D.L.: Estimating medical image registration error and confidence: a taxonomy and scoping review. Med. Image Anal. 102531 (2022)
5. Shams, R., Xiao, Y., Hébert, F., Abramowitz, M., Brooks, R., Rivaz, H.: Assessment of rigid registration quality measures in ultrasound-guided radiotherapy. IEEE Trans. Med. Imaging **37**(2), 428–437 (2017)
6. Saygili, G.: Local-search based prediction of medical image registration error. In: Medical Imaging,: Image Perception, Observer Performance, and Technology Assessment, vol. 10577, pp. 327–332. SPIE 2018 (2018)
7. Sokooti, H., Saygili, G., Glocker, B., Lelieveldt, B.P., Staring, M.: Quantitative error prediction of medical image registration using regression forests. Med. Image Anal. **56**, 110–121 (2019)
8. Sokooti, H., Saygili, G., Glocker, B., Lelieveldt, B.P.F., Staring, M.: Accuracy estimation for medical image registration using regression forests. In: Ourselin, S., Joskowicz, L., Sabuncu, M.R., Unal, G., Wells, W. (eds.) MICCAI 2016. LNCS, vol. 9902, pp. 107–115. Springer, Cham (2016). https://doi.org/10.1007/978-3-319-46726-9_13
9. Eppenhof, K.A., Pluim, J.P.: Error estimation of deformable image registration of pulmonary CT scans using convolutional neural networks. J. Med. Imaging **5**(2), 024 003–024 003 (2018)
10. Sokooti, H., Yousefi, S., Elmahdy, M.S., Lelieveldt, B.P., Staring, M.: Hierarchical prediction of registration misalignment using a convolutional LSTM: application to chest CT scans. IEEE Access **9**, 62 008–62 020 (2021)
11. Fonov, V.S., Dadar, M., Adni, T.P.-A.R.G., Collins, D.L.: DARQ: deep learning of quality control for stereotaxic registration of human brain MRI to the T1w MNI-ICBM 152 template. Neuroimage **257**, 119266 (2022)

12. Bierbrier, J., Eskandari, M., Di Giovanni, D.A., Collins, D.L.: Towards estimating MRI-Ultrasound registration error in image-guided neurosurgery. IEEE Transactions on Ultrasonics, Ferroelectrics, and Frequency Control (2023)
13. Yang, J., Li, C., Dai, X., Gao, J.: Focal modulation networks. Adv. Neural Inf. Process. Syst. **35**, 4203–4217. Curran Associates Inc. (2022)
14. Vaswani, A., et al.: Attention is all you need. In: Advances in Neural Information Processing Systems (NeurIPS), vol. 30 (2017)
15. Gal, Y., Ghahramani, Z.: Dropout as a Bayesian approximation: representing model uncertainty in deep learning. In: International Conference on Machine Learning, pp. 1050–1059. PMLR (2016)
16. Xiao, Y., Fortin, M., Unsgård, G., Rivaz, H., Reinertsen, I.: Retrospective evaluation of cerebral tumors (RESECT): a clinical database of pre-operative MRI and intra-operative ultrasound in low-grade glioma surgeries. Med. Phys. **44**(7), 3875–3882 (2017)
17. Lotfi, T., Tang, L., Andrews, S., Hamarneh, G.: Improving probabilistic image registration via reinforcement learning and uncertainty evaluation. In: Wu, G., Zhang, D., Shen, D., Yan, P., Suzuki, K., Wang, F. (eds.) MLMI 2013. LNCS, vol. 8184, pp. 187–194. Springer, Cham (2013). https://doi.org/10.1007/978-3-319-02267-3_24
18. Abdar, M., et al.: A review of uncertainty quantification in deep learning: techniques, applications and challenges. Inf. Fusion **76**, 243–297 (2021)

Optical Ultrasound Imaging for Endovascular Repair of Abdominal Aortic Aneurysms: A Pilot Study

Callum Little[1] (ID), Shaoyan Zhang[1] (ID), Richard Colchester[1] (ID), Sacha Noimark[1] (ID), Sunish Mathews[1] (ID), Edward Zhang[1] (ID), Paul Beard[1] (ID), Malcolm Finlay[2] (ID), Tara Mastracci[2] (ID), Roby Rakhit[3] (ID), and Adrien Desjardins[1] (✉) (ID)

[1] University College London, London, UK
a.desjardins@ucl.ac.uk
[2] Barts Heart Centre, London, UK
[3] Royal Free Hospital, London, UK

Abstract. An abdominal aortic aneurysm (AAA) is a persistent localized dilatation of the aorta to more than 1.5 times the expected diameter, which may lead to rupture with resultant high mortality. Endovascular repair (EVAR) of AAAs is a minimally invasive procedure that involves the peripheral delivery of one or more covered endografts to the aneurysmal segment, via a catheter-based system. A particularly challenging group of patients to treat are those in which the aneurysmal sac extends proximally to include the origin of the renal arteries (15% of all AAAs). To maintain the patency of renal side branches in these "complex" cases, *in situ* fenestration (ISF) of endografts during AAA procedures has been proposed. The challenges addressed in this study were a) to develop an endovascular imaging system for visualizing side branches beyond deployed endografts and thereby to determine the locations for ISF; b) to obtain an initial assessment of the clinical utility of this system. Here, all-optical ultrasound (OpUS) imaging with a fiber optic transducer was used for real-time guidance, wherein ultrasonic pulses are generated in nanocomposite coatings via the photoacoustic effect and received optically using a Fabry-Perot cavity. These custom OpUS transducer components were integrated into a steerable sheath (6 Fr) that also included a separate optical fiber for delivering laser pulses for fenestrating the endograft. In an ex-vivo model, it was shown that OpUS imaging extended through the endograft and underlying aortic tissue, and permitted aortic side-branch visualization. During an EVAR procedure in a porcine model *in vivo*, an aortic side branch was visualized with OpUS imaging after the endograft was deployed and optical fenestration of the stent graft was successfully performed. This study showed that OpUS is a promising modality for guiding EVAR and could find particularly utility with identifying aortic side branches for ISF during treatment of complex AAAs.

Keywords: Optical ultrasound imaging · abdominal aortic aneurysm · endovascular repair

H. Greenspan et al. (Eds.): MICCAI 2023, LNCS 14228, pp. 699–707, 2023.
https://doi.org/10.1007/978-3-031-43996-4_67

1 Introduction

Intervention options for the treatment of abdominal aortic aneurysms (AAAs) include open surgery or endovascular repair (EVAR). A minimally invasive procedure, EVAR involves the peripheral delivery of one or more covered endografts to the aneurysmal segment via a catheter-based system. EVAR has been demonstrated to be superior to open surgery with regards to early mortality [1, 2]. A particularly challenging group of patients to treat are those in which the aneurysmal sac extends proximally to include the origin of the renal arteries. These are known as juxtarenal or "complex" AAAs and account for roughly 15% of all AAAs [3].

With complex AAA repair, it is of paramount importance to maintain the patency of major aortic side branches to allow for end-organ perfusion. Traditionally, open surgery was preferred for these complex cases; however, with the advent of endografts that have pre-made openings (fenestrations) in the graft material, these patients can be treated using fenestrated endovascular repair (FEVAR). The concept of *in situ* fenestration (ISF) for endografts during FEVAR has been proposed as an alternative to the use of pre-fenestrated endografts. With ISF, the fenestrations are generated within the aorta following deployment of the endograft. Potential benefits of this approach include both greater anatomical conformity and reduced device cost. Once ISF has been performed, a wire is passed through the fenestration into the side-branch. The fenestration is then dilated with incrementally sized non-compliant or cutting balloons and secured with a stent that maintains communication between the aorta and side-branch [4]. Several fenestration methods have been proposed, of which mechanical puncture and thermal ablation of stent material are the most prominent. Thermal ablation methods for ISF are attractive as there is no requirement to exert penetrative force on the endograft material. Instead, the material is vaporized through the use of radiofrequency (RF) [5] or laser energy, in which 94% success was achieved in 16 patients with no major complications [6].

A key challenge with ISF that we address here is to visualize aortic side branches in order to precisely identify the locations for fenestration. Pre-procedural imaging in concert with X-ray fluoroscopy can be used for guidance [7]; however, distortion of the aorta due to deployment of the endograft limits the accuracy of this technique [8]. All-optical ultrasound (OpUS) is an emerging imaging modality that involves the optical generation and reception of ultrasound [9], which can be performed with fiber optics that are readily integrated into medical devices [10]. OpUS is a potentially attractive option for guiding ISF for several reasons: the depth penetration achieved for OpUS (>1 cm) is relevant to aneurysmal vessels, high bandwidths enable high resolution imaging (*ca.* 50–100 μm), and a blood-free environment is not required for image acquisition. The feasibility of visualizing aortic side branches with OpUS has previously been demonstrated [9, 11] and intracardiac imaging *in vivo* has been performed [10]. In this study, we explore the application of OpUS to EVAR and ISF, by addressing two open questions: does this technique permit visualization of side branches through endograft material, and can the optical transducers be integrated into a steerable sheath with forward imaging that also permits optical fenestration?

2   Methods

2.1   Device Design

The custom device comprised two primary fiber optic components which were integrated within a catheter: 1) an OpUS transmitter/receiver pair for visualizing side branches; 2) an ISF fiber for optical fenestration. These optical fibers were secured together using heat shrink and housed within a 6 Fr [inner diameter (ID): 2 mm] catheter (Multipurpose-1, Cordis, Santa Clara, CA, USA), leaving a 0.7 mm diameter channel to allow a 0.014″ or 0.018″ guidewire to be delivered through a generated fenestration. The catheter was delivered to the endovascular stent graft using a 7 Fr ID steerable sheath (Tourguide, Medtronic, USA), a current standard that allowed for bi-directional 180°/90° deflection with manual control and thereby allowed the catheter to be appropriately positioned to detect side-branches behind the endovascular stent (Fig. 1).

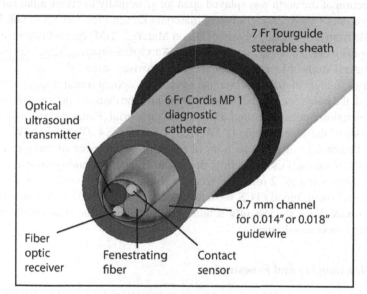

Fig. 1. Custom device for performing all-optical ultrasound imaging and fenestration during endovascular repair (EVAR), shown schematically. The device comprised an optical ultrasound transmitter and receiver for imaging, and a third optical fiber to deliver light for *in situ* fenestration (ISF) of endograft stent material. A custom fiber optic contact sensor was not used in this study. This device was enclosed by heat shrink and positioned within two off-the-shelf, commercially-available devices: a diagnostic catheter and a steerable sheath.

The OpUS transmitter comprised a custom nanocomposite material for photoacoustic generation of ultrasound, with optical absorption of excitation light by a near-infrared absorbing dye embedded within a medical-grade elastomeric host [12] that was applied to a 400 μm circular surface following the design of Colchester et al. [13]. The OpUS receiver comprised a plano-concave microresonator positioned at the distal end of a fiber. These fiber optic OpUS components were interrogated by a custom console that delivered pulsed excitation multi-mode light to the transmitter (wavelength: 1064 nm; pulse duration: 2 ns; pulse energy: 20.1 µJ; repetition rate: 100 Hz) and wavelength-tunable continuous wave single-mode light to the receiver (1500–1600 nm; 4 mW). This custom console, previously described in detail [9], comprised software for real-time processing and display of concatenated A-scans as M-mode OpUS images. The ISF fiber for fenestration had an inner diameter of 400 µm and transmitted light from a laser unit that permitted control of the pulse duration and power (Diomed D15 Diode, Diomed Holdings, USA; wavelength: 808 nm; maximum power: 4W). This wavelength was chosen to minimize damage to the underlying vascular tissues [14].

2.2 Benchtop Imaging and Fenestration

To assess whether OpUS permits visualization of side branches through endograft material, a benchtop model with ex-vivo swine aorta (Medmeat Ltd, United Kingdom) was used. A section of the aorta was splayed open longitudinally to create a flat surface and secured to a cork board. Endograft stent material (Zenith Flex® AAA, Cook Medical, USA) made from a synthetic polyester Dacron Material (DM) (polyethylene terephthalate) was positioned above the aortic sample. An OpUS imaging scan was performed with 600 lateral steps of length 50 µm using a motorised stage [9].

To test the effects of different laser parameters on optical fenestration, the following were varied: the pulse energy, the pulse duration, and the distance between the distal end of the fenestration optical fiber and the endograft material. Fenestrations with different combinations of these parameters (2.5W, 0.5s; 1.8W, 0.5s; 4.2W, 0.5s; 4.2W, 1s; 4.2W, 5s) were attempted in different sections of aortic tissue. A piece of endograft material was secured above each tissue section; the distance from the endograft material to the tissue was approximately 2 mm. The tissue and endograft material were immersed in anticoagulated sheep blood (TCS Biosciences, Buckingham, UK). A fenestration was deemed to have been successfully achieved if adjacent tissue was visible when the surrounding blood was drained.

2.3 *In Vivo* Imaging and Fenestration

To obtain a preliminary validation of the system's potential for guiding EVAR and ISF, OpUS imaging and endograft fenestration were performed in a clinically realistic environment using an *in vivo* porcine model. The primary objectives of this experiment were a) to visualize an aortic side branch behind a deployed endograft; b) to perform optical fenestration of the endograft. All procedures on animals were conducted in accordance with U.K. Home Office regulations and the Guidance for the Operation of Animals (Scientific Procedures) Act (1986). The protocol for this study was reviewed and ratified by the local animal welfare ethical review board.

Following successful anesthesia, transcutaneous ultrasound (US) was used to gain percutaneous access to both common femoral arteries; 8 Fr vascular sheaths were inserted bilaterally. Both fluoroscopic acquisition and digital subtraction angiography with an iodine-based contrast agent were performed to create a road-map for further stages of the experiment. The ISF device was inserted via the Right Femoral Artery (RFA) and directed with the steerable catheter into the aorta. The inner catheter of the ISF device was advanced into the aorta.

For the EVAR procedure, a non-fenestrated iliac endograft was used (Zenith Spiral-Z, Cook Medical, USA; proximal/distal diameter: 13 mm/11 mm; length: 129 mm). The choice of an iliac endograft originally for human patients was for the comparatively smaller dimensions of the swine aorta. The sheath in the RFA was up-sized to a 14 Fr in order to accommodate the delivery system. The stent was deployed across the renal artery bifurcation. The ISF device was inserted into the lumen of the endograft and OpUS imaging was performed with a longitudinal pullback across the covered left renal artery. Once positioned over the desired location, a single pulse from the 808 nm fenestration laser (power: 4.2 W; pulse duration: 5 s) was delivered to the endograft to create a fenestration. A coronary guidewire (0.018″; Balanced Middleweight Universal, Abbott, USA) was then inserted through the ISF device and through the fenestration into the side-branch. Expansion of the fenestration in the endograft was then performed with a series of three balloons (2 mm non-compliant, 3 mm non-compliant and 3 mm cutting) to create a fenestration large enough to accommodate a stent. Following successful ISF, the procedure was ended and the ISF device was removed.

3 Results

3.1 Benchtop Imaging and Fenestration

With OpUS imaging of the paired endograft and underlying aortic tissue, the stent graft manifested as an echo-bright linear structure (Fig. 2). Aortic tissue beneath the graft could be imaged to a depth >6 mm from the transducer surface (>3 mm from the tissue surface). The pullback revealed the side-branch within the aorta, which was apparent as a well demarcated area with signal drop-out, consistent with an absence of tissue.

During benchtop attempts, a fenestration was successfully created with a pulse duration of 0.5 s and an optical power of 1.8 W when the distal end of the fiber was in contact with the endograft material. This combination of pulse duration and optical power corresponded to the smallest energy that yielded fenestration with this device. There was no macroscopic tissue damage with these parameters. The size of the fenestration increased with both higher power outputs and longer laser pulse durations. When the distal end of the optical fiber was recessed a distance of 3 mm from the endograft material, both a higher power output and longer pulse duration were required to generate a fenestration. No macroscopic tissue damage was identified during any of the fenestration experiments. As a control, the optical fiber was placed directly on the tissue with no stent material. In this configuration, a power of 4.2 W and a pulse duration of 5 s resulted in a macroscopic thermal ablation of the tissue.

3.2 *In Vivo* Imaging and Fenestration

The renal bifurcation was readily observed with fluoroscopic guidance and X-ray contrast injections. Due to the small diameter of the swine aorta relative to that of an adult human, it proved challenging to achieve sufficient bending of the steerable sheath whilst

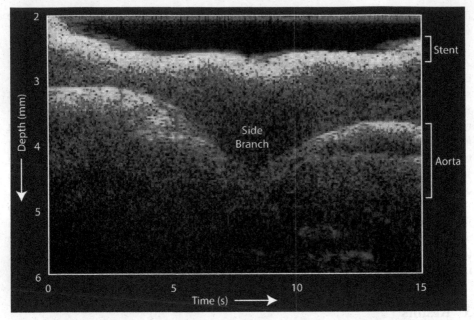

Fig. 2. All optical ultrasound (OpUS) imaging of the endograft paired with underlying aortic tissue, performed on the benchtop. The stent endograft material was strongly hyperechoic. The underlying aorta is apparent as a 2–3 mm thick region. The side-branch presented as an absence of signal, corresponding to a discontinuity in the aortic tissue.

maintaining a gap between the imaging device and the aortic tissue. In this configuration, bend-induced losses in the fiber optic receiver significantly lowered the signal-to-noise (SNR) ratio relative to that obtained during benchtop imaging. Nonetheless, a manual pullback with imaging performed at a slightly non-perpendicular angle relative to the aortic surface proved feasible. With OpUS imaging during this pullback, the endograft material presented as a strongly hyperechoic region and the underlying aorta could be imaged to a depth >3 mm. A side branch presented as a hypoechoic region consistent with an absence of tissue within the renal vessel lumen (Fig. 3a).

For ISF, the laser output parameters that were chosen (4.2 W, 5 s) were intentionally higher than the minimum values observed in benchtop experiments. Following the ISF, the guidewire was passed into the renal artery. A post-mortem dissection confirmed a successful optical fenestration (Fig. 3b). Passage of the guidewire through the fenestration in the endograft material was confirmed with fluoroscopic imaging performed *in vivo* (Fig. 3c). No tearing in the endograft material that may have arisen from balloon-induced expansion was apparent (Fig. 3b).

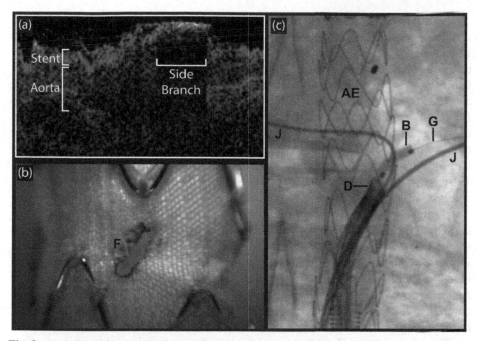

Fig. 3. (a) All-Optical Ultrasound (OpUS) imaging obtained from a manual pullback across a renal bifurcation, formed from concatenated A-scans. The endograft material presents as a hyperechoic region above the underlying aortic tissue; a side branch that manifested as an absence of signal. (b) With optical microscopy of the endograft performed post autopsy, a fenestration (F) is apparent. (c) Fluoroscopic imaging immediately after optical *in situ* fenestration (ISF) showing successful insertion of the guidewire (G) into the renal artery. AE: aortic endograft; D: custom device for optical ultrasound imaging and fenestration; B: balloon inflated to dilated the fenestration; J: J-tip 0.035" wires that mark the location of the renal arteries.

4 Discussion

A key observation of this study is that OpUS offers direct visualization of aortic tissue beneath an AAA endograft, and detection of vascular side-branches. This capability is important for appropriate ISF positioning, thereby maintaining side-branch patency. The ability to detect a side-branch immediately before performing fenestration with the same device could reduce the risk of inappropriate positioning and the subsequent need for conversion to open surgery. In current clinical practice, there is no device that allows for direct imaging of aortic side branches with the added capability of achieving endograft fenestration.

Current methods for generating ISF, although promising, are not guided by intravascular imaging. With these methods, there is a risk of creating a fenestration that is not aligned with the side-branch, leading to a potential endoleak, and a risk of causing injury to underlying tissue structures. The potential cost implication of ISF is significant. The average costs of pre-fenestrated and patient-specific grafts are ca. £15,400 and £24,000 (GBP), respectively; the cost of a non-fenestrated aortic endograft that could be used for ISF is ca. £6,000.

Several improvements to the device can be envisaged. An increase in the lateral OpUS imaging resolution could be effected with greater US beam collimation, for instance with the use of a larger-diameter transmitting surface; however, this change would likely result in increased device size and reduced flexibility. Improvements in the SNR of OpUS images could potentially be achieved with depth-dependent frequency filtering and with the application of deep-learning methods [15]. In future iterations, imaging and ISF could be performed with a single optical fiber [16, 17], thereby reducing device size, increasing device flexibility/deliverability and potentially reducing device cost. Using the same ultrasonic receiver used for reception during imaging, it could be possible to use ultrasonic tracking to localize the device relative to an external imaging device [15, 18, 19], thereby reducing the dependency on fluoroscopic image guidance and corresponding contrast injections.

During the *in vivo* procedure in this study, the power of the fenestration laser was chosen to be higher than the minimum value observed in the benchtop component of this study to increases the chances of success. It remains to be seen whether a lower output setting may be viable in this context, with the presence of device motion and flowing blood and with pathological changes in the arterial wall. In future studies, real-time feedback from OpUS imaging could be incorporated to determine the power and pulse duration of the fenestrating laser; a similar probe could be used for cardiac ablation, for instance in the context of treatment of atrial fibrillation [20].

This study presented a novel imaging probe for performing interventional image guidance for endovascular procedures. We conclude that OpUS imaging is a promising modality for guiding EVAR; in concert with and optical fenestration, it could find particularly utility with identifying aortic side branches and performing ISF during treatment of complex AAAs.

Acknowledgements. This work was supported by the Wellcome/EPSRC Centre for Interventional and Surgical Sciences, which is funded by grants from the Wellcome (203145/Z/16/Z) and the Engineering and Physical Sciences Research Council (ESPRC; NS/A000050/1). Additional funding was provided by the National Institute for Health Research UCL Biomedical Research Centre, the Royal Academy of Engineering (RF/201819/18/125), the Wellcome (Innovative Engineering for Health award WT101957), and by the EPSRC (Healthcare Technologies Challenge Award: EP/N021177/1).

References

1. Prinssen, M., et al.: A randomized trial comparing conventional and endovascular repair of abdominal aortic aneurysms. N. Engl. J. Med. **351**, 1607–1618 (2004)
2. Greenhalgh, R.M.: Comparison of endovascular aneurysm repair with open repair in patients with abdominal aortic aneurysm (EVAR trial 1), 30-day operative mortality results: randomised controlled trial. Lancet **364**, 843–848 (2004)
3. Jongkind, V., et al.: Juxtarenal aortic aneurysm repair. J. Vasc. Surg. **52**, 760–767 (2010)
4. Eadie, L.A., Soulez, G., King, M.W., Tse, L.W.: Graft durability and fatigue after in situ fenestration of endovascular stent grafts using radiofrequency puncture and balloon dilatation. Eur. J. Vasc. Endovasc. Surg. **47**, 501–508 (2014)

5. Numan, F., Arbatli, H., Bruszewski, W., Cikirikcioglu, M.: Total endovascular aortic arch reconstruction via fenestration in situ with cerebral circulatory support: an acute experimental study. Interact. Cardiovasc. Thorac. Surg. **7**, 535–538 (2008)
6. le Houérou, T., et al.: In situ antegrade laser fenestrations during endo-vascular aortic repair. Eur. J. Vasc. Endovasc. Surg. **56**, 356–362 (2018)
7. Bailey, C.J., Edwards, J.B., Giarelli, M., Zwiebel, B., Grundy, L., Shames, M.: Cloud-based fusion imaging improves operative metrics during fenestrated endovascular aneurysm repair. J Vasc Surg. **77**, 366–373 (2023)
8. Koutouzi, G., Sandström, C., Roos, H., Henrikson, O., Leonhardt, H., Falkenberg, M.: Orthogonal rings, fiducial markers, and overlay accuracy when image fusion is used for EVAR guidance. Eur. J. Vasc. Endovasc. Surg. **52**, 604–611 (2016)
9. Colchester, R.J., Zhang, E.Z., Mosse, C.A., Beard, P.C., Papakonstantinou, I., Desjardins, A.E.: Broadband miniature optical ultrasound probe for high resolution vascular tissue imaging. Biomed. Opt. Express. **6**, 1502–1511 (2015)
10. Finlay, M.C., et al.: Through-needle all-optical ultrasound imaging in vivo: a preclinical swine study. Light Sci. Appl. **6**, e17103–e17103 (2017)
11. Alles, E.J., Noimark, S., Zhang, E., Beard, P.C., Desjardins, A.E.: Pencil beam all-optical ultrasound imaging. Biomed. Opt. Express **7**, 3696–3704 (2016)
12. Lewis-Thompson, I., Zhang, S., Noimark, S., Desjardins, A.E., Colchester, R.J.: PDMS composites with photostable NIR dyes for multi-modal ultra-sound imaging. MRS Adv. **7**, 499–503 (2022)
13. Colchester, R.J., Alles, E.J., Desjardins, A.E.: A directional fibre optic ultra-sound transmitter based on a reduced graphene oxide and polydimethylsiloxane composite. Appl. Phys. Lett. **114**, 113505 (2019)
14. Piazza, R., et al.: In situ diode laser fenestration: an ex-vivo evaluation of irradiation effects on human aortic tissue. J. Biophotonics. **12**, e201900032 (2019)
15. Maneas, E., et al.: Enhancement of instrumented ultrasonic tracking images using deep learning. Int. J. Comput. Assist. Radiol. Surg. **18**, 395–399 (2023)
16. Colchester, R.J., Zhang, E.Z., Beard, P.C., Desjardins, A.E.: High-resolution sub-millimetre diameter side-viewing all-optical ultrasound transducer based on a single dual-clad optical fibre. Biomed. Opt. Express **13**, 4047–4057 (2022)
17. Noimark, S., et al.: Polydimethylsiloxane composites for optical ultrasound generation and multimodality imaging. Adv. Funct. Mater. **28**, 1704919 (2018)
18. Mathews, S.J., et al.: Ultrasonic needle tracking with dynamic electronic focusing. Ultrasound Med. Biol. **48**, 520–529 (2022)
19. Xia, W., West, S.J., Mari, J.-M., Ourselin, S., David, A.L., Desjardins, A.E.: 3D ultrasonic needle tracking with a 1.5D transducer array for guidance of fetal interventions. In: Ourselin, S., Joskowicz, L., Sabuncu, M.R., Unal, G., Wells, W. (eds.) MICCAI 2016. LNCS, vol. 9900, pp. 353–361. Springer, Cham (2016). https://doi.org/10.1007/978-3-319-46720-7_41
20. Zhang, S., Zhang, E.Z., Beard, P.C., Desjardins, A.E., Colchester, R.J.: Dual-modality fibre optic probe for simultaneous ablation and ultrasound imaging. Commun. Eng. **1**, 20 (2022)

Self-supervised Sim-to-Real Kinematics Reconstruction for Video-Based Assessment of Intraoperative Suturing Skills

Loc Trinh[1], Tim Chu[1], Zijun Cui[1], Anand Malpani[2], Cherine Yang[3], Istabraq Dalieh[4], Alvin Hui[5], Oscar Gomez[1], Yan Liu[1], and Andrew Hung[3(✉)]

[1] University of Southern California, Los Angeles, USA
{loctrinh,tnchu,zijuncui,gomez,yanliu.cs}@usc.edu
[2] Mimic Technologies Inc., Seattle, USA
malpani.anand.89@gmail.com
[3] Department of Urology, Cedars-Sinai Medical Center, Los Angeles, USA
{cherine.yang,andrew.hung}@cshs.org
[4] Boston University Henry M. Goldman School of Dental Medicine, Boston, USA
idalieh@bu.edu
[5] Western University of Health Sciences, Pomona, USA
alvin.hui@westernu.edu

Abstract. Suturing technical skill scores are strong predictors of patient functional recovery following robot-assisted radical prostatectomy (RARP), but manual assessment of these skills is a time and resource-intensive process. By automating suturing skill scoring through computer vision methods, we can significantly reduce the burden on healthcare professionals and enhance the quality and quantity of educational feedbacks. Although automated skill assessment on simulated virtual reality (VR) environments have been promising, applying vision methods to live ('real') surgical videos has been challenging due to: 1) the lack of kinematic data from the da Vinci® surgical system, a key source of information for determining the movement and trajectory of robotic manipulators and suturing needles, and 2) the lack of training data due to the labor-intensive task of segmenting and scoring individual stitches from live videos. To address these challenges, we developed a self-supervised pre-training paradigm whereby sim-to-real generalizable representations are learned without requiring any live kinematic annotations. Our model is based on a masked autoencoder (MAE), termed as *LiveMAE*. We augment live stitches with VR images during pre-training and require LiveMAE to reconstruct images from both domains while also predicting the corresponding kinematics. This process learns a visual-to-kinematic mapping that seeks to locate the positions and orientations of surgical manipulators and needles, deriving "kinematics" from live videos without requiring supervision. With

Supplementary Information The online version contains supplementary material available at https://doi.org/10.1007/978-3-031-43996-4_68.

an additional skill-specific finetuning step, LiveMAE surpasses supervised learning approaches across 6 technical skill assessments, ranging from 0.56–0.84 AUC (0.70–0.91 AUPRC), with particular improvements of 35.78% in AUC for *wrist rotation* skills and 8.7% for *needle driving* skills. Mean-squared error for test VR kinematics was as low as 0.045 for each element of the instrument poses. Our contributions provide the foundation to deliver personalized feedback to surgeons training in VR and performing live prostatectomy procedures.

Keywords: Vision transformers · Masked autoencoders · Self-supervised learning · sim-to-real generalization · Suturing skill assessment

1   Introduction

Previous studies have shown that surgeon performance directly affects patient clinical outcomes [1,2,13,14]. In one instance, manually rated suturing technical skill scores were the strongest predictors of patient continence recovery following a robot-assisted radical prostatectomy compared to other objective measures of surgeon performance [3]. Ultimately, the value of skill assessment is not only in its ability to predict surgical outcomes, but also in its function as formative feedback for training surgeons. The need to automate skills assessment is readily apparent, especially since manual assessments by expert raters are subjective, time-consuming, and unscalable [4,5]. View Fig. 1 for problem setup.

Preliminary work has shown favorable results for automated skill assessments on simulated VR environments, demonstrating the benefits of machine learning (ML) methods. ML approaches for automating suturing technical skills leveraged instrument kinematic (motion-tracking) data as the sole input to recurrent networks have been able to achieve effective area-under-ROC-curve (AUC), up to 0.77 for skill assessment in VR sponge suturing exercises [6]. Multi-modality approaches that fused information from both kinematics and video modalities have demonstrated increased performance over uni-modal approaches in both VR sponge and tube suturing exercises, reaching up to 0.95 AUC [7].

Despite recent advances, automated skill assessment in live scenarios is still a difficult task due to two main challenges: 1) the lack of kinematic data from the da Vinci® system, and 2) the lack of training data due to the labor-intensive labeling task. Unlike simulated VR environments where kinematic data can be readily available, current live surgical systems do not output motion-tracking data, which is a key source of information for determining the movement and trajectory of robotic manipulators and suturing needles. Moreover, live surgical videos do not have a clear and painted target area for throwing stitches, unlike VR videos, which makes the task additionally difficult. On the other hand, due to the labor-intensive task of segmenting and scoring individual stitches from each surgical video, the quantity of available and labeled training data is quite low, rendering traditional supervised learning approaches ineffective.

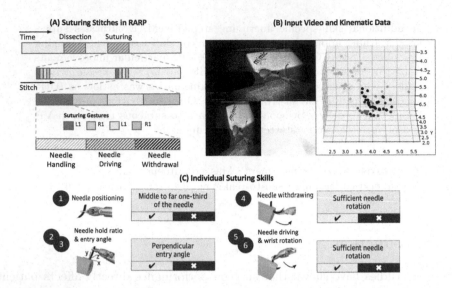

Fig. 1. Suturing skill assessment. (a) The suturing step of RARP is composed of multiple stitches, each of which can also be broken down into three sub-phases (needle handling, driving, and withdrawal). (b) Input video and kinematics data for each VR suturing exercise. Live data do not have kinematics. Colors indicate different instruments such as left/right manipulators and needle/targets. (c) Each sub-phrase can be divided into specific EASE skills [8] and assessed for their quality (low vs high).

To address these challenges, we propose LiveMAE which learns sim-to-real generalizable representations without requiring any live kinematic annotations. Leveraging available video and sensor data from previous VR studies, LiveMAE can map from surgical images to instrument kinematics and derive surrogate "kinematic" automatically by learning to reconstruct images from both VR and live stitches while also predicting the corresponding VR kinematics. This creates a shared encoded representation space between the two visual domains while using available kinematic data from only one domain, the VR domain. Moreover, our pre-training strategy is not skill-specific which brings a bonus in improving data efficiency. LiveMAE enjoys up to six times more training data across the six suturing skills seen in Fig. 1c, especially when we further break down video clips and kinematic sequences into multiple (image, kinematic) pairs.

Overall, our main contributions include:

1. We propose LiveMAE which learns sim-to-real generalizable representations without requiring any live kinematic annotations.
2. We design a pre-training paradigm that increases the number of effective training samples significantly by combining data across suturing skills.
3. We conduct rigorous evaluations to verify the effectiveness of LiveMAE on surgical data collected and labeled across multiple institutions and surgeons. Finetuning on suturing skill assessment tasks yields better performance on 5/6 skills on live surgical videos compared to supervised learning baselines.

2 Methodology

Masked autoencoding is a method for self-supervised pre-training of Vision Transformers (ViTs [12]) on images. It has demonstrated the capability to learn efficient and useful visual representations for downstream tasks such as image classification and segmentation. Our model builds on top of mask autoencoders (MAEs) and we provide a preliminary intro for MAE in Appendix 1.1.

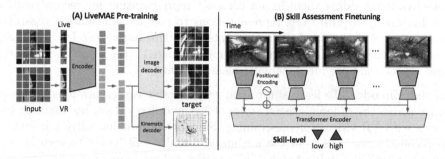

Fig. 2. LiveMAE Overview. (a) Pre-training with a shared encoder between Live and VR images and a kinematic decoder for predicting instrument kinematics. (b) Skill-specific finetuning for suturing skill assessment using pre-trained LiveMAE mapping.

The input to our system contains both VR and live surgical data. VR data for a suturing skill s is defined as $D_s^{VR} = \{(x_i, k_i, y_i)\}_{i=0}^{N_s}$ consisting of segmented video clips $x_i \in \mathbb{R}^{F \times H \times W \times 3}$, aligned kinematic sequence $k_i \in \mathbb{R}^{F \times 70}$, and EASE technical skill score $y_i \in \{0, 1\}$ for non-ideal vs ideal performance. F denotes the number of frames in the video clip. Live data for s is similarly $D_s^L = \{(x_i, y_i)\}_{i=0}^{M_s}$, except there are no aligned kinematics. Kinematic data has 70 features tracking 10 instruments of interest, each pose contains 3 elements for coordinates and 4 elements for quarternions. There are six technical skill labels, see Fig. 1c.

2.1 LiveMAE

Since D_s^L lacks kinematic information that is crucial for suturing skill assessment, we propose LiveMAE to automatically derive "kinematics" from live videos that can be helpful for downstream prediction. Specifically, we aim to learn a mapping $\phi : \mathbb{R}^{H \times W \times 3} \to \mathbb{R}^{70}$ from images to instrument kinematics using available video and sensor data from D_s^{VR}, and subsequently utilizing that mapping ϕ on live videos. Although the visual style between VR and live surgical videos can differ, this mapping is possible since we know that both simulated VR and live instruments share the exact same dimensions and centered coordinate frames. Our method builds on top of MAE and has three main components: a kinematic decoder, a shared encoder, and an expanded training set.

Kinematic Decoder. For mapping from a surgical image to the instrument kinematics, we propose an additional kinematic output head along with a corresponding self-supervised task of reconstructing kinematics from masked input. See Fig. 2a. The kinematic decoder is also a lightweight series of Transformer blocks that takes in a full sequence of both the (i) encoded visible patches, and (ii) learnable mask tokens. The last layer of the decoder is a linear projection whose number of output channels equals to 70, the dimension of the kinematic data. Similar to the image reconstruction task, which aims to learn visual concepts and semantics by encoding them into a compact representation for reconstruction, we additionally require these representations to contain information regarding possible poses of the surgical instruments. The kinematic decoder also has a reconstruction loss, which computes the mean squared error (MSE) between the reconstructed and original kinematic measurements.

Shared Encoder. To learn sim-to-real generalizable representations that generalize across the different visual styles of VR and live videos, we augment live images with VR videos for pre-training. Since we do not have live kinematics, the reconstruction loss from the kinematic decoder will be set to zero for live samples within a training batch. This creates a shared encoded representation space between the two visual domains such that visual concepts and semantics about manipulators and suturing needles can be shared between them. Moreover, as we simultaneously train the kinematic reconstruction task, we are learning a mapping that can generalize to live videos, since two similar positioning in either VR or live should have similar corresponding kinematics.

Expanded Training Set. Since we have limited surgical data, and the mapping from image to instrument kinematics is not specific to any one suturing skill, we can combine visual and kinematic data across different skills during pre-training. Specifically, we pre-train the model on all data combined across 6 suturing skills to help learn the mapping. In addition, we can further break down video clips and kinematic sequences into $F * (N_s + M_s)$ (image, kinematics) pairs to increase the effective training set size without needing heavy data augmentations. These two key facts provide is a unique advantage over traditional supervised learning, since training each skill assessment task required the full video clip to learn temporal signature along with skill-specific scorings.

Finetuning of LiveMAE for Skill Assessment After pre-training, we discard the image decoder and only use the pathway from the encoder to the kinematic decoder as our mapping ϕ. See Fig. 2b. We applied ϕ to our live data D_s^L and extract a surrogate kinematic sequence for each video clip. The extracted kinematics are embedded by a linear projection with added positional embeddings and processed with a lightweight sequential DistilBERT model. We append a linear layer on top of the pooled output from DistilBERT for classification. We finetune the last layer of ϕ and the sequential model with the cross-entropy loss using a small learning rate, e.g. 1e−5.

3   Experiments and Results

Datasets. We utilize a previously validated suturing assessment tool (EASE [8]) to evaluate the robotic suturing skill in both VR and live surgery. We collected 156 VR videos and 54 live surgical videos from 43 residents, fellows, and attending urologic surgeons in this 5-center multi-institutional study. VR suturing exercises were completed on the Surgical Science^{TM} Flex VR simulator and live surgical videos of surgeons performing the vesico-urethra anastomosis (VUA) step of a RARP were recorded. Each video was split into stitches, ($n = 3448$) total, and each stitch was segmented into sub-phrases with 6 binary assessment labels (low vs. high skill). See data breakdown and processing in Appendix 1.2.

Table 1. Suturing skill assessments on VR data. Boldfaced denotes best and $\pm$ are standard deviations across 5 held-out institutions.

Modality	Model	Repositions		HoldRatio		HoldAngle	
		AUC	AUPRC	AUC	AUPRC	AUC	AUPRC
Kinematics	LSTM	0.808 ± 0.02	0.888 ± 0.03	0.567 ± 0.09	0.859 ± 0.06	0.469 ± 0.06	0.804 ± 0.07
	Transformer	**0.852 ± 0.03**	**0.916 ± 0.04**	**0.652 ±0.07**	**0.895 ±0.05**	0.457 ±0.08	0.796 ± 0.06
Video	ConvLTSM	0.715 ± 0.06	0.840 ± 0.04	0.587 ± 0.07	0.883 ± 0.05	0.552 ± 0.02	0.831 ± 0.05
	ConvTransformer	0.842 ± 0.02	0.912 ± 0.03	0.580 ± 0.03	0.880 ± 0.05	**0.560 ±0.06**	**0.837 ±0.06**
Video + Kin.	AuxTransformer	0.851 ± 0.02	0.912 ± 0.04	0.597 ± 0.07	0.886 + 0.05	0.557 ± 0.03	0.842 ± 0.06
Modality	Model	DrivingSmoothness		WristRotation		WristRotationNW	
		AUC	AUPRC	AUC	AUPRC	AUC	AUPRC
Kinematics	LSTM	0.851 ± 0.06	0.953 ± 0.03	0.615 ± 0.07	0.894 ± 0.03	0.724 ± 0.04	0.942 ± 0.03
	Transformer	**0.878 ±0.06**	**0.963 ±0.02**	**0.640 ±0.08**	**0.899 ±0.02**	**0.725 ± 0.03**	**0.942 ±0.03**
Video	ConvLTSM	0.851 ± 0.05	0.938 ± 0.02	0.636 ± 0.10	0.897 ± 0.03	0.661 ± 0.03	0.934 ± 0.02
	ConvTransformer	0.858 ± 0.04	0.956 ± 0.02	0.634 ± 0.10	0.895 ± 0.04	0.700 ± 0.06	0.937 ± 0.02
Video + Kin	AuxTransformer	0.868 ± 0.06	0.963 ± 0.02	0.649 ± 0.10	0.898 ± 0.04	0.675 ± 0.05	0.935 ± 0.02

Metrics and Baselines. Across the five institutions, we use 5-fold cross-validation to evaluate our model, training and validating on data from 4 institutions while testing on the 5th held-out institution. This allows us to test for generalization on unseen cases across both surgeons and medical centers. We measure and report the mean $\pm$ std. dev. for the two metrics: (1) Area-under-the-ROC curve (AUC) and (2) Area-under-the-PR curve (AUPRC) for the 5 test folds.

To understand the benefits of each data modality, we compare LiveMAE against 3 setups: (1) train/test using only kinematics, (2) train/test using only videos, and (3) train using kinematic and video data while testing only on video (no live kinematics). For kinematics-only baselines, we use two sequential models (1) LSTM recurrent model [9], and (2) DistillBERT transformer-based model [10]. For video-only baselines, we used two models based on pre-trained CNNs (3) ConvLSTM and (4) ConvTransformer. Both used pre-trained AlexNet to extract visual and flow features from the penultimate layer for each frame. The features are then flattened as used as input vectors to the sequential model (1) and (2). For a multi-modal baseline, we compare against recent work, AuxTransformer [7], which uses kinematics as privileged data in the form of an auxiliary loss during training. Unlike our method, they have additional kinematic supervision for the live video domain which we do not have.

3.1   Understanding the Benefits of Kinematics

Table 1 presents automated suturing assessment results for each technical skill on VR data from unseen surgeons across the 5 held-out testing institutions. We make 3 key observations: (1) we successfully reproduced assessment performance seen in previous works and showed that sequential models trained on kinematic-only data often achieve the best results (outperforming video and multi-modal on 5/6 skills with high mean AUCs (0.652–0.878) and AUPRC (0.895–0.963). (2) Vision model trained on video-only data can help with skill assessment, especially in certain skills such as *needle hold angle* where the angle between the needle tip and the target tissue (largely responsible for high/low score) is better represented visually, opposed kinematic poses. (3) Lastly, we demonstrated the benefits of using kinematics data as supervisory signals during training, which yields improved performance on video-only baselines where kinematic data are not available during testing, seen with AuxTranformer's numbers. Overall, kinematics provide a wealth of clean motion signals that is essential for skill assessment, which helps to inspire LiveMAE for assessment in live videos.

Table 2. Suturing skill assessment on Live data. Boldfaced denotes best and ± are standard deviations across 5 held-out institutions.

Modality	Model	Repositions		HoldRatio		HoldAngle	
		AUC	AUPRC	AUC	AUPRC	AUC	AUPRC
Video	ConvTransformer	0.822 ± 0.02	0.905 ± 0.02	**0.564 ±0.10**	**0.697 ±0.15**	0.489 ± 0.06	0.813 ± 0.03
Video + Kin.	AuxTransformer	0.831 ± 0.02	0.900 ± 0.01	0.466 ± 0.06	0.631 ± 0.19	0.505 ± 0.02	0.805 ± 0.06
	AuxTransformer-FT	0.828 ± 0.02	0.897 ± 0.01	0.472 ± 0.06	0.630 ± 0.19	0.499 ± 0.05	0.790 ± 0.07
	(Ours) LiveMAE	0.832 ± 0.03	0.911 ± 0.03	0.430 ± 0.05	0.5930 ± 0.20	**0.550 ±0.10**	**0.844 ±0.07**
	(Ours) LiveMAE-FT	**0.837 ±0.01**	**0.912 ±0.02**	0.474 ± 0.03	0.610 ± 0.20	0.489 ± 0.04	0.822 ± 0.04

Modality	Model	DrivingSmoothness		WristRotation		WristRotationNW	
		AUC	AUPRC	AUC	AUPRC	AUC	AUPRC
Video	ConvTransformer	0.667 ± 0.07	0.894 ± 0.06	0.435 ± 0.06	0.649 ± 0.05	0.445 ± 0.01	0.702 ± 0.05
Video + Kin.	AuxTransformer	0.502 ± 0.03	0.830 ± 0.07	0.519 ± 0.04	0.708 ± 0.04	0.519 ± 0.04	0.708 ± 0.04
	AuxTransformer-FT	0.483 ± 0.04	0.810 ± 0.10	0.517 ± 0.04	0.707 ± 0.04	0.520 ± 0.08	0.753 ± 0.06
	(Ours) LiveMAE	0.683 ± 0.08	0.878 ± 0.08	0.543 ± 0.13	0.721 ± 0.10	0.486 ± 0.12	0.723 ± 0.12
	(Ours) LiveMAE-FT	**0.725 ±0.12**	**0.903 ±0.06**	**0.562 ±0.08**	**0.733 ±0.08**	**0.634 ±0.06**	**0.826 ±0.01**

3.2   Evaluation of LiveMAE on Live Videos

Quantitative Results. Table 2 presents automated suturing assessment results for each technical skill on Live data across the 5 held-out institutions. We make 3 key observations: (1) Skill assessment on live stitch using LiveMAE or LiveMAE-finetuned often achieves the best results (outperforming supervised baselines and AuxTransformer with mean AUCs (0.550–0.837) and AUPRC (0.733–0.912) with particular improvements of 35.78% in AUC for wrist rotation skills and 8.7% for needle driving skill. (2) LiveMAE can learn generalizable representations from VR to Live using its shared encoder and kinematic mapping, achieving reasonable performance even without fine-tuning in the *needle repositioning, hold angle* and *wrist rotation* skills. (3) Clinically, we observe that VR data can directly help with live skill assessment, especially in certain skills such as *wrist rotation* and *wrist rotation withdrawal* (+35.78% increase in AUC), where medical students confirmed that the rotation motions (largely responsible for high/low score) are more pronounced in VR suturing videos and less so in Live videos due to how manipulators are visualized in the simulation. Hence training with VR data can help to teach LiveMAE of the desired assessment procedure that is not as clear in Live data and supervised training paradigm. Overall, LiveMAE contributes positively to the task of automated skill assessment, especially in live scenarios where it is not possible to obtain kinematics from the da Vinci® surgical system.

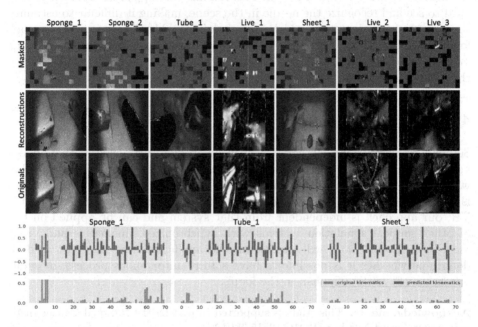

Fig. 3. Visualizations of reconstructed images and kinematics. Images for different exercises {sponge, tube, sheet} and live videos are presented. (a) Masked inputs and reconstructed images vs original images for held-out VR and live samples. (b) Predicted and original 70 kinematic features for the 4 VR samples. The bottom row plots the absolute difference. MSE for held-out VR kinematics are 0.045 ± 0.001.

Qualitative results. We visualized both reconstructed images and kinematics from masked inputs in Fig. 3. Top row of Fig. 3 a shows the 75% masked image where only 1/4 of the visible patches are input into the model. The block patterns are input patches to LiveMAE that were not masked. The middle row shows the image's visual reconstruction vs. the original images (last row). We observe that LiveMAE can pick out and reconstruct the positioning of the manipulators quite well. It also does a good job at reconstructing the target tissue, especially in *Tube1* and *Sheet1*. However, we also observe very small reconstruction artifacts in darker/black regions. This can be attributed to the training data, which sometimes contain all black borders that were not cropped out, yielding the confusion between black borders in live videos and black manipulators in the VR videos. In Fig. 3b, we plot in the top row the original and predicted kinematics of the VR samples in blue and orange, respectively. The bottom row plots their absolute difference. LiveMAE does well in predicting kinematics from unseen samples, especially in *Sheet1* where it gets both positioning and orientations correctly for all instruments of interest, off by at most 0.2. In *Sponge1* and *Tube1*, we notice it does a poor job at estimating poses for some of the instruments, namely the orientation of the needle and target positions (index 4–7, 60–70) in *Sponge1* and the needle orientation (index 4–7) in *Tube1*. This can happen in cases where it is hard to see and recognize the needle in the scene, making it difficult to estimate the exact needle orientation, which may explain LiveMAE's poorer performance for the skill *Needle hold ratio.* and presents a promising direction for future work in diving deeper into CV models to segment out instruments of interest since they can be easily ignored.

4   Conclusion

Self-supervised learning methods, as utilized in our work, showed that video-based evaluation of suturing technical skills in live surgical videos is achievable with robust performance across multiple institutions. Although current work is limited to using VR data from one setup, namely Surgical Science^{TM} Flex VR, our approach is independent from that system and can be applied on top of other surgical simulation systems with synchronized kinematics and video recordings. Future work will expand on the applications we demonstrated to determine whether it is possible to have a fully autonomous process, or semi-autonomously with a "human-in-the-loop".

Acknowledgements. This study is supported in part by the National Cancer Institute under Award Number 1RO1CA251579-01A1.

References

1. Birkmeyer, J.D., et al.: Surgical skill and complication rates after bariatric surgery. N. Engl. J. Med. **369**(15), 1434–1442 (2013). https://doi.org/10.1056/NEJMsa130062
2. Hung, A.J., et al.: A deep-learning model using automated performance metrics and clinical features to predict urinary continence recovery after robot-assisted radical prostatectomy. BJU Int. **124**(3), 487–495 (2019). https://doi.org/10.1111/bju.14735
3. Trinh, L., et al.: Survival analysis using surgeon skill metrics and patient factors to predict urinary continence recovery after robot-assisted radical prostatectomy. Eur. Urol. Focus. **S2405–4569**(21), 00107–00113 (2021). https://doi.org/10.1016/j.euf.2021.04.001

4. Chen, J., et al.: Objective assessment of robotic surgical technical skill: a systematic review. J. Urol. **201**(3), 461–469 (2019). https://doi.org/10.1016/j.juro.2018.06.078

5. Lendvay, T.S., White, L., Kowalewski, T.: Crowdsourcing to assess surgical skill. JAMA Surg. **150**(11), 1086–1087 (2015). https://doi.org/10.1001/jamasurg.2015.2405

6. Hung, A.J., et al.: Road to automating robotic suturing skills assessment: battling mislabeling of the ground truth. Surgery **S0039–6060**(21), 00784–00794 (2021). https://doi.org/10.1016/j.surg.2021.08.014

7. Hung, A.J., Bao, R., Sunmola, I.O., Huang, D.A., Nguyen, J.H., Anandkumar, A.: Capturing fine-grained details for video-based automation of suturing skills assessment. Int. J. Comput. Assist. Radiol. Surg. **18**(3), 545–552 (2023). Epub 2022 Oct 25. PMID: 36282465; PMCID: PMC9975072. https://doi.org/10.1007/s11548-022-02778-x

8. Sanford, D.I., et al.: Technical skill impacts the success of sequential robotic suturing substeps. J. Endourol. **36**(2), 273–278 (2022). PMID: 34779231; PMCID: PMC8861914. https://doi.org/10.1089/end.2021.0417

9. Graves, A.: Long short-term memory. In: Supervised Sequence Labelling with Recurrent Neural Networks. Studies in Computational Intelligence, vol. 385, pp. 37–45. Springer, Heidelberg (2012). https://doi.org/10.1007/978-3-642-24797-2_4

10. Sanh, V., Debut, L., Chaumond, J., Wolf, T.: DistilBERT, a distilled version of BERT: smaller, faster, cheaper and lighter. arXiv preprint arXiv:1910.01108 (2019)

11. He, K., Chen, X., Xie, S., Li, Y., Dollár, P., Girshick, R.: Masked autoencoders are scalable vision learners. In: Proceedings of the IEEE/CVF Conference on Computer Vision and Pattern Recognition, pp. 16000–16009 (2022)

12. Dosovitskiy A, et al.:. An image is worth 16x16 words: transformers for image recognition at scale. arXiv preprint arXiv:2010.11929 (2020)

13. Balvardi, S., et al.: The association between video-based assessment of intraoperative technical performance and patient outcomes: a systematic review. Surg. Endosc. **36**(11), 7938–7948 (2022). Epub 2022 May 12. PMID: 35556166. https://doi.org/10.1007/s00464-022-09296-6

14. Fecso, A.B., Szasz, P., Kerezov, G., Grantcharov, T.P.: The effect of technical performance on patient outcomes in surgery: a systematic review. Ann Surg. **265**(3), 492–501 (2017). PMID: 27537534. https://doi.org/10.1097/SLA.0000000000001959

Cascade Transformer Encoded Boundary-Aware Multibranch Fusion Networks for Real-Time and Accurate Colonoscopic Lesion Segmentation

Ao Wang[1]([✉]), Ming Wu[1]([✉]), Hao Qi[1], Wenkang Fan[1], Hong Shi[3]([✉]),
Jianhua Chen[3], Sunkui Ke[4], Yinran Chen[1], and Xiongbiao Luo[1,2]([✉])

[1] Department of Computer Science and Technology,
Xiamen University, Xiamen, China
awang.xmu@gmail.com, xiongbiao.luo@gmail.com, wuming@stu.xmu.edu.cn
[2] National Institute for Data Science in Health and Medicine,
Xiamen University, Xiamen, China
[3] Fujian Cancer Hospital, Fujian Medical University Cancer Hospital, Fuzhou, China
endoshihong@hotmail.com
[4] Zhongshan Hospital, Xiamen University, Xiamen 361004, China

Abstract. Automatic segmentation of colonoscopic intestinal lesions is essential for early diagnosis and treatment of colorectal cancers. Current deep learning-driven methods still get trapped in inaccurate colonoscopic lesion segmentation due to diverse sizes and irregular shapes of different types of polyps and adenomas, noise and artifacts, and illumination variations in colonoscopic video images. This work proposes a new deep learning model called cascade transformer encoded boundary-aware multibranch fusion networks for white-light and narrow-band colorectal lesion segmentation. Specifically, this architecture employs cascade transformers as its encoder to retain both global and local feature representation. It further introduces a boundary-aware multibranch fusion mechanism as a decoder that can enhance blurred lesion edges and extract salient features, and simultaneously suppress image noise and artifacts and illumination changes. Such a newly designed encoder-decoder architecture can preserve lesion appearance feature details while aggregating the semantic global cues at several different feature levels. Additionally, a hybrid spatial-frequency loss function is explored to adaptively concentrate on the loss of important frequency components due to the inherent bias of neural networks. We evaluated our method not only on an in-house database with four types of colorectal lesions with different pathological features, but also on four public databases, with the experimental results showing that our method outperforms state-of-the-art network models. In particular, it can improve the average dice similarity coefficient and intersection over union from (84.3%, 78.4%) to (87.0%, 80.5%).

A. Wang and M. Wu—Shows the equally contributed authors.
X. Luo and H. Shi are the corresponding authors.

1   Introduction

Colorectal cancer (CRC) is the third most commonly diagnosed cancer but ranks second in terms of mortality worldwide [11]. Intestinal lesions, particularly polyps and adenomas, are usually developed to CRC in many years. Therefore, diagnosis and treatment of colorectal polyps and adenomas at their early stages are essential to reduce morbidity and mortality of CRC. Interventional colonoscopy is routinely performed by surgeons to visually examine colorectal lesions. However these lesions in colonoscopic images are easily omitted and wrongly classified due to limited knowledge and experiences of surgeons. Automatic and accurate segmentation is a promising way to improve colorectal examination.

Many researchers employ U-shaped network [7,13,18] for colonoscopic polyp segmentation. ResUNet++ [7] combines residual blocks and atrous spatial pyramid pooling and Zhao et al. [18] designed a subtraction unit to generate the difference features at multiple levels and constructed a training-free network to supervise polyp-aware features. Unlike a family of U-Net driven segmentation methods, numerous papers have been worked on boundary constraints to segment colorectal polyps. Fan et al. [2] introduced PraNet with reverse attention to establish the relationship between boundary cues from global feature maps generated by a parallel partial decoder. Both polyp boundary-aware segmentation methods work well but still introduce much false positive. Based on PraNet [2] and HardNet [1], Huang et al. [6] removed the attention mechanism and replaced Res2Net50 by HardNet to build HardNet-MSEG that can achieve faster segmentation. In addition, Kim et al. [9] modified PraNet to construct UACANet with parallel axial attention and uncertainty augmented context attention to compute uncertain boundary regions. Although PraNet and UACANet aim to extract ambiguous boundary regions from both saliency and reverse saliency features, they simply set the saliency score to 0.5 that cannot sufficiently detect complete boundaries to separate foreground and background regions. More recently, Shen et al. [10] introduced task-relevant feature replenishment networks for cross-center polyp segmentation, while Tian et al. [12] combined transformers and multiple instance learning to detect polyps in a weakly supervised way.

Unfortunately, limited field of view and illumination variations usually result in insufficient boundary contrast between intestinal lesions and their surrounding tissues. On the other hand, various polyps and adenomas with different pathological features have similar visual characteristics to intestinal folds. To address these issues mentioned above, we explore a new deep learning architecture called cascade transformer encoded boundary-aware multibranch fusion (CTBMF) networks with cascade transformers and multibranch fusion for polyp and adenoma segmentation in colonoscopic white-light and narrow-band video images. Several technical highlights of this work are summarized as follows. First, we construct cascade transformers that can extract global semantic and subtle boundary features at different resolutions and establish weighted links between global semantic cues and local spatial ones for intermediate reasoning, providing long-range dependencies and a global receptive field for pixel-level segmentation. Next, a hybrid spatial-frequency loss function is defined to compensate for loss features in the spatial domain but available in the frequency domain. Additionally, we built a new colonoscopic lesion image database and will make it publicly available, while this work also conducts a thorough evaluation and comparison on our new database and four publicly available ones (Fig. 2).

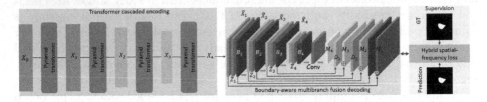

Fig. 1. CTBMF consists of cascade transformers, boundary-aware multibranch fusion, and hybrid spatial-frequency loss.

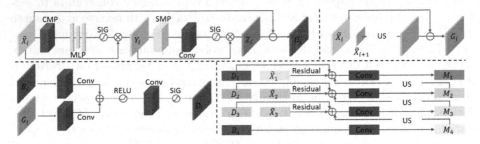

Fig. 2. The boundary-aware multibranch fusion decoder employs the boundary-aware attention module to compute B_i, G_i, and D_i and introduces residual multibranch fusion to calculate M_i.

2   Approaches

This section details our CTBMF networks that can refine inaccurate lesion location, rough or blurred boundaries, and unclear textures. Figure 1 illustrates the encoder-decoder architecture of CTMBF with three main modules.

2.1   Transformer Cascade Encoding

This work employs a pyramid transformer [15] to build a transformer cascaded encoder. Let X_0 and X_i be input patches and the feature map at stage i, respectively. Overlapping patch embedding (OPE) separates an image into fixed-size patches and linearly embeds them into tokenized images while making adjacent windows overlap by half of a patch. Either key K_i or value V_i is the input sequence of linear spatial reduction (LSR) that implements layer normalization (LN) and average pooling (AP) to reduce the input dimension:

$$\text{LSR}(K_i) = \text{AP}(\text{Reshape}(\text{LN}(K_i \oplus \Omega(K_i), R_i)W_{K_i}) \quad (1)$$

where $\Omega(\cdot)$ denotes the output parameters of position embedding, $\oplus$ is the element-wise addition, W_{K_i} indicates the parameters that reduces the dimension of K_i or V_i, and R_i is the reduction ratio of the attention layers at stage i. As the output of LSR is fed into multihead attention, we can obtain attention feature map A_i^j from head j ($j = 1, 2, \cdots, N$, N is the head number of the attention layer) at stage i:

$$A_i^j = \text{Attention}(QW_{Q_i}^j, \text{LSR}(K_i)W_{K_i}^j, \text{LSR}(V_i)W_{V_i}^j) \quad (2)$$

where Attention($\cdot$) is calculated as the original transformer [14]. Subsequently, the output $\text{LSRA}(Q_i, K_i, V_i)$ of LSRA is

$$\text{LSRA}(Q_i, K_i, V_i) = (A_i^1 \odot \cdots A_i^j \odot \cdots A_i^N)W_{A_i} \quad (3)$$

where $\odot$ is the concatenation and W_{A_i} is the linear projection parameters. Then, $\text{LSRA}(Q_i, K_i, V_i)$ is fed into convolutional feed-forward (CFF):

$$\text{CFF}(Q_i, K_i, V_i) = \text{FC}(\text{GELU}(\text{DC}(\text{FC}(\text{LN}(\Psi))))) \quad (4)$$

$$\Psi = ((Q_i, K_i, V_i) \oplus \Omega(Q_i, K_i, V_i)) \oplus \text{LSRA}(Q_i, K_i, V_i) \quad (5)$$

where DC is a 3×3 depth-wise convolution [5] with padding size of between the fully-connected (FC) layer and the Gaussian error linear unit (GELU) [4] in the feed-forward networks. Eventually, the output feature map X_i of the pyramid transformer at stage i can be represented by

$$X_i = \text{Reshape}(\Psi \oplus \text{CFF}(Q_i, K_i, V_i)) \quad (6)$$

2.2   Boundary-Aware Multibranch Fusion Decoding

Boundary-Aware Attention Module. Current methods [2,9] detect ambiguous boundaries from both saliency and reverse-saliency maps by predefining a saliency score of 0.5. Unfortunately, a predefined score cannot distinguish foreground and background of different colonoscopic lesions [3]. Based on [17], this work explores an effective boundary-aware attention mechanism to adaptively extract boundary regions.

Given the feature map X_i with semantic cues and rough appearance details, we perform convolution (Conv) on it and obtain $\tilde{X}_i = \text{Conv}(X_i)$, which is further augmented by channel and spatial attentions. The channel attention performs channel maxpooling (CMP), multilayer perceptron (MLP), and sigmoid (SIG) to obtain the intermediate feature map Y_i:

$$Y_i = \tilde{X}_i \otimes \text{SIG}(\text{MLP}(\text{CMP}(\tilde{X}_i))) \quad (7)$$

where $\otimes$ indicates the elementwise product. Subsequently, the detail enhanced feature map Z_i of the channel-spatial attention is

$$Z_i = Y_i \otimes \mathrm{SIG}(\mathrm{Conv}(\mathrm{SMP}(Y_i))) \tag{8}$$

where SMP indicates spatial maxpooling. We subtract the feature map $\tilde{X}_i$ from the enhanced map Z_i to obtain the augmented boundary attention map B_i, and also establish the correlation between the neighbor layers X_{i+1} and X_i to generate multilevel boundary map G_i:

$$B_i = Z_i \ominus \tilde{X}_i, i = 1, 2, 3, 4 \qquad G_i = \tilde{X}_i \ominus \mathrm{US}(\tilde{X}_{i+1}), i = 1, 2, 3 \tag{9}$$

where $\ominus$ and US indicate subtraction and upsampling.

Residual Multibranch Fusion Module. To highlight salient regions and suppress task-independent feature responses (e.g., blurring), we linearly aggregate B_i and G_i to generate discriminative boundary attention map D_i:

$$D_i = \mathrm{SIG}(\mathrm{Conv}(\mathrm{RELU}(\mathrm{Conv}(B_i) \oplus \mathrm{Conv}(G_i)))), \tag{10}$$

where $i = 1, 2, 3$ and RELU is the rectified linear unit function.

We obtain the fused feature representation map M_i $(i = 1, 2, 3, 4)$ from the elementwise addition or summation of M_{i+1}, D_i, and the residual feature $\tilde{X}_i$ by

$$M_i = \mathrm{Conv}(\tilde{X}_i \oplus D_i \oplus \mathrm{US}(M_{i+1})), i = 1, 2, 3 \qquad M_4 = \mathrm{Conv}(B_4) \tag{11}$$

Eventually, the output M_1 of the boundary-aware multibranch fusion decoder is represented by the following equation:

$$M_1 = \mathrm{Conv}(\tilde{X}_1 \oplus D_1 \oplus \mathrm{US}(M_2)) \tag{12}$$

which precisely combines global semantic features with boundary or appearance details of colorectal lesions.

2.3   Hybrid Spatial-Frequency Loss

This work proposes a hybrid spatial-frequency loss function $\mathcal{H}_L$ to train our network architecture for colorectal polyp and adenoma segmentation:

$$\mathcal{H}_L = \mathcal{S}_L + \mathcal{F}_L \tag{13}$$

where $\mathcal{S}_L$ and $\mathcal{F}_L$ are a spatial-domain loss and a frequency-domain loss to calculate the total difference between prediction P and ground truth G, respectively. The spatial-domain loss $\mathcal{S}_L$ consists of a weighted intersection over union loss and a weighted binary cross entropy loss [16].

The frequency-domain loss $\mathcal{F}_L$ can be computed by [8]

$$\mathcal{F}_L = \lambda \frac{1}{WH} \sum_{u=0}^{W-1} \sum_{v=0}^{H-1} \gamma(u,v)|\mathcal{G}(u,v) - \mathcal{P}(u,v)|^2 \tag{14}$$

where $W \times H$ is the image size, λ is the coefficient of $\mathcal{F}_L$, $\mathcal{G}(u, v)$ and $\mathcal{P}(u, v)$ are a frequency representation of ground truth G and prediction P using 2-D discrete Fourier transform. $\gamma(u, v)$ is a spectrum weight matrix that is dynamically determined by a non-uniform distribution on the current loss of each frequency.

3   Experiments

Our clinical in-house colonoscopic videos were acquired from various colonoscopic procedures under a protocol approved by the research ethics committee of the university. These white-light and narrow-band colonoscopic images contain four types of colorectal lesions with different pathological features classified by surgeons: (1) 268 cases of hyperplastic polyp, (2) 815 cases of inflammatory polyp, (3) 1363 cases of tubular adenoma, and (4) 143 cases of tubulovillous adenoma. Additionally, four public datasets including Kvasir, ETIS-LaribPolypDB, CVC-ColonDB, and CVC-ClinicDB were also used to evaluate our network model.

We implemented CTBMF on PyTorch and trained it with a single NVIDIA RTX3090 to accelerate the calculations for 100 epochs at mini-batch size 16. Factors λ (Eq. (14)) were set to 0.1. We employ the stochastic gradient descent

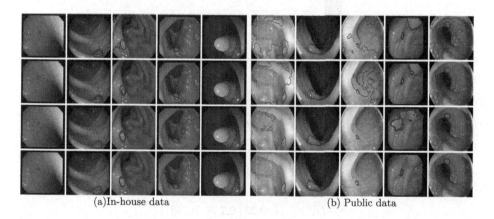

(a)In-house data   (b) Public data

Fig. 3. Visual comparison of the segmentation results of using the four different methods tested on those in-house and public datasets. *Green* and *blue* show ground truth and prediction. (Color figure online)

Table 1. Results and computational time of using five databases(our in-house and four public databases)

Average	DSC	IoU	F_β	In-house	Public	Average
PraNet [2]	0.813	0.751	0.795	28.7 FPS	30.5 FPS	29.6 FPS
HardNet [6]	0.830	0.764	0.812	39.2 FPS	39.9 FPS	39.5 FPS
UACANet [9]	0.843	0.784	0.824	16.5 FPS	16.7 FPS	16.6 FPS
CTBMF (Ours)	**0.870**	**0.805**	**0.846**	**33.1 FPS**	**33.6 FPS**	**33.4 FPS**

algorithm to optimize the overall parameters with an original learning rate of 0.0001 for cascade transformer encoding and 0.05 for other parts and use warm-up and linear decay strategies to adjust it. The momentum and weight decay were set as 0.9 and 0.0005. Further, we resized input images to 352×352 for training and testing and the training time was nearly 1.5 h to achieve the convergence. We employ three metrics to evaluate the segmentation: Dice similarity coefficient (DSC), intersection over union (IoU), and weighted F-measure (F_β).

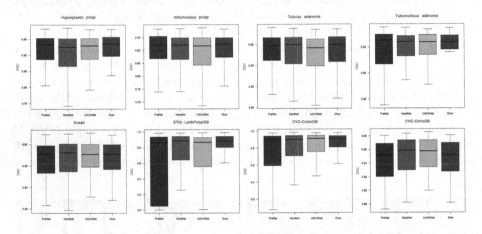

Fig. 4. DSC boxplots of using the four methods evaluated on our in-house and publicly available databases

Table 2. Public data segmented results of our ablation study

Modules	DSC	IoU	F_β
D_1	0.681	0.593	0.634
D_2	0.822	0.753	0.798
D_3	0.820	0.748	0.793
Residual	0.828	0.757	0.802
$\mathcal{F}_L$	0.834	0.764	0.804

4   Results and Discussion

Figure 3 visually compares the segmentation results of the four methods tested on our in-house and public databases. Our method can accurately segment polyps in white-light and narrow-band colonoscopic images under various scenarios, and CTBMF can successfully extract small, textureless and weak boundary and colorectal lesions. The segmented boundaries of our method are sharper and clear than others especially in textureless lesions that resemble intestinal lining.

Figure 4 shows the DSC-boxplots to evaluate the quality of segmented polyps and adenomas, which still demonstrate that our method works much better than the others. Figure 5 displays the enhanced feature maps using the boundary-aware attention module. Evidently, small and weak-boundary or textureless lesions can be enhanced with good boundary feature representation.

Table 1 summarizes the quantitative results in accordance with the three metrics and computational time of four methods. Evidently, CTBMF generally works better than the compared methods on the in-house database with four types of colorectal lesions. Furthermore, we also summarizes the average three metrics computed from all the five databases (the in-house dataset and four public datasets). Our method attains much higher average DSC and IoU of (0.870, 0.805) than the others on the five databases.

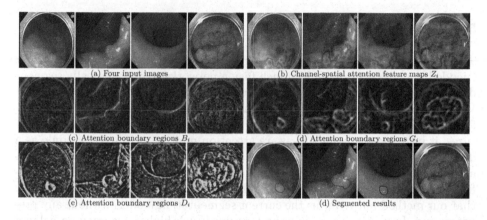

(a) Four input images   (b) Channel-spatial attention feature maps Z_i

(c) Attention boundary regions B_i (d) Attention boundary regions G_i

(e) Attention boundary regions D_i (d) Segmented results

Fig. 5. Effectiveness of the boundary-aware multibranch fusion decoder generated various boundary-aware feature maps

We performed an ablation study to evaluate the effectiveness of each module used in CTBMF. The baseline is the standard version of cascade pyramid transformers. Modules D_1, D_2, D_3, residual connections, and frequency loss $\mathcal{F}_L$ are gradually added into the baseline, evaluating the effectiveness of each module and comparing the variants with each other. We tested these modules on the four public databases. Table 2 shows all the ablation study results. Each module can improve the segmentation performance. Particularly, the boundary-aware attention module critically improves the average DSC, IoU, and F_β.

Our method generally works better than the other three methods. Several reasons are behind this. First, the cascade-transformer encoder can extract local and global semantic features of colorectal lesions with different pathological characteristics due to its pyramid representation and linear spatial reduction attention. While the pyramid operation extracts multiscale local features, the attention mechanism builds global semantic cues. Both pyramid and attention strategies facilitate the representation of small and textureless intestinal lesions in encoding, enabling to characterize the difference between intestinal folds (linings) and subtle-texture polyps or small adenomas. Next, the boundary-aware attention mechanism drives the multibranch fusion, enhancing the representation of intestinal lesions in weak boundary and nonuniform lighting. Such a mechanism first extracts the channel-spatial attention feature map, from which subtracts the current pyramid transformer's feature map to enhance the boundary information. Also, the multibranch fusion generates multilevel boundary maps by subtracting the next pyramid transformer's upsampling output from the current pyramid transformer's output, further improving the boundary contrast. Additionally, the hybrid spatial-frequency loss was also contributed to the improvement of colorectal lesion segmentation. The frequency-domain information can compensate loss feature information in the spatial domain, leading to a better supervision in training.

5   Conclusion

This work proposes a new deep learning model of cascade pyramid transformer encoded boundary-aware multibranch fusion networks to automatically segment different colorectal lesions of polyps and adenomas in colonoscopic imaging. While such an architecture employs simple and convolution-free cascade transformers as an encoder to effectively and accurately extract global semantic features, it introduces a boundary-aware attention multibranch fusion module as a decoder to preserve local and global features and enhance structural and boundary information of polyps and adenomas, as well as it uses a hybrid spatial-frequency loss function for training. The thorough experimental results show that our method outperforms the current segmentation models without any preprocessing. In particular, our method attains much higher accuracy on colonoscopic images with small, illumination changes, weak-boundary, textureless, and motion blurring lesions, improving the average dice similarity coefficient and intersection over union from (89.5%, 84.1%) to (90.3%, 84.4%) on our in-house database, from (78.9%, 72.6%) to (83.4%, 76.5%) on the four public databases, and from (84.3%, 78.4%) to (87.0%, 80.5%) on the five databases.

Acknowledgements. This work was supported partly by the National Natural Science Foundation of China under Grants 61971367 and 82272133, the Natural Science Foundation of Fujian Province of China under Grant 2020J01004, and the Fujian Provincial Technology Innovation Joint Funds under Grant 2019Y9091.

References

1. Chao, P., Kao, C.Y., Ruan, Y.S., Huang, C.H., Lin, Y.L.: Hardnet: a low memory traffic network. In: Proceedings of the IEEE/CVF International Conference on Computer Vision, pp. 3552–3561 (2019)
2. Fan, D.-F., et al.: PraNet: parallel reverse attention network for polyp segmentation. In: Martel, A.L., et al. (eds.) MICCAI 2020. LNCS, vol. 12266, pp. 263–273. Springer, Cham (2020). https://doi.org/10.1007/978-3-030-59725-2_26

3. Guo, X., Yang, C., Liu, Y., Yuan, Y.: Learn to threshold: thresholdnet with confidence-guided manifold mixup for polyp segmentation. IEEE Trans. Med. Imaging **40**(4), 1134–1146 (2020)
4. Hendrycks, D., Gimpel, K.: Gaussian error linear units (gelus). arXiv preprint arXiv:1606.08415 (2016)
5. Howard, A., et al.: Efficient convolutional neural networks for mobile vision. arXiv preprint arXiv:1704.04861 (2017)
6. Huang, C.H., Wu, H.Y., Lin, Y.L.: HarDNet-MSEG: a simple encoder-decoder polyp segmentation neural network that achieves over 0.9 mean dice and 86 fps. arXiv preprint arXiv:2101.07172 (2021)
7. Jha, D., et al.: ResUNet++: an advanced architecture for medical image segmentation. In: 2019 IEEE International Symposium on Multimedia (ISM), pp. 225–2255. IEEE (2019)
8. Jiang, L., Dai, B., Wu, W., Loy, C.C.: Focal frequency loss for image reconstruction and synthesis. In: Proceedings of the IEEE/CVF International Conference on Computer Vision, pp. 13919–13929 (2021)
9. Kim, T., Lee, H., Kim, D.: UACANet: uncertainty augmented context attention for polyp segmentation. In: Proceedings of the 29th ACM International Conference on Multimedia, pp. 2167–2175 (2021)
10. Shen, Y., Lu, Y., Jia, X., et al.: UACANet: uncertainty augmented context attention for polyp segmentation. In: Medical Image Computing and Computer Assisted Intervention (MICCAI), pp. 599–608 (2022)
11. Sung, H., et al.: Global cancer statistics 2020: GLOBOCAN estimates of incidence and mortality worldwide for 36 cancers in 185 countries. CA Cancer J. Clin. **71**(3), 209–249 (2021)
12. Tian, Y., Pang, G., Liu, F., et al.: Contrastive transformer-based multiple instance learning for weakly supervised polyp frame detection. In: Wang, L., Dou, Q., Fletcher, P.T., Speidel, S., Li, S. (eds.) Medical Image Computing and Computer Assisted Intervention – MICCAI 2022. MICCAI 2022. LNCS, vol. 13433, pp. 88–98. Springer, Cham (2022). https://doi.org/10.1007/978-3-031-16437-8_9
13. Tomar, N.K., et al.: DDANet: dual decoder attention network for automatic polyp segmentation. In: Del Bimbo, A., et al. (eds.) ICPR 2021. LNCS, vol. 12668, pp. 307–314. Springer, Cham (2021). https://doi.org/10.1007/978-3-030-68793-9_23
14. Vaswani, A., et al.: Attention is all you need. Adv. Neural Inf. Process. Syst. **30** (2017)
15. Wang, W., Xie, E., Li, X., et al.: PVT v2: improved baselines with pyramid vision transformer. Comput. Vis. Media **8**, 415–424 (2022)
16. Wei, J., Wang, S., Huang, Q.: F^{3}net: fusion, feedback and focus for salient object detection. In: Proceedings of the AAAI Conference on Artificial Intelligence, vol. 34, pp. 12321–12328 (2020)
17. Woo, S., Park, J., Lee, J.-Y., Kweon, I.S.: CBAM: convolutional block attention module. In: Ferrari, V., Hebert, M., Sminchisescu, C., Weiss, Y. (eds.) ECCV 2018. LNCS, vol. 11211, pp. 3–19. Springer, Cham (2018). https://doi.org/10.1007/978-3-030-01234-2_1
18. Zhao, X., Zhang, L., Lu, H.: Automatic polyp segmentation via multi-scale subtraction network. In: de Bruijne, M., et al. (eds.) MICCAI 2021. LNCS, vol. 12901, pp. 120–130. Springer, Cham (2021). https://doi.org/10.1007/978-3-030-87193-2_12

Learnable Query Initialization for Surgical Instrument Instance Segmentation

Rohan Raju Dhanakshirur[1]([⊠]), K. N. Ajay Shastry[1], Kaustubh Borgavi[1],
Ashish Suri[2], Prem Kumar Kalra[1], and Chetan Arora[1]

[1] Indian Institute of Technology Delhi, New Delhi, India
rohanrd@sit.iitd.ac.in, rohanrd28296@gmail.com
[2] AIIMS, New-Delhi, India

Abstract. Surgical tool classification and instance segmentation are crucial for minimally invasive surgeries and related applications. Though most of the state-of-the-art for instance segmentation in natural images use transformer-based architectures, they have not been successful for medical instruments. In this paper, we investigate the reasons for the failure. Our analysis reveals that this is due to incorrect query initialization, which is unsuitable for fine-grained classification of highly occluded objects in a low data setting, typical for medical instruments. We propose a class-agnostic Query Proposal Network (QPN) to improve query initialization inputted to the decoder layers. Towards this, we propose a deformable-cross-attention-based learnable Query Proposal Decoder (QPD). The proposed QPN improves the recall rate of the query initialization by 44.89% at 0.9 IOU. This leads to an improvement in segmentation performance by 1.84% on Endovis17 and 2.09% on Endovis18 datasets, as measured by ISI-IOU. The source code can be accessed at https://aineurosurgery.github.io/learnableQPD.

1 Introduction

Background: Minimally invasive surgeries (MIS) such as laparoscopic and endoscopic surgeries have gained widespread popularity due to the significant reduction in the time of surgery and post-op recovery [16,21]. Surgical instrument instance segmentation (SIIS) in these surgeries opens up the doors to increased precision and automation [18]. However, the problem is challenging due to the lack of large-scale well-annotated datasets, occlusions of tool-tip (the distinguishing part of the surgical instrument), rapid changes in the appearance, reflections due to the light source of the endoscope, smoke, blood spatter etc. [5].

The Challenge: Most modern techniques for SIIS [11,18,31] are multi-stage architectures, with the first stage generating region proposals (rectilinear boxes)

Supplementary Information The online version contains supplementary material available at https://doi.org/10.1007/978-3-031-43996-4_70.

and the second stage classifying each proposal independently. Unlike natural images, rectilinear bounding boxes are not an ideal choice for medical instruments, which are long, thin, and often visible diagonally in a bounding box. Thus, the ratio of the visible tool area to the area of the bounding box is highly skewed in medical scenarios. E.g., the ratio is 0.45 for the Endovis17 dataset and 0.47 for Endovis18, the two popular MIS datasets. In contrast, it is 0.60 for the MS-COCO dataset of natural images. The ratio is important because lower numbers imply more noise due to background and a more difficult classification problem.

Current Solution Strategy: Recently, S3Net [4] adapted the MaskRCNN [14] backbone to propose a 3-stage architecture. Their third stage implements hard attention based on the predicted masks from the second stage and re-classifies the proposals. The hard attention avoids distraction due to the presence of large background regions in the proposal boxes, allowing them to outperform all the previous state of the art for medical instruments or natural images.

Our Observation: In the last few years, attention-based transformer architectures have outperformed CNN architectures for many computer-vision-based tasks. Recent transformer-based object detection models implement deformable attention [6,19,33] which predicts sampling points to focus attention on the fine-grained features in an image. One expects that this would allow transformer architectures to concentrate only on the tool instead of the background, leading to high accuracy for medical instrument instance segmentation. However, in our experiments, as well as the ones reported by [4], this is not observed. We investigate the reasons and report our findings on the probable causes. We also propose a solution strategy to ameliorate the problem. Our implementation of the strategy sets up a new state of the art for the SIIS problem.

Contributions: (1) We investigate the reason for the failure of transformer-based object detectors for the medical instrument instance segmentation tasks. Our analysis reveals that incorrect query initialization is to blame. We observe that recall of an instrument based on the initialized queries is a lowly 7.48% at 0.9 IOU, indicating that many of the relevant regions of interest do not even appear in the initialized queries, thus leading to lower accuracy at the last stage. **(2)** We observe that CNN-based object detectors employ a non-maximal suppression (NMS) at the proposal stage, which helps spread the proposal over the whole image. In contrast in transformer-based detection models, this has been replaced by taking the highest confidence boxes. In this paper, we propose to switch back to NMS-based proposal selection in transformers. **(3)** The NMS uses only bounding boxes and does not allow content interaction for proposal selection. We propose a Query Proposal Decoder block containing multiple layers of self-attention and deformable cross-attention to perform region-aware refinement of the proposals. The refined proposals are used by a transformer-based decoder backbone for the prediction of the class label, bounding box, and segmentation mask. **(4)** We show an improvement of 1.84% over the best-performing SOTA technique on the Endovis17 and 2.09% on the Endovis18 dataset as measured by ISI-IOU.

2   Related Work

Datasets: Surgical tool recognition and segmentation has been a well-explored research topic [11]. Given the data-driven nature of recent computer vision techniques, researchers have also proposed multiple datasets for the problem. Twinanda et al. [28] have proposed an 80-video dataset of cholecystectomy surgeries, with semantic segmentation annotation for 7 tools. Al Hajj et al. [1] proposed a 50-video dataset of phacoemulsification cataract surgeries with 21 tools annotated for bounding boxes. Cadis [12] complements this dataset with segmentation masks. Ross et al. [25] in 2019 proposed a 30-video dataset corresponding to multiple surgeries with one instrument, annotated at image level for its presence. Endovis datasets, Endovis 2017 (EV17) [3] and Endovis 2018 (EV18) [2] have gained popularity in the recent past. Both of them have instance-level annotations for 7 tools. EV17 is a dataset of 10 videos of the Da Vinci robot, and EV18 is a dataset of 15 videos of abdominal porcine procedures.

Instance Segmentation Techniques for Medical Instruments: Multiple attempts have been made to perform instance segmentation using these datasets. [11] and [18] use a MaskRCNN-based [14] backbone pre-trained on natural images and perform cross-domain fine-tuning. Wang et al. [31] assign categories to each pixel within an instance and convert the problem into a pixel classification. They then use the ResNet backbone to solve the problem. [29] modified the MaskRCNN architecture and proposed a Sample Consistency Network to bring closer the distribution of the samples at training and test time. Ganea et al. [10] use the concept of few-shot learning on top of MaskRCNN to improve the performance. Wentao et al. [8] add a mask prediction head to YoLo V3 [23] All these algorithms use CNN-based architectures, with ROI-Align, to crop the region of interest. Since the bounding boxes are not very tight in surgical cases due to the orientation of the tools, a lot of background information is passed along with the tool, and thereby the classification performance is compromised. Baby et al. [4] use a third-stage classifier on top of MaskRCNN to correct the misclassified masks and improve the performance.

Transformer-based Instance Segmentation for Natural Images: On the other hand, transformer-based instance segmentation architectures [6,13,17,19] generate sampling points to extract the features and thereby learn more localised information. This gives extra leverage to transformer architectures to perform better classification. [17] propose the first transformer-based end-to-end instance segmentation architecture. They predict low-level mask embeddings and combine them to generate the actual masks. [13] learn the location-specific features by providing the information on position embeddings. [6] uses a deformable-multi-head-attention-based mechanism to enrich the segmentation task. Mask DINO [19] utilizes better positional priors as originally proposed in [33]. They also perform box refinement at multiple levels to obtain the tight instance mask. In these architectures, the query initialization is done using the *top-k* region proposals based on their corresponding classification score. Thus ambiguity in the classification results in poor query initialization, and thereby the entire mask corresponding to that instance is missed. This leads to a significant reduction in the recall rate of these models.

3   Proposed Methodology

Backbone Architecture: We utilize Mask DINO [19] as the backbone architecture for our model. As illustrated in Fig. 1, it uses ResNet [15] as the feature extractor to generate a multi-scale feature map at varying resolutions. These feature maps are then run through an encoder to generate an enhanced feature map with the same resolution as the original feature map. The enhanced feature maps are used to generate a set of region proposals. Then, the region proposals are sorted based on their classification logit values. Mask DINO uses a d dimensional query vector to represent an object's information. The top n_q generated region proposals are used to initialize a set of n_q queries. These queries are then passed through a series of decoder layers to obtain a set of refined queries. These refined queries are used for the purpose of detection, classification, and segmentation.

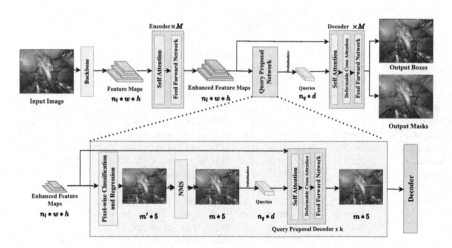

Fig. 1. Proposed Query Proposal Network

Problems with Mask DINO: The model predicts various outputs using the queries initialized with the top n_q region proposals. Our analysis reveals that most false negative outputs are due to the queries initialized with significantly fewer to no region proposals corresponding to the missed objects. During the initialization procedure, Mask DINO sorts region proposals based on their classification logit values. The surgical instruments resemble one another and have few distinguishing features. In case of label confusion, which happens often in medical instruments, the encoder outputs a proposal with low confidence. When sorted, these low-confidence proposals get deleted. Hence, in the proposed architecture we argue for class-independent proposal selection, which does not rely on the classification label or its confidence at this stage.

Query Proposal Network: Spatial Diversification of the Proposals: The proposed Query Proposal Network (QPN) is shown in Fig. 1. QPN takes the enhanced feature maps as input and performs a pixel-wise classification and regression to obtain the initial region proposals. These initial region proposals undergo Non-Maximal Suppression (NMS) in order to remove the duplicates, but more importantly, output the proposal boxes which are spread all through the image. This is important because, given the complexity of the classification in medical instruments, we do not wish to overly rely on the classification label and would rather explore more regions of interest in the decoder. Hence, we choose top k proposals based on the NMS instead of the label confidence.

Table 1. The comparison of the proposed methodology against the other SOTA architectures for the Endovis 17 [3] dataset. Here rows 4–17 are obtained from [4]. Note that [34] is a video instance segmentation paper and uses video information to perform segmentation.

Method	Conference	Instrument Classes IOU							Ch. IOU	ISI. IOU	MC. IOU
		BF	PF	LND	VS	GR	MCS	UP			
Dataset EV17											
SimCaL [30]	ECCV20	39.44	38.01	46.74	16.52	1.90	1.98	13.11	49.56	45.71	23.78
CondInst [27]	ECCV20	44.29	38.03	47.38	24.77	4.51	15.21	15.67	59.02	52.12	27.12
BMaskRCNN [7]	ECCV20	32.89	32.82	41.93	12.66	2.07	1.37	14.43	49.81	38.81	19.74
SOLO [31]	NeurIPS20	22.05	23.17	41.07	7.68	0.00	11.29	4.60	35.41	33.72	15.79
ISINET [11]	MICCAI20	38.70	38.50	50.09	27.43	2.01	28.72	12.56	55.62	52.20	28.96
SCNet [29]	AAAI21	43.96	29.54	48.75	22.89	1.19	4.90	14.47	48.17	46.92	25.98
MFTA [10]	CVPR21	31.16	35.07	39.9	12.05	2.28	6.08	11.61	46.16	41.77	20.27
Detectors [22]	CVPR21	48.54	34.36	49.72	20.33	2.04	8.92	10.58	50.93	47.38	24.93
Orienmask [8]	ICCV21	40.42	28.78	44.48	12.11	3.91	15.18	12.32	42.09	39.27	23.22
QueryInst [9]	ICCV21	20.87	12.37	46.75	10.48	0.52	0.39	4.58	33.59	33.06	15.32
FASA [32]	ICCV21	20.13	18.81	39.12	8.34	0.68	2.17	3.46	34.38	29.67	13.24
Mask2Former [6]	CVPR22	19.60	20.22	45.44	11.95	0.00	1.48	22.10	40.39	39.84	17.78
TraSeTR* [34]	ICRA22	45.20	**56.70**	55.80	38.90	11.40	31.30	18.20	60.40	65.20	36.79
S3Net [4]	WACV23	**75.08**	54.32	61.84	35.50	27.47	43.23	28.38	72.54	71.99	46.55
Mask DINO [19]	Archive 22	64.14	45.29	78.96	59.35	21.90	36.70	30.72	75.96	77.63	43.86
Proposed architecture		70.61	45.84	**80.01**	**63.41**	**33.64**	**66.57**	**35.28**	**77.8**	**79.58**	**49.92**

Query Proposal Network: Content-based Inter-proposal Interaction: The top k region proposals from NMS are used to initialize the queries for the proposed Query Proposal Decoder (QPD). Note that the NMS works only on the basis of box coordinates and does not take care of content embeddings into account. We try to make up for the gap through the QPD module. The QPD module consists of self-attention, cross-attention and feed-forward layers. The self-attention layer of QPD allows the queries to interact with each other and the duplicates are avoided. We use the standard deformable cross-attention module as proposed in [35]. Unlike traditional cross-attention-based mechanisms, this method attends only to a fixed number of learnable key points around the middlemost pixel of every region proposal (query) irrespective of the size of the feature maps. This allows us to achieve better convergence in larger feature maps. Thus, the cross-attention layer allows the interaction of queries with the enhanced feature maps from the encoder and feature representation for each query is obtained. The feed-forward layer refines the queries based on the feature representation obtained in the previous layer. We train the QPD layers using the mask and bounding box loss as is common in transformer architectures [19]. Note that no classification loss is back-propagated. This allows the network to perform query refinement irrespective of the classification label and retains queries corresponding to the object instances, which were omitted due to ambiguity in classification. The queries outputted by the QPD module are used to initialize the standard decoder network of Mask DINO.

Table 2. The comparison of the proposed methodology against the other SOTA architectures for the Endovis 18 [2] dataset. Here rows 4–17 are obtained from [4] Note that [34] is a video instance segmentation paper and uses video information to perform segmentation.

Method	Conference	Instrument Classes IOU							Ch. IOU	ISI. IOU	MC. IOU
		BF	PF	LND	SI	CA	MCS	UP			
Dataset EV18											
SimCaL [30]	ECCV20	73.67	40.35	5.57	0.00	0.00	89.84	0.00	68.56	67.58	29.92
Condlnst [27]	ECCV20	77.42	37.43	7.77	43.62	0.00	87.8	0.00	72.27	71.55	36.29
BMaskRCNN [7]	ECCV20	70.04	28.91	9.97	45.01	4.28	86.73	3.31	68.94	67.23	35.46
SOLO [31]	NeurIPS20	69.46	23.92	2.61	36.19	0.00	87.97	0.00	65.59	64.88	31.45
ISINET [31]	MICCAI20	73.83	48.61	30.98	37.68	0.00	88.16	2.16	73.03	70.97	40.21
SCNet [29]	AAAI21	78.40	47.97	5.22	29.52	0.00	86.69	0.00	71.74	70.99	35.40
MFTA [10]	CVPR21	71.00	31.62	3.93	43.48	9.90	87.77	3.86	69.20	67.97	35.94
Detectors [22]	CVPR21	73.94	46.85	0.00	0.00	0.00	79.92	0.00	66.69	65.06	28.67
Orienmask [8]	ICCV21	68.95	38.66	0.00	31.25	0.00	91.21	0.00	67.69	66.77	32.87
Querylnst [9]	ICCV21	74.13	31.68	2.30	0.00	0.00	87.28	0.00	66.44	65.82	27.91
FASA [32]	ICCV21	72.82	37.64	5.62	0.00	0.00	89.02	1.03	68.31	66.84	29.45
Mask2Former [6]	CVPR22	69.35	24.13	0.00	0.00	0.00	89.96	10.29	65.47	64.69	27.67
TraSeTR* [34]	ICRA22	76.30	53.30	46.5	40.6	13.90	86.3	17.5	76.20		47.77
S3Net [4]	WACV23	77.22	50.87	19.83	50.59	0.00	92.12	7.44	75.81	74.02	42.58
Mask DINO [19]	Archive 22	82.35	57.67	0.83	60.46	0.00	90.73	0.00	75.63	76.39	41.73
Proposed architecture		82.8	60.94	19.96	49.70	0.00	93.93	0.00	77.77	78.43	43.84

Implementation Details: We train the proposed architecture using three kinds of losses, classification loss $\mathcal{L}_{cls}$, box regression loss $\mathcal{L}_{box}$, and the mask prediction loss $\mathcal{L}_{mask}$. We use focal loss [20] as $\mathcal{L}_{cls}$. We use ℓ_1 and GIOU [24] loss for $\mathcal{L}_{box}$. For $\mathcal{L}_{mask}$, we use cross entropy and IOU (or dice) loss. The total loss is:

$$\mathcal{L}_{total} = \lambda_1 \mathcal{L}_{cls} + \lambda_2 \ell_1 + \lambda_3 \mathcal{L}_{giou} + \lambda_4 \mathcal{L}_{ce} + \lambda_5 \mathcal{L}_{dice} \tag{1}$$

Through hyper-parameter tuning, we set $\lambda = [0.19, 0.24, 0.1, 0.24, 0.24]$. We use a batch size of 8. The initial learning rate is set to 0.0001, which drops by 0.1 after every 20 epochs. We set 0.9 as the Nesterov momentum coefficient. We train the network for 50 epochs on a server with 8 NVidia A100, 40 GB GPUs. Besides QPN we use the exact same architecture as proposed in MaskDINO [19]. However, we perform transfer learning using the MaskDINO pre-trained weights, and therefore, we do not use "GT+noise" as an input to the decoder.

4   Results and Discussion

Evaluation Methodology: We demonstrate the performance of the proposed methodology on two benchmark Robot-assisted endoscopic surgery datasets Endovis 2017 [3] (denoted as EV17), and Endovis 2018 [2] (denoted as EV18) as performed by [4]. EV17 is a dataset of 10 videos (1800 images) obtained from the Da Vinci robotic system with 7 instruments. We adopt the same four-fold cross-validation strategy as shown by [26] for the EV17 dataset. EV18 is a real

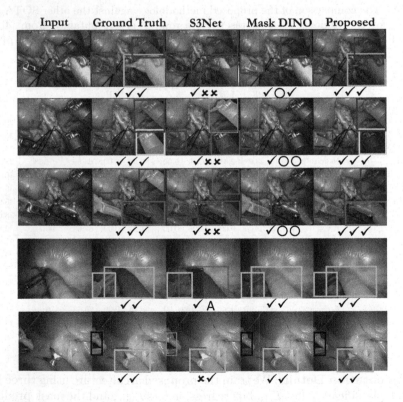

Fig. 2. The qualitative analysis of instance segmentation by the proposed methodology against the other SOTA algorithms: Here, ✓ indicates that the instrument is classified and segmented correctly. O indicates missed instance, × indicates incorrect classification and A indicates ambiguous instance.

surgery dataset of 15 videos containing 1639 training and 596 validation images, with 7 instruments. [4] corrected the misclassified ground truth labels and added instance-level annotations to this dataset. We evaluate the performance of the proposed algorithm using challenge IOU as proposed by [3], as well as ISI IOU and Mean class IOU as reported in [11]. We also report per instrument class IOU as suggested by [4].

Quantitative Results on EV17, EV18, and Cadis: The results of our technique for the EV17 dataset are shown in Table 1 and for the EV18 dataset are shown in Table 2. It can be observed that the proposed technique outperforms the best-performing SOTA methods by 1.84% in terms of challenge IOU for EV17 and 1.96% for EV18 datasets. Due to the paucity of space, we show the performance of the proposed methodology on the Cadis [12] dataset in the supplementary material. The qualitative analysis of instance segmentation by the proposed methodology against the other SOTA algorithms is shown in Fig. 2. We demonstrate improved performance in the occluded and overlapping cases. We observe a testing speed of 40 FPS on a standard 40GB Nvidia A100 GPU cluster.

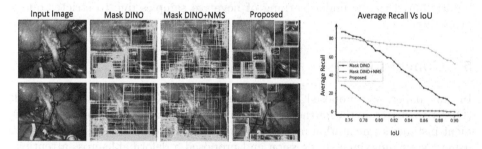

Fig. 3. Proposed improvement in the Query Initialization. Left: two sample images and their corresponding query initialization by different architectures. Right: shows the recall rate at various IOU thresholds

Table 3. Ablation analysis on EV18 dataset to demonstrate the importance of each block in the proposed architecture and to check for the most optimal number of Query Proposal Decoder (QPD) layers.

Configuration	Instrument Classes IOU							Ch. IOU	ISI. IOU	MC. IOU
	BF	PF	LND	SI	CA	MCS	UP			
Vanilla MaskDINO	82.35	57.67	0.83	60.46	0.00	90.73	0.00	75.63	76.39	41.73
Mask DINO + 1 QPD	82.41	66.61	0.00	40.57	0.00	89.65	0.00	75.58	76.28	39.89
Mask DINO + 2 QPD	83.71	60.92	10.55	50.4	0.00	92.22	4.84	76.67	77.51	43.23
Mask DINO + 3 QPD	76.24	61.26	7.06	48.91	0.00	92.30	0.00	74.11	75.00	40.83
Mask DINO + 4 QPD	75.43	45.67	2.67	48.31	0.00	94.23	0.00	71.97	73.46	38.04
Mask DINO + 5 QPD	78.25	27.39	0.92	47.96	0.00	93.89	0.00	71.38	71.79	35.49
Mask DINO + Random Input to NMS	81.46	51.73	25.36	40.73	0.00	92.73	0.00	76.19	76.60	41.70
Mask DINO + Encoder Input to NMS	82.05	63.02	15.33	50.61	0.00	88.71	0.00	76.21	76.64	42.82
Mask DINO + NMS + 2 QPD (Proposed)	82.80	60.94	19.96	49.70	0.00	93.93	0.00	77.77	78.43	43.84

Evidence of Query Improvement: The improvement in the query initialization due to the proposed architecture is demonstrated in Fig. 3. Here, we mark the centre of initialized query boxes and generate the scatter plot. The average recall rate for the EV18 dataset at 0.9 IOU for the top 300 queries in vanilla Mask DINO is 7.48%. After performing NMS, the recall rate decreases to 0.00%, but the queries get diversified. After passing the same through the proposed Query Proposal Decoder (QPD), the recall rate is observed to be 52.38%. The increase in recall rate to 52.38% indicates successful learning of QPD. It indicates that the proposed architecture is able to cover more objects in the initialized queries thereby improving the performance at the end. While the diversity of the queries is important, it is also important to ensure that the queries with higher confidence are near the ground truth to ensure better learning. Random initialization for NMS is observed with sub-optimal performance and is shown in Table 3.

Ablation Study: We perform an ablation on the EV18 dataset to show the importance of each block in the proposed methodology. The results of the same are summarised in Table 3. We also experiment with the number of layers in QPD. We have used the best-performing 2-layer QPD architecture for all other experiments. The reduction in performance with more QPD layers can be attributed to the under-learning of negative samples due to stricter query proposals

5   Conclusion

In this paper, we proposed a novel class-agnostic Query Proposal Network (QPN) to better initialize the queries for a transformer-based surgical instrument instance segmentation model. Towards this, we first diversified the queries using the non-maximal suppression and proposed a deformable-cross-attention-based learnable Query Proposal Decoder (QPD). On average, the proposed QPN improved the recall rate of the query initialization by 52.38% at 0.9 IOU. The improvement translates to an improved ISI-IOU of 1.84% and 2.09% in the publicly available Endovis 2017 and Endovis 2018 datasets, respectively.

Acknowledgements. We thank Dr Britty Baby for her assistance. We also thank DBT, Govt of India, and ICMR, Govt of India, for the funding under the projects BT/PR13455/CoE/34/24/2015 and 5/3/8/1/2022/MDMS(CARE) respectively.

References

1. Al Hajj, H., et al.: CATARACTS: challenge on automatic tool annotation for cataract surgery. MIA **52**, 24–41 (2019)
2. Allan, M., et al.: 2018 robotic scene segmentation challenge. arXiv preprint arXiv:2001.11190 (2020)
3. Allan, M., et al.: 2017 robotic instrument segmentation challenge. arXiv preprint arXiv:1902.06426 (2019)

4. Baby, B., et al.: From forks to forceps: a new framework for instance segmentation of surgical instruments. In: WACV, pp. 6191–6201 (2023)
5. Bouget, D., Allan, M., Stoyanov, D., Jannin, P.: Vision-based and marker-less surgical tool detection and tracking: a review of the literature. MIA **35**, 633–654 (2017)
6. Cheng, B., Choudhuri, A., Misra, I., Kirillov, A., Girdhar, R., Schwing, A.G.: Mask2former for video instance segmentation. arXiv preprint arXiv:2112.10764 (2021)
7. Cheng, T., Wang, X., Huang, L., Liu, W.: Boundary-preserving mask R-CNN. In: Vedaldi, A., Bischof, H., Brox, T., Frahm, J.-M. (eds.) ECCV 2020. LNCS, vol. 12359, pp. 660–676. Springer, Cham (2020). https://doi.org/10.1007/978-3-030-58568-6_39
8. Du, W., Xiang, Z., Chen, S., Qiao, C., Chen, Y., Bai, T.: Real-time instance segmentation with discriminative orientation maps. In: ICCV, pp. 7314–7323 (2021)
9. Fang, Y., et al.: Instances as queries. In: ICCV, pp. 6910–6919 (2021)
10. Ganea, D.A., Boom, B., Poppe, R.: Incremental few-shot instance segmentation. In: CVPR, pp. 1185–1194 (2021)
11. González, C., Bravo-Sánchez, L., Arbelaez, P.: ISINet: an instance-based approach for surgical instrument segmentation. In: Martel, A.L., et al. (eds.) MICCAI 2020. LNCS, vol. 12263, pp. 595–605. Springer, Cham (2020). https://doi.org/10.1007/978-3-030-59716-0_57
12. Grammatikopoulou, M., et al.: Cadis: cataract dataset for image segmentation. arXiv preprint arXiv:1906.11586 (2019)
13. Guo, R., Niu, D., Qu, L., Li, Z.: Sotr: segmenting objects with transformers. In: ICCV, pp. 7157–7166 (2021)
14. He, K., Gkioxari, G., Dollár, P., Girshick, R.: Mask R-CNN. In: ICCV, pp. 2961–2969 (2017)
15. He, K., Zhang, X., Ren, S., Sun, J.: Deep residual learning for image recognition. In: Proceedings of the IEEE Conference on Computer Vision and Pattern Recognition, pp. 770–778 (2016)
16. Himal, H.: Minimally invasive (laparoscopic) surgery. Surg. Endosc. Interv. Tech. **16**, 1647–1652 (2002)
17. Hu, J., et al.: ISTR: end-to-end instance segmentation with transformers. arXiv preprint arXiv:2105.00637 (2021)
18. Kong, X., et al.: Accurate instance segmentation of surgical instruments in robotic surgery: model refinement and cross-dataset evaluation. IJCARS **16**(9), 1607–1614 (2021)
19. Li, F., Zhang, H., Liu, S., Zhang, L., Ni, L.M., Shum, H.Y., et al.: Mask DINO: towards a unified transformer-based framework for object detection and segmentation. arXiv preprint arXiv:2206.02777 (2022)
20. Lin, T.Y., Goyal, P., Girshick, R., He, K., Dollár, P.: Focal loss for dense object detection. In: ICCV, pp. 2980–2988 (2017)
21. Westebring-van der Putten, E.P., Goossens, R.H., Jakimowicz, J.J., Dankelman, J.: Haptics in minimally invasive surgery-a review. Minim. Invasive Therapy Allied Technol. **17**(1), 3–16 (2008)
22. Qiao, S., Chen, L.C., Yuille, A.: Detectors: detecting objects with recursive feature pyramid and switchable atrous convolution. In: CVPR, pp. 10213–10224 (2021)
23. Redmon, J., Farhadi, A.: Yolov3: an incremental improvement. arXiv preprint arXiv:1804.02767 (2018)

24. Rezatofighi, H., Tsoi, N., Gwak, J., Sadeghian, A., Reid, I., Savarese, S.: Generalized intersection over union: a metric and a loss for bounding box regression. In: CVPR, pp. 658–666 (2019)
25. Ross, T., et al.: Comparative validation of multi-instance instrument segmentation in endoscopy: results of the ROBUST-MIS 2019 challenge. MIA **70**, 101920 (2021)
26. Shvets, A.A., Rakhlin, A., Kalinin, A.A., Iglovikov, V.I.: Automatic instrument segmentation in robot-assisted surgery using deep learning. In: 2018 17th ICMLA, pp. 624–628. IEEE (2018)
27. Tian, Z., Shen, C., Chen, H.: Conditional convolutions for instance segmentation. In: Vedaldi, A., Bischof, H., Brox, T., Frahm, J.-M. (eds.) ECCV 2020. LNCS, vol. 12346, pp. 282–298. Springer, Cham (2020). https://doi.org/10.1007/978-3-030-58452-8_17
28. Twinanda, A.P., Shehata, S., Mutter, D., Marescaux, J., De Mathelin, M., Padoy, N.: EndoNet: a deep architecture for recognition tasks on laparoscopic videos. IEEE TMI **36**(1), 86–97 (2016)
29. Vu, T., Kang, H., Yoo, C.D.: SCNet: training inference sample consistency for instance segmentation. In: AAAI, vol. 35, pp. 2701–2709 (2021)
30. Wang, T., et al.: The devil is in classification: a simple framework for long-tail instance segmentation. In: Vedaldi, A., Bischof, H., Brox, T., Frahm, J.-M. (eds.) ECCV 2020. LNCS, vol. 12359, pp. 728–744. Springer, Cham (2020). https://doi.org/10.1007/978-3-030-58568-6_43
31. Wang, X., Zhang, R., Kong, T., Li, L., Shen, C.: SOLOv2: dynamic and fast instance segmentation. ANIPS **33**, 17721–17732 (2020)
32. Zang, Y., Huang, C., Loy, C.C.: FASA: feature augmentation and sampling adaptation for long-tailed instance segmentation. In: ICCV, pp. 3457–3466 (2021)
33. Zhang, H., et al.: DINO: detr with improved denoising anchor boxes for end-to-end object detection. In: 11th ICLR (2022)
34. Zhao, Z., Jin, Y., Heng, P.A.: Trasetr: track-to-segment transformer with contrastive query for instance-level instrument segmentation in robotic surgery. In: ICRA, pp. 11186–11193. IEEE (2022)
35. Zhu, X., Su, W., Lu, L., Li, B., Wang, X., Dai, J.: Deformable detr: deformable transformers for end-to-end object detection. arXiv preprint arXiv:2010.04159 (2020)

Correction to: Semantic Segmentation of Surgical Hyperspectral Images Under Geometric Domain Shifts

Jan Sellner, Silvia Seidlitz, Alexander Studier-Fischer,
Alessandro Motta, Berkin Özdemir, Beat Peter Müller-Stich,
Felix Nickel, and Lena Maier-Hein

Correction to:
Chapter 59 in: H. Greenspan et al. (Eds.): *Medical Image Computing and Computer Assisted Intervention – MICCAI 2023*, LNCS 14228, https://doi.org/10.1007/978-3-031-43996-4_59

The updated version of this chapter can be found at
https://doi.org/10.1007/978-3-031-43996-4_59

Author Index

H. Greenspan et al. (Eds.): MICCAI 2023, LNCS 14228, pp. 739–743, 2023.
https://doi.org/10.1007/978-3-031-43996-4

Printed in the United States
by Baker & Taylor Publisher Services.

Printed in the United States
by Baker & Taylor Publisher Services